HIGH SPEED RAIL PLANNING, POLICY, AND ENGINEERING, VOLUME III

HIGH SPEED RAIL PLANNING, POLICY, AND ENGINEERING, VOLUME III

System Operations

TERRY L. KOGLIN, PE

MOMENTUM PRESS
ENGINEERING

High Speed Rail Planning, Policy, and Engineering, Volume III: System Operations

First published in 2017 by
Momentum Press®, LLC
222 East 46th Street, New York, NY 10017
www.momentumpress.net

ISBN-13: 978-1-60650-983-8 (print)
ISBN-13: 978-1-60650-984-5 (e-book)

Momentum Press Transportation Engineering Collection

Cover and interior design by S4Carlisle Publishing Service Ltd.
Chennai, India

First edition: 2017

10 9 8 7 6 5 4 3 2 1

Printed in the United States of America

ABSTRACT

Volume III of *High-Speed Rail Planning, Policy, and Engineering—* Operations explores the high-speed operations of a hypothetical reconstruction of a former railroad main line between Chicago and New York. The former Pennsylvania Railroad main line between New York and Chicago, via Trenton, Harrisburg, Pittsburgh, Canton, and Fort Wayne, is studied in its existing condition and under various phases of rehabilitation and reconstruction. Operation of high-speed passenger and freight trains under various scenarios of reconstruction of the aforementioned rail line is studied. The possibility of long-distance commuter operations is investigated. Cost analysis, marketing, track maintenance, and equipment maintenance for a proposed high-speed rail system are discussed.

KEYWORDS

Railroad, High speed, Passenger, Freight, Trailer-on-Flatcar (TOFC), Signals, Controls, New York, Philadelphia, Pittsburgh, Canton, Fort Wayne, Chicago, Terminals, Locomotive, Diesel, Electric, Crews, Double Track, Multiple Track, Passing Track, Service, Schedule, Amtrak, Privatization, Trackage Rights, Superelevation, Positive Train Control, Highway Trailers, No Train Horn, Maintenance, Cab Car, Federal Railroad Administration, Surface Transportation Board, Grade Crossing, Drone Technology, Cab Signals, Automatic Train Stop, Automatic Train Control, Cost Analysis, Marketing, Demand Pricing, Competition, Fixed Cost, Variable Cost, Incremental Cost, Drayage, Reliability

CONTENTS

List of Figures

ACKNOWLEDGMENTS

Peter F. Chang, PE has provided valuable connections to photographers of overseas high-speed rail trains and equipment. He also provided critical review of the contents. Rick Harnish of the Midwest High Speed Rail Association provided valuable criticism and constructive suggestions toward the content and format of the book. Gordon Davids, PE, who retired from the Federal Railroad Administration, pointed out several aspects of FRA policy and also provided a critical view of parts of the text including an opinion on some innovative concepts presented in the text. Maria Grazia Bruschi, PE assisted in formation of the basic plan of the book, guidance on SI nomenclature and criticism of the contents. Dave Peterson was kind enough to review an early draft. Dale Muellerleile, PE provided a thorough review of the manuscript and offered many helpful suggestions based on his experience in high-speed rail and railroad engineering in general. Maria and Dale also allowed access to their extensive photograph files. I thank you all for assisting in what was a very, very difficult project in a very controversial and opinion-filled field. In spite of many suggestions, the radical ideas as well as the rest of the content of this volume are mine and mine alone.

Most artwork and drawings were prepared by Lucas Chase of R H Batterman.

CHAPTER 1

INTRODUCTION

High-speed passenger and freight trains are in operation in many countries around the world. Most of them are operated as a public service and are not self-sustaining. Trains are ideal vehicles for moving large numbers of people, or large volumes of goods, rapidly and inexpensively. In order for trains to perform their services efficiently, they must be operated sensibly. To maximize service while minimizing cost, the high-speed rail system must be competently managed, with efficient use of labor and equipment, without compromising quality of service.

A high-speed rail system is a very large investment. Railroads, unlike other transportation systems, require an exclusive right-of-way and roadway, or at least one with severe restrictions on access, the full length of the intended service corridor. In order to justify the investment in such a system, it must be extremely productive. A high-speed passenger rail system must carry large numbers of people day in and day out. The high-speed rail system must attract large numbers of passengers and keep attracting them for repeat trips as often as they travel. The system must operate efficiently so that it can generate net income, over and above its operating costs, and preferably need not be required to rely on subsidies to keep in operation.

Even a totally privately funded railroad line, high speed or not, must accommodate itself to political realities. There is government involvement in railroad operations in the United States at many different levels. This involvement varies with the political winds, but railroad operators must be aware of and adhere to government regulations. Even private, for profit, freight-only railroads are subject to government restrictions. Space does not allow complete discussion of these restrictions, but some mention will be made in appropriate areas.*1

The high-speed rail system must deliver the best possible service at the least possible cost, within practical limitations. In order to do this, all the personnel within the organization must be dedicated to performing

their jobs to the best of their abilities. In order that this may be the case, management must provide the best possible working environment for all employees, on trains and off, giving them the tools they need to do their jobs.

A true high-speed rail system will have no interface with other railroads other than interchange tracks for delivery of equipment, materials, and supplies. A possible exception would be the use of common terminals for high-speed and other passenger trains. It may be necessary to transfer passenger cars to and from other railroads to provide for extensions of transportation beyond the high-speed system, unless or until the system reaches its full practical extent. It may be advantageous to transfer high-speed freight traffic, as described in Volume 2, between railroads to provide nationwide service. Any such arrangement should be considered temporary, to be replaced by complete service by the high-speed system, once it has expanded and developed to cover the entire market.

This book considers only high-speed steel-wheel-on-steel-rail systems. The definition of "high-speed" is somewhat flexible. Nothing in this book constitutes a manual or textbook for design, construction, or operation of a high-speed rail system. Any cost estimates given are very approximate. In North America, the American Railway Engineering and Maintenance-of-Way Association (AREMA) performs the function of providing guidelines for engineering aspects of railroad design and construction. The AREMA *Manual for Railway Engineering* consists of 21 chapters, in four volumes, on the various aspects of railway engineering and has a special Chapter 17 entitled "High Speed Rail Systems". Volume 1 (Track), Volume 2 (Structures), Volume 3 (Infrastructure), and Volume 4 (System Management) all contain materials essential to the design, construction, and maintenance of most of the fixed property aspects of a railroad system. The Manual is revised annually by special committees that consist of experts on the aspects of their particular chapters in the Manual so that the information contained in the Manual is as up-to-date as is practical. Various portions of the Manual will be referenced in these volumes where appropriate. The Association of American Railroads (AAR) serves the function of a joint committee on operations and other aspects of railroading that benefit from coordination. Most railroads in the United States are members of this group. The AAR has taken a neutral position regarding the use of freight railroad property for high-speed passenger rail operations while insisting that any such usage is not to impede the rail freight operations of its members.

With few exceptions, all dimensions in this volume are in English units. Some SI conversions are given in parentheses. A comprehensive system of conversion between English and SI units, as published by AREMA, is provided as an appendix to Volume 1.

OPERATIONS GOALS

The proposed high-speed rail system, as described in Volume 2, is intended to provide the best possible quality of rail passenger and trailer-on-flatcar (TOFC) service between Metropolitan Chicago and Metropolitan New York City. The high-speed passenger operations would be retail, sold directly to the consumer or via travel agents and other outlets. The freight service, consisting of highway trailers carried on specially equipped platforms, similar to conventional TOFC equipment but aerodynamically enclosed and capable of high speed operation, would be sold primarily on a wholesale basis, relying on the multitude of entities of various kinds that specialize in direct customer contact for the services to be provided.

Long-distance business travelers typically prefer to leave their place of origin early in the morning and arrive at their destination in time for meetings in midmorning. If it is necessary to arrive earlier, these travelers fly the evening before and book a hotel room. A high-speed passenger rail system as presently constituted, with scheduled trains operating at 200 mph (322 km/hr) maximum speed so that overall 100 mph (161 km/hr) or higher average speeds can be maintained, can duplicate these travel arrangements up to a distance of 500 to 600 miles (800 to 965 km). Alternately, the high-speed passenger trains can provide complete hotel services aboard the train, so that the travel and accommodation services are combined. Such accommodations had been provided on trains in the past, and could easily be furnished on a high-speed train. This would in many cases allow business travelers to depart their homes late in the evening and yet arrive refreshed at early morning meetings in distant cities.

Trains have tremendous capacity to modify operations to suit the needs of the traveling public. Through trains can be nonstop, serving end point terminals only, or can connect major terminals with busier intermediate stations. Equipment utilization should be maximized so that trains operate filled or nearly so over as much of the route as possible. A long train on which most of the cars are empty for a great part of the trip wastes resources. Trains, ships, and buses, as well as airplanes, operate unprofitably unless a sufficient number of their seats are filled. Amtrak and public

agency commuter train operations typically are not ultimately responsible for financing the purchase of their rolling stock and do not consider capital costs, including purchase of equipment, as part of their budgets. Not being responsible for paying for their cars, they do not concern themselves with efficiency in their use. On the other hand, rather than paying the expense, which would come from their operating budgets, of having the workers available to add or remove cars from trains, they often run 10- or 12-car trains with only as few as two cars used by passengers during a significant portion of, or perhaps the entire, trip. The rail cars that are used are very expensive, and an operator responsible for paying for them should get the most productive use out of them.

As mentioned above, rail travel requires an exclusive right-of-way, or one that is limited to certain types of traffic. Experiments, such as in Great Britain after "privatizing" British Rail, with multiple independent users of a single rail line have found that conflicts can occur. A high-speed rail line would of necessity have to be restricted to only high-speed trains or be operated under very strict rules. In the interest of safety, all traffic on the line would have to be under a single control. A railroad then becomes a monopoly of sorts, as opposed to airports and air traffic controllers, and highways, which are open to all. Antimonopoly sentiment is another reason why some people are opposed to government funding of a rail system. Amtrak is required to provide practically unrestricted access to its tracks; the Northeast Corridor hosts trains operated by MARC, Septa, New Jersey Transit, the Long Island Rail Road, Norfolk Southern, and others in addition to Amtrak's own trains.

CHAPTER 2

BEGINNING OPERATIONS

The proposed system anticipates two types of traffic: high-speed passenger trains and high-speed freight trains consisting of highway trailers on flatcars modified for efficiency in high-speed service. In beginning service, it may be expeditious initially to operate combined trains, containing passengers and trailer-on-flat-car (TOFC) in the same train.

The proposed system could begin by taking over operations of an existing carrier, or piggybacking its operations on to those of an existing carrier. This should eliminate, or at least minimize, difficulties due to complaints of other carriers objecting to the instigation of the proposed service. In the United States, government restrictions on entry of new providers are much relaxed today from what they were under the now-defunct Interstate Commerce Commission, but the Surface Transportation Board still has some say in the matter. Shippers, carriers, and other entities are constantly petitioning Congress for special consideration in such matters so that the rules could change at any time, and any entity considering commencement of such services should pay close attention to the current political scene. There is currently (May 2016) much agitation over the intention of a private entity to build a new freight-only bypass around Chicago.*1

If operations of the proposed system are begun immediately upon organization, it may be necessary to use existing trackage belonging to others. Segments of rail lines that are to be used temporarily, for initial operations only, will be used as is, and speed limits will be as existing, or perhaps a little faster depending on track condition, signals in use, and the willingness of the owner of the trackage to give consideration to the proposed service. At locations where the proposed system is to continue permanent operations, improvements will eventually be made, as described in Volume 2. These improvements will be performed gradually, as funds are available, with operating times reduced as the improvements are made.

Negotiating trackage rights or haulage rights with a railroad or railroads forming part of a connection between the terminals may be difficult.

Both CSX and Norfolk Southern (NS) have very busy main lines between the New York area and Chicago. They also heavily market their trailer and container train service in that market. They are not likely to welcome an independent operation that does not fit into their main line operations as to speed and schedule. A railroad line not performing a service similar to the proposed highway trailer carriage would likely be less averse to providing tracks for the proposed initial operations. Both Canadian Pacific (CP) and Canadian National (CN) provide rail service east from Chicago. CN has its own tracks from Chicago to Niagara Falls, and also to points east of Lake Ontario, with roundabout independent connections to New York City. CP actually goes from Chicago to the New York City area, but uses trackage rights for the Chicago–Detroit segment and Buffalo–New York City. Other routes might be possible, such as Norfolk Southern's comparatively lightly used former NKP–Erie route from Chicago to the metropolitan New York City area. This route could be preferred if it is believed that operations must start immediately, and it is not feasible to begin operations over the route described in detail below.

INITIAL ROUTE

There are eight basic sections of railroad that are intended to be used initially for the beginning operations of the proposed service. These are as follows: (1) Chicago, Fort Wayne and Eastern between Liverpool, Indiana and Crestline, Ohio; (2) NS secondary line between Crestline and Alliance, Ohio; (3) another NS secondary line between Alliance, Ohio and Rochester, Pennsylvania via East Liverpool; (4) NS main line between Rochester and Wilmerding, Pennsylvania via downtown Pittsburgh; (5) NS main line between Wilmerding and Harrisburg; (6) Amtrak between Harrisburg and Glen Loch, Pennsylvania; (7) NS between Glen Loch and Morrisville, Pennsylvania; and (8) Amtrak from Morrisville to the Jersey Meadows. All or part of these track segments except number 4, from Rochester to Wilmerding, are to be part of the proposed permanent high-speed system. Further details of existing conditions can be found in Volume 2.

CHICAGO, FORT WAYNE AND EASTERN, LIVERPOOL–CRESTLINE

This is a regional railroad operating over a portion of the former Pennsylvania Railroad main line between Chicago and Pittsburgh. It operates fairly

extensive local freight service and also allows CSX and NS trains to use its tracks. There are several trains each day operating over the length of the line intended to be used for the proposed high-speed service. The line is no longer signaled, and so the maximum speed allowable will be 59 mph for the proposed TOFC trains, which are allowed by the Federal Railroad Administration (FRA) to operate at passenger train speeds. The trains must be rigidly scheduled, and must not suffer delays due to interference with other rail traffic. The existing track will be used, with lining, leveling, and maintenance to allow the 59 mph maximum speed per FRA regulations in unsignaled territory.*2 A trackage rights agreement will have to be negotiated with the present operator, with a high-enough usage fee to allow acceptance of the proposed traffic on their railroad, with strict performance assurances including a high penalty for failure to allow the proposed trains to operate unhindered through the length of the line on a strict, mutually agreed upon schedule. It is intended to take control of operations on this line as soon as is possible. This can be accomplished by obtaining ownership of the line or by a lease or other such arrangement.

NORFOLK SOUTHERN, CRESTLINE–ALLIANCE

This secondary line has some traffic and is signaled. It has multiple tracks on some stretches. The existing tracks will be used at a negotiated rate to provide acceptable revenue to NS while guaranteeing on-schedule performance of the proposed service. Speed will be maximum 70 mph unless NS will allow higher speed operation. It is intended to take control of operations on this line between Crestline and Fairhope (a few miles west of Alliance) as soon as is possible. This can be accomplished by obtaining ownership of the line or by a lease or other such arrangement.

NORFOLK SOUTHERN, ALLIANCE–ROCHESTER

This secondary line is largely unsignaled and has a 30 mph speed limit between Alliance and Bayard, Ohio. Speed limits vary up to 50 mph further east as far as Rochester, Pennsylvania. This line has been reduced to single track between Alliance and Yellow Creek, Ohio. It once carried very heavy iron ore and coal traffic, but this has been much reduced. The existing track will be used, with initial lining and leveling. The tracks will be used at a negotiated rate to provide acceptable revenue to NS while

guaranteeing on-schedule performance of the proposed service. It is intended to take control of operations on this line between Bayard, Ohio, and Midland, Pennsylvania, as soon as possible. This can be accomplished by obtaining ownership of the line or by a lease or other such arrangement. Eventually, most of this existing line segment would be bypassed by new construction, but control of the line would expedite matters.

NORFOLK SOUTHERN, ROCHESTER–WILMERDING

This is a very busy section of track, passing alongside NS's main classification yard for the Pittsburgh area, at Conway. It also passes through the main Amtrak Pittsburgh station. It has a few sharp curves that will be almost impossible to eliminate. It will be bypassed as soon as is feasible, as discussed in Volume 2.

NORFOLK SOUTHERN, WILMERDING–HARRISBURG

This section of heavily used main line once had four tracks throughout most of its length, but has been reduced to two tracks except at a few isolated locations. It will be very difficult to obtain permission to operate the proposed trains on the existing tracks. Maximum speed limit on this line is presently 70 mph, except for a 75 mph speed limit for a short distance west of Harrisburg. This is a very busy stretch of railroad, so that on-time operation will be very difficult to achieve as long as track is shared with NS and Amtrak trains.

AMTRAK, HARRISBURG–GLEN LOCH

This section has been rebuilt for 110 mph maximum speeds. Initial operations will be on existing track with no modifications. There are two tracks from Harrisburg to Thorndale, and three tracks from Thorndale, where SEPTA train service to Philadelphia terminates, to Glen Loch. There are a few late night SEPTA and Amtrak trains between Thorndale and Glen Loch, but tight scheduling will avoid interferences with the proposed service. NS schedules very few freight trains over this line, but they may operate at the same time as the proposed high-speed trains; provision must

be made in operating agreements to avoid interference. There should be little or no difficulty in coming to agreement with Amtrak for trains operating over this line on trackage rights.

NORFOLK SOUTHERN, GLEN LOCH TO MORRISVILLE

This line has been reduced to single track over most of its length, and signals have been removed. The speed limit on this line was 50 mph, but trains should be able to operate at 59 mph over most of the length of the line unless NS restricts speed to a lower level. The existing track will be used at a negotiated rate to provide acceptable revenue to NS while guaranteeing on-schedule performance of the proposed service.

AMTRAK, MORRISVILLE–JERSEY MEADOWS

Amtrak has a four-track high-speed line in place between Morrisville and the Jersey Meadows. There will be no difficulty with track capacity at night, when the proposed trains will operate, but severe congestion will occur on the line on weekday mornings, between New Brunswick and Jersey Meadows due to the presence of commuter trains heading into New York City. The problem is particularly severe between Kearney and New York City, where the greatest amount of train traffic occurs because of additional trains entering the line. There is also a passenger station at Secaucus, and only two tracks are available over the Hackensack River in the middle of the Meadows, adding to congestion. The trains of the proposed service must vacate these tracks before 7 a.m. to minimize delays.

CHAPTER 3

Initial Operations

Initial operation would be as detailed in Appendix B0. The modified route used, following Norfolk Southern from Beaver to Wilmerding, is a heavily used railroad so that it will be difficult to add the trains of the proposed service to it without resulting in delays to the proposed trains. This segment of rail line is used by the proposed system only temporarily and does not warrant construction of benefits to make operation faster or on-time operation easier. Three or more tracks are available on this route from Rochester, Pennsylvania to Wilmerding, and so delays due to conflicting traffic may not be intolerable. While this segment is used by the proposed system, extra time might be scheduled for trips so that overall on-time operation can be maintained despite delays in this area.

It may be necessary at certain locations to operate trains at restricted speeds, lower than the calculated maximum speeds shown in Appendix B due to the existing level of track maintenance, existing superelevation amounts, and existing signaling, or lack thereof. There are likely to be community concerns about new operations on trackage that may not have seen any trains for some time, or no trains operating at high speeds, or more trains than were operated for some time previous to the introduction of the new service. It may be possible to alleviate these concerns by providing fencing and extra protection at crossings. "No train horn" rules might be put into effect at appropriate locations. The proposed trains can normally operate at slower than the maximum speed over the final few 100 miles of the route but operate faster, up to the maximum allowable, when necessary to allow for compensation for delays here or elsewhere. Early arrivals can occur if delays are less than anticipated, but may cause conflicts because of capacity constraints at terminals. Again, even in the early stages of operation, the key element of the service is reliability. Overall schedule time will not be as fast as possible. It will be as fast as can be consistently, reliably maintained.

TERMINALS

Any abandoned or lightly used rail yard can be quickly adapted into a temporary trailer-on-flatcar (TOFC) terminal. Obtaining the property on a short-term lease should not be difficult. After a few stub-end tracks are fitted with ramps, a few pieces of equipment to move trailers on and off the trains are obtained, some sort of clerical billing office is established, and rudimentary servicing and repair facilities have been set up, the terminal will be ready. One terminal somewhere in the Chicago area and one in the New York area, with connections to the rail network, will allow operations to begin. Passengers, if carried at this stage, will be unlikely to be willing to board and leave trains in the environment that the trailer terminal is likely to be in, but there are many existing passenger train stations in operation in both the Chicago area and New York–Northern New Jersey. The passenger cars can be brought to and taken from the trailer terminal to and from the passenger station being used, and connected and disconnected to the freight portion of the train at the trailer terminal. It might be possible to use Chicago Union Station (CUS) initially as the passenger terminal at the west end of the line; connection can be made via existing trackage, Tolleston–Kensington–CUS. At the east end, Secaucus Transfer would be an excellent terminal for the passenger portion of the initial trains, if the trains of the proposed service cannot be operated directly into Pennsylvania Station in New York City (PSNY). Depending on location of terminals, trains could begin with trailers and pick up the loaded passenger cars before proceeding or vice versa, with the reverse procedure on returning.

The initial New York area TOFC terminal would have to be located away from the Northeast Corridor, at least the portion of it between Rahway and the North River tunnels, and a convenient point would have to be found to disconnect the passenger portion of the proposed train from its TOFC portion. There are several locations where a temporary intermodal yard for use of the proposed TOFC trains could be sited, such as Greenville yard, Port Reading, the former CNJ Elizabeth Yard, and others. Minimizing expense and complications of construction, a yard presently accessible off the Northeast Corridor would be best, southwest of but as close as practical to New York City. To minimize difficulties inherent with interface with conflicting Northeast Corridor commuter traffic, the TOFC trains should exit the Northeast Corridor south/west of Metropark station. The freight yard located alongside the Northeast Corridor between Edison and Metuchen would serve. It is not long enough to provide for

loading and unloading a full-length TOFC train without breaking the train in two (dividing the train into two parts), but the initial operations will call for circus-style loading, so that if the TOFC train is long it would be expeditious to break it up anyway. The location has easy access to the New Jersey Turnpike and other Interstate highways. It could serve as a permanent location for a central-northern New Jersey TOFC terminal. The initial TOFC operation would consist of standard intermodal flatcars equipped for circus-style loading and unloading. A few of the existing tracks in the yard can be isolated for use for the proposed service and equipped with ramps at the south/west end. If there is insufficient space at Liverpool to construct an adequate temporary TOFC terminal for circus-style loading and unloading trailers, another temporary location must be found. Several abandoned or relatively lightly used rail freight yards exist in the Chicago area. Gibson, near Liverpool, with access to the Chicago, Fort Wayne and Eastern line would be a good choice. It is on the Tolleston–Kensington–Chicago line that connects the proposed permanent Liverpool, Indiana trailer terminal location with CUS.

Unloading and loading of trailers can be done with conventional equipment. One loading track could be used for more than one cut of cars in a train, and expedited and nonexpedited trailers can be segregated into different groups for loading and unloading.

The least expensive mode of trailer loading and unloading would be circus-style operation, with a ramp or ramps at the end of stub track(s) arranged so that trailers can be driven on and off the conventional intermodal railcars. This type of facility could be placed anywhere near the desired terminal location provided sufficient space is available for driving on and off the ramp, sufficient trailer parking area exists, and access to the main railroad track and local highways is convenient. No special pavement is needed such as must be supplied for heavy side-loading equipment. Clerical duties can be performed from a temporary structure located at the highway entrance to the facility. Initial operations would be as described in the following sections:

INITIAL TRAINS

Assuming that CUS and PSNY are chosen as the initial passenger terminals for the proposed service, and passengers during the initial stages of operations are carried on combined passenger-TOFC trains, the trains would be configured as follows:

Trains will be powered by diesel-electric locomotives. Locomotive axle loading must be sufficient to allow the efficient use of diesel-electric locomotives to pull the trains. There will be only a small advantage in having passenger trains in multiple-unit form with consequent lower axle loading; therefore, passenger trains will consist of unpowered passenger cars pulled by diesel-electric locomotives. Motive power would be distributed, located at each end of the train. Diesel-electric locomotive power units would consist of one end unit at front of the train and one at the rear of the train, with additional units included as necessary to provide sufficient motive power for the train. The passenger cars would be adjacent to the western group of locomotive power units. The passenger cars would be push-pull type, with a control cab unit placed as the easternmost passenger car for control of the eastbound train when operating passenger-only. The TOFC cars would be placed between the passenger cars and the eastern group of locomotive power units. Separable couplings would connect the easternmost passenger car to the westernmost TOFC car. Throughout most of the route, during initial operations, trains will not operate at over 70 mph. Motive power requirements will thus be rather modest as compared with trains of the fully operational high-speed system. Motive power will be largely dependent on the size of the train. If a train has 10 passenger cars and 200 highway trailer carriers and is fully loaded, six (6) locomotive power units at 4,000 horsepower apiece should be sufficient for the train. These locomotive units could be distributed, with two units at the west end of the train and four at the east end. One of the units at the west end of the train could be a diesel-electric/electric locomotive so that the passenger portion of the train could proceed into Penn Station New York while powered by a straight electric locomotive, with the diesel engine or engines shut down. This transition could occur at or near Secaucus Transfer station in the Jersey Meadows.

All cars and motive power for initial operations would be standard, mass-produced, conventional railroad equipment as manufactured for use in North America. Used equipment, if available and in acceptable condition, would be used. A private operator, short of capital, must obtain rolling stock by some means. Leasing is a very attractive alternative to purchase and is frequently used by privately owned freight railroads.

To make up an eastbound combined passenger–TOFC train, the passenger cars and western end group of locomotive power units will be placed at CUS for passenger boarding. At the same time, highway trailers will be in the process of being loaded onto the TOFC cars at the terminal in Liverpool (or elsewhere). The passenger portion of the train will take

approximately 26 minutes to travel a little over 28 miles from CUS to Liverpool. The TOFC portion of the train should be completely loaded and ready for departure by the time the passenger portion of the train arrives at Liverpool. If the loaded passenger portion of the train leaves CUS at 4 p.m., it would arrive at Liverpool at 4:26 p.m. Allowing 4 minutes to make connections and test the train, the complete train should depart Liverpool for the east at 4:30 p.m. The eastbound combined passenger–TOFC train would proceed to Edison, New Jersey, where it would stop with the passenger cars at the north/eastbound NJ Transit passenger platform at 8:30 a.m. (9:30 Eastern Time). The TOFC portion and the eastern locomotive power units would disconnect from the passenger cars and proceed into the yard. The passenger cars, propelled by means of the locomotive power units at the west end of the train and controlled from the easternmost "cab" car, would then proceed to Secaucus Transfer or Penn Station New York. The TOFC portion of the train would, in the meantime, be placed on the unloading tracks in the yard, broken up into as many sections as necessary on as many unloading/loading tracks as needed.

With the timing shown for the unimproved route as operated in Appendix B0, the proposed trains would take approximately 16¼ hours to run from Liverpool to the Jersey Meadows. Combined passenger–TOFC trains departing Liverpool in the early to midevening would then arrive at Jersey Meadows in midmorning, interfering to some extent with rush hour commuter traffic into PSNY. This would be unacceptable under the initial conditions of operation. After both the passenger and freight sections of the eastbound train are unloaded, they would be taken to designated service/storage areas to be prepared for a westbound trip. The TOFC portion would, if possible, remain in the yard, presumably at Edison, for servicing and storage until it was time to reload the train with westbound trailers. The passenger portion could be taken to Sunnyside Yard in New York City, if an arrangement could be made to service and store the passenger part of the train there. Other locations are possible.

The westbound train would operate in similar fashion to the eastbound. At the proper time, the passenger portion of the train will be returned to Penn Station New York or Secaucus Transfer for loading, and then will proceed west to Edison, stopping at the west/southbound NJ Transit passenger platform. The TOFC part of the westbound train will have commenced loading earlier and be ready to move by the time the passenger portion approaches Edison. The TOFC portion of the train will pull out of the yard from the north/east end and proceed to Metuchen,

New Jersey, where it will work its way through the existing crossovers and onto Track Number Three, where it will wait for the westbound passenger portion of the train to pass it on Track Number Four. After the westbound passenger portion passes through Metuchen, the TOFC portion will proceed south/west, crossing over to Track Number Four, where it will follow the passenger portion of the train, and couple onto it after the passenger portion stops on Track Number Four at the west/southbound platform at Edison Station.

The complete train will then proceed west to Liverpool, Indiana, where the passenger and TOFC portions will again separate. The TOFC portion of the train will remain at Liverpool (or proceed to Gibson or another site where the temporary TOFC terminal is located) for unloading while the passenger portion of the train will proceed to CUS and the passengers will disembark. If the westbound TOFC portion of the train must proceed more than a short distance to its terminal after separating from the passenger portion of the train, the use of a suitable cab or control car will be advisable, located at the west end of the TOFC part of the train.

INTERCITY OPERATIONS

Assuming that schedules can be held per the operating times shown in Appendix B0 for initial operations between Liverpool and Edison, with delay times as shown for railroad crossings in Ohio and Indiana, crew change points can be set up so that each crew works an eight hour day and layovers are minimized. With only the minimal improvements made so that the proposed trains can run over the entire route as proposed, with as many crews as possible operating four hours in one direction and then returning on a four-hour trip in the other direction, crew change points would be near:

Upper Sandusky, Ohio
Beaver, Pennsylvania
Vineyard, Pennsylvania, a short distance east of Mt Union

Operating only a few overnight trains under initial conditions, there is little opportunity for 8- to 12-hour round trips for the crews. Most crews would be required to operate eight-hour shifts, lay over for minimum eight hours of rest, and then return to their point of origin on an eight-hour shift. If daylight trains were added, passenger-only or otherwise, there would be

more opportunity for out-and-back crew operations. Trains will require at least two labor shifts to make a one-way trip until considerable improvements are made, and so some means of crew change en route will be necessary. It is unlikely that trains under initial conditions will hold rigidly to any schedule, and so picking a precise crew change point will be difficult (see Figure 3-3-1). Productivity would be low, as is seen in Figure 2-5-1, because of the inability to keep the locomotives and cars in constant operation. Only one trip a day would be made by trains. Terminal operations will also likely be less efficient than with expanded, higher speed service.

On whatever existing intercity route might be available, pilots would be needed initially on the trains for each segment operated within a particular jurisdiction not under the control of the proposed system. This would add substantially to operating labor costs. Eventually, if operation on the tracks of other railroads continued for some time, employees of the proposed high-speed service could become qualified so that the pilots would no longer be necessary. A simple solution would be to place dormitory cars on the trains and have an off-duty crew ride in these cars and take over duties when the first crew's time has expired. This arrangement has been proposed in the past for freight service*1 and is commonly used for certain on-train passenger-service personnel. The crews, on duty and off, can transfer between trains at the meeting points so that no overnight out-of-town stays are necessary. Having a meeting point for an eastbound and a westbound train will mean having a range of possible meets, each of which will have to have a side track or second main track available for the trains to pass. This will mean delays unless a suitable arrangement can be made with the railroad over which the trains are operating when they meet. It is very unlikely that a great deal of business will be obtained via the initial manner of operation, but if the service is priced low enough, there will be a few customers. It will give company personnel an opportunity to become familiar with an admittedly rough version of the proposed service, and it will allow marketing staff to gauge the reception that the service as envisioned will receive. It will demonstrate to potential investors, other financial entities, and potential customers that the proposed system is in fact a bona fide transportation provider.

SCHEDULE INITIAL OPERATIONS

ALL TIMI

MILES	WESTBOUND READ DOWN TRAIN NUMBER 1	3	5	7	STATION	EASTBOUND READ UP TRAIN NUMBER 2	4	6	8
-3.38					PENN STATION NEW YORK E				
-3.12					PENN STATION NEW YORK W				
0.00					EAST MEADOWS	6:00	7:00	8:00	9:00
2.63	16:00	17:00	18:00	19:00	WEST MEADOWS	5:33	6:33	7:33	8:33
6.69	16:05	17:05	18:05	19:05	NEWARK	5:28	6:28	7:28	8:28
21.54	16:17	17:17	18:17	19:17	METROPARK	5:17	6:17	7:17	8:17
29.97	16:24	17:24	18:24	19:24	NEW BRUNSWICK	5:12	6:12	7:12	8:12
45.04	16:31	17:31	18:31	19:31	PRINCETON JUNCTION	5:04	6:04	7:04	8:04
53.10	16:36	17:36	18:36	19:36	TRENTON	4:57	5:57	6:57	7:57
75.57	16:54	17:54	18:54	19:54	YORK ROAD	4:39	5:39	6:39	7:39
101.43	17:03	18:03	19:03	20:03	GLEN LOCH	4:17	5:17	6:17	7:17
121.14	17:26	18:26	19:26	20:26	PARKESBURG	4:03	5:03	6:03	7:03
143.52	17:45	18:45	19:45	20:45	LANCASTER	3:44	4:44	5:44	6:44
180.53	18:10	19:10	20:10	21:10	HARRISBURG STATE TOWER	3:15	4:15	5:15	6:15
195.89	18:31	19:31	20:31	21:31	DUNCANNON	2:56	3:56	4:56	5:56
241.96	19:19	20:19	21:19	22:19	LEWISTOWN	2:08	3:08	4:08	5:08
277.60	19:55	20:55	21:55	22:55	HUNTINGDON	1:33	2:33	3:33	4:33
311.01	20:37	21:37	22:37	23:37	ALTOONA	0:50	1:50	2:50	3:50
349.14	21:31	22:31	23:31	0:31	JOHNSTOWN	23:54	0:54	1:54	2:54
363.66	21:51	22:51	23:51	0:51	CONPITT JUNCTION	23:37	0:37	1:37	2:37
372.09	22:01	23:01	0:01	1:01	TORRANCE	23:25	0:25	1:25	2:25
385.77	22:15	23:15	0:15	1:15	LATROBE	23:13	0:13	1:13	2:13
394.36	22:21	23:21	0:21	1:21	GREENSBURG	23:08	0:08	1:08	2:08
412.02	22:37	23:37	0:37	1:37	TURTLE CREEK (WILMERDING)	22:49	23:49	0:49	1:49
425.20	22:57	23:57	0:57	1:57	PITTSBURGH STATION SQUARE	22:28	23:28	0:28	1:28
445.41	23:16	0:16	1:16	2:16	ALIQUIPPA	22:14	23:14	0:14	1:14
449.79	23:19	0:19	1:19	2:19	MONACA	22:07	23:07	0:07	1:07
460.81	23:34	0:34	1:34	2:34	MIDLAND	21:54	22:54	23:54	0:54
498.05	0:38	1:38	2:38	3:38	KENSINGTON	21:02	22:02	23:02	0:02
534.62	1:27	2:27	3:27	4:27	CANTON	20:18	21:18	22:18	23:18
543.95	1:38	2:38	3:38	4:38	MASSILLON	20:06	21:06	22:06	23:06
568.34	2:01	3:01	4:01	5:01	WOOSTER LIBERTY AVENUE	19:44	20:44	21:44	22:44
608.83	2:37	3:37	4:37	5:37	MANSFIELD	19:09	20:09	21:09	22:09
634.14	3:08	4:08	5:08	6:08	BUCYRUS	18:36	19:36	20:36	21:36
693.97	3:48	4:48	5:48	6:48	LIMA CSX CROSSING	17:58	18:58	19:58	20:58
752.88	4:29	5:29	6:29	7:29	FORT WAYNE - NS CROSSING	17:20	18:20	19:20	20:20
860.67	5:36	6:36	7:36	8:36	VALPARAISO - CN CROSSING	16:14	17:14	18:14	19:14
871.06	5:46	6:46	7:46	8:46	LIVERPOOL EAST	16:00	17:00	18:00	19:00
873.67	6:00	7:00	8:00	9:00	LIVERPOOL WEST				
					CHICAGO UNION STATION E				
					CHICAGO UNION STATION W				

Figure 3-3-1. Schedule of initial operations

CHAPTER 4

Established System Operation

After business has built up sufficiently, separate freight and passenger trains will be operated. The line shall have been improved so that trains operating between Chicago and New York can be competitive with air travel, as indicated in Volume 2. Trains will have aerodynamically designed cab units facing outward at each end. Additional locomotive power units without cabs, with streamlined sides but flush ends, will be added as necessary. The passenger trains will operate between Chicago Union Station and New York Penn Station.

PERMANENT ROUTE

The temporary initial operation on Norfolk Southern (NS) between Beaver and Wilmerding will be terminated, and operation on CSX from Beaver to Homestead will commence. This, too, is a very busy single-track railroad. A lease will be negotiated for a portion of the right-of-way over the full distance to be used, and new track for the sole use of the proposed service will be installed. Until signals are provided, speed will be limited to 59 mph. If necessary, CSX can be offered limited trackage rights for some of their trains to operate on the new track, subject to control by the proposed service and operation of trains by crews furnished by the proposed service. After high-speed operation commences, axle loading will be restricted.

NS will be used from Homestead to Wilmerding via Port Perry, connecting to the NS line to Harrisburg via Greensburg and Latrobe. This segment carries heavy local and through traffic. Operation here will be at slow speed. The existing tracks will be used at a negotiated rate to provide acceptable revenue to NS while guaranteeing on-schedule performance of the proposed service. This segment will eventually be replaced with a new line to be constructed when economically feasible (see Volume 2).

NORFOLK SOUTHERN, WILMERDING–HARRISBURG

A lease will be negotiated for part of the right-of-way over the length of this line and new track installed for the exclusive use of the proposed service. Amtrak and possibly certain NS trains could be allowed trackage rights on this new line.

At some locations where existing trackage is to be used, the traffic of the owning railroad may be heavy enough, and of certain composition, as to not allow the desired superelevation to be installed to allow the proposed trains to operate at the desired speeds. All these undesirable conditions will be eliminated by further development of the proposed system. Locations of unreliability are addressed first. These include railroad grade crossings and heavily used track owned and operated upon by others.

Once the system has established control of its complete permanent route, serious high-speed operation can commence, with dedicated, specially designed high-speed equipment, as described in Volume 2, including enclosed, aerodynamically designed cars carrying the highway trailers. The proposed line will not be electrified initially or in the future unless or until the need for motive power is greatly increased because of higher speeds and larger volumes of traffic, and electrification becomes economical. There are limitations to just how fast the trains can travel. Without signals of some sort, the trains cannot exceed 59 mph. Without cab signals, automatic train stop, automatic train control, or positive train control (PTC) (which will eventually be mandatory in all cases), the trains cannot exceed 79 mph. The distance on the proposed system, from Liverpool, Indiana to the Jersey Meadows, is now slightly over 871 miles. At a constant 79 mph, the trip time between Liverpool and Meadows would be about 11 hours. As seen in Appendix B2, the trip time for a trailer-on-flatcar (TOFC) train between these points, with improved but not modified right-of-way, traveling at 120 mph maximum speed, is about 10.7 hours. The distance from Penn Station New York to Chicago Union Station over the proposed route is approximately 905 miles. At 79 mph, this trip would take about 11½ hours. Operating on the improved but not modified right-of-way, as detailed in Appendix B3, with optimum signaling and curve superelevations, and with rectification of highway but not railroad grade crossings, the trip time for a high-speed passenger-only train between Chicago Union Station and Penn Station in New York City would be less than 7½ hours, with trains traveling at a maximum speed of 200 mph (see Figure 3-4-1). Note that productivity of equipment and labor will not be as high as shown in Figure 2-13-1, unless traffic can be found for trains operating continuously around the clock and track capacity allows passenger and freight trains to operate at the same time.

SCHEDULE OF OPERATIONS ALL TIMES EASTERN STANDARD EXCEPT AS NOTED

INTERMEDIATE PHASE - PASSENGER AND FREIGHT 24 HOUR CLOCK ALL TIMES TO NEAREST MINUTE

IDENTIFICATION OF STATION DOES NOT INDICATE STOP AT THAT STATION

ALL TRAINS NON-STOP BETWEEN CHICAGO/LIVERPOOL AND MEADOWS/PENN STATION

	TRAIN NUMBER							TRAIN NUMBER					
	WESTBOUNC READ DOWN						STATION				EASTBOUND READ UP		
	1	3	5	101	103	105		2	4	6	102	104	106
	PASSENGER			FREIGHT				PASSENGER			FREIGHT		
MILES	FIRST CLASS			SECOND CLASS			PENN STATION NEW YORK	FIRST CLASS			SECOND CLASS		
-0.26				-	-	-	EAST END PLATFORM - ARRIVE	12:15	13:15	14:15	-	-	-
0.00	7:00	8:00	9:00	-	-	-	DEPART - WEST END PLATFORM	12:13	13:13	14:13	-	-	-
3.38	7:03	8:03	9:03	-	-	-	EAST MEADOWS - ARRIVE	12:10	13:10	14:10	4:30	5:30	6:30
6.01	7:05	8:05	9:05	19:00	20:00	21:00	DEPART - WEST MEADOWS	12:08	13:08	14:08	4:03	5:03	6:03
10.07	7:10	8:10	9:10	19:05	20:05	21:05	NEWARK	12:07	13:07	14:07	4:01	5:01	6:01
24.92	7:12	8:12	9:12	19:12	20:12	21:12	METROPARK	12:02	13:02	14:02	3:54	4:54	5:54
33.35	7:15	8:15	9:15	19:16	20:16	21:16	NEW BRUNSWICK	11:59	12:59	13:59	3:48	4:48	5:48
48.42	7:20	8:20	9:20	19:24	20:29	21:29	PRINCETON JUNCTION	11:54	12:54	13:54	3:40	4:40	5:40
58.10	7:23	8:23	9:23	19:29	20:32	21:32	TRENTON	11:52	12:52	13:52	3:36	4:36	5:36
78.95	7:30	8:30	9:30	19:41	20:41	21:41	YORK ROAD	11:44	12:44	13:44	3:24	4:24	5:24
104.81	7:39	8:39	9:39	19:53	20:53	21:53	GLEN LOCH	11:35	12:35	13:35	3:11	4:11	5:11
124.23	7:45	8:45	9:45	20:04	21:04	22:04	PARKESBURG	11:30	12:30	13:30	3:00	4:00	5:00
146.12	7:54	8:54	9:54	20:16	21:16	22:16	LANCASTER	11:21	12:21	13:21	2:47	3:47	4:47
183.13	8:07	9:07	10:07	20:36	21:36	22:36	HARRISBURG STATE TOWER	11:07	12:07	13:07	2:27	3:27	4:27
198.50	8:15	9:15	10:15	20:49	21:49	22:49	DUNCANNON	10:58	11:58	12:58	2:13	3:13	4:13
228.65	8:31	9:31	10:31	21:10	22:10	23:10	PORT ROYAL	10:43	11:43	12:43	1:51	2:51	3:51
244.10	8:39	9:39	10:39	21:24	22:24	23:24	LEWISTOWN	10:34	11:34	12:34	1:38	2:38	3:38
279.72	8:58	9:58	10:58	21:50	22:50	23:50	HUNTINGDON	10:16	11:16	12:16	1:14	2:14	3:14
305.13	9:15	10:15	11:15	22:16	23:16	0:16	FOSTORIA	9:58	10:58	11:58	0:48	1:48	2:48
313.14	9:18	10:18	11:18	22:22	23:22	0:22	ALTOONA	9:56	10:56	11:56	0:44	1:44	2:44
346.03	9:33	10:33	11:33	22:45	23:45	0:45	CONEMAUGH	9:41	10:41	11:41	0:19	1:19	2:19
346.88	9:33	10:33	11:33	22:47	23:47	0:47	JOHNSTOWN	9:40	10:40	11:40	0:18	1:18	2:18
359.00	9:38	10:38	11:38	22:55	23:55	0:55	CONPITT JUNCTION	9:35	10:35	11:35	0:11	1:11	2:11
367.44	9:44	10:44	11:44	23:03	0:03	1:03	TORRANCE	9:29	10:29	11:29	0:02	1:02	2:02
381.12	9:51	10:51	11:51	23:14	0:14	1:14	LATROBE	9:23	10:23	11:23	23:52	0:52	1:52
389.71	9:54	10:54	11:54	23:18	0:18	1:18	GREENSBURG	9:20	10:20	11:20	23:48	0:48	1:48
407.37	10:02	11:02	12:02	23:28	0:28	1:28	TURTLE CREEK (WILMERDING)	9:13	10:13	11:13	23:36	0:36	1:36
420.55	10:11	11:11	12:11	23:48	0:48	1:48	PITTSBURGH STATION SQUARE	9:03	10:03	11:03	23:21	0:21	1:21
439.81	10:20	11:20	12:20	23:58	0:58	1:58	ALIQUIPPA	8:54	9:54	10:54	23:08	0:08	1:08
444.93	10:22	11:22	12:22	0:02	1:02	2:02	MONACA	8:52	9:52	10:52	23:05	0:05	1:05
458.26	10:27	11:27	12:27	0:07	1:07	2:07	MIDLAND	8:48	9:48	10:48	23:00	0:00	1:00
494.50	10:53	11:53	12:53	0:45	1:45	2:45	KENSINGTON	8:24	9:24	10:24	22:21	23:21	0:21
519.36	11:01	12:01	13:01	0:58	1:58	2:58	CANTON	8:15	9:15	10:15	22:08	23:08	0:08
528.01	11:06	12:06	13:06	1:04	2:04	3:04	MASSILLON	8:09	9:09	10:09	22:01	23:01	0:01
553.08	11:18	12:18	13:18	1:22	2:22	3:22	WOOSTER LIBERTY AVENUE	7:57	8:57	9:57	21:43	22:43	23:43
593.57	11:54	12:54	13:54	1:50	2:50	3:50	MANSFIELD - CSX CROSSING	7:37	8:37	9:37	21:15	22:15	23:15
618.87	12:03	13:03	14:03	2:03	3:03	4:03	BUCYRUS - NS CROSSING	7:29	8:29	9:29	21:01	22:01	23:01
635.83	12:07	13:07	14:07	2:11	3:11	4:11	UPPER SANDUSKY	7:23	8:23	9:23	20:53	21:53	22:53
678.71	12:20	13:20	14:20	2:33	3:33	4:33	LIMA - CSX -CROSSING	7:11	8:11	9:11	20:32	21:32	22:32
737.61	12:39	13:39	14:39	3:02	4:02	5:02	FORT WAYNE - NS CROSSING	6:53	7:53	8:53	20:01	21:01	22:01
845.41	13:12	14:12	15:12	3:56	4:56	5:56	VALPARAISO - CN CROSSING	6:20	7:20	8:20	19:07	20:07	21:07
855.79	13:15	14:15	15:15	4:02	5:02	6:02	LIVERPOOL EAST - EST - DEPART	6:17	7:17	8:17	19:00	20:00	21:00
							LIVERPOOL EAST - CST	-	-	-	18:00	19:00	20:00
857.95	13:16	14:16	15:16	4:30	5:30	6:30	ARRIVE - LIVERPOOL WEST - EST	6:16	7:16	8:16	-	-	-
	12:16	13:16	14:16	3:30	4:30	5:30	LIVERPOOL WEST - CST	-	-	-	-	-	-
886.21	13:31	14:31	15:31	-	-	-	CHICAGO UNION STATION E - DE	6:00	7:00	8:00	-	-	-
	12:31	13:31	14:31				CHICAGO UNION STATION E - DE	5:00	6:00	7:00	-	-	-
886.46	13:35	14:35	15:35	-	-	-	ARRIVE - CHICAGO UNION STATI	-	-	-	-	-	-
	12:35	13:35	14:35	-	-	-	ARRIVE - CHICAGO UNION STATI	-	-	-	-	-	-

Figure 3-4-1. Schedule of operations on existing right-of-way with high-speed equipment

PERMANENT TERMINAL OPERATIONS

With dedicated high-speed equipment operating in and out of permanent terminals, as described in Volume 2, all TOFC trains will be unloaded immediately upon arrival at their terminals. Trailers will be available to consignees immediately upon unloading. Special expedited service will be available so that certain designated trailers can be unloaded first and available for quick pickup. At the originating terminal, loading slots can be reserved for late-arriving trailers to be loaded on trains just before departure. These special services will warrant extra charges. All terminals will be open 24 hours per day, seven days per week, for drop off and pickup of trailers.

EMPTY TRAILERS

The return of empty trailers can present problems. Presumably the number of empty trailers transported can be minimized by aggressive solicitation of backhauls. Empties can be carried on regular TOFC trains as space permits, but this can delay train departures. Special empty-trailer-only TOFC trains can be run at times of low demand such as mid-day, operated on a low-priority basis to avoid delaying more remunerative traffic.

ESTABLISHED PERMANENT OPERATIONS

Once permanent terminals have been constructed and the desired permanent rail line has been placed in service, early-stage full high-speed TOFC operations can begin. Enclosed trailer-carrying cars, as described in Volume 2, will be used. The Jersey Meadows TOFC terminal will be short on space, and so the most efficient layout possible will be made there, and the Liverpool terminal will be designed to accommodate the trains to and from the east. An ideal location for the Meadows terminal will be on the east bank of the Hackensack River, to the south of the Amtrak main line. If mixed trains are still operated at this stage, an eastbound train will be able to drop its passenger cars at Secaucus Transfer for forwarding to Penn Station New York by an electric locomotive.

The Liverpool TOFC terminal will have a similar configuration to that at Meadows, with the same orientation of loading and unloading tracks. From Liverpool, the passenger cars of a westbound mixed train can

be forwarded to Chicago Union Station. The passenger parts of eastbound trains can be dispatched to Liverpool from Chicago Union Station. Trains that combine freight and passenger equipment would have the passenger sections separating from the freight sections and continuing on to and back from the freight terminal in push-pull mode, then reconnecting to a full train before departure.

Trains that are freight-only or passenger-only can have locomotive units provided at each end of the trains in distributed mode, with streamlining at the end locomotive units, so that no switching of locomotive units is required and good aerodynamic design is achieved at both ends of each train.

Trains will operate on a very strict schedule. Operations at terminals will also be strictly controlled for maximum speed and efficiency. Teamsters will be encouraged to bring their trailers into the departing terminal somewhat ahead of the cutoff time, but trailers will be accepted up until just before it is too late to load before the train departs. The final trailer can be placed on the final TOFC car just a few minutes before departure. A reserve fee can be charged for the final trailer or the final few trailers, to assure space on the train in spite of arriving just before departure. This will provide as much time as possible for the trailers to travel from the origin point to the terminal in spite of late departures from points of origin. Discounts can be offered to encourage earlier arrival and late departure of trailers. With arriving TOFC trains, rush loads can be identified and the cut or cuts of cars that contain them can have priority in unloading, at a premium fee.

Passenger trains and TOFC trains will not be operating at the same times over the same parts of the railroad so that there will be no conflicts with track operations that might force slower trains to be passed by faster ones. Passenger trains will be largely limited to daylight operations, and TOFC trains will be overnight.

The operating times for trains, as shown in Appendix B, are extremely tight. What would normally be considered excess motive power is applied so that trains can accelerate rapidly out of slow sections of line, and high amounts of braking capacity are provided to allow trains to operate at full speed as far as possible toward a slow section and then decelerate quickly to the required reduced speed through the restricted area. This will require very frequent application and release of brakes so that substantial amounts of compressed air will be used. Some locations of slow allowable speed and locations of unrestricted speed are very short; the train engineer will be expected to rapidly adjust the controls to make the trip in the minimum time. These operational conditions will become less severe as speed-restricting locations are eliminated or improved over time, as shown in Volume 2.

On the road, once overall trip times are improved to 12 hours or less, a train crew will be able to travel half way to the other terminal and return without running out of time. When practical, trains will meet at some point near midway between the terminals and exchange crews so that on-train employees can go on and off duty at the same place. A full double-track section of line will be in place at midpoint between terminals and will be extended as far as necessary so that each eastbound train can meet its westbound counterpart there.

Essentially nonstop express passenger trains and "local" passenger trains making intermediate stops will not make stops at the freight terminals, but will make one suburban stop outside the metropolitan areas departing and arriving. Possible stations for these stops will be Whiting, Indiana and Iselin (Metropark), New Jersey. The "local" trains will stop at such intermediate points as Valparaiso and Fort Wayne, Indiana, Lima, Mansfield and Canton, Ohio, Pittsburgh, Altoona, Harrisburg and King of Prussia (Philadelphia) Pennsylvania, and Trenton and New Brunswick in New Jersey. Other stops at other locations can be made if demand warrants. Passenger trains, both express and local, can make an additional stop on the "far" side of the metropolitan areas. For instance, a train can originate at Elgin, Illinois, make a major stop at Chicago Union Station, stop again at Hammond–Whiting in Indiana, operate express nonstop to Iselin, New Jersey, make its major stop at Penn Station in New York City, and then make a final stop at New Rochelle, New York. These suburban stations, for eastbound and westbound operations, can be selected for ease of automobile access, availability of parking, or convenience of train servicing, as well as potential ridership.

One difficulty with high-speed trains, particularly operating on a hybrid or blended system, is maintaining observance of the track ahead. Blocked grade crossings on segments operating at or less than 125 mph, and other locations with curvature or hills, can provide a situation wherein the train driver is unable to stop a train before striking an obstruction on the tracks, or running into track damage such as a "sun kink." Drone technology might be applied to resolve this issue, by having a second engineer or "copilot" operating a drone ahead of the train and observing video transmitted from a camera on the drone. The drone could perhaps be automated to follow the track without direct human guidance, with a visual screen observable by the train engineer (train driver).

With complete reconstruction of the existing right-of-way and several bypass segments built to allow 200 mph passenger train operation over the majority of the route, the system can be operated to provide faster service than air

flights, when terminal and other air service delays are considered. As discussed in Volume 2, the line relocation work would include, in addition to the work indicated in Appendix B, the Pavonia Cutoff and the Sam Rea Line, revised from its original conception (see Figure 3-4-2). Productivity and efficiency will not be as high as shown in Figure 1-10-2, unless trains can be operated 24 hours per day and passengers can be found to fill them.

Trains scheduled for trip times in excess of the maximum allowable service hours will stop at midway, in the neighborhood of Pittsburgh to exchange crews between opposing trains. This should occur only for passenger trains during the earlier stages of operation, before the system has been extensively rebuilt. Exchange of crews will be standard procedure for TOFC trains at mid-route between Meadows and Liverpool.

Operating crews will consist of two persons qualified to act as train drivers or engineers, both on duty within the control cab of the locomotive leading the train. Passenger cars will have attendants capable of performing trainman's duties as well as providing services to the passengers. Communication with the operating personnel in the locomotive cab will be available in case the need arises. One technician will be aboard the train, competent to resolve any minor issue with the locomotives or the cars. Diesel locomotives will carry sufficient fuel to make a complete one-way journey without refueling. Standard locomotives in the United States today have this capacity, but maximum fuel consumption must be accurately determined so that fuel tanks can be provided with ample capacity.

Operating rules will be similar to the consolidated rules currently followed by most Class 1 railroads, but with simplifications commensurate with the segregation of operations of the proposed system from the North American railroad network.

The Federal Railroad Administration (FRA) is the administrative agency of the U.S. government that is responsible for administering the railroad safety rules legislated by the U.S. Congress. The FRA publishes standards for various aspects of railroad construction and operation that are direct factors in the cost of high-speed rail construction and operation. Automatic signal systems are required on railroads over which freight trains operate in excess of 49 mph and passenger trains operate in excess of 59 mph. Cab signals, automatic train stop, automatic train control, or PTC are required where trains operate in excess of 79 mph. After December, 2018, PTC will be required on all passenger-carrying railroads, including high-speed ones. There is a provision in the FRA regulations excluding "A railroad that operates only on track inside an installation that is not part of the general railroad system of transportation."*1

SCHEDULE OF OPERATIONS — ALL TIMES EASTERN STANDARD EXCEPT AS NOTED

HIGH SPEED PASSENGER TRAINS — 24 HOUR CLOCK — ALL TIMES TO THE NEAREST MINUTE

IDENTIFICATION OF STATION DOES NOT INDICATE STOP AT THAT STATION

ALL TRAINS NON-STOP BETWEEN CHICAGO/LIVERPOOL AND MEADOWS/PENN STATION

WESTBOUND READ DOWN — EASTBOUND READ UP

MILES	1	3	5	7	9	11	STATION*	2	4	6	8	10	12
	FIRST CLASS						PENN STATION NEW YORK	FIRST CLASS					
-0.26				-	-	-	EAST END PLATFORM - ARRIVE	9:59	10:59	11:59	12:59	13:59	14:59
0.00	5:00	6:00	7:00	8:00	9:00	10:00	DEPART - WEST END PLATFORM	9:59	10:59	11:59	12:59	13:59	14:59
3.38	5:02	6:02	7:02	8:02	9:02	10:02	EAST MEADOWS - ARRIVE	9:56	10:56	11:56	12:56	13:56	14:56
6.01	5:03	6:03	7:03	8:03	9:03	10:03	DEPART - WEST MEADOWS	9:55	10:55	11:55	12:55	13:55	14:55
10.07	5:05	6:05	7:05	8:05	9:05	10:05	NEWARK	9:54	10:54	11:54	12:54	13:54	14:54
24.92	5:10	6:10	7:10	8:10	9:10	10:10	METROPARK	9:49	10:49	11:49	12:49	13:49	14:49
33.35	5:13	6:13	7:13	8:13	9:13	10:13	NEW BRUNSWICK	9:46	10:46	11:46	12:46	13:46	14:46
48.42	5:18	6:18	7:18	8:18	9:18	10:18	PRINCETON JUNCTION	9:41	10:41	11:41	12:41	13:41	14:41
58.10	5:21	6:21	7:21	8:21	9:21	10:21	TRENTON	9:38	10:38	11:38	12:38	13:38	14:38
78.95	5:27	6:27	7:27	8:27	9:27	10:27	YORK ROAD	9:31	10:31	11:31	12:31	13:31	14:31
104.81	5:36	6:36	7:36	8:36	9:36	10:36	GLEN LOCH	9:23	10:23	11:23	12:23	13:23	14:23
124.23	5:42	6:42	7:42	8:42	9:42	10:42	PARKESBURG	9:17	10:17	11:17	12:17	13:17	14:17
146.12	5:50	6:50	7:50	8:50	9:50	10:50	LANCASTER	9:09	10:09	11:09	12:09	13:09	14:09
183.13	6:03	7:03	8:03	9:03	10:03	11:03	HARRISBURG STATE TOWER	8:56	9:56	10:56	11:56	12:56	13:56
198.45	6:09	7:09	8:09	9:09	10:09	11:09	(DUNCANNON)	8:50	9:50	10:50	11:50	12:50	13:50
222.88	6:18	7:18	8:18	9:18	10:18	11:18	PORT ROYAL	8:40	9:40	10:40	11:40	12:40	13:40
237.46	6:23	7:23	8:23	9:23	10:23	11:23	LEWISTOWN	8:36	9:36	10:36	11:36	12:36	13:36
259.71	6:30	7:30	8:30	9:30	10:30	11:30	PETERSBURG	8:29	9:29	10:29	11:29	12:29	13:29
275.38	6:34	7:34	8:34	9:34	10:34	11:34	BELLWOOD	8:24	9:24	10:24	11:24	12:24	13:24
285.51	6:37	7:37	8:37	9:37	10:37	11:37	DEAN	8:21	9:21	10:21	11:21	12:21	13:21
298.80	6:41	7:41	8:41	9:41	10:41	11:41	ST BENEDICT	8:17	9:17	10:17	11:17	12:17	13:17
305.38	6:43	7:43	8:43	9:43	10:43	11:43	ALVERDA	8:16	9:16	10:16	11:16	12:16	13:16
311.95	6:45	7:45	8:45	9:45	10:45	11:45	CHERRY HILL MANOR	8:14	9:14	10:14	11:14	12:14	13:14
318.54	6:47	7:47	8:47	9:47	10:47	11:47	U.S. 119	8:12	9:12	10:12	11:12	12:12	13:12
325.11	6:49	7:49	8:49	9:49	10:49	11:49	THOMAS BRIDGE	8:10	9:10	10:10	11:10	12:10	13:10
331.68	6:51	7:51	8:51	9:51	10:51	11:51	IDAHO	8:08	9:08	10:08	11:08	12:08	13:08
338.26	6:53	7:53	8:53	9:53	10:53	11:53	COCHRANS MILLS	8:06	9:06	10:06	11:06	12:06	13:06
344.83	6:55	7:55	8:55	9:55	10:55	11:55	LOOKABOUGH CORNERS	8:04	9:04	10:04	11:04	12:04	13:04
351.40	6:57	7:57	8:57	9:57	10:57	11:57	EKASTOWN	8:02	9:02	10:02	11:02	12:02	13:02
357.97	6:59	7:59	8:59	9:59	10:59	11:59	CHERRY VALLEY	8:00	9:00	10:00	11:00	12:00	13:00
371.12	7:03	8:03	9:03	10:03	11:03	12:03	CRANBERRY	7:57	8:57	9:57	10:57	11:57	12:57
380.38	7:06	8:06	9:06	10:06	11:06	12:06	MONACA	7:54	8:54	9:54	10:54	11:54	12:54
421.23	7:18	8:18	9:18	10:18	11:18	12:18	KENSINGTON	7:42	8:42	9:42	10:42	11:42	12:42
444.69	7:25	8:25	9:25	10:25	11:25	12:25	CANTON	7:35	8:35	9:35	10:35	11:35	12:35
456.01	7:28	8:28	9:28	10:28	11:28	12:28	MASSILLON	7:31	8:31	9:31	10:31	11:31	12:31
480.60	7:35	8:35	9:35	10:35	11:35	12:35	NEW PITTSBURG	7:26	8:26	9:26	10:26	11:26	12:26
487.17	7:37	8:37	9:37	10:37	11:37	12:37	ENGLAND	7:25	8:25	9:25	10:25	11:25	12:25
493.71	7:39	8:39	9:39	10:39	11:39	12:39	MILTON	7:23	8:23	9:23	10:23	11:23	12:23
500.26	7:40	8:40	9:40	10:40	11:40	12:40	AMOY	7:21	8:21	9:21	10:21	11:21	12:21
506.79	7:42	8:42	9:42	10:42	11:42	12:42	TOLEDO JUNCTION	7:19	8:19	9:19	10:19	11:19	12:19
513.35	7:44	8:44	9:44	10:44	11:44	12:44	VERNON	7:18	8:18	9:18	10:18	11:18	12:18
519.90	7:45	8:45	9:45	10:45	11:45	12:45	LIBERTY	7:16	8:16	9:16	10:16	11:16	12:16
526.45	7:47	8:47	9:47	10:47	11:47	12:47	BUCYRUS	7:14	8:14	9:14	10:14	11:14	12:14
533.00	7:49	8:49	9:49	10:49	11:49	12:49	NEVADA	7:13	8:13	9:13	10:13	11:13	12:13
539.54	7:50	8:50	9:50	10:50	11:50	12:50	U.S. 23	7:10	8:10	9:10	10:10	11:10	12:10
541.74	7:51	8:51	9:51	10:51	11:51	12:51	UPPER SANDSUSKY	7:09	8:09	9:09	10:09	11:09	12:09
570.24	7:58	8:58	9:58	10:58	11:58	12:58	ADA	7:02	8:02	9:02	10:02	11:02	12:02
584.61	8:02	9:02	10:02	11:02	12:02	13:02	LIMA	6:58	7:58	8:58	9:58	10:58	11:58
643.52	8:18	9:18	10:18	11:18	12:18	13:18	FORT WAYNE	6:41	7:41	8:41	9:41	10:41	11:41
748.10	8:45	9:45	10:45	11:45	12:45	13:45	VALPARAISO	6:14	7:14	8:14	9:14	10:14	11:14
761.70	8:49	9:49	10:49	11:49	12:49	13:49	LIVERPOOL EAST - EST - DEPART	6:10	7:10	8:10	9:10	10:10	11:10
	5:00	6:00	7:00	8:00	9:00	10:00	LIVERPOOL EAST - CST	6:00	7:00	8:00	9:00	10:00	11:00
763.86	8:50	9:50	10:50	11:50	12:50	13:50	ARRIVE - LIVERPOOL WEST - EST	6:09	7:09	8:09	9:09	10:09	11:09
	5:00	6:00	7:00	8:00	9:00	10:00	LIVERPOOL WEST - CST	5:09	6:09	7:09	8:09	9:09	10:09
792.12	8:59	9:59	10:59	11:59	12:59	13:59	CHICAGO UNION STATION E - DEPART - EST	6:00	7:00	8:00	9:00	10:00	11:00
	7:59	8:59	9:59	10:59	11:59	12:59	CHICAGO UNION STATION E - DEPART - CST	5:00	6:00	7:00	8:00	9:00	10:00
792.37	8:59	9:59	10:59	11:59	12:59	13:59	ARRIVE - CHICAGO UNION STATION W - EST	-	-	-	-	-	-
	7:59	8:59	9:59	10:59	11:59	12:59	ARRIVE - CHICAGO UNION STATION W - CST	-	-	-	-	-	-

Figure 3-4-2. Schedule of high-speed passenger-only operations

CHAPTER 5

High-Speed Rail Scheduling

Optimum passenger transit time between New York and Chicago would be four hours. Present nonstop air schedules advertise two-hour trips. Adding airport entry before the flight and departure after the flight, flight delays, and security checks, compared with what would be experienced when taking the train, would make the air travel time the equivalent of at least three hours; four hours would be more the norm.

For a passenger train to travel the 896.44 miles (1439.8 km) by rail on the proposed route as it now exists (see Appendix B1) between Union Station in Chicago and Penn Station in New York City in four hours, the average train speed would have to be 225 miles per hour. This speed, 360 Km/hr, is a more or less everyday speed for a TGV-type train, but much faster than that needed for the optimum high-speed freight train. Operation of many passenger trains at speeds in excess of 200 mph while carrying large volumes of trailer-on-flatcar (TOFC) traffic at considerably slower speeds would require separate tracks or coordinated schedules for freight and passenger trains. High-speed trailer on flatcar trains of suitable configuration can be operated on the same tracks as 200 mph (320 km/hr) passenger trains up to a certain level of traffic. Some carload freight could be carried on these trains, but double stack container trains, coal trains, and trains carrying bulk grain and other such commodities should not be operated on high-speed tracks regardless of the speed at which they operate. After reconstruction for high-speed passenger trains is complete, the distance on the line between New York and Chicago would be 800 miles or less (see Volume 2). This could be traversed in four hours at a 200 mph average speed. At an intermediate stage of reconstruction of the proposed line, a typical schedule for a high-speed, 5½-hour passenger train at a maximum speed of 200 mph (320 km/hr) and an 8½-hour freight train at a maximum speed of 120 mph (193 km/hr), with trains running on two tracks, might be something like that shown in Figure 3-5-1.

SCHEDULE OF OPERATIONS — INTERMEDIATE PHASE - PASSENGER AND FREIGHT

ALL TIMES EASTERN STANDARD EXCEPT AS NOTED — 24 HOUR CLOCK — ALL TIMES TO THE NEAREST MINUTE

IDENTIFICATION OF STATION DOES NOT INDICATE STOP AT THAT STATION — ALL TRAINS NON-STOP BETWEEN CHICAGO/LIVERPOOL AND MEADOWS/PENN STATION

MILES	TRAIN NUMBER WESTBOUND 1	3	5	301	303	101	103	READ DOWN 105	STATION	TRAIN NUMBER EASTBOUND 2	4	6			102	104	READ UP 106
	PASSENGER FIRST CLASS			FREIGHT THIRD CLASS		FREIGHT SECOND CLASS				PASSENGER FIRST CLASS			FREIGHT THIRD CLASS		FREIGHT SECOND CLASS		
									PENN STATION NEW YORK								
-0.26									EAST END PLATFORM - ARRIVE	12:15	13:15	14:15					
0.00	7:00	8.00	9:00						DEPART - WEST END PLATFORM	12:13	13:13	14:13					
3.38	7:03	8:03	9:03						EAST MEADOWS - ARRIVE	12:10	13:10	14:10	20:30	21:30	4:30	5:30	6:30
6.01	7:05	8:05	9.05	11:00	12:00	19:00	20:00	21:00	DEPART - WEST MEADOWS	12:08	13:08	14:08	20:03	21:03	4:03	5.03	6.03
10.07	7:10	8:10	9:10	11:05	12:05	19:05	20:05	21:05	NEWARK	12:07	13:07	14:07	20:01	21:01	4:01	5:01	6:01
24.92	7:12	8.12	9:12	11:12	12:12	19:12	20:12	21.12	METROPARK	12:02	13:02	14:02	19:54	20:54	3:54	4:54	5:54
33.35	7:15	8.15	9:15	11:16	12:16	19:16	20.16	21:16	NEW BRUNSWICK	11:59	12:59	13.59	19:48	20:48	3:48	4:48	5:48
48.42	7:20	8:20	9:20	11:24	12:29	19:24	20:29	21:29	PRINCETON JUNCTION	11:54	12:54	13:54	19:40	20:40	3.40	4:40	5:40
58.10	7:23	8:23	9.23	11.29	12:32	19:29	20:32	21:32	TRENTON	11:52	12:52	13:52	19:36	20:36	3:36	4.36	5:36
104.81	7:39	8:39	9:39	11:53	12:53	19:53	20.53	21:53	GLEN LOCH	11:35	12:35	13:35	19.11	20:11	3:11	4:11	5:11
146.12	7:54	8.54	9:54	12:16	13:16	20:16	21:16	22:16	LANCASTER	11:21	12:21	13:21	18:47	19:47	2:47	3:47	4:47
183.13	8:07	9:07	10:07	12:36	13:36	20:36	21:36	22:36	HARRISBURG STATE TOWER	11:07	12:07	13:07	18:27	19:27	2:27	3:27	4:27
244.10	8:39	9:39	10:39	13:24	14:24	21:24	22:24	23:24	LEWISTOWN	10:34	11:34	12.34	17:38	18:38	1:38	2:38	3:38
279.72	8:58	9:58	10:58	13:50	14:50	21:50	22:50	23:50	HUNTINGDON	10:16	11:16	12:16	17:14	18:14	1:14	2:14	3.14
313.14	9:18	10:18	11:18	14:22	15:22	22:22	23:22	0:22	ALTOONA	9:56	10:56	11:56	16:44	17:44	0:44	1:44	2:44
346.88	9:33	10:33	11:33	14:47	15:47	22:47	23:47	0.47	JOHNSTOWN	9.40	10:40	11:40	16:18	17:18	0:18	1:18	2:18
381.12	9.51	10:51	11:51	15:14	16:14	23:14	0:14	1:14	LATROBE	9:23	10:23	11:23	15:52	16:52	23:52	0:52	1:52
420.55	10:11	11:11	12:11	15:48	16:48	23:48	0:48	1:48	PITTSBURGH STATION SQUARE	9:03	10:03	11:03	15:21	16:21	23.21	0.21	1:21
444.93	10:22	11:22	12.22	16.02	17:02	0:02	1:02	2:02	MONACA	8:52	9:52	10:52	15:05	16:05	23:05	0:05	1:05
519.36	11:01	12:01	13:01	16:58	17:58	0:58	1:58	2:58	CANTON	8:15	9:15	10:15	14:08	15.08	22:08	23:08	0:08
528.01	11:06	12:06	13:06	17:04	18.04	1.04	2.04	3:04	MASSILLON	8.09	9:09	10:09	14:01	15:01	22:01	23:01	0:01
553.08	11:18	12:18	13:18	17:22	18:22	1:22	2:22	3:22	WOOSTER LIBERTY AVENUE	7:57	8:57	9.57	13:43	14:43	21:43	22.43	23:43
593.57	11:54	12:54	13:54	17:50	18:50	1:50	2:50	3:50	MANSFIELD - CSX CROSSING	7:37	8:37	9:37	13:15	14:15	21:15	22:15	23:15
618.87	12:03	13:03	14:03	18.03	19:03	2:03	3:03	4:03	BUCYRUS - NS CROSSING	7:29	8:29	9:29	13:01	14:01	21:01	22:01	23:01
635.83	12:07	13:07	14:07	18:11	19:11	2:11	3:11	4:11	UPPER SANDUSKY	7:23	8:23	9:23	12:53	13:53	20:53	21:53	22:53
676.71	12:20	13:20	14:20	18:33	19:33	2:33	3:33	4:33	LIMA - CSX - CROSSING	7:11	8.11	9:11	12:32	13.32	20:32	21:32	22:32
737.61	12:39	13:39	14:39	19:02	20:02	3.02	4.02	5:02	FORT WAYNE - NS CROSSING	6:53	7:53	8:53	12:01	13:01	20:01	21.01	22:01
845.41	13:12	14:12	15:12	19:56	20:56	3:56	4:56	5:56	VALPARAISO - CN CROSSING	6:20	7:20	8:20	11:07	12:07	19:07	20:07	21:07
855.79	13:15	14:15	15:15	20:02	21:02	4:02	5:02	6:02	LIVERPOOL EAST - EST - DEPART	6:17	7:17	8.17	11:00	12:00	19:00	20:00	21:00
									LIVERPOOL EAST - CST				10:00	11:00	18:00	19:00	20:00
857.95	13:16	14:16	15:16	20:30	21:30	4:30	5:30	6:30	ARRIVE - LIVERPOOL WEST - EST	6:16	7.16	8:16					
	12:16	13:16	14:16	19:30	20:30	3:30	4:30	5:30	LIVERPOOL WEST - CST								
886.21	13:31	14:31	15:31						CHICAGO UNION STATION E - DEPART - EST	6:00	7:00	8:00					
	12:31	13:31	14:31						CHICAGO UNION STATION E - DEPART - CST	5:00	6:00	7:00					
886.46	13.35	14.35	15.35						ARRIVE - CHICAGO UNION STATION W - EST								
	12:35	13:35	14:35						ARRIVE - CHICAGO UNION STATION W - CST								

Figure 3-5-1. Schedule of fully developed operations

OPERATION SCHEDULES

To minimize costs, the minimum number of trains will be operated to satisfy customer's needs. High quality of service is the basic modus operandi of the system so that as many trains as necessary to satisfy customers will be operated. There may be many travel times that will be satisfactory to passengers so that different schedules will be run. Those departure and arrival times that do not generate a reasonable amount of business will be dropped, while popular schedules will be served by trains as long as are necessary to accommodate the public, but rather than have passenger trains become too long for efficient operation, second, third, or fourth *sections*, as many as are necessary, can be operated. Each section will have the suitable number of locomotive power units to carry it across the system on the scheduled time. Each succeeding section will follow its leader on the closest possible time within the limitations of the signal, control, and positive train control systems employed. At certain times of year, heavier passenger traffic than normal can be anticipated; every effort will be made to ensure that the maximum number of cars and locomotive units are available for service and that, preferably, no units are undergoing repair at that time. Other sources of suitable equipment will be sought out to satisfy the need if a sufficient number of units is not available within the system.

It is presumed that most TOFC customers will be satisfied with overnight service so that, theoretically, once train terminal-to-terminal times are brought down to satisfactory numbers, a single train of whatever length is necessary to handle the number of trailers being carried could be operated. It would most likely be necessary to break the TOFC train up into multiple sections for operating convenience. There may also be a sufficient number of customers who desire the minimum trip time for their shipment, so that holding a train to fill it out becomes undesirable. As this is likely to be the case, fewer, shorter, scheduled trains will be operated so that delay time at the originating terminal is minimized. TOFC trains carrying only empty trailers could be operated on daytime schedule slots not needed for passenger trains.

Passenger revenue can be increased by running faster passenger-only trains for the length of the line, perhaps with stops at major intermediate points or with express and local service. Daytime is more likely to be the preferred time for operating passenger trains, whereas the preferred time for operating freight trains is at night, providing overnight service between terminals. Dispatching 24 overnight freight-only TOFC trains each way, at 15-minute intervals between 6:15 p.m. and midnight, leaves 18 hours of the day for dispatching additional passenger or other trains. This should be adequate, at least initially, to carry the available traffic.

With the fully developed system, to avoid the faster passenger trains having to pass slower TOFC freight trains on the single- or double-tracked railroad, the high-speed passenger train should not depart an end terminal before about 3 hours after the last TOFC freight train has departed.

Boarding of passenger trains could be allowed after 8 p.m. for people reserving sleeping compartments for overnight travel on any trains departing in the early morning hours. At the other end of the day, passenger trains could depart end terminals as late as 6 p.m. and arrive at the other end of the line by 10 p.m., providing about the same convenience in scheduling as flying.

Other fast passenger trains could operate at other times during the day. Shorter runs could be made similar to the shorter TOFC freight runs suggested. An intermediate station at Lima, Ohio, about 70 miles (110 km) south of Toledo and 70 miles (110 km) north of Dayton on I-75 could serve those areas easily for TOFC traffic, but might be too far away to pick up a large share of the passenger traffic. Similarly, a station near Mansfield on I-71 could serve Cleveland and Columbus and a station at Canton on I-77 could serve Cleveland and Akron. Intermodal freight service to Baltimore and Washington could be provided from a station at Harrisburg, Pennsylvania, via I-83, but might not prove feasible for passenger traffic. Most promising intermediate passenger stops would be Fort

Wayne, Canton, Pittsburgh, and Philadelphia. All these would be added without increasing the investment in main line trackage or right-of-way; only additional terminals and trainsets would be required.

Long-distance commuting could be established with trains operating, for example, from central Pennsylvania to New York City on commuter schedules. Trains could originate at Altoona, Pennsylvania, stopping at Huntingdon, Lewistown, and Millersburg and operating nonstop from there to Pennsylvania Station in New York City. Trains could operate to arrive in NYC at approximately 7:30, 8:00, and 8:30 in the morning, departing from Penn Station for Altoona at 4:45, 5:15, and 5:45 p.m. (see Figure 3-5-2). Productivity of equipment and labor would be low, as it will be unlikely to keep the trains in profitable operation around the clock. More likely, the trains will make one round trip per day and layover during off-peak hours, and crews will have to be paid a full day's wages for about 4 hours' work.

SCHEDULE OF OPERATIONS ALL TIMES EASTERN STANDARD
HIGH SPEED COMMUTER PASSENGER TRAINS STANDARD CLOCK
IDENTIFICATION OF STATION DOES NOT INDICATE STOP AT THAT STATION
ALL TRAINS NON-STOP BETWEEN MILLERSTOWN AND PENN STATION NEW YORK

	WESTBOUND	READ DOWN					EASTBOUND				READ UP
	TRAIN NUMBER					STATION	TRAIN NUMBER	EASTBOUND			READ UP
MILES	101	103	105	109	111		102	106	108	112	116
	FIRST CLASS					PENN STATION NEW YORK	FIRST CLASS				
-0.26				-	-	EAST END PLATFORM - ARRIVE	7:25 AM	8:10 AM	8:25 AM	8:55 AM	9:25 AM
0.00	4:45 PM	5:00 PM	5:15 PM	5:45 PM	6:00 PM	DEPART - WEST END PLATFORM	7:25 AM	8:10 AM	8:25 AM	8:55 AM	9:25 AM
3.38	4:47 PM	5:02 PM	5:17 PM	5:47 PM	6:02 PM	EAST MEADOWS - ARRIVE	7:22 AM	8:07 AM	8:22 AM	8:52 AM	9:22 AM
6.01	4:48 PM	5:03 PM	5:18 PM	5:48 PM	6:03 PM	DEPART - WEST MEADOWS	7:21 AM	8:06 AM	8:21 AM	8:51 AM	9:21 AM
9.85	4:50 PM	5:05 PM	5:20 PM	5:50 PM	6:05 PM	NEWARK	7:21 AM	8:06 AM	8:21 AM	8:51 AM	9:21 AM
24.80	4:55 PM	5:10 PM	5:25 PM	5:55 PM	6:10 PM	METROPARK	7:16 AM	8:01 AM	8:16 AM	8:46 AM	9:16 AM
32.65	4:58 PM	5:13 PM	5:28 PM	5:58 PM	6:13 PM	NEW BRUNSWICK	7:14 AM	7:59 AM	8:14 AM	8:44 AM	9:14 AM
48.42	5:02 PM	5:17 PM	5:32 PM	6:02 PM	6:17 PM	PRINCETON JUNCTION	7:09 AM	7:54 AM	8:09 AM	8:39 AM	9:09 AM
58.10	5:05 PM	5:20 PM	5:35 PM	6:05 PM	6:20 PM	TRENTON	7:06 AM	7:51 AM	8:06 AM	8:36 AM	9:06 AM
78.95	5:12 PM	5:27 PM	5:42 PM	6:12 PM	6:27 PM	YORK ROAD	6:57 AM	7:42 AM	7:57 AM	8:27 AM	8:57 AM
104.81	5:20 PM	5:35 PM	5:50 PM	6:20 PM	6:35 PM	GLEN LOCH	6:49 AM	7:34 AM	7:49 AM	8:19 AM	8:49 AM
124.23	5:26 PM	5:41 PM	5:56 PM	6:26 PM	6:41 PM	PARKESBURG	6:43 AM	7:28 AM	7:43 AM	8:13 AM	8:43 AM
146.12	5:35 PM	5:50 PM	6:05 PM	6:35 PM	6:50 PM	LANCASTER	6:35 AM	7:20 AM	7:35 AM	8:05 AM	8:35 AM
183.13	5:47 PM	6:02 PM	6:17 PM	6:47 PM	7:02 PM	HARRISBURG STATE TOWER	6:22 AM	7:07 AM	7:22 AM	7:52 AM	8:22 AM
210.74	5:58 PM	6:13 PM	6:28 PM	6:58 PM	7:13 PM	MILLERSTOWN	6:11 AM	6:56 AM	7:11 AM	7:41 AM	8:11 AM
222.88	6:03 PM	6:18 PM	6:33 PM	7:03 PM	7:18 PM	PORT ROYAL	6:06 AM	6:51 AM	7:06 AM	7:36 AM	8:06 AM
237.46	6:09 PM	6:24 PM	6:39 PM	7:09 PM	7:24 PM	LEWISTOWN (STATE COLLEGE)	6:00 AM	6:45 AM	7:00 AM	7:30 AM	8:00 AM
259.71	6:18 PM	6:33 PM	6:48 PM	7:18 PM	7:33 PM	PETERSBURG (HUNTINGDON)	5:50 AM	6:35 AM	6:50 AM	7:20 AM	7:50 AM
275.38	6:23 PM	6:38 PM	6:53 PM	7:23 PM	7:38 PM	BELLWOOD (ALTOONA)	5:45 AM	6:30 AM	6:45 AM	7:15 AM	7:45 AM

Figure 3-5-2. Commuter train schedule, Central Pennsylvania–New York City

TOFC freight trains between New York and Chicago would optimally depart after the day shift at factories is ended and arrive in early morning before the start of the workday.

The most important single aspect of the proposed service is reliability. It is most important that the advertised departure and arrival times be met every time a scheduled train operates. The trains should operate only at the fastest overall trip time that can be reliably maintained day after day. For TOFC traffic, the schedules must include the time to load and unload trailers, and this must also be guaranteed and reliably performed on a regular basis.

SCHEDULING

Timing is critical for the success of a high-speed rail system. Operating at high speed between point of origin and destination will be to no avail if train arrivals and departures are erratic. Consistent adherence to schedules is important for both freight and passenger high-speed operations.

To arrive at a schedule, the quickest time that can reliably be made between origin and destination must be determined. To be high-speed, the trains must be high speed. It is important to be on time at departure and arrival at all stations, particularly at the origin and destination terminals. No train should ever leave a terminal or station before its scheduled time. There are situations when this might be permissible, but they are few and should be clearly announced beforehand, preferably by notation in published schedules. All trains should arrive at destination terminals and intermediate station stops, if any, on or before the published schedule time. There should be absolutely no exceptions to this rule. All staff personnel should be acutely aware of this rule, and any activity short of an emergency that results in failure to adhere to this rule should be grounds for dismissal.

CHAPTER 6

Cost Analysis and Marketing

It is assumed that the proposed service will have the lowest out-of-pocket cost per unit of transportation of any comparable service by any mode. This can be achieved initially because of the simplicity of operations, on a single rail line with only two terminals, one in the metropolitan New York City area and the other in the Chicago area. Management and supervision will be fairly uncomplicated so that costs can be strictly controlled. The highest utilization must be made of equipment and labor. Labor must be "direct" as much as possible. The time of train operating crews can easily be designated to certain product. The time for nonoperating personnel must be noted indicating which trains they were doing work for—billing, loading, unloading, maintenance, and any other clerical or other laboring function that can be traced to services provided.

MARKETING

To determine the optimum schedule of trains, both passenger and freight (trailer-on-flatcar [TOFC]), existing traffic patterns between New York and Chicago should be carefully studied. If capacity warrants, intermediate routes could also be considered, but would also require study. These could include Chicago–Philadelphia, Chicago–Pittsburgh, Pittsburgh–New York, Fort Wayne–Pittsburgh, Fort Wayne–Philadelphia, Fort Wayne–New York, and others. Short TOFC routes and passenger routes could be established as justified or on an experimental basis.

Pricing of services is a critical element of the marketing process. Pricing of freight (trailer) transportation will be an entirely different operation from passenger services. Both will be highly responsive to demand, but passenger sales will be largely automated via a website, whereas freight sales will be handled personally.

Demand pricing will be the rule. The low cost of providing the service will allow extremely low pricing to meet almost any competition but allow prices to rise when demand warrants it. The passenger website will maintain real-time inventory of available seats and evaluate rate of inquiry and response to adjust prices accordingly.

Trailer pricing will be based on weight of shipment. A single trailer will occupy a platform completely regardless of the trailer length or volume, and so the size of the trailer is irrelevant in cost of providing the service. Weight is a factor, however, as it has some effect on the amount of fuel required to propel the train and in establishing the number of locomotive units required to pull it. A 200-car train carrying only trailers is assumed to weigh, at maximum, slightly over 13,000 tons. At this loading, the actual payload inside each trailer would average only about 25 tons so that the net cargo weight of a trailer train would be 5,000 tons, or only about 38% of the total weight. The trailers would each be weighed upon arrival at the entry terminal, primarily to isolate extremely overloaded trailers. A small per-cwt charge would be added to the basic per-trailer fee, and an additional surcharge for special handling, if desired, would be added to arrive at the total price for trailer transport. For instance, at Phase 5 (see Volume 2) a typical trailer charge for carriage between New York (Meadows) and Chicago (Liverpool) would be $2,000, consisting of the following(see Figure 3-6-1):

Base charge	$1,200
Lading(including trailer) 40,000 lbs @ $1.50/cwt	$600
Expedited unloading	$200
Total price	$2,000

Figure 3-6-1. Typical TOFC charge

Pricing and service can go explicitly hand in hand with the proposed TOFC service. All trains will travel as rapidly as possible with the desired reliability between the initial permanent TOFC terminal at Liverpool, Indiana, and the permanent initial metropolitan NYC terminal in the Jersey Meadows. Terminal operations will also be expedited, with free access at all times for delivery and pickup of trailers. Special expedited TOFC terminal service can be offered, for an extra fee. At the originating terminal, loading slots can be reserved for late-arriving trailers so that a rush shipment can arrive at the terminal just prior to train departure and still make the trip. At the destination terminal, certain trailers can be flagged for immediate unloading and placement for departure on the highway. These services can be priced at what the market will bear so that the number of trailers receiving expedited service does not exceed the capacity of the terminal operations to provide that service.

The service will be value priced, charging as much as the traffic will bear consistent with maximization of net revenue. In other words, charging $2,000 per trailer and carrying 10,000 trailers per day at a profit per trailer of $600 for a net profit of $6,000,000 per day is better than charging $6,000 per trailer and carrying 1,000 trailers per day at a profit of $4,600 per trailer for a net profit of $4,600,000 per day. Similar figures apply to passenger transport.

The important thing is to know what the costs of providing service are so that price can be cut as necessary to meet competition, but a loss is avoided. As indicated elsewhere, the proposed system should be the lowest-cost provider of services so that any competitor who undercuts the lowest prices the proposed service can offer will be losing money and cannot sustain such operations indefinitely.

Carriage of empty trailers could be free or priced so that empty hauls are minimized and trains are as full of loaded trailers as possible. Empty trailers could be carried such as space is available, or on daytime TOFC trains carrying only empty trailers.

The possibility of competition cannot be overlooked. In spite of the cost advantage expected to be retained by the proposed system, there may be some trucking companies or airlines willing to lose money to maintain or gain market share. In addition, the expected high net revenue from the proposed system may encourage others to develop a similar system. For instance, a direct New York–Chicago route could be developed of about the same length as the proposed, utilizing parts of various existing or abandoned lines, shown in Appendix A of Volume I, with alternate connections. One such line would not be as convenient as the proposed line for intermediate traffic, but would serve Cleveland and Toledo directly. It would be more expensive to upgrade to 120 mph operation and much more expensive to provide four-hour passenger rail service between New York and Chicago. It is not likely that any of these suggested attempts at competition would be successful, as long as the proposed system had sufficient working capital.

The proposed system will be able to accurately determine cost of producing service because of simple operations. There will be three clearly defined areas of cost, namely, fixed costs, variable costs, and incremental costs, which are discussed as follows:

FIXED COSTS

The cost of providing the railroad will be a fixed number. This will involve ownership costs of the right-of-way and track, and any structures needed. The cost of overall executive management will also be fixed, at least over the short term.

VARIABLE COSTS

There will be a certain cost incurred for each train that is operated. This cost will be a fixed amount, including operating personnel and minimal equipment, and minimum fuel and supplies for a trip. There will be a certain minimum number of personnel at each terminal to provide basic services such as ticket sales, trailer loading and unloading, and clerical activities.

INCREMENTAL COST

Many costs will vary with the amount of traffic. These will include the cost of additional equipment for longer trains and additional staff for terminals to handle additional traffic.

SALES

The freight business could be operated on a wholesale basis. There are many various forwarders, trucking companies, and related entities out there with their pulses on the market that can handle local drayage. Shippers can deliver trailer loads to the TOFC terminal themselves, and receivers can pick the trailers up for final delivery. Third-party providers can take responsibility for the complete movement, using the proposed service as a part of that movement. Railroads have traditionally been very poor marketers, and little success has been had in railroads trying to provide complete door-to-door freight service. On the contrary, cooperative service between railroads and various trucking and package delivery firms has been successful.

To guarantee rapid, reliable transportation from origin to destination, it may be necessary for the proposed system to place the entire movement under its responsibility. To assure success, it may be necessary to have the local drivers become permanent full-time employees of the proposed service.

It is intended for the proposed high-speed TOFC service not to provide road drayage of trailers, but this can be provided, for a fee, if demand warrants it.

Passenger marketing and sales will be a completely separate operation from trailer transportation. There will be a few common elements, such as demand pricing and cost analysis, but efforts on the passenger side will be retail-oriented. A website will allow customers to purchase tickets directly. Ticket prices will vary so that gross income is maximized. Customers will be encouraged to fill potentially empty seats by discounts that can be obtained via various "hoops" that will be developed. Advance purchase will be encouraged by price reductions, and last-second sales will be very deeply discounted. Prices will vary seasonally with demand. But purchase the few days before or on the day of transportation will normally be full price.

CHAPTER 7

MAINTENANCE

TRACK MAINTENANCE

Track must be maintained at the quality level commensurate with acceptable ride comfort for passengers at the speed the passenger trains will operate, and at a quality level to avoid damage to freight being carried at the speeds the trailer-on-flatcar trains will be operating. The passenger train's comfort level and speed will be the deciding factor. While airline passengers readily accept a bumpy ride on frequent occasions, intercity rail passengers expect a smooth ride that will allow them to walk comfortably around the train, eat and drink without fear of spillage, and work at their seats and be able to write legibly and perform other tasks without difficulty. Passenger train travel should always be smooth enough to avoid anyone suggesting that seat belts might be a good idea.

Track maintenance must also include adequate inspection and preventive maintenance procedures so that the likelihood of a derailment or other accident due to a broken rail or other such failure is reduced to infinitesimal proportions. There are many means of nondestructive track inspection. Track rails under load experience severe strain. Rails have been tested for years with equipment of the type developed by Sperry in the 1920s. More recent nondestructive rail testing is performed by ultrasonic, magnetic induction, and eddy current methods.

Time has to be provided for safe access to the tracks for maintenance and inspection. This will require gaps in scheduled train service. This can avoid accidents. There are various systems available to help prevent accidents due to conflicts between maintenance, repair, and operations. These are not 100% reliable. Education and supervision must be of the highest order to minimize unfortunate incidents such as the fatal collision between a revenue train and a track maintenance vehicle that occurred on Amtrak's Northeast Corridor in 2016.

Track maintenance at highway crossings is difficult. Maintaining track alignment at highway crossings when rail speeds are as high as 125 mph is much more difficult. Use of the Railroad Highway Crossing (RHC) described in Volume I* will allow for better and easier maintenance of the track through the crossing by exposing the track, when the RHC is in the raised position, to complete track maintenance in identical fashion to the track outside the crossing (see Figure 3-7-1). There should be little difficulty in obtaining permission from the Federal Railroad Administration (FRA) for use of the RHC at highway crossings for rail speeds not exceeding 110 mph, such as in the early stages of operation between Liverpool and Crestline, when speed limit will be 59 mph due to the absence of signaling. Once acceptability of the RHC is proven under these conditions, test applications of the RHC at crossings with higher rail speeds can be attempted, up to the present FRA maximum of 125 mph for track with at-grade highway crossings. Then, a case can be made for allowing the RHC on FRA Class 8 and Class 9 track, by using the argument that (1) the RHC is not an at-grade crossing and (2) that the RHC effectively prevents automobile–train collisions at crossings so equipped.

EQUIPMENT MAINTENANCE

High-speed train equipment for the proposed service must be maintained at the highest levels to assure reliability of operation. No train delays due to equipment failure can be tolerated. Preventive maintenance procedures must be developed for all rolling stock so that all wearing parts and parts subject to fatigue or other time-related failure are replaced before they fail. Careful inspections must be made of each train before departure from its originating terminal to assure that all critical components are in proper operating condition.

Fueling and maintenance can be performed while the trains are loading and unloading so that quick turnarounds can be made and equipment utilization is maximized.

Arrangements will be made along the route to provide service in case of breakdown of equipment. No location along the proposed line will be more than one hour away from this necessary emergency equipment for immediate use in case the need arises. Every effort will be made, in such cases, to quickly repair the defect and continue with the entire train. No car or locomotive will be set out unless repair is either impossible or will delay the train for an unacceptable length of time. Setout may mean removing the

Figure 3-7-1. Track maintenance at rail highway crossing

offending piece of equipment from the track and setting it on the ground. If it is a loaded freight car, the trailer will be removed and taken to its destination by alternate means as expeditiously as possible. Contract services such as Hulcher* can be used to provide rapid response at any location. Special equipment such as helicopters might be used to advantage for rapid access for personnel and small equipment. It must be remembered that the chief mission of the proposed system is the provision of reliable transportation so that delays must be minimized at any practical cost.

APPENDIX A

BIBLIOGRAPHY

There have been hundreds of books, papers, reports, newspaper and magazine articles, and other documents written and published over the last several years on the subject of high-speed rail systems in the United States. Some of the technical publications have been produced by private contractors and published by government agencies. Some books, documents and papers on the topic are as follows:
Recent Publications

The Economics and Politics of High-Speed Rail: Lessons from Experiences Abroad

By Daniel Albalate and Germa Bel, Lexington Books 2014. A critical view of the high-speed rail experience in Europe and Asia. Compares economics-oriented and social policy –oriented master plans and questions the rationality of large investments in high-speed rail for little practical benefit.

High Speed Rail Systems: Impacts on Mobility, on Tourism and on Mobile Workers

By Francesca Pagliara, Lambert Academic Publishing 2014. Considers effects of availability of high-speed passenger rail systems on work practices.

The Economics of Investment in High-Speed Rail

By OECD, International Transport Forum 2014

High Speed Rail: Background and Issues

Jonathan S. Fischer, editor, Nova Science Publishers 2013. Compendium of statements by and to federal agencies on subject of high-speed passenger rail.

The Development of High Speed Rail in the United States: Issues and Recent Events

By David Randall Peterman, John Frittelli and William J. Mallett, Congressional Research Service 2012. Overview of high-speed passenger rail activities in the United States, including active and potential Federal initiatives.

Fast Trains – America's High Speed Future

By Emmy Louie and Nancy Bolts, FasTrack Communications 2012. Mainly a review of high speed rail in other countries with discussion on application to the U.S.

High-Speed Rail: International lessons for U.S. Policy Makers

By Petra Todorovich, Daniel Schned and Robert Lane. Lincoln Institute of Land Policy 2011. Discusses high-speed rail activities around the world and their relevance to the United States.

High Speed Trains

By Peter Clark Rosenberg Publishing Pty Ltd 2011. Compilation of data on various high-speed rail lines around the world.

Transportation Research Board

The Transportation Research Board provides funding for studies and publications on various transportation issues. For instance, "An Analysis

of Proposed U.S. House of Representatives Actions and Their Impact on Public Transportation" (September 2011), including high-speed rail passenger systems, such as produced by the American Public Transit Association; "Hearings on the Future of Intercity Rail" (Sept 2011) report on a U.S. Senate subcommittee's hearings. More technical titles include "Rail Track Maintenance Planning An Assessment Model" "Asphalt Trackbed Technology Development The First 20 Years".

High-Speed Passenger Trains

Federal Railroad Administration 2010. This is an Ebook. It is a digest of federal publications on high-speed rail.

Illinois High-Speed Rail Four-Quadrant Gate Reliability Assessment

U.S. Department of Transportation 2009. A partially theoretical evaluation of the effectiveness of 4-quadrant gates at highway-rail at-grade crossings on a moderately high-speed passenger rail corridor.

High Speed Rail (HSR) in the United States

David Randall Peterman, John Freittelli and William J. Mallett, Congressional Research service 2009 preliminary version of *The Development of High Speed Rail in the United States: Issues and Recent Events.*

Transport Revolutions

By Richard Gilbert and Anthony Perl, New Society Publishers 2007. This book stresses the replacement of internal combustion engine transport with electric motor driven transportation, which is most easily accomplished by means of railroads. It largely discusses other modes of transport.

High-Speed Rail Projects in the United States: Identifying the Elements of Success Part 2

By the Mineta Transportation Institute, 2006. Discusses high speed rail developments in Chicago, Pennsylvania, and the Northeast Corridor (Amtrak). Looks at ways that appear to have possibility of success in furthering high speed rail passenger transportation.

High-Speed Rail Projects in the United States: Identifying the Elements of Success

By the Mineta Transportation Institute, 2005. Describes the status of and activity in various potential high speed rail corridors in the United States.

Impact of Blocked Highway Rail Grade Crossings On Emergency Response Services

U.S. Department of Transportation Federal Railroad Administration 2006. Provides suggestions for alleviating difficulties of slow freight trains passing through communities.

End of the Line: The Failure of Amtrak Reform and the Future of America's Passenger Trains

By Joseph Vranich, AEI Press 2004. A rather negative critique of Amtrak, but neither the first nor the last.

Modern Trains and Splendid Stations: Architecture, Design, and Rail Travel for the Twenty-First Century

By the Art Institute of Chicago, Martha Thorne editor, Merrell Publishers 2001. Mainly a discussion of train stations but mentions high-speed trains in Europe, Japan and the United States.

New Departures: Rethinking Rail Passenger Policy in the Twenty-First Century

By Anthony Perl, University Press of Kentucky 2001. Looks at the need for high-speed rail and North American public policy changes, and discusses the obstacles to these changes.
Other Books, Papers and Publications

Supertrains

By Joseph Vranich, St Martin's Press, New York, 1991. Discusses worldwide developments in high-speed passenger trains, including magnetic levitation. Acts as cheerleader for development of high speed rail systems in the United States. This book does not get into technical details on high-speed trains, nor does it provide details on a proposed system.

Europe's High Speed Trains: A Study in Geo-Economics

Mitchell P. Strohl, 1992. Discusses high speed passenger rail systems in Europe and elsewhere. Provides some engineering information, but is primarily a guide for Americans as to what has been done elsewhere. Also includes an extensive bibliography.

High Speed Ground Transportation Systems I/ Proceedings of the First International Conference on High Speed Ground Transportation (HSGT) Systems

Edited by Murthy V.A. Bondata and Roger L. Wayson, American Society of Civil Engineers (ASCE), New York 1992. Proceedings of First International Conference on High Speed Ground Transportation (HSGT) Systems, held in Orlando, Florida in October 1992. Papers discuss many attempts at building high speed passenger rail systems around the United States; also various experiences in high speed rail in other countries; provides suggestions for technical guidelines for high speed rail systems; other papers present other high speed rail topics. This is a collection of independent papers and not a comprehensive, organized text on the subject.

Financing a High Speed Rail Project with Freight Revenues

T.L. Koglin, International Conference on High Speed Ground Transportation (HSGT) Systems, Orlando, Florida, 1992. Proposed high speed passenger/freight rail system (not published in proceedings for above) (see appendix D).

High Speed Rail Project for New York City,

T.L. Koglin, International Association for Bridge and Structural Engineering, Congress on Urban Infrastructure, Lucerne, Switzerland, October 2000. Describes freight line, potentially high speed, to carry goods into New York City from the west (see Appendix C).

Northeast Corridor Improvement Project – Redirection Study

(1979) By United States Department of Transportation. Looks at results of NECIP as of the Implementation Master Plan of 1977 compared to directives of Congress in the 4R act and concludes that certain mandated goals, such as 2 hour 40 minute trip time between New York and Washington, with intermediate stops in Baltimore, Wilmington, Philadelphia, Trenton and Newark, cannot be met with existing constraints of time and money, and provides suggested modifications.

Standard and Important References; not all pertain directly or exclusively to high speed rail systems

FRA Standards

The Federal Railroad Administration (FRA) administers federal law on high-speed trains and railroads in general. It provides many documents that guide constructors and operators of rail systems, including high-speed rail. Specifically, their track and signaling handbooks are essential. The FRA is constantly doing research and evaluation of new technologies, and updates their standards in CFR Title 49 accordingly.

AREMA Manual

The American Railway Engineering and Maintenance-of-Way Association *Manual for Railway Engineering* contains sections on specific requirements for high-speed rail systems as well as a chapter on high-speed rail. The Manual is published in several volumes and provides complete comprehensive information and guidelines on all aspects of engineering of freight and passenger rail systems, including right of way, track, and structures, as well as operational characteristics. The chapter on high-speed rail systems provides specific details of engineering requirements for high-speed trains, as well as references to other relevant chapters. The Manual is revised annually and is in the process of expanding its high-speed rail coverage.

Hay, William W. *Railroad Engineering*, John Wiley & Sons (New York, 2nd Edition, 1982). Standard, definitive text on railroad engineering. Includes chapters on all elements of engineering applicable to railroad design. Discusses high-speed rail systems. Provides data for calculating power requirements for railroad trains, a factor of weight and speed, including factors for different degrees of aerodynamic soundness. Unfortunately somewhat dated as many advances in technology have been made in the last 30 years, as well as in research and development on high-speed train operation.

Seventh Report on the High Speed Ground Transportation Act of 1965 and the Railroad Technology Program, U.S. Department of Transportation, Federal Railroad Administration, 1973. Discussion of various federally funded programs on railroad technology, including high-speed passenger trains, such as the Northeast Corridor Metroliners and Turbotrains, and mentions foreign advances such as the French TGV and Japanese Bullet trains. Discusses cost-benefits of electrified operations.

References

Armstrong, John, *All About Signals*, Kalmbach Publishing, Milwaukee, 1957. Detailed discussion of railroad signals and operation, vis-à-vis 1950's technology.

Ballon, Hilary, *New York's Pennsylvania Stations*, Norton, New York, 2002. History of construction of Penn Station New York and discussion of present plans for reconstruction of same.

Barriger, John W., *The Pittsburgh & Lake Erie Railroad*, address to New York Society of Security Analysts, December 1958. Detailed discussion of condition of Pittsburgh & Lake Erie Railroad, a key component suitable for use as a high speed railroad to the west from Pittsburgh, Pennsylvania.

Benson, Lee, *Merchants, Farmers, & Railroads*, Harvard University Press, 1955. Background on hostility of the public to railroads, particularly the monopolistic tendencies of trunk line railroads in the latter half of the nineteenth century, and development of railroad regulation. Considers return on invested capital and watered stock; Public vs private discussions. While concerned primarily with transportation of freight, monopolies and public vs private discussions may be applicable to twenty-first century high-speed passenger rail systems.

Bezilla, Michael, *Electric Traction on the Pennsylvania Railroad 1895-1968*. Pennsylvania State University Press, University Park, Pennsylvania, 1980. Detailed discussion of electric railroad locomotive technology as applied on the Pennsylvania Railroad.

Burgess, George H. and Kennedy, Miles C, *Centennial History of The Pennsylvania Railroad Company*, PRR Philadelphia 1949. Comprehensive history of the Pennsylvania Railroad and railroads it absorbed. Written under contract by engineers employed by Coverdale & Colpitts, a consulting firm noted for its investigations of various railroads, particularly in reference to determining their financial viability. The book covers construction and financing in particular detail. Much of the information is taken directly from an earlier version also published by the railroad. Publication of this book was financed by the Pennsylvania Railroad and it appears that no attempt was made to be completely unbiased, but many valuable facts are disclosed, as it were, for which historians should be grateful, although most of the material in the book could be found in original form if one were to take the time and effort to investigate and delve through the various repositories of the records of the Pennsylvania Railroad, its predecessors and its corporate successors.

Consolidated Rail Corporation (Conrail), Engineering Department, *Maintenance Program and Track Chart*(s), various dates and volumes. Detailed descriptions of the railroad fixed plant, showing track alignment, curves and grades; distances between points, junctions and crossings, at grade, overhead, and undergrade, with highways, Conrail branches, and other railroads, relevant structures, and bridges and culverts over waterways of various descriptions. Also shows rail

weights, ages, in-place treatments (grinding) and whether welded or bolted connections; ballast type and age, tie replacement history, allowable speed per all relevant factors including condition of track, tonnage carried, signals and detection devices, and presence of buried cables. Each division in each region has a book, updated as required. Conrail was notorious for its management initiatives, so these books have had a history of reissuance due to modification and renaming of divisions and regions, as well as removal of various pieces of trackage due to abandonment or sale. These are private documents issued by the railroad for internal use only. They were intended to be up-to-date documents describing the physical plant, and fulfill that purpose, with minor errors and occasional reference to obsolete names of streets, other organizations, and other secondary or informational items. They usually do not show certain details, such as track configuration at junctions, or speed limits through those junctions, nor are crossovers and such always indicated where multiple tracks exist.

Coyle, J.B., et al, *Transportation*, West Publishing, St Paul/Minneapolis, 4th ed 1994. Chapters on transportation and the economy, demand for transportation, regulation, transportation policy, as applied to railroads and other modes, and other transportation subjects.

Cunningham, John T, *Railroads of New Jersey the Formative Years*, Afton Publishing, Andover, NJ 1997. History of New Jersey railroads, including those that formed part of what were original lines of what is now the Northeast Corridor.

Derleth, August, *The Milwaukee Road*, Creative Age Press, New York, 1948. History of the Chicago, Milwaukee, St Paul and Pacific Railroad, including its high-speed passenger operations of the 1930's.

Diehl, Lorraine B, *The Late, Great Pennsylvania Station*, Stephen Greene Press, Lexington MA, 1987 (originally published by American Heritage Press, 1985). Inspiration, design, construction and destruction of Penn Station New York, with plans and profiles of original construction from Transactions, ASCE, Vol LXVIII, No. 1150 by Charles Raymond, chairman of the board of engineers in charge of the project.

Jonnes, Jill, *Conquering Gotham*, Viking Press, London, 2007. Popular press history of conceptualization and construction of Penn Station New York.

Klein, Maury *Unfinished Business: The Railroad in American Life* University Press of New England, 1994. Mr. Klein is the author of many more or less scholarly books on railroads, particularly railroad history. This book is a compilation of articles and presentations on several

railroad subjects. His chapter on high-speed trains and their lack of implementation in the United States touches on several relevant topics.

Klohn, Charles H, Editor, *Joint Usage of Utility and Transportation Corridors*, American Society of Civil Engineers, New York, 1981

Kurtz, C.M. *Track and Turnout Engineering*, Simmons-Boardman Publishing Corporation, New York, 1945 (and later editions). A standard handbook for detailing railroad track, with detailed discussion on curves: easements, superelevation, vertical curves; and turnouts, crossings, yard layouts, and extensive data tables.

Lyon, Peter, *To Hell in a Day Coach*, J. B. Lippincott, Philadelphia, 1968. Popular press, unsympathetic exposition of the public's attitude about railroads pre-Amtrak, pre-Conrail & pre-deregulation.

Meeks, Carroll L.V. *The Railroad Station An Architectural History*, Yale University Press, New Haven, 1956. More interested in the appearance of a station than its functionality, but considers utility.

Middleton, William D. *Manhattan Gateway New York's Pennsylvania Station*, Kalmbach Publishing, Waukesha, Wisconsin 1996. History of New York Penn Station and its approaches and ancillary facilities, including Hell Gate Bridge.

Middleton, William D. *When the Steam Railroads Electrified, 2nd edition*, Indiana University Press, 1974, 2001. Description in some detail of the conception, design, development, financing, construction, operation and, where it occurred, demise of main line railroad electrification projects in the United States and Latin America

National Railroad Passenger Corporation (Amtrak) *Track Charts* See Consolidated Rail Corporation

New Jersey Driver's Manual, State of New Jersey Department of Law and Public Safety, Division of Motor Vehicles, 1989 edition. Official manual for motorists, includes one paragraph (typical of most states) on railroad grade crossings.

Patton, Spiro G, *Charles Ellet, Jr and the Canal vs. Railroad Controversy*, Proceedings of the Canal history and Technology Symposium, 1983. Discussion of the railroad usurpage of canal transport, comparable to the later replacement of rails by air and highway transport, and what could occur as high-speed trains take over from airways and highways.

Penn Central Corporation (and other names) *Track Charts* See Consolidated Rail Corporation

Pittsburgh & Lake Erie Railroad, Office of Chief Engineer, *Track Chart*(s) See Consolidated Rail Corporation. Similar to CR charts but simpler in presentation as is the case with most smaller railroads.

Railway Age, Simmons-Boardman Publishing, July 1, 1957. Articles on railroad vs airlines passenger travel advertising budgets and attitude of railroad managements on passenger travel advertising.

Roberts, Charles, et al *Triumph,* Barnard Roberts 1998-2010. Several volumes, titled Triumph I to Triumph IX, semi-scholarly history of the Pennsylvania Railroad, including details on the United States' Northeast Corridor rail line from before Pennsylvania Railroad through Amtrak. Also historical coverage of Pennsylvania Railroad line from Philadelphia to Pittsburgh that forms part of the proposed high-speed line of Part 2 of this book. Each volume contains images of much archival material showing construction reconstruction, proposed modifications to the rail line, and other information.

Solomon, Brian, *Bullet Trains* MBI 2001. Pictures, descriptions, definitions and histories of high-speed steel-wheel-on-steel-rail systems worldwide.

Solomon, Brian *Railroad Signaling*, MBI 2003. A history of signal systems on railroads, but covers in some detail modern signal systems up to but not including Positive Train Control.

U.S. Department of Transportation, Federal Highway Administration, *Manual on Uniform Traffic Control Devices*, U.S. Government Printing Office, Washington D.C. updated periodically. Part VIII covers railroad-highway grade crossings

United States Department of Transportation, Federal Railroad Administration

Several volumes of studies have been prepared under the aegis of the FRA. These include

Fire Safety of Passenger Trains (December 1993)

Recommended Emergency Preparedness Guidelines for Passenger Trains (December 1993)

Safety of High Speed Ground Transportation Systems

An Overview of Biological Effects and Mechanisms Relevant to EMF Exposures from Mass Transit and Electric Rail Systems (August 1993)

Broadband Magnetic Fields: Their Possible Role in EMF-Associated Bioeffects (August 1993)

Collision Avoidance and Accident Survivability 4 volumes (March 1993)

Comparison of Magnetic and Electric Fields of Conventional and Advanced Electrified Transportation Systems (August 1993)

EMF Exposure Environments Summary
Report (August 1993)
Potential Health Effects of Low Frequency Electromagnetic Fields Due to Maglev and Other Electric Rail Systems (August 1993)
Railroad Safety Statistics 2004 Annual Report
Review of Existing EMF Guidelines, Standards and Regulations (August 1993)
The Biological Effects of Maglev Magnetic Field Exposures (August 1993)
Work Breakdown Structure (Nov 1993)
Safety of Vital Control and Communication Systems in Guided Ground Transportation ((May 1993)

Westing, Fred, *Penn Station its Tunnels and Side Rodders*, Superior Publishing, Seattle, 1978. Largely a reprint of a report on the original construction of Penn Station New York and its connections, edited by William Couper, published by Isaac H Blanchard company, New York, 1912. Also provides information on development of electric locomotives that were to be used in propelling trains into and out of Pennsylvania Station, New York City.

In addition to the above publications, many other books have appeared from time to time on the subject of high-speed passenger rail or magnetic levitation systems. Also, many hundreds of pamphlets, papers, articles and other documents have been produced on high-speed passenger rail, in print and online. There are too many for them all to be listed here. A few are mentioned in the footnotes. A great deal is accessible on-line, including many printed papers and articles. These can be easily searched.

On –Line railroad news publications have, almost daily, items on high-speed rail developments around the world. Some of these have been cited in the footnotes.

APPENDIX B

Operating Spreadsheets, freight Meadows to Liverpool, passenger New York Penn Station to Chicago Union Station, various states of reconstruction.

These tables of distance, speed and time are simplified to reduce their size. Some simplifications of calculations have also been made, but without producing results that deviate substantially from what would be experienced by actual operations under the conditions given. Decelerations of special high-speed equipment are assumed to take place at 5 mph/second; conventional equipment is assumed to decelerate at the rate of 2 mph/second. Trains are required to reduce fully to the restricted speed before entering any restriction. The effect of gradient on accelerations and decelerations is ignored. This could result in substantial errors under some conditions, such as when ascending or descending steep inclines on long tangents. In the situations analyzed, the net result of ignoring grades is a very small change in overall trip times; they would be a few seconds longer than shown. This is primarily due to the large amount of restrictive curvature in mountainous or hilly territory, reducing the maximum allowable speeds where steep gradients exist so that more locomotive horsepower is available for acceleration, as well as the high horsepower per ton ratios and large amounts of braking power available on the trains modeled in the calculations. Reserve horsepower is available to maintain allowable speed when ascending most steep grades, and to accelerate rapidly when necessary, as the speed at these locations is much reduced below the maximum for the system due to curvature or other restrictions. The tight curvature encountered at most steep downhill locations on the existing alignment also provides additional deceleration capability in these cases due to curve resistance. On the improved alignments, while the curvature is reduced, the gradients are as well so that the effect of gradient is minimal. There are fewer locations of acceleration and deceleration with the improved alignments, reducing any cumulative error substantially. Rotational inertia of wheels and axles and any other inertial components of trains other than the total mass of the moving train are ignored. It is assumed that this simplification results in much less than a 5 percent reduction in acceleration rates.

Accelerations are calculated to nearest integral mph in 1 second increments for what might be considered underpowered (freight) trains and in ¼ second increments for what might be considered overpowered (passenger) trains. Exact accelerations could be computed within the tables, but

the large amount of computation involved would make the spreadsheets extremely complicated. Limited time and space for this project prevented pursuing this further, especially considering the small improvement in accuracy that would be achieved. Transition curvature into and out of curves and other transitions into or out of speed restrictions are ignored. Abrupt accelerations or decelerations would be undesirable for freight and passengers, in the interest of simplicity as well as to produce minimum overall trip times by maximized acceleration and deceleration rates are assumed. The transition segments entering and leaving curves can be used to reduce the abruptness of accelerations and decelerations without sacrificing time or violating the curve speed limits. For compound curves, only the most restrictive part of a curve is considered in determining allowable speed through the curve. In the tables, the entire distance of the curve is traversed at the most restrictive speed. Accelerations over very short distances, followed immediately or nearly immediately by speed restrictions, are generally not included in the tables as the resulting reduction in overall trip times would be nearly nil and these extremely short bursts of acceleration, followed by very quick application of brakes, would be almost impossible for a train driver to accomplish considering the response time of the equipment. The response time factor would also prevent instantaneous changes from acceleration at the rates indicated, or deceleration, to a constant speed. This is also ignored. Superelevation of curves is 3 inches for unmodified track carrying conventional equipment and 6-1/2 inches for track carrying high-speed equipment unless otherwise specified in the tables. All allowable curve speeds are calculated per FRA specifications. At some locations the calculated maximum curve speed for a significant distance is high enough to increase the allowable speed sufficiently to allow an upgrade of the segment of track to a higher class. The higher track classification is implemented unless stated otherwise.

The type of equipment in use is critical in determining allowable speed on curves. For a curve on which conventional equipment is allowed 71 mph, a high-speed TOFC train is allowed 95 mph and a high-speed passenger train is allowed 117 mph, with superelevation as appropriate for each type of equipment. This is per FRA allowances for special equipment under special conditions, particularly with the height of center of gravity.

Each table lists highway grade crossings, at their approximate locations. All known road crossings are listed. "Private" indicates a farm or other crossing not open to the public. These crossings are eliminated or otherwise specially protected when grade separation occurs such as at major crossings of other railroads, and where improvements are made to

allow trains to travel at speeds in excess of 125 mph.

Table B0 is based on existing conditions using existing trackage except some improved track lining and leveling on sections to be used in permanent route of high-speed trains in the proposed system. Maximum speeds are per currently allowable by owning railroads, providing a 16-hour trip.

Table B1 is calculated for a maximum speed of 120 miles per hour (193 km/hr) using conventional TOFC equipment. Table B1 is for unreconstructed track except for reconfigured junctions and lining and leveling track, with signaling upgraded where necessary to allow maximum speed for conventional equipment, with superelevation per existing. With appropriate track rehabilitation and appropriate signals and other appurtenances, but no realignments or other improvements; 14 hour trip.

Tables B2, B4 and B6 have been prepared for special high-speed TOFC (Trailer On Flat Car) equipment ("flatcar" is a misnomer as the "flatcars" are fully enclosed with roofs, sides and ends) as described in Volume 2 with a maximum height of center of gravity above top of rail of 97 inches, allowed higher speed per FRA as discussed elsewhere; Table B2 is calculated over line rebuilt from Table B1 right of way, with FRA Class 7 track where needed; maximum speed remains 120 mph (193 km/hr) with high speed equipment.

All reconstruction indicated in Volume 2 is reflected in these tables at the proper phases of the work. Track rebuilt at curves for 120 mph maximum speed for high-speed TOFC equipment can sustain high speed tilting passenger equipment speeds of 147 mph, primarily due to the lower center of gravity. The same track, with the same curvature, will only allow a maximum speed of 89 mph with conventional equipment, due to much more conservative FRA standards, which consider, in part, variable condition and design of equipment, variable heights of centers of gravity, and variable braking capacity.

Wherever little or no extra cost is involved in the reconstruction to make the curves acceptable for 200 mph instead of 147 mph for the special passenger equipment, the tables reflect that change. All highway grade crossings are removed at any location where the track alignment is altered, by closure of the crossing or grade separation. Highway grade crossing locations shown in the tables are approximate.

Even with simplifications in operation per above, it may be difficult to operate the trains with the frequent precise accelerations and decelerations shown in the table, particularly for the 200 mph passenger trains.

B2 Unreconfigured, but with adjustment of superelevations and transitions at curves, appropriate track rehabilitation and appropriate signals and other appurtenances so that maximum speed with high-speed intermodal equipment of 120 MPH, for 11 hour trips, can be attained.

B4 Line reconstructed at various locations for 10 hour TOFC freight trip with high speed equipment at 120 MPH maximum..

B6 Reconstructed to 9-1/4 hour freight trip at 120 MPH maximum speed for high speed TOFC freight trains with further line improvements resulting in reduced trip times

Table B3 is for 200 mph passenger trains with Class 9 track where needed with rail and highway grade crossings closed, grade separated or otherwise remediated. The line is reconstructed without realignment to allow maximum speed per curvature restrictions with high speed tilting passenger equipment. Tables B3, B5 and B7 have been prepared for special tilting passenger equipment with effective center of gravity 60 inches above top of rail.Unreconfigured, but with adjustment of superelevations and transitions at curves, appropriate track rehabilitation and appropriate signals and other appurtenances, optimized, maximum speed with high-speed tilting passenger equipment; maximum speed 200 MPH, 7-3/4 hour trip New York Penn Station –Chicago Union Station

B5 Reconstructed for 7-1/4 hour passenger trip at 200 MPH maximum speed Table B5 has line reconstructed per Table B4 but calculated for a maximum passenger train speed of 200 mph on Class 9 track where allowed, for trains consisting of locomotives plus tilt body passenger equipment.

B7 Reconstructed to 6-1/2 hour passenger trip at 200 MPH maximum speed for high speed passenger trains with further line improvements resulting in reduced trip times.

B8 Reconstructed for 4 hour passenger trip at 235 MPH maximum speed.

The tables B0 [[RNGU]]through B8 are very complicated, and are fed by tables of allowable speed on curves, acceleration rates, and horsepower requirements for trains that are also lengthy and complicated. Every effort has been made to eliminate mathematical errors in these tables, but some small errors may be present. The results of any individual errors of any

magnitude should be readily apparent and are believed to have been corrected. Any small errors would have little effect on the end result, which is the time required to travel between Chicago and New York. Such errors would be random, and as such could be negative or positive in their effect, and may very well add up to zero and have no effect on the overall travel time. The author would, however, appreciate being made aware of any such errors found by the reader. The simplifications and assumptions in some aspects of theoretically perfect operations, to make the development of these tables manageable, have been fully explained here and elsewhere and should be taken into account when reviewing these tables.

Appendix B0 - Unreconstructed Existing Right of Way Existing Right-of-Way as available. 120 mph maximum with conventional equipment. Trains freight only or mixed freight and passenger, Northern New Jersey to Northwest Indiana. Trains consist of 200 TOFC platforms maximum, with passenger equipment as needed.
FRA Class 7 track where available.
FRA Pre-Positive Train control rules apply as of June 2016.
Motive power 20 units maximum diesel electric, 3,440 P for traction each. Lead unit compatible with Amtrak east of Harrisburg.
Braking/Deceleration rate = 2 mph/second.
Assumed E_a = 4 inches, E_u = 3 inches Signals per existing. Additional train controls added per FRA requirements. Curves L = left, R = right, per facing west bound. Normal curve speed reduced 20% at reverse curves including up to 600 feet between curves. Speed not to exceed existing allowable per latest railroad information.
Center of gravity of moving equipment assumed to be 120 inches above top of rail. Train length = 11,000 feet
See Appendix B1 for right of way characteristics 875 miles, North Jersey to Northwest Indiana

Westbound trip 16.394 hours
Eastbound trip 16.181 hours

READ DOWN WESTBOUND TRAFFIC — APPENDIX 60 (8 ZEROS) STATION — READ UP EASTBOUND TRAFFIC

PAGE 1 COLUMN 1

MILES	DISTANCE, ft	HWY GRADE CROSSING	MAX MPH	Δ DISTANCE, feet	SPEED, mph	Δt, hours	TIME, hours	STATION	TIME, hours	Δt, hours	Δ DISTANCE, feet	SPEED, mph	MILES
0								PENNSYLVANIA STATION NYC					40.82
3.36								E MEADOWS					41.59
4.00	3274												42.59
4.79	4187												43.39
4.79								EAST END	16.0818	0.0000		0	44.39
4.85	850							LAUTENBERG PLATFORM	16.3016	0.0080	1300 D	60	45.39
4.95								AT SECAUCUS TRANSFER	16.3034	0.0070	2230	60	46.39
5.00	266							SLOW UP	16.3184	0.0184	5191	60	47.39
5.04	188								16.3010				48.39
5.04								WEST END	16.3200				48.45
5.45	2191			5464 A 61		0.0089	0.0286		16.3030	0.0075	2377	60	49.59
5.73	1479		137				0.0268 R		16.2925	0.0047	1479	60	50.39
5.85	589						0.0286		16.2878	0.0019	589	60	51.39
6.04	1024						0.0286 A		16.2860	0.0026	838	60	51.71
6.04							0.0288	W MEADOWS	16.2883	0.0000			51.93
6.14	528			320	60	0.0010	0.0200		16.2833	0.0017	528	60	52.59
6.20	330		60	330	60	0.0016	0.0307	HACKENSACK R	16.2817	0.0019	330	60	53.39
6.25							0.0307		16.2808	0.0004	135 D 85		54.39
6.20							0.0307		16.2802	0.0010	314	60	55.39
7.04	4407			4407	60	0.0139	0.0446		16.2798	0.0147	4214 A 44		56.39
7.42	2022			2022	60	0.0084	0.0510	HUDSON	16.2645	0.0079	1762	44	57.20
8.07	3443		141	3443	60	0.0109	0.0616	R HUDSON	16.2570	0.0148	3443	44	57.35
8.31	1273		141	1129	60	0.0036	0.0654	R	16.2421	0.0085	1273	44	57.46
8.41	487			3442 A 60		0.0100	0.0704		16.2387	0.0021	487	44	56.13
9.17	4035					0.0200	0.0754		16.2348	0.0174	4035	44	58.27
[illegible]	[illegible]	[illegible]	[illegible]	[illegible]	[illegible]	[illegible]	[illegible]	[illegible]	[illegible]	[illegible]	[illegible]	[illegible]	[illegible]

PAGE 1 COLUMN 2

MILES	DISTANCE, ft	HWY GRADE CROSSING	MAX MPH	Δ DISTANCE, feet	SPEED, mph	Δt, hours	TIME, hours	STATION	TIME, hours	Δt, hours	Δ DISTANCE, feet	SPEED, mph
[illegible]	[illegible]	[illegible]	[illegible]	[illegible]	[illegible]	[illegible]	[illegible]	[illegible]	[illegible]	[illegible]	[illegible]	[illegible]

READ DOWN WESTBOUND TRAFFIC — APPENDIX 9D (B 25803) STATION — READ UP EASTBOUND TRAFFIC

PAGE 2 COLUMN 1

MILES	DISTANCE ft	HWY GRADE CROSSINGS	MAX MPH	DISTANCE feet	SPEED mph	ET hours	TIME hours	STATION
91.00	1400		92	1400	59	0.0045	1.2787	LYRAMBO
91.05	280		92	280	59	0.0009	1.2796	L
91.32	1400			1400	59	0.0045	1.2840	
92.00	3600		99	3600	59	0.0116	1.2869	RHENDERSON
92.11	580		93	580	59	0.0019	1.2875	R
93.00	4700			4700	59	0.0151	1.3107	
93.17	880			880	59	0.0028	1.3154	
93.51	1800		100	1800	59	0.0058	1.3212	L
94.00	2600			2592	59	0.0083	1.3295	KING of PRUSSIA
95.00	5280			5280	59	0.0169	1.3465	
96.00	5280			5280	59	0.0169	1.3634	
96.11	580			580	59	0.0019	1.3653	
96.41	1600		175	1600	59	0.0051	1.3704	R
97.00	3100			3100	59	0.0100	1.3803	
98.00	5280			5280	59	0.0169	1.3972	NORRISTOWN VILLE
98.53	2800			2800	59	0.0090	1.4063	
98.85	1692		141	1692	59	0.0054	1.4117	R
99.00	798		141	798	59	0.0026	1.4142	R
99.15	750			750	59	0.0025	1.4168	
99.84	2800		141	2800	59	0.0083	1.4251	L
99.85	1113			1113	59	0.0036	1.4287	
100.00	787			787	59	0.0025	1.4312	
100.74	3909			3909	59	0.0125	1.4437	
100.85	580			580	59	0.0019	1.4456	
100.89	211			211	59	0.0007	1.4463	
101.00	580		245	580	59	0.0019	1.4481	R
101.18	880		245	880	59	0.0031	1.4513	R
101.85	3806			3806	59	0.0110	1.4528	
102.00	792			792	59	0.0025	1.4621	
102.00					59	0.0020	1.4651	
102.21	1100			1100	59	0.0035	1.4686	
102.32	580		173	580	59	0.0019	1.4705	L
102.32							1.4705	
102.78	2440			2442	59	0.0078	1.4780	
102.85	370			370	59	0.0012	1.4795	
103.00	786			788	59	0.0025	1.4820	
103.21	1100	2801	181	1100	59	0.0036	1.4858	R
103.39	950			950	59	0.0031	1.4887	
103.72	1737			1737	59	0.0056	1.4942	
103.85	692		170	692	59	0.0022	1.4965	L
104.00	786		140	786	59	0.0025	1.4990	L
104.15	800			405	59	0.0013	1.5003	
104.15		8843 45.96				0.0020	1.5002	W HENDERSON
104.20	265			18	59	0.0001	1.5001	
104.20				830 D	59	0.0024	1.5027	AMTRAK WB TR & OHIO
104.45	1315	1575	42	1313	42	0.0059	1.5086	R
104.50	280			281	42	0.0042	1.5098	
104.59		UG 4.35		927	42	0.0024	1.5122	R RAVINE ROAD
104.64	780			728 D	42	0.0011	1.5133	
104.72	403		34	277	34	0.0016	1.5148	R
104.74	101			277 D	34	0.0018	1.5166	
104.79	274		20	274	20	0.0028	1.5194	L REVERSE CURVE
104.84	267		34	267	20	0.0024	1.5218	L
104.84						0.0020	1.5218	GLEN LOCH 30 MPH PERMIT
104.84						0.0000	1.5218	
104.84						0.0000	1.5218	
104.84			40			0.0000	1.5218	XT SWITCH
105.03	980			980	20	0.0093	1.5311	SPEED LIMIT 20 MPH TO HERE
105.32	1524		181	1524	20	0.0144	1.5455	L
105.55	1252			1252	20	0.0119	1.5574	T
105.68	673			673	20	0.0064	1.5637	T
106.00	1190		130	1190	20	0.0113	1.5750	L
106.13	1181		172	1181	20	0.0113	1.5863	R REVERSE CURVE
106.55	2226			2226	20	0.0211	1.6074	T
106.55						0.0000	1.6074	
107.55	5280			1707	20	0.0129	1.6187	
107.55				884 A	20	0.0096	1.6209	T
107.79	1230	6840 26.26				0.0000	1.6209	U-GRADE - ABANDONED
108.56	4050		206	10673 A	57	0.0352	1.6568	L
108.71	845		258			0.0000	1.6566	L
109.30	3221			896	56	0.0015	1.6600	T
109.30				927	0.76	0.0028	1.6609	
109.55	1314		57	1014	57	0.0040	1.6666	R
109.55						0.0020	1.6858	RADFORD HILLS FIRST CHECK
109.74	870		298	870	57	0.0032	1.6680	R
109.85	1024			1024	57	0.0034	1.6733	T
110.30	1640		45	1640	57	0.0064	1.6797	
110.30							1.6797	
110.55	1347			1347	57	0.0045	1.6840	T
110.76	2173			1173	57	0.0039	1.6881	T
111.17	2080		76	2080	57	0.0069	1.6886	R
111.17							1.6886	
111.55	2027			2027	57	0.0067	1.7017	T
111.76	1100			639	57	0.0015	1.7030	T
111.76	4			1878 A	57	0.0055	1.7085	DOWNINGTOWN
112.12	1777		107	894	63	0.0028	1.7111	L
112.37	1946			1046	63	0.0031	1.7142	T
112.55	1283		107	948	63	0.0010	1.7152	R
112.76	2106		107	5308 A	63	0.0144	1.7297	R
113.21	2376			0	76	0.0000	1.7297	T
113.55	1795		107	8311 A	76	0.0147	1.7444	L
113.64	466		107	16375 A	86	0.0311	1.7756	L
113.78	721					0.0000	1.7758	T
113.99	1089		100			0.0000	1.7753	R
113.99							1.7756	
114.19	1087		200			0.0000	1.7756	R
114.19							1.7752	
114.38	1085					0.0000	1.7755	T
114.55	880		141			0.0000	1.7755	T
114.60	216					0.0000	1.7756	T
114.74	734		116			0.0000	1.7756	L
115.42	3610			0		0.0000	1.7756	T
115.55	700		113			0.0000	1.7755	T
115.59	200		113			0.0000	1.7755	T
115.74	784		113	0	100	0.0000	1.7756	R REVERSE CURVE
116.43	3607						1.7756	T
116.55	480		400	0	100	0.0000	1.7758	R
117.02	2433		100	0	100	0.0000	1.7758	R
117.13	633			0	100	0.0000	1.7795	T
117.50	1914		100	0	100	0.0000	1.7796	L
117.55	251			251	100	0.0008	1.7781	T
117.55					100	0.0000	1.7781	COATESVILLE
117.98	2240			2240	100	0.0042	1.7808	T
118.55	3040		141	3040	100	0.0058	1.7861	L
118.63	434		141	414	100	0.0038	1.7869	L
118.91	1485			1485	100	0.0027	1.7896	T
119.38	2485		116	2485	100	0.0047	1.7943	R
119.55	932			932	100	0.0014	1.7964	T
120.30	3951			149	100	0.0003	1.7964	T
120.55	1246		114	20986 A	100	0.0411	1.8375	L
120.65	485		114			0.0000	1.8376	T
120.92	1436					0.0000	1.8375	T

PAGE 2 COLUMN 2

READ DOWN WESTBOUND TRAFFIC — APPENDIX 8C (9 ZERO) STATION — READ UP EASTBOUND TRAFFIC

PAGE 3 COLUMN 1

MILES	DISTANCE, ft	HWY GRADE CROSSING	MAX MPH	d DISTANCE, feet	SPEED, mph	dt, hours	TIME, hours	STATION	TIME, hours	dt, hours	d DISTANCE, feet	SPEED, mph
157.37	1045		80	1045	80	0.0022	2.2856 L		14.0847	0.0027	1041	80
157.56	990			990	80	0.0021	2.2870 T		14.0820	0.0021	990	80
158.55	5280			5280	89	0.0112	2.2989		14.0808	0.0112	5286	89
158.70	778	XING 79.1		778	89	0.0017	2.3006	MOORES MILL ROAD	14.0486	0.0017	778	89
158.90	1056		80	1056	89	0.0022	2.3028 R		14.0470	0.0022	1056	89
159.31	2168			2168	89	0.0046	2.3074 T		14.0447	0.0007	336 D 94-	
159.55	1278		122	1278	89	0.0027	2.3101 R		14.0440	0.0022	1092	94
159.65	500		122	500	89	0.0011	2.3112 R		14.0418	0.0000	0	
160.42	4057			4057	89	0.0088	2.3196 T		14.0418	0.0000	0	
160.55	723		116	723	89	0.0015	2.3213 R		14.0418	0.0000		
160.71	834	XING	116	834	89	0.0015	2.3231 R	MARKET ST MT JOY	14.0418	0.0000		
161.11	2112			1440	89	0.0081	2.3302 T		14.0418	0.0000		
161.55	2324		100			0.0000	2.3382 R		14.0418	0.0384	18080 A 71-	
161.55				17852 A 89-		0.0236	2.3617		14.0064	0.0000		
161.59	165		100			0.0000	2.3617 R		14.0054	0.0000		
162.34	3960					0.0000	2.3617 T		14.0054	0.0017	616	71
162.55	1155		141			0.0000	2.3617 L		14.0036	0.0031	1156	71
162.78	1210		141			0.0000	2.3617 L		14.0007	0.0002	1210	71
163.26	2530			0	100	0.0000	2.3617 T		13.9975	0.0067	2530	71
163.55	1540		113	0	100	0.0000	2.3617 L		13.9907	0.0041	1546	71
163.80	1278		112	1339	100	0.0026	2.3643 L		13.9880	0.0034	1276	71
164.30	2640			1323	100	0.0029	2.3672 T		13.9832	0.0071	2664	71
164.30				1818 D 100		0.0040	2.3812		13.9701	0.0000		
164.55	1334		71	1334	71	0.0036	2.3848 R		13.9761	0.0038	1334	71
164.89	1779		71	1779	71	0.0047	2.3895 R	ELIZABETHTOWN	13.9725	0.0047	1779	71
165.03	723			723	71	0.0019	2.3714 T		13.9678	0.0019	723	71
165.32	1536		100	1586	71	0.0042	2.3756 L		13.9659	0.0042	1558	71
165.44	611			611	71	0.0016	2.3772 T		13.9617	0.0016	611	71
165.55	611		71	611	71	0.0016	2.3788 L		13.9601	0.0018	611	71
165.74	1000		71	1000	71	0.0027	2.3815 L		13.9585	0.0027	1000	71
165.74							2.3815		13.9558	0.0005	1401 D 96-	
166.55	4280			4280	71	0.0114	2.3929 T		13.9524	0.0006	258	96
167.27	3779			3779	71	0.0101	2.4030 T		13.9519	0.0000	0	
167.52	1334		141	1334	71	0.0036	2.4066		13.9519	0.0000		
167.55	167			167	71	0.0004	2.4070 T		13.9519	0.0000		
168.55	5280			1440	71	0.0038	2.4108		13.9519	0.0000	0	
168.55	5280			21157 A 71		0.0484	2.4572		13.9519	0.0000		
168.55						0.0000	2.4572 T		13.9519	0.0528	21982 A 57-	
170.33	4062			0		0.0000	2.4572 T		13.8904	0.0017	521	57
170.55	1156		98	0	98	0.0000	2.4572 L		13.8877	0.0040	1156	57
170.71	825		98		98	0.0000	2.4572 L		13.8837	0.0027	825	57
171.20	2588				98	0.0000	2.4572 T		13.8810	0.0068	2585	57
171.55	1870		100		98	0.0000	2.4572 R		13.8824	0.0062	1829	57
171.85	1584		100		98	0.0000	2.4572 R		13.8762	0.0052	1556	57
172.31	2445			194	98	0.0004	2.4576 T		13.8710	0.0091	2445	57
172.51				2030 D 98		0.0057	2.4633		13.8629	0.0000		
172.56	1278		57	1278	57	0.0042	2.4676 R		13.8620	0.0042	1278	57
172.91	1870		57	1800	57	0.0062	2.4738 R		13.8580	0.0062	1870	57
172.91							2.4738	REVERSE CURVE	13.8504	0.0013	408 D 86-	
173.49	3080		98	3080	57	0.0102	2.4840 L		13.8512	0.0077	2674	86
173.55	330			330	57	0.0011	2.4851 T		13.8436	0.0032	1307 D 89-	
173.55						0.0000	2.4851	REMALTON - ROYALTON BRANCH	13.8403	0.0004	391	89
174.15	3168			3168	57	0.0106	2.4956 T		13.8399	0.0043	2000	89
174.29	723		162	723	57	0.0024	2.4980 L		13.8356	0.0015	723	89
174.37	389			389	57	0.0013	2.4993 T		13.8341	0.0008	389	89
174.52	834		346	834	57	0.0028	2.5021 L		13.8338	0.0018	834	89
174.54	111			111	57	0.0004	2.5025 T		13.8315	0.0002	111	89
174.55	56		111	56	57	0.0002	2.5027 L		13.8313	0.0001	56	89
174.55						0.0000	2.5027	MIDDLETOWN	13.8311	0.0000		
174.85	1556		89	1556	57	0.0052	2.5078 L		13.8311	0.0033	1556	89
174.92	388			389	57	0.0013	2.5091	REVERSE CURVE	13.8278	0.0011	548 D 97-	
175.12	1056		108	364	57	0.0012	2.5103 R		13.8267	0.0018	899	97
175.32	1056			2718 A 57		0.0068	2.5187 T		13.8250	0.0021	1056	97
175.52	1056		115	86	66	0.0002	2.5189 L		13.8229	0.0021	1056	97
175.55	167			167	66	0.0005	2.5194 T		13.8208	0.0103	5296	97
175.85	1584			150	66	0.0003	2.5197 T		13.8105	0.0000	0	97
176.00	778		346			0.0000	2.5197 L		13.8105	0.0000	0	97
176.46	2446			13182 A 66-		0.0317	2.5514 T	FLYOVER	13.8106	0.0000	0	97
176.55	500		97			0.0000	2.5514 R		13.8105	0.0000	0	97
176.85	1584		97	0	98	0.0000	2.5514 R		13.8105	0.0000	0	97
177.45	3168			0	98	0.0000	2.5514 T		13.8106	0.0000	0	97
177.55	556		224			0.0000	2.5514 L		13.8106	0.0000	0	
177.75	1045		224	0	97	0.0000	2.5514 L		13.8106	0.0000		
178.49	3905			0	97	0.0000	2.5514 T		13.8105	0.0000	0	
178.55	330		185	0	97	0.0000	2.5514 R		13.8105	0.0000		
179.01	2385		185	6478 A 98		0.0172	2.5686 R		13.8105	0.0000		
179.55	2865					0.0000	2.5686 T		13.8106	0.0481	21346 A 97-	
180.55	5280						2.5686		13.7625			
180.55				4336 A 97		0.0085	2.5769	STEELTON	13.7625	0.0000		
181.18	3300			7056 A 104		0.0131	2.5900 T		13.7625	0.0235	8541 A 35-	
181.52	1825		186	0	104	0.0000	2.5900 R		13.7391	0.0008	153	35
181.55	165			165	104	0.0003	2.5903 L		13.7383	0.0000	0	35
181.55				485	104	0.0008	2.5911		13.7383	0.0022	365 A 30-	
182.00	2334			289 D 100		0.0008	2.5919 T		13.7361	0.0029	458	30
182.24	1276		106	1276	100	0.0024	2.5940 R		13.7327	0.0001	1270	30
182.46	1167			1167	100	0.0022	2.5963 T		13.7291	0.0074	1187	30
182.53	388		122	220	100	0.0004	2.5967 L		13.7177	0.0025	368	30
182.55	111			0	100	0.0000	2.5967 T		13.7153	0.0007	111	30
182.89	1760			19	100	0.0000	2.5967 T		13.7146	0.0111	1760	30
182.89				2021 D 100		0.0046	2.6013		13.7036	0.0000		
183.01	660		67	660	67	0.0019	2.6032 R		13.7036	0.0042	860	30
183.55	2860			2860	67	0.0081	2.6112 T		13.6993	0.0181	2860	30
183.89	1778			581	67	0.0018	2.6129 T		13.6812	0.0112	1778	30
183.89				1197 D 67		0.0044	2.6173		13.6700	0.0000		
183.94	269		35	269	35	0.0015	2.6188 L		13.6700	0.0017	266	30
183.95	54		35	54	35	0.0003	2.6191 L	HARRISBURG	13.6683	0.0005	54	30
183.99	215			96	35	0.0005	2.6196 T		13.6680	0.0014	215	30
143.05	1595			1595	71	0.0043	2.6239 T		14.2062	0.0030	1595	106
183.99				119 D 36		0.0007	2.6203	REVERSE CURVE T	13.6666	0.0000		
184.06	377		30	377	30	0.0024	2.6227 R		13.6666	0.0004	377	30
184.18	590			563	30	0.0037	2.6264	REVERSE CURVE	13.6642	0.0006	98 D 34-	
184.18						0.0000	2.6264		13.6637	0.0028	488	34
184.31	700		38	700	30	0.0044	2.6308 L		13.6609	0.0039	700	34
184.45	754			754	30	0.0048	2.6356 T		13.6570	0.0042	754	34
184.55	539			539	30	0.0034	2.6390 T		13.6505	0.0030	539	34
184.60	216			218	30	0.0014	2.6404 T		13.6498	0.0012	218	34
184.63	163		45	163	30	0.0010	2.6414 R		13.6486	0.0009	163	34
184.76	707			707	30	0.0045	2.6459	MUST HAVE AIR COMPATIBLE LC	13.6477	0.0030	707	34
184.76	0	HERR STREET	10		30	0.0000	2.6459	WEST END AMTRAK - EAST END P	13.6437	0.0000	0	
185.02	1361			1361	30	0.0086	2.6545	MAXIMUM SPEED 75 MPH	13.6407	0.0076	1361	34
185.10	435		34	435	30	0.0027	2.6572 L		13.6362	0.0024	435	34
185.10	0			0	30	0.0000	2.6572	REVERSE CURVE	13.6337	0.0030	810 D 58-	
185.18	435		80	435	30	0.0027	2.6599 R		13.6304	0.0001	38	58
185.31	653			658	30	0.0041	2.6641 T		13.6303	0.0008	250	58
185.43	660		87	660	30	0.0042	2.6708 R		13.6295	0.0032	985	58
185.55	337			327	30	0.0021	2.6723 T		13.6263	0.0011	327	58
185.68	701			701	30	0.0044	2.6767 T		13.6252	0.0108	3312	58
185.72	162		56	160	30	0.0010	2.6776 L		13.6144	0.0033	987 A 54-	
186.55	4428			2272	30	0.0143	2.6921 T		13.6111	0.0104	2958	54
186.55				284 A 30-		0.0017	2.6938		13.6007	0.0086	2179 A 41-	
187.55	5282			5186	34	0.0289	2.7227 T		13.5921	0.0062	1129	41
187.55				3885 A 34		0.0156	2.7382		13.5869	0.0000		

PAGE 3 COLUMN 2

MILES	DISTANCE, ft	HWY GRADE CROSSING	MAX MPH	d DISTANCE, feet	SPEED, mph	dt, hours	TIME, hours	STATION	TIME, hours	dt, hours	d DISTANCE, feet	SPEED, mph
188.43	4736			2846	58	0.0060	2.7469 T		13.5609	0.0219	4736	41
188.45				164 D 58		0.0009	2.7474		13.5450	0.0000		
188.48	163		54	163	54	0.0006	2.7480 R		13.5450	0.0008	163	41
188.51	163			163	54	0.0006	2.7486	REVERSE CURVE	13.5443	0.0009	163	41
188.55	216		54	216	54	0.0008	2.7493 L		13.5436	0.0010	218	41
188.64	457			457	54	0.0015	2.7508 T		13.5425	0.0020	427	41
188.88	1267		67	1267	54	0.0044	2.7553 L		13.5405	0.0059	1257	41
189.42	2897			2446	54	0.0086	2.7638 T		13.5347	0.0134	2897	41
189.42				458 D 64		0.0018	2.7656		13.5413	0.0000		
189.47	219		41	219	41	0.0010	2.7667 L		13.5413	0.0010	219	41
189.57	547		41	547	41	0.0025	2.7692 L	ROCKVILLE	13.5403	0.0025	547	41
189.61	216		41	218	41	0.0010	2.7702 L		13.5378	0.0010	218	41
189.61						0.0000	2.7702		13.5367	0.0031	639 D 62	
190.32	3756			3756	41	0.0174	2.7875 T		13.5337	0.0095	3150	63
190.32						0.0000	2.7875		13.5242	0.0044	1423 A 58	
190.32						0.0000	2.7875	SUSQUEHANNA RIVER	13.5198	0.0007	203	58
190.50	925		80	925	41	0.0043	2.7918 R		13.5191	0.0090	925	58
190.57	381			381	41	0.0016	2.7930 T		13.5101	0.0012	381	58
190.74	926			916	41	0.0042	2.7978	MAXIMUM SPEED LIMIT 70 MPH	13.5148	0.0008	263 A 57	
191.29	2909		82	2909	41	0.0134	2.8112 R		13.5140	0.0037	1120	57
191.57	1455			1455	41	0.0067	2.8180 T		13.5103	0.0040	1456	57
191.57				638	41	0.0030	2.8210		13.5065	0.0000		
192.10	2797					0.0000	2.8210 T		13.5065	0.0098	2797	57
192.59	2581		83	4589 A 41-		0.0164	2.8374 L		13.4962	0.0086	2581	57
192.65	335		63	1194	63	0.0036	2.8410 L		13.4876	0.0011	335	57
192.83	930			202 D 80		0.0007	2.8417 T		13.4865	0.0032	930	57
192.91	391		58	391	58	0.0013	2.8429 R		13.4833	0.0013	391	57
193.03	671			671	58	0.0022	2.8451 T		13.4820	0.0032	671	57
193.05	70			28	58	0.0001	2.8452		13.4788	0.0023	700	57
193.05		XING 114.25		42 D 58		0.0001	2.8454	PRIVATE	13.4775	0.0000		
193.27	1200		57	1200	57	0.0040	2.8494 L		13.4775	0.0030	1180	57
193.27						0.0000	2.8494		13.4736	0.0003	85 D 59	
193.53	1370			1370	57	0.0046	2.8539		13.4733	0.0041	1285	59
194.08	2900			2800	57	0.0096	2.8635		13.4692	0.0080	2800	59
194.30	1176		108	1176	57	0.0039	2.8675 R		13.4599	0.0026	1176	59
194.53	1200	XING 115.32		1200	57	0.0040	2.8714	PRIVATE	13.4581	0.0026	1200	59
195.38	4500	XING 115.90		4354	57	0.0145	2.8859	PRIVATE	13.4523	0.0074	2313	59
195.53	791		58	311 A 57		0.0017	2.8876 R		13.4449	0.0000		
195.74	1094		59	1510	59	0.0048	2.8924 R		13.4449	0.0181	4427 A 30	
196.31	3000			3000	59	0.0096	2.9021		13.4268	0.0166	2939	30
196.55	1200		66	1200	59	0.0039	2.9060 R		13.4101	0.0078	1200	30
197.10	3000		80	3000	59	0.0096	2.9155 R		13.4026	0.0189	3000	30
197.23	665			665	59	0.0021	2.9177		13.3836	0.0042	666	30
197.38	800		65	800	59	0.0026	2.9202 R		13.3794	0.0051	800	30
197.53	800		67	800	59	0.0026	2.9228 R		13.3744	0.0091	800	30
197.74	1100		67	960	60	0.0030	2.9259 R		13.3653	0.0089	1100	30
197.89	800			4	59	0.0000	2.9259		13.3624	0.0051	800	30
197.89				946 D 56		0.0040	2.9298		13.3575	0.0000		
198.10	1100		30	1100	30	0.0089	2.9386 L		13.3573	0.0069	1103	30
198.10							2.9386		13.3504	0.0004	69 D 33	
198.15	270			270	30	0.0017	2.9403	TRC TT MPH35	13.3500	0.0012	201	33
198.30	800		33	800	30	0.0051	2.9436 R		13.3488	0.0046	800	33
198.30							2.9436		13.3442	0.0051	1387 D 70	
198.51	1100			1100	30	0.0069	2.9505		13.3391	0.0003	101	70
198.53	100			100	30	0.0006	2.9512		13.3388	0.0000		70
199.31	4098		78	4098	30	0.0259	2.9770 R	DUNCANNON	13.3388	0.0103	3800	70
199.40	500			500	30	0.0032	2.9802		13.3285	0.0014	500	70
199.86	2402			2402	30	0.0152	2.9954		13.3272	0.0065	2402	70
200.42	3000		88	1730	30	0.0109	3.0063 L		13.3207	0.0081	3000	70
200.67	1300		80	165 A 30		0.0011	3.0074 L		13.3126	0.0035	1300	70
201.03	1880			864	33	0.0051	3.0125		13.3091	0.0051	1880	70
201.27	1300		100	7945 A 33		0.0278	3.0402 R		13.3040	0.0036	1300	70
201.42	800						3.0402		13.3004	0.0022	800	70
201.58	800			0			3.0402		13.2982	0.0022	800	70
201.83	1886	XING 125.104		201	70	0.0006	3.0409 L	AQUEDUCT	13.2961	0.0241	8018	70
202.43	2800	XING 125.39		2800	70	0.0070	3.0479	PRIVATE	13.2720	0.0011	407 A 60	
203.43	5285	XING 125.80		5286	70	0.0143	3.0622	PRIVATE	13.2709	0.0013	456	69
204.43	5279			5279	70	0.0143	3.0765		13.2696	0.0145	5279	69
204.65	1200			1200	70	0.0032	3.0797		13.2551	0.0035	1200	69
205.05	2083	XING 126.70		2083	70	0.0056	3.0854 L	PRIVATE	13.2516	0.0057	2085	69
205.39	1800			1800	70	0.0046	3.0902		13.2461	0.0049	1800	69
205.42	182			131	70	0.0004	3.0906		13.2412	0.0006	182	69
205.42		XING 127.30		51 D 70		0.0001	3.0907	PRIVATE	13.2407	0.0001	770	69
205.98	3000		69	3000	69	0.0082	3.0980 L		13.2386	0.0081	2769 A 61	
206.22	1200			1200	69	0.0033	3.1023		13.2306	0.0021	681	61
206.39	900		73	900	69	0.0025	3.1047 R		13.2266	0.0028	900	61
206.53	752		108	752	69	0.0021	3.1068 R		13.2257	0.0025	792	61
206.74	1100		69	1100	69	0.0030	3.1098 R		13.2233	0.0034	1100	61
206.97	1200		94	1200	69	0.0033	3.1131 R		13.2199	0.0037	1200	61
207.20	1200	XING 126	75	1200	69	0.0033	3.1164 R	PRIVATE	13.2162	0.0037	1200	61
207.42	1200			1200	69	0.0033	3.1197		13.2125	0.0037	1200	61
207.63	1100			1100	69	0.0030	3.1227		13.2087	0.0034	1100	61
207.93	1600	XING 126	69	1600	69	0.0044	3.1271 L	PRIVATE	13.2053	0.0050	1600	61
208.16	1287			906	69	0.0025	3.1296		13.2004	0.0040	1287	61
208.16				381 D 65		0.0011	3.1307		13.1964	0.0000		
208.42	1350		61	1300	61	0.0040	3.1347 L		13.1964	0.0040	1300	61
208.60	930		61	930	61	0.0029	3.1376 L		13.1923	0.0029	930	61
208.60							3.1376		13.1894	0.0008	185 D 65	
209.43	4400		97	4400	61	0.0137	3.1513 L		13.1889	0.0123	4215	65
209.68	1313			1313	61	0.0041	3.1554 T	201.6	13.1766	0.0038	1313	65
210.40	3800		71	3800	61	0.0118	3.1672 R		13.1728	0.0111	3800	65
211.01	3200		65	1447	61	0.0045	3.1718 R		13.1617	0.0093	3200	65
211.01							3.1718		13.1524	0.0007	248 D 70	
211.20	1024			1204 A 65		0.0000	3.1707		13.1517	0.0021	777	70
211.37	900		67	2343	65	0.0068	3.1826 L		13.1496	0.0024	900	70
211.47	500		97	500	65	0.0015	3.1840 L		13.1472	0.0014	500	70
211.73	1400			1400	65	0.0041	3.1880 T	NEWPORT	13.1458	0.0038	1400	70
212.07	1800	XING 130	97	1800	65	0.0052	3.1933 R	PRIVATE	13.1420	0.0049	1800	70
212.26	1002			1000	65	0.0029	3.1962		13.1371	0.0027	1000	70
212.33	373	XING 131	97	373	65	0.0011	3.1973 R	PRIVATE	13.1344	0.0061	2245	70
212.80	2478		97	2478	65	0.0072	3.2045 R		13.1284	0.0022	828 A 66	
213.33	2800			1525	65	0.0044	3.2089		13.1261	0.0072	2598	66
213.49	825			1882 A 68		0.0053	3.2142		13.1189	0.0026	825	66
214.32	4400		106	7693	70	0.0208	3.2350 L		13.1163	0.0123	4400	66
214.92	3178			101 D 70		0.0003	3.2353		13.1044	0.0086	2178	68
215.32	2120		68	2100	68	0.0058	3.2412 L		13.0958	0.0025	898	68
215.66	1800		78	1800	68	0.0050	3.2460 L		13.0930	0.0084	3153 A 58	
215.74	400			400	68	0.0011	3.2473 T	HC MPH70(TT)	13.0836	0.0008	284	58
215.67	719		77	719	68	0.0020	3.2493 R		13.0827	0.0023	719	58
216.31	2300			2300	68	0.0064	3.2557		13.0804	0.0075	2300	58
216.84	2800			2800	68	0.0078	3.2635		13.0729	0.0091	2800	58
216.98	713		94	713	68	0.0020	3.2655 R		13.0637	0.0023	713	58
217.32	1800			1800	68	0.0050	3.2706 T	MILLERSTOWN	13.0614	0.0059	1800	58
217.77	2414		176	1952	68	0.0054	3.2759 L		13.0556	0.0078	2414	58
217.77				462 D 58		0.0014	3.2773		13.0477	0.0000		
218.30	2800		58	2800	58	0.0091	3.2865 L		13.0477	0.0091	2800	58
218.66	1900			1900	58	0.0062	3.2927		13.0385	0.0017	563 D 70	
218.83	902	XING 140	114	902	58	0.0029	3.2956 R	PRIVATE	13.0368	0.0004	146	70
219.08	1300			1300	58	0.0042	3.2998		13.0365	0.0117	5082 A 58	
219.31	1200		115	1200	58	0.0039	3.3038 R		13.0246	0.0021	651	58
219.33	100		116	100	58	0.0003	3.3041 R		13.0227	0.0003	100	58
219.97	3400			3400	58	0.0111	3.3152		13.0224	0.0111	3400	58

READ DOWN WESTBOUND TRAFFIC — APPENDIX B0 (B ZERO) STATION — READ UP EASTBOUND TRAFFIC

PAGE 4 COLUMN 1

MILES	DISTANCE, ft	HWY GRADE CROSSING	MAX MPH	∆DISTANCE, feet	SPEED, mph	∆T, hours	TIME, hours	STATION	TIME, hours	∆T, hours	∆DISTANCE, feet	SPEED, mph
220.28	1620		65	1620	58	0.0052	3.3206	L THURWARD CURVE	13.0113	0.0065	1620	58
220.33	249		86	249	58	0.0006	3.3213	L	13.0050	0.0006	249	58
220.89	3000			329	58	0.0011	3.3224		13.0052	0.0088	3000	58
221.03	700		121	3962	A 58	0.0117	3.3340	R	12.9964	0.0023	700	56
221.12	500					0.0000	3.3340		12.9931	0.0016	500	58
221.27	800		85	146	70	0.0004	3.3344	R	12.9915	0.0028	600	56
221.27				583	D 70	0.0017	3.3361		12.9889	0.0000		
221.48	1100		58	1100	58	0.0036	3.3397	L	12.9889	0.0056	1100	56
221.48							3.3397		12.9833	0.0001	40	D 56
221.59	600			600	58	0.0020	3.3417	REVERSE CURVE	12.9851	0.0018	567	58
221.78	1000		74	1000	58	0.0033	3.3449	R THOMPSONTOWN	12.9833	0.0032	1000	58
221.84	282			282	58	0.0009	3.3458	T RC MPH58	12.9801	0.0009	282	58
222.05	1100		58	1100	58	0.0036	3.3494	L	12.9792	0.0035	1100	58
222.05							3.3494		12.9757	0.0015	520	D 70
222.27	1200			1200	58	0.0039	3.3534		12.9742	0.0018	680	70
222.31	184			184	58	0.0006	3.3540		12.9728	0.0005	184	70
222.52	1100		87	1100	58	0.0036	3.3576	R	12.9718	0.0030	1100	70
223.27	4000			4000	58	0.0131	3.3706		12.9689	0.0108	4000	70
224.16	4700			1534	58	0.0050	3.3756		12.9580	0.0104	3828	70
224.26	506	XING	79	256	A 58	0.0008	3.3765	R PRIVATE	12.9477	0.0000		70
224.37	496		79	2724	59	0.0087	3.3852	R	12.9477	0.0106	3704	A 59
225.28	4800			3704	A 59	0.0108	3.3960		12.9368	0.0088	3080	59
225.91	3340			5712	70	0.0155	3.4115		12.9270	0.0107	3340	59
226.29	2000		121	2000	70	0.0054	3.4169	L	12.9163	0.0064	2000	59
226.31	100		121	100	70	0.0003	3.4172	L	12.9099	0.0003	100	59
226.78	2500			1900	70	0.0051	3.4223		12.9096	0.0080	2500	59
226.76				80	70	0.0002	3.4225		12.9015	0.0000		
226.78				520	D 70	0.0015	3.4241		12.9015	0.0000		
227.26	2500		59	2500	59	0.0080	3.4321	R TUSCARORA	12.9015	0.0080	2500	59
227.34	465		59	465	59	0.0015	3.4336	R	12.8936	0.0015	465	59
227.34		XING 145.6				0.0000	3.4336	PRIVATE	12.8920	0.0014	489	D 69
227.34						0.0000	3.4336		12.8906	0.0004	100	69
227.34							3.4336		12.8903	0.0040	1475	A 66
227.84	2800			2800	59	0.0083	3.4419		12.8862	0.0016	580	66
228.25	2200		72	2200	59	0.0071	3.4490	L	12.8848	0.0064	2200	66
228.65	2100		72	2100	59	0.0067	3.4557	L	12.8782	0.0061	2100	66
228.91	1360			1360	59	0.0045	3.4601		12.8720	0.0039	1360	66
229.17	1400		80	1400	59	0.0045	3.4645	R	12.8681	0.0041	1400	66
229.46	1500			1390	59	0.0045	3.4689	150	12.8640	0.0006	210	66
229.46	0		80	1822	A 59	0.0050	3.4744	R PORT ROYAL	12.8634	0.0089	2827	A 56
229.46				238	66	0.0007	3.4751		12.8545	0.0005	87	55
229.82	1900			0		0.0000	3.4751		12.8542	0.0076	2044	A 44
230.15	1767		86	1767	66	0.0051	3.4803	R	12.8466	0.0020	480	44
230.20	270		66	270	66	0.0008	3.4810	R	12.8445	0.0044	940	A 36
231.15	5000			5000	66	0.0146	3.4956		12.8400	0.0204	3888	36
231.68	2800	XING 152.1		2800	66	0.0080	3.5036	PRIVATE	12.8197	0.0147	2800	36
231.83	800		71	360	66	0.0010	3.5046	L RC	12.8049	0.0042	800	36
231.83				440	D 66	0.0014	3.5062		12.8007	0.0000		
232.18	1731		56	1731	56	0.0060	3.5122	R MPH 50	12.8007	0.0091	1731	36
232.23	400			1	56	0.0000	3.5122		12.7916	0.0021	400	36
232.29				390	D 56	0.0015	3.5137		12.7895	0.0000		
232.50	1400		44	1165	44	0.0050	3.5187	R RC	12.7895	0.0074	1400	36
232.50				235	D 44	0.0011	3.5198		12.7822	0.0000		
232.61	600		36	600	36	0.0032	3.5230	L MPH40	12.7827	0.0032	600	36
232.61		XING 153.00					3.5230	JUNIATA STREET	12.7790	0.0019	406	D 46
233.03	2210			2210	36	0.0116	3.5346	T MIFFLIN	12.7772	0.0070	1805	49
233.15	600		49	600	36	0.0032	3.5378	L	12.7702	0.0023	600	49
233.15							3.5378		12.7679	0.0007	189	D 54
233.39	1300		78	1300	36	0.0068	3.5446	L MPH 45	12.7672	0.0039	1111	54
233.49	500			500	36	0.0026	3.5472	T RC	12.7633	0.0018	500	54
233.80	1680		54	1680	36	0.0088	3.5561	R	12.7618	0.0058	1680	54
233.80							3.5561		12.7557	0.0013	366	D 63
233.94	700			700	36	0.0037	3.5598		12.7544	0.0008	259	63
234.15	1100		71	1100	36	0.0058	3.5655	L	12.7536	0.0086	2845	A 53
234.77	3205		71	2910	36	0.0153	3.5808	L	12.7450	0.0065	1605	53
235.00	1200			1760	A 36	0.0078	3.5886		12.7386	0.0043	1200	53
235.15	800		71	825	49	0.0024	3.5910	R	12.7343	0.0029	800	53
235.30	800		71	425	49	0.0018	3.5927	R	12.7314	0.0029	800	53
235.51	1100			984	A 49	0.0036	3.5963		12.7286	0.0039	1100	53
235.84	700		35	1191	54	0.0042	3.6005	R	12.7246	0.0025	700	53
236.15	2886			2885	54	0.0086	3.6090		12.7221	0.0086	2826	53
236.51	1900			1822	54	0.0064	3.6165	T DENHOLM	12.7126	0.0068	1900	53
236.51				39	D 54	0.0001	3.6166		12.7067	0.0000		
236.87	1900		53	1900	53	0.0066	3.6232	L	12.7067	0.0068	1900	53
236.87							3.6232		12.6999	0.0030	1258	D 79
237.00	700			700	53	0.0025	3.6257		12.6963	0.0005	203	79
237.15	781		85	781	53	0.0028	3.6285	L	12.6948	0.0000	0	
237.26	581		85	581	53	0.0021	3.6306	L	12.6948	0.0014	581	79
238.02	4000			4000	53	0.0143	3.6449		12.6934	0.0053	2218	79
238.15	700		95	700	53	0.0025	3.6474	R	12.6881	0.0000	0	
238.20	236		95	235	53	0.0008	3.6482	R	12.6881	0.0000	0	
238.56	1900			1900	53	0.0068	3.6550		12.6881	0.0000	0	
238.78	1200		97	1200	53	0.0043	3.6593	L	12.6881	0.0000	0	
239.14	1900			900	53	0.0032	3.6625	CP HAWSTONE	12.6881	0.0000	0	
239.14				5184	A 53	0.0158	3.6784	180	12.6881	0.0258	8202	A 56
239.29	800					0.0000	3.6784	1281.6	12.6625	0.0026	748	56
239.49	1014		121			0.0000	3.6784	L	12.6599	0.0053	1437	A 48
240.15	3600			1168	70	0.0032	3.6816		12.6547	0.0085	1843	48
240.30	800			800	70	0.0022	3.6838		12.6462	0.0032	800	48
240.58	1471		115	1471	70	0.0040	3.6876	L	12.6430	0.0056	1471	48
240.90	1700			1700	70	0.0046	3.6924		12.6382	0.0067	1700	48
241.17	1400		95	1400	70	0.0038	3.6962	R	12.6325	0.0055	1400	48
241.36	1000		95	1000	70	0.0027	3.6988	R	12.6270	0.0039	1000	48
241.51	800			113	70	0.0003	3.6992		12.6230	0.0032	800	48
241.51				688	D 70	0.0021	3.7013		12.6199	0.0000		
241.81	1586		55	1586	56	0.0056	3.7067	L	12.6199	0.0063	1586	48
241.92	600			338	56	0.0012	3.7079	REVERSE CURVE	12.6136	0.0024	606	48
241.92				264	D 56	0.0010	3.7089		12.6113	0.0000		
242.11	1000		48	1000	48	0.0039	3.7128	R	12.6113	0.0032	806	48
242.26	800		48	800	48	0.0032	3.7160	R	12.6081	0.0083	1796	A 33
242.66	2114			2114	48	0.0065	3.7245	SHAWNEE	12.5997	0.0075	1309	33
243.10	2300		80	2300	48	0.0091	3.7334	R	12.5922	0.0132	2300	33
243.23	700		80	700	48	0.0028	3.7361	R	12.5790	0.0040	700	33
243.53	1600			1600	48	0.0063	3.7424		12.5750	0.0092	1600	33
243.78	1291		57	1291	48	0.0051	3.7475	L	12.5658	0.0074	1291	33
243.95	900			900	48	0.0036	3.7511	0	12.5584	0.0052	900	33
244.10	800		56	800	48	0.0032	3.7543	R	12.5532	0.0046	800	33
244.25	800		58	800	48	0.0032	3.7574	R	12.5487	0.0046	800	33
244.42	900		62	495	48	0.0020	3.7594	R	12.5441	0.0052	900	33
244.50	400			380	48	0.0014	3.7608	T RC MPH35	12.5389	0.0023	400	33
244.50				446	D 33	0.0021	3.7629		12.5366	0.0000		
244.71	1100		33	1100	33	0.0063	3.7692	L	12.5366	0.0063	1100	33
244.71							3.7692		12.5303	0.0021	446	D 48
244.91	1054			1054	33	0.0060	3.7752		12.5282	0.0024	609	48
245.10	1000		48	1000	33	0.0057	3.7810	L	12.5258	0.0039	1000	48
245.38	1500		48	1500	33	0.0086	3.7896	L LEWISTOWN	12.5219	0.0059	1500	48
245.38							3.7896		12.5159	0.0054	1931	D 87
245.99	3200			3200	33	0.0184	3.8079		12.5105	0.0028	1269	87
246.09	562		87	562	33	0.0032	3.8112	R	12.5076	0.0012	562	87
247.02	4907		87	3684	33	0.0211	3.8323	R	12.5066	0.0119	5455	87
247.12	500		87			0.0000	3.8323	R	12.4947	0.0000		
247.55	2300			1790	A 33	0.0083	3.8406		12.4947	0.0000		
247.72	672		141	1795	40	0.0068	3.8476	R	12.4947	0.0106	3604	A 56
248.02	1800			6148	A 48	0.0194	3.8670		12.4838	0.0042	1220	55
248.27	1300	XING 169.40					3.8670	GRANVILLE	12.4796	0.0045	1300	55
248.77	2680		97	778	70	0.0021	3.8691	L	12.4752	0.0002	2680	56
249.02	1300			1300	70	0.0035	3.8726		12.4699	0.0045	1300	55
249.57	4500	XING 170.1		3813	70	0.0103	3.8829	PRIVATE	12.4614	0.0136	4500	56
249.87		XING 170.[illegible]		858	D 70	0.0021	3.8850	ANDERSON	12.4499	0.0000		
250.03	771		85	771	[illegible]	[illegible]	3.8877	L	12.4490	0.0027	771	55
250.38	1900		55	1900	56	0.0066	3.8942	L	12.4463	0.0065	1900	56

READ DOWN WESTBOUND TRAFFIC — APPENDIX B0 (B ZERO) STATION — READ UP EASTBOUND TRAFFIC

PAGE 4 COLUMN 2

MILES	DISTANCE, ft	HWY GRADE CROSSING	MAX MPH	∆DISTANCE, feet	SPEED, mph	∆T, hours	TIME, hours	STATION	TIME, hours	∆T, hours	∆DISTANCE, feet	SPEED, mph
250.38							3.8942		12.4368	0.0011	346	D 63
250.51	700			700	56	0.0024	3.8966		12.4356	0.0011	354	63
250.78	1000		63	1300	56	0.0045	3.9011	R	12.4346	0.0039	1300	63
250.78							3.9011		12.4307	0.0006	191	D 67
250.98	1000		77	1200	56	0.0041	3.9052	R	12.4301	0.0029	1000	67
251.02	179	XING 172.20		179	56	0.0006	3.9058	PRIVATE	12.4275	0.0005	179	67
251.54	2774	XING 173.36		2774	55	0.0086	3.9154	PRIVATE	12.4268	0.0078	2774	67
252.02	2500		95	2500	55	0.0086	3.9240	R	12.4189	0.0071	2500	67
252.24	1174		71	1174	55	0.0040	3.9281	R	12.4118	0.0032	1174	67
252.77	2800			1173	55	0.0040	3.9321		12.4085	0.0079	2800	67
253.02	1300		67	2170	A 55	0.0066	3.9380	R	12.4006	0.0037	1300	67
253.02		XING 174.64			69	0.0000	3.9380	HORNINGFORD	12.3969	0.0003	100	D 69
253.43	2168		173	1334	A 69	0.0036	3.9426	R	12.3967	0.0057	2068	69
254.01	3100	XING 175.33		4681	67	0.0133	3.9562	PRIVATE	12.3910	0.0085	3100	69
254.46	2378			2379	67	0.0067	3.9629		12.3825	0.0191	6907	69
255.01	2900	XING 176	69	2900	67	0.0082	3.9711	L SWIGARTS	12.3634	0.0022	798	A 67
256.02	5266	XING 177.16		463	67	0.0013	3.9724	PRIVATE	12.3611	0.0079	2800	67
256.05	191	XING 177.61		798	A 67	0.0022	3.9746	STATE HWY 103 McVEYTOWN	12.3532	0.0005	191	67
256.56	2700		100	6985	66	0.0190	3.9937	L	12.3527	0.0076	2700	67
257.00	2300		69	2814	69	0.0077	4.0014	L MATAWANNA	12.3450	0.0065	2300	67
257.00				407	A 69	0.0011	4.0025		12.3386	0.0000		
257.02	100			1906	70	0.0052	4.0077		12.3386	0.0003	100	67
257.57	2900			161	D 70	0.0004	4.0081		12.3382	0.0082	2900	67
257.89	1700		67	1700	67	0.0048	4.0129	R	12.3300	0.0048	1700	67
258.02	680			680	67	0.0019	4.0148		12.3252	0.0001	50	D 69
258.98	5100			5100	67	0.0144	4.0293		12.3251	0.0047	1674	66
259.01	148	XING 18[illegible]	75	148	67	0.0004	4.0297	R PRIVATE	12.3204			
259.16	800		75	800	67	0.0023	4.0319	R RYDE	12.3204			
259.24	400			400	67	0.0011	4.0331		12.3204			
259.39	800		95	800	67	0.0023	4.0353	R	12.3204			
259.53	711		68	711	67	0.0020	4.0373	R	12.3204	0.0244	8687	A 36
260.02	2600			2591	67	0.0067	4.0440		12.2960	0.0140	2578	36
261.01	5227			396	A 67	0.0011	4.0451	T VINEYARD	12.2820	0.0283	5272	35
261.61	3200			7018	68	0.0196	4.0647	T RC & CURVE	12.2536	0.0173	3000	35
261.61				1246	D 68	0.0046	4.0693		12.2365	0.0000		
261.83	1137		35	1137	35	0.0062	4.0754	R RC	12.2365	0.0062	1137	35
261.83						0.0000	4.0754		12.2303	0.0001	26	D 36
262.00	900		36	900	35	0.0049	4.0803	L	12.2302	0.0046	874	36
262.17	917		36	917	35	0.0050	4.0852	L	12.2256	0.0048	917	36
262.17						0.0000	4.0852		12.2207	0.0047	1321	D 70
262.53	1900		100	1900	35	0.0103	4.0955	L	12.2160	0.0019	579	70
262.72	1000			1000	35	0.0054	4.1009		12.2145	0.0027	1000	70
262.85	700		71	700	35	0.0038	4.1047	R	12.2116	0.0019	700	70
263.01	800		82	800	35	0.0043	4.1091	R	12.2099	0.0022	800	70
263.10	500		100	500	35	0.0027	4.1118	R	12.2077	0.0014	500	70
263.87	4065			4065	35	0.0221	4.1339		12.2063	0.0111	4065	70
264.01	700		173	198	35	0.0011	4.1349	R	12.1955	0.0019	700	70
264.55	2874		173	58	A 35	0.0003	4.1352	R	12.1934	0.0078	2874	70
265.01	2400		234	1764	36	0.0093	4.1445	R	12.1856	0.0065	2400	70
265.44	2280		234	7893	A 36	0.0264	4.1709	R	12.1791	0.0062	2280	70
266.01	3000			1546	70	0.0042	4.1751		12.1730	0.0081	3000	70
266.44	2300			2300	70	0.0062	4.1813		12.1648	0.0062	2300	70
266.79	1834		82	1834	70	0.0050	4.1863	L NEWTON HAMILTON	12.1586	0.0050	1834	70
266.98	1000			1000	70	0.0027	4.1890		12.1537	0.0054	1983	70
267.05	402			402	70	0.0011	4.1900		12.1483	0.0042	1487	A 60
267.91	4500		89	4500	70	0.0122	4.2022	R	12.1441	0.0069	2412	60
268.42	2700		89	2700	70	0.0073	4.2095	R	12.1372	0.0077	2700	69
268.91	2591			2591	70	0.0070	4.2165		12.1294	0.0072	2504	68
268.91				3086	70	0.0084	4.2249		12.1222	0.0108	3457	A 54
269.53	3267			199	D 70	0.0006	4.2256		12.1114	0.0031	838	A 50
269.91	2000	XING 191	66	2000	66	0.0057	4.2312	L PRIVATE	12.1084	0.0041	1085	50
270.12	1100		60	1100	66	0.0032	4.2344	L MT UNION	12.1042	0.0042	1100	50
270.21	477			477	66	0.0014	4.2357	REVERSE CURVE	12.1000	0.0018	477	50
270.70	2600		74	2600	66	0.0075	4.2432	R	12.0982	0.0095	2600	50
271.01	1600		74	1600	66	0.0046	4.2478	R	12.0864	0.0061	1600	50
271.32	1662			1662	66	0.0048	4.2526		12.0823	0.0084	1660	50
271.70	2000		82	1472	66	0.0042	4.2568	L	12.0760	0.0076	2000	50
271.70				528	D 66	0.0017	4.2585		12.0684	0.0000		
271.76	300		54	0	54	0.0000	4.2585	L	12.0684	0.0011	300	50
271.79	148			285	54	0.0010	4.2595	T RC	12.0672	0.0006	148	50
271.79				158	D 54	0.0006	4.2601	MAPLETON	12.0667	0.0000		
272.70	4800		50	4800	50	0.0182	4.2783	R MPH80	12.0667	0.0182	4800	50
272.92	1200		50	1200	50	0.0045	4.2828	R	12.0485	0.0045	1200	50
272.92							4.2828		12.0440	0.0026	829	D 66
272.92							4.2828		12.0415	0.0002	71	69
273.36	2300			2300	50	0.0087	4.2915		12.0411	0.0038	1400	69
273.64	1477		80	1477	50	0.0056	4.2971	R	12.0373	0.0041	1477	69
273.70	300			300	50	0.0011	4.2983		12.0332	0.0001	51	D 70
274.08	2000			2000	50	0.0076	4.3068		12.0331	0.0061	2249	70
274.69	3246		100	3246	50	0.0123	4.3181	L	12.0270	0.0088	3246	70
274.90	1085		100	1085	50	0.0041	4.3223	L	12.0182	0.0029	1085	70
275.20	1600			592	50	0.0022	4.3245		12.0153	0.0142	5288	70
275.46	1400		87	2903	A 56	0.0092	4.3337	L	12.0010	0.0000	0	70
275.69	1200		89	705	69	0.0019	4.3356	L MILL CREEK	12.0010	0.0033	1205	A 67
276.22	2790	XING 197	82	407	A 69	0.0011	4.3367	L CASSVILLE	11.9977	0.0018	627	67
276.69	2500			4885	70	0.0132	4.3499		11.9962	0.0071	2500	67
276.73	178			178	70	0.0005	4.3504		11.9892	0.0031	1110	67
277.39	3600		141	3600	70	0.0095	4.3599	R	11.9860	0.0117	3744	A 54
277.89	1600			1600	70	0.0043	4.3642		11.9744	0.0111	3177	54
278.20	2700			2700	70	0.0073	4.3715		11.9632	0.0017	468	A 52
278.20				151	D 70	0.0004	4.3719		11.9616	0.0000		
278.43	1179	XING 195	67	1179	67	0.0033	4.3753	R JACOBS	11.9616	0.0024	840	52
278.69	1400			1249	67	0.0035	4.3788		11.9592	0.0051	1400	52
279.70	5267			4821	67	0.0136	4.3927		11.9541	0.0193	5267	52
279.70				577	D 67	0.0016	4.3945		11.9348	0.0000		
279.94	1300		54	1300	54	0.0048	4.3991	L	11.9348	0.0047	1300	52
279.99	243			243	54	0.0009	4.3999	REVERSE CURVE	11.9300	0.0009	243	52
280.39	2100		61	2022	54	0.0071	4.4070	R	11.9291	0.0076	2100	52
280.39				78	D 54	0.0003	4.4073	REVERSE CURVE	11.9215	0.0000		
280.67	1500		52	1500	52	0.0055	4.4127	L	11.9215	0.0055	1500	52
280.86	1000	XING 202	52	1000	52	0.0036	4.4164	L MPH 80 4TH STREET	11.9160	0.0036	1000	52
281.01	800			800	52	0.0029	4.4193	T HUNTINGDON	11.9124	0.0029	800	52
281.20	1000		121	1000	52	0.0036	4.4229	R	11.9095	0.0036	1000	52
281.33	700			705	52	0.0025	4.4255		11.9058	0.0025	700	52
281.62	1500		67	1500	52	0.0055	4.4310	R	11.9033	0.0055	1500	52
281.66	239			239	52	0.0009	4.4318		11.8978	0.0009	239	52
282.27	3197			3197	52	0.0118	4.4436		11.8969	0.0118	3197	52
282.51	1300		107	1300	52	0.0047	4.4482	R	11.8853	0.0047	1300	52
282.67	800		52	800	52	0.0029	4.4511	R	11.8806	0.0029	800	52
282.82	800		52	800	52	0.0029	4.4540	R	11.8777	0.0029	800	52
282.87	300			300	52	0.0011	4.4551	T DEER	11.8747	0.0011	300	52
283.06	1000		132	1000	52	0.0036	4.4588	L RC MPH 55	11.8736	0.0066	1748	52
283.23	900		107	900	52	0.0033	4.4621	L	11.8671	0.0028	731	A 48
283.54	1800			1800	52	0.0058	4.4679		11.8643	0.0039	978	48
283.66	670		98	670	52	0.0024	4.4703	R	11.8605	0.0026	670	48
283.72	277		98	277	52	0.0010	4.4713	R	11.8578	0.0011	277	48
284.66	5000			5000	52	0.0182	4.4895		11.8568	0.0197	5000	48
284.78	800			800	52	0.0027	4.4917		11.8370	0.0024	800	48
284.98	1100		52	653	52	0.0024	4.4941	L	11.8347	0.0043	1100	48
285.17	977		68	977	52	0.0036	4.4977	L	11.8303	0.0036	977	48
285.43	1400	XING 205.6		1400	52	0.0056	4.5031	T WARRIOR RIDGE	11.8285	0.0056	1400	46
285.43				147	D 52	0.0006	4.5037		11.8208	0.0000		
285.66	1200		46	1200	46	0.0047	4.5084	R	11.8209	0.0047	1200	46
285.91	1300		48	1300	48	0.0051	4.5135	R	11.8162	0.0051	1300	48
285.91							4.5135		11.8111	0.0000	224	D 54
285.92	67			67	48	0.0003	4.5138	REVERSE CURVE	11.8102	0.0002	63	54
286.15	1200		54	1200	48	0.0047	4.5186		11.8100	0.0035	1000	54
286.15							4.5186		11.8066	0.0008	231	D 80
286.38	1200		60	1200	48	0.0047	4.5234		11.8067	0.0030	540	60
286.36							4.5234		11.8027	0.0004	135	D 65
286.66	1900			1900	46	0.0059	4.5285		[illegible]	[illegible]	[illegible]	[illegible]

READ DOWN WESTBOUND TRAFFIC

APPENDIX B (3 ZERO) STATION

READ UP EASTBOUND TRAFFIC

PAGE 5
COLUMN 1

PAGE 5
COLUMN 2

READ DOWN WESTBOUND TRAFFIC — APPENDIX 80 (B ZERO) STATION — READ UP EASTBOUND TRAFFIC

MILES	DISTANCE, ft	HWY GRADE CROSSING	MAX MPH	d DISTANCE, feet	SPEED, mph	dT, hours	TIME, hours	PAGE 6 COLUMN 1	TIME, hours	dT, hours	d DISTANCE, feet	SPEED, mph	MILES	DISTANCE, ft	HWY GRADE CROSSING	MAX MPH	d DISTANCE, feet	SPEED, mph	dT, hours	TIME, hours	PAGE 6 COLUMN 2	TIME, hours	dT, hours	d DISTANCE, feet	SPEED, mph
344.83	199			199	32	0.0012	5.8382	RC	10.4479	0.0010	199	36	370.82	1019		71	1919	44	0.0070	6.4477	L	9.8980	0.0079	1819	39
345.05	600		30	600	32	0.0036	5.8417	R	10.4468	0.0063	1578	36	370.98	1900		71	1900	44	0.0082	6.4500	L BOLIVAR	9.8620	0.0090	1900	39
345.20	800		60	800	32	0.0047	5.8485	R	10.4385	0.0022	599 A	31	371.15	600			600	44	0.0084	6.4594		9.8428	0.0039	600	39
345.36	870			870	32	0.0051	5.8510		10.4363	0.0015	299	31	371.30	900		68	900	44	0.0039	6.4630	R	9.8389	0.0044	900	39
345.44	400		56	400	32	0.0026	5.8540	L	10.4346	0.0024	400	31	371.38	1161		71	1058	44	0.0048	6.4678	R	9.8346	0.0057	1161	39
345.53	800		60	600	32	0.0047	5.8587	L	10.4320	0.0046	800	31	371.53				125 D	44	0.0006	6.4683		9.8288	0.0000		
345.72	700		56	700	32	0.0041	5.8628	L MINERAL POINT	10.4271	0.0043	700	31	371.67	1900		40	1500	40	0.0085	6.4768	R	9.8288	0.0087	1800	39
345.86	700			700	32	0.0041	5.8670		10.4220	0.0043	700	31	371.98	569			540	40	0.0026	6.4794	RC	9.8200	0.0028	569	39
345.97	600		38	600	32	0.0036	5.8705	R RC	10.4186	0.0037	600	31	371.98				20 D	40	0.0001	6.4790		9.8175	0.0000		
346.05	400			400	32	0.0024	5.8729	T TT MPH 35	10.4148	0.0024	400	31	372.53	2800		39	2900	39	0.0141	6.4837	L	9.8175	0.0141	2800	39
346.20	800		47	800	32	0.0047	5.8776	L	10.4125	0.0040	800	31	372.60	400		39	400	39	0.0019	6.4956	L	9.8032	0.0019	400	39
346.33	700		47	700	32	0.0041	5.8819	L	10.4076	0.0043	700	31	372.90							6.4956		9.8013	0.0017	306 D	51
346.59	1400		36	1400	32	0.0083	5.8905	L	10.4030	0.0086	1400	31	372.72	831			831	39	0.0031	6.4987		9.7996	0.0009	285	51
346.63	200			200	32	0.0012	5.8913	REVERSE CURVE	10.3948	0.0012	200	31	372.97	1300		71	1300	39	0.0063	6.5050	R	9.7987	0.0048	1300	51
346.75	600		36	600	32	0.0047	5.8960	R	10.3935	0.0040	800	31	373.16	1000			1000	39	0.0049	6.5098		9.7939	0.0037	1000	51
346.84	500			131	32	0.0008	5.8968	RC	10.3868	0.0016	300	31	373.46	1600		100	1600	39	0.0078	6.5176	R	9.7902	0.0043	1171	51
347.14	1600		39	300 A	32	0.0017	5.8984	L	10.3898	0.0086	1800	31	373.51	264		100	264	39	0.0013	6.5189	R	9.7858	0.0033	888 A	46
347.20	286			1755	36	0.0080	5.9077		10.3770	0.0017	286	31	374.04	2800			2800	39	0.0138	6.5326		9.7825	0.0108	2687	46
347.24	215			215	36	0.0011	5.9086	RC	10.3753	0.0013	215	31	374.36	1700		54	1700	39	0.0083	6.5407	R	9.7718	0.0070	1700	46
347.39	800		38	877	36	0.0038	5.9124	R	10.3740	0.0049	800	31	374.45	900		61	500	39	0.0024	6.5432	R	9.7646	0.0021	500	46
347.39				123 D	36	0.0007	5.9101		10.3691	0.0000			374.61	900			860	38	0.0039	6.5470		9.7626	0.0032	860	46
347.54	800		31	800	31	0.0049	5.9179	R	10.3691	0.0049	800	31	375.16	2900		51	405	39	0.0020	6.5490	R	9.7596	0.0119	2900	46
347.62	400			400	31	0.0024	5.9204	RC	10.3642	0.0024	400	31	375.37	1118			1797 A	39	0.0075	6.5565		9.7473	0.0046	1118	46
347.75	700		34	700	31	0.0043	5.9247	L	10.3618	0.0043	700	31	375.46	500		86	2317	51	0.0086	6.5651	R	9.7427	0.0021	500	46
347.88	700		30	700	31	0.0043	5.9289	L	10.3575	0.0043	700	31	375.51	244		66	445	51	0.0017	6.5668	R TORRANCE	9.7407	0.0010	244	46
347.90	100			150	31	0.0006	5.9296	RC	10.3532	0.0006	100	31	375.62	800			221	51	0.0008	6.5676	E END 2 TR (SIDING)	9.7397	0.0023	800	46
348.07	900		33	900	31	0.0056	5.9350	R	10.3526	0.0056	900	31	375.82				178 D	51	0.0007	6.5683	REVERSE CURVE	9.7372	0.0000		
348.20	700		31	700	31	0.0043	5.9393	R	10.3471	0.0043	700	31	376.00	2000		40	2000	46	0.0082	6.5765	L	9.7372	0.0087	2000	46
348.28	377		31	377	31	0.0023	5.9416	R	10.3428	0.0023	377	31	376.23	1200		71	1200	46	0.0049	6.5815	L	9.7290	0.0049	1200	46
348.28							5.9416	AO	10.3405	0.0008	150 D	37	376.38	700		67	700	46	0.0029	6.5843	L	9.7240	0.0028	700	46
348.52	1300			1300	31	0.0079	5.9496		10.3397	0.0059	1150	37	376.46	500		53	500	46	0.0021	6.5864	L	9.7211	0.0021	500	46
348.75	1200		42	1200	31	0.0073	5.9569	R	10.3338	0.0061	1200	37	376.54	436		53	436	46	0.0018	6.5882	L	9.7191	0.0018	436	46
348.98	1200			1200	31	0.0073	5.9642		10.3277	0.0061	1200	37	377.30	4000			4000	46	0.0165	6.6047	DRAVS	9.7173	0.0165	4000	46
349.20	1200		44	1200	31	0.0073	5.9716	L	10.3215	0.0061	1200	37	377.43	700		46	700	46	0.0029	6.6075	R	9.7008	0.0036	700	46
349.39	1000		58	1000	31	0.0061	5.9777	L	10.3154	0.0051	1000	37	377.54	800		46	600	46	0.0025	6.6100	R	9.6979	0.0025	600	46
349.66	1400		37	1400	31	0.0086	5.9862	L	10.3102	0.0072	1400	37	377.54							6.6100		9.6955	0.0015	415 D	57
349.70	200			200	31	0.0012	5.9874	RC	10.3031	0.0010	200 D	44	377.58	177			177	46	0.0007	6.6107	RC	9.6939	0.0002	82	57
349.89	1000		44	1000	31	0.0061	5.9935	R	10.3021	0.0043	892	44	378.11	2800		57	2800	46	0.0115	6.6223	L	9.6937	0.0083	2800	57
349.89					31	0.0000	5.9936		10.2978	0.0010	244 D	51	378.11		XING 301 S					6.6223	T 868 PA	9.6854	0.0014	955 D	70
349.93	248		51	248	31	0.0015	5.9951	R	10.2969	0.0000	2	51	378.41	1800			1800	46	0.0066	6.6289		9.6839	0.0027	995	70
350.05	500		144	600	31	0.0037	5.9987	R	10.2969	0.0022	600	51	378.43	100	XING 302	71	100	46	0.0004	6.6293	R PRIVATE	9.6800	0.0003	100	70
350.20	800			800	31	0.0049	6.0036		10.2948	0.0030	800	51	378.67	1289	XING 302	71	1289	46	0.0053	6.6346	R PRIVATE	9.6801	0.0035	1289	70
350.60	2100			854	31	0.0052	6.0088	T CONEMAUGH TOWER	10.2917	0.0039	1060	51	379.43	4000	XING 303	60	4000	46	0.0165	6.6511	T 889 PA	9.6772	0.0108	4000	70
350.69	500		50	0	31	0.0000	6.0088	R	10.2877	0.0000			380.45	5278			1034	46	0.0043	6.6553		9.6664	0.0143	5278	70
350.71	100			499 A	31	0.0026	6.0118	RC	10.2877	0.0075	1797 A	39	381.00	3000			2087 A	46	0.0081	6.6634		9.6521	0.0081	3000	70
350.94	1200		124	2567	37	0.0131	6.0247	L	10.2802	0.0052	1085	39	381.03	162		102	770	57	0.0026	6.6650	L	9.6440	0.0004	162	70
351.17	1200			1200	37	0.0061	6.0309		10.2751	0.0058	1200	39	381.35	1700			4215 A	57	0.0125	6.6764		9.6435	0.0046	1700	70
351.26	476		66	476	37	0.0024	6.0333	L	10.2692	0.0023	476	38	381.45	400		71	2314	70	0.0083	6.6847	R	9.6389	0.0011	400	70
351.31	300			300	37	0.0015	6.0348	RC	10.2669	0.0015	300	38	381.62	1000		71	1020	70	0.0027	6.6874	R WILLWOOD	9.6378	0.0027	1000	70
351.45	700		102	700	37	0.0036	6.0384	R	10.2655	0.0034	700	38	381.81	1000			1000	70	0.0027	6.6901		9.6351	0.0027	1000	70
351.62	900		63	900	37	0.0046	6.0431	R	10.2621	0.0044	900	38	381.93	677		109	677	70	0.0018	6.6919	L	9.6324	0.0018	677	70
351.83	1100		55	1100	37	0.0056	6.0487	R	10.2577	0.0060	1100	38	382.07	700		141	700	70	0.0019	6.6938	L	9.6306	0.0019	700	70
352.01	1000			654 A	37	0.0039	6.0526		10.2524	0.0049	1000	39	382.43	1900			1900	70	0.0051	6.6990		9.6287	0.0051	1900	70
352.20	1000		51		44	0.0000	6.0526	R	10.2475	0.0049	1000	38	382.52	483			483	70	0.0013	6.7003		9.6236	0.0013	483	70
352.41	1081		56			0.0000	6.0526	R	10.2426	0.0052	1081	39	382.65	700		100	700	70	0.0019	6.7022	R	9.6223	0.0019	700	70
352.58	900			1167 A	44	0.0047	6.0573	T JOHNSTOWN	10.2375	0.0036	600	38	383.45	4150			4150	70	0.0111	6.7133		9.6204	0.0111	4100	70
352.67	600		152	2480	51	0.0091	6.0664	R	10.2338	0.0026	600	38	383.97	2887			2887	70	0.0078	6.7211	T DERRY	9.6093	0.0078	2887	70
352.82	800			404	51	0.0015	6.0679	W END 3 TR	10.2307	0.0033	671	38	384.43	2400	XING 304	100	2400	70	0.0065	6.7276	R VALLEY STREET	9.6015	0.0065	2400	70
352.82				386 D	51	0.0017	6.0696	E END 1 TR SANG HOLLOW	10.2274	0.0000			384.43	287		100	287	70	0.0008	6.7283	R	9.5950	0.0008	287	70
352.86	200		39	200	39	0.0010	6.0706	R	10.2274	0.0060	898 A	28	385.43	5000			5000	70	0.0135	6.7419		9.5942	0.0136	5000	70
353.05	1000			1000	39	0.0049	6.0755		10.2224	0.0009	433	28	385.56	700			700	70	0.0019	6.7438		9.5807	0.0019	700	70
353.24	1000		60	1000	39	0.0048	6.0802	L	10.2195	0.0069	1070	28	386.01	2368		71	2368	70	0.0064	6.7501	R BRADENVILLE	9.5738	0.0064	2368	70
353.32	438		135	438	39	0.0021	6.0823	L	10.2127	0.0030	438	28	386.31	1600			1600	70	0.0043	6.7545		9.5724	0.0043	1600	70
353.70	2000			2000	39	0.0097	6.0920		10.2066	0.0135	2000	28	387.13	4300			4300	70	0.0116	6.7661		9.5661	0.0116	4300	70
353.85	800		115	800	39	0.0039	6.0959	R	10.1962	0.0054	800	28	387.31	989		100	989	70	0.0027	6.7688	L	9.5564	0.0027	989	70
354.23	2000		84	2000	39	0.0097	6.1056	R	10.1908	0.0135	2000	28	388.31	5266		100	5266	70	0.0142	6.7830	L	9.5537	0.0142	5266	70
354.52	1529		70	1529	39	0.0074	6.1131	R	10.1773	0.0103	1529	28	388.42	800		100	800	70	0.0016	6.7846	L	9.5396	0.0016	800	70
354.67	800			800	39	0.0039	6.1170		10.1670	0.0054	800	28	388.18	4000			4000	70	0.0108	6.7955	T LATROBE	9.5378	0.0108	4000	70
354.90	1200		50	1200	39	0.0058	6.1228	R	10.1616	0.0081	1200	28	389.51	700		200	700	70	0.0019	6.7974	R	9.5270	0.0019	700	70
355.05	800			230	39	0.0011	6.1239	DORNICK	10.1534	0.0064	960	28	390.28	5112		200	5112	70	0.0138	6.8112	R	9.5252	0.0138	5112	70
355.05				297	39	0.0014	6.1254		10.1480	0.0000			390.70	2200		104	2200	70	0.0060	6.8171	R BEATTY	9.5113	0.0060	2200	70
355.05				270 D	38	0.0015	6.1269		10.1480	0.0000			391.12	2225		89	2225	70	0.0060	6.8232	R	9.5054	0.0060	2225	70
355.24	1000		28	1000	28	0.0068	6.1336	L	10.1480	0.0068	1000	28	391.15	177		89	177	70	0.0006	6.8236	R	9.4990	0.0005	177	70
355.41	869		28	869	28	0.0060	6.1397	L	10.1413	0.0060	869	28	392.12	5100			5100	70	0.0138	6.8374		9.4989	0.0138	5100	70
355.41							6.1397		10.1352	0.0026	88 D	32	393.12	5278			5278	70	0.0143	6.8517		9.4851	0.0143	5278	70
355.45	200			200	28	0.0014	6.1410	RC	10.1347	0.0007	112	32	393.38	1400			1400	70	0.0038	6.8555		9.4708	0.0038	1400	70
355.64	1000		32	1000	28	0.0066	6.1478	R	10.1340	0.0059	1000	32	393.78	2074		141	2074	70	0.0056	6.8611	L	9.4670	0.0056	2074	70
355.64							6.1478		10.1281	0.0022	469 D	45	394.12	1800			1800	70	0.0049	6.8660	DONOHOE	9.4614	0.0049	1800	70
355.81	900		141	900	28	0.0061	6.1539	R	10.1259	0.0017	431	45	394.65	2800			2800	70	0.0076	6.8736		9.4565	0.0076	2800	70
356.00	1000		46	1000	28	0.0068	6.1606	R	10.1242	0.0030	1000	48	394.80	1039		82	1339	70	0.0036	6.8772	L	9.4489	0.0036	1338	70
356.00							6.1606		10.1202	0.0024	109 D	51	395.34	2300		62	2300	70	0.0062	6.8834	L	9.4453	0.0062	2300	70
356.03	200			200	28	0.0014	6.1620	REVERSE CURVE	10.1198	0.0000	81	51	395.47	682			682	70	0.0018	6.8853		9.4391	0.0018	682	70
356.18	800		56	800	28	0.0054	6.1674	L	10.1196	0.0090	800	51	395.81	1800		82	1800	70	0.0049	6.8901	R	9.4373	0.0049	1800	70
356.20	100			100	28	0.0007	6.1681		10.1105	0.0004	100	51	395.88	400			400	70	0.0011	6.8912		9.4324	0.0011	400	70
356.51	1600			1600	28	0.0108	6.1789		10.1101	0.0059	1600	51	396.00	600			600	70	0.0016	6.8928		9.4313	0.0016	600	70
356.81	1600		78	1600	28	0.0108	6.1897	R	10.1102	0.0059	1600	51	396.28	1500		189	1500	70	0.0041	6.8969	L	9.4297	0.0041	1500	70
356.96	800			800	28	0.0054	6.1951		10.1043	0.0030	800	51	396.55	1397		245	1397	70	0.0038	6.9007	L	9.4256	0.0038	1397	70
357.08	600		130	600	28	0.0041	6.1992	R	10.1013	0.0022	600	51	396.89	1800			1800	70	0.0049	6.9056		9.4218	0.0049	1800	70
357.15	400		54	400	28	0.0027	6.2019	R	10.0991	0.0015	400	51	397.78	4700			4700	70	0.0127	6.9183	T GREENSBURG	9.4170	0.0127	4700	70
357.20	282			282	28	0.0019	6.2038	RC	10.0976	0.0010	282	51	397.89	579		141	579	70	0.0016	6.9198	R	9.4043	0.0016	579	70
357.34	700		61	700	28	0.0047	6.2085	L	10.0965	0.0026	700	51	398.15	1400		141	1400	70	0.0038	6.9236	R	9.4027	0.0080	2235	70
357.34							6.2085	WEST END JOHNSTOWN	10.0939	0.0015	458 D	62	398.15				0			6.9236		9.3948	0.0011	407 A	68
358.09	4000			818	28	0.0063	6.2141	SC	10.0924	0.0106	3644	62	398.45	1594		71	1594	70	0.0043	6.9279	R	9.3936	0.0010	354	68
358.20	586		141	286 A	28	0.0017	6.2157	L	10.0818	0.0093	2711	62	398.89	2300			2300	70	0.0062	6.9342		9.3946	0.0063	2300	68
358.25	257		141	900	32	0.0053	6.2211	L	10.0725	0.0078	2589 A	53	399.21	1700		141	1700	70	0.0046	6.9388	L	9.3882	0.0047	1700	68
359.20	5000			34	32	0.0002	6.2213		10.0656	0.0027	745	53	399.82	3000			3000	70	0.0067	6.9474		9.3836	0.0085	3000	68
359.85	300			1847 A	32	0.0068	6.2280		10.0629	0.0020	600	53	399.88	335		200	335	70	0.0009	6.9483	L	9.3748	0.0006	335	68
360.11	4000		62	53	48	0.0002	6.2301	L	10.0600	0.0140	4000	53	399.94	311		200	311	70	0.0008	6.9492	L	9.3739	0.0009	311	60
360.14	200			560 A	48	0.0022	6.2323	RC SANG HOLLOW	10.0457	0.0007	200	53	400.47	2800			2749	70	0.0074	6.9566		9.3730	0.0077	2800	60
360.18	237		64	8502	51	0.0261	6.2584	R	10.0450	0.0006	237	53	400.47				51 D	70	0.0001	6.9567		9.3653	0.0000		
360.19	20		64	2751 A	51	0.0082	6.2656	R	10.0441	0.0001	20	53	400.64	900		60	900	60	0.0025	6.9592	R	9.3653	0.0053	1846	60
360.95	4000			5920	62	0.0183	6.2819		10.0441	0.0143	4000	53	400.66	100			100	60	0.0003	6.9595		9.3600	0.0106	3865 A	53
361.14	1000		92	821	62	0.0018	6.2850	L	10.0296	0.0026	1000	53	401.66	5905			5905	68	0.0146	6.9740	T RADEBAUGH	9.3494	0.0026	802	58
361.14				360 D	62	0.0015	6.2850	L SEWARD	10.0062	0.0000			401.88	1700			1700	39	0.0047	6.9787		9.3468	0.0015	698	58
361.71	2896		53	2896	53	0.0107	6.2957	E END 2 TR SANG HOLLOW	10.0062	0.0107	2896	53	402.67	3588		100	3588	68	0.0098	6.9885	L JEANNETTE	9.3203	0.0089	1971 A	49
362.09	2000			2000	53	0.0071	6.3029		10.0155	0.0024	787 D	75	403.16	2600		100	300	69	0.0008	6.9894	L	9.3134	0.0002	46	49
362.13	214			214	53	0.0008	6.3037	T	10.0131	0.0038	1447	70	403.46	1592			407 A	59	0.0011	6.9905		9.3132	0.0062	1592	49
363.07	5000		141	5000	53	0.0179	6.3215	R	10.0092	0.0135	5000	70	403.46				2913	70	0.0079	6.9984		9.3120	0.0000		
363.50	2286		141	2286	53	0.0081	6.3296	R	9.9957	0.0061	2286	70	403.46				563 D	70	0.0017	7.0001		9.3120	0.0000		
364.07	3000			1521	55	0.0054	6.3351	W END 2 TR SANG HOLLOW	9.9896	0.0061	3000	70	403.67	1100		58	1100	58	0.0036	7.0037	R	9.3120	0.0043	1100	49
364.48	2139			8184 A	53	0.0158	6.3506	T NEW FLORENCE	9.9815	0.0056	2139	70	403.77	582		58	582	58	0.0019	7.0056	R	9.3078	0.0022	582	49
365.04	3000		89	1434	70	0.0039	6.3548	R	9.9757	0.0081	3000	70	404.46	3800	XING 328 6		3247	58	0.0106	7.0161	WRIGLEYS	9.3056	0.0139	3800	49
365.42	2000		89	2000	70	0.0054	6.3602	R	9.9675	0.0054	2000	70	404.46				353 D	58	0.0013	7.0173		9.2917	0.0000		
365.80	2000			2000	70	0.0054	6.3656		9.9621	0.0054	2000	70	404.65	1000		49	1000	49	0.0039	7.0212	L	9.2917	0.0039	1000	49
365.96	820		141	820	70	0.0022	6.3678	L	9.9567	0.0004	142	70	404.66	100			100	49	0.0004	7.0216		9.2878	0.0004	100	49
366.11	824		69	824	70	0.0022	6.3700	L	9.9563	0.0000		70	404.82	800			800	49	0.0031	7.0247		9.2874	0.0031	800	49
366.87	4000			4000	70	0.0108	6.3809		9.9563	0.0000		70	404.99	900		49	900	49	0.0035	7.0282	R	9.2843	0.0035	900	49
366.87					70	0.0000	6.3809	200	9.9563	0.0128	4218 A	57	405.23	1300			1300	49	0.0050	7.0332	T MANOR	9.2800	0.0050	1300	49
367.07	1056			1056	70	0.0029	6.3837	T CONPITT JCT	9.9438	0.0004	737	57	405.44	1105		49	1105	49	0.0043	7.0375	L	9.2758	0.0043	1105	49
367.13	300	XING 250		300	70	0.0008	6.3845	PRIVATE	9.9414	0.0000			405.44							7.0375		9.2718	0.0028	815 D	70
367.20	400		113	400	70	0.0011	6.3856	R REVERSE CURVE	9.9414	0.0084	2038 A	44	405.67	1200			1200	49	0.0046	7.0421		9.2687	0.0008	354	70
367.28	400		113	400	70	0.0011	6.3867	L MPH 50	9.9317	0.0007	158	44	405.93	1400			1400	49	0.0054	7.0475	T GRAFTON	9.2679	0.0038	1400	70
367.79	2720			2720	70	0.0074	6.3941		9.9312	0.0117	2720	44	406.66	3810		71	3610	49	0.0147	7.0622	R	9.2641	0.0103	3810	70
368.71	4687		71	4292	70	0.0114	6.4055	R	9.9195	0.0208	4837	44	407.55	4720		135	4590	49	0.0177	7.0800	R [illegible]	9.2538	0.0108	4000	70
368.71				905 D	70	0.0018	6.4073		9.8987	0.0148	3351	44	407.62	376		135	5833 A	48	0.0186	7.0986	R	9.2430	0.0010	376	70
369.28	3000		57	2700	57	0.0090	6.4163	R	9.8841	0.0022	494 A	40	408.47	4500			0			7.0986		9.2420	0.0122	4500	70
369.33	275			94	57	0.0003	6.4166	RC	9.8819	0.0107	2250	40	408.53	285			0			7.0986		9.2298	0.0008	285	70
369.33				481 D	57	0.0016	6.4181	INDIA	9.8713	0.0000			408.79	1400		100	756	70	0.0021	7.1006	R LARIMER	9.2290	0.0058	1400	70
369.61	1500		44	1500	44	0.0065	6.4246	L	9.8713	0.0006	118 A	39	409.47	3600			3600	70	0.0097	7.1104		9.2232	0.0097	3600	70
369.99	2000		44	2005	44	0.0086	6.4335	L LOCKPORT	9.8707	0.0026	531	39	408.53	285			285	70	0.0008	7.1111		9.2155	0.0008	285	70
370.32	1700			1700	44	0.0073	6.4406	[illegible]	9.8681	0.0048	1700	39	408.70	900		90	900	70	0.0024	7.1136	L ARDARA	9.2147	0.0080	2200	70

PAGE 7 COLUMN 1

				READ DOWN WESTBOUND TRAFFIC				APPENDIX 9B (B ZERO) STATION	READ UP EASTBOUND TRAFFIC			
MILES	DISTANCE, ft	HWY GRADE CROSSING	MAX MPH	Δ DISTANCE, feet	SPEED, mph	ΔT, hours	TIME, hours		TIME, hours	ΔT, hours	Δ DISTANCE, feet	SPEED, mph
409.81	600		100	600	70	0.0016	7.1162 L		9.2088	0.0053	1887 A	65
410.48	3500			3500	70	0.0095	7.1247		9.2035	0.0027	918	65
410.85	2000			2000	70	0.0054	7.1301		9.2008	0.0058	2000	65
411.37	2749		100	2749	70	0.0074	7.1375 R		9.1950	0.0080	2749	65
411.46	455		105	455	70	0.0012	7.1387 R	CP TRAFF	9.1870	0.0020	460	65
411.46				0	70	0.0000	7.1387	TRAFFORD	9.1840	0.0000		
412.38	4800			4858	70	0.0126	7.1513		9.1842	0.0161	4741 A	45
412.38				249 C	70	0.0007	7.1520		9.1681	0.0000		
412.90	2700		66	2700	66	0.0079	7.1598 L		9.1681	0.0032	771	45
413.42	2743		74	2743	66	0.0080	7.1678 L		9.1648	0.0182	4314	45
413.42							7.1676		9.1467	0.0086	1571	45
413.66	1300		74	1300	66	0.0038	7.1718 L		9.1401	0.0022	305 A	41
414.10	2300			2300	66	0.0067	7.1783	PITCAIRN YARD	9.1378	0.0070	1524	41
414.36	1500		71	1500	66	0.0044	7.1827	R WALL	9.1309	0.0068	1500	41
414.64	1351			1351	66	0.0039	7.1866		9.1239	0.0067	1351	41
414.64				0			7.1866		9.1176	0.0030	647	41
414.83	1005			220	65	0.0007	7.1873		9.1147	0.0017	348 A	35
414.83				807 D	62	0.0028	7.1901		9.1130	0.0002	36	38
415.10	1407		45	1407	45	0.0059	7.1960 R		9.1128	0.0070	1407	38
415.31	1115			1115	45	0.0047	7.2007	WILMERDING	9.1058	0.0051	1032	38
415.34	185		66	185	45	0.0007	7.2014 L		9.1006			
415.34						0.0000	7.2014	CP WING	9.1006	0.0025	488 A	33
415.53	1004		90	1004	45	0.0042	7.2056 L		9.0981	0.0046	784	32
415.63	515			515	45	0.0020	7.2076		9.0936	0.0030	515	33
415.88	1364			1258	45	0.0053	7.2131		9.0907	0.0072	1364	33
415.88	0			126 D	45	0.0008	7.2138		9.0827	0.0000	0	33
416.16	1411		41	1411	41	0.0065	7.2201 L		9.0827	0.0081	1411	33
416.61	2402			2402	41	0.0111	7.2312		9.0746	0.0138	2402	33
416.74	671			671	41	0.0031	7.2343		9.0609	0.0039	671	33
416.91	887		58	800	41	0.0037	7.2380 R		9.0570	0.0051	887	33
416.91				87 D	41	0.0004	7.2385		9.0519	0.0000		33
417.08	887		58	887	38	0.0044	7.2428 R		9.0519	0.0051	887	33
417.20	634			634	38	0.0032	7.2460		9.0468	0.0036	634	33
417.38	966		58	806	38	0.0042	7.2502 L		9.0432	0.0055	966	33
417.38	0			130 D	38	0.0007	7.2509		9.0376	0.0000	0	33
417.47	460			460	33	0.0029	7.2538		9.0376	0.0026	458	33
417.57	549		33	549	33	0.0032	7.2567 R		9.0350	0.0032	549	33
417.58	58			58	33	0.0003	7.2570		9.0319	0.0003	30 D	39
417.60	116		36	116	33	0.0007	7.2577 L		9.0316	0.0007	124	35
417.60					33	0.0000	7.2577		9.0309	0.0006	109 D	36
417.68	424		44	424	33	0.0024	7.2601 L		9.0303	0.0015	315	39
418.11	2251			2251	33	0.0129	7.2730		9.0288	0.0159	2011	39
418.26	794		45	794	33	0.0046	7.2776 R		9.0179	0.0039	784	38
418.56	1665			1665	33	0.0097	7.2873		9.0140	0.0082	1595	39
418.60	106		42	106	33	0.0006	7.2879 R		9.0058	0.0025	106	39
418.73	667		40	667	33	0.0038	7.2917 R		9.0033	0.0032	667	39
418.87	757			757	33	0.0043	7.2960		9.0021	0.0037	757	39
419.00	667		26	667	33	0.0038	7.2998 L		8.9984	0.0032	667	38
419.40	2113			2113	33	0.0121	7.3120		8.9952	0.0014	323 D	46
419.61	1107		50	1107	33	0.0064	7.3183 L		8.9938	0.0112	2897	49
419.61	0			259	33	0.0015	7.3198		8.9826			
419.70	500		50	189 A	33	0.0011	7.3209 L		8.9826	0.0019	500	49
419.73	156			363 A	35	0.0018	7.3226		8.9807	0.0008	156	49
420.00	1428		54	1228	38	0.0059	7.3286 R		8.9801	0.0054	1405	49
420.28	1484			1484	38	0.0072	7.3360		8.9748	0.0057	1464	49
420.60	1719		49	1719	38	0.0083	7.3444 R		8.9690	0.0066	1719	49
420.61	26			29	38	0.0001	7.3445		8.9622	0.0019	575 D	63
420.61	0				38	0.0000	7.3445	SWISSVALE	8.9603	0.0004	137	63
421.04	2288			2489	38	0.0121	7.3566		8.9599	0.0046	1605	63
421.27	1220		71	1490 A	38	0.0061	7.3627 L		8.9551	0.0018	589	63
421.60	1702			1292	49	0.0050	7.3677		8.9538			63
421.60	0				49	0.0000	7.3677		8.9535	0.0150	4342 A	43
422.07	2480			2489	48	0.0096	7.3773		8.9380	0.0021	480	43
422.35	1509		92	1509	48	0.0058	7.3831 L		8.9359	0.0218	4058	43
422.68	1721		85	1721	48	0.0067	7.3898 L		8.9141	0.0006	125 A	42
422.68					49	0.0000	7.3898	WILKINSBURG	8.9139	0.0113	2499	42
423.57	4720			44	49	0.0032	7.3900		8.9032	0.0047	841 A	34
423.57				3304 A	49	0.0114	7.4014	EAST LIBERTY	8.8981	0.0090	1621	34
424.06	2589			3072	65	0.0092	7.4106		8.8890	0.0017	264 A	30
424.06				777 D	62	0.0028	7.4134		8.8874	0.0012	188	30
424.27	1140		43	1140	43	0.0050	7.4184 L		8.8862	0.0072	1140	30
424.58	1604			1604	43	0.0071	7.4255		8.8790	0.0101	1604	30
424.66	560			560	43	0.0026	7.4279		8.8688	0.0035	562	30
424.87	1001		100	1001	43	0.0044	7.4323 L		8.8653	0.0063	1001	30
425.07	1022		57	1228	43	0.0054	7.4377 L		8.8590	0.0056	1027	30
425.11	238			31 D	42	0.0001	7.4379		8.8526	0.0015	236	30
425.49	1918		42	1918	42	0.0086	7.4465 R		8.8510	0.0121	1918	30
425.58	530			530	42	0.0024	7.4489		8.8389	0.0033	530	30
425.67	507		30	489	42	0.0035	7.4519 R		8.8356	0.0032	507	30
425.75	395			223 D	40	0.0011	7.4531		8.8324	0.0024	388	30
425.86	612		34	1811	34	0.0051	7.4651 L		8.8298	0.0039	612	30
426.11	1285			94 D	34	0.0006	7.4657		8.8281	0.0082	1285	30
426.25	776		30	778	30	0.0048	7.4695 R		8.8179	0.0046	778	30
426.35	491			491	30	0.0031	7.4717		8.8130	0.0061	974	30
426.58	1200		30	1200	30	0.0076	7.4795 L		8.8069	0.0018	283 A	25
426.58	0				30	0.0000	7.4795		8.8049	0.0036	464	25
426.72	752		37	752	30	0.0047	7.4842 L		8.8014	0.0057	752	25
426.95	1196			1196	30	0.0075	7.4918		8.7957	0.0091	1196	25
427.11	860		67	860	30	0.0054	7.4971 L		8.7867	0.0065	862	25
427.57	2456			2456	30	0.0155	7.5126		8.7802	0.0180	2450	25
427.77	1061			1061	30	0.0066	7.5195		8.7610	0.0080	1061	25
427.82	264		71	264	30	0.0017	7.5209 R		8.7536	0.0020	264	25
428.37	3064			3064	30	0.0044	7.5403		8.7516	0.0030	3065	25
428.57	0			101 D	30	0.0007	7.5400	PITTSBURGH PENN STATION	8.7216			
428.77	1016		25	1016	25	0.0077	7.5537 R		8.7218	0.0077	1016	25
428.77	0						7.5537		8.7139	0.0014	320 D	35
428.77	0						7.5537	ALLEGHENY RIVER	8.7126	0.0005	58	35
428.90	675		50	675	25	0.0051	7.5588 L		8.7122	0.0022	400	36
428.96	328			328	25	0.0025	7.5613		8.7100	0.0018	328	36
429.12	837		27	837	25	0.0063	7.5677 L		8.7083	0.0045	837	35
429.17	261			261	25	0.0020	7.5697		8.7037	0.0014	261	36
429.38	1145		42	1145	25	0.0087	7.5783 R		8.7023	0.0060	1145	36
429.60	1124			1124	25	0.0085	7.5869		8.6963	0.0061	1124	35
429.65	263		50	263	25	0.0022	7.5891 R		8.6900	0.0016	293	36
429.73	417		35	417	25	0.0032	7.5922 R	FEDERAL STREET	8.6885	0.0023	417	36
429.81	422		45	422	25	0.0032	7.5954 R		8.6862	0.0023	422	35
429.89	412		35	412	25	0.0031	7.5985 R		8.6839	0.0022	412	36
429.92	153			153	25	0.0012	7.5997	REVERSE CURVE	8.6817	0.0006	106 D	38
430.00	417		39	417	25	0.0032	7.6029 L		8.6811	0.0022	481	39
430.08	422		66	422	25	0.0032	7.6061 L		8.6789	0.0014	320 D	46
430.35	1465			1465	25	0.0111	7.6171		8.6775	0.0025	646	49
430.47	626		57	626	25	0.0046	7.6218 R		8.6750	0.0024	626	49
430.48	53			53	25	0.0004	7.6223 RC		8.6726	0.0002	53	49
430.56	422		57	422	25	0.0032	7.6255 L		8.6724	0.0016	422	49
430.65	476			476	25	0.0036	7.6291		8.6707	0.0016	476	49
430.65	0				25	0.0000	7.6291	JCT CORR WB	8.6690	0.0000	0	49
430.77	620			620	25	0.0047	7.6338		8.6690	0.0024	620	49
430.85	440		60	437	25	0.0033	7.6371 R		8.6665	0.0017	440	49
430.89	228			115 A	29	0.0008	7.6379 RC		8.6648	0.0009	228	49
431.02	680		56	799	27	0.0056	7.6435 L		8.6640	0.0026	685	49
431.24	1113			927	27	0.0065	7.6500		8.6612	0.0043	1118	49
431.36	638		40	558 A	27	0.0030	7.6530 R		8.6570	0.0025	636	49
431.60	1287			1561	33	0.0084	7.6618		8.6545	0.0026	916 D	70
431.63	168		97	189	38	0.0010	7.6628 R		8.6516	0.0039	1456	70
431.82	1013		97	1015	38	0.0055	7.6680 R		8.6477	0.0027	1015	70
432.05	1210			796	39	0.0041	7.6724		8.6449	0.0033	1210	70
432.15	501		71	383 A	35	0.0019	7.6743 L		8.6417	0.0014	501	70
432.41	1396			487	39	0.0009	7.6752		8.6403	0.0036	1396	70
432.48	373		100	1435 A	38	0.0061	7.6814 L		8.6368	0.0005	285	70
432.66	917			1919	49	0.0063	7.6876		8.6359			
432.69	175			175	49	0.0007	7.6863		8.6359			

PAGE 7 COLUMN 2

				READ DOWN WESTBOUND TRAFFIC				APPENDIX 9B (B ZERO) STATION	READ UP EASTBOUND TRAFFIC			
MILES	DISTANCE, ft	HWY GRADE CROSSING	MAX MPH	Δ DISTANCE, feet	SPEED, mph	ΔT, hours	TIME, hours		TIME, hours	ΔT, hours	Δ DISTANCE, feet	SPEED, mph
432.81	818		72	618	48	0.0024	7.6887 L		8.6359	0.0072	2520 A	68
433.10	1534			1534	49	0.0059	7.6946		8.6287	0.0025	940	68
433.21	606		64	605	49	0.0023	7.6969 L		8.6261	0.0005	96	63
433.43	1166			1166	49	0.0045	7.7015		8.6259			63
433.48							7.7015	JCT CORR EB	8.6259			68
433.51	414		74	36	48	0.0001	7.7016 R		8.6259	0.0079	2410 A	54
433.63	637			5053 A	49	0.0186	7.7222		8.6181	0.0011	319	54
433.65	100		160				7.7222 R		8.6170	0.0004	108	54
433.91	1382		130				7.7222 R		8.6166	0.0045	1382	54
433.99	385						7.7222 RC		8.6117	0.0014	385	54
434.11	670		68				7.7222 L		8.6104	0.0025	670	54
434.68	2805			440	70	0.0013	7.7235		8.6080	0.0105	2805	54
434.72	343			1584	70	0.0043	7.7278	AVALON	8.5995	0.0012	343	54
435.02	1582		110	541 D	70	0.0016	7.7288 L		8.5968	0.0059	1582	54
435.22	1000		65	1060	63	0.0032	7.7320 L		8.5912	0.0037	1000	54
435.32	500			506	63	0.0015	7.7335 RC		8.5875	0.0018	500	54
435.53	1107		70	1107	63	0.0033	7.7368 R		8.5857	0.0039	1107	54
435.59	327			327	63	0.0010	7.7378 RC		8.5819	0.0011	327	54
435.66	348		60	340	63	0.0010	7.7388 L		8.5807	0.0012	348	54
435.66	0			366 D	63	0.0011	7.7400	EMSWORTH	8.5796	0.0000	0	54
435.75	491		54	491	54	0.0017	7.7416 L		8.5796	0.0017	491	54
435.81	312			312	54	0.0011	7.7428 RC		8.5778	0.0022	727 D	70
435.93	671		80	671	54	0.0024	7.7452 R		8.5756	0.0007	258	70
436.21	1489			1492	54	0.0053	7.7504		8.5749	0.0040	1489	70
436.29	402		122	402	54	0.0014	7.7518 R		8.5709	0.0011	402	70
436.65	1808			1666	54	0.0060	7.7584		8.5698	0.0031	1806	70
436.66	53		141	53	54	0.0002	7.7586 L		8.5667	0.0001	53	70
436.93	1431		141	1431	54	0.0050	7.7637 L		8.5666	0.0039	1431	70
437.14	1140		194	1140	54	0.0040	7.7677 L		8.5627	0.0031	1140	70
437.65	2703			2703	54	0.0093	7.7771		8.5576	0.0058	2137	70
437.65	0						7.7771		8.5518	0.0042	1497 A	68
438.25	3150			935	54	0.0033	7.7804		8.5476	0.0064	2219	66
438.45	1073		100	4949 A	54	0.0160	7.7964 R		8.5413	0.0035	1073	66
438.65	1062						7.7964		8.5380	0.0030	1062	66
438.74	464						7.7964		8.5352	0.0013	464	66
438.84	536		60	19	70	0.0001	7.7965 L		8.5338	0.0015	536	66
439.00	853			815	70	0.0022	7.7987		8.5323	0.0025	853	66
439.18	960		100	960	70	0.0026	7.8013 R		8.5296	0.0029	960	66
439.68	2512			2510	70	0.0068	7.8070		8.5272	0.0072	2512	66
439.68	0			1182	70	0.0031	7.8102		8.5200	0.0000	0	66
439.91	1239			109 D	70	0.0005	7.8107		8.5200	0.0039	1239	66
440.08	887		68	887	66	0.0025	7.8132 L		8.5161	0.0026	887	66
440.17	472			472	66	0.0014	7.8146 RC		8.5136	0.0014	472	66
440.34	892		66	892	66	0.0026	7.8172 R		8.5122	0.0026	892	66
440.42	401			431	66	0.0012	7.8184 RC		8.5096	0.0004	109 D	70
440.46	319		96	319	66	0.0009	7.8194 L		8.5091	0.0015	561	70
440.54	319		80	316	66	0.0009	7.8203 L		8.5076	0.0009	319	70
440.66	644			644	66	0.0018	7.8221		8.5067	0.0017	644	70
440.70	211			211	66	0.0006	7.8227	SEWICKLEY	8.5050	0.0006	211	70
440.79	475		141	475	66	0.0014	7.8241 R		8.5044	0.0015	475	70
440.91	644			644	66	0.0024	7.8265		8.5031	0.0023	644	70
441.22	1403		102	1403	66	0.0040	7.8305 R		8.5006	0.0038	1403	70
441.42	1050		80	1050	66	0.0030	7.8335 R		8.4970	0.0028	1050	70
441.58	833		75	835	66	0.0024	7.8359 R		8.4942	0.0023	833	70
441.71	723		90	725	66	0.0021	7.8380 R		8.4919	0.0020	723	70
441.78	369			369	66	0.0011	7.8391		8.4900	0.0010	369	70
441.96	924			924	66	0.0027	7.8417		8.4890	0.0025	924	70
442.05	489		141	489	66	0.0014	7.8431 L		8.4865	0.0013	489	70
442.34	1525			1525	66	0.0044	7.8475		8.4852	0.0041	1525	70
442.43	489		141	441	66	0.0013	7.8488 L		8.4810	0.0013	489	70
442.78	1729			1407 A	66	0.0042	7.8530		8.4797	0.0047	1729	70
442.80	214			466	70	0.0013	7.8543		8.4750	0.0006	214	70
442.90	985		82	985	70	0.0028	7.8571 R		8.4746	0.0026	985	70
443.22	1733			1233	70	0.0033	7.8604		8.4718	0.0018	1233	70
443.44	1180		87	1180	70	0.0032	7.8636 L		8.4685	0.0032	1180	70
443.66	1340			1340	70	0.0036	7.8670		8.4653	0.0054	1880	70
443.69	0						7.8670		8.4599	0.0064	2280 A	64
444.66	5187			3197	70	0.0141	7.8811		8.4536	0.0068	2283	64
445.30	3282		178	3282	70	0.0089	7.8900 R		8.4468	0.0097	3282	64
445.60	1588		82	1588	70	0.0043	7.8943 R		8.4371	0.0047	1588	64
445.68	423		194	425	70	0.0011	7.8954 R		8.4324	0.0013	423	64
446.03	789		194	789	70	0.0021	7.8976 R	AMBRIDGE	8.4311	0.0023	789	64
446.15	1669		115	2342	70	0.0063	7.9039 R		8.4288	0.0050	1669	64
446.33	948			295 D	82	0.0008	7.9047		8.4238	0.0028	948	64
446.59	1372		64	1372	64	0.0041	7.9088 R		8.4210	0.0041	1372	64
446.66	385			385	64	0.0011	7.9099 RC		8.4169	0.0005	194 D	68
446.66	0						7.9099		8.4164	0.0008	171	68
446.69	158			158	64	0.0005	7.9103		8.4159	0.0004	158	68
446.99	1583		66	1583	64	0.0047	7.9150 L		8.4154	0.0044	1583	68
447.33	1810			1810	64	0.0054	7.9204		8.4110	0.0002	101 D	70
447.64	1647		135	1647	64	0.0049	7.9252 R		8.4108	0.0091	3356	70
447.65	53			53	64	0.0002	7.9254		8.4017	0.0001	53	70
447.94	1502			1502	64	0.0044	7.9298		8.4015	0.0041	1502	70
448.13	1012		200	1012	64	0.0030	7.9328 L		8.3975	0.0027	1012	70
448.45	1716			1716	64	0.0051	7.9379		8.3947	0.0046	1716	70
448.57	585		96	585	64	0.0017	7.9397 L		8.3901	0.0016	585	70
448.66	479			479	64	0.0014	7.9411		8.3885	0.0013	478	70
448.79	724			95	64	0.0003	7.9415		8.3872	0.0020	724	70
448.82	661		88	1451 A	64	0.0042	7.9458 L		8.3852	0.0018	661	70
449.41	2812			635	64	0.0018	7.9475		8.3834	0.0071	2812	70
449.50	463		129	609 A	56	0.0022	7.9496 L		8.3764	0.0013	463	70
449.63	666			2148	70	0.0058	7.9554		8.3751	0.0018	666	70
449.65	126		158	126	70	0.0003	7.9557 R		8.3733	0.0003	126	70
449.73	423		158	423	70	0.0011	7.9568 R	EAST CONWAY	8.3730	0.0011	423	70
449.90	886			588	70	0.0025	7.9592		8.3719	0.0027	985	70
450.08	945		106	945	70	0.0026	7.9617 R		8.3684	0.0042	1487 A	66
450.26	965		104	965	70	0.0026	7.9643 R		8.3650	0.0008	202	66
450.56	1607			1607	70	0.0043	7.9687		8.3641	0.0001	50	66
450.66	517		116	517	70	0.0014	7.9701 R		8.3640			
450.66	0						7.9701	CP BADEN	8.3640	0.0083	2718 A	57
451.17	2679		115	2679	70	0.0072	7.9773 R		8.3556	0.0056	2025	57
451.66	2605			2605	70	0.0070	7.9844		8.3489	0.0087	2605	57
451.66	0						7.9844		8.3402	0.0022	662	57
451.74	415			418	70	0.0011	7.9856		8.3380	0.0017	464 A	55
451.90	838		74	838	70	0.0023	7.9878 R		8.3364	0.0003	85	55
452.08	942		50	942	70	0.0025	7.9903 R		8.3361	0.0064	1713 A	48
452.28	1075		73	830	70	0.0024	7.9926 R		8.3297	0.0013	320	46
452.29	26			189 D	65	0.0006	7.9933 RC		8.3284	0.0001	26	46
452.55	1391		80	1391	66	0.0039	7.9972 L		8.3283	0.0058	1391	46
452.64	471			471	66	0.0014	7.9986		8.3227	0.0019	471	46
452.72	450			450	66	0.0013	7.9999		8.3207	0.0019	450	46
452.80	403		68	374	66	0.0011	8.0009 R		8.3188	0.0017	403	46
452.87	377			406 D	66	0.0013	8.0022 RC		8.3172	0.0016	377	46
453.10	1220		87	1220	57	0.0041	8.0063 L		8.3157	0.0050	1220	46
453.30	1057		113	1057	57	0.0035	8.0098 L		8.3106	0.0044	1057	46
453.46	837		96	837	57	0.0028	8.0126 L		8.3063	0.0034	837	46
453.57	576		90	576	57	0.0019	8.0145 L		8.3025	0.0024	576	46
453.65	416		113	416	57	0.0014	8.0158 L		8.3006	0.0017	416	46
453.78	686		112	686	57	0.0023	8.0181 L		8.2987	0.0028	686	46
453.96	997		71	82 D	57	0.0003	8.0184 L		8.2959	0.0041	997	46
454.10	691		58	681	56	0.0024	8.0208 L		8.2918	0.0028	691	46
454.17	369		58	389	56	0.0013	8.0221 L		8.2890	0.0015	369	46
454.20	158			158	56	0.0005	8.0226 RC		8.2874	0.0007	158	46
454.24	211		92	1150	56	0.0040	8.0266 R		8.2868	0.0009	211	46
454.31	369			383 D	56	0.0013	8.0279 RC		8.2859	0.0015	369	46
454.39	422		48	422	46	0.0017	8.0296 L		8.2844	0.0017	422	46
454.47	420		56	422	46	0.0017	8.0313 L		8.2827	0.0006	141 D	55
454.56	475			475	46	0.0020	8.0333	JCT BAYARD BRANCH	8.2821	0.0028	759	50
454.65	505		100	528	46	0.0022	8.0354 L	MAXIMUM SPEED LIMIT 90	8.2792	0.0020	525	50
454.92	1417			1417	46	0.0058	8.0413		8.2773	0.0054	1417	50
455.08	795		60	735	46	0.0030	8.0443 L		8.2719	0.0028	735	50

PAGE 8
COLUMN 1

PAGE 8
COLUMN 2

READ DOWN WESTBOUND TRAFFIC — APPENDIX 80 (S ZERO) STATION — READ UP EASTBOUND TRAFFIC

PAGE 9 COLUMN 1

MILES	DISTANCE, ft	HWY GRADE CROSSING	MAX MPH	DISTANCE, feet	SPEED, mph	ET, hours	TIME, hours	STATION	TIME, hours	ET, hours	DISTANCE, feet	SPEED, mph
507.49	3752			3752	40	0.0178	9.3312		8.9872	0.0297	3752	30
507.66	900			900	40	0.0040	9.3354	MPH 40	8.9705	0.0057	900	30
507.96	1600	XING	100	1600	40	0.0076	9.3430	R US HWY 30	8.9679	0.0101	1600	30
508.11	800			800	40	0.0038	9.3468	E ROCHESTER	8.9577	0.0051	800	30
508.51	2075		80	2075	40	0.0098	9.3569	L	8.9527	0.0131	2075	30
508.58	300		60	300	40	0.0014	9.3580	L	8.9396	0.0019	300	30
508.70	714			714	40	0.0034	9.3614	BAYARD BRANCH	8.9377	0.0045	714	30
508.70		JCT 54.9		267 D	40	0.0014	9.3628	JCT MAHONING SECTY	8.9332	0.0000	0	30
509.14	2320		30	2320	30	0.0146	9.3775	R BAYARD TT 30 MPH	8.9332	0.0146	2320	30
509.14	3	XING 54.7		3	30	0.0000	9.3775	RC BAYARD ROAD	8.9186	0.0000	3	30
509.25	600		115	600	30	0.0038	9.3813	L	8.9186	0.0038	600	30
510.14	4683	XING 58.9		4683	30	0.0296	9.4108	RINKER RD	8.9146	0.0296	4683	30
510.51	1946			1946	30	0.0123	9.4231		8.8852	0.0123	1946	30
510.64	700		141	700	30	0.0044	9.4275	L	8.8729	0.0044	700	30
510.79	785	XING 38.5		785	30	0.0050	9.4325	SANDY SPRING	8.8685	0.0050	785	30
510.89	545		46	545	30	0.0034	9.4359	L	8.8635	0.0034	545	30
510.91	120			120	30	0.0008	9.4367		8.8601	0.0008	120	30
511.00	480		67	480	30	0.0030	9.4397	R	8.8593	0.0030	480	30
511.14	700			700	30	0.0044	9.4441		8.8563	0.0044	700	30
511.29	800			800	30	0.0051	9.4492		8.8519	0.0051	800	30
511.50	1100		86	1100	30	0.0019	9.4561	R	8.8468	0.0069	1100	30
512.14	3382	XING 57.9		3382	30	0.0214	9.4729	SR 212	8.8399	0.0200	3382	30
513.14	5280			5280	30	0.0334	9.5068		8.8069	0.0334	5280	30
513.32	990			990	30	0.0061	9.5119		8.8065	0.0061	990	30
513.51	1010		100	1010	30	0.0064	9.5183	R	8.8005	0.0064	1010	30
513.92	2147			2147	30	0.0136	9.5319		8.7941	0.0136	2147	30
514.12	1090		100	1080	30	0.0069	9.5387	L	8.7806	0.0069	1090	30
514.35	1200		100	1200	30	0.0076	9.5463	L	8.7767	0.0200	1200	30
515.30	4984			4984	30	0.0315	9.5777		8.7567	0.0315	4084	30
516.11	4300			4300	30	0.0271	9.6048		8.7250	0.0271	4300	30
516.22	598		99	598	30	0.0037	9.6084	R	8.6981	0.0037	598	30
516.30	398			398	30	0.0025	9.6111		8.6943	0.0025	398	30
517.28	5200			5200	30	0.0328	9.6439		8.6918	0.0328	5200	30
517.30	78		67	78	30	0.0005	9.6444	R	8.6590	0.0005	78	30
517.34	240		57	240	30	0.0015	9.6460	L REVERSE CURVE	8.6585	0.0015	240	30
517.67	1728	XING 60.19		1728	30	0.0109	9.6569	PRIVATE	8.6570	0.0109	1728	30
517.81	720	XING 60.1	100	720	30	0.0046	9.6614	L PRIVATE	8.6461	0.0046	720	30
518.30	2600			2600	30	0.0164	9.6778	TIMETABLE 30 MPH	8.6416	0.0164	2600	30
518.53	1248			1248	30	0.0079	9.6857	BAYARD - ALLIANCE	8.6251	0.0136	1248	30
518.74	1100		100	1100	30	0.0069	9.6926	L	8.6115	0.0068	1100	30
519.29	2900			2900	30	0.0183	9.7109		8.6013	0.0183	2900	30
519.59	1600		100	1600	30	0.0101	9.7210	L	8.5829	0.0101	1600	30
520.30	3712	XING 66.17		3712	30	0.0229	9.7435	MAHONING AVE	8.5728	0.0234	3717	30
521.24	5000	XING 66.4		5000	30	0.0316	9.7751	SUMMIT STREET	8.5494	0.0316	5000	30
521.30	274	XING	71	274	30	0.0017	9.7768	L BROADWAY	8.5179	0.0017	274	30
521.30			50				9.7768	L JCT ALLIANCE	8.5101	0.0008	587 D	30
521.41	600		50	600	30	0.0038	9.7806	MAXIMUM SPEED LIMIT 70 S	8.5133	0.0004	13	30
521.58	920	XING	127	920	30	0.0058	9.7864	L FREEDOM	8.5133	0.0026	690 D	70
521.75	880	XING	100	880	30	0.0056	9.7920	L MECHANIC AVENUE	8.5105	0.0005	920	70
522.10	1860			1860	30	0.0117	9.8037	MPH 60 ALLI-FAIRH (OR)	8.5080	0.0050	1860	70
522.50	2090	XING 64.0		2090	30	0.0132	9.8169	LINCOLN AVE	8.5020	0.0067	2090	70
522.72	1184		100	1164	30	0.0073	9.8245	L	8.4970	0.0031	1164	70
523.11	2070	XING 64.7		2070	30	0.0131	9.8373	KEITER'S CROSSING	8.4942	0.0144	5025	70
523.66	2915	XING 65.5		1210	30	0.0086	9.8463	PRIVATE	8.4798	0.0022	809 A	68
524.11	2340		81	2343 A	30	0.0108	9.8571	L	8.4776	0.0055	1193	68
524.26	800		81	2346 A	50	0.0021	9.8592	L	8.4742	0.0022	800	68
525.11	4481	XING 86.5		4174	60	0.0132	9.8785	FRESHLY RD	8.4720	0.0125	4481	68
525.21	546			546	60	0.0017	9.8801		8.4595	0.0015	546	68
525.38	900		115	900	60	0.0025	9.8826	L	8.4580	0.0022	800	68
525.98	3160			3180	60	0.0100	9.8926		8.4558	0.0089	3180	68
526.11	780	XING	88	780	60	0.0025	9.8951	L PRIVATE	8.4469	0.0021	780	68
526.34	1200		88	1200	60	0.0038	9.8989	L	8.4448	0.0033	1200	68
526.70	1900	XING 58.4		1900	60	0.0060	9.9049	BEECH RD	8.4415	0.0053	1900	68
526.87	900		100	900	60	0.0028	9.9077	L	8.4362	0.0025	900	68
527.11	1300	XING 88.9		1300	60	0.0041	9.9118	PRIVATE	8.4337	0.0036	1300	68
527.34	1210		88	1210	60	0.0038	9.9156	R	8.4300	0.0034	1210	68
527.60	1342	XING	100	1342	60	0.0042	9.9198	R EASTON ST	8.4267	0.0005	101 D	70
527.60		XING 90.15			60	0.0000	9.9199	PRIVATE	8.4264	0.0034	1241	70
528.60	5311	XING 90.4		5311	60	0.0168	9.9366	PRIVATE	8.4230	0.0144	5311	70
528.61	60	XING 90.6		60	60	0.0002	9.9369	PARKS ROAD	8.4087	0.0002	60	70
528.79	940		107	940	60	0.0030	9.9398	R	8.4085	0.0025	940	70
528.93	750	XING 91.2		750	60	0.0024	9.9422	PARIS AVENUE	8.4060	0.0020	780	70
529.29	1880		70	1880	60	0.0059	9.9481	R	8.4039	0.0051	1880	70
529.52	1244			1244	60	0.0039	9.9520		8.3988	0.0001	52 D	71
529.60	400		141	400	60	0.0013	9.9533	R	8.3987	0.0097	2620	71
529.81	1106		141	1106	60	0.0035	9.9568	R EISHELMAN ROAD	8.3900	0.0011	414 A	70
530.37	2960			2960	60	0.0094	9.9662	PRIVATE	8.3808	0.0074	2744	70
530.51	1206		60	1206	60	0.0040	9.9702	R MAPLE GROVE	8.3834	0.0035	1284	70
530.66	269		65	269	60	0.0008	9.9710	R	8.3800	0.0022	909 A	68
531.03	1924	XING 93.3		1924	60	0.0061	9.9771	MESSIE ROAD	8.3778	0.0038	1368	68
531.25	1160		160	1130	60	0.0037	9.9808	L	8.3736	0.0030	1180	68
531.61	1960			1960	60	0.0059	9.9867		8.3707	0.0020	727	68
531.94	1755	XING 94.0		1755	60	0.0056	9.9923	NICKEL PLATE ROAD	8.3686	0.0000	0	68
532.18	1280		70	1280	60	0.0040	9.9963	L	8.3686	0.0156	4678 A	50
532.60	2227	XING 94.9		2227	60	0.0070	10.0033	CHAPPEL STREET	8.3526	0.0054	1418	50
532.85	1330			1330	60	0.0042	10.0075	MAIN STREET LOUISVILLE	8.3474	0.0050	1326	50
533.12	1430		88	1430	60	0.0045	10.0120	R	8.3424	0.0054	1430	50
533.61	2580	XING 95.5		2580	60	0.0081	10.0201	PRIVATE	8.3370	0.0097	2586	50
533.80	989			989	60	0.0031	10.0232	FAIRHOPE	8.3273	0.0037	988	50
534.01	1120		65	2852	60	0.0080	10.0312	R	8.3235	0.0042	1120	50
534.42	2155			403 D	50	0.0014	10.0326		8.3193	0.0082	2150	50
534.51	500	JCT 0	30	500	50	0.0019	10.0356	MP 95 JCT FT W LINE	8.3111	0.0019	500	50
534.51		OPFAIRHOPE				0.0000	10.0356	JCT BAYARD RR CP FAIRHO	8.3092	0.0000	0	
534.58	376		50	378	50	0.0014	10.0369	10.94 MILES LONGER VIA AL	8.3092	0.0014	378	50
534.68	500	XING 96.9		4076 A	50	0.0156	10.0506	BECK ROAD	8.3078	0.0008	480 D	70
534.88							10.0528		8.3050	0.0001	20	70
534.79	600	XING 97.13			68	0.0202	10.0508	BROADWAY	8.3050	0.0006	200	70
534.85	300		90		68	0.0000	10.0528	L	8.3044	0.0008	300	70
534.90	300		93		68	0.0000	10.0506	L	8.3036	0.0008	300	70
535.24	1758			0	68	0.0000	10.0520		8.3028	0.0047	1755	70
535.45	1100	XING	98	0	68	0.0000	10.0508	L CARNAHAN AVENUE	8.2981	0.0030	1100	70
535.69	1300	XING 100		800 A	68	0.0022	10.0566	8TH STREET NE	8.2951	0.0236	8865	70
536.69	5298	XING 100.77		5353	70	0.0145	10.0905	6TH STREET NE	8.2716	0.0022	809 A	68
537.69	5280	XING 100		5280	70	0.0143	10.0856	5TH STREET	8.2694	0.0066	2364	68
538.68	5238	XING 100		5298	70	0.0142	10.0976	ROWLAND AVENUE	8.2628	0.0146	5236	68
538.98	1600	XING 101		1800	70	0.0049	10.1023	2ND AVENUE NE	8.2482	0.0045	1800	68
538.99		XING 101		0		0.0187	10.1169	TUSC AVE	8.2437			
538.99		XING	90	1996	70	0.0048	10.1236	WALE WANDLE 3RD STREE	8.2437	0.0167	RR CROSSIN	
539.33	1800	XING 105		101 D	70	0.0003	10.1236	CANTON	8.2271	0.0050	1400	68
539.58	1400	XING	84	1400	68	0.0038	10.1277	R MCKINLEY	8.2220	0.0039	1400	68
539.71	616	XING 102		610	68	0.0017	10.1294	CLEVELAND AVENUE	8.2181	0.0003	101 D	70
539.83	634	XING 102		634	68	0.0018	10.1312	HIGH STREET	8.2179	0.0026	843	70
540.70	4600	XING 102		4600	68	0.0128	10.1440	PROSPECT AVENUE	8.2153	0.0130	4860	70
541.71	5308	XING 102		5150	68	0.0144	10.1583	DUEBER AVENUE	8.2023	0.0132	4674	70
541.71		XING 103		800 A	68	0.0020	10.1605	CLARENDON AVENUE	8.1891	0.0022	800 A	68
541.71		XING 103.67				0.0000	10.1606	MARYLAND AVENUE	8.1869	0.0065	2316	68
541.99	1500	XING	79	841	70	0.0023	10.1626	SHAFF RD	8.1806	0.0098	2967 A	61
542.73	3780			3981	70	0.0108	10.1736		8.1708	0.0007	200	61
543.39	3800	XING 106		3000	70	0.0089	10.1824	PERRY DRIVE	8.1728	0.0112	3600	61
543.54	800		141	800	70	0.0027	10.1850	L	8.1617	0.0087	3139	61
543.71	887			887	70	0.0024	10.1873		8.1519	0.0017	505 A	58
543.86	800			498	70	0.0013	10.1880		8.1503	0.0026	762	58
543.88				101 D	70	0.0005	10.1890		8.1477			
544.14	1500	XING	98	1500	68	0.0042	10.1927	L GENOA STREET	8.1477	0.0050	2718 A	46
544.73	2861			2800	68	0.0078	10.2005		8.1390	0.0067	1638	46
544.75	182			52	68	0.0001	10.2017		8.1313	0.0008	192	46
544.75				321 D	82	0.0010	10.2027		8.1305			
545.01	1400		61	1400	61	0.0043	10.2070	R	8.1305	0.0025	304 A	42
545.33	1700			1700	61	0.0063	10.2125		8.1280	0.0032	704	42
545.41	400		100	400	61	0.0012	10.2136	L	8.1248	0.0018	400	42
545.21	1600			1600	61	0.0040	10.2166		8.1230	0.0072	1600	42

READ DOWN WESTBOUND TRAFFIC — APPENDIX 80 (S ZERO) STATION — READ UP EASTBOUND TRAFFIC

PAGE 9 COLUMN 2

MILES	DISTANCE, ft	HWY GRADE CROSSING	MAX MPH	DISTANCE, feet	SPEED, mph	ET, hours	TIME, hours	STATION	TIME, hours	ET, hours	DISTANCE, feet	SPEED, mph
545.94	1187			1187	61	0.0037	10.2202		6.1150	0.0054	1187	42
546.03	600		100	600	61	0.0016	10.2237	L	6.1106	0.0023	600	42
546.16	700			612	61	0.0019	10.2256		6.1082	0.0032	700	42
546.19				88 D	61	0.0003	10.2259		6.1051			
546.47	1600	XING 105	59	1600	59	0.0051	10.2310	R TT MPH 60 16TH STREET	6.1051	0.0072	1600	42
546.71	1300			1300	59	0.0042	10.2352	REED	6.0978	0.0056	1300	43
546.63	600			100	59	0.0003	10.2365		6.0920	0.0027	600	42
546.83				901 D	56	0.0016	10.2373		6.0893			
547.19	1909		46	1909	46	0.0079	10.2452	L	6.0866	0.0058	1909	40
547.28	600			371	46	0.0015	10.2467	REVERSE CURVE	6.0807	0.0023	500	42
547.28				129 D	46	0.0006	10.2473		6.0784			
547.70	2200		42	2200	42	0.0089	10.2572	R	6.0784	0.0089	2200	42
547.78	400		42	400	42	0.0018	10.2590	R	6.0696	0.0018	400	42
547.79							10.2590		6.0687	0.0008	950 D	50
547.98	1100			1100	42	0.0050	10.2640	MASSILLON	6.0633	0.0004	160	66
548.12	700		100	700	42	0.0032	10.2671	R	6.0629	0.0020	700	68
548.46	1800			1800	42	0.0081	10.2752		6.0608	0.0052	1800	68
548.66	1061		68	1061	42	0.0048	10.2800	L	6.0567	0.0030	1061	68
548.66					BSC	0.0000	10.2800	MACE TOWER	6.0527	RR CROSSING		
548.74	400			400	42	0.0018	10.2818		6.0527	0.0011	400	68
549.15	2200	XING 111.38		2200	42	0.0099	10.2918	EARL ROAD	6.0516	0.0063	2200	68
549.47	1700		66	1700	42	0.0077	10.2994	L TT MPH 70/60/50	6.0452	0.0049	1700	66
549.47							10.2994		6.0404	0.0006	195 D	70
549.58	564	XING 111.90		594	42	0.0027	10.3021	RC BOYDS CRN	6.0398	0.0011	395	70
549.72	700		75	700	40	0.0032	10.3053	R	6.0387	0.0018	700	70
549.89	900		75	745	42	0.0034	10.3088	R	6.0368	0.0024	900	70
549.83				1218 A	42	0.0050	10.3136		6.0344	0.0000		70
550.10	1100	XING 112.98		4291 A	50	0.0139	10.3275	BEAUMONT AVENUE	6.0344	0.0030	1100	70
550.53	2300		79	0	50	0.0000	10.3275	L	6.0314	0.0062	2300	70
550.72	1000			0	50	0.0000	10.3275		6.0252	0.0027	1000	70
550.76	187			0	50	0.0000	10.3275		6.0225	0.0005	187	70
551.02	1400		79			0.0000	10.3275	R	6.0220	0.0038	1400	70
551.72	3700			3452	66	0.0099	10.3374		6.0182	0.0100	3700	70
552.50	4500	XING 114.05		1487 A	66	0.0042	10.3416	KENYON AVENUE	6.0082	0.0111	4100	70
552.69	1000		98	4487	70	0.0121	10.3537	L	5.9971	0.0027	1000	70
552.72	135			135	70	0.0004	10.3541		5.9944	0.0004	135	70
553.72	5282	XING 115.18		5282	70	0.0143	10.3684	SR 93	5.9940	0.0143	5282	70
554.47	4000	XING 116.15		4000	70	0.0108	10.3792	CLERMONT	5.9797	0.0108	4000	70
554.63	800	XING 116	100	800	70	0.0022	10.3814	L ALABAMA AVE.	5.9688	0.0022	800	70
554.72	511	XING 117.00		511	70	0.0014	10.3827	RAILROAD STREET N LAWR	5.9667	0.0014	511	70
555.73	5298			5296	70	0.0143	10.3971	NORTH LAWRENCE	5.9654	0.0143	5298	70
556.74	5377	XING 119.28		5377	70	0.0145	10.4116	ANDERSON ROAD	5.9510	0.0145	5377	70
557.46	3800	XING 119.84		3800	70	0.0103	10.4219	PRIVATE	5.9365	0.0103	3800	70
557.73	1422	XING 119	86	1422	70	0.0038	10.4258	L PRIVATE	5.9262	0.0038	1422	70
557.75	98	XING 120	98	98	70	0.0003	10.4260	LMP 120 PRIVATE	5.9224	0.0003	98	70
558.74	5200	XING 120.99		5200	70	0.0141	10.4401	SR 94	5.9221	0.0141	5200	70
559.74	5297	XING 121.82		5287	70	0.0143	10.4544	TANNERVILLE	5.9080	0.0143	5287	70
560.46	3800			3800	70	0.0103	10.4647		5.8937	0.0103	3800	70
560.63	900		87	900	70	0.0024	10.4671	R	5.8834	0.0024	900	70
560.74	599	XING 123.82		598	70	0.0016	10.4688	WALNUT	5.8810	0.0016	599	70
561.03	1500	XING 123.86		1500	70	0.0041	10.4728	MARKET STREET	5.8793	0.0041	1500	70
561.75	3789		94	3788	70	0.0103	10.4831	L	5.8753	0.0100	3789	70
562.24	2600		100	2600	70	0.0070	10.4901	L ORRVILLE	5.8650	0.0070	2600	70
562.75	2750			2700	70	0.0073	10.4974	NLE OHIO	5.8580	0.0073	2700	70
563.03	1500	XING 125.11		1500	70	0.0041	10.5015	CROWN HILL	5.8507	0.0041	1500	70
563.30	1400	XING 125	71	1400	70	0.0038	10.5053	R SCHOOLHOUSE ROAD	5.8466	0.0038	1400	70
563.56	1400		108	1400	70	0.0038	10.5090	R	5.8429	0.0038	1400	70
563.75	1028		87	1028	70	0.0028	10.5118	R	5.8391	0.0028	1028	70
564.00	1300	XING 126	105	1300	70	0.0035	10.5153	R MUSKRAT ROAD	5.8365	0.0035	1300	70
564.26	1365	XING 122	98	1365	70	0.0037	10.5190	R BACK ORRVILLE	5.8328	0.0060	2221	70
564.76	2600	XING 127	60	2600	70	0.0070	10.5261	ROHRER ROAD	5.8268	0.0194	6148 A	40
565.76	5302	XING 128.06		5302	70	0.0143	10.5404	PRIVATE	5.8073	0.0089	2477	40
565.80	200	XING 108.77		200	70	0.0005	10.5410	APPLEGATE	5.7976	0.0008	300	48
566.76	5100		93	5100	70	0.0138	10.5548	L	5.7967	0.0201	5150	48
567.07	1600			1600	70	0.0043	10.5591		5.7766	0.0019	477	48
567.24	900		95	900	70	0.0024	10.5615	L MPH 50	5.7747	0.0018	274 A	45
567.37	700	XING 129.57		700	70	0.0019	10.5634	CO RD 75	5.7729			
567.56	1000		104	1000	70	0.0027	10.5661	R	5.7729	0.0078	1872	45
567.67	600			600	70	0.0016	10.5677		5.7649	0.0026	600	45
567.87				960 D	70	0.0031	10.5706		5.7624			
567.77	518		46	516	48	0.0020	10.5728	L SMITHVILLE	5.7624	0.0088	2321	45
567.98	1123		48	1123	48	0.0044	10.5773	L	5.7536	0.0022	506 A	41
568.74	4000			4000	48	0.0159	10.5931		5.7504	0.0121	2615	41
568.78	187			167	48	0.0007	10.5938	TT MPH 35	5.7383	0.0009	187	41
569.08	1600		50	1600	48	0.0063	10.6001	R	5.7375	0.0074	1600	41
569.32	1300			1196	48	0.0047	10.6048		5.7301	0.0060	1300	41
569.33				100 D	48	0.0004	10.6065		5.7241			
569.74	2200		45	2200	45	0.0093	10.6145	L	5.7241	0.0102	2200	41
569.76	96		48	96	45	0.0004	10.6149	L	5.7139	0.0005	96	41
569.91	800			800	45	0.0034	10.6183		5.7134	0.0037	800	41
570.22	1600		49	1600	45	0.0067	10.6250	R	5.7098	0.0074	1600	41
570.33	600			474	45	0.0020	10.6270		5.7024	0.0028	600	41
570.58				126 D	45	0.0006	10.6276		5.6996			
570.73	2100		41	2100	41	0.0097	10.6373	R	5.6996	0.0067	2100	41
570.86	700	XING	41	700	41	0.0032	10.6405	R GEYGERS CHAPEL RD	5.6929	0.0032	700	41
570.86							10.6406	EAST END 2 TRACKS	5.6897	0.0004	94 D	44
570.90	230			230	41	0.0011	10.6416	REVERSE CURVE	5.6892	0.0026	143	44
571.40	2600	XING 137	44	2600	41	0.0120	10.6536	L	5.6856	0.0112	2600	44
571.62	1200			1200	41	0.0055	10.6592		5.6744	0.0062	1200	44
571.76	700		45	700	41	0.0032	10.6624	L	5.6683	0.0030	700	44
571.89	700		40	700	41	0.0032	10.6656	L	5.6663	0.0030	700	44
571.93	218			218	41	0.0010	10.6666	REVERSE CURVE	5.6632	0.0009	218	44
572.02	500		44	500	41	0.0023	10.6689	R	5.6620	0.0022	500	44
572.02							10.6689		5.6602	0.0017	440 D	56
572.46	2300			2300	41	0.0106	10.6795		5.6585	0.0060	1985	56
572.61	800		57	600	41	0.0037	10.6833	L	5.6525	0.0027	800	56
572.73	600			600	41	0.0028	10.6860		5.6498	0.0020	600	56
572.82	500			500	41	0.0023	10.6883		5.6475	0.0017	500	56
573.05	1207	XING 135	56	646	41	0.0030	10.6913	L WOOSTER LIBERTY AVEN	5.6458	0.0041	1207	56
573.05		XING 135.24					10.6913	SOUTH STREET	5.6417	0.0019	647 D	77
573.26	1100			375 A	41	0.0017	10.6930		5.6397	0.0012	453	70
573.33	400		77	754	44	0.0032	10.6961	R	5.6385	0.0011	400	70
573.64	1600			1600	44	0.0069	10.7030		5.6374	0.0049	1800	70
573.75	600	XING 136	82	600	44	0.0026	10.7056	L PRIVATE WEST END STR	5.6326	0.0016	600	70
573.76	86	XING 136	62	68	44	0.0005	10.7059	L PRAIRIE ROAD	5.6309	0.0002	86	70
574.65	4700			2779	44	0.0120	10.7178		5.6308	0.0127	4700	70
574.78	700			2089 A	44	0.0086	10.7265		5.6180	0.0019	700	70
574.90	600		105	952	58	0.0032	10.7296	R	5.6161	0.0016	600	70
574.90	38			36	58	0.0001	10.7297	REVERSE CURVE	5.6145	0.0001	38	70
575.02	600		105	600	58	0.0020	10.7318	L OK	5.6144	0.0016	600	70
575.74	3800			900	58	0.0030	10.7361		5.6128	0.0103	3800	70
576.59	4500			4464 A	58	0.0135	10.7496		5.6025	0.0122	4500	70
576.73	714		70	3690	70	0.0096	10.7561	R KILLBUCK	5.5904	0.0019	714	70
577.48	4000			4000	70	0.0108	10.7669		5.5884	0.0108	4000	70
577.69	1100		100	1100	70	0.0030	10.7719	R FISH	5.5776	0.0030	1100	70
577.72	134	XING 140.10		134	70	0.0004	10.7723	MESSNER ROAD	5.5746	0.0348	12877	70
577.72					70	0.0000	10.7723	MP 140	5.5398	0.0081	2815 A	62
578.74	5406	XING 141.85		5405	70	0.0146	10.7869	WILLOW ROAD	5.5317	0.0005	151	62
579.75	5382	XING 142.64		5362	70	0.0145	10.8014	FORCE ROAD	5.5313	0.0053	1649 A	56
580.75	5030			5252	70	0.0142	10.8156		5.5260	0.0009	276	56
581.12	2000	XING 143.3		2000	70	0.0054	10.8209	CEMETERY ROAD	5.5251	0.0039	1102 A	51
581.35	1216		138	1216	70	0.0033	10.8242	L	5.5212	0.0017	466	51
581.73	2000			2000	70	0.0054	10.8296		5.5194	0.0074	2000	51
581.96	1200	XING 144.181		1200	70	0.0032	10.8329	L WOOD	5.5120	0.0045	1200	51
582.09	700	XING 144.45		700	70	0.0019	10.8348	ROBINSON	5.5075	0.0026	700	51
582.41	1700	XING 144	70	1700	70	0.0046	10.8394	R MARKET STREET	5.5048	0.0063	1700	51
582.60	1000	XING 144.62		1000	70	0.0027	10.8421	SHREVE MAIN STREET	5.4986	0.0037	1000	51
582.75	758		100	368	70	0.0010	10.8431	R	5.4949	0.0028	753	51
582.75				387 D	70	0.0011	10.8442		5.4921			
582.99	1300		82	1300	62	0.0040	10.8481	R OK	5.4921	0.0048	1300	51
583.09	500			740	62	0.0037	10.8489	REVERSE CURVE	5.4873	0.0019	500	51

READ DOWN WESTBOUND TRAFFIC — APPENDIX B0 (B ZERO) STATION — READ UP EASTBOUND TRAFFIC

PAGE 10 COLUMN 1

MILES	DISTANCE, ft	HWY GRADE CROSSING	MAX MPH	d DISTANCE, feet	SPEED, mph	dT, hours	TIME, hours	STATION	TIME, hours	dT, hours	d DISTANCE, feet	SPEED, mph	MILES
583.09				200	D 82	0.0008	10.8497		5.4854				619.36
583.35	1378		58	1180	58	0.0040	10.8937	L	5.4854	0.0061	1378	51	619.49
583.38				196	D 82	0.0007	10.8644		5.4693				619.74
583.61	1400		51	1400	51	0.0052	10.8596	L	5.4808	0.0052	1400	51	619.65
583.61							10.8680		5.4751	0.0046	238	D 82	620.03
583.71	500			500	51	0.0019	10.8815	RC	5.4745	0.0009	262	57	620.03
583.75	200		57	200	51	0.0007	10.8822	R	5.4734	0.0007	200	57	620.03
584.05	1601	XING	57	1601	51	0.0058	10.8801	R BROWNS	5.4727	0.0063	1601	57	620.03
584.75	3700	XING 147		3700	51	0.0137	10.8819	PRIVATE	5.4874	0.0016	808	D 70	620.86
584.75							10.8818		5.4030	0.0064	3005	70	621.79
585.58	4392	XING 146		4392	51	0.0163	10.8982	SCRUBY'S	5.4577	0.0118	4282	70	622.87
585.92	1800	XING 148.2		867	51	0.0023	10.9004	BIG PRAIRIE RTE 226	5.4454	0.0049	1800	70	623.88
586.09	867		100			0.0000	10.9004	L	5.4406	0.0024	867	70	624.60
586.75	3640			1981	A 51	0.0047	10.9052		5.4381	0.0087	3096	70	625.83
586.81	277			950	57	0.0032	10.9083		5.4284	0.0022	809	A 68	626.23
587.05	1300		100	4215	A 67	0.0126	10.9208	L	5.4282	0.0025	900	68	626.23
587.39	1800		77	2441	70	0.0086	10.9274	L	5.4257	0.0058	1800	68	626.33
587.53	700			700	70	0.0019	10.9285		5.4187	0.0019	700	68	626.50
587.75	1200		88	1200	70	0.0032	10.9328	R	5.4187	0.0029	1000	68	626.50
587.98	1100		92	1100	70	0.0030	10.9358	R	5.4140	0.0031	1100	68	626.54
588.04	400			400	70	0.0011	10.9368		5.4109	0.0017	604	68	626.54
588.28	1300	XING	70	1300	70	0.0035	10.9401	PLAKEVILLE GILLS ROAD	5.4082	0.0118	3600	A 56	626.65
588.76	2500	XING 150.1		3069	70	0.0100	10.9502	STATE ROUTE 179	5.3975	0.0029	836	56	626.65
589.00	1300			101	D 70	0.0005	10.9504		5.3944	0.0046	1300	50	626.61
589.23	1217		98	1217	68	0.0034	10.9535	L	5.3899	0.0039	1079	A 50	626.88
589.76	2800			2800	56	0.0078	10.9616		5.3860	0.0066	1796	56	627.01
590.76	5281			5401	68	0.0150	10.9767		5.3792	0.0200	5281	50	627.06
590.90	766			596	D 62	0.0016	10.9786		5.3592	0.0027	766	50	627.18
591.49	3122		56	3122	56	0.0108	10.9892		5.3566	0.0118	3122	50	627.89
591.51	92			92	56	0.0003	10.9895	REVERSE CURVE	5.3447	0.0003	92	50	626.19
591.51		XING 163.1		195	D 50	0.0007	10.9902	R TR 565	5.3444	0.0000			626.30
591.77	1368		50	1368	50	0.0052	10.9954	L	5.3444	0.0052	1368	50	626.32
591.77	18			18	50	0.0001	10.9956		5.3392	0.0015	448	D 61	626.47
591.90	712			713	50	0.0027	10.9982		5.3377	0.0009	281	61	628.47
592.08	921		61	921	50	0.0035	11.0017	L	5.3368	0.0029	921	61	629.15
592.23	804		64	804	50	0.0030	11.0047	L	5.3339	0.0004	135	D 64	629.28
592.77	2850			2850	50	0.0108	11.0155		5.3335	0.0104	3517	64	629.45
593.66	4700	XING 165.1		4700	50	0.0178	11.0333	TR 6	5.3231	0.0139	4700	64	630.28
593.77	554	XING	67	554	50	0.0021	11.0354	R LOUDONVILLE JEFFERSO	5.3092	0.0016	554	64	631.27
594.00	2848	XING	84	442	50	0.0017	11.0371	R SPRING STREET	5.3076	0.0084	2646	64	632.27
594.30				2612	A 50	0.0088	11.0460		5.2991	0.0006	296	D 70	633.27
594.78	2500			0	91	0.0000	11.0460	CP MOHICAN	5.2985	0.0060	2208	70	634.27
595.61	4400			916	A 91	0.0026	11.0487		5.2823	0.0119	4400	70	635.23
595.78	888	XING	77	6476	84	0.0192	11.0679	L PRIVATE	5.2804	0.0024	888	70	636.27
595.90	616		77	610	84	0.0018	11.0697	L	5.2780	0.0017	616	70	637.27
596.79	4700			2607	84	0.0077	11.0774		5.2764	0.0127	4700	70	636.26
597.03	1305	XING	80	2200	A 84	0.0064	11.0838	L E PERRY PRIVATE	5.2686	0.0035	1303	70	638.47
597.79	4000			5146	70	0.0139	11.0977		5.2801	0.0158	4000	70	638.86
597.98	1000	XING 160.1		1000	70	0.0027	11.1004	BRIDGE STREET	5.2483	0.0027	1000	70	639.85
598.79	4264	XING 104		4264	70	0.0115	11.1120	R PRIVATE	5.2468	0.0115	4264	70	638.85
599.37	3100		96	3100	70	0.0084	11.1203	R PERRYSVILLE	5.2351	0.0143	5271	70	639.85
599.80	2280	XING 162.1		2280	70	0.0061	11.1265	PRIVATE	5.2208	0.0092	3176	A 61	638.65
600.58	4100	XING 162.x		4100	70	0.0111	11.1376	TOWNSHIP ROAD 1057	5.2116	0.0031	1012	61	639.05
600.81	1197		98	1197	70	0.0032	11.1408	L	5.2085	0.0037	1197	61	639.05
600.90	500		98	505	70	0.0014	11.1421	L	5.2046	0.0024	781	61	639.15
601.09	1000			1000	70	0.0027	11.1448		5.2024	0.0050	1533	A 56	639.17
601.39	1600		72	1600	70	0.0043	11.1492	L	5.1974	0.0028	506	56	639.26
601.66	1390	XING 163.1		1390	70	0.0038	11.1529	COUNTY ROAD 1075	5.1946	0.0045	1390	56	639.26
601.81	800	XING	96	800	70	0.0022	11.1551	L PRIVATE	5.1898	0.0028	800	56	639.47
601.96	800		96	800	70	0.0022	11.1573	L	5.1871	0.0026	800	56	640.26
602.47	2700			2288	70	0.0061	11.1634		5.1843	0.0068	2700	55	641.26
602.47				432	D 72	0.0013	11.1647		5.1750	0.0000			642.26
602.68	1104		81	1104	81	0.0034	11.1681	R	5.1750	0.0038	1104	55	643.29
602.81	700			700	81	0.0022	11.1703		5.1712	0.0024	700	55	644.28
603.32	2700			2446	81	0.0076	11.1779		5.1688	0.0096	2700	55	645.29
603.32				268	D 61	0.0008	11.1787		5.1592	0.0000			646.16
603.61	1500		55	1500	55	0.0052	11.1839	L	5.1596	0.0052	1500	55	647.28
603.81						0.0000	11.1839		5.1544	0.0057	537	D 67	647.28
603.82	1102			1102	55	0.0038	11.1876		5.1527	0.0018	565	67	647.28
604.16	1800			1800	55	0.0062	11.1938		5.1511	0.0061	1800	67	647.28
604.31	800		69	800	55	0.0028	11.1966	L	5.1460	0.0023	800	67	648.31
604.51	1067	XING 168.x		1067	55	0.0037	11.2003	PRIVATE	5.1437	0.0030	1067	67	648.31
604.74	1200		69	1200	55	0.0041	11.2044	R	5.1407	0.0034	1200	67	648.31
604.81	400			400	55	0.0014	11.2058		5.1373	0.0011	400	67	649.31
604.97	850			850	55	0.0028	11.2085		5.1362	0.0023	850	67	650.31
605.10	700		72	700	55	0.0024	11.2109	R	5.1339	0.0020	700	67	651.31
605.65	2900	XING 167.x		2900	55	0.0100	11.2209	PRIVATE	5.1320	0.0082	2900	67	652.31
605.72	400	XING	67	231	55	0.0008	11.2217	R LUCAS PERRYSVILLE	5.1238	0.0011	400	67	653.31
605.81	457			1660	A 55	0.0058	11.2276		5.1226	0.0013	457	67	654.31
605.87	295						11.2276		5.1213	0.0008	295	67	654.46
606.04	900		67				11.2276	L	5.1206	0.0025	900	67	655.31
606.04		XING 168.3					11.2276	UNION STREET	5.1180	0.0004	101	D 70	655.31
606.04		XING 168.48					11.2276	CHURCH	5.1175	0.0025	649	70	655.31
606.43	2110	XING 168.x		2033	67	0.0057	11.2333	LUCAS HANLEY BRETZ	5.1103	0.0006	1100	70	655.31
606.81	2000		60	2000	67	0.0057	11.2390	R	5.1120	0.0054	2000	70	655.31
607.23	2200		100	2200	67	0.0062	11.2452	R	5.1089	0.0060	2200	70	655.31
607.51	1500			1500	67	0.0042	11.2494	CP LUCAS	5.1009	0.0041	1500	70	655.77
607.64	693		82	693	67	0.0020	11.2514	R	5.0969	0.0019	693	70	655.77
607.78	700		72	700	67	0.0020	11.2534	R	5.0950	0.0019	700	70	655.80
607.82	200			205	67	0.0006	11.2539		5.0931	0.0006	200	70	655.80
608.31	2600			1808	67	0.0061	11.2600		5.0929	0.0070	2600	70	655.90
608.74	2300		71	1205	A 67	0.0035	11.2634	L	5.0855	0.0062	2300	70	655.69
608.74				1867	70	0.0051	11.2675		5.0793	0.0000		70	656.31
608.81	377	XING 171.x		377	70	0.0010	11.2685	PRIVATE	5.0793	0.0050	377	70	657.31
609.81	5281	XING 171.x		5281	70	0.0143	11.2826	BITTS ROAD	5.0763	0.0143	5281	70	658.28
610.76	5000			5000	70	0.0136	11.2963		5.0640	0.0135	5000	70	658.29
610.81	279		113	279	70	0.0008	11.2971	L	5.0504	0.0008	279	70	659.29
610.87	300	XING	113	300	70	0.0008	11.2979	URC ILLINOIS AVENUE	5.0497	0.0008	300	70	660.29
610.96	481		113	481	70	0.0013	11.2992	RTT MPH 70	5.0489	0.0013	481	70	661.29
611.81	4600			4500	70	0.0122	11.3113		5.0476	0.0127	4500	70	662.29
612.19	2000			2000	70	0.0054	11.3168		5.0354	0.0054	2000	70	663.29
612.31	600		67	600	70	0.0016	11.3184	R	5.0350	0.0016	600	70	664.29
612.42	600	XING	102	600	70	0.0016	11.3200	R 6TH STREET	5.0284	0.0016	600	70	665.30
612.55	600	XING	82	600	70	0.0016	11.3216	R 5TH STREET	5.0267	0.0016	600	70	666.29
612.82	1512	XING 175.x		1512	70	0.0041	11.3257	MANSFIELD ORANGE STRE	5.0251	0.0041	1512	70	667.29
613.54	3800	XING 175.x		3800	70	0.0103	11.3360	SURREY ROAD	5.0210	0.0105	3800	70	667.29
613.54			90		70	0.0000	11.3360	R&O XING	5.0108	0.0000		70	667.29
613.60	300	XING	141	300	70	0.0008	11.3368	MAIN STREET	5.0108	0.0000	67	70	667.29
613.84	1274	XING 175.x		1274	70	0.0034	11.3403	LW MANS LONGVIEW AVEN	5.0108	0.0000			667.29
614.01	900			900	70	0.0024	11.3427		5.0108	0.0117	3960	A 58	667.29
614.08	400		141	400	70	0.0011	11.3438	L	4.9989	0.0000			667.29
614.32	1268			1268	70	0.0034	11.3472		4.9989	0.0004	113	58	667.29
614.67	1800	XING	90	1800	70	0.0049	11.3521	L EMP STL MILL	4.9985	0.0059	1800	56	667.29
614.64	900	XING 177.x		900	70	0.0024	11.3545	PRIVATE	4.9927	0.0028	900	58	667.29
615.48	3300			3300	70	0.0089	11.3634		4.9897	0.0108	3300	58	668.30
615.84	1967		100	1967	70	0.0054	11.3688	L	4.9789	0.0086	1967	58	668.30
616.39	2900			2900	70	0.0062	11.3750		4.9725	0.0095	2900	58	669.30
616.39				37	70	0.0001	11.3751		4.9630	0.0000			670.30
616.39				563	D 70	0.0017	11.3768		4.9630	0.0000			671.29
616.60	1100		58	1100	58	0.0036	11.3804	R	4.9630	0.0036	1150	56	672.30
616.60						0.0000	11.3804		4.9594	0.0017	583	D 70	673.29
616.71	593			593	58	0.0019	11.3823	REVERSE CURVE	4.9577	0.0014	530	70	673.29
616.84	700		80	700	58	0.0023	11.3846	L	4.9563	0.0006	200	70	674.56
617.03	997		80	997	58	0.0033	11.3879	L	4.9556	0.0027	997	70	674.58
617.18	800			800	58	0.0026	11.3906		4.9531	0.0022	800	70	674.56
617.29	600		133	600	58	0.0020	11.3924	L	4.9509	0.0016	600	70	674.58
617.43	700	XING 179.x		700	58	0.0023	11.3947	LEPPO ROAD	4.9489	0.0041	1504	70	674.58
617.58	800		104	800	58	0.0028	11.3975	R	4.9452	0.0011	407	A 69	674.58
617.82	1300	XING 180.x		1300	58	0.0047	11.4016	SPRING MILL	4.9441	0.0004	368	69	674.58
618.11	1510		100	1510	58	0.0049	11.4065	L	4.9410	0.0041	1510	69	675.29
618.65	3000			3000	58	0.0098	11.4163		4.9375	0.0107	3000	69	676.29

PAGE 10 COLUMN 2

DISTANCE, ft	HWY GRADE CROSSING	MAX MPH	d DISTANCE, feet	SPEED, mph	dT, hours	TIME, hours	STATION	TIME, hours	dT, hours	d DISTANCE, feet	SPEED, mph
2700			3662	A 58	0.0017	11.4280		4.9268	0.0074	2700	69
701		70	1568	70	0.0043	11.4323	L	4.9194	0.0019	701	69
1500	XING 181.94		51	D 70	0.0001	11.4324	PRIVATE	4.9175	0.0056	1300	69
800		68	800	69	0.0018	11.4341	L	4.9139	0.0016	800	68
907	XING 182	69	907	69	0.0026	11.4368	L ROCK ROAD	4.9122	0.0025	907	69
						11.4380		4.9097	0.0014	540	D 79
						11.4356		4.9080	0.0318	13240	79
						11.4380		4.8765	0.0214	7935	A 60
4400	XING 183.86		4400	60	0.0121	11.4487	PRIVATE	4.8562	0.0005	147	60
4912	XING 184.72		4912	66	0.0136	11.4622	SHELBY ROAD	4.8547	0.0050	1507	A 54
5710			1686	69	0.0046	11.4668		4.8497	0.0027	760	54
5295	XING 186.49		407	A 69	0.0011	11.4679	HOBBINS ROAD	4.8469	0.0022	617	A 51
5300			14219	70	0.0385	11.5064		4.8447	0.0031	844	51
5302	XING 188.07		5302	70	0.0143	11.5207	COUNTY LINE	4.8416	0.0197	5302	51
1800	XING 188.18		1325	70	0.0036	11.5243	BUCYRUS STREET	4.8219	0.0067	1800	51
			477	D 70	0.0014	11.5257	WEST END 2 TRACKS	4.8152			
564	XING 190	60	554	60	0.0017	11.5274	R WASHINGTON	4.8152	0.0021	564	51
900			900	60	0.0025	11.5300	CRESTLINE	4.8132	0.0030	900	51
		90	44	D 60	0.0001	11.5301	CP CREST E4 XING	4.8026	RR CROSSING 90 MPH		
200			49	59	0.0002	11.5303	MAXIMUM SPEED LIMIT 90	4.8098	0.0007	200	51
			207	D 56	0.0007	11.5310		4.8091			
600		54	635	54	0.0017	11.5327	R	4.8081	0.0022	600	51
				54	0.1333	11.6660	RR XING DELAY	4.8006	0.1333		
600			116	D 54	0.0004	11.6664		4.6735	0.0030	800	51
400		51	400	51	0.0015	11.6679	R	4.6705	0.0015	400	51
700		51	700	51	0.0026	11.6705	R	4.6691	0.0026	700	51
204			204	51	0.0008	11.6713	REVERSE CURVE	4.6665	0.0008	156	D 56
400		56	400	51	0.0015	11.6727	L	4.6659	0.0016	440	55
4000			4005	51	0.0149	11.6876		4.6644	0.0138	4005	55
1800			1800	51	0.0069	11.6935		4.6506	0.0064	1800	55
600		57	600	51	0.0022	11.6958	L	4.6451	0.0021	600	55
78			78	51	0.0008	11.6961	REV CUR TT40MPH	4.6420	0.0005	78	55
800		55	800	51	0.0030	11.6990	R	4.6427	0.0026	800	55
					0.0000	11.6990		4.6400	0.0008	167	D 58
3600	XING 191.00		3818	51	0.0128	11.7118	SR 598	4.6394	0.0110	3430	58
700	XING 191.130		867	A 51	0.0031	11.7144	R CRESTLINE ROAD	4.6284	0.0022	700	58
900	XING 192.133		1625	55	0.0063	11.7207	R BIDDLE ROAD	4.6262	0.0029	900	58
4360	XING 193.30		4360	55	0.0151	11.7364	BECK ROAD	4.6233	0.0141	4360	59
5102	XING 194.62		620	55	0.0021	11.7378	MAIN STREET	4.6092	0.0197	5102	58
5286	XING 194.71				0.0000	11.7378	WESTERN AVENUE	4.5925	0.0170	5286	59
5264	XING 195.46		6032	A 55	0.0298	11.7656	OLENTANGY ROAD	4.5758	0.0170	5064	58
5269	XING 195.87		11291	59	0.0301	11.7985	PRIVATE	4.5586	0.0170	5289	59
5082	XING 196.26		5092	59	0.0163	11.8156	PRIVATE	4.5418	0.0308	9813	59
5470	XING 197.05		5470	59	0.0176	11.8334	OWR LEESVILLE	4.5108	0.0128	3640	A 40
5275	XING 197.87		5275	59	0.0189	11.8504	PRIVATE	4.4874	0.0127	2681	40
5254	XING 197.78		5254	59	0.0168	11.8672	PRIVATE	4.4847	0.0248	5064	40
1100	XING 196.00		1100	59	0.0035	11.8707	PRIVATE	4.4599	0.0052	1100	40
985	XING 198.96		985	59	0.0031	11.8738	R ALBRIGHT	4.4547	0.0040	985	40
1000	XING 200.08		310	59	0.0010	11.8748	WHETSTONE STREET	4.4501	0.0047	1000	40
				59	0.0667	11.9415	RR XING DELAY	4.4454	0.0667		
		90	890	D 56	0.0026	11.9441	COLSAN NEW XING	4.3787	0.0000	0	40
	XING 200.79			40		11.9441	LANE STREET BUCYRUS	4.3767	0.0000	0	40
1100	XING 20X	63	1100	40	0.0052	11.9494	LTT 40 MPH WALNUT STREE	4.3787	0.0052	1100	40
	XING 200.00					11.9494	ALLEY - TRACK IN GALEN S	4.3735	0.0000	0	
500	XING 200.97		500	40	0.0024	11.9517	N SANDUSKY STATE HWY	4.3736	0.0024	500	40
100		110	100	40	0.0006	11.9522	L TRACK IN GALEN STREET	4.3711	0.0006	960	D 59
500	XING 201.10		500	40	0.0024	11.9546	RC TT 40 SPRING STREET	4.3686	0.0000	10	59
120 201 2 TRACK IN TREET WALNUT						11.9546	SANDUSKY RIVER	4.3684			
1100	XING 201.104		1100	40	0.0052	11.9598	R SEARS STREET	4.3684	0.0052	1000	59
4188	XING 201.46		4188	40	0.0198	11.9796	MANSFIELD STREET	4.3632	0.0134	4188	59
5286	XING 202.18		5286	40	0.0250	12.0046	McCRACKEN	4.3519	0.0170	5286	59
5080	XING 203.86		476	40	0.0020	12.0066	SIMMS CR	4.3348	0.0169	5280	59
5590	XING 204.70		3643	A 40	0.0133	12.0200	PRIVATE	4.3179	0.0173	5360	59
5272	XING 204.91		11971	59	0.0384	12.0584	GLENVILLE CR	4.3006	0.0169	5272	59
5286	XING 205.50		5298	59	0.0170	12.0754	PRIVATE	4.2837	0.0170	5298	59
4595	XING 206.13		4595	59	0.0147	12.0902	STRIBS CR KN	4.2680	0.0147	4595	59
5938	XING 207.10		5938	59	0.0190	12.1092	MARION-MELMO	4.2518	0.0190	5928	59
	XING 207.70			59	0.0000	12.1092	PRIVATE	4.2329	0.0000		59
	XING 208.13			59	0.0000	12.1092	COUNTY LINE	4.2328	0.0000		59
	XING 208.69			59	0.0000	12.1092	PRIVATE	4.2329	0.0000		59
5440	XING 209.02		5440	59	0.0175	12.1267	NEVADA AYERS STREET	4.2328	0.0175	5440	59
	XING 209.10			59	0.0000	12.1267	MORRISON STREET	4.2154	0.0000		59
	XING 209.27			59	0.0000	12.1267	GOODWRIAD STREET	4.2154	0.0000		59
5270	XING 210.01		5270	59	0.0169	12.1436	EDENVILLE ROAD	4.2154	0.0169	5270	59
5090	XING 211.00		5090	59	0.0170	12.1606	PRIVATE	4.1985	0.0170	5290	59
5288	XING 212.12		5289	59	0.0170	12.1775	DOUGLAS RD	4.1815	0.0170	5280	59
5269	XING 213.25		5269	59	0.0169	12.1944	PRIVATE	4.1645	0.0169	5269	59
5280	XING 213.80		5280	59	0.0169	12.2114	ROCK RUN ROAD	4.1476	0.0169	5280	59
5264	XING 214.37		5264	59	0.0170	12.2284	PRIVATE	4.1307	0.0170	5264	59
787	XING 214.143		787	59	0.0025	12.2309	L WILLIAMS CR	4.1137	0.0025	787	59
4500	XING 215.60		4500	59	0.0144	12.2453	PRIVATE	4.1112	0.0144	4500	59
	XING 215.70			59	0.0000	12.2453	PRIVATE	4.0967	0.0000		
	XING 216.10			59	0.0000	12.2453	PRIVATE	4.0967	0.0000		
	XING 216.31			59	0.0000	12.2453	KOINS RD-TRACKS STRAIG	4.0967	0.0000		
	XING 216.59			59	0.0000	12.2453	S 5TH STREET	4.0967	0.0000		
	XING 216.90			59	0.0000	12.2453	S SANDUSKY	4.0967	0.0000		
2440	XING 217.06		2440	59	0.0079	12.2532	UPPER SANDUSKY 7TH ST	4.0967	0.0079	2440	59
	XING 217.20			59	0.0000	12.2532	EIGHTH STREET	4.0888	0.0000		59
160	XING 217.34		160	59	0.0005	12.2537	HAZEL STREET	4.0889	0.0005	160	59
	XING 217.44			59	0.0000	12.2537	S WARPOLE	4.0884	0.0000		59
		80		59	0.0500	12.3037	USAND OSC XING	4.0884	0.0500		59
479		62	479	59	0.0015	12.3052	L	4.0384	0.0015	479	59
2200	XING 218.18		2200	59	0.0071	12.3125	TWP ROAD	4.0369	0.0071	2200	59
5281	XING 219.19		5281	59	0.0170	12.3292	WHITE CR	4.0298	0.0170	5281	59
5185	XING 219.90		5185	59	0.0166	12.3458	PRIVATE	4.0128	0.0166	5185	59
	XING 220.20			59	0.0000	12.3458	MP 220 PRIVATE	3.9962	0.0000		59
5281	XING 220.73		5281	59	0.0170	12.3628	WILL CR	3.9962	0.0170	5281	59
5281	XING 220.89		5281	59	0.0170	12.3798	MIGRET CR	3.9792	0.0170	5281	59
5283	XING 221.38		5283	59	0.0170	12.3967	PRIVATE	3.9623	0.0170	5283	59
5280	XING 222.25		5280	59	0.0169	12.4137	GAMBER CR	3.9453	0.0169	5280	59
5279	XING 220.80		5279	59	0.0169	12.4306	PRIVATE	3.9284	0.0169	5279	59
6060	XING 223.54		5282	59	0.0170	12.4476	KRAUS CR	3.9114	0.0170	5282	59
5264	XING 223.60		5264	59	0.0170	12.4646	PRIVATE	3.8945	0.0170	5284	59
5275	XING 223.90		5275	59	0.0169	12.4815	PRIVATE	3.8775	0.0169	5275	59
5279	XING 224.26		5279	59	0.0169	12.4984	MAIN STREET SP 506	3.8606	0.0169	5279	59
	XING 224.30			59	0.0000	12.4984	ALLEY STREET	3.8436	0.0000		59
	XING 225.80			59	0.0000	12.4984	LILES CR	3.8436	0.0000		59
	XING 226.64			59	0.0000	12.4984	HELLER CR	3.8436	0.0000		59
	XING 227.20			59	0.0000	12.4984	PRIVATE	3.8436	0.0000		59
	XING 227.50			59	0.0000	12.4984	PRIVATE	3.8436	0.0000		59
	XING 227.80			59	0.0000	12.4984	PRIVATE	3.8436	0.0000		59
	XING 227.94			59	0.0000	12.4984	ANGLING ROAD	3.8436	0.0000		59
	XING 228.38			59	0.0000	12.4984	COUNTY LINE	3.8436	0.0000		59
	XING 228.90			59	0.0000	12.4984	LOUISA STREET	3.8436	0.0000		59
5298	XING 228.97		5298	59	0.0170	12.5154	FOREST MARY STREE	3.8436	0.0170	5298	59
	XING 229.21			59	0.0000	12.5154	MP 230 COMLY STREET	3.8266	0.0000		59
5275	XING 229.34		5275	59	0.0169	12.5324	MARTIN STREET SR	3.8266	0.0169	5275	59
5273	XING 229.56		5273	59	0.0169	12.5493		3.8097	0.0169	5273	59
5278	XING 229.75		5278	59	0.0169	12.5662	CAMPBELL STREET	3.7928	0.0169	5278	59
5269	XING 230.57		5269	59	0.0170	12.5832	MERLIN CR	3.7758	0.0170	5269	59
5268	XING 231.50		5268	59	0.0169	12.6001	PRIVATE	3.7588	0.0169	5268	59
5278	XING 231.70		5278	59	0.0169	12.6171	PRIVATE	3.7419	0.0169	5278	59
1500	XING 234.11		1500	59	0.0048	12.6218	GROAT ROAD	3.7250	0.0048	1500	59
	XING 234.59			59	0.0000	12.6219	WYKES CR	3.7202	0.0000		59
	XING 235.10			59	0.0000	12.6218	PRIVATE	3.7202	0.0000		59
	XING 236.08			59	0.0000	12.6219	WALNUT STREET	3.7202	0.0000		59
	XING 236.14			59	0.0000	12.6219	MAIN STREET	3.7202	0.0000		59
		90		59	0.0333	12.6552	DUNKIRK NYC XING	3.7202	0.0000		59
	XING 236.99			59	0.0000	12.6552	RTE 135	3.6869	0.0000		59
3771	XING 237.61		3771	59	0.0121	12.6673	HOPPS ROAD	3.6869	0.0121	3771	59
5276	XING 238.71		5278	59	0.0169	12.6843	WAYNE STREET	3.6747	0.0169	5276	59

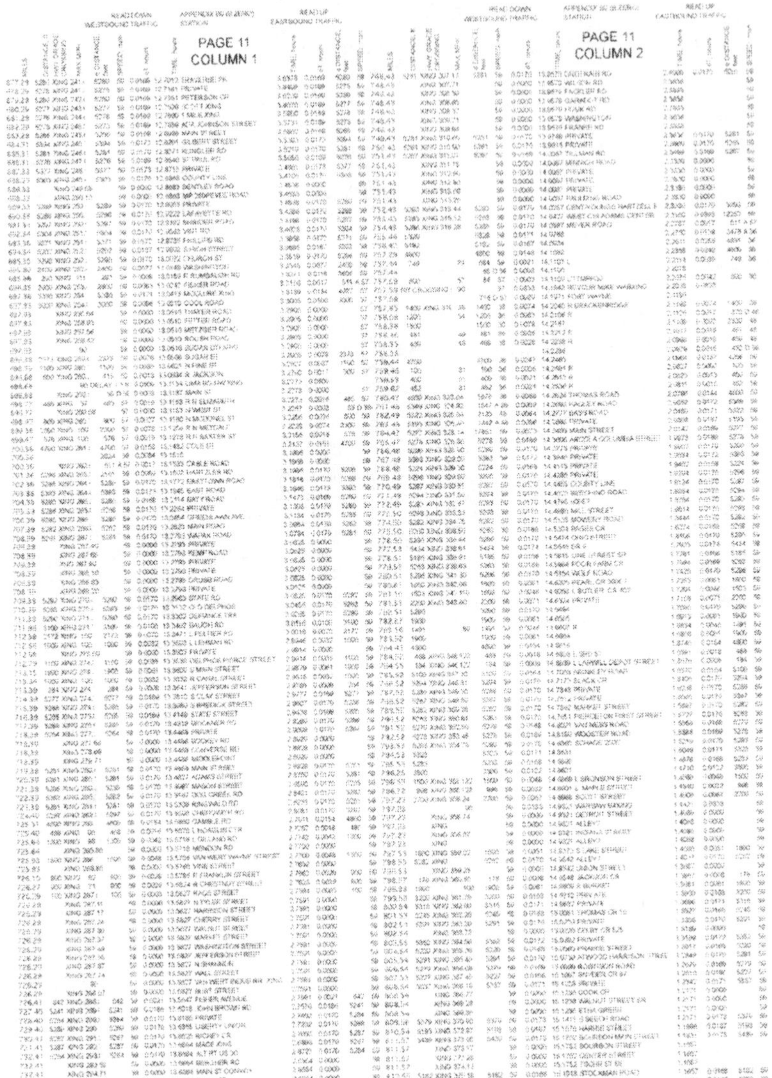
READ DOWN
WESTBOUND TRAFFIC
READ UP
EASTBOUND TRAFFIC
PAGE 11
COLUMN 1
READ DOWN
WESTBOUND TRAFFIC
READ UP
EASTBOUND TRAFFIC
PAGE 11
COLUMN 2

READ DOWN WESTBOUND TRAFFIC — APPENDIX B0 (B ZERO) STATION — READ UP EASTBOUND TRAFFIC

PAGE 12 COLUMN 1

MILES	DISTANCE, ft	HWY GRADE CROSSING	MAX MPH	d DISTANCE, feet	SPEED, mph	dT, hours	TIME, hours	STATION	TIME, hours	dT, hours	d DISTANCE, feet	SPEED, mph
829.54	5268	XING 388.		5268	59	0.0169	15.4798	PRIVATE	0.8780	0.0169	5268	59
830.55	5298	XING 390.		5298	59	0.0170	15.4968	UNION ROAD	0.8611	0.0170	5298	59
831.55	5298			5298	59	0.0170	15.5138	DONALDSON	0.8441	0.0170	5298	59
832.57	5411	XING 393.		5411	59	0.0174	15.5312	PRIVATE	0.8271	0.0174	5411	59
833.55	5164	XING 394.		5164	59	0.0166	15.5477	SR 23	0.8097	0.0166	5164	59
834.55	5285	XING 395.		5285	59	0.0170	15.5647	WALSH CR	0.7932	0.0170	5285	59
835.55	5288	XING 396.		5288	59	0.0170	15.5817	750E	0.7762	0.0170	5288	59
836.55	5285	XING 398.		5285	59	0.0170	15.5987	STARKE STREET	0.7592	0.0170	5285	59
837.55	5279	XING 398.		5279	59	0.0169	15.6158	HAMLET JEFFERSON STREI	0.7423	0.0169	5279	59
838.56	5301	XING 399.		5301	59	0.0170	15.6326	500E	0.7253	0.0170	5301	59
839.56	5282	XING 401.		5282	59	0.0170	15.6496	300E	0.7083	0.0170	5282	59
840.56	5282	XING 401.		5282	59	0.0170	15.6665	PRIVATE	0.6913	0.0170	5282	59
841.56	5270	XING 402.		5270	59	0.0169	15.6834	PRIVATE	0.6744	0.0169	5270	59
842.56	5287	XING 403.		5287	59	0.0170	15.7004	50E	0.6575	0.0170	5287	59
843.56	5292	XING 405.		5292	59	0.0170	15.7174	100W	0.6405	0.0170	5292	59
844.56	5280	XING 405.		5280	59	0.0169	15.7344	1400S	0.6235	0.0169	5280	59
845.59	5422	XING 406.		5422	59	0.0174	15.7518	SR 39	0.6066	0.0174	5422	59
846.54	5045	XING 407.		5045	59	0.0162	15.7679	LONG LANE	0.5892	0.0162	5045	59
846.54		XING 407.70			59	0.0000	15.7679	PRIVATE	0.5730	0.0000		59
847.38	4400	XING 408.		4400	59	0.0141	15.7821	THOMPSON STREET	0.5730	0.0141	4400	59
847.38		XING 408.74			59	0.0000	15.7821	OHIO STREET	0.5588	0.0000		59
847.38			90		59	0.0000	15.7821	HANNA C&O XING	0.5588	0.0000		59
847.56	966	XING 408.		966	59	0.0031	15.7852	450W	0.5588	0.0031	966	59
848.56	5281	XING 409.		5281	59	0.0170	15.8021	PRIVATE	0.5557	0.0170	5281	59
848.56		XING 410.35			59	0.0000	15.8021	MP 410 SCHULL 600W	0.5388	0.0000		59
849.56	5285	XING 411.		5285	59	0.0170	15.8191	FARRESTER ROAD	0.5388	0.0170	5285	59
850.56	5273	XING 411.		5273	59	0.0169	15.8360	TILDON 750W	0.5218	0.0169	5273	59
851.73	6182	XING 412.		6182	59	0.0198	15.8559	PRIVATE	0.5049	0.0198	6182	59
852.56	4367	XING 413.		4367	59	0.0140	15.8699	KLEINS 900W	0.4850	0.0140	4367	59
853.56	5280	XING 414.		5280	59	0.0169	15.8868	N MAIN STREET	0.4710	0.0169	5280	59
854.56	5272	XING 414.		5272	59	0.0169	15.9038	WANATAH S ILLINOIS	0.4541	0.0169	5272	59
855.55	5251	XING 415.		5251	59	0.0169	15.9206	S WASHINGTON	0.4372	0.0169	5251	59
856.55	5293	XING 415.		5293	59	0.0170	15.9376	LINCOLN STREET	0.4203	0.0170	5293	59
857.55	5270	XING 415.		5270	59	0.0169	15.9545	1100W	0.4033	0.0169	5270	59
858.16	3200	XING 416.		3200	59	0.0103	15.9648	OSBORN CR-CO	0.3864	0.0103	3200	59
858.16		XING 417.78					15.9648	FISHER 575E	0.3761			
858.16		XING 419.08					15.9648	DOLSON 450E	0.3761			
858.27	570	XING 133		570	59	0.0018	15.9666	R BICKEL 400E	0.3761	0.0018	570	59
858.55	1500	XING 420.		1500	59	0.0048	15.9714	MP 420 PRIVATE	0.3743	0.0048	1500	59
859.55	5280	XING 420.		5280	59	0.0169	15.9884	MONTDALE ROAD	0.3695	0.0169	5280	59
860.55	5263	XING 421.		5263	59	0.0169	16.0053	PRIVATE	0.3525	0.0169	5263	59
860.68	700	XING 422.		700	59	0.0022	16.0075	CEMETERY ROAD	0.3356	0.0022	700	59
861.47	4200	XING 100		4200	59	0.0135	16.0210	R GREENWICH STREET	0.3334	0.0135	4200	59
861.55	394	XING 423.		394	59	0.0013	16.0223	AXE STREET	0.3199	0.0013	394	59
862.08	2800	XING 423.		2800	59	0.0090	16.0313	FRANKLIN STREET	0.3186	0.0090	2800	59
862.17	486	XING 100		486	59	0.0016	16.0328	L VALPARAISO WASHINGTO	0.3096	0.0016	486	59
862.55	2000	XING 423.		2000	59	0.0064	16.0392	LAFAYETTE STREET	0.3081	0.0064	2000	59
863.55	5279	XING 423.		5279	59	0.0169	16.0562	NAPOLEON STREET	0.3017	0.0169	5279	59
864.55	5274	XING 424.		5274	59	0.0169	16.0731	STEVENS XING	0.2847	0.0169	5274	59
865.38	4400	XING 425.		4400	59	0.0141	16.0872	150 W	0.2678	0.0141	4400	59
865.38			90		59	0.0667	16.1539	VALPO GTW XING	0.2537	0.0667		59
865.38		XING 426.93			59	0.0000	16.1539	250W	0.1870	0.0000		59
865.55	892	XING 427.		892	59	0.0029	16.1568	400W	0.1870	0.0029	892	59
866.55	5299	XING 428.		5299	59	0.0170	16.1738	EAST SALT CR	0.1841	0.0170	5299	59
867.57	5340	XING 428.		5340	59	0.0171	16.1909	W SALT CR	0.1671	0.0171	5340	59
868.55	5206	XING 430.		5206	59	0.0167	16.2076	PRIVATE	0.1500	0.0167	5206	59
869.55	5291	XING 430.		5291	59	0.0170	16.2246	WHEELER PARK AVE	0.1333	0.0170	5291	59
870.56	5289	XING 431.		5289	59	0.0170	16.2416	GOTT RD 7TH	0.1163	0.0170	5289	59
871.54	5212	XING 432.		5212	59	0.0167	16.2583	TREAGER RD	0.0993	0.0167	5212	59
872.56	5354	XING 433.		5354	59	0.0172	16.2755	COUNTY LINE	0.0826	0.0172	5354	59
873.56	5316	XING 434.		5316	59	0.0171	16.2926	HOBART ILLINOIS STREET	0.0654	0.0171	5316	59
873.70	700	XING 434.		700	59	0.0022	16.2948	LINDA STREET	0.0483	0.0022	700	59
873.70		XING 434.82			59	0.0000	16.2948	CLEVELAND STREET	0.0461	0.0000		
873.86	844	XING 71		844	59	0.0027	16.2975	R LAKE PARK AVE	0.0461	0.0027	844	59
874.56	3700	XING 435.		3700	59	0.0119	16.3094	WISCONSIN STREET	0.0434	0.0000	0	
874.56					59	0.0000	16.3094		0.0434	0.0156	4658	59
875.56	5320			5320	59	0.0171	16.3265		0.0278	0.0000	0	
875.76	1060			1060	59	0.0034	16.3299		0.0278	0.0278	5222	A 0-59
875.76		XING 437.16			59	0.0000	16.3299	LIVERPOOL RD E END LIV T EAST END LIVERPOOL TERMINAL				
876.56	4200	XING 438.		4200	59	0.0135	16.3434	INDIANA AVENUE				
876.67	600	XING 439.		600	59	0.0019	16.3453	VIRGINIA STREET				
876.67					59	0.0000	16.3453	I-65				
876.89	1122			1122	59	0.0036	16.3489					
876.89					59	0.0000	16.3489	I-80				
877.57	3600			3600	59	0.0116	16.3605					
877.93	1897		D 11	2984	59	0.0096	16.3700					
878.38	2363			1276	D 59	0.0062	16.3782					
878.38					0		16.3782	WEST END LIVERPOOL TERMINAL				

Appendix B1 - Unreconstructed Right of Way 120 mph maximum with conventional equipment. FRA Class 7 track where allowed per existing curvature. Trains freight only or mixed freight and passenger, Northern New Jersey to Northwest Indiana. Trains consist of 200 TOFC platforms maximum, with passenger equipment as needed.
FRA Pre-Positive Train control rules apply as of June 2016.
Motive power 20 units maximum diesel electric, 6,000 hp each. Lead unit compatible with Amtrak east of Harrisburg.
Braking/Deceleration rate = 2 mph/second.
Assumed E_a = 4 inches, E_u = 3 inches
Signals reactivated or reinstalled where required for higher speeds. Additional train controls added per FRA requirements. Curves L = left, R = right, per facing west bound. Normal curve speed reduced 20% at reverse curves including up to 600 feet between curves.
Center of gravity of moving equipment assumed to be 120 inches above top of rail. Train length = 11,000 feet
Highway crossing locations are approximate.
Distance Jersey Meadows to Liverpool IN 875.1 miles

Time Westbound 13.611 hours
Time Eastbound 13.709 hours

READ DOWN WESTBOUND TRAFFIC — APPENDIX B1 STATION — READ UP EASTBOUND TRAFFIC

MILES	DISTANCE FT	HWY GRADE CROSSING	MAX MPH	DISTANCE	SPEED	ST. HOURS	TIME HOURS	PAGE 1 COLUMN 1	TIME HOURS	ST. HOURS	DISTANCE	SPEED
0.00								E MEADOWS	13.7141	0.0000		0
0.62	3274								13.7141	0.0083	1320	D 80
0.62									13.7058	0.0065	2044	80
1.62	5281								13.6993	0.0164	5191	80
2.07	2377								13.6829	0.0075	2377	80
2.35	1479		137					R	13.6754	0.0047	1479	80
2.46	589						0.0000		13.6708	0.0019	589	80
2.62	838								13.6689	0.0028	838	80
2.62								W MEADOWS	13.6663	0.0000		
2.77	526			3407	A 84	0.0203	0.0203		13.6663	0.0017	526	80
2.78	330		60		60	0.0000	0.0203	HACKENSACK R	13.6646	0.0010	330	80
2.78							0.0203		13.6636	0.0017	561	D 72
2.78						0.0000	0.0203		13.6619	0.0002	80	72
3.62	4409			1858	60	0.0058	0.0261		13.6617	0.0128	4006	A 44
4.00	2022			2022	60	0.0064	0.0325	HUDSON	13.6489	0.0076	1762	44
4.65	3443		141	3443	60	0.0109	0.0434	R HUDSON	13.6413	0.0148	3443	44
4.89	1273		141	796	60	0.0025	0.0459	R	13.6265	0.0055	1273	44
4.99	487			3207	A 60	0.0089	0.0548		13.6210	0.0021	487	44
5.75	4036					0.0000	0.0548		13.6189	0.0174	4036	44
5.75				384	26	0.0010	0.0558		13.6015	0.0000		
5.75				1408	D 26	0.0044	0.0602		13.6015			
5.98	1206		44	1206	44	0.0052	0.0654	L HARRISON	13.6015	0.0052	1206	44
5.98							0.0654		13.5964	0.0021	987	D 59
6.05	371			371	44	0.0016	0.0670	REVERSE CURVE	13.5943	0.0002	86	59
6.23	955		66	955	44	0.0041	0.0711	R	13.5941	0.0000	0	59
6.28	265			265	44	0.0011	0.0722	REVERSE CURVE	13.5941	0.0000	0	59
6.38	531		98	531	44	0.0023	0.0745	L	13.5941	0.0010	300	59
6.46	424			424	44	0.0018	0.0763		13.5931	0.0014	424	59
6.68	1168		141	1168	44	0.0050	0.0814	L NEWARK	13.5917	0.0037	1168	59
6.73	265			265	44	0.0011	0.0825		13.5880	0.0006	265	59
6.85	637		59	637	44	0.0027	0.0853	L	13.5871	0.0020	637	59
6.85					44		0.0853		13.5851	0.0032	1189	D 82
6.96	584			584	44	0.0025	0.0878		13.5819	0.0013	584	82
7.09	686			686	44	0.0030	0.0907		13.5806	0.0016	686	82
7.20	581		52	581	44	0.0025	0.0932	R HUNTER	13.5790	0.0013	581	82
7.96	4013			4533	44	0.0195	0.1127		13.5776	0.0025	1201	D 10
8.16	1054					0.0000	0.1127		13.5751	0.0006	300	100
8.51	1832		100	1687	A 44	0.0058	0.1186	L	13.5745	0.0306	13164	A 57
8.99	2563			2929	59	0.0094	0.1280		13.5440	0.0013	391	57
9.98	5262			5119	A 59	0.0136	0.1416		13.5427	0.0022	639	A 52
10.13	1002						0.1416		13.5405	0.0001	38	52
10.56	2004		165			0.0000	0.1416	R	13.5404	0.0073	2004	52
10.99	2267						0.1416		13.5331	0.0083	2267	52
11.01	105			7791	A 82	0.0161	0.1577		13.5249	0.0004	105	52
11.16	742		165			0.0000	0.1577	L	13.5245	0.0027	742	52
11.98	4389					0.0000	0.1577		13.5216	0.0160	4389	52
12.06	434			1134	100	0.0021	0.1598		13.5056	0.0016	434	52
12.06				2475	D 16	0.0060	0.1658		13.5042			
12.23	867		57	830	57	0.0028	0.1686	L ELIZABETH	13.5042	0.0032	867	52
12.26	163			6	57	0.0000	0.1686	REVERSE CURVE	13.5010	0.0006	163	52
12.26				200	D 57	0.0007	0.1693	TT 55 MPH	13.5004			
12.70	2330		52	2330	57	0.0085	0.1777	R	13.5004	0.0085	2330	52
12.70							0.1777	ELMORA	13.4918	0.0086	2774	D 11
12.98	1517			1517	57	0.0056	0.1833	E END 6 TRACKS	13.4833	0.0009	540	114
13.96	5283			5253	57	0.0192	0.2025		13.4824	0.0047	2807	A 11
14.99	5239			4200	57	0.0153	0.2178		13.4777	0.0085	4968	111
15.99	5258			21565	A 52	0.0467	0.2645		13.4692	0.0078	4554	111
16.13	774					0.0000	0.2645		13.4615	0.0000		
16.36	1187		111			0.0000	0.2645	R	13.4615	0.0000		
16.74	2014			1808	A 11	0.0031	0.2675		13.4615	0.0000		
16.99	1342		200			0.0000	0.2675	L	13.4615	0.0000		
17.13	695					0.0000	0.2675		13.4615	0.0000		
17.30	909		135			0.0000	0.2675	R	13.4615	0.0000		
17.49	1016					0.0000	0.2675		13.4615	0.0000		
17.52	160					0.0000	0.2675		13.4615	0.0000		
17.56	214					0.0000	0.2675		13.4615	0.0000		
17.75	1016		200			0.0000	0.2675	L	13.4615	0.0000		
17.98	1230					0.0000	0.2675		13.4615	0.0000		
18.23	1295					0.0000	0.2675		13.4615	0.0244	13044	A 89
18.57	1781		113			0.0000	0.2675	R	13.4370	0.0028	1295	59
18.64	378					0.0000	0.2675	REVERSE CURVE	13.4343	0.0008	378	59
18.74	540		132			0.0000	0.2675	L	13.4335	0.0038	1804	59
18.99	1296					0.0000	0.2675		13.4296	0.0056	2509	A 82
19.58	3108			2092	113	0.0035	0.2710		13.4241	0.0015	630	82
19.69	621		115	457	113	0.0005	0.2715	R	13.4226	0.0014	621	82
19.81	621			164	D 11	0.0003	0.2721		13.4212	0.0014	621	82
19.97	828		111	826	111	0.0014	0.2735	L	13.4197	0.0006	256	82
20.00	158		111	1159	111	0.0020	0.2755	L	13.4191	0.0019	627	A 79
20.41	2158			1613	D 11	0.0031	0.2786		13.4172	0.0049	2092	79
20.68	1421		88	1421	88	0.0030	0.2816	R	13.4123	0.0034	1421	79
20.77	474			474	88	0.0010	0.2826	REVERSE CURVE	13.4089	0.0011	474	79
20.98	1106		88	1105	88	0.0024	0.2849	L	13.4077	0.0026	1106	79
21.28	1602		88	1602	88	0.0034	0.2883	L ISELIN	13.4051	0.0038	1602	79
21.41	701		88	262	88	0.0006	0.2889	L	13.4012	0.0017	701	79
21.54	681			439	D 86	0.0010	0.2899	METROPARK	13.3996	0.0016	681	79
21.80	1361		82	1361	82	0.0031	0.2930	R	13.3979	0.0006	246	79
21.99	995			818	82	0.0019	0.2949		13.3973	0.0064	2506	A 69
22.10	598			177	D 82	0.0004	0.2953		13.3909	0.0006	292	69
22.53	2264		79	2284	79	0.0066	0.3008	L	13.3904	0.0053	2284	69
22.72	979			979	79	0.0023	0.3031		13.3841	0.0027	979	69
23.01	1522		88	1522	79	0.0036	0.3068	R	13.3814	0.0042	1522	69
23.46	2398		88	2398	79	0.0057	0.3125	R	13.3772	0.0066	2398	69
23.98	2756			2756	79	0.0066	0.3191		13.3707	0.0076	2756	69
24.15	880			317	79	0.0008	0.3199		13.3631	0.0024	886	69
24.15				543	D 79	0.0014	0.3213		13.3607			
24.41	1397		88	1397	88	0.0038	0.3251	R S METUCHEN	13.3607	0.0038	1397	69
24.41							0.3251		13.3569	0.0071	3534	D 12
24.81	2095			2095	88	0.0058	0.3309		13.3498	0.0001	47	120
24.98	913		141	913	88	0.0025	0.3334	L	13.3497	0.0000	0	120
25.09	573		141	573	88	0.0016	0.3349	L	13.3497	0.0000	0	120
25.27	937			937	88	0.0026	0.3375		13.3497	0.0015	937	120
25.62	1874		104	1574	68	0.0061	0.3477	L	13.3483	0.0030	1874	120
25.98	1874			1574	68	0.0061	0.3478		13.3453	0.0030	1874	120
26.76	4135			2784	88	0.0075	0.3553		13.3424	0.0085	4135	120
26.96	1074		141	13519	A 69	0.0289	0.3842	R	13.3358	0.0017	1074	120
26.98	107					0.0000	0.3842		13.3341	0.0002	107	120
27.98	5273					0.0000	0.3842		13.3340	0.0083	5273	120
28.15	875					0.0000	0.3842		13.3258	0.0014	875	120
28.65	2676		139	15220	A 10	0.0256	0.4098	L	13.3243	0.0042	2676	120
28.97	1698			0	120	0.0000	0.4098		13.3200	0.0027	1698	120
29.05	420			0	120	0.0000	0.4098		13.3174	0.0007	420	120
29.26	1103		137	0	120	0.0000	0.4098	L	13.3167	0.0017	1103	120
29.97	3727			0	120	0.0000	0.4098	NEW BRUNSWICK	13.3150	0.0059	3727	120
30.97	5290				120	0.0000	0.4098		13.3091	0.0083	5290	120
31.54	3029				120	0.0000	0.4098		13.3007	0.0048	3029	120
31.96	2208		124	1421	120	0.0022	0.4120	L	13.2959	0.0035	2208	120

READ DOWN WESTBOUND TRAFFIC — APPENDIX B1 STATION — READ UP EASTBOUND TRAFFIC

MILES	DISTANCE FT	HWY GRADE CROSSING	MAX MPH	DISTANCE	SPEED	ST. HOURS	TIME HOURS	PAGE 1 COLUMN 2	TIME HOURS	ST. HOURS	DISTANCE	SPEED
32.59	5404			5404	120	0.0065	0.4206		13.2925	0.0065	5404	120
33.97	5202			5202	120	0.0062	0.4287		13.2839	0.0039	2454	120
34.98	5303			5303	120	0.0064	0.4371		13.2801	0.0161	9803	A 11
35.97	5259			5259	120	0.0063	0.4454		13.2639	0.0059	3457	111
36.97	5285			5285	120	0.0063	0.4537		13.2580	0.0090	5285	111
37.05	430			430	120	0.0007	0.4544	DEANS	13.2490	0.0007	430	111
37.28	1183		165	1068	120	0.0017	0.4561	L	13.2483	0.0020	1183	111
37.40	645			762	D 12	0.0013	0.4574		13.2463	0.0011	645	111
37.97	3013		111	3013	111	0.0051	0.4625	R	13.2452	0.0061	3013	111
38.97	5253			5253	111	0.0090	0.4715	MONMOUTH JUNCTIO	13.2400	0.0013	762	D 12
39.97	5314			5314	111	0.0091	0.4806		13.2388	0.0155	9805	120
40.97	5287			433	111	0.0007	0.4813		13.2233	0.0083	5282	120
41.97	5286			9863	A 11	0.0161	0.4974		13.2150	0.0083	5286	120
42.97	5284			5566	120	0.0088	0.5062		13.2066	0.0083	5284	120
43.97	5276			5276	120	0.0083	0.5145		13.1983	0.0083	5276	120
44.97	5274			5274	120	0.0083	0.5228		13.1900	0.0083	5274	120
45.03	317			317	120	0.0005	0.5233		13.1816	0.0005	317	120
45.97	4965			4968	120	0.0078	0.5312	PRINCETON JUNCTIO	13.1811	0.0078	4968	120
46.97	5268			5268	120	0.0083	0.5395		13.1733	0.0083	5268	120
47.97	5296			5295	120	0.0084	0.5478		13.1650	0.0011	714	120
48.29	1671			1671	120	0.0026	0.5505		13.1639	0.0000		120
48.51	1148		158	1148	120	0.0018	0.5523	LAWRENCE STATION	13.1639	0.0161	9853	A 11
48.97	2452			2453	120	0.0039	0.5561		13.1477	0.0000		
49.98	5291			5291	120	0.0084	0.5645		13.1477	0.0082	4614	111
50.97	5277			5277	120	0.0083	0.5728		13.1395	0.0056	3290	A 10
51.97	5280			5280	120	0.0083	0.5812		13.1340	0.0000		
52.97	5278			5278	120	0.0083	0.5895		13.1340	0.0438	18072	A 44
53.78	4273			4273	120	0.0067	0.5962		13.0901	0.0000		
53.97	990		173	990	120	0.0016	0.5978	L	13.0901	0.0000		
54.07	532		173	3172	120	0.0050	0.6028	L	13.0901	0.0141	3274	44
54.72	3402			762	D 12	0.0013	0.6040	TRENTON	13.0760	0.0025	535	A 36
54.20	691		111	1009	111	0.0017	0.6058	R	13.0732	0.0003	93	36
54.97	638			320	D 11	0.0006	0.6063	A	13.0729	0.0034	635	36
55.58	3236		107	2948	107	0.0052	0.6115	L	13.0695	0.0170	3236	36
55.88	1600			0	107	0.0000	0.6115	MORRISVILLE PA	13.0525	0.0084	1600	36
55.98	500		141	0	107	0.0000	0.6115	R	13.0441	0.0026	500	36
56.19	1100				107	0.0000	0.6115		13.0414	0.0056	1100	36
56.19				3483	D 16	0.0085	0.6203		13.0357			
56.34	800		44	800	44	0.0034	0.6237	R	13.0357	0.0042	800	36
56.39	300			300	44	0.0013	0.6250	REVERSE CURVE	13.0314	0.0015	300	36
56.58	980		46	980	44	0.0042	0.6292	L	13.0299	0.0052	980	36
56.64	300		46	300	44	0.0013	0.6305	L	13.0247	0.0016	300	36
56.66	100			100	44	0.0004	0.6310	REVERSE CURVE	13.0231	0.0005	100	36
56.81	800		46	800	44	0.0034	0.6344	R	13.0226	0.0042	800	36
56.92	580			345	44	0.0015	0.6359		13.0184	0.0031	580	36
56.90				235	D 44	0.0011	0.6370		13.0153			
57.06	700		36	700	36	0.0037	0.6407	R	13.0153	0.0037	700	36
57.06							0.6407		13.0117	0.0006	111	D 46
57.14	500			500	36	0.0026	0.6433	REVERSE CURVE	13.0111	0.0018	389	40
57.33	1000		50	1000	36	0.0053	0.6486	L	13.0093	0.0047	1000	40
57.39	300			300	36	0.0016	0.6502	REVERSE CURVE	13.0045	0.0014	300	40
57.58	1000		113	1000	36	0.0053	0.6554	R	13.0031	0.0047	1000	40
57.79	1100			1100	36	0.0058	0.6612		12.9984	0.0052	1100	40
58.02	1200		57	1200	36	0.0063	0.6675	R	12.9932	0.0057	1200	40
58.07	300			300	36	0.0016	0.6691	REVERSE CURVE	12.9875	0.0014	300	40
58.22	600		48	800	36	0.0042	0.6733	L	12.9861	0.0038	800	40
58.22							0.6733		12.9823	0.0021	523	D 55
58.58	1880			1880	36	0.0099	0.6832		12.9802	0.0047	1053	55
58.41	4380			2920	36	0.0154	0.6986		12.9755	0.0151	4380	55
58.56	500		58			0.0000	0.6986	R	12.9604	0.0028	500	55
58.56	0			115	A 38	0.0005	0.6991	W YARD	12.9577			55
59.58	100			2245	40	0.0106	0.7097		12.9577	0.0003	100	55
59.60	100			100	40	0.0005	0.7102		12.9573	0.0003	100	55
59.69	500		55	500	40	0.0024	0.7126	R	12.9570	0.0017	500	55
59.69							0.7126		12.9553	0.0015	483	D 66
59.69	0						0.7126		12.9537	0.0006	206	66
59.69	0						0.7126		12.9530	0.0044	1446	A 57
60.58	4680			3240	40	0.0153	0.7279		12.9487	0.0084	2540	57
61.15	3000			1405	A 40	0.0028	0.7307		12.9403	0.0100	3000	57
61.56	2280		80	5312	55	0.0183	0.7490	R	12.9303	0.0076	2280	57
61.71	840		80	680	55	0.0023	0.7513	R	12.9227	0.0028	680	57
62.15	2500			380	55	0.0012	0.7526	REVERSE CURVE	12.9206	0.0080	2500	57
62.15				1598	A 55	0.0053	0.7579		12.9122			
62.16				41	56	0.0001	0.7580		12.9122			
62.15				406	D 56	0.0013	0.7592		12.9122			
62.56	2100		57	2100	57	0.0070	0.7662	L WOODBOURNE	12.9122	0.0070	2100	57
62.56							0.7662		12.9052	0.0054	2168	D 96
63.58	5280			5280	657	0.0015	0.7677		12.8998	0.0006	298	96
64.00	2200			2200	57	0.0073	0.7750		12.8992	0.0119	5604	A 82
64.24	1280		141	1280	57	0.0043	0.7793	R	12.8872	0.0015	640	82
64.33	500			500	57	0.0017	0.7809		12.8858	0.0012	500	82
64.58	1300		173	1300	57	0.0043	0.7853	R	12.8846	0.0030	1300	82
64.73	700		173	440	57	0.0015	0.7867	R	12.8816	0.0018	700	82
65.38	3300			5273	A 57	0.0150	0.8017		12.8798	0.0076	3300	82
65.56	1200		100			0.0000	0.8017	R	12.8722	0.0028	1200	82
65.71	680		100	2120	A 82	0.0047	0.8065	R	12.8694	0.0016	680	82
66.20	2600			253	83	0.0005	0.8070		12.8678	0.0060	2600	82
66.20				374	D 86	0.0008	0.8078		12.8618			
66.45	1300		82	1300	82	0.0030	0.8108	L	12.8618	0.0030	1300	82
66.58	700			700	82	0.0016	0.8124	HOLLAND	12.8588	0.0016	700	82
66.67	500			500	82	0.0012	0.8136		12.8572	0.0012	500	82
66.81	700		100	700	82	0.0016	0.8152	R	12.8561	0.0016	700	82
67.11	1560			1560	82	0.0036	0.8189		12.8544	0.0036	1560	82
67.31	1100		82	1100	82	0.0025	0.8214	L	12.8508	0.0025	1100	82
67.58	1400			1400	82	0.0032	0.8246		12.8480	0.0032	1400	82
67.60	80			80	82	0.0002	0.8248		12.8450	0.0002	80	82
67.98	2100		115	2100	82	0.0049	0.8297	R	12.8448	0.0049	2100	82
68.31	1700			1700	82	0.0039	0.8336		12.8400	0.0039	1700	82
68.48	900	XING	100	900	82	0.0021	0.8357	L BUSTLETON	12.8361	0.0021	900	82
68.52	500			500	82	0.0012	0.8369		12.8340	0.0012	500	82
68.97	2080			2080	82	0.0048	0.8416		12.8328	0.0048	2080	82
69.50	2800		82	2800	82	0.0065	0.8481	R STREET RD	12.8280	0.0065	2800	82
69.58	400						0.8481		12.8215	0.0047	2488	D 11
70.28	3660			24003	A 82	0.0322	0.8803		12.8168	0.0001	59	116
70.52	1300		122				0.8803	L	12.8167	0.0000		
70.58	300	XING	15.95				0.8803	COUNTY LINE	12.8167	0.0000		
71.58	5280						0.8803		12.8167	0.0219	12593	A 10
72.58	5280			0		0.0000	0.8803		12.7948	0.0021	1120	100
73.58	5280			581	119	0.0009	0.8813		12.7927	0.0100	5280	100
74.45	4600			1526	D 11	0.0026	0.8839		12.7827	0.0087	4600	100
74.58	660		100	680	100	0.0013	0.8852	L	12.7740	0.0013	680	100
74.86	1480		100	1480	100	0.0028	0.8880	L	12.7727	0.0028	1480	100
75.56	3700			3700	100	0.0070	0.8950	YORK RD	12.7699	0.0013	690	D 10
75.58	100		141	100	100	0.0002	0.8952	L	12.7686	0.0015	882	100
76.26	3580		141	3580	100	0.0068	0.9020	L	12.7671	0.0108	5688	A 10
76.58	1700			1700	100	0.0032	0.9052		12.7562	0.0029	1520	100

APPENDIX B1 — PAGE 2 COLUMN 1

MILES	DISTANCE, ft	HWY GRADE CROSSINGS	MAX MPH	DISTANCE, feet	SPEED, mph	ET, hours	TIME, hours	STATION	TIME, hours	ET, hours	DISTANCE, feet	SPEED, mph
READ DOWN	WESTBOUND TRAFFIC								READ UP	EASTBOUND TRAFFIC		
76.77	980			980	100	0.0019	0.9071		12.7534	0.0019	980	100
77.37	3200		141	940	100	0.0018	0.9088	R	12.7515	0.0061	3200	100
77.58	1100			5988	A 10	0.0108	0.9197		12.7454	0.0021	1100	100
78.38	4200			882	109	0.0015	0.9212		12.7434	0.0057	3009	100
78.38				690	D 10	0.0013	0.9225		12.7377			
78.56	1080		100	1080	100	0.0020	0.9245	R	12.7377	0.0000	0	100
78.73	800		100	800	100	0.0015	0.9260	R'CAMP HILL	12.7377	0.0000	6	100
79.26	2800			2800	100	0.0053	0.9313		12.7377	0.0000	6	100
79.58	1680		115	1680	100	0.0032	0.9345	L	12.7377	0.0181	7791	A 82
79.90	1700		115	1700	100	0.0032	0.9377	L	12.7218	0.0034	1460	82
80.58	3580			3580	100	0.0068	0.9445		12.7182	0.0083	3560	82
80.58				2395	A 10	0.0044	0.9489		12.7099			
81.58	5280			2065	104	0.0038	0.9577		12.7099	0.0122	5280	82
81.71	680			1500	D 10	0.0031	0.9558		12.6977	0.0016	680	82
82.11	2100		82	2100	82	0.0049	0.9606	R'STENTON	12.6961	0.0049	2100	82
82.58	2500			2500	82	0.0058	0.9664		12.6913	0.0044	2300	D 11
83.58	5280			5280	82	0.0122	0.9786		12.6869	0.0009	540	114
84.58	5280	XING29.18		3220	82	0.0074	0.9860	CONSHOHOCKEN	12.6860	0.0000		
85.58	5280	XING 29.5		0		0.0000	0.9860	GRAVERS	12.6860	0.0347	18366	A 82
85.58	0					0.0000	0.9860	MP 30	12.6512	0.0042	1840	82
85.99	2180			13779	A 82	0.0269	1.0130		12.6470	0.0050	2180	82
86.47	2500		110	0		0.0000	1.0130	L	12.6420	0.0058	2500	82
86.58	600			612	109	0.0011	1.0140	EARNEST	12.6362	0.0014	600	82
87.31	3880			2109	D 10	0.0038	1.0178		12.6346	0.0090	3880	82
87.58	1400		82	1400	82	0.0032	1.0210	L'RAMBO	12.6258	0.0032	1400	82
87.83	280		82	280	82	0.0006	1.0217	L	12.6226	0.0006	280	82
87.90	1400			1400	82	0.0032	1.0249		12.6219	0.0015	706	D 93
88.58	3600		93	3600	82	0.0083	1.0332	R'HENDERSON	12.6204	0.0087	4294	93
88.69	580		93	580	82	0.0013	1.0345	R	12.6117	0.0012	580	93
89.58	4700			4700	82	0.0109	1.0454		12.6105	0.0010	495	D 10
89.75	880			720	82	0.0017	1.0471		12.6095	0.0007	385	100
90.09	1800		100	1791	A 82	0.0161	1.0632	L	12.6088	0.0034	1800	100
90.58	2600			2769	100	0.0052	1.0684	KING of PRUSSIA	12.6054	0.0028	1913	D 12
91.58	5280			5280	100	0.0100	1.0784		12.6026	0.0099	6267	120
92.58	5280			3120	100	0.0059	1.0843		12.5927	0.0083	5253	120
92.89	580			17615	A 10	0.0300	1.1143		12.5644	0.0000	0	120
92.99	1600		173	0	120	0.0000	1.1143	R	12.5644	0.0000	0	120
93.58	3100			0	120	0.0000	1.1143		12.5644	0.0000		
94.58	5280			0	120	0.0000	1.1143	HOWELLVILLE	12.5644	0.0000		
95.11	2800			0	120	0.0000	1.1143		12.5644	0.0175	10663	A 10
95.43	1692		141	0	120	0.0000	1.1143	R	12.5669			
95.58	788		141	385	120	0.0006	1.1149	R	12.5669			
95.73	780			780	120	0.0012	1.1160		12.5669			
96.22	2600		141	2600	120	0.0041	1.1203	L	12.5669			
96.43	1113			1113	120	0.0018	1.1220		12.5669			
96.58	787			787	120	0.0012	1.1233		12.5669			
97.32	3909			3909	120	0.0062	1.1294		12.5669			
97.43	580			580	120	0.0009	1.1304		12.5669			
97.47	211			211	120	0.0003	1.1307		12.5669			
97.58	580		245	580	120	0.0009	1.1316	R	12.5669			
97.77	980		245	980	120	0.0015	1.1331	R	12.5669	0.0488	21651	A 42
98.43	3508			3508	120	0.0055	1.1387		12.5180	0.0066	1458	42
98.58	792			792	120	0.0013	1.1399		12.5115	0.0036	792	42
98.79	1100			1100	120	0.0017	1.1417		12.5079			
98.90	580		173	580	120	0.0009	1.1426	L	12.5079	0.0044	627	A 20
99.36	2442			2442	120	0.0039	1.1464		12.5034	0.0060	638	20
99.38		EAST END GLEN LOCH					1.1464	STREAM 2.07	12.4974			
99.43	370			370	120	0.0006	1.1470		12.4974	0.0035	370	20
99.58	788			788	120	0.0012	1.1483		12.4939	0.0075	788	20
99.79	1100		141	1100	120	0.0017	1.1500	R	12.4864	0.0104	1100	20
99.97	963			550	120	0.0009	1.1500		12.4760	0.0091	963	20
100.30	1737			0	120	0.0000	1.1509	RTE 352	12.4669	0.0164	1737	20
100.30				843	D 12	0.0014	1.1523		12.4506	0.0000		
100.43	692		110	0	110	0.0000	1.1523	L	12.4505	0.0066	692	20
100.58	788		110		120	0.0000	1.1523	L	12.4439	0.0075	788	20
100.73	802	XING 45.18			120	0.0000	1.1523	W'HENDERSON	12.4384	0.0078	802	20
100.78	245			44	110	0.0001	1.1523		12.4288	0.0023	245	20
100.78				3790	D 11	0.0094	1.1618	AMTRAK WB TR 4 OHD	12.4266	0.0000		
101.03	1313	1573	42	1313	42	0.0058	1.1677	R	12.4266	0.0124	1313	20
101.08	260			260	42	0.0012	1.1689		12.4141	0.0025	260	20
101.08		ug 0.35		527	42	0.0024	1.1712	R RAVINE ROAD	12.4116	0.0000		
101.22	750			223	D 42	0.0011	1.1724		12.4116	0.0071	750	20
101.30	453		34	277	34	0.0015	1.1739	R	12.4045	0.0043	453	20
101.32	101			277	D 34	0.0019	1.1758		12.4002	0.0010	101	20
101.38	274		20	274	20	0.0026	1.1784	L REVERSE CURVE	12.3993	0.0026	274	20
101.42	257		31	257	20	0.0024	1.1809	L	12.3967	0.0015	206	D 31
101.42						0.0000	1.1809	GLEN LOCH 30 MPH OF	12.3952	0.0003	51	31
101.42						0.0000	1.1809	JCT SWITCH	12.3948	0.0000	0	31
101.61	980			980	20	0.0093	1.1902	T	12.3948	0.0000	0	0
101.90	1524		133	1524	20	0.0144	1.2046	L	12.3948	0.0000		
102.14	1252			1252	20	0.0119	1.2164	T	12.3948	0.0000		31
102.26	673			673	20	0.0064	1.2228	T	12.3948	0.0013	234	D 40
102.49	1190		130	1190	20	0.0113	1.2341	L	12.3936	0.0000		40
102.71	1191		122	1191	20	0.0113	1.2454	R REVERSE CURVE	12.3936	0.0074	2585	D 93
103.14	2226			2226	20	0.0211	1.2664	T	12.3862	0.0006	224	93
103.14						0.0000	1.2664		12.3858	0.0272	9619	A 57
104.14	5280			1707	20	0.0114	1.2778		12.3586	0.0055	1594	57
104.14				484	A 20	0.0036	1.2814	T	12.3531	0.0000		
104.37	1230	OHD 28.38				0.0000	1.2814	LO-GRADE - ABANDONE	12.3531	0.0041	1230	57
105.14	4050		258	10453	A 31	0.0114	1.2925	L	12.3490	0.0135	4050	57
105.30	845		258			0.0000	1.2925	L	12.3356	0.0028	845	57
105.91	3221			203	90	0.0004	1.2933	T	12.3327	0.0107	3221	57
105.91				1779	D 90	0.0046	1.2978		12.3220			
106.14	1214		57	1214	57	0.0040	1.3019	R	12.3220	0.0040	1214	57
106.14						0.0000	1.3019	BRADFORD HILLS- EAS	12.3180	0.0008	264	D 63
106.32	970		203	970	57	0.0032	1.3051	R	12.3171	0.0021	706	63
106.51	1024			1024	57	0.0034	1.3085	T	12.3150	0.0031	1024	63
106.88	1940		63	1940	57	0.0064	1.3149	L	12.3119	0.0058	1940	63
106.88							1.3149		12.3061	0.0018	583	D 76
107.14	1347			1347	57	0.0045	1.3194	T	12.3043	0.0017	584	76
107.36	1173			1173	57	0.0039	1.3233	T	12.3026	0.0029	1173	76
107.75	2080		76	2080	57	0.0069	1.3302	R	12.2987	0.0052	2080	76
107.75							1.3302		12.2945	0.0033	1549	D 10
108.14	2027			2027	57	0.0067	1.3370	T	12.2912	0.0009	478	100
108.38	1202			438	57	0.0015	1.3384		12.2902	0.0023	1202	100
108.38	0			879	A 57	0.0026	1.3412	DOWNINGTOWN	12.2880	0.0000	0	
108.70	1777		100	1661	83	0.0058	1.3462	L	12.2880	0.0034	1777	100
108.90	1046			1046	83	0.0031	1.3493		12.2846	0.0020	1046	100
109.14	1256		107	348	83	0.0010	1.3504	R	12.2826	0.0024	1255	100
109.35	1109		107	2757	A 63	0.0075	1.3579	R	12.2802	0.0021	1108	100
109.80	2376			1843	76	0.0046	1.3625	T	12.2781	0.0045	2376	100
110.14	1796		107	3091	A 76	0.0072	1.3697	L	12.2736	0.0034	1796	100
110.22	466		107	6437	A 86	0.0131	1.3828	L	12.2702	0.0009	466	100
110.38	725					0.0000	1.3828	T	12.2694	0.0014	725	100
110.57	1087		100			0.0000	1.3828	R	12.2680	0.0021	1057	100
110.57							1.3828		12.2659	0.0003	148	D 10

APPENDIX B1 — PAGE 2 COLUMN 2

MILES	DISTANCE, ft	HWY GRADE CROSSINGS	MAX MPH	DISTANCE, feet	SPEED, mph	ET, hours	TIME, hours	STATION	TIME, hours	ET, hours	DISTANCE, feet	SPEED, mph
			READ DOWN	WESTBOUND TRAFFIC					READ UP	EASTBOUND TRAFFIC		
110.77	1087		200			0.0000	1.3828	R	12.2657	0.0010	548	102
110.77							1.3828		12.2646	0.0022	1186	A 10
110.97	1035					0.0000	1.3828	T	12.2624	0.0005	240	100
111.14	880		141			0.0000	1.3828	L	12.2620	0.0017	880	100
111.18	216					0.0000	1.3828	T	12.2603	0.0004	216	100
111.32	754		115			0.0000	1.3828	L	12.2599	0.0014	754	100
112.00	3610			5988	A 10	0.0108	1.3936	T	12.2585	0.0068	3610	100
112.14	700		113			0.0000	1.3936	L	12.2516	0.0013	700	100
112.18	209		113			0.0000	1.3936	L	12.2503	0.0004	209	100
112.32	784		113	541	109	0.0009	1.3945	R REVERSE CURVE	12.2499	0.0015	784	100
113.01	3607			600	D 10	0.0013	1.3958	T	12.2484	0.0068	3607	100
113.14	680		100	680	100	0.0013	1.3971	R	12.2416	0.0013	680	100
113.60	2433		100	2433	100	0.0046	1.4017	R	12.2403	0.0046	2433	100
113.71	621			621	100	0.0012	1.4029		12.2357	0.0012	621	100
114.08	1915		100	1915	100	0.0036	1.4065	L	12.2345	0.0036	1915	100
114.14	311			311	100	0.0006	1.4071		12.2309	0.0021	1183	D 11
114.14					100	0.0000	1.4071	COATESVILLE	12.2288	0.0027	1015	115
114.56	2240			2240	100	0.0042	1.4113	T	12.2261	0.0017	1008	A 15
115.14	3040		141	3040	100	0.0058	1.4171	L	12.2245	0.0030	1786	114
115.21	414		141	414	100	0.0008	1.4179	L	12.2215	0.0007	414	114
115.49	1449			1449	100	0.0027	1.4206	T	12.2206	0.0024	1449	114
115.96	2485		115	2485	100	0.0047	1.4253	R	12.2164	0.0041	2485	114
116.14	932			932	100	0.0018	1.4271	T	12.2143	0.0015	932	114
116.58	3934			129	100	0.0002	1.4273	T	12.2127	0.0066	3934	114
117.14	1346		114	10569	A 10	0.0186	1.4459	L	12.2062	0.0022	1346	114
117.23	485		114			0.0000	1.4459	L	12.2040	0.0008	485	114
117.50	1455					0.0000	1.4459	T	12.2032	0.0018	1068	114
117.89	2047		118			0.0000	1.4459	R	12.2014	0.0000	0	114
118.14	1293			6067	114	0.0101	1.4560	T	12.2014	0.0006		
119.14	5280					0.0000	1.4560		12.2014	0.0000		
119.62	2536					0.0000	1.4560	T	12.2014	0.0000		
119.80	984		162	4427	A 11	0.0072	1.4632	L	12.2014	0.0000		
120.14	1760			2619	A 11	0.0042	1.4674	T	12.2014	0.0419	21282	A 71
121.14	5280			2589	120	0.0041	1.4715	PARKESBURG	12.1595	0.0000		
121.23	496			495	120	0.0042	1.4756	T	12.1595	0.0018	684	71
121.47	1265		141	1265	120	0.0020	1.4776	R	12.1576	0.0072	2451	A 57
122.14	3520			3520	120	0.0056	1.4832	T	12.1504	0.0015	440	57
122.90	4015			3990	120	0.0063	1.4895	T	12.1489	0.0133	4015	57
123.04	770		141	0	120	0.0000	1.4895	R	12.1356	0.0026	770	57
123.14	495			0	120	0.0000	1.4895	T	12.1330	0.0016	495	57
123.54	2145			3	120	0.0000	1.4895	T	12.1314	0.0071	2145	57
123.54				3432	D 12	0.0068	1.4963		12.1243	0.0000		
124.14	3135		71	2478	71	0.0066	1.5029	R	12.1243	0.0104	3135	57
124.14				657	D 71	0.0019	1.5049	ATGLEN	12.1139	0.0000		
124.29	834		57	834	57	0.0028	1.5075	R	12.1139	0.0028	834	57
124.39	500			500	57	0.0017	1.5093	REVERSE CURVE	12.1111	0.0062	1878	57
124.80	2168		80	2168	57	0.0072	1.5165	L	12.1048	0.0031	883	A 50
124.87	389			389	57	0.0013	1.5175	REVERSE CURVE	12.1018	0.0012	316	50
125.14	1389		80	1389	57	0.0046	1.5224	R	12.1006	0.0053	1389	50
125.34	1056		80	1056	57	0.0035	1.5259	R CHRISTIANA	12.0953	0.0040	1056	50
125.95	3224			3224	57	0.0107	1.5366	T	12.0913	0.0122	3224	50
126.14	1000		80	1000	57	0.0033	1.5399	L	12.0791	0.0038	1006	50
126.24	550		90	550	57	0.0018	1.5418	L	12.0753	0.0021	556	50
126.34	550			468	57	0.0016	1.5433	REVERSE CURVE	12.0733	0.0021	550	50
126.34				82	D 62	0.0003	1.5436		12.0712	0.0000		
126.74	2090		55	2090	55	0.0072	1.5508	R	12.0712	0.0079	2090	50
126.90	825			633	55	0.0022	1.5530	T	12.0633	0.0031	825	50
126.90				193	D 55	0.0007	1.5537		12.0601			50
127.14	1265		50	1265	50	0.0048	1.5585	L GAP	12.0601	0.0048	1265	50
127.45	1650		50	1650	50	0.0063	1.5647	L	12.0553	0.0063	1650	50
127.80	1870		162	1870	50	0.0071	1.5718	L	12.0491	0.0071	1870	50
128.14	1760			1760	50	0.0067	1.5785	T	12.0420	0.0067	1760	50
128.20	330			330	50	0.0013	1.5797		12.0353	0.0013	330	50
128.59	2090		50	2090	50	0.0079	1.5876	R	12.0341	0.0079	2090	50
128.59							1.5876		12.0262	0.0026	829	D 69
128.89	1540			1540	50	0.0058	1.5935	T	12.0236	0.0020	711	69
129.14	1320		69	1320	50	0.0050	1.5985	L	12.0216	0.0036	1320	69
129.33	1045		69	1045	50	0.0040	1.6024	L	12.0180	0.0029	1045	69
129.33							1.6024	0	12.0151	0.0044	1995	D 10
129.81	2530			2530	50	0.0095	1.6120	T	12.0106	0.0003	164	101
129.81							1.6120		12.0103	0.0011	591	A 10
130.14	1705		141	1705	50	0.0065	1.6185	L	12.0092	0.0028	1485	100
130.61	2530			2530	50	0.0096	1.6281	T	12.0064	0.0048	2530	100
130.79	935		141	330	50	0.0013	1.6293	L	12.0016	0.0018	935	100
131.14	1815			2903	A 50	0.0092	1.6385	T	11.9998	0.0023	1198	100
131.94	4235			1002	69	0.0028	1.6412	T	11.9975	0.0000		
132.14	1045		100	11124	A 69	0.0244	1.6657	R	11.9976			
132.14						0.0000	1.6657		11.9976	0.0161	7791	A 82
132.75	3245		100			0.0000	1.6657	R	11.9815	0.0031	1351	82
133.14	2035					0.0000	1.6657	T	11.9783	0.0047	2035	82
133.51	1980					0.0000	1.6657	T	11.9736	0.0028	1232	82
133.79	1485		122	1486	100	0.0028	1.6685	L	11.9708	0.0083	3333	A 69
134.14	1815			2879	100	0.0055	1.6739	T	11.9625	0.0020	715	69
134.56	2334			1201	D 10	0.0025	1.6764	T	11.9605	0.0064	2334	69
135.06	2501		82	2501	82	0.0058	1.6822	L	11.9541	0.0069	2501	69
135.14	445			445	82	0.0010	1.6832	T	11.9472	0.0012	445	69
135.64	2640	XING 59.1		2640	82	0.0061	1.6893	IRISHTOWN ROAD	11.9460	0.0072	2640	69
135.83	1045		89	1045	82	0.0024	1.6917	R	11.9388	0.0029	1045	69
136.08	1320			600	82	0.0014	1.6931		11.9359	0.0036	1320	69
136.08				720	D 82	0.0018	1.6949		11.9323			
136.14	275		69	275	69	0.0008	1.6957	R	11.9323	0.0008	275	69
136.74	3190		69	3190	69	0.0088	1.7044	R	11.9315	0.0088	3190	69
137.09	1870			1870	69	0.0051	1.7096		11.9228	0.0003	103	D 71
137.14	220		71	220	69	0.0006	1.7102	L	11.9225	0.0053	1587	71
137.59	2390		71	2390	69	0.0066	1.7167	L BIRD IN HAND	11.9172	0.0064	2390	71
137.80	1112			1112	69	0.0031	1.7198		11.9108	0.0040	1818	D 10
138.14	1779		100	1779	69	0.0049	1.7247	L	11.9068	0.0020	1073	100
138.25	605		100	605	69	0.0017	1.7263	L	11.9047	0.0011	605	100
139.06	4235			3024	69	0.0063	1.7346		11.9036	0.0080	4235	100
139.05				411	A 69	0.0011	1.7358		11.8966	0.0000		
139.14	440		141	440	71	0.0012	1.7369	L	11.8966	0.0008	440	100
139.33	1045		141	1045	71	0.0028	1.7397	L	11.8947	0.0020	1045	100
139.64	1595			1595	71	0.0043	1.7440	T	11.8928	0.0030	1595	100
139.84	1595		100	989	71	0.0026	1.7466	R	11.8897	0.0030	1595	100
139.94							1.7466		11.8867	0.0006	299	D 10
140.14	1045			10713	A 71	0.0233	1.7699	T	11.8862	0.0007	386	104
140.14						0.0000	1.7699		11.8855	0.0044	2395	A 10
140.94	4224					0.0000	1.7699	T	11.8810	0.0074	3914	100
141.14	1056		173			0.0000	1.7699	R	11.8736	0.0020	1056	100
141.64	2640		173	0		0.0000	1.7699	R	11.8716	0.0050	2640	100
142.14	2640					0.0000	1.7699	T	11.8666	0.0006	342	100
142.19	275			3781	A 10	0.0069	1.7769	T	11.8660	0.0000		
142.28	466		113			0.0000	1.7769	R	11.8660	0.0000		
142.36	440					0.0000	1.7769	REVERSE CURVE	11.8660	0.0053	2718	A 95
142.55	990		113	539	106	0.0010	1.7778	L	11.8607	0.0001	55	95
142.60	275			453	D 10	0.0003	1.7787	T	11.8606	0.0006	275	95

READ DOWN WESTBOUND TRAFFIC — APPENDIX B1 STATION — READ UP EASTBOUND TRAFFIC

PAGE 3 COLUMN 1

MILES	DISTANCE	HWY GRADE CROSSING	MAX MPH	DISTANCE	SPEED	ET HOURS	TIME HOURS	STATION	TIME HOURS	ET HOURS	DISTANCE	SPEED
142.79	990		100	990	100	0.0019	1.7806	L	11.8800	0.0020	990	95
143.14	1815					0.0000	1.7806	T	11.8061	0.0036	1815	95
143.52	2006			3821	100	0.0072	1.7878		11.8544	0.0040	2006	95
144.14	3274			3274	100	0.0062	1.7940	LANCASTER STATION	11.8504	0.0065	3274	95
144.14				2000	100	0.0033	1.7978		11.8439	0.0000		
144.63	2585			228	100	0.0004	1.7982	T	11.8439	0.0052	2585	95
144.63				358	D 10	0.0007	1.7989	DILLERVILLE - COLA BF	11.8386	0.0000		
145.14	2695		95	2695	96	0.0054	1.8043	R	11.8386	0.0054	2695	95
145.33	1045		95	1045	95	0.0021	1.8064	R	11.8334	0.0021	1045	95
146.14	4235			4235	96	0.0084	1.8148	T	11.8313	0.0007	358	D 10
146.14						0.0000	1.8148		11.8306	0.0073	3878	100
146.79	3465			3465	95	0.0069	1.8217	T	11.8233	0.0066	3465	100
147.14	1815		100	1815	95	0.0036	1.8253	L	11.8167	0.0034	1815	100
147.18	220		100	220	95	0.0004	1.8258	L	11.8132	0.0004	220	100
147.82	3410	XING	71.6	1265	95	0.0025	1.8283	E PETER ROAD	11.8128	0.0048	2543	100
148.14	1650		100	13287	A 95	0.0239	1.8522	L	11.8080	0.0001	2133	A 94
148.28	660		100	0		0.0000	1.8522	L	11.8019	0.0001	44	94
148.54	1485			0		0.0000	1.8522	T	11.8018	0.0030	1485	94
148.93	2035		100			0.0000	1.8522	R	11.7988	0.0041	2035	94
149.14	1100					0.0000	1.8522	T	11.7947	0.0022	1101	94
150.14	5280					0.0000	1.8522		11.7925	0.0106	5280	94
150.14				590	114	0.0010	1.8532		11.7819	0.0000		
150.34	1056			1525	D 11	0.0028	1.8560	T	11.7819	0.0021	1056	94
150.84	2668		94	2668	94	0.0054	1.8613	L	11.7798	0.0054	2668	94
150.99	778			778	94	0.0016	1.8629	T	11.7744	0.0016	778	94
151.14	778		100	778	94	0.0016	1.8645	R	11.7728	0.0029	1434	94
151.34	1056		100	1056	94	0.0021	1.8666	R	11.7699	0.0044	2149	A 89
152.04	3724	XING	75.8	3724	94	0.0075	1.8741	PRIVATE	11.7655	0.0042	1975	89
152.14	500		224	500	94	0.0010	1.8751	L	11.7613	0.0011	500	89
152.64	2668		224	2668	94	0.0054	1.8805	L	11.7602	0.0057	2668	89
153.14	2612			1496	94	0.0030	1.8835	T	11.7545	0.0058	2612	89
153.14				3133	A 94	0.0061	1.8896		11.7490	0.0000		
153.22	440					0.0000	1.8896	T	11.7490	0.0008	440	89
153.45	1210		100	466	100	0.0009	1.8905	L	11.7480	0.0026	1210	89
153.75	1595			782	D 10	0.0015	1.8920	T	11.7455	0.0034	1595	89
153.95	1045		89	1045	89	0.0022	1.8942	L	11.7421	0.0022	1045	89
154.14	990			990	89	0.0021	1.8963	T	11.7398	0.0021	990	89
155.14	5280			5280	89	0.0112	1.9076		11.7377	0.0112	5280	89
155.28	778	XING	79.1	778	89	0.0017	1.9092	MOORES MILL ROAD	11.7265	0.0017	778	89
155.48	1056		89	1056	89	0.0022	1.9115	R	11.7248	0.0022	1056	89
155.89	2168			2168	89	0.0046	1.9161	T	11.7226	0.0026	1372	D 10
156.14	1278		122	1278	89	0.0027	1.9188	R	11.7200	0.0004	734	100
156.23	500		122	500	89	0.0011	1.9199	R	11.7196	0.0000	0	
157.00	4067			4067	89	0.0086	1.9285	T	11.7196	0.0103	5100	A 10
157.14	723		118	723	89	0.0016	1.9301	R	11.7093	0.0000		
157.29	834	XING	118	834	89	0.0018	1.9318	R MARKET ST MT JOY	11.7093	0.0000		
157.69	2112			1440	89	0.0031	1.9349	T	11.7093	0.0000		
158.14	2334		100			0.0000	1.9349	R	11.7093	0.0233	10713	A 71
158.14				11124	A 89	0.0106	1.9454		11.6859	0.0000		
158.17	165		100				1.9454		11.6859			
158.92	3960					0.0000	1.9454	T	11.6859	0.0017	619	71
159.14	1155		141			0.0000	1.9454	L	11.6843	0.0031	1155	71
159.36	1210		141			0.0000	1.9454	L	11.6812	0.0032	1210	71
159.84	2530			902	100	0.0017	1.9472	T	11.6780	0.0067	2530	71
160.14	1540		113	1540	100	0.0029	1.9501	L	11.6712	0.0041	1540	71
160.38	1278		112	440	100	0.0008	1.9509	L	11.6671	0.0034	1278	71
160.38				1188	A 10	0.0022	1.9531		11.6637	0.0000		
160.88	2668			354	102	0.0007	1.9538	T	11.6637	0.0071	2668	71
160.88				1966	D 10	0.0043	1.9581		11.6566	0.0000		
161.14	1334		71	1334	71	0.0036	1.9616	R ELIZABETHTOWN	11.6566	0.0036	1334	71
161.47	1779		71	1779	71	0.0047	1.9664	R	11.6530	0.0047	1779	71
161.61	723			723	71	0.0019	1.9683	T	11.6483	0.0019	723	71
161.90	1556		100	1556	71	0.0042	1.9725	L	11.6464	0.0042	1556	71
162.02	611			611	71	0.0016	1.9741	T	11.6422	0.0016	611	71
162.14	611		71	611	71	0.0016	1.9757	L	11.6406	0.0016	611	71
162.33	1000		71	1000	71	0.0027	1.9784	L	11.6390	0.0027	1000	71
162.33							1.9784		11.6363	0.0066	2689	D 11
163.14	4280			4280	71	0.0114	1.9898	T	11.6307	0.0001	86	111
163.85	3779			3779	71	0.0101	1.9999	T	11.6306	0.0014	810	A 11
164.10	1334		141	1334	71	0.0036	2.0035		11.6292	0.0000		
164.14	167			167	71	0.0004	2.0039	T	11.6292	0.0000		
165.14	5280			1440	71	0.0038	2.0077		11.6292	0.0147	8114	A 98
166.14	5280			12501	A 71	0.0267	2.0344		11.6145	0.0000		
166.14						0.0000	2.0344	T	11.6145	0.0283	12000	A 57
166.14				333	103	0.0006	2.0350		11.5861	0.0000		
166.91	4082			389	D 10	0.0007	2.0357	T	11.5861	0.0017	521	57
167.14	1196		98	1196	98	0.0023	2.0380	L	11.5844	0.0040	1196	57
167.29	825		98		98	0.0000	2.0380	L	11.5804	0.0027	825	57
167.78	2585				98	0.0000	2.0380	T	11.5777	0.0086	2585	57
168.14	1870		100		98	0.0000	2.0380	R	11.5691	0.0062	1870	57
168.43	1556		100		98	0.0000	2.0380	R	11.5629	0.0052	1556	57
168.89	2445			6661	98	0.0134	2.0515	T	11.5577	0.0061	2445	57
168.89				2330	D 98	0.0057	2.0572		11.5496	0.0000		
169.14	1278		57	1278	57	0.0042	2.0614	R	11.5496	0.0042	1278	57
169.49	1870		57	1870	57	0.0062	2.0676	R	11.5453	0.0062	1870	57
169.49							2.0676	REVERSE CURVE	11.5391	0.0013	406	D 66
170.07	3080		66	3080	57	0.0102	2.0779	L	11.5379	0.0077	2674	66
170.14	330			330	57	0.0011	2.0789	T	11.5302	0.0032	1307	D 88
170.14						0.0000	2.0789	ROYALTON - ROYALTO	11.5270	0.0000	11	89
170.74	3168			3168	57	0.0106	2.0895	T	11.5270	0.0046	2100	89
170.87	723		152	723	57	0.0024	2.0919	L	11.5224	0.0015	723	89
170.95	389			389	57	0.0013	2.0932	T	11.5208	0.0008	389	89
171.10	834		340	834	57	0.0028	2.0959	L	11.5200	0.0018	834	89
171.13	111			111	57	0.0004	2.0963	T	11.5182	0.0002	111	89
171.14	56		111	56	57	0.0002	2.0965	L	11.5180	0.0001	58	89
171.43	1556		89	1556	57	0.0052	2.1017	L MIDDLETOWN	11.5179	0.0033	1556	89
171.50	389			389	57	0.0013	2.1030	REVERSE CURVE	11.5145	0.0011	546	D 97
171.70	1056		108	364	57	0.0012	2.1042	R	11.5134	0.0018	899	97
171.90	1056			1446	A 57	0.0044	2.1086	T	11.5117	0.0021	1056	97
172.10	1056		115	1358	66	0.0039	2.1125	L	11.5096	0.0021	1056	97
172.14	167			167	66	0.0005	2.1130	T	11.5076	0.0003	167	97
172.43	1556			109	66	0.0003	2.1133	T	11.5072	0.0030	1556	97
172.58	778		346			0.0000	2.1133	L	11.5042	0.0015	778	97
173.04	2445			6436	A 66	0.0158	2.1289	T FLYOVER	11.5027	0.0048	2445	97
173.14	500		97			0.0000	2.1289	R	11.4979	0.0010	500	97
173.43	1556		97	290	89	0.0006	2.1295	R	11.4969	0.0030	1556	97
174.03	3168			441	89	0.0009	2.1304	T	11.4939	0.0018	987	D 11
174.14	658		224			0.0000	2.1304	L	11.4921	0.0024	1421	110
174.14				3549	A 88	0.0077	2.1376		11.4898	0.0000		
174.33	1045		224	779	97	0.0015	2.1392	L	11.4896	0.0000		
175.07	3906			3905	97	0.0076	2.1468	T	11.4896	0.0125	6962	A 10
175.14	330		133	330	97	0.0006	2.1474	R	11.4771	0.0000		
175.14						0.0000	2.1474		11.4771	0.0033	1733	A 97
175.59	2396		133	1996	97	0.0039	2.1513	R	11.4738	0.0000		
176.14	2885					0.0000	2.1513	T	11.4738	0.0222	9790	A 67

PAGE 3 COLUMN 2

MILES	DISTANCE	HWY GRADE CROSSING	MAX MPH	DISTANCE	SPEED	ET HOURS	TIME HOURS	STATION	TIME HOURS	ET HOURS	DISTANCE	SPEED
177.14	5280					0.0000	2.1513		11.4516	0.0047	1673	67
177.14				1733	A 97	0.0033	2.1547	STEELTON	11.4468	0.0000		
177.78	3300					0.0000	2.1547	T	11.4468	0.0131	3625	A 35
178.10	1815		165	12593	A 10	0.0218	2.1766	R	11.4338	0.0007	136	35
178.14	168			0	0	0.0000	2.1766	T	11.4331	0.0009	165	35
178.14				585	116	0.0010	2.1776		11.4322	0.0014	238	A 30
178.58	2334			1267	D 11	0.0022	2.1798	T	11.4308	0.0029	459	30
178.62	1278		100	1278	100	0.0024	2.1822	R	11.4279	0.0081	1278	30
179.04	1167			1167	100	0.0022	2.1844	T	11.4198	0.0074	1167	30
179.11	389		122	220	100	0.0034	2.1848	L	11.4124	0.0025	389	30
179.14	111			0	100	0.0000	2.1848	T	11.4100	0.0007	111	30
179.47	1760			19	100	0.0000	2.1849	T	11.4093	0.0111	1760	30
179.47				2021	D 10	0.0046	2.1894		11.3982	0.0000		
179.59	660		67	660	67	0.0019	2.1913	R	11.3982	0.0042	660	30
180.14	2880			2880	67	0.0081	2.1994	T	11.3840	0.0181	2880	30
180.47	1778			581	67	0.0016	2.2010	T	11.3760	0.0112	1778	30
180.47				1197	D 10	0.0044	2.2055		11.3647	0.0000		
180.52	260		35	260	35	0.0015	2.2069	L HARRISBURG	11.3647	0.0017	260	30
180.53	54		35	54	35	0.0003	2.2072	L	11.3630	0.0003	54	30
180.57	215			96	35	0.0005	2.2077	T	11.3627	0.0014	215	30
180.57				119	D 35	0.0007	2.2084	REVERSE CURVE	11.3613	0.0000		
180.65	377		30	377	30	0.0024	2.2108	R	11.3613	0.0024	377	30
180.76	590			592	30	0.0037	2.2146	REVERSE CURVE	11.3590	0.0006	94	D 34
180.76						0.0000	2.2146		11.3584	0.0028	496	34
180.89	700		36	700	30	0.0044	2.2190	L	11.3556	0.0039	700	34
181.03	754			754	30	0.0048	2.2237	T	11.3517	0.0042	754	34
181.14	539			539	30	0.0034	2.2271	T	11.3475	0.0030	539	34
181.18	218			218	30	0.0014	2.2285	T	11.3445	0.0012	218	34
181.21	163		45	163	30	0.0010	2.2296	R	11.3433	0.0009	163	34
181.34	707			707	30	0.0045	2.2340	T	11.3424	0.0038	707	34
181.34	0	HERR STREET		10	30	0.0000	2.2340	WEST END AMTRAK	11.3385	0.0000	0	
181.60	1361			1361	30	0.0086	2.2426	T	11.3385	0.0070	1361	34
181.68	435		34	435	30	0.0027	2.2454	L	11.3309	0.0024	435	34
181.68	0			0	30	0.0000	2.2454	REVERSE CURVE	11.3286	0.0033	810	D 58
181.76	435		80	435	30	0.0027	2.2481	R	11.3251	0.0001	28	58
181.89	653			653	30	0.0041	2.2522	T	11.3250	0.0008	250	58
182.07	980		67	980	30	0.0062	2.2584	R	11.3242	0.0032	980	58
182.13	327			327	30	0.0021	2.2605	T	11.3210	0.0011	327	55
182.27	701			701	30	0.0044	2.2649	T	11.3199	0.0122	3220	58
182.30	162		58	162	30	0.0010	2.2659	L	11.3078	0.0019	576	A 54
183.14	4419			2272	30	0.0143	2.2803	T	11.3058	0.0138	3946	54
183.14				186	A 30	0.0011	2.2814		11.2920	0.0047	1189	A 41
184.14	5281			5282	34	0.0194	2.3008	T	11.2873	0.0052	1129	41
184.14				2209	A 34	0.0089	2.3197		11.2821	0.0000		
185.03	4736			1085	56	0.0034	2.3231	T	11.2821	0.0219	4736	41
185.03				184	D 58	0.0005	2.3237		11.2602	0.0000		
185.06	163		54	163	54	0.0006	2.3243	R	11.2602	0.0006	163	41
185.09	163			163	54	0.0006	2.3249	REVERSE CURVE	11.2594	0.0006	163	41
185.14	218		54	218	54	0.0008	2.3256	L	11.2587	0.0010	218	41
185.22	437			437	54	0.0015	2.3271	T	11.2577	0.0020	437	41
185.46	1257		57	1257	54	0.0044	2.3316	L	11.2567	0.0058	1257	41
186.00	2897			2444	54	0.0086	2.3401	T	11.2498	0.0134	2897	41
186.00				453	D 54	0.0018	2.3419		11.2365	0.0000		
186.05	219		41	219	41	0.0010	2.3429	L	11.2365	0.0010	219	41
186.15	547		41	547	41	0.0025	2.3455	L ROCKVILLE	11.2355	0.0026	547	41
186.19	218		41	218	41	0.0010	2.3465	L	11.2329	0.0010	218	41
186.19						0.0000	2.3465		11.2319	0.0031	839	D 63
186.90	3756			3756	41	0.0174	2.3638	T	11.2289	0.0118	3863	63
186.90						0.0000	2.3638		11.2172	0.0022	710	A 58
186.90						0.0000	2.3638	SUSQUEHANNA RIVER	11.2150	0.0008	287	58
187.08	925		82	925	41	0.0043	2.3681	R	11.2141	0.0030	925	58
187.15	381			381	41	0.0018	2.3698	T	11.2111	0.0012	381	58
187.32	916			916	41	0.0042	2.3741	T	11.2086	0.0006	169	A 57
187.87	2900		82	2909	41	0.0134	2.3875	R	11.2087	0.0037	1120	57
188.15	1456			1456	41	0.0067	2.3942	T	11.2055	0.0049	1456	57
188.15				668	41	0.0030	2.3973		11.2007	0.0000		
188.68	2797					0.0000	2.3973	T	11.2007	0.0093	2797	57
189.17	2581		63	2475	A 41	0.0089	2.4062	L	11.1914	0.0086	2581	57
189.23	335		63	5580	63	0.0198	2.4269	L	11.1828	0.0011	335	57
189.41	950			222	D 63	0.0007	2.4266	T	11.1817	0.0032	950	57
189.49	391		58	391	58	0.0013	2.4279	R	11.1786	0.0013	391	57
189.61	671			671	58	0.0022	2.4301	T	11.1773	0.0022	671	57
189.63	70			28	58	0.0001	2.4302		11.1750	0.0023	700	57
189.63		XING	114	42	D 58	0.0001	2.4303	PRIVATE	11.1727	0.0000		
189.85	1200		57	1200	57	0.0040	2.4343	L	11.1727	0.0039	1163	57
189.85						0.0000	2.4343		11.1689	0.0003	55	D 59
190.11	1370			1370	57	0.0046	2.4389		11.1686	0.0041	1265	59
190.66	2900			2900	57	0.0096	2.4485		11.1645	0.0060	2900	59
190.69	1176		108	1176	57	0.0039	2.4524	R	11.1551	0.0038	1176	59
191.11	1200	XING	115.	1200	57	0.0040	2.4564	PRIVATE	11.1514	0.0039	1200	59
191.96	4500	XING	115.5	4354	57	0.0145	2.4709	PRIVATE	11.1475	0.0137	4364	59
192.11	791		58	254	A 57	0.0008	2.4717	R	11.1338	0.0000		
192.32	1084		58	1767	59	0.0057	2.4774	R	11.1338	0.0104	2476	A 30
192.89	3000			3000	59	0.0096	2.4870		11.1236	0.0166	2635	30
193.12	1200		89	1200	59	0.0039	2.4909	R	11.1069	0.0078	1200	30
193.88	3000		89	3000	59	0.0096	2.5005	R	11.0983	0.0189	3000	30
193.81	665			665	59	0.0021	2.5026		11.0804	0.0042	665	30
193.96	800		65	800	59	0.0026	2.5052	R	11.0762	0.0051	800	30
194.11	800		32	800	59	0.0026	2.5078	R	11.0712	0.0051	800	30
194.32	1100		67	960	59	0.0030	2.5108	R	11.0661	0.0068	1100	30
194.47	800			4	59	0.0000	2.5108		11.0592	0.0051	800	30
194.47				946	D 66	0.0040	2.5149		11.0541	0.0000		
194.68	1100		30	1100	30	0.0069	2.5218	L	11.0541	0.0069	1100	30
194.68							2.5218		11.0472	0.0004	68	D 33
194.73	270			270	30	0.0017	2.5236	FRC 17 MPHGS	11.0467	0.0012	201	33
194.88	800		33	800	30	0.0051	2.5286	R	11.0456	0.0046	800	30
194.88							2.5286		11.0410	0.0060	1719	D 76
195.09	1100			1100	30	0.0069	2.5355		11.0350	0.0004	179	76
195.11	100			100	30	0.0006	2.5361		11.0346	0.0000	0	76
196.89	4098		70	4098	30	0.0259	2.5620	R DUNCANNON	11.0346	0.0095	3400	76
196.89							2.5620		11.0251	0.0014	594	D 66
196.98	500			500	30	0.0032	2.5652		11.0247	0.0000	8	66
196.44	2402			2402	30	0.0152	2.5803		11.0247	0.0051	2302	86
197.00	3000		86	1730	30	0.0109	2.5912	L	11.0196	0.0066	3000	86
197.25	1300		86	139	A 30	0.0008	2.5921	L	11.0130	0.0029	1300	86
197.61	1880			931	33	0.0053	2.5974		11.0102	0.0015	738	D 97
197.85	1300		100	5719	A 33	0.0189	2.6163	R	11.0086	0.0002	115	97
198.00	800			3091	A 76	0.0072	2.6235		11.0084	0.0081	3938	A 88
198.16	800			3690	58	0.0081	2.6317		11.0004	0.0125	5249	A 70
198.51	1896	XING	104	6437	A 36	0.0131	2.6447	L 'AQUEDUCT	10.9879	0.0104	3851	70
199.01	2600	XING	123	1136	A 10	0.0022	2.6469	PRIVATE	10.9774	0.0006	204	A 69
200.01	5265	XING	120 60			0.0000	2.6469	PRIVATE	10.9769	0.0013	456	69
201.01	5279			6	102	0.0000	2.6469		10.9756	0.0146	5279	69
201.24	1200			399	102	0.0007	2.6477		10.9611	0.0030	1200	69
201.24				2018	D 10	0.0044	2.6521		10.9578	0.0000		

READ DOWN WESTBOUND TRAFFIC — APPENDIX B STATION — PAGE 4 COLUMN 1

MILES	DISTANCE	XING	MAX SPD	DISTANCE	SPEED	ET HOURS	CUM HOURS	
201.63	2083	XING	70	2083	70	0.0056	2.6578	L PRIVATE
201.97	1800			1800	70	0.0049	2.6626	
202.01	182			131	70	0.0004	2.6630	
202.01		XING 127...		51	D 85	0.0001	2.6631	PRIVATE
202.57	3000		69	3000	69	0.0082	2.6714	L
202.80	1200			1200	69	0.0033	2.6746	
202.97	900		73	900	69	0.0025	2.6771	R
203.11	752		103	752	69	0.0021	2.6792	R
203.32	1100		69	1100	69	0.0030	2.6822	R
203.55	1200		94	1200	69	0.0033	2.6855	R
203.78	1200	XING	73	1200	69	0.0033	2.6888	R PRIVATE
204.00	1200			1200	69	0.0033	2.6921	
204.21	1100			1100	69	0.0030	2.6951	
204.52	1600	XING	69	1600	69	0.0044	2.6995	L PRIVATE
204.76	1287			906	69	0.0025	2.7020	
204.76				381	D 69	0.0011	2.7031	
205.01	1300		61	1300	61	0.0040	2.7071	L
205.18	930		61	930	61	0.0029	2.7100	L
205.18							2.7100	
206.01	4400		97	4400	61	0.0137	2.7237	L
206.26	1313			1313	61	0.0041	2.7278	T 201.6
206.98	3800		71	3800	61	0.0118	2.7396	R
207.59	3200		65	1467	61	0.0046	2.7442	R
207.59							2.7442	
207.78	1024			548	A 61	0.0019	2.7461	
207.96	800		97	2068	65	0.0087	2.7548	L
208.06	500		97	500	65	0.0015	2.7563	L
208.31	1400			1400	65	0.0041	2.7604	T NEWPORT
208.59	1500	XING	97	1500	65	0.0052	2.7656	R PRIVATE
208.64	1000			1000	65	0.0029	2.7686	
208.91	373	XING	97	373	65	0.0011	2.7696	R PRIVATE
209.38	2478		97	2478	65	0.0072	2.7768	R
209.91	2800			1525	65	0.0044	2.7812	
210.07	825			7822	A 85	0.0186	2.7998	
210.90	4400		106	446	92	0.0009	2.8008	L
211.50	3176			1408	D 92	0.0033	2.8041	
211.90	2100		68	2100	68	0.0058	2.8100	L
212.24	1800		78	1800	68	0.0050	2.8150	L
212.37	400			400	68	0.0011	2.8161	T RC MPH [illegible]
212.46	719		77	719	68	0.0020	2.8181	R
212.89	2300			2300	68	0.0064	2.8245	
213.42	2800			2800	68	0.0078	2.8323	
213.56	713		84	713	68	0.0020	2.8343	R
213.90	1800			1800	68	0.0050	2.8393	T MILLERSTOWN
214.36	2414		178	1952	68	0.0054	2.8447	L
214.36				462	D 68	0.0014	2.8461	
214.90	2800		58	2800	58	0.0091	2.8553	L
215.24	1800			1800	58	0.0060	2.8615	
215.42	932	XING	154	932	58	0.0029	2.8644	R PRIVATE
215.66	1300			1300	58	0.0042	2.8687	
215.89	1200		115	1200	58	0.0039	2.8726	R
215.91	100		116	100	58	0.0003	2.8729	R 211.6
216.56	3400			3400	58	0.0111	2.8840	
216.86	1620		86	1620	58	0.0053	2.8893	L OUTWARD CURVE
216.91	249		83	249	58	0.0008	2.8901	L
217.47	3000			329	58	0.0011	2.8912	
217.61	700		121	3467	A 58	0.0087	2.9006	R
217.70	500					0.0000	2.9006	
217.86	800		85	320	76	0.0008	2.9017	R
217.86				884	D 76	0.0025	2.9042	
218.66	1100		58	1100	58	0.0036	2.9078	L
218.66							2.9078	
218.77	600			600	58	0.0020	2.9098	REVERSE CURVE
218.96	1000		74	1000	58	0.0032	2.9130	R THOMPSONTOWN
219.02	282			262	58	0.0009	2.9140	TRC MPH55
219.23	1100		55	1100	58	0.0036	2.9175	L
219.23							2.9175	
219.45	1200			1200	58	0.0039	2.9215	
219.48	184			184	58	0.0006	2.9221	
219.69	1100		97	1100	58	0.0036	2.9257	R
219.85	4000			4000	56	0.0131	2.9387	
220.74	4700			1534	56	0.0050	2.9437	
220.86	596	XING	79	55	A 56	0.0003	2.9440	R PRIVATE
220.95	496		79	2297	59	0.0094	2.9533	R
221.86	4800			4292	A 69	0.0117	2.9650	
222.49	3340			7564	79	0.0191	2.9841	
222.87	2000		121	388	A 79	0.0014	2.9855	L
222.89	100		121			0.0000	2.9855	L
223.36	2500					0.0000	2.9855	
223.36				23	41	0.0001	2.9856	
223.36				1129	D 51	0.0031	2.9886	
223.84	2500		69	2500	69	0.0080	2.9966	R TUSCARORA
223.93	465		59	465	69	0.0015	2.9981	R
223.93		XING 148.5				0.0000	2.9981	PRIVATE
223.93						0.0000	2.9981	
223.93							2.9981	
224.42	2600			2600	59	0.0083	3.0066	
224.83	2200		72	2200	60	0.0071	3.0136	L
225.23	2100		72	2100	60	0.0067	3.0203	L
225.49	1350			1350	59	0.0043	3.0246	
225.75	1400		60	1400	59	0.0045	3.0291	R
226.04	1500			1360	56	0.0043	3.0334	
226.04	0		80	1788	A 56	0.0069	3.0404	R PORT ROYAL
226.04				57	60	0.0002	3.0406	
226.40	1900			197	D 61	0.0006	3.0411	
226.73	1767		66	1767	65	0.0051	3.0463	R
226.78	270		65	270	65	0.0008	3.0471	R
227.73	5000			5000	65	0.0146	3.0618	
228.29	2900	XING 152		2900	65	0.0087	3.0698	PRIVATE
228.41	800		71	360	65	0.0010	3.0708	L RC
228.41				440	D 66	0.0014	3.0722	
228.74	1731		95	1731	55	0.0060	3.0782	R TT MPH 55
228.82	400			1	55	0.0000	3.0782	
228.82				399	D 55	0.0015	3.0797	
229.08	1400		44	1185	44	0.0050	3.0847	RRC
229.08				235	D 44	0.0011	3.0858	
229.19	600		35	600	36	0.0032	3.0890	L TT MPH40
229.19		XING 153.90					3.0890	JUNIATA STREET
229.61	2210			2210	36	0.0116	3.1006	T MIFFLIN
229.73	600		49	600	36	0.0032	3.1038	L
229.73							3.1038	
229.97	1300		76	1300	36	0.0068	3.1106	L TT MPH 45
230.07	500			500	36	0.0026	3.1133	TRC
230.39	1680		54	1680	36	0.0088	3.1221	R

READ UP EASTBOUND TRAFFIC

CUM HOURS	ET HOURS	DISTANCE	SPEED	MILES	DISTANCE	XING
10.9678	0.0057	2083	69	230.39		
10.9621	0.0049	1650	69	230.52	700	
10.9472	0.0005	182	69	230.73	1100	
10.9467	0.0058	2105	69	231.35	3295	
10.9409	0.0042	1434	A 61	231.58	1200	
10.9367	0.0021	691	61	231.73	800	
10.9347	0.0028	900	61	231.86	700	
10.9319	0.0023	752	61	232.09	1200	
10.9296	0.0034	1100	61	232.22	700	
10.9261	0.0037	1200	61	232.73	2695	
10.9224	0.0037	1200	61	232.98	1500	
10.9187	0.0037	1200	61	233.08		
10.9149	0.0034	1100	61	233.45	1900	
10.9115	0.0050	1600	61	233.45		
10.9066	0.0040	1287	41	233.58	700	
10.9026	0.0000			233.73	781	
10.9026	0.0040	1300	61	233.84	581	
10.8985	0.0029	930	61	234.60	4000	
10.8956	0.0006	185	D 10	234.73	700	
10.8951	0.0123	4215	65	234.78	235	
10.8828	0.0038	1313	65	235.14	1900	
10.8790	0.0111	3600	65	235.36	1200	
10.8679	0.0097	3200	65	235.72	1900	
10.8584	0.0039	1620	D 93	235.72		
10.8547	0.0006	274	93	235.88	800	
10.8541	0.0000			236.07	1014	
10.8541	0.0000			236.73	3500	
10.8541	0.0000			236.86	800	
10.8541	0.0000			237.14	1471	
10.8541	0.0000			237.46	1750	
10.8541	0.0000			237.75	1400	
10.8541	0.0181	7750	A 68	237.94	1000	
10.8361	0.0072	2599	68	238.09	800	
10.8288	0.0023	825	68	238.09		
10.8265	0.0123	4400	68	238.39	1356	
10.8143	0.0088	3176	68	238.50	600	
10.8054	0.0088	2316	68	238.50		
10.7988	0.0050	1670	A 58	238.69	1000	
10.7938	0.0006	194	58	238.84	801	
10.7930	0.0023	719	58	239.24	2114	
10.7907	0.0075	2300	58	239.66	2300	
10.7831	0.0091	2800	58	239.81	700	
10.7740	0.0023	713	58	240.11	1600	
10.7717	0.0059	1800	58	240.36	1291	
10.7658	0.0079	2414	58	240.53	900	
10.7579	0.0000			240.85	800	
10.7579	0.0081	2800	58	240.83	900	
10.7485	0.0025	884	D 76	241.00	900	
10.7463	0.0008	320	76	241.00		
10.7455	0.0091	3467	A 58	241.08	400	
10.7367	0.0021	571	58	241.09		
10.7317	0.0003	100	58	241.29	1100	
10.7334	0.0111	3400	58	241.29		
10.7225	0.0053	1620	58	241.49	1054	
10.7190	0.0008	249	58	241.68	1000	
10.7160	0.0098	3000	58	241.96	1500	
10.7064	0.0023	706	58	241.96		
10.7041	0.0016	500	58	242.57	3200	
10.7024	0.0026	800	58	242.67	562	
10.6998	0.0000			243.60	4900	
10.6998	0.0036	1100	58	243.70	540	
10.6962	0.0001	43	D 69	244.13	2300	
10.6961	0.0018	657	69	244.30	870	
10.6943	0.0032	1000	59	244.60	1600	
10.6911	0.0009	282	59	244.85	1300	XING 169.40
10.6902	0.0035	1100	59	245.35	2680	
10.6867	0.0028	1012	D 79	245.60	1300	
10.6839	0.0006	168	79	246.45	1502	XING 176
10.6834	0.0004	184	79	246.45		XING 176
10.6830	0.0026	1100	79	246.50	271	
10.6804	0.0096	4000	79	246.98	1600	
10.6708	0.0078	3240	79	246.98		
10.6630	0.0000			247.06	700	
10.6630	0.0117	4292	A 59	247.34	1300	
10.6513	0.0098	3066	59	247.34		
10.6415	0.0107	3340	59	247.98	1700	
10.6308	0.0064	2000	59	247.90	170	XING 172
10.6244	0.0007	150	59	248.12	2774	XING 173
10.6240	0.0080	2500	59	248.60	2500	
10.6160	0.0000			248.82	1174	
10.6160	0.0000			249.05	2800	
10.6160	0.0080	2500	69	249.60	1300	
10.6080	0.0015	465	69	249.80		XING 174
10.6065	0.0018	624	D 72	250.01	2168	
10.6047	0.0001	29	72	250.59	3100	XING 175
10.6046	0.0039	1408	A 65	251.06	2379	
10.6008	0.0015	550	65	251.59	2500	XING
10.5960	0.0064	2250	65	252.80	3200	XING 177
10.5927	0.0061	2150	65	252.83	191	XING 177
10.5866	0.0039	1350	65	253.14	2700	
10.5827	0.0041	1400	65	253.58	2300	
10.5786	0.0045	1536	65	253.56		
10.5741	0.0047	1501	A 55	253.90	100	
10.5694	0.0036	1022	55	254.15	2500	
10.5659	0.0042	1049	A 44	254.47	1700	
10.5617	0.0035	814	44	254.66	830	
10.5582	0.0028	686	A 36	256.56	3100	
10.5554	0.0204	3889	36	256.58	148	XING
10.5351	0.0147	2800	36	256.74	600	
10.5203	0.0042	800	36	256.82	400	
10.5161	0.0000			256.97	800	
10.5161	0.0091	1731	36	258.11	711	
10.5070	0.0021	400	36	256.60	2600	
10.5049	0.0000			257.58	5222	
10.5049	0.0074	1400	36	258.19	3200	
10.4975	0.0000			258.19		
10.4975	0.0032	600	36	258.41	1137	
10.4944	0.0015	405	D 49	258.41		
10.4929	0.0070	1605	49	258.58	900	
10.4859	0.0023	400	49	258.75	917	
10.4836	0.0007	199	D 54	258.75		
10.4829	0.0039	1111	54	259.11	1900	
10.4787	0.0018	500	54	259.30	1000	
10.4769	0.0098	1680	54	259.44	700	

READ DOWN WESTBOUND TRAFFIC — APPENDIX B STATION — PAGE 4 COLUMN 2

MAX SPD	DISTANCE	SPEED	ET HOURS	CUM HOURS	
				3.1221	
	700	36	0.0037	3.1258	
71	1100	36	0.0058	3.1316	L
71	2910	36	0.0153	3.1469	L
	967	A 36	0.0064	3.1533	
41	1280	49	0.0054	3.1556	R
11	425	49	0.0016	3.1600	R
	928	A 49	0.0019	3.1620	
98	1647	54	0.0058	3.1680	R
	1305	54	0.0046	3.1726	
	3212	54	0.0113	3.1838	T DENHOLM
	38	D 54	0.0001	3.1840	
53	1600	53	0.0058	3.1905	L
				3.1905	
	700	53	0.0025	3.1930	
95	781	53	0.0025	3.1955	L
56	581	53	0.0021	3.1981	L
	4000	53	0.0143	3.2124	
56	700	53	0.0025	3.2149	R
96	235	53	0.0008	3.2158	R
	1900	53	0.0068	3.2227	
97	1200	53	0.0043	3.2268	L
	900	53	0.0032	3.2301	CP HAWSTONE
	10001	A 93	0.0209	3.2570	
			0.0000	3.2570	
121			0.0000	3.2570	L
			0.0000	3.2570	
			0.0000	3.2570	
115			0.0000	3.2570	L
			0.0000	3.2570	
96			0.0000	3.2570	R
95	350	95	0.0006	3.2576	R
	0	95	0.0000	3.2576	
	2200	D 95	0.0056	3.2632	
65	1536	55	0.0055	3.2688	L
	334	56	0.0012	3.2700	REVERSE CURVE
	264	D 56	0.0010	3.2709	
46	1000	48	0.0039	3.2749	R
48	800	48	0.0032	3.2780	R
	2114	48	0.0083	3.2864	SHAWNEE
60	2300	48	0.0091	3.2955	R
80	700	48	0.0028	3.2982	R
	1600	48	0.0063	3.3045	
57	1291	48	0.0051	3.3096	L
	900	48	0.0036	3.3132	
58	800	48	0.0032	3.3163	R
58	800	48	0.0032	3.3195	R
82	495	45	0.0020	3.3214	R
	143	A 48	0.0006	3.3220	
	145	50	0.0006	3.3226	TRC TT MPH50
	514	D 50	0.0024	3.3249	
53	5100	33	0.0063	3.3312	L
				3.3312	
	1054	33	0.0060	3.3372	
48	1000	33	0.0057	3.3430	L
48	1500	33	0.0086	3.3516	L LEWISTOWN
				3.3516	
	2250	33	0.0164	3.3700	
87	562	33	0.0032	3.3732	R
87	3664	33	0.0211	3.3943	R
87			0.0000	3.3943	R
	1577	A 33	0.0060	3.3993	
141	2477	48	0.0098	3.4091	R
	6114	A 48	0.0219	3.4311	
				3.4311	GRANVILLE
97				3.4311	L
	2511	A 87	0.0052	3.4363	
	34	93	0.0001	3.4364	PRIVATE
	2062	D 93	0.0053	3.4417	ANDERSON
55	771	55	0.0026	3.4443	L
56	1900	56	0.0065	3.4509	L
				3.4509	
	700	56	0.0024	3.4533	
50	1300	53	0.0046	3.4578	R
				3.4578	
77	1200	55	0.0041	3.4619	R
	179	55	0.0006	3.4625	PRIVATE
	2774	56	0.0096	3.4721	PRIVATE
96	2500	55	0.0086	3.4807	R
71	1174	55	0.0040	3.4847	R
	1173	55	0.0045	3.4868	
67	1326	A 55	0.0036	3.4924	R
	874	63	0.0026	3.4950	HORNINGFORD
173	702	A 63	0.0022	3.4972	R
	5403	67	0.0154	3.5126	PRIVATE
	2379	67	0.0067	3.5190	
99	2800	67	0.0082	3.5275	L SWIGARTS
177	453	67	0.0013	3.5268	PRIVATE
177	389	A 67	0.0011	3.5299	S H 103 MCVEYTOWN
106	2334	56	0.0091	3.5380	L
89	2314	59	0.0077	3.5575	L MATAWANNA
	1832	A 69	0.0050	3.5625	
	26	77	0.0001	3.5629	
	126	D 77	0.0014	3.5643	
67	1700	67	0.0046	3.5690	R
	686	67	0.0019	3.5709	
	5150	67	0.0144	3.5853	
75	148	67	0.0004	3.5858	R PRIVATE
75	820	67	0.0023	3.5880	R RYDE
	400	67	0.0011	3.5892	
56	800	67	0.0023	3.5914	R
48	711	67	0.0020	3.5934	R
	2361	67	0.0067	3.6001	
	198	A 67	0.0006	3.6007	T VINEYARD
	7217	68	0.0201	3.6208	TRC & CURVE
	1246	D 68	0.0048	3.6253	
36	1137	35	0.0062	3.6315	RRC
			0.0000	3.6315	
36	900	35	0.0048	3.6364	L
36	917	36	0.0049	3.6413	L
			0.0000	3.6413	
100	1900	35	0.0103	3.6516	L
	1000	35	0.0054	3.6570	
71	700	35	0.0038	3.6608	R

READ UP EASTBOUND TRAFFIC

CUM HOURS	ET HOURS	DISTANCE	SPEED
10.4710	0.0021	677	D 69
10.4690	0.0002	89	69
10.4687	0.0078	2525	A 53
10.4609	0.0065	1805	53
10.4545	0.0043	1200	53
10.4502	0.0029	800	53
10.4473	0.0029	800	53
10.4445	0.0039	1100	53
10.4406	0.0025	700	53
10.4381	0.0096	2695	53
10.4284	0.0068	1900	53
10.4216	0.0000		
10.4214	0.0068	1900	53
10.4148	0.0058	2279	D 56
10.4090	0.0010	510	96
10.4080	0.0000	0	
10.4080	0.0000	0	
10.4080	0.0000	0	
10.4080	0.0000	0	
10.4080	0.0000	0	
10.4080	0.0000	0	
10.4080	0.0000	0	
10.4080	0.0000	0	
10.4080	0.0261	10693	A 55
10.3819	0.0049	1427	55
10.3770	0.0028	759	A 48
10.3742	0.0065	1643	48
10.3677	0.0032	800	48
10.3646	0.0058	1471	48
10.3587	0.0067	1700	48
10.3520	0.0056	1400	48
10.3465	0.0039	1000	48
10.3426	0.0032	800	48
10.3394	0.0000		
10.3394	0.0063	1536	48
10.3332	0.0024	609	48
10.3308	0.0000		
10.3308	0.0060	1528	48
10.3248	0.0090	3017	A 33
10.3158	0.0035	1209	33
10.3123	0.0132	2300	33
10.2990	0.0040	700	33
10.2950	0.0092	1600	33
10.2858	0.0074	1291	33
10.2784	0.0052	900	33
10.2732	0.0046	800	33
10.2687	0.0046	800	33
10.2641	0.0052	900	33
10.2588	0.0000		
10.2588	0.0023	400	33
10.2566	0.0000		
10.2564	0.0063	1100	33
10.2505	0.0021	446	D 46
10.2487	0.0024	609	48
10.2456	0.0039	1000	48
10.2418	0.0058	1500	48
10.2360	0.0054	1931	D 67
10.2306	0.0028	1270	67
10.2278	0.0012	562	67
10.2266	0.0036	1104	67
10.2231	0.0000		
10.2231	0.0000		
10.2231	0.0192	7366	A 65
10.2039	0.0042	1226	56
10.1997	0.0045	1300	55
10.1952	0.0087	2680	55
10.1866	0.0046	1300	55
10.1818	0.0150	4500	55
10.1668	0.0000		
10.1668	0.0027	771	55
10.1634	0.0066	1600	55
10.1568	0.0011	348	D 62
10.1557	0.0011	364	63
10.1547	0.0036	1300	63
10.1507	0.0006	191	D 67
10.1502	0.0029	1009	67
10.1473	0.0006	179	67
10.1468	0.0078	2774	67
10.1390	0.0071	2500	67
10.1318	0.0033	1174	67
10.1286	0.0079	2800	67
10.1207	0.0037	1300	67
10.1170	0.0053	700	D 69
10.1167	0.0057	2088	69
10.1111	0.0085	3100	69
10.1026	0.0202	7366	68
10.0820	0.0011	300	A 67
10.0812	0.0079	2300	67
10.0733	0.0005	191	67
10.0727	0.0076	2700	67
10.0651	0.0065	2300	67
10.0586	0.0000		
10.0586	0.0003	100	67
10.0583	0.0082	2900	67
10.0501	0.0046	1700	67
10.0455	0.0001	50	D 68
10.0460	0.0133	4499	68
10.0318			
10.0318			
10.0318			
10.0318			
10.0318	0.0136	3820	A 35
10.0182	0.0140	2578	35
10.0042	0.0083	5277	36
9.9960	0.0173	3300	36
9.9587	0.0000		
9.9587	0.0062	1137	35
9.9526	0.0001	26	D 36
9.9524	0.0046	874	36
9.9478	0.0048	917	36
9.9430	0.0046	1373	D 71
9.9381	0.0014	537	71
9.9367	0.0027	1000	71
9.9341	0.0019	700	71

READ DOWN WESTBOUND TRAFFIC — APPENDIX B1 STATION — READ UP EASTBOUND TRAFFIC

PAGE 5 COLUMN 1

MILES	DISTANCE, ft	HWY GRADE CROSSING	MAX MPH	Δ DISTANCE, feet	SPEED, mph	dT, hours	TIME, hours	STATION	TIME, hours	dT, hours	Δ DISTANCE, feet	SPEED, mph
259.44						0.0000	3.6608		9.9322	0.0015	617	D 82
259.59	800		82	800	35	0.0043	3.6651	R	9.9307	0.0004	183	82
259.68	500		100	500	35	0.0027	3.6678	R	9.9302	0.0021	985	D 97
260.46	4085			4085	35	0.0221	3.6899		9.9282	0.0002	97	97
260.59	700		173	198	35	0.0011	3.6910	R	9.9280	0.0128	5058	A 82
261.13	2874		173	52	A 35	0.0003	3.6913	R	9.9152	0.0024	1020	82
261.58	2400		234	1785	38	0.0093	3.7006	R	9.9128	0.0055	2400	82
262.02	2260		234	4383	A 36	0.0150	3.7156	R	9.9073	0.0053	2280	82
262.59	3000			2922	A 71	0.0072	3.7228		9.9020	0.0069	3000	82
263.02	2300			3301	A 82	0.0072	3.7300		9.8951	0.0053	2300	82
263.02				382	91	0.0008	3.7308		9.8898	0.0000		
263.02				571	D 91	0.0013	3.7320		9.8898	0.0000		
263.37	1834		82	1834	82	0.0042	3.7363	L'NEWTON HAMILTON	9.8898	0.0032	1397	82
263.56	1000			1000	82	0.0023	3.7386		9.8866	0.0000		
263.64	402			402	82	0.0009	3.7395		9.8866	0.0100	3927	A 66
264.49	4500		89	4500	82	0.0104	3.7499	R	9.8766	0.0069	2412	66
265.00	2700		89	2700	82	0.0062	3.7561	R	9.8696	0.0077	2700	66
265.49	2591			2386	82	0.0055	3.7617		9.8619	0.0129	4486	66
265.49				2120	A 82	0.0047	3.7664	190	9.8490	0.0000		
265.49				128	68	0.0003	3.7667		9.8490	0.0058	1853	A 54
266.11	3297			1242	D 68	0.0031	3.7697		9.8432	0.0017	456	A 50
266.49	2000	XING	68	2000	66	0.0067	3.7755	L PRIVATE	9.8415	0.0041	1093	50
266.70	1100		66	1100	66	0.0032	3.7786	L'MT UNION	9.8374	0.0042	1100	50
266.79	477			477	66	0.0014	3.7800	REVERSE CURVE	9.8332	0.0018	477	50
267.28	2600		74	2600	66	0.0075	3.7875	R	9.8314	0.0098	2600	50
267.59	1600		74	1600	66	0.0046	3.7920	R	9.8216	0.0061	1600	50
267.91	1682			1682	66	0.0048	3.7968		9.8155	0.0064	1682	50
268.26	2000		82	1472	66	0.0042	3.8011	L	9.8091	0.0076	2000	50
268.28				528	D 66	0.0017	3.8028		9.8016	0.0000		
268.34	300		54	0	54	0.0000	3.8028	L	9.8016	0.0011	300	50
268.37	148			295	54	0.0010	3.8038	TRC	9.8004	0.0006	148	50
268.37				153	D 54	0.0006	3.8044	MAPLETON	9.7989	0.0000		
269.28	4800		50	4800	50	0.0182	3.8225	R'MPH50	9.7989	0.0182	4800	50
269.51	1200		50	1200	50	0.0045	3.8271	R	9.7817	0.0045	1200	50
269.51							3.8271		9.7771	0.0028	829	D 69
269.51							3.8271		9.7745	0.0002	71	69
269.94	2300			2300	50	0.0087	3.8358		9.7743	0.0036	1400	69
270.22	1477		69	1477	50	0.0056	3.8414	R	9.7704	0.0041	1477	69
270.28	300			300	50	0.0011	3.8425		9.7664	0.0018	720	D 82
270.66	2000			2000	50	0.0076	3.8501		9.7646	0.0036	1580	82
271.27	3246		100	3246	50	0.0123	3.8624	L	9.7609	0.0075	3246	82
271.48	1085		100	1085	50	0.0041	3.8665	L	9.7534	0.0025	1085	82
271.78	1600			592	50	0.0022	3.8687		9.7509	0.0037	1000	82
272.04	1400		67	2903	A 50	0.0092	3.8779	L	9.7472	0.0026	1136	82
272.27	1200		89	705	69	0.0019	3.8798	L'MILL CREEK	9.7446	0.0094	3732	A 67
272.80	2790	XING	82	3333	A 69	0.0083	3.8882	L CASSVILLE	9.7352	0.0015	522	67
273.27	2500			1957	82	0.0045	3.8927		9.7337	0.0071	2500	67
273.31	178			178	82	0.0004	3.8931		9.7266	0.0079	2808	67
273.97	3500		141	3500	82	0.0081	3.9012	R	9.7187	0.0064	2048	A 54
274.27	1600			1000	82	0.0037	3.9049		9.7123	0.0120	3411	54
274.78	2700			2700	82	0.0062	3.9111		9.7003	0.0008	232	A 52
274.78				820	D 82	0.0021	3.9132		9.6995	0.0000		
275.01	1179	XING	67	1179	67	0.0033	3.9165	R JACOBS	9.6995	0.0024	860	52
275.27	1400			581	67	0.0016	3.9182		9.6971	0.0051	1400	52
276.28	5297			4720	67	0.0133	3.9315		9.6920	0.0193	5297	52
276.28				577	D 66	0.0018	3.9333		9.6727	0.0000		
276.52	1300		54	1300	54	0.0046	3.9379	L	9.6727	0.0047	1300	52
276.57	243			243	54	0.0009	3.9387	REVERSE CURVE	9.6680	0.0009	243	52
276.97	2100		61	2022	54	0.0071	3.9458	R	9.6671	0.0076	2100	52
276.97				78	D 54	0.0003	3.9461	REVERSE CURVE	9.6594	0.0000		
277.25	1500		52	1500	52	0.0055	3.9516	L	9.6594	0.0055	1500	52
277.44	1000	XING	52	1000	52	0.0036	3.9552	L'MPH 55 4TH STREET	9.6540	0.0036	1000	52
277.59	800			800	52	0.0029	3.9581	T'HUNTINGDON	9.6503	0.0029	800	52
277.78	1000		121	1000	52	0.0036	3.9618	R	9.6474	0.0036	1000	52
277.91	700			700	52	0.0025	3.9643		9.6438	0.0025	700	52
278.20	1500		87	1500	52	0.0055	3.9698	R	9.6412	0.0055	1500	52
278.24	239			239	52	0.0009	3.9707		9.6358	0.0009	239	52
278.85	3197			3197	52	0.0116	3.9823		9.6349	0.0116	3197	52
279.09	1300		107	1300	52	0.0047	3.9870	R	9.6233	0.0047	1300	52
279.25	800		52	800	52	0.0029	3.9900	R	9.6185	0.0029	800	52
279.40	800		52	800	52	0.0029	3.9929	R	9.6156	0.0029	800	52
279.45	300			300	52	0.0011	3.9940	T'DEER	9.6127	0.0011	300	52
279.64	1000		132	1000	52	0.0036	3.9976	L'RC MPH 55	9.6116	0.0079	2157	52
279.81	900		107	900	52	0.0033	4.0009	L	9.6037	0.0014	367	A 48
280.12	1600			1600	52	0.0058	4.0067		9.6024	0.0039	976	48
280.24	670		98	670	52	0.0024	4.0091	R	9.5985	0.0025	670	48
280.30	277		98	277	52	0.0010	4.0102	R	9.5959	0.0011	277	48
281.24	5000			5000	52	0.0182	4.0284		9.5948	0.0197	5000	48
281.36	600			600	52	0.0022	4.0306		9.5750	0.0024	600	48
281.57	1100		52	653	52	0.0024	4.0329	L	9.5727	0.0043	1100	48
281.75	977		68	977	52	0.0036	4.0365	L	9.5683	0.0039	977	48
282.02	1400	XING 206.[illegible]		1700	52	0.0062	4.0427	T'WARRIOR RIDGE	9.5645	0.0055	1400	48
282.02				147	D 52	0.0006	4.0432		9.5590	0.0000		
282.24	1200		48	1200	48	0.0047	4.0480	R	9.5590	0.0047	1200	48
282.49	1300		48	1300	48	0.0051	4.0531	R	9.5542	0.0051	1300	48
282.49							4.0531		9.5491	0.0008	224	D 54
282.51	87			87	48	0.0003	4.0534	REVERSE CURVE	9.5483	0.0002	63	54
282.73	1200		54	1200	48	0.0047	4.0582	L	9.5480	0.0035	1000	54
282.73							4.0582		9.5445	0.0006	251	D 60
282.96	1200		60	1200	48	0.0047	4.0629	L	9.5437	0.0030	949	60
282.96							4.0629		9.5407	0.0004	135	D 63
283.24	1500			1500	48	0.0059	4.0688		9.5403	0.0041	1365	63
283.47	1190			1190	48	0.0047	4.0735		9.5362	0.0036	1190	63
283.86	2000		93	2000	48	0.0079	4.0814	L PETERSBURG	9.5326	0.0060	2000	63
284.25	2100		63	2100	48	0.0083	4.0897	L	9.5266	0.0063	2100	63
284.25							4.0897		9.5203	0.0026	1010	D 82
284.42	900			900	48	0.0036	4.0933		9.5176	0.0004	167	82
284.77	1885		67	823	48	0.0032	4.0965	R	9.5173	0.0000		
285.25	2500			598	A 48	0.0022	4.0987		9.5173	0.0194	6719	A 44
285.25		XING 210.[illegible]		668	54	0.0024	4.1011	PRIVATE 210	9.4978	0.0000		
285.25	5289	XING 210.[illegible]		836	A 54	0.0050	4.1061	PRIVATE	9.4978	0.0115	2678	44
287.16	4800			364	60	0.0011	4.1073		9.4863	0.0207	4800	44
287.22	322		94	450	A 80	0.0014	4.1087	R	9.4656	0.0114	2653	44
287.22				6340	63	0.0191	4.1277		9.4542	0.0000		
287.22				5848	A 63	0.0147	4.1425		9.4542	0.0000		
287.60	2000	XING	94	46	86	0.0001	4.1426	R'BARREE	9.4542	0.0014	309	A 40
287.83	1200			2002	D 86	0.0068	4.1494		9.4528	0.0027	560	40
288.13	1600		44	1600	44	0.0069	4.1553	L	9.4502	0.0076	1600	40
288.21	415		44	415	44	0.0018	4.1571	L	9.4426	0.0020	415	40
288.36	805			805	44	0.0035	4.1606	L MPH 50WB	9.4406	0.0036	805	40
288.47	600			600	44	0.0026	4.1631	L	9.4368	0.0028	600	40
288.89	2190		63	2190	44	0.0094	4.1725		9.4340	0.0104	2190	40
289.04	810			810	44	0.0035	4.1760	TUNNEL - NORTH REBI	9.4236	0.0038	810	40
289.16	630			630	44	0.0027	4.1787	R	9.4198	0.0030	630	40
289.31	800			800	44	0.0034	4.1822	T'SPRUCE CREEK	9.4168	0.0036	800	40
289.50	1000		58	1000	44	0.0043	4.1865	R	9.4130	0.0047	1000	40

PAGE 5 COLUMN 2

MILES	DISTANCE, ft	HWY GRADE CROSSING	MAX MPH	Δ DISTANCE, feet	SPEED, mph	dT, hours	TIME, hours	STATION	TIME, hours	dT, hours	Δ DISTANCE, feet	SPEED, mph
289.69	1000		62	1000	44	0.0043	4.1908	R	9.4083	0.0188	3550	40
289.80	590			467	44	0.0020	4.1928	REVERSE CURVE	9.3915	0.0019	379	A 34
289.80				123	D 53	0.0006	4.1934		9.3895	0.0000		
290.14	1800		40	1800	40	0.0085	4.2019	L	9.3896	0.0092	1658	34
290.18	200		106	200	40	0.0009	4.2028	L	9.3803	0.0006	97	A 32
290.58	2100			2100	40	0.0099	4.2126		9.3797	0.0000	6	32
290.83	1300		46	1300	40	0.0062	4.2189	L	9.3797	0.0077	1300	32
291.03	1077			1077	40	0.0051	4.2240		9.3720	0.0064	1077	32
291.16	700		77	700	40	0.0033	4.2273	R	9.3656	0.0041	700	32
291.37	1100		41	1100	40	0.0052	4.2325	R	9.3615	0.0155	2624	32
291.54	900		63	900	40	0.0043	4.2368	R	9.3459	0.0006	91	A 30
291.74	1062	XING 216.[illegible]		1062	40	0.0050	4.2418	PEMBERTON	9.3454	0.0022	347	30
291.95	1100		100	1100	40	0.0052	4.2470	L	9.3432	0.0069	1100	30
292.14	1000			1000	40	0.0047	4.2518		9.3362	0.0063	1000	30
292.33	1000			461	40	0.0022	4.2540		9.3299	0.0063	1000	30
292.33				182	A 40	0.0008	4.2548		9.3236	0.0000		
292.33				103	43	0.0005	4.2553		9.3236	0.0000		
292.33				254	D 43	0.0013	4.2565	UNION FURNACE	9.3236	0.0000		
292.57	1300		34	1300	34	0.0072	4.2637	L	9.3236	0.0082	1300	30
292.66	453			405	34	0.0023	4.2660	T'RC MPH 35	9.3154	0.0028	453	30
292.65				48	D 34	0.0003	4.2663		9.3126	0.0000		
292.89	1200		32	1200	32	0.0071	4.2734	R	9.3126	0.0076	1200	30
293.15	1400			1400	32	0.0083	4.2817		9.3050	0.0088	1400	30
293.59	2300			2300	32	0.0136	4.2953		9.2961	0.0145	2300	30
293.65	300		105	300	32	0.0018	4.2971	R	9.2816	0.0019	300	30
293.76	600			555	32	0.0033	4.3003	REVERSE URVE	9.2797	0.0038	600	30
293.76				45	D 32	0.0003	4.3006		9.2759	0.0000		
293.80	200		30	200	30	0.0013	4.3019	L	9.2759	0.0013	200	30
293.80							4.3019		9.2747	0.0003	45	D 32
294.10	1600		33	1600	30	0.0101	4.3120	R RC	9.2744	0.0092	1555	32
294.15	247			247	30	0.0016	4.3135		9.2652	0.0015	247	32
294.22	400			400	30	0.0025	4.3161		9.2637	0.0024	400	32
294.41	1000		70	1000	30	0.0063	4.3224	L	9.2614	0.0069	1000	32
294.68	1400			1400	30	0.0088	4.3312	T'BIRMINGHAM	9.2554	0.0083	1400	32
295.14	2467		94	2467	30	0.0156	4.3468	L	9.2472	0.0146	2467	32
295.31	900		94	900	30	0.0057	4.3525	L	9.2326	0.0063	900	32
295.50	1000			1000	30	0.0063	4.3588		9.2272	0.0059	1000	32
295.67	900		33	900	30	0.0057	4.3645	R	9.2213	0.0053	900	32
295.73	300			300	30	0.0019	4.3664	REVERSE URVE	9.2160	0.0018	300	32
295.94	1085		32	766	30	0.0050	4.3713	L	9.2142	0.0064	1085	32
295.94							4.3713		9.2078	0.0003	48	D 34
296.15	1100			91	A 30	0.0006	4.3719		9.2075	0.0057	1015	34
296.30	800			2106	32	0.0125	4.3843		9.2019	0.0046	800	34
296.49	1000		43	1000	32	0.0059	4.3903	L	9.1974	0.0056	1000	34
296.54	300			300	32	0.0018	4.3920	REVERSE CURVE	9.1918	0.0017	300	34
296.75	1100	XING	44	1100	32	0.0065	4.3986	R PRIVATE	9.1902	0.0061	1100	34
296.90	800			800	32	0.0047	4.4033		9.1840	0.0045	800	34
297.14	1262		65	1262	32	0.0075	4.4108	L'TTMPH 35	9.1796	0.0070	1262	34
297.30	860		34	860	32	0.0051	4.4158	L'TYRONE	9.1725	0.0048	860	34
297.30		XING 222.18					4.4158	WILSON ROAD	9.1678	0.0021	457	D 49
297.40	500			500	32	0.0030	4.4188	RC	9.1657	0.0002	44	49
297.63	1200		56	1200	32	0.0071	4.4259	R	9.1655	0.0046	1200	49
297.84	1100		49	1100	32	0.0065	4.4324	R	9.1609	0.0043	1100	49
297.84							4.4324		9.1566	0.0001	36	D 50
297.95	600			600	32	0.0036	4.4360	RC	9.1565	0.0023	600	50
298.14	1000		50	378	32	0.0022	4.4382	L	9.1542	0.0038	1000	50
298.19	273		50	97	A 32	0.0006	4.4388	L	9.1504	0.0010	273	50
298.36	900			1698	34	0.0095	4.4482		9.1494	0.0006	153	D 54
298.68	1700		97	1700	34	0.0095	4.4577	R	9.1488	0.0086	2447	54
299.14	2400			2400	34	0.0134	4.4711		9.1402	0.0084	2400	54
299.52	2000			1327	34	0.0074	4.4785		9.1318	0.0070	2000	54
299.88	1900		54	1121	A 34	0.0050	4.4835	L	9.1248	0.0067	1900	54
299.88		XING 224.[illegible]		1452	49	0.0056	4.4891	PRIVATE	9.1181	0.0001	40	D 55
299.99	600	XING 225.[illegible]		227	49	0.0009	4.4899	T'TTMPH 60 RC PRIVA	9.1180	0.0019	560	55
300.14	782	XING	56	72	A 49	0.0003	4.4902	R TIPTON ROAD	9.1161	0.0027	782	55
300.42	1473		56	1801	50	0.0066	4.4970	R	9.1134	0.0051	1473	55
301.14	3800			456	A 50	0.0017	4.4987	T'TIPTON	9.1083	0.0131	3800	55
302.14	5276			8444	54	0.0296	4.5283		9.0952	0.0182	5276	55
302.95	4300		55	160	A 54	0.0006	4.5289	L	9.0771	0.0148	4300	55
302.96		XING 227.93					4.5289	PA RTE 502	9.0622	0.0004	124	D 58
303.01	300			5373	55	0.0033	4.5322	T'FOSTORIA RC	9.0618	0.0006	176	58
303.07	358		58	358	55	0.0012	4.5334	R'TTMPH 60	9.0613	0.0012	358	58
303.07							4.5334		9.0601	0.0019	867	D 72
303.53	2409		72	2409	55	0.0083	4.5417	R	9.0581	0.0046	1742	72
303.53							4.5417		9.0536	0.0036	1621	D 98
304.06	2800			2800	55	0.0096	4.5514		9.0500	0.0005	257	98
305.06	5293	XING 230.[illegible]		5133	55	0.0177	4.5681	PRIVATE	9.0495	0.0000		
306.59	2800			416	A 55	0.0014	4.5704		9.0495	0.0211	9551	A 71
306.84	1276		97	242	58	0.0117	4.5821	L	9.0283	0.0020	742	71
306.06	1200			2481	A 58	0.0072	4.5893		9.0264	0.0032	1200	71
306.06				8771	A 72	0.0194	4.6088		9.0232	0.0000		
307.06	5258			974	97	0.0019	4.6107		9.0232	0.0140	5258	71
307.78	3800			1802	D 97	0.0036	4.6143	T'ANTIS	9.0091	0.0101	3800	71
308.02	1300		71	1300	71	0.0035	4.6178	R	8.9990	0.0035	1300	71
308.06	202			202	71	0.0005	4.6183	T'MPH60(TT)	8.9955	0.0011	440	D 79
308.08	106			106	71	0.0003	4.6186	RC	8.9944	0.0005	202	79
308.58	2600		101	2600	71	0.0069	4.6255	L	8.9939	0.0111	4117	A 60
309.05	2500			2500	71	0.0067	4.6322		8.9828	0.0021	651	60
309.09	200			200	71	0.0005	4.6327		8.9808	0.0006	200	60
309.56	2500		126	2500	71	0.0067	4.6394	R	8.9801	0.0079	2500	60
310.04	2549			2549	71	0.0068	4.6462		8.9722	0.0080	2549	60
310.65	3200			341	71	0.0009	4.6471		8.9642	0.0101	3200	60
310.78	700		173	3285	A 71	0.0081	4.6552	L	8.9541	0.0022	700	60
311.01	1200			263	83	0.0006	4.6558	T'TTMPH30	8.9519	0.0073	2324	60
311.01				1206	D 83	0.0032	4.6590	ALTOONA	8.9446	0.0000		
311.04	142		60	142	60	0.0004	4.6594	R	8.9446	0.0000		
311.13	520		60	520	60	0.0016	4.6611	R	8.9446	0.0075	2025	A 41
312.04	4800			4800	60	0.0152	4.6762	T'ALTO	8.9371	0.0330	7149	41
312.76	3787			3787	60	0.0120	4.6882	T'SLOPE	8.9040	0.0025	487	A 33
313.03	1400		82	1400	60	0.0044	4.6926	L BRICKYARD	8.9015	0.0112	1947	33
313.29	1400	XING	82	351	60	0.0011	4.6937	L COBURN ROAD	8.8904	0.0008	139	A 30
313.69	2100			1815	A 60	0.0053	4.6990	T'TTMPH 35	8.8895	0.0081	1278	30
313.69				154	70	0.0004	4.6994		8.8815	0.0000		
313.69				1180	D 70	0.0040	4.7034		8.8815	0.0000		
313.97	1500		41	1500	41	0.0069	4.7103	R	8.8815	0.0095	1500	30
314.04	372			372	41	0.0017	4.7121		8.8720	0.0023	372	30
314.34	1574			1574	41	0.0073	4.7193		8.8696	0.0178	2821	30
314.53	1000		60	1000	41	0.0046	4.7239	R	8.8518	0.0014	209	A 26
314.74	1100			1100	41	0.0051	4.7290	T'TTMPH35	8.8504	0.0025	337	26
314.91	900		45	900	41	0.0042	4.7332	L	8.8480	0.0066	900	26
315.04	700			700	41	0.0032	4.7364		8.8414	0.0051	700	26
315.19	800			583	41	0.0027	4.7391		8.8363	0.0058	800	26
315.19				217	D 41	0.0011	4.7402		8.8305	0.0000		26
315.48	1500		33	1500	33	0.0086	4.7488	R	8.8305	0.0109	1500	26
315.53	276			207	33	0.0012	4.7500	T'TTMPH 35 RC	8.8196	0.0020	276	26
315.53		UG 240.69		69	D 33	0.0004	4.7504	SCOTTS RUN	8.8176	0.0000		

READ DOWN WESTBOUND TRAFFIC								APPENDIX B: STATION	READ UP EASTBOUND TRAFFIC						
MILES	DISTANCE ft	HWY GRADE CROSSING	MAX MPH	DISTANCE ft	SPEED mph	ET hours	TIME hours	PAGE 6 COLUMN 1	TIME hours	ET hours	DISTANCE ft	SPEED mph	MILES	DISTANCE ft	HWY GRADE CROSSING
318.76	1200		30	1200	30	0.0076	4.7580	L	8.8176	0.0087	1200	26	340.44	900	
318.04	1500			1500	30	0.0095	4.7675		8.8088	0.0109	1500	26	340.59	800	
316.19	800			800	30	0.0051	4.7725		8.7979	0.0056	800	26	340.74	800	
316.34	800		100	800	30	0.0051	4.7776	L	8.7921	0.0058	800	26	340.74		
316.52	900			900	30	0.0057	4.7833		8.7862	0.0068	900	26	340.76	200	
316.63	600		60	600	30	0.0036	4.7871	R	8.7797	0.0044	600	26	340.93	600	
316.74	600		42	600	30	0.0038	4.7908	R	8.7753	0.0044	600	26	341.14	1100	
316.76	87			5	30	0.0000	4.7908	RC	8.7709	0.0006	87	28	341.14		
316.76				82 D	36	0.0006	4.7914		8.7703	0.0000			341.31	900	
316.93	900		26	900	26	0.0066	4.7980	L	8.7703	0.0068	900	26	341.31		
317.04	600		33	602	26	0.0044	4.8024	L'HORSESHOE	8.7638	0.0044	600	26	341.48	900	
317.21	900		26	900	28	0.0066	4.8089	L	8.7594	0.0066	900	26	341.51	199	
317.21					26		4.8089		8.7528	0.0015	310 D	39	341.63	800	
317.28	152			182	28	0.0013	4.8102	RC	8.7510	0.0049	1001	39	341.70	500	
317.32	400		57	400	28	0.0029	4.8131	R	8.7462	0.0017	451 A	34	341.94	670	
317.42	500			500	28	0.0036	4.8168		8.7445	0.0027	461	34	342.02	400	
317.57	800		45	800	26	0.0066	4.8226	R	8.7418	0.0025	364 A	26	342.17	800	
317.66	500		100	500	26	0.0036	4.8263	R	8.7393	0.0003	45	26	342.30	700	
317.78	500		38	600	26	0.0044	4.8306	R	8.7389	0.0044	600	26	342.44	700	
317.93	800		43	800	26	0.0058	4.8365	R	8.7346	0.0058	800	26	342.56	600	
318.04	600			600	26	0.0044	4.8408		8.7287	0.0044	600	26	342.63	400	
318.23	1000			1000	26	0.0073	4.8481		8.7244	0.0073	1000	26	342.78	900	
318.40	900		71	900	26	0.0066	4.8547	L'MG	8.7171	0.0066	900	26	342.91	700	
318.72	1670			1670	26	0.0122	4.8668		8.7106	0.0122	1670	26	343.18	1400	
319.04	1700		71	1700	26	0.0124	4.8792	R	8.6984	0.0124	1700	26	343.21	200	
319.25	1100		47	1100	26	0.0080	4.8872	R	8.6860	0.0080	1100	26	343.36	800	
319.36	600			248	26	0.0018	4.8890		8.6780	0.0044	600	26	343.40	300	
319.57	1100		141	394 A	26	0.0025	4.8915	R	8.6736	0.0080	1100	26	343.72	1000	
319.72	800		34	1767	34	0.0098	4.9014	R	8.6656	0.0058	800	26	343.76	218	
319.74	85				34	0.0000	4.9014	REVERSE CURVE	8.6597	0.0006	85	26	343.82	215	
319.74				176 D	34	0.0011	4.9025		8.6591	0.0000			343.97	800	
319.93	1000		26	1000	26	0.0073	4.9098	L	8.6591	0.0073	1000	26	343.97		
319.93							4.9098		8.6518	0.0001	19 D	27	344.12	800	
319.97	200			200	26	0.0015	4.9112	RC	8.6517	0.0013	181	27	344.20	400	
320.06	500		42	500	26	0.0036	4.9148	R	8.6504	0.0036	500	27	344.33	200	
320.16	500		42	500	26	0.0036	4.9185	R	8.6468	0.0035	500	27	344.46	200	
320.18	149			149	26	0.0011	4.9196	T'RC	8.6434	0.0010	149	27	344.48	100	
320.37	1000		26	1000	26	0.0073	4.9269	L'MPH 30(TT)	8.6424	0.0070	1000	27	344.65	900	
320.49	600			600	26	0.0044	4.9313		8.6354	0.0042	600	27	344.79	700	
320.62	700		71	700	26	0.0051	4.9364	R	8.6312	0.0049	700	27	344.86	377	
320.77	800		41	800	26	0.0058	4.9422	R	8.6262	0.0056	800	27	344.86		
320.92	800			802	26	0.0066	4.9480		8.6206	0.0056	800	27	345.10	1300	
321.02	400		63	400	26	0.0029	4.9509	R	8.6150	0.0028	400	27	345.33	1200	
321.04	200		42	200	26	0.0015	4.9524	R	8.6122	0.0014	200	27	345.56	1200	
321.08	211			211	26	0.0015	4.9539	RC	8.6108	0.0015	211	27	345.78	1200	
321.32	1300		27	1300	26	0.0095	4.9634	L	8.6093	0.0091	1300	27	345.87	1000	
321.32							4.9634		8.6002	0.0015	262 D	38	346.24	1400	
321.42	500			500	26	0.0036	4.9670	T'BENNY	8.5987	0.0012	238	38	346.28	200	
321.51	500		141	500	26	0.0036	4.9707	L'EB -2.27% per cr b4pbs	8.5975	0.0025	500	38	346.42	1000	
321.63	600			600	26	0.0044	4.9750	T(lt 1)	8.5950	0.0030	600	38	346.42		
322.04	2200		100	2040	26	0.0149	4.9899	R	8.5920	0.0110	2200	38	346.51	246	
322.16	607		100	787 A	26	0.0003	4.9902	R	8.5811	0.0030	607	38	346.63	600	
322.27	600		71	600	27	0.0042	4.9944	R	8.5780	0.0030	600	38	346.78	600	
322.44	900			900	27	0.0063	5.0007		8.5750	0.0045	900	38	347.18	2126	
323.05	3200			3200	27	0.0224	5.0232	T'TUNNELS	8.5705	0.0150	3200	38	347.22	500	
323.12	400			400	27	0.0028	5.0260	T'TUNNELS	8.5548	0.0020	400	38	347.29	100	
323.26	700			700	27	0.0049	5.0309	GALLITZEN	8.5526	0.0035	700	35	347.52	1200	
323.48	1200		38	793	27	0.0056	5.0364	L@UNT@ARE END 4 T	8.5491	0.0060	1200	38	347.75	1200	
323.48							5.0364		8.5431	0.0010	213 D	45	347.84	476	
323.72	1246			1175 A	27	0.0035	5.0400	EAST END 5 TR	8.5422	0.0043	1033	45	347.90	300	
323.93	1100		71	1575	38	0.0076	5.0479	R	8.5378	0.0040	1100	45	348.03	700	
324.27	1800		71	1802	38	0.0090	5.0569	R	8.5337	0.0075	1800	45	348.20	900	
324.40	700			700	38	0.0035	5.0604		8.5256	0.0029	700	45	348.41	1100	
324.67	1400		45	1400	38	0.0070	5.0673	L	8.5227	0.0060	1400	45	348.66	1000	
324.67							5.0673		8.5166	0.0060	2067 D	86	348.79	1000	
324.92	1338			1338	38	0.0067	5.0740		8.5106	0.0010	450	86	348.99	1061	
325.92	5304			3416	38	0.0170	5.0910	WAWEST END 5 TRACKS	8.5098	0.0119	5047 A	71	349.14	800	
326.38	2400			490 A	38	0.0022	5.0932	T'CRESSON	8.4979	0.0039	1453	71	349.25	600	
326.67	1536		141	5334	45	0.0224	5.1157	R	8.4940	0.0041	1536	71	349.40	800	
326.90	1200			422	45	0.0018	5.1175		8.4899	0.0032	1200	71	349.40		
327.90	5311	XING 253		5480 A	45	0.0164	5.1338	RISSING ROAD	8.4867	0.0142	5311	71	349.44	200	
327.90				227	78	0.0006	5.1344		8.4725	0.0000			349.60	1000	
327.90				382 D	78	0.0010	5.1354		8.4725	0.0000			349.62	1000	
328.19	1500			1500	71	0.0040	5.1394		8.4725	0.0040	1500	71	349.90	438	
328.60	2162		71	2162	71	0.0058	5.1452	L	8.4685	0.0058	2162	71	350.28	2000	
328.83	1500			1500	71	0.0040	5.1492		8.4628	0.0040	1500	71	350.43	800	
329.07	982			982	71	0.0026	5.1518		8.4588	0.0026	982	71	350.81	2000	
329.24	900		71	900	71	0.0024	5.1542	R'LILLY	8.4562	0.0024	900	71	351.10	1529	
329.24					71	0.0000	5.1542		8.4538	0.0024	991 D	88	351.25	800	
329.77	2800			2800	71	0.0075	5.1616		8.4514	0.0007	311	88	351.48	1200	
330.11	1794			1794	71	0.0048	5.1664		8.4507	0.0119	5042 A	71	351.63	800	
330.49	2000		100	2000	71	0.0053	5.1718	R	8.4388	0.0126	4726	71	351.63		
330.64	800			800	71	0.0021	5.1739	CASSANDRA	8.4262	0.0011	411 A	69	351.63		
331.49	4519			4519	71	0.0121	5.1859		8.4250	0.0012	425	69	351.82	1000	
332.37	4631			4631	71	0.0124	5.1983		8.4238	0.0127	4631	69	351.99	889	
332.52	800			800	71	0.0021	5.2004		8.4112	0.0022	800	69	351.99		
332.62	500		71	500	71	0.0013	5.2016	R	8.4090	0.0014	500	69	352.00	200	
332.71	500			500	71	0.0013	5.2031	T'PORTAGE	8.4076	0.0014	500	69	352.27	1000	
332.98	1400		80	1400	71	0.0037	5.2068	R	8.4062	0.0038	1400	69	352.27		
333.23	1346		89	2643	71	0.0071	5.2139	R	8.4024	0.0037	1346	69	352.38	900	
333.50	1400			100 D	71	0.0003	5.2142		8.3987	0.0038	1400	69	352.56	1000	
333.76	1400		69	1400	69	0.0038	5.2180	R	8.3949	0.0038	1400	69	352.58		
334.03	1400			1400	69	0.0038	5.2219		8.3910	0.0003	100 D	71	352.61	200	
334.11	426		71	426	69	0.0012	5.2230	L	8.3907	0.0046	1723	71	352.77	800	
334.52	2200		71	2200	69	0.0060	5.2291	L	8.3861	0.0058	2200	71	352.78	100	
335.02	2626			2626	69	0.0072	5.2363	WILMORE	8.3803	0.0022	827 D	87	353.08	1500	
335.81	3100			3100	69	0.0085	5.2448	WEST END POP RUN	8.3780	0.0000	121	87	353.39	1500	
335.94	1274		141	1248	69	0.0034	5.2482	R	8.3778	0.0253	8875 A	38	353.54	600	
336.56	3381		141	411 A	69	0.0011	5.2493	R	8.3525	0.0047	935	38	353.66	600	
336.67	1500			3016	71	0.0096	5.2590		8.3478	0.0075	1500	38	353.73	400	
336.99	665			7188 A	71	0.0164	5.2753		8.3404	0.0033	665	38	353.79	282	
337.48	2600		155			0.0000	5.2753	L	8.3371	0.0130	2600	38	353.92	700	
337.79	1600					0.0000	5.2753	SUMMERHILL	8.3241	0.0060	1600	38	353.92		
338.28	2600					0.0000	5.2753		8.3181	0.0226	4527	38	354.68	4600	
338.49	1100		86	116	93	0.0002	5.2756	R	8.2935	0.0003	56 A	37	354.78	566	
338.49				2642 D	93	0.0076	5.2832	E END MIN POINT	8.2933	0.0006			354.83	257	
338.68	1000		38	1000	38	0.0050	5.2882	R	8.2933	0.0006	116	37	355.78	5000	
338.73	300			300	38	0.0015	5.2897	REVERSE CURVE	8.2927	0.0016	300	37	356.63	300	
338.78	248		41	248	38	0.0012	5.2909	L	8.2911	0.0013	243	37	356.69	4000	
338.89	600		41	600	38	0.0030	5.2939	L	8.2899	0.0058	1136	37	356.73	200	
339.05	800		80	800	38	0.0040	5.2979	L	8.2841	0.0017	306 A	32	356.77	237	
339.42	2000			2000	38	0.0100	5.3079	T'T'MPH 40	8.2824	0.0116	1984	32	356.77	20	
339.78	1865		91	1865	38	0.0093	5.3172	R'SOUTH FORK	8.2708	0.0110	1865	32	357.53	4600	
340.13	1871			1871	38	0.0093	5.3265	T-H850-@SUM(C734..C	8.2598	0.0111	1871	32	357.72	1000	
340.28	800		98	800	38	0.0040	5.3305	L	8.2487	0.0047	800	32	357.72		

READ DOWN WESTBOUND TRAFFIC					APPENDIX B: STATION	READ UP EASTBOUND TRAFFIC			
MAX MPH	DISTANCE ft	SPEED mph	ET hours	TIME hours	PAGE 6 COLUMN 2	TIME hours	ET hours	DISTANCE ft	SPEED mph
59	800	38	0.0042	5.3345	L	8.2439	0.0047	800	32
46	800	38	0.0040	5.3385	L'MPH 35	8.2392	0.0047	800	32
	773	38	0.0038	5.3423		8.2345	0.0047	800	32
	28 D	38	0.0001	5.3425		8.2297	0.0000		
37	200	37	0.0010	5.3435	R	8.2297	0.0012	200	32
37	800	37	0.0041	5.3476	R	8.2286	0.0047	800	32
48	974	37	0.0050	5.3526	R REV CURVE	8.2238	0.0065	1100	32
	127 D	44	0.0007	5.3533		8.2173	0.0000		
32	900	32	0.0053	5.3586	L'MPH 30	8.2173	0.0053	900	32
				5.3586		8.2120	0.0006	100 D	36
46	900	32	0.0053	5.3638	L	8.2114	0.0042	800	36
	199	32	0.0012	5.3651	RC	8.2072	0.0010	199	36
39	800	32	0.0036	5.3686	R	8.2062	0.0091	1726	36
60	800	32	0.0047	5.3734	R	8.1971	0.0017	245 A	31
	870	32	0.0051	5.3785		8.1954	0.0018	299	31
56	400	32	0.0024	5.3809	L	8.1936	0.0024	400	31
63	800	32	0.0047	5.3856	L	8.1912	0.0049	800	31
53	700	32	0.0041	5.3898	L'MINERAL POINT	8.1863	0.0043	700	31
	700	32	0.0041	5.3939		8.1820	0.0043	700	31
38	600	32	0.0036	5.3975	R'RC	8.1777	0.0037	600	31
	400	32	0.0024	5.3998	T'T'TMPH 35	8.1741	0.0024	400	31
47	800	32	0.0047	5.4046	L	8.1716	0.0049	800	31
47	700	32	0.0041	5.4087	L	8.1667	0.0043	700	31
36	1400	32	0.0083	5.4170	L	8.1624	0.0086	1400	31
	200	32	0.0012	5.4182	REVERSE CURVE	8.1539	0.0012	200	31
36	800	32	0.0047	5.4229	R	8.1527	0.0049	800	31
	131	32	0.0008	5.4237	RC	8.1478	0.0018	300	31
35	180 A	32	0.0011	5.4248	L	8.1459	0.0038	1600	31
	1866	36	0.0098	5.4346		8.1362	0.0017	298	31
	215	36	0.0011	5.4357	RC	8.1344	0.0013	215	31
36	677	36	0.0036	5.4393	R	8.1331	0.0045	800	31
	123 D	38	0.0007	5.4399		8.1282	0.0000		
31	800	31	0.0049	5.4448	R	8.1282	0.0049	800	31
	400	31	0.0024	5.4473	RC	8.1233	0.0024	400	31
34	700	31	0.0043	5.4516	L	8.1209	0.0043	700	31
32	700	31	0.0043	5.4556	L	8.1168	0.0043	700	31
	100	31	0.0006	5.4564	RC	8.1123	0.0006	100	31
33	900	31	0.0056	5.4619	R	8.1117	0.0056	900	31
31	700	31	0.0043	5.4662	R	8.1062	0.0043	700	31
31	377	31	0.0023	5.4685	R	8.1020	0.0023	377	31
				5.4685	AC	8.0996	0.0008	150 D	37
	1300	31	0.0079	5.4765		8.0988	0.0059	1150	37
42	1200	31	0.0073	5.4838	R	8.0929	0.0061	1200	37
	1200	31	0.0073	5.4911		8.0868	0.0061	1200	37
44	1200	31	0.0073	5.4985	L	8.0806	0.0061	1200	37
58	1000	31	0.0061	5.5046	L	8.0745	0.0051	1000	37
37	1400	31	0.0086	5.5131	L	8.0694	0.0072	1400	37
	200	31	0.0012	5.5143	RC	8.0622	0.0010	298 D	44
44	1000	31	0.0061	5.5205	R	8.0612	0.0043	992	44
		31	0.0000	5.5205		8.0570	0.0010	344 D	51
51	246	31	0.0015	5.5220	R	8.0560	0.0000	2	51
144	600	31	0.0037	5.5256	R	8.0560	0.0022	600	51
	800	31	0.0049	5.5305		8.0538	0.0030	800	51
	854	31	0.0052	5.5357	T'CONEMAUGH TOWE	8.0508	0.0069	1859	51
60	0	31	0.0000	5.5357	R	8.0439	0.0000		
	352 A	31	0.0019	5.5377	RC	8.0439	0.0042	998 A	39
124	2714	37	0.0139	5.5516	L	8.0397	0.0052	1063	39
	1206	37	0.0061	5.5577		8.0345	0.0058	1200	39
98	476	37	0.0024	5.5601	L	8.0287	0.0023	476	39
	300	37	0.0015	5.5617	RC	8.0264	0.0015	300	39
102	700	37	0.0036	5.5653	R	8.0250	0.0034	700	39
63	900	37	0.0046	5.5699	R	8.0216	0.0044	900	39
53	1100	37	0.0056	5.5756	R	8.0172	0.0053	1100	39
	479 A	37	0.0022	5.5777		8.0119	0.0049	1000	39
51	279	44	0.0012	5.5789	R	8.0070	0.0049	1000	39
58			0.0000	5.5789	R	8.0021	0.0052	1061	39
	531 A	44	0.0025	5.5814	T'JOHNSTOWN	7.9970	0.0039	800	39
152	3072	51	0.0114	5.5928	R	7.9931	0.0029	600	39
	404	51	0.0015	5.5943	W END 3 TR	7.9902	0.0046	930	39
	398 D	51	0.0017	5.5960	E END 1 TR SANG HO	7.9857	0.0000		
39	200	39	0.0010	5.5970	R	7.9857	0.0036	837 A	28
	1000	39	0.0049	5.6018		7.9821	0.0029	433	28
60	1000	39	0.0049	5.6067	L	7.9791	0.0068	1000	28
135	438	39	0.0021	5.6088	L	7.9724	0.0030	438	28
	2000	39	0.0097	5.6185		7.9694	0.0135	2000	28
115	800	39	0.0039	5.6224	R	7.9559	0.0054	800	28
64	2000	39	0.0097	5.6321	R	7.9506	0.0135	2000	28
70	1529	39	0.0074	5.6396	R	7.9369	0.0103	1529	28
	800	39	0.0039	5.6434		7.9266	0.0054	800	28
50	1200	39	0.0058	5.6493	R	7.9212	0.0061	1200	28
	233	39	0.0011	5.6504	DORNOCK	7.9131	0.0054	800	28
	297	39	0.0014	5.6518		7.9077	0.0000		
	270 D	39	0.0015	5.6534		7.9077	0.0000		
28	1000	26	0.0068	5.6601	L	7.9077	0.0068	1000	28
26	889	26	0.0060	5.6661	L	7.9008	0.0060	889	28
				5.6661		7.8949	0.0006	88 D	32
	200	26	0.0014	5.6675	RC	7.8943	0.0007	112	32
32	1000	28	0.0068	5.6742	R	7.8937	0.0059	1000	32
				5.6742		7.8877	0.0022	469 D	48
141	900	26	0.0061	5.6803	R	7.8856	0.0017	431	48
45	1000	26	0.0068	5.6871	R	7.8838	0.0039	1000	48
				5.6871		7.8799	0.0004	109 D	51
	200	28	0.0014	5.6885	REVERSE CURVE	7.8795	0.0003	91	51
56	800	28	0.0054	5.6939	L	7.8791	0.0030	800	51
	100	28	0.0007	5.6945		7.8761	0.0004	100	51
	1600	26	0.0108	5.7054		7.8758	0.0059	1500	51
75	1600	26	0.0108	5.7162	R	7.8698	0.0059	1500	51
	800	26	0.0054	5.7216		7.8639	0.0030	800	51
136	600	28	0.0041	5.7257	R	7.8609	0.0022	600	51
54	400	28	0.0027	5.7284	R	7.8587	0.0015	400	51
	282	28	0.0019	5.7303	RC	7.8572	0.0010	282	51
51	700	28	0.0047	5.7350	L	7.8562	0.0026	700	51
				5.7350	WEST END JOHNSTO	7.8536	0.0016	456 D	62
	818	28	0.0056	5.7406	SO	7.8520	0.0106	3544	62
141	219 A	28	0.0014	5.7419	L	7.8412	0.0083	2730	62
141	257	32	0.0015	5.7434	L	7.8329	0.0064	2380 A	53
	724	32	0.0043	5.7477		7.8265	0.0027	743	53
	1125 A	32	0.0053	5.7530		7.8238	0.0029	800	53
62	775	48	0.0031	5.7561	L	7.8210	0.0143	4000	53
	291 A	48	0.0011	5.7572	RC SANG HOLLOW	7.8067	0.0007	200	53
84	6791	51	0.0252	5.7824	R	7.8060	0.0008	237	53
84	1507 A	51	0.0050	5.7874	R	7.8051	0.0001	20	53
	6575	62	0.0201	5.8075		7.8050	0.0143	4000	53
62	621	62	0.0019	5.8094	L	7.7907	0.0036	1000	53
	380 D	62	0.0013	5.8106	L'SEWARD	7.7872	0.0000		

READ DOWN WESTBOUND TRAFFIC — APPENDIX B1 STATION — READ UP EASTBOUND TRAFFIC

PAGE 7 COLUMN 1

MILES	DISTANCE	HWY GRADE CROSSING	MAX MPH	DISTANCE	SPEED	ST. HOURS	TIME HOURS	STATION	TIME HOURS	ST. HOURS	DISTANCE	SPEED
358.29	2996		53	2996	53	0.0107	5.8213	E END 2 TR SANG HOL	7.7872	0.0107	2996	53
358.87	2000			2000	53	0.0071	5.8285		7.7765	0.0054	2074	D 92
358.71	214			214	53	0.0008	5.8293	T	7.7710	0.0007	351	92
358.71							5.8293		7.7703	0.0025	1194	A 89
359.85	5000		141	5000	53	0.0179	5.8471	R	7.7678	0.0077	3586	89
360.06	2265		141	2265	53	0.0081	5.8552	R	7.7600	0.0046	2265	89
360.65	3000			1521	53	0.0054	5.8607	W END 2 TR SANG HOL	7.7554	0.0064	3000	69
361.08	2139			5367	A 53	0.0217	5.8823	T NEW FLORENCE	7.7490	0.0046	2139	89
361.62	3000		89			0.0000	5.8823	R	7.7444	0.0023	1089	89
362.00	2000		89	251	89	0.0005	5.8829	R	7.7421	0.0000	0	
362.38	2000			2006	89	0.0043	5.8871		7.7421	0.0000		
362.54	820		141	820	89	0.0017	5.8888	L	7.7421	0.0128	5431	A 71
362.89	824		89	824	89	0.0018	5.8906	L	7.7293	0.0057	2124	71
363.45	4000			4000	89	0.0085	5.8991		7.7237	0.0007	267	71
363.45						0.0000	5.8991	290	7.7230	0.0072	2451	A 57
363.65	1056			1056	89	0.0022	5.9014	T CONRAIL JCT	7.7157	0.0064	1929	57
363.71	300	XING 290		300	89	0.0006	5.9020	PRIVATE	7.7093	0.0000		
363.78	400		113	400	89	0.0009	5.9029	R REVERSE CURVE	7.7093	0.0050	1348	A 44
363.86	400		113	400	89	0.0009	5.9037	L MPH 50	7.7043	0.0007	169	44
364.37	2720			1664	89	0.0035	5.9072		7.7036	0.0117	2720	44
364.37				1056	D 89	0.0025	5.9097		7.6919	0.0000		
365.29	4837		71	4186	71	0.0112	5.9209	R	7.6919	0.0208	4837	44
365.28				657	D 71	0.0019	5.9228		7.6711	0.0005		
365.86	3000		57	2700	57	0.0090	5.9318	R	7.6711	0.0167	3876	44
365.91	275			94	57	0.0003	5.9321	RC	7.6544	0.0109	2311	40
365.91				481	D 57	0.0018	5.9339	INDIA	7.6435	0.0000		
366.20	1500		44	1500	44	0.0065	5.9404	L	7.6435	0.0003	58	A 39
366.57	2000		44	2000	44	0.0086	5.9490	L LOCKPORT	7.6432	0.0026	531	39
366.90	1700			1700	44	0.0073	5.9563	END TRNS	7.6406	0.0083	1700	39
367.20	1619		71	1619	44	0.0070	5.9633	L	7.6323	0.0079	1619	39
367.58	1900		71	1900	44	0.0082	5.9715	L BOLIVAR	7.6245	0.0092	1900	39
367.71	800			800	44	0.0034	5.9749		7.6153	0.0039	800	39
367.88	900		58	900	44	0.0039	5.9788	R	7.6114	0.0044	900	39
368.11	1181		71	1058	44	0.0046	5.9833	R	7.6070	0.0057	1181	38
368.11				123	D 44	0.0006	5.9839		7.6013	0.0000		
368.46	1800		40	1800	40	0.0085	5.9924	R	7.6013	0.0087	1800	39
368.56	569			540	40	0.0026	5.9950	RC	7.5926	0.0028	569	39
368.56				29	D 40	0.0001	5.9951		7.5898	0.0000		
369.11	2900		39	2900	39	0.0141	6.0092	L	7.5898	0.0141	2900	39
369.18	400		39	400	39	0.0019	6.0111	L	7.5757	0.0019	400	39
369.18							6.0111		7.5737	0.0017	396	D 51
369.30	631			631	39	0.0031	6.0142		7.5721	0.0009	236	51
369.55	1300		71	1300	39	0.0063	6.0205	R	7.5712	0.0048	1300	51
369.74	1000			1000	39	0.0049	6.0254		7.5664	0.0037	1000	51
370.04	1600		100	1600	39	0.0078	6.0331	R	7.5627	0.0057	1529	51
370.09	264		100	264	39	0.0013	6.0344	R	7.5570	0.0019	498	A 48
370.62	2800			2800	39	0.0136	6.0480		7.5550	0.0109	2637	46
370.84	1700		54	1700	39	0.0083	6.0563	R	7.5442	0.0070	1700	46
371.04	500		61	500	39	0.0024	6.0587	R	7.5372	0.0021	500	46
371.19	800			800	39	0.0039	6.0626		7.5351	0.0033	800	46
371.74	2900		51	405	39	0.0020	6.0646	R	7.5318	0.0119	2900	46
371.96	1119			898	A 33	0.0042	6.0687		7.5199	0.0046	1119	46
372.04	500		66	3116	51	0.0116	6.0803	R	7.5153	0.0021	500	46
372.09	244		86	244	51	0.0009	6.0812	R TORRANCE	7.5132	0.0010	244	46
372.20	600			422	51	0.0016	6.0828	E END 3 TR (SIDING)	7.5122	0.0025	600	46
372.20				178	D 51	0.0007	6.0835	REVERSE CURVE	7.5097	0.0000		
372.58	2000		46	2000	46	0.0082	6.0917	L	7.5097	0.0082	2000	46
372.81	1200		71	1200	46	0.0049	6.0966	L	7.5015	0.0049	1200	46
372.94	700		67	700	46	0.0029	6.0995	L	7.4966	0.0029	700	46
373.04	500		53	500	46	0.0021	6.1016	L	7.4937	0.0021	500	46
373.12	436		53	436	46	0.0018	6.1034	L	7.4916	0.0018	436	46
373.88	4000			4000	46	0.0165	6.1199	GRAYS	7.4898	0.0165	4000	46
374.01	700		46	700	46	0.0029	6.1227	R	7.4734	0.0029	700	46
374.12	600		46	600	46	0.0025	6.1252	R	7.4705	0.0025	600	46
374.12							6.1252		7.4680	0.0015	415	D 57
374.16	177			177	46	0.0007	6.1259	RC	7.4665	0.0002	62	57
374.69	2800		57	2800	46	0.0115	6.1374	L	7.4663	0.0083	2500	57
374.69		XING 301 8					6.1374	T 589 PA	7.4580	0.0019	657	D 71
374.98	1600			1600	46	0.0066	6.1440		7.4560	0.0025	943	71
375.01	100	XING	71	100	46	0.0004	6.1444	R PRIVATE	7.4535	0.0003	100	71
375.25	1289	XING	71	1289	46	0.0053	6.1498	R PRIVATE	7.4532	0.0034	1289	71
375.25							6.1498		7.4498	0.0013	498	D 82
375.25							6.1498		7.4486	0.0007	314	86
375.25							6.1498		7.4478	0.0058	2326	A 71
376.01	4000	XING 303		4000	46	0.0165	6.1662	T 888 PA	7.4420	0.0023	865	71
377.01	5278			1034	46	0.0043	6.1705		7.4397	0.0141	5278	71
377.58	3000			1213	A 46	0.0044	6.1749		7.4256	0.0080	3000	71
377.61	162		122	1764	57	0.0058	6.1808	L	7.4176	0.0004	162	71
377.93	1700			2451	A 57	0.0072	6.1880		7.4172	0.0045	1700	71
378.01	400		71	4078	71	0.0109	6.1989	R	7.4126	0.0011	400	71
378.20	1000		71	1000	71	0.0027	6.2016	R MILLWOOD	7.4116	0.0027	1000	71
378.20							6.2016		7.4089	0.0033	1461	D 95
378.39	1000			1000	71	0.0027	6.2042		7.4056	0.0008	376	95
378.51	677		109	677	71	0.0018	6.2060	L	7.4048	0.0000		
378.65	700		141	700	71	0.0019	6.2079	L	7.4048	0.0000		
379.01	1900			1900	71	0.0051	6.2130		7.4048	0.0000		
379.10	483			483	71	0.0013	6.2142		7.4048	0.0000		
379.23	700		100	700	71	0.0019	6.2161	R	7.4048	0.0000		
380.01	4100			4100	71	0.0109	6.2271		7.4048	0.0181	7995	A 71
380.56	2887			1440	71	0.0038	6.2309	T DERRY	7.3868	0.0070	2813	71
381.01	2400	XING	100			0.0000	6.2309	R VALLEY STREET	7.3798	0.0064	2400	71
381.06	287		100			0.0000	6.2309	R	7.3734	0.0008	287	71
382.01	5000			7965	A 71	0.0181	6.2489		7.3726	0.0133	5000	71
382.14	700			378	95	0.0008	6.2497		7.3593	0.0019	700	71
382.14				1461	D 95	0.0030	6.2500		7.3574	0.0000		
382.58	2359		71	2359	71	0.0063	6.2563	R BRADENVILLE	7.3574	0.0063	2359	71
382.58							6.2563		7.3511	0.0040	1818	D 100
382.89	1600			1600	71	0.0043	6.2606		7.3471	0.0002	82	100
383.71	4300			4300	71	0.0115	6.2721		7.3469	0.0075	4000	100
383.88	989		100	989	71	0.0026	6.2777	L	7.3394	0.0019	989	100
384.88	5268		100	4111	71	0.0110	6.2887	L	7.3375	0.0030	1006	100
385.01	600		100	0		0.0000	6.2887	L	7.3345	0.0106	5282	A 89
385.76	4000			10713	A 71	0.0247	6.3134	T LATROBE	7.3239	0.0064	2988	89
385.90	700		200			0.0000	6.3134	R	7.3176	0.0015	700	89
386.86	5112		200	2292	109	0.0043	6.3177	R	7.3161	0.0109	5112	89
387.25	2200		104	782	D 100	0.0016	6.3193	R BEATTY	7.3052	0.0047	2200	89
387.70	2225		89	2225	89	0.0047	6.3240	R	7.3005	0.0047	2225	89
387.74	177		89	177	89	0.0004	6.3244	R	7.2958	0.0109	5100	89
388.70	5100			5100	89	0.0109	6.3352		7.2849	0.0056	2509	A 82
389.70	5278			5278	89	0.0112	6.3465		7.2793	0.0068	2926	82
389.97	1400			622	89	0.0013	6.3478		7.2726	0.0032	1400	82
390.36	2074		141	5873	A 89	0.0117	6.3594	L	7.2693	0.0048	2074	82
390.70	1800			304	101	0.0006	6.3600	DONOHOE	7.2645	0.0042	1800	82
391.23	2800			1275	D 101	0.0026	6.3627		7.2604	0.0066	2800	82
391.46	1339		82	1339	82	0.0031	6.3658	L	7.2539	0.0031	1339	82

PAGE 7 COLUMN 2

MILES	DISTANCE	HWY GRADE CROSSING	MAX MPH	DISTANCE	SPEED	ST. HOURS	TIME HOURS	STATION	TIME HOURS	ST. HOURS	DISTANCE	SPEED
391.92	2300		82	2300	82	0.0028	6.3711	L	7.2508	0.0053	2300	82
392.05	682			682	82	0.0015	6.3726		7.2455	0.0016	682	82
392.39	1800		82	1800	82	0.0042	6.3768	R	7.2439	0.0031	1800	D 10
392.46	400			400	82	0.0009	6.3777		7.2409	0.0007	387	104
392.56	600			600	82	0.0014	6.3791		7.2402	0.0044	2385	A 10
392.86	1500		185	1500	82	0.0035	6.3826	L	7.2357	0.0000		
393.13	1397		245	1397	82	0.0032	6.3858	L	7.2357	0.0000		
393.47	1800			1800	82	0.0042	6.3900		7.2357	0.0000		
394.36	4700			4700	82	0.0109	6.4008	T GREENSBURG	7.2357	0.0000		
394.47	579		141	578	82	0.0013	6.4021	R	7.2357	0.0000		
394.73	1400		141	783	82	0.0018	6.4040	R	7.2357	0.0000		
394.73				617	D 82	0.0015	6.4055		7.2357	0.0244	1124	A 69
395.04	1684		71	1594	71	0.0043	6.4097	R	7.2113	0.0010	384	69
395.47	2300			2300	71	0.0061	6.4158		7.2103	0.0063	2300	69
395.79	1700		141	1700	71	0.0045	6.4204	L	7.2040	0.0047	1700	69
396.40	3200			3200	71	0.0085	6.4289		7.1993	0.0088	3200	69
396.46	335		290	335	71	0.0009	6.4298	L	7.1905	0.0009	335	69
396.52	311		290	311	71	0.0008	6.4307	L	7.1896	0.0008	311	69
397.05	2800			2697	71	0.0072	6.4379		7.1888	0.0077	2800	69
397.05				103	D 85	0.0003	6.4381		7.1811	0.0000		
397.22	900		69	900	69	0.0025	6.4406	R	7.1811	0.0080	2912	69
397.24	100			100	69	0.0003	6.4409		7.1731	0.0056	1971	A 58
398.29	5306			5306	69	0.0146	6.4554	T RADEBAUGH	7.1676	0.0050	1522	58
398.57	1700			1700	69	0.0047	6.4601		7.1626	0.0220	6738	58
399.25	3586		100	3586	69	0.0098	6.4700	L JEANNETTE	7.1406	0.0039	1154	A 49
399.74	2600		100	309	69	0.0008	6.4708	L	7.1367	0.0002	48	49
400.04	1597			2738	A 69	0.0069	6.4777		7.1365	0.0062	1592	49
400.04				31	40	0.0001	6.4778		7.1303	0.0000		
400.04				1113	D 80	0.0031	6.4809		7.1303	0.0000		
400.25	1100		58	1100	58	0.0036	6.4845	R	7.1303	0.0043	1100	49
400.36	562		58	562	58	0.0013	6.4858	R	7.1261	0.0022	562	49
401.04	3600	XING 328		3247	58	0.0106	6.4969	WRIGLEYS	7.1239	0.0139	3600	49
401.04				353	D 58	0.0013	6.4982		7.1100	0.0000		
401.23	1000		49	1000	49	0.0039	6.5020	L	7.1100	0.0039	1000	49
401.25	100			100	49	0.0004	6.5024		7.1061	0.0004	100	49
401.40	800			800	49	0.0031	6.5055		7.1057	0.0031	800	49
401.57	900		49	900	49	0.0035	6.5090	R	7.1027	0.0035	900	49
401.81	1300			1300	49	0.0050	6.5140	L MANOR	7.0992	0.0050	1300	49
402.02	1105		49	1105	49	0.0043	6.5183	L	7.0942	0.0043	1105	49
402.02							6.5183		7.0899	0.0031	968	D 71
402.28	1200			1200	49	0.0046	6.5229		7.0868	0.0006	232	71
402.52	1400			1400	49	0.0054	6.5283	T GRAFTON	7.0862	0.0037	1400	71
403.24	3810		71	3810	49	0.0147	6.5430	R	7.0825	0.0102	3810	71
403.24							6.5430		7.0723	0.0029	1255	D 92
403.24							6.5430		7.0694	0.0012	562	92
403.24							6.5430		7.0682	0.0050	2349	A 86
404.13	4720		135	4590	48	0.0177	6.5608	R IRWIN	7.0632	0.0012	554	86
404.20	378		135	3386	A 48	0.0106	6.5713	R	7.0620	0.0008	376	86
405.05	4500			1620	71	0.0043	6.5757		7.0612	0.0099	4500	86
405.11	286			1404	71	0.0037	6.5794		7.0513	0.0006	286	86
405.37	1400		100	4278	A 71	0.0103	6.5897	R LARIMER	7.0506	0.0031	1400	86
406.06	3600					0.0000	6.5897		7.0476	0.0055	2494	86
406.11	285					0.0000	6.5897		7.0421	0.0000		
406.28	906		86	790	86	0.0017	6.5914	L ARDARA	7.0421	0.0000		
406.39	600		105	600	86	0.0013	6.5927	L	7.0421	0.0136	5473	A 65
407.06	3500			3500	86	0.0077	6.6005		7.0285	0.0027	918	65
407.44	2000			2000	86	0.0044	6.6048		7.0258	0.0058	2009	65
407.96	2749		100	2749	86	0.0061	6.6109	R	7.0200	0.0080	2749	65
408.04	433		100	433	86	0.0010	6.6119	R CP TRAFF	7.0119	0.0148	5093	65
408.04				1718	86	0.0038	6.6157	TRAFFORD	6.9971	0.0000		
408.04				1549	A 86	0.0033	6.6190		6.9971	0.0000		
408.97	4600			212	90	0.0004	6.6194		6.9971	0.0086	2534	A 45
408.97				1421	D 90	0.0035	6.6229		6.9885	0.0000		
409.48	2700		65	2730	65	0.0073	6.6303	L	6.9885	0.0017	405	45
410.00	2743		74	2743	65	0.0080	6.6386	L	6.9868	0.0206	4885	45
410.24	1300		74	1300	65	0.0038	6.6425	L	6.9662	0.0022	496	A 38
410.68	2300			2300	65	0.0067	6.6493	PITCAIRN YARD	6.9640	0.0048	968	38
410.96	1500		71	1500	65	0.0044	6.6536	R WALL	6.9592	0.0128	2589	38
411.22	1361			1351	65	0.0039	6.6576		6.9464	0.0017	313	A 33
411.48	1400			593	65	0.0017	6.6593		6.9446	0.0072	1261	33
411.48				807	D 65	0.0028	6.6621		6.9375	0.0008	139	A 30
411.81	1700		45	1700	45	0.0072	6.6692	R	6.9367	0.0106	1679	30
412.01	1093			1093	45	0.0046	6.6738	WILMERDING	6.9261	0.0069	1093	30
412.15	748		60	748	45	0.0031	6.6770	L PRR Branch conn to rw	6.9192	0.0047	748	30
412.21	300			200	45	0.0008	6.6778		6.9145	0.0128	2034	30
412.46	1300			31	45	0.0001	6.6779	340.3 Jct Port Perry Br	6.9018	0.0014	206	A 26
412.46				1370	D 54	0.0063	6.6842		6.9002	0.0000		
412.66	1080		38	1060	38	0.0054	6.6896	L	6.9002	0.0032	440	26
412.74	400		38	400	38	0.0020	6.6916	L	6.8970	0.0029	400	26
413.15	2200			2200	38	0.0110	6.7025	EB MORE TO EAST	6.8941	0.0180	2200	26
413.23	405		63	400	38	0.0020	6.7045	L	6.8761	0.0029	400	26
413.31	400			270	38	0.0014	6.7059		6.8732	0.0029	400	26
413.31				130	D 32	0.0007	6.7066		6.8723	0.0000		
413.35	400		33	400	33	0.0023	6.7089	L WESTINGHOUSE BR	6.8723	0.0029	400	26
413.44	300			931	33	0.0053	6.7142	RC	6.8694	0.0022	300	26
413.57	700		33	69	D 33	0.0004	6.7146	R	6.8672	0.0051	700	26
413.66	480		30	480	30	0.0030	6.7176	R	6.8621	0.0035	480	26
413.78	600		31	600	30	0.0038	6.7214	R	6.8586	0.0044	600	26
413.87	500			500	30	0.0032	6.7246	RC	6.8542	0.0036	500	26
413.96	480		36	480	30	0.0030	6.7276	L	6.8506	0.0035	480	26
414.09	700			700	30	0.0044	6.7320		6.8471	0.0051	700	26
414.19	500		45	500	30	0.0032	6.7352	L	6.8420	0.0036	500	26
414.66	2500			2418	30	0.0153	6.7506		6.8383	0.0182	2500	26
414.66				82	D 30	0.0004	6.7510	PT PERRY TUNNEL	6.8201	0.0000		
414.76	524		26	524	26	0.0038	6.7548	R	6.8201	0.0038	524	26
414.76						0.0000	6.7548	PORT PERRY	6.8163	0.0007	40	D 28
414.76							6.7548		6.8160	0.0001	10	28
414.80	200			200	26	0.0015	6.7563	REVERSE CURVE	6.8159	0.0010	150	28
414.87	400		28	400	26	0.0029	6.7592	L	6.8149	0.0027	400	28
414.97							6.7592		6.8122	0.0031	629	D 50
415.08	1100			1100	26	0.0080	6.7672		6.8092	0.0018	471	50
415.23	800		109	800	26	0.0058	6.7730	R	6.8074	0.0030	800	50
415.37	700		60	700	26	0.0051	6.7781	R	6.8044	0.0027	700	50
415.53	858		60	864	26	0.0063	6.7844	R	6.8017	0.0032	858	50
415.74	1100			1100	26	0.0080	6.7924		6.7985	0.0042	1100	50
415.76	100		86	100	26	0.0007	6.7931	R RANKIN BRIDGE	6.7943	0.0004	100	50
415.89	700		86	700	26	0.0051	6.7982	R	6.7939	0.0027	700	50
416.04	800			800	26	0.0058	6.8041		6.7913	0.0030	800	50
416.08	180		100	180	26	0.0013	6.8054	L	6.7882	0.0111	180	50
416.26	1000			1000	26	0.0073	6.8127		6.7771	0.0038	1000	50
416.53	1400		100	1400	26	0.0102	6.8229	L	6.7683	0.0053	1400	50
416.76	1200			1200	26	0.0087	6.8316		6.7630	0.0045	1200	50
416.87	580			248	26	0.0018	6.8334		6.7585	0.0022	580	50
416.94	386		57	217	26	0.0016	6.8350	L CONNECTION ALIGN	6.7563	0.0015	386	50
417.18	1264			28	A 26	0.0006	6.8356		6.7548	0.0048	1264	50

MILES	DISTANCE, ft	HWY GRADE CROSSING	MAX MPH	DISTANCE, feet	SPEED, mph	ET, hours	TIME, hours	STATION (APPENDIX B1) PAGE 8 COLUMN 1	TIME, hours	ET, hours	DISTANCE, feet	SPEED, mph
			READ DOWN WESTBOUND TRAFFIC						READ UP EASTBOUND TRAFFIC			
417.25	364		100				6.8355	R PRIOR CONNECTION	6.7500	0.0014	364	50
417.52	1422			522	28	0.0035	6.8391		6.7487	0.0064	1422	50
417.75	1254		71	1487	A 28	0.0072	6.8463	L	6.7433	0.0048	1254	50
418.27	2746			2746	50	0.0104	6.8567	L	6.7385	0.0104	2746	50
418.27		P&LE PRR					6.8567	AMITY STREET	6.7281			
418.28	46		50	46	50	0.0002	6.8568	L	6.7281	0.0002	46	50
418.43	758			3478	50	0.0132	6.8700		6.7279	0.0029	758	50
418.46	152		91	152	50	0.0006	6.8706	L EASTWARD TRACK	6.7251	0.0066	152	50
418.46	19		91	19	50	0.0001	6.8707		6.7245	0.0001	19	50
418.49	149			149	50	0.0006	6.8712		6.7244	0.0006	149	50
418.51	129		69	129	50	0.0005	6.8717	R EASTWARD TRACK	6.7239	0.0006	129	50
418.65	703			703	50	0.0027	6.8744		6.7234	0.0027	703	50
418.83	1000			1000	50	0.0038	6.8782		6.7207	0.0038	1000	50
419.02	1000			1000	50	0.0038	6.8820		6.7169	0.0038	1000	50
419.15	680			660	50	0.0025	6.8845		6.7131	0.0025	660	50
419.21	340			340	50	0.0013	6.8858		6.7106	0.0013	340	50
419.40	1000			1000	50	0.0038	6.8895		6.7093	0.0038	1000	50
419.51	540			540	50	0.0020	6.8916		6.7056	0.0020	540	50
419.59	480		53	480	50	0.0018	6.8934	R	6.7035	0.0017	460	50
419.71	596		53	596	50	0.0023	6.8957	R	6.7018	0.0023	566	50
419.78	415			415	50	0.0016	6.8972		6.6995	0.0016	415	50
419.90	632			632	50	0.0024	6.8996		6.6979	0.0024	632	50
419.97	368		84	368	50	0.0014	6.9010	R	6.6956	0.0014	368	50
420.08	553		84	553	50	0.0021	6.9031		6.6942	0.0021	553	50
420.12	233			233	50	0.0009	6.9040		6.6921	0.0009	233	50
420.15	135		95	135	50	0.0005	6.9045	R	6.6912	0.0005	135	50
420.16	79		100	79	50	0.0003	6.9046	R	6.6907	0.0003	79	50
420.26	499		76	499	50	0.0019	6.9067	R	6.6904	0.0019	499	50
420.31	305		67	305	50	0.0012	6.9079	R	6.6885	0.0012	305	50
420.35	196			196	50	0.0007	6.9086		6.6873	0.0007	196	50
420.40	266			266	50	0.0010	6.9096		6.6866	0.0010	266	50
420.50	535		60	530	50	0.0020	6.9116	R	6.6856	0.0020	539	50
420.54	195			195	50	0.0007	6.9124		6.6835	0.0010	275	D 57
420.72	940			940	50	0.0036	6.9159		6.6826	0.0029	860	57
420.73	60		87	60	50	0.0002	6.9162	R	6.6797	0.0002	60	57
420.92	1000			1000	50	0.0038	6.9200		6.6795	0.0033	1000	57
421.08	840			840	50	0.0032	6.9231	R	6.6762	0.0006	167	57
421.11	160			160	50	0.0006	6.9237		6.6756	0.0031	863	A 50
421.16	260			260	50	0.0010	6.9247		6.6726	0.0009	230	50
421.30	734		138	734	50	0.0028	6.9275	R	6.6717	0.0007	173	50
421.32	97		138	97	50	0.0004	6.9279	R	6.6710	0.0028	672	A 42
421.45	706			706	50	0.0027	6.9306		6.6683	0.0031	692	42
421.49	197			197	50	0.0007	6.9313	BECKS RUN	6.6651	0.0008	188	42
421.49							6.9313		6.6643	0.0006	120	A 40
421.57	421			421	50	0.0016	6.9329		6.6637	0.0015	310	40
421.68	579		57	579	50	0.0022	6.9351	R	6.6623	0.0027	579	40
421.69	62			62	50	0.0002	6.9353	R	6.6595	0.0003	82	40
421.81	652		71	652	50	0.0025	6.9378		6.6592	0.0031	852	40
421.87	286			286	50	0.0011	6.9389	L	6.6561	0.0014	286	40
421.91	239			239	50	0.0009	6.9398	L	6.6548	0.0011	239	40
422.02	580		100	580	50	0.0021	6.9419	L	6.6537	0.0026	580	40
422.06	211		71	211	50	0.0008	6.9427	L	6.6511	0.0010	211	40
422.06	37		71	37	50	0.0001	6.9428	L	6.6501	0.0002	37	40
422.15	439			439	50	0.0017	6.9445		6.6499	0.0021	439	40
422.24	524			524	50	0.0020	6.9465		6.6478	0.0025	524	40
422.36	634			634	50	0.0024	6.9489		6.6453	0.0030	634	40
422.42	269		56	269	50	0.0010	6.9499	L	6.6423	0.0013	269	40
422.43	97		100	97	50	0.0004	6.9502	L	6.6410	0.0005	97	40
422.44	44			44	50	0.0002	6.9504		6.6406	0.0016	330	40
422.49	273		71	273	50	0.0010	6.9514	L	6.6390	0.0022	430	A 33
422.62	683			494	50	0.0019	6.9533		6.6366	0.0014	240	33
422.62				863	A 50	0.0031	6.9564		6.6354	0.0000		
422.66	217						6.9564		6.6354	0.0012	217	33
422.78	619		71	142	57	0.0005	6.9568	L	6.6342	0.0038	619	33
422.81	154			164	57	0.0005	6.9574		6.6306	0.0009	164	33
422.93	630			630	57	0.0021	6.9596		6.6297	0.0036	630	33
423.00	370		71	370	57	0.0012	6.9607	L	6.6261	0.0021	370	33
423.04	175			175	57	0.0006	6.9613		6.6239	0.0010	175	33
423.15	822			550	57	0.0018	6.9631		6.6229	0.0036	622	33
423.19	203					0.0000	6.9631		6.6194	0.0012	203	33
423.19				36	57	0.0001	6.9632		6.6182	0.0000		
423.20	36			275	D 57	0.0010	6.9642		6.6182	0.0002	36	33
423.25	297		50	297	50	0.0011	6.9653	L	6.6180	0.0017	297	33
423.33	385		100	385	50	0.0015	6.9668	L	6.6163	0.0022	385	33
423.38	282			0	50	0.0000	6.9668		6.6141	0.0016	282	33
423.38				123	50	0.0005	6.9673		6.6125	0.0000		
423.40	111			270	D 50	0.0011	6.9684		6.6125	0.0006	111	33
423.51	556		42	556	42	0.0025	6.9709	R	6.6118	0.0032	556	33
423.57	333			333	42	0.0015	6.9724	REVERSE	6.6086	0.0019	333	33
423.57				51	42	0.0002	6.9726	STATION 29+75 P&LE	6.6067	0.0000		
423.59	111			60	D 42	0.0003	6.9729	CURVE	6.6067	0.0006	111	33
423.73	720		40	720	40	0.0034	6.9763	L	6.6061	0.0041	720	33
423.76	194			194	40	0.0009	6.9772		6.6020	0.0011	194	33
423.93	890			899	40	0.0043	6.9815		6.6005	0.0052	899	33
423.95	101		138	101	40	0.0005	6.9819	L	6.5957	0.0006	101	33
423.97	103		138	103	40	0.0005	6.9824	L	6.5951	0.0006	103	33
424.07	535			535	40	0.0025	6.9850	REVERSE CURVE	6.5945	0.0031	535	33
424.12	262		138	262	40	0.0012	6.9862	R	6.5914	0.0015	262	33
424.14	100			100	40	0.0005	6.9867		6.5899	0.0006	100	33
424.15	45			45	40	0.0002	6.9869	REVERSE CURVE	6.5894	0.0003	45	33
424.20	268			268	40	0.0013	6.9882		6.5891	0.0015	268	33
424.24	189		113	189	40	0.0009	6.9891	L	6.5876	0.0011	189	33
424.33	498			495	40	0.0024	6.9914	REVERSE CURVE	6.5866	0.0028	498	33
424.37	185			185	40	0.0009	6.9923		6.5836	0.0011	185	33
424.49	651		64	651	40	0.0031	6.9954	R	6.5826	0.0037	651	33
424.52	164			164	40	0.0008	6.9962	REVERSE CURVE	6.5788	0.0009	164	33
424.55	147			147	40	0.0007	6.9968		6.5779	0.0008	147	33
424.60	285		40	285	40	0.0013	6.9982	L	6.5770	0.0018	285	33
424.66	303			187	D 42	0.0010	6.9992	REVERSE CURVE	6.5754	0.0017	303	33
424.69	156		33	156	33	0.0009	7.0001	R	6.5737	0.0009	156	33
424.71	109		53	109	33	0.0006	7.0007	R	6.5728	0.0019	411	D 47
424.73	122		53	122	33	0.0007	7.0014		6.5708	0.0000		47
424.74	11		53	11	33	0.0001	7.0015	LIBERTY BRIDGE	6.5708	0.0006	0	47
424.79	270		47	270	33	0.0015	7.0030	R	6.5708	0.0004	101	47
424.90	600			600	33	0.0034	7.0064		6.5704	0.0004	107	D 50
424.97	350		100	350	33	0.0020	7.0085	R	6.5700	0.0032	843	50
425.14	914			914	33	0.0052	7.0137		6.5668	0.0035	914	50
425.16	86			86	33	0.0005	7.0142	SMITHFIELD STREET	6.5634	0.0003	86	50
425.20	220		71	220	33	0.0013	7.0155	R	6.5630	0.0008	220	50
425.28	450			450	33	0.0026	7.0180	RC P&LE STATION	6.5622	0.0017	450	50
425.32	200		50	200	33	0.0011	7.0192	L	6.5606	0.0008	200	50
425.33	50			50	33	0.0003	7.0195		6.5597	0.0002	50	50
425.33				33			7.0195		6.5595	0.0010	275	D 57
425.52	1000			1000	33	0.0057	7.0252		6.5586	0.0027	825	57
425.66	750			750	33	0.0043	7.0295		6.5568	0.0022	850	57

MILES	DISTANCE, ft	HWY GRADE CROSSING	MAX MPH	DISTANCE, feet	SPEED, mph	ET, hours	TIME, hours	STATION (APPENDIX B1) PAGE 8 COLUMN 2	TIME, hours	ET, hours	DISTANCE, feet	SPEED, mph
			READ DOWN WESTBOUND TRAFFIC						READ UP EASTBOUND TRAFFIC			
425.71	250		100	250	33	0.0014	7.0310	R FORT PITT BRIDGE	6.5537	0.0008	250	57
425.79	400		100	400	33	0.0023	7.0332	R	6.5528	0.0013	400	57
425.90	600			600	33	0.0034	7.0367		6.5515	0.0020	600	57
426.09	1000			1000	33	0.0057	7.0424		6.5495	0.0036	1043	57
426.13	200			200	33	0.0011	7.0436		6.5460	0.0014	407	A 54
426.20	400		58	400	33	0.0023	7.0459	L	6.5447	0.0005	150	54
426.26	300		58	300	33	0.0017	7.0476	L	6.5441	0.0011	300	54
426.28	100			100	33	0.0006	7.0482		6.5431	0.0004	100	54
426.40	650			650	33	0.0037	7.0519		6.5427	0.0023	650	54
426.47	350		82	350	33	0.0020	7.0539	R	6.5406	0.0012	350	54
426.49	100		82	100	33	0.0006	7.0545	R	6.5392	0.0004	100	54
426.66	900		103	900	33	0.0052	7.0596	R WEST END BRIDGE	6.5389	0.0032	900	54
426.85	1000		103	618	33	0.0035	7.0632	R	6.5357	0.0035	1000	54
426.87	100		103				7.0632		6.5322	0.0004	100	54
427.04	900			938	A 33	0.0044	7.0676		6.5319	0.0032	900	54
427.17	700			282	A 47	0.0011	7.0688		6.5287	0.0025	700	54
427.19	130		160	1108	50	0.0042	7.0729	R	6.5262	0.0005	130	54
427.23	170		160	170	50	0.0006	7.0736	R	6.5256	0.0006	170	54
427.41	1000		160	834	50	0.0032	7.0768	R	6.5252	0.0035	1000	54
427.58	850		92	863	A 50	0.0031	7.0798	R	6.5217	0.0030	850	54
427.60	150			303	57	0.0010	7.0808	REVERSE CURVE	6.5187	0.0005	150	54
427.62	100			100	57	0.0003	7.0811		6.5182	0.0004	100	54
427.71	450		57	450	57	0.0015	7.0826	L	6.5178	0.0015	450	54
427.79	450			450	57	0.0015	7.0841		6.5163	0.0016	450	54
427.98	1000			1000	57	0.0033	7.0875		6.5147	0.0035	1000	54
428.10	600			600	57	0.0020	7.0895		6.5112	0.0021	600	54
428.17	400		66	400	57	0.0013	7.0908	L	6.5091	0.0014	400	54
428.21	190			190	57	0.0006	7.0914	REVERSE CURVE	6.5077	0.0007	190	54
428.26	280			138	57	0.0005	7.0919	CORLISS	6.5070	0.0009	260	54
428.26				122	D 57	0.0004	7.0923		6.5061	0.0000		54
428.36	550		54	550	54	0.0019	7.0942	R	6.5061	0.0019	550	54
428.47	550		54	550	54	0.0019	7.0961	R	6.5042	0.0019	550	54
428.47							7.0961		6.5022	0.0004	122	D 57
428.55	450			450	54	0.0016	7.0977		6.5018	0.0011	328	57
428.74	1000	XING 3 4		1000	54	0.0035	7.1012	RIVER AVENUE	6.5007	0.0033	1000	57
428.93	1000			1000	54	0.0035	7.1047	WB MORE TO WEST	6.4974	0.0033	1000	57
429.12	1000			1000	54	0.0035	7.1082		6.4941	0.0033	1000	57
429.20	450			450	54	0.0016	7.1098		6.4907	0.0015	450	57
429.31	550			550	54	0.0019	7.1118		6.4893	0.0018	550	57
429.42	600			600	54	0.0021	7.1139		6.4874	0.0020	600	57
429.50	400		80	400	54	0.0014	7.1153	R	6.4854	0.0013	400	57
429.52	100		80	100	54	0.0004	7.1156	R	6.4841	0.0003	100	57
429.56	250			250	54	0.0009	7.1165	REVERSE CURVE	6.4838	0.0008	250	57
429.68	600		80	600	54	0.0021	7.1186	L	6.4829	0.0020	600	57
429.69	50			50	54	0.0002	7.1188		6.4809	0.0002	50	57
429.88	1000			1000	54	0.0035	7.1223		6.4808	0.0033	1000	57
430.07	1000			1000	54	0.0035	7.1258		6.4775	0.0033	1000	57
430.11	220			220	54	0.0008	7.1266		6.4741	0.0007	220	57
430.21	550			550	54	0.0019	7.1285		6.4734	0.0016	550	57
430.28	230		57	230	54	0.0008	7.1293	L DAVIS SCHOOL	6.4716	0.0008	230	57
430.38	650		57	650	54	0.0023	7.1316	L	6.4708	0.0022	650	57
430.38					54	0.0000	7.1316		6.4687	0.0019	857	D 71
430.45	350			300	54	0.0011	7.1326		6.4667	0.0002	80	71
430.63	1000			600	54	0.0021	7.1347		6.4666	0.0016	600	71
430.82	1000			407	A 54	0.0014	7.1361	NEVILLE ISLAND	6.4649	0.0027	1000	71
430.94	600			1643	57	0.0065	7.1416		6.4622	0.0016	600	71
430.98	300		71	300	57	0.0010	7.1426	R	6.4606	0.0008	300	71
430.99							7.1426		6.4598	0.0036	1602	D 97
431.01	100			100	57	0.0003	7.1429		6.4562	0.0003	139	97
431.09	400			400	57	0.0013	7.1442		6.4559	0.0000		
431.20	600			600	57	0.0020	7.1462		6.4559	0.0000		
431.38	1000			1000	57	0.0033	7.1495		6.4559	0.0000		
431.44	240			240	57	0.0008	7.1503		6.4559	0.0000		
431.58	780		96	780	57	0.0025	7.1529	L	6.4559	0.0000		
431.59	50		98	50	57	0.0002	7.1530	L	6.4559	0.0000		
431.77	950			950	57	0.0032	7.1562		6.4559	0.0000		
431.96	1000			1000	57	0.0033	7.1595		6.4559	0.0000		
432.11	800			800	57	0.0027	7.1622		6.4559	0.0000		
432.15	200		113	200	57	0.0007	7.1628	L	6.4559	0.0000		
432.21	300		113	300	57	0.0010	7.1638	L	6.4559	0.0000		
432.33	640		113	640	57	0.0021	7.1660	L	6.4559	0.0000		
432.34	60			60	57	0.0002	7.1662		6.4559	0.0000		
432.42	450			450	57	0.0015	7.1677	REVERSE CURVE	6.4559	0.0000		
432.48	300		113	200	57	0.0007	7.1683	R	6.4559	0.0000		
432.52	210						7.1683		6.4559	0.0000		
432.53	40		113				7.1683	L	6.4559	0.0000		
432.72	1000		113	2451	A 57	0.0072	7.1755	L	6.4559	0.0161	7459	A 77
432.85	700		113				7.1755	L	6.4398	0.0015	600	77
432.90	250						7.1755		6.4383	0.0006	250	77
432.91	50		113	5566	A 71	0.0192	7.1947	R	6.4377	0.0001	50	77
432.96	300		113				7.1947	R	6.4376	0.0007	300	77
433.10	700						7.1947		6.4368	0.0017	700	77
433.20	550						7.1947		6.4351	0.0014	550	77
433.29	450						7.1947		6.4338	0.0011	450	77
433.48	1000					0.0000	7.1947		6.4327	0.0025	1000	77
433.66	1000						7.1947		6.4302	0.0025	1000	77
433.82	800						7.1947		6.4277	0.0020	800	77
433.85	200		100				7.1947	R	6.4258	0.0005	200	77
434.04	1000		100				7.1947	R	6.4253	0.0025	1000	77
434.18	700		100				7.1947	R	6.4228	0.0017	700	77
434.20	120	XING 8.95					7.1947	PRIVATE?	6.4211	0.0003	120	77
434.23	180						7.1947		6.4208	0.0004	180	77
434.42	1000						7.1947		6.4204	0.0025	1000	77
434.61	1000	XING 8.4					7.1947	LENORA AVENUE	6.4179	0.0025	1000	77
434.80	1000			238	96	0.0005	7.1952		6.4154	0.0025	1000	77
434.82	100			1205	D 38	0.0026	7.1978		6.4130	0.0002	100	77
434.97	550		77	550	77	0.0014	7.1992	L	6.4127	0.0014	550	77
434.99	350			350	77	0.0009	7.2000		6.4114	0.0028	1276	D 97
435.01	90			90	77	0.0002	7.2003		6.4086	0.0007	336	97
435.18	910			910	77	0.0022	7.2025		6.4080	0.0000	0	97
435.36	950	XING 10.3		950	77	0.0023	7.2048	BROADWAY STREET	6.4080	0.0000	0	97
435.37	50		113	50	77	0.0001	7.2050	L CORAOPOLIS	6.4080	0.0000	0	97
435.41	200	XING	113	200	77	0.0005	7.2054	L MULBERRY STREET	6.4080	0.0000	0	97
435.50	500	XING 16.5		500	77	0.0017	7.2067	RC MILL STREET	6.4080	0.0000	0	97
435.56	300		113	300	77	0.0007	7.2074	R	6.4080	0.0000	0	97
435.62	300			300	77	0.0007	7.2081		6.4080	0.0000	0	97
435.66	250		113	250	77	0.0006	7.2088	R	6.4080	0.0000	0	97
435.75	450			450	77	0.0011	7.2099	REVERSE CURVE	6.4080	0.0000	0	97
435.76	50			50	77	0.0001	7.2100		6.4080	0.0000	0	97
435.81	300		113	300	77	0.0007	7.2107	L	6.4080	0.0000	0	97
435.89	400			200	77	0.0010	7.2117		6.4080	0.0000	0	97
435.91	100	XING	113	100	77	0.0002	7.2120	R MAIN STREET	6.4080	0.0081	3938	A 83
435.94	150			150	77	0.0004	7.2123		6.3999	0.0006	350	88
436.04	550			550	77	0.0014	7.2137					
436.09	250		141	250	77	0.0006	7.2143	R	6.3891	0.0005	250	88

READ DOWN WESTBOUND TRAFFIC — APPENDIX B1 STATION — READ UP EASTBOUND TRAFFIC — READ DOWN WESTBOUND TRAFFIC — APPENDIX B1 STATION — READ UP EASTBOUND TRAFFIC

MILES	DISTANCE, ft	HWY GRADE CROSSING	MAX MPH	DISTANCE, feet	SPEED, mph	dT, hours	TIME, hours	PAGE 9 COLUMN 1		TIME, hours	dT, hours	DISTANCE, feet	SPEED, mph	MILES	DISTANCE, ft	HWY GRADE CROSSINGS	MAX MPH	DISTANCE, feet	SPEED, mph	dT, hours	TIME, hours	PAGE 9 COLUMN 2	TIME, hours	dT, hours	DISTANCE, feet	SPEED, mph
436.13	200	XING	10.9	200	77	0.0005	7.2148	WATT STREET		6.3986	0.0004	200	88	447.77	450			0	98	0.0000	7.3474		6.2591	0.0000		70
436.20	400			400	77	0.0010	7.2158			6.3982	0.0009	400	88	447.87	550		100	0	98	0.0000	7.3474	R	6.2591	0.0000		70
436.32	600			600	77	0.0015	7.2172			6.3973	0.0013	600	88	447.90	150		100	77	98	0.0001	7.3476	R	6.2591	0.0000		70
436.51	1000			1000	77	0.0025	7.2197			6.3960	0.0022	1000	88	447.90				1673	D 98	0.0036	7.3513		6.2591	0.0000		
436.64	700	XING	11.4	700	77	0.0017	7.2214	PRIVATE?		6.3939	0.0015	700	88	447.95	300		71	300	71	0.0008	7.3521	R	6.2591	0.0000		70
436.70	300		98	300	77	0.0007	7.2222	R		6.3924	0.0006	300	88	447.97	100			100	71	0.0003	7.3524		6.2591	0.0000		70
436.88	1000		98	1000	77	0.0025	7.2246	R		6.3917	0.0008	390	88	448.01	200		71	200	71	0.0005	7.3529	R	6.2591	0.0000		70
436.93	250	XING	98	250	77	0.0006	7.2252	R		6.3909	0.0000	0	88	448.06	250			250	71	0.0007	7.3536		6.2591	0.0000		70
437.07	750	XING	11.8	400	77	0.0010	7.2262	PRIVATE?		6.3909	0.0000	0	88	448.08	100			100	71	0.0003	7.3538		6.2591	0.0000		70
437.20	650					0.0000	7.2262			6.3909	0.0000		88	448.09	80		100	80	71	0.0002	7.3540	R	6.2591	0.0000		70
437.20				5104	A 77	0.0114	7.2376		12	6.3909	0.0053	2359	A 81	448.20	540			540	71	0.0014	7.3555		6.2591	0.0000		70
437.26	350					0.0000	7.2376			6.3856	0.0006	251	81	448.25	280			280	71	0.0007	7.3562		6.2591	0.0000		70
437.39	650					0.0000	7.2376			6.3850	0.0015	850	81	448.37	850			598	71	0.0016	7.3578		6.2591	0.0000		70
437.45	350		122			0.0000	7.2376	R		6.3835	0.0006	350	81	448.37				57	D 71	0.0001	7.3580		6.2591	0.0000		
437.62	900		122			0.0000	7.2376	R		6.3829	0.0021	900	81	448.44	350		70	350	70	0.0009	7.3589	L	6.2591	0.0061	2075	A 58
437.64	100		112			0.0000	7.2376	R		6.3808	0.0002	100	81	448.63	1000		70	1000	70	0.0027	7.3616	L COLONA	6.2530	0.0067	2046	58
437.83	1000		112			0.0000	7.2376	R		6.3806	0.0023	1000	81	448.80	900		70	900	70	0.0024	7.3641	L	6.2464	0.0007	216	58
437.95	600		112				7.2376	R STOOPS FERRY		6.3782	0.0014	600	81	448.82	100			100	70	0.0003	7.3643		6.2456	0.0000		
438.02	400			182	92	0.0004	7.2380			6.3768	0.0009	400	81	449.01	1000			1000	70	0.0027	7.3670		6.2456	0.0000		
438.06	200			264	D 92	0.0006	7.2385			6.3759	0.0005	200	81	449.07	350			350	70	0.0009	7.3680		6.2456	0.0000		
438.20	740		88	740	88	0.0016	7.2401	L		6.3754	0.0017	740	81	449.20	650		79	650	70	0.0018	7.3697	L	6.2456	0.0000		
438.21	60		88	60	88	0.0001	7.2403	L		6.3737	0.0001	60	81	449.21	80		79	80	70	0.0002	7.3700	L	6.2456	0.0000		
438.32	600		88	600	88	0.0013	7.2416	L		6.3736	0.0014	600	81	449.22	70		79	70	70	0.0002	7.3701	L	6.2456	0.0156	2738	A 22
438.40	400			400		0.0000	7.2416			6.3722	0.0009	450	81	449.38	850			850	70	0.0023	7.3724		6.2301	0.0172	2000	22
438.58	950			950	88	0.0020	7.2436			6.3712	0.0022	950	81	449.57	1000			1000	70	0.0027	7.3752		6.2129	0.0086	1000	22
438.59	50		89	50	88	0.0001	7.2437	L		6.3690	0.0001	50	81	449.76	1000	XING	82	1000	70	0.0027	7.3779	L 4TH STREET	6.2043	0.0086	1000	22
438.73	750		89	750	88	0.0016	7.2453	L		6.3689	0.0003	149	81	449.75	100		82	100	70	0.0003	7.3781	L MONACA	6.1957	0.0009	100	22
438.78	250			250	88	0.0005	7.2458			6.3685	0.0000	0	81	449.95	900			900	70	0.0024	7.3806		6.1948	0.0077	900	22
438.97	1000			1000	88	0.0022	7.2480			6.3686	0.0000	0	81	450.07	600			400	70	0.0011	7.3816		6.1871	0.0052	600	22
439.16	1000			1000	88	0.0022	7.2502			6.3685	0.0000			450.07					70	0.0000	7.3816	OHIO RIVER BRANCH	6.1819	0.0000		22
439.19	180			180	88	0.0004	7.2506			6.3685	0.0000			450.14	400			37	70	0.0001	7.3817		6.1819	0.0034	400	22
439.19						0.0000	7.2506		14	6.3685	0.0028	1400	A 78	450.14				563	D 84	0.0017	7.3834		6.1784	0.0000		22
439.35	820			820	88	0.0018	7.2523			6.3658	0.0006	250	78	450.19	250		58	250	58	0.0008	7.3842	R	6.1784	0.0022	250	22
439.54	1000			1516	88	0.0033	7.2556			6.3652	0.0024	1000	78	450.33	750		58	750	58	0.0024	7.3867	R	6.1763	0.0065	750	22
439.72	950	XING	14.5	434	D 88	0.0010	7.2566	PRIVATE		6.3627	0.0023	950	78	450.35	80		58	80	58	0.0003	7.3869	R	6.1698	0.0007	80	22
439.73	50		81	50	81	0.0001	7.2567	R GLENWILLARD		6.3604	0.0001	50	78	450.41	320			320	58	0.0010	7.3880		6.1691	0.0026	320	22
439.91	1000		81	1000	81	0.0023	7.2590	R		6.3603	0.0024	1000	78	450.41					58	0.0000	7.3880	OHIO RIVER S BANK	6.1664	0.0000		22
439.93	100		81	100	81	0.0002	7.2592	R		6.3579	0.0002	100	78	450.52	600			600	58	0.0020	7.3899		6.1664	0.0052	600	22
440.10	900			900	81	0.0021	7.2613			6.3576	0.0022	900	78	450.65	700			700	58	0.0023	7.3922		6.1612	0.0050	700	22
440.20	500			500	81	0.0012	7.2625			6.3564	0.0012	500	78	450.65					58	0.0000	7.3922	OHIO RIVER N BANK	6.1562	0.0000		22
440.29	460			460	81	0.0011	7.2636			6.3542	0.0011	460	78	450.71	300			300	58	0.0010	7.3932		6.1562	0.0026	300	22
440.29	40		141	40	81	0.0001	7.2637	R		6.3531	0.0001	40	78	450.79	400		59	400	58	0.0013	7.3945	R	6.1526	0.0034	400	22
440.48	1000		141	1000	81	0.0023	7.2660	R		6.3530	0.0024	1000	78	450.90	600		59	600	58	0.0020	7.3965	R	6.1492	0.0052	600	22
440.57	450		141	450	81	0.0011	7.2671	R		6.3506	0.0011	450	78	450.94	200		59	200	58	0.0007	7.3971	R	6.1440	0.0017	200	22
440.67	550			550	81	0.0013	7.2684			6.3495	0.0013	550	78	451.01	400			340	58	0.0011	7.3982	BEAVER	6.1423	0.0034	400	22
440.86	1000	XING	15.6	1000	81	0.0023	7.2707	PRIVATE		6.3482	0.0024	1000	78	451.09	400		141	0	58	0.0000	7.3982	R	6.1388	0.0034	400	22
441.05	1000			1000	81	0.0023	7.2730			6.3457	0.0024	1000	78	451.20	600			4	58	0.0000	7.3982		6.1354			
441.19	750			750	81	0.0018	7.2748	COUNTY LINE ALLE-BE		6.3433	0.0018	750	78	451.20				1056	D 58	0.0050	7.4032	JCT BEAVER VALLEY	6.1354			
441.24	250			825	81	0.0019	7.2767			6.3415	0.0006	250	78	451.32	600		22	600	22	0.0052	7.4084	L	6.1354	0.0052	600	22
441.38	750			175	D 81	0.0004	7.2771	SOUTH HEIGHTS		6.3409	0.0018	750	78	451.43	600		130	600	22	0.0052	7.4136	R REVERSE CURVE	6.1302	0.0017	246	D 34
441.43	250		78	250	78	0.0006	7.2777	R		6.3390	0.0006	250	78	452.24	4300			4300	22	0.0370	7.4506	5TH AVE - TRACK IN S	6.1286	0.0240	4300	34
441.57	750		78	750	78	0.0018	7.2796	R		6.3384	0.0018	750	78	452.32	400		42	400	22	0.0034	7.4540	L	6.1046	0.0022	400	34
441.62	250			250	78	0.0006	7.2802			6.3366	0.0049	2451	D 11	452.41	500			500	22	0.0043	7.4583	RC	6.1024	0.0028	500	34
441.81	1000			1000	78	0.0024	7.2826			6.3315	0.0027	1596	113	452.45	200	XING	42	200	22	0.0017	7.4601	R TUSCARAWAS RD	6.0996	0.0011	200	34
442.00	1000			1000	78	0.0024	7.2850			6.3291	0.0000		113	452.77	1700			1700	22	0.0146	7.4747		6.0985	0.0095	1700	34
442.19	1000			1000	78	0.0024	7.2875			6.3291	0.0000			452.96	1000	XING		1000	22	0.0086	7.4833	WALNUT LANE	6.0890	0.0056	1000	34
442.21	120			120	78	0.0003	7.2878			6.3291	0.0000			453.00	200		34	200	22	0.0017	7.4850	R	6.0834	0.0011	200	34
442.35	880			880	78	0.0021	7.2899			6.3291	0.0000			453.29	1500		61	1500	22	0.0129	7.4979	L REVERSE CURVE	6.0823	0.0014	298	D 44
442.51	700			700	78	0.0017	7.2916			6.3291	0.0000			453.29		XING				0.0000	7.4979	SPRING LANE	6.0809	0.0052	1214	44
442.57	300		141	300	78	0.0007	7.2923	R		6.3291	0.0000			453.50	1600			600	22	0.0052	7.5031		6.0757	0.0069	1600	44
442.76	1000		141	1000	78	0.0024	7.2947	R		6.3291	0.0000			453.68	500		74	535	A 22	0.0036	7.5067	L	6.0688	0.0022	500	44
442.93	940		141	940	78	0.0023	7.2970	R		6.3291	0.0000			453.76	400			400	34	0.0022	7.5090		6.0667	0.0017	400	44
442.95	60			60	78	0.0001	7.2972			6.3291	0.0000			453.85	500			500	34	0.0028	7.5117		6.0650	0.0022	500	44
443.13	1000			1000	78	0.0024	7.2996			6.3291	0.0000			453.95	500		44	500	34	0.0028	7.5145	L	6.0628	0.0022	500	44
443.21	380			380	78	0.0009	7.3005			6.3291	0.0000			454.01	300			300	34	0.0017	7.5162		6.0607	0.0006	135	D 48
443.21	20			20	78	0.0000	7.3006			6.3291	0.0000			454.01		XING					7.5162	PA HWY 68	6.0601	0.0007	165	48
443.27	300		113	300	78	0.0007	7.3013	L		6.3291	0.0000			454.04	200			200	34	0.0011	7.5173		6.0594	0.0006	200	48
443.32	300			300	78	0.0007	7.3020	REVERSE CURVE		6.3291	0.0000			454.12	400		48	400	34	0.0022	7.5196	R	6.0587	0.0016	400	48
443.33	40			40	78	0.0001	7.3021	WEST ECONOMY		6.3291	0.0000			454.18	300			300	34	0.0017	7.5212		6.0571	0.0040	1329	D 77
443.37	200		113	200	78	0.0005	7.3026	R		6.3291	0.0000			454.18							7.5212	VANPORT C&P JCT	6.0530	0.0001	46	77
443.51	760			760	78	0.0018	7.3045			6.3291	0.0000			454.25	375			375	34	0.0021	7.5233		6.0529	0.0000		77
443.65	720			720	78	0.0017	7.3062			6.3291	0.0000			454.38	700	XING	3.04	700	34	0.0039	7.5272	PRIVATE	6.0529	0.0000	0	77
443.69	200		113	30	78	0.0001	7.3063	L		6.3291	0.0000			454.82	2300		89	2300	34	0.0128	7.5400	L	6.0529	0.0057	2300	77
443.70	80					0.0000	7.3063	REVERSE CURVE		6.3291	0.0000			454.98	879			879	34	0.0049	7.5449	WB MORE TO WEST	6.0473	0.0022	879	77
443.73	150					0.0000	7.3063			6.3291	0.0000			455.10	800		113	1511	34	0.0084	7.5533	L MPH50(TT)	6.0451	0.0015	800	77
443.79	300		113			0.0000	7.3063	R		6.3291	0.0000			455.25	800		113	688	A 34	0.0033	7.5567	R RC	6.0436	0.0020	800	77
443.89	550					0.0000	7.3063			6.3291	0.0169	9670	A 10	455.36	600			0	44	0.0000	7.5567		6.0417	0.0015	600	77
444.08	1000					0.0000	7.3063			6.3121	0.0000			455.53	900		104	701	44	0.0030	7.5597	R	6.0402	0.0022	900	77
444.19	590					0.0000	7.3063			6.3121	0.0000			455.89	1900			1900	44	0.0082	7.5678		6.0380	0.0047	1900	77
444.19				8964	A 78	0.0189	7.3252		19	6.3121	0.0000			456.06	900		79	746	44	0.0032	7.5711	R	6.0333	0.0022	900	77
444.27	380					0.0000	7.3252			6.3121	0.0000			456.25	978			340	A 44	0.0014	7.5724		6.0311	0.0024	978	77
444.27	30		141			0.0000	7.3252	L		6.3121	0.0000			456.36	600			560	48	0.0022	7.5747		6.0287	0.0015	600	77
444.46	1000		141			0.0000	7.3252	L ALIQUIPPA		6.3121	0.0000			456.49	700		77	4976	A 48	0.0147	7.5894	R	6.0272	0.0013	525	77
444.53	350		141			0.0000	7.3252	L		6.3121	0.0000			456.64	781		112		77	0.0000	7.5894	R	6.0259	0.0103	3718	A 59
444.65	650					0.0000	7.3252			6.3121	0.0000			457.25	3200				77	0.0000	7.5894		6.0156	0.0014	438	59
444.82	900					0.0000	7.3252			6.3121	0.0000			457.85	3200	XING	6.18	4700	77	0.0116	7.6009	PRIVATE	6.0142	0.0103	3200	59
444.84	100		100			0.0000	7.3252	R		6.3121	0.0000			458.08	1173		100	1173	77	0.0029	7.6038	R E IND	6.0040	0.0112	3491	59
444.95	600		100			0.0000	7.3252	R		6.3121	0.0000			458.25	900			900	77	0.0022	7.6060	INDUSTRY	5.9928	0.0008	254	A 57
445.03	400					0.0000	7.3252			6.3121	0.0022	1162	A 98	458.64	2100	XING	7.85	1746	77	0.0043	7.6103	WABASH STREET	5.9919	0.0014	428	57
445.20	900					0.0000	7.3252			6.3099	0.0000			458.85	1069		93	2132	A 77	0.0050	7.6153	L W IND	5.9905	0.0036	1069	57
445.22	100					0.0000	7.3252			6.3099	0.0000			459.25	2100	XING	8.08	100	84	0.0001	7.6155	PRIVATE	5.9869	0.0070	2100	57
445.27	250					0.0000	7.3252			6.3099	0.0000			459.25				1311	D 84	0.0035	7.6189		5.9799	0.0000		
445.36	600		100	136	100	0.0003	7.3254	L		6.3099	0.0000			459.40	800		59	800	59	0.0026	7.6215	L	5.9799	0.0027	800	57
445.41	150			150	100	0.0003	7.3257	WEST ALIQUIPPA		6.3099	0.0000			459.74	1800			1800	59	0.0058	7.6273		5.9773	0.0060	1800	57
445.60	1000			1000	100	0.0019	7.3276			6.3099	0.0000			459.96	1100		155	1100	59	0.0035	7.6308	L	5.9713	0.0037	1100	57
445.79	1000			1000	100	0.0019	7.3295			6.3099	0.0211	9561	A 71	460.19	1283			1283	59	0.0041	7.6349		5.9676	0.0043	1283	57
445.98	1000			1000	100	0.0019	7.3314			6.2888	0.0030	1121	71	460.25	300		70	300	59	0.0010	7.6359		5.9634	0.0010	300	57
446.13	800			800	100	0.0015	7.3329			6.2858	0.0021	800	71	460.42	900		70	900	59	0.0029	7.6388	R	5.9624	0.0030	900	57
446.16	200		100	200	100	0.0004	7.3333	L		6.2837	0.0017	640	71	460.65	1200			1115	59	0.0036	7.6424		5.9594	0.0040	1200	57
446.20	210			210	100	0.0004	7.3337	WB MORE TO WEST		6.2820	0.0006	207	A 70	460.65				85	D 56	0.0003	7.6426		5.9554	0.0000		
446.32	590		100	590	100	0.0011	7.3348	L		6.2814	0.0004	150	70	460.85	1100		57	1100	57	0.0037	7.6463	R	5.9554	0.0037	1100	57
446.35	200		122	200	100	0.0004	7.3352	L		6.2810	0.0005	200	70	460.85							7.6463		5.9517	0.0010	311	D 64
446.47	600		122	600	100	0.0011	7.3363	L		6.2805	0.0016	600	70	460.85						0.0000	7.6463		5.9508	0.0001	20	64
446.54	400			400	100	0.0008	7.3371			6.2790	0.0011	400	70	460.85							7.6463		5.9507	0.0033	1065	A 57
446.73	1000			1000	100	0.0019	7.3390	HOPEWELL TOWNSHIP		6.2778	0.0027	1000	70	461.02	900			900	57	0.0030	7.6493		5.9474	0.0008	247	A 55
446.75	100			100	100	0.0002	7.3392			6.2751	0.0003	100	70	461.25	1175		89	1175	57	0.0039	7.6532	R	5.9465	0.0015	432	55
446.92	900		138	900	100	0.0017	7.3409	L		6.2748	0.0024	900	70	462.25	5278		75	5278	57	0.0175	7.6707	R MIDLAND	5.9450	0.0253	7345	55
446.98	300		138	655	100	0.0012	7.3421	L		6.2723	0.0008	300	70	462.30	290		114	290	57	0.0010	7.6717	R	5.9197	0.0014	392	A 52
447.07	500			145	D 1[illegible]	0.0003	7.3424	REVERSE CURVE		6.2715	0.0014	500	70	463.04	3900			3900	57	0.0130	7.6846		5.9184	0.0069	1682	52
447.11	200		98	200	98	0.0004	7.3428	R		6.2702	0.0005	200	70	463.25	1100		74	1100	57	0.0037	7.6883	L	5.9115	0.0036	896	A 42
447.20	480		98	480	98	0.0009	7.3437	R		6.2696	0.0013	480	70	463.25				82	D 57	0.0003	7.6886		5.9079	0.0000		
447.30	520		98	520	98	0.0010	7.3447	R		6.2683	0.0014	520	70	463.42	900		56	900	56	0.0031	7.6917	L MPH45	5.9079	0.0043	953	42
447.32	80		98	80	98	0.0002	7.3449	R		6.2669	0.0002	80	70	463.63	1100			1100	55	0.0038	7.6955		5.9036	0.0050	1100	42
447.49	920			920	98	0.0018	7.3466			6.2667	0.0025	920	70	463.74	600		100	600	55	0.0021	7.6975	L	5.8986	0.0027	600	42
447.68	1000			400	98	0.0008	7.3474			6.2642	0.0051	1879	70	463.89	800			800	55	0.0028	7.7003	MIDL YD LEAD	5.8959	0.0036	800	42

READ DOWN WESTBOUND TRAFFIC — APPENDIX B1 STATION — READ UP EASTBOUND TRAFFIC

PAGE 10 COLUMN 1

MILES	DISTANCE	WAYSIDE/XING	MAX MPH	DISTANCE	SPEED MPH	ET HOURS	TIME HOURS	STATION	TIME HOURS	ET HOURS	DISTANCE	SPEED MPH
464.01	600		76	600	56	0.0021	7.7024	L	5.8823	0.0027	600	42
464.11	569			569	56	0.0020	7.7043	WB MORE TO WEST	5.8896	0.0026	569	42
464.25	700		100	700	56	0.0024	7.7067	L	5.8870	0.0032	700	42
464.80	2900			2782	55	0.0096	7.7163	7.7792 18862	5.8839	0.0131	2900	42
464.80				116	D 55	0.0004	7.7187	5.869154895	5.8708	0.0000		
464.95	800		52	800	52	0.0029	7.7196	L'OKWB	5.8708	0.0036	800	42
465.17	900	XING 13.9		900	52	0.0033	7.7229	CHURCH STREET	5.8672	0.0041	900	42
465.25	678		80	678	52	0.0025	7.7254	RMPH45	5.8631	0.0031	678	42
465.28	200		80	200	52	0.0007	7.7261	R	5.8601	0.0009	200	42
465.32	200			118	52	0.0004	7.7265	MPH 45	5.8592	0.0008	200	42
465.32				345	D 52	0.0014	7.7279	REVERSE CURVE	5.8583	0.0000		
465.44	631		42	631	42	0.0026	7.7305	LRC	5.8583	0.0028	631	42
465.44							7.7308	LITTLE BEAVER	5.8554	0.0058	1940	D 84
465.50	300			300	42	0.0014	7.7321		5.8496	0.0004	161	84
465.61	600	XING	113	600	42	0.0027	7.7348	R' REVERSE PRIVATE	5.8492	0.0000		
465.73	600		113	600	42	0.0027	7.7375	L' CURVE	5.8492	0.0106	4265	A 68
466.24	2700			2700	42	0.0122	7.7497	OKWB OHIO/PA LINE	5.8387	0.0114	3374	A 42
467.25	5332	XING 15.1		5332	42	0.0240	7.7738	PUTNAM STREET	5.8273	0.0006	124	42
467.32	400	XING 15.2		400	42	0.0018	7.7756	MYRTLE STREET	5.8267	0.0018	400	42
467.37	258	XING	110	256	42	0.0012	7.7767	L WALTERS STREET	5.8249	0.0012	258	42
467.47	500		113	500	42	0.0023	7.7790	R' RC RC	5.8238	0.0023	500	42
467.67	1100	XING 16.5		312	42	0.0014	7.7804	MPH50(TT) VIRGINIA ST	5.8215	0.0050	1100	42
467.86	1000		96				7.7804	L	5.8165	0.0045	1000	42
468.07	1100		105			0.0000	7.7804	L	5.8121	0.0050	1100	42
468.24	900			6668	A 42	0.0197	7.8001		5.8071	0.0041	900	42
468.73	2600					0.0000	7.8001	US 30	5.8030	0.0117	2600	42
468.92	1000		107			0.0000	7.8001	L	5.7913	0.0045	1000	42
469.00	400			64	81	0.0001	7.8003	RC	5.7868	0.0018	400	42
469.00				710	D 81	0.0018	7.8021		5.7860	0.0000		
469.09	500		68	506	68	0.0014	7.8035	R	5.7850	0.0023	500	42
469.19	500		84	0	58	0.0000	7.8035	RMPH50(TT)	5.7827	0.0023	500	42
469.25	370			71	62	0.0002	7.8036		5.7805	0.0014	320	42
469.31	300				62	0.0000	7.8036		5.7790	0.0014	300	42
469.31				1049	D 64	0.0038	7.8073		5.7777	0.0000		
469.34	200	XING	42	200	42	0.0009	7.8082	R PRIVATE	5.7777	0.0008	200	42
469.34							7.8082		5.7768	0.0004	96	D 45
469.48	700			700	42	0.0032	7.8113		5.7764	0.0025	604	45
469.52	252	XING	45	252	42	0.0011	7.8125	R BROADWAY	5.7738	0.0011	252	45
469.52							7.8125		5.7726	0.0003	67	D 47
469.60	400	XING 18.4		400	42	0.0018	7.8143	TT MPH 35 WASHINGTON	5.7725	0.0000		47
469.71	600	XING	47	600	42	0.0027	7.8170	R MARKET STREET	5.7725	0.0024	600	47
469.71		XING 18.60					7.8170	JACKSON STREET	5.7701	0.0006	220	D 53
469.53	500			600	42	0.0027	7.8197	E LIVERPOOL	5.7692	0.0014	380	53
469.94	600		68	600	42	0.0027	7.8224	R	5.7679	0.0021	600	53
470.05	600			600	42	0.0027	7.8251	MPH35	5.7657	0.0021	600	53
470.17	600		71	600	42	0.0027	7.8278	R	5.7636	0.0021	600	53
470.24	400			400	42	0.0018	7.8296		5.7615	0.0014	400	53
470.30	300			300	42	0.0014	7.8309		5.7600	0.0011	300	53
470.40	500		121	500	42	0.0023	7.8332	L	5.7590	0.0018	500	53
470.57	900			900	42	0.0041	7.8373		5.7572	0.0032	900	53
470.66	500		53	500	42	0.0023	7.8395	L	5.7540	0.0018	500	53
470.66							7.8396		5.7522	0.0019	816	D 57
470.77	600		71	600	42	0.0027	7.8422	L	5.7502	0.0002	64	67
470.96	1100			1100	42	0.0050	7.8472		5.7500	0.0000	1000	67
471.15	900		70	900	42	0.0041	7.8512	L	5.7494	0.0026	906	67
471.19	215			215	42	0.0010	7.8522		5.7469	0.0065	2311	67
471.76	3000			1233	42	0.0056	7.8578		5.7403	0.0044	1472	A 58
471.99	1200		71				7.8578	R	5.7359	0.0021	632	58
472.24	1337			266	A 42	0.0011	7.8593	MPH 50(TT)	5.7338	0.0087	2671	58
472.51	1400			897	45	0.0025	7.8618		5.7251	0.0036	1032	A 50
472.85	1800		71	138	A 45	0.0006	7.8624	L	5.7215	0.0032	834	50
473.02	900		67	865	47	0.0035	7.8659	L	5.7183	0.0034	900	50
473.23	1131		76	660	A 47	0.0025	7.8684	L'OKWB	5.7149	0.0043	1131	50
473.38	800			4340	53	0.0155	7.8839		5.7107	0.0030	806	50
473.56	900		180	2126	A 53	0.0067	7.8906	R	5.7076	0.0034	906	50
473.69	700			2857	67	0.0081	7.8986		5.7042	0.0027	706	50
473.80	600		82	500	67	0.0017	7.9003	L	5.7016	0.0023	600	50
473.95	800			368	67	0.0011	7.9014	MPH 50(TT)	5.6993	0.0036	800	50
473.95				413	D 67	0.0013	7.9026		5.6953	0.0000	0	50
474.19	1256		58	1256	58	0.0041	7.9067	R	5.6953	0.0056	1486	50
474.25	300			300	58	0.0010	7.9077		5.6906	0.0000		
474.34	500			500	58	0.0016	7.9094		5.6906	0.0050	1171	A 34
474.51	900		71	900	58	0.0029	7.9123	R	5.6854	0.0017	298	34
474.77	1380	XING 23.4		1063	56	0.0035	7.9158	16TH STREET	5.6837	0.0077	1380	34
474.77				317	D 56	0.0011	7.9169	WELLSVILLE	5.6760	0.0000		
474.94	900		50	900	50	0.0034	7.9203	L	5.6760	0.0050	900	34
475.08	700			700	50	0.0027	7.9229		5.6710	0.0039	700	34
475.21	700		88	700	50	0.0027	7.9256	L	5.6671	0.0039	705	34
475.25	200			200	50	0.0006	7.9263		5.6632	0.0011	200	34
475.40	800			800	50	0.0030	7.9294		5.6621	0.0045	800	34
475.57	900		60	900	50	0.0034	7.9328	L	5.6576	0.0050	900	34
475.83	1400			1400	50	0.0053	7.9381	WB MORE TO WEST	5.6526	0.0078	1400	34
476.02	1000		50	1000	50	0.0038	7.9419	L	5.6448	0.0056	1000	34
476.25	1222			1222	50	0.0046	7.9465		5.6392	0.0068	1222	34
476.35	500			500	50	0.0019	7.9484		5.6324	0.0028	500	34
476.48	600		80	450	50	0.0017	7.9501	L'OKWB	5.6297	0.0033	600	34
476.54	400			57	50	0.0002	7.9503	REVERSE CURVE	5.6263	0.0022	400	34
476.54				493	D 50	0.0022	7.9525		5.6241	0.0000		
476.67	700		34	700	34	0.0038	7.9564	R	5.6241	0.0039	700	34
476.67							7.9564		5.6202	0.0006	106	D 38
476.80	700	XING 25.7		700	34	0.0038	7.9603	TT MPH 35 PRIVATE	5.6196	0.0030	594	38
476.96	800		35	550	34	0.0045	7.9648	L YELLOW CREEK OKV	5.6167	0.0006	500	38
477.07	600			600	34	0.0033	7.9681	MPH 35 REVERSE CUR	5.6161	0.0003	57	D 40
477.25	952		40	980	34	0.0053	7.9734	R OKWB	5.6158	0.0071	1496	40
477.52	1400		40	1400	34	0.0078	7.9812	R OKWB	5.6088	0.0066	1400	40
477.72	1100			1100	34	0.0061	7.9874		5.6021	0.0007	156	D 45
477.82	496		71	495	34	0.0026	7.9901	L	5.6014	0.0102	2423	45
478.08	1400		71	1400	34	0.0076	7.9979	L	5.5912	0.0090	906	A 29
478.22	700			700	34	0.0039	8.0018		5.5862	0.0006	130	29
478.40	1000			1000	34	0.0056	8.0074		5.5854	0.0065	1000	29
478.63	1200		56	1200	34	0.0067	8.0141	L	5.5789	0.0078	1200	29
478.86	1200			653	34	0.0036	8.0177		5.5710	0.0078	1200	29
478.86		XING 27.8		264	A 34	0.0014	8.0191	DANDOS CROSSING	5.5632	0.0000		
479.27	2195		53	1236	38	0.0033	8.0224	R	5.5632	0.0143	2195	29
479.43	500		68	115	A 38	0.0006	8.0230	R	5.5488	0.0052	800	29
479.43				1927	45	0.0091	8.0321		5.5436	0.0000		
479.58	800		47	500	45	0.0038	8.0358	R	5.5436	0.0052	800	29
479.58				110	45	0.0006	8.0364		5.5384	0.0000		
479.73	800		45	375	A 45	0.0017	8.0381	RC	5.5384	0.0052	800	29
480.03	1600	XING 28.6		1915	45	0.0081	8.0462	HAMMONDSVILLE WAS	5.5332	0.0104	1600	29
480.16	700		73	841	45	0.0035	8.0497	L	5.5227	0.0046	700	29
480.27	575		47	434	D 45	0.0022	8.0519	L	5.5181	0.0038	575	29
480.44	900		29	900	29	0.0059	8.0578	L	5.5144	0.0059	900	29
480.49	225			225	28	0.0015	8.0593	REVC	5.5085	0.0001	22	D 30

PAGE 10 COLUMN 2

MILES	DISTANCE	WAYSIDE/XING	MAX MPH	DISTANCE	SPEED MPH	ET HOURS	TIME HOURS	STATION	TIME HOURS	ET HOURS	DISTANCE	SPEED MPH
480.49				5		0.0000	8.0593		5.5084	0.0013	203	30
480.70	1100		30	1100	29	0.0072	8.0664	R	5.5071	0.0069	1100	30
480.87	900			907	29	0.0059	8.0723		5.5001	0.0010	172	D 37
481.13	1400		43	1400	29	0.0091	8.0815	L	5.4992	0.0109	2120	37
481.28	800	XING 30.0		800	29	0.0052	8.0867	EAST AVENUE	5.4883	0.0041	800	37
481.47	1000		46	1000	29	0.0065	8.0932	R	5.4842	0.0051	1000	37
481.66	1100			1100	29	0.0072	8.1004	IRONDALE	5.4791	0.0056	1100	37
481.85	1300		70	1300	29	0.0085	8.1089	L	5.4734	0.0067	1300	37
482.17	1300		50	1300	29	0.0085	8.1174	L	5.4668	0.0067	1300	37
482.28	900			900	29	0.0039	8.1217		5.4601	0.0142	2783	37
482.34	300			300	29	0.0020	8.1237		5.4458	0.0006	107	A 36
482.74	2100		37	905	29	0.0064	8.1296	R	5.4451	0.0006	100	36
482.77	145			44	A 29	0.0003	8.1299	REVC	5.4448	0.0006	145	36
483.07	1600		37	1281	30	0.0081	8.1380	L'OKWB	5.4440	0.0084	1600	36
483.30	1200		61	397	A 30	0.0022	8.1402	L	5.4356	0.0063	1200	36
483.60	1600			3638	37	0.0202	8.1604	NEW SALISBURY	5.4293	0.0084	1600	36
483.77	900		68	900	37	0.0046	8.1650	R	5.4209	0.0047	900	36
483.90	700	XING	40	700	37	0.0036	8.1686	PRIVATE	5.4162	0.0037	700	36
484.13	1200		56	1200	37	0.0061	8.1747	L	5.4125	0.0063	1200	36
484.29	856			855	37	0.0044	8.1791		5.4062	0.0046	856	36
484.46	900			900	37	0.0046	8.1837		5.4017	0.0047	900	36
484.83	900		60	900	37	0.0046	8.1883	L	5.3969	0.0047	900	36
484.80	900			873	37	0.0045	8.1928		5.3922	0.0047	900	36
484.80				27	D 37	0.0001	8.1929		5.3875	0.0000		
485.05	1300		36	1300	36	0.0068	8.1997	R	5.3875	0.0068	1300	36
485.06						0.0000	8.1997		5.3806	0.0019	441	D 50
485.05						0.0000	8.1997		5.3787	0.0054	1415	50
486.09	100			190	36	0.0010	8.2007	REVC	5.3733	0.0050	672	A 42
486.22	700		50	700	36	0.0037	8.2044	L	5.3683	0.0050	1110	42
486.27	300			300	36	0.0016	8.2060		5.3633	0.0014	200	A 37
486.81	2800			2800	36	0.0147	8.2207		5.3619	0.0100	1945	37
486.01	1100		73	1101	36	0.0058	8.2265	R	5.3520	0.0005	55	A 36
486.29	1471			1471	36	0.0077	8.2342		5.3514	0.0033	633	36
486.39	500			500	36	0.0026	8.2368		5.3481	0.0026	500	36
486.56	900	XING	56	900	36	0.0047	8.2415	L MCGARY	5.3454	0.0047	900	36
487.03	2500	XING 35.7		2520	36	0.0132	8.2548	NOLAN	5.3407	0.0132	2500	36
487.27	1267		114	539	36	0.0028	8.2576	R	5.3276	0.0067	1267	36
487.29	100			1932	A 36	0.0076	8.2654		5.3209	0.0005	100	36
487.29				52	57	0.0002	8.2656		5.3204	0.0000		36
487.61	1700	XING 36.1		545	D 57	0.0021	8.2676	PRIVATE	5.3204	0.0089	1700	36
487.82	1100		42	1100	42	0.0050	8.2726	R	5.3114	0.0058	1100	36
487.88	300	XING 36.5		155	42	0.0007	8.2733	REV C HAYTI	5.3056	0.0016	300	36
487.88				145	D 40	0.0007	8.2740		5.3041	0.0000		36
488.24	1900		37	1900	37	0.0097	8.2837	L'ROGERS	5.3041	0.0100	1900	36
488.20	100			72	37	0.0004	8.2841	RC	5.2941	0.0005	100	36
488.28				27	D 37	0.0001	8.2842		5.2935	0.0000		36
488.27	100		36	100	36	0.0005	8.2848	R	5.2935	0.0005	100	36
488.55	1431		36	1431	36	0.0075	8.2923	R	5.2930	0.0075	1431	36
488.56		XING 37.20				0.0000	8.2923	MAIN STREET	5.2865	0.0000	111	D 40
488.64	500			500	36	0.0026	8.2949	RC SALINEVILLE	5.2848	0.0018	389	40
488.79	900		41	800	36	0.0042	8.2991	L	5.2831	0.0038	600	40
488.96	900			900	36	0.0047	8.3038		5.2793	0.0043	900	40
489.13	900		71	900	36	0.0047	8.3086	L	5.2750	0.0043	900	40
489.28	800			800	36	0.0042	8.3126		5.2708	0.0038	800	40
489.57	1500			1500	36	0.0078	8.3207		5.2670	0.0071	1500	40
489.70	700	XING	51	700	36	0.0037	8.3244	R FOUNDRY HILL RD	5.2598	0.0033	700	40
489.76	300			306	36	0.0016	8.3260	REVC	5.2566	0.0014	300	40
489.67	600		57	600	36	0.0032	8.3291	L	5.2552	0.0028	600	40
489.96	400			400	36	0.0021	8.3312	REVC	5.2523	0.0019	400	40
490.14	1000		40	1000	36	0.0053	8.3365	R	5.2504	0.0047	1000	40
490.16	116			116	36	0.0006	8.3371	REV C	5.2457	0.0009	188	D 46
490.27	600		46	600	36	0.0032	8.3403	L	5.2449	0.0022	527	46
490.43	1100		46	1100	36	0.0058	8.3460	L'MPH 35	5.2427	0.0045	1100	46
490.48		XING 38.33			36	0.0000	8.3460	DOBSON	5.2382	0.0010	264	D 53
490.52	208			208	36	0.0011	8.3471	REV C	5.2372	0.0002	54	53
490.95	700		60	576	36	0.0030	8.3502	R	5.2370	0.0021	600	53
490.80	800		56			0.0000	8.3502	R	5.2348	0.0073	2031	53
490.97	900			277	A 36	0.0014	8.3516		5.2276	0.0042	1060	A 42
491.28	1600		113	3147	40	0.0149	8.3665	L'MP	5.2234	0.0010	219	42
491.37	500		113	500	40	0.0024	8.3688	L40	5.2224	0.0064	1191	42
491.50	700			700	40	0.0033	8.3721		5.2171	0.0014	290	A 37
491.66	800	XING	50	800	40	0.0038	8.3759	R HULL ROAD	5.2157	0.0027	519	37
491.77	600			800	40	0.0028	8.3788	REVERSE CURVE	5.2130	0.0031	600	37
492.26	2615		57	2378	40	0.0112	8.3900	L'MPH 35(TT)	5.2099	0.0134	2613	37
492.41	786		57	0		0.0000	8.3900	L'SUMMITVILLE DIV P	5.1966	0.0038	768	37
492.56	800			442	A 40	0.0019	8.3920		5.1926	0.0041	800	37
492.56				1363	46	0.0056	8.3976		5.1885	0.0000		
492.88	1700		53	11	46	0.0000	8.3976	R	5.1885	0.0067	1700	37
493.09	500			728	A 46	0.0028	8.4004		5.1798	0.0031	800	37
493.11	600		81	2161	53	0.0077	8.4081	R	5.1768	0.0031	800	37
493.26	800			700	53	0.0026	8.4106		5.1737	0.0041	800	37
493.32	300			57	53	0.0001	8.4107		5.1696	0.0016	300	37
493.32				383	D 53	0.0015	8.4122		5.1681	0.0000		
493.57	1300		42	1300	42	0.0059	8.4181	L	5.1681	0.0067	1300	37
493.64	400			347	42	0.0015	8.4196	REV CUR TT MPH 35	5.1614	0.0020	400	37
493.64				58	D 42	0.0003	8.4199		5.1594	0.0000		
493.79	800		37	800	37	0.0041	8.4240	R OK	5.1594	0.0041	800	37
493.79						0.0000	8.4240	REVERSE CURVE	5.1553	0.0004	85	D 40
493.93	700		40	700	37	0.0036	8.4276	L	5.1549	0.0029	615	40
493.93						0.0000	8.4276		5.1519	0.0043	1262	D 71
494.04	600			600	37	0.0031	8.4306	SUMMITVILLE	5.1476	0.0001	48	71
494.10	310		71	310	37	0.0016	8.4322	L'W END 2 TRACKS	5.1475	0.0000	0	71
494.25	800			800	37	0.0041	8.4363	MPH 40(TT)	5.1475	0.0011	400	71
494.26	32			32	37	0.0002	8.4365		5.1464	0.0001	32	71
494.35	500		71	500	37	0.0026	8.4390	R	5.1464	0.0010	300	71
495.26	4800	XING 43.4		4800	37	0.0246	8.4636	PRIVATE	5.1452	0.0279	10433	71
495.26						0.0000	8.4636		5.1172	0.0006	207	A 70
496.28	5280			3258	37	0.0167	8.4803		5.1166	0.0060	2206	70
496.98	3800	XING 46.1		170	A 37	0.0005	8.4811		5.1107	0.0067	2075	A 58
497.19	1100		71	536	40	0.0025	8.4836	L	5.1040	0.0002	60	58
497.27	450			12437	A 40	0.0136	8.4972		5.1038	0.0015	460	58
498.26	5200			383	71	0.0010	8.4983		5.1023	0.0170	5200	58
498.45	1000			52	D 71	0.0001	8.4984		5.0853	0.0033	1000	58
498.60	800		70	800	70	0.0023	8.5006	L	5.0821	0.0026	800	58
499.26	3480			2917	70	0.0079	8.5085		5.0794	0.0114	3480	58
498.28				563	D 70	0.0017	8.5102		5.0681			
498.48	1200		58	1206	58	0.0039	8.5141	L'KENSINGTON	5.0681	0.0039	1200	58
498.48						0.0000	8.5141		5.0642	0.0018	615	D 71
499.08	1000			1000	58	0.0032	8.5173	MPH 40(TT)	5.0624	0.0010	385	71
499.85	900		71	900	58	0.0029	8.5203	R	5.0613	0.0024	900	71
500.15	1600			1600	58	0.0052	8.5256		5.0589	0.0043	1600	71
500.25	524		71	524	58	0.0017	8.5272	R	5.0547	0.0014	524	71
500.28	185		71	185	58	0.0006	8.5278	R	5.0533	0.0005	185	71
500.28						0.0000	8.5278		5.0525	0.0038	1673	D 98

READ DOWN WESTBOUND TRAFFIC — APPENDIX B1 STATION — READ UP EASTBOUND TRAFFIC

PAGE 11 COLUMN 1

MILES	DISTANCE, ft	HWY GRADE CROSSING	MAX MPH	DISTANCE, feet	SPEED, mph	ΔT, hours	TIME, hours	STATION	TIME, hours	ΔT, hours	DISTANCE, feet	SPEED, mph
501.25	5100			5100	58	0.0167	8.5445	MP	5.0490	0.0006	260	98
502.10	4500	XING 51.1		1691	58	0.0056	8.5500	BROGAN ROAD 50	5.0485			
502.16	300	XING 122				0.0000	8.5500	L LIPPINCOT ROAD	5.0485	0.0153	7223	A 80
502.25	476	XING 51.8		2262	A 56	0.0067	8.5567	PRIVATE	5.0332			
503.26	5342	XING 52.2		1927	71	0.0051	8.5618	PRIVATE	5.0332	0.0225	7001	A 30
503.41	800					0.0000	8.5618		5.0107	0.0023	361	30
503.51	500		200			0.0000	8.5618	R	5.0085	0.0032	500	30
504.22	3752			10713	A 71	0.0130	8.5757		5.0053	0.0237	3752	30
504.39	900					0.0000	8.5757	MPH 40	4.9816	0.0057	900	30
504.69	1800	XING 100		1037	100	0.0020	8.5777	R US HWY 30	4.9759	0.0101	1600	30
504.84	800			1320	D 10	0.0028	8.5804	E ROCHESTER	4.9658	0.0061	800	30
505.23	2073		80	1070	80	0.0025	8.5830	L	4.9608	0.0131	2073	30
505.29	300		80	1760	D 80	0.0056	8.5885	L	4.9477	0.0019	300	30
505.43	714			0	40	0.0000	8.5885	BAYARD BRANCH	4.9458	0.0045	714	30
505.43		JCT 54.6		257	D 40	0.0014	8.5899	JCT MAHONING SECO	4.9413	0.0000	0	30
505.87	2320		30	2320	30	0.0146	8.6046	R BAYARD	4.9413	0.0146	2320	30
505.87		XING 54.71				0.0000	8.6046	BAYARD ROAD	4.9267	0.0000		30
505.87	3			3	30	0.0000	8.6046	RC JCT TUSC SEC	4.9267	0.0000	3	30
505.98	600		113	600	30	0.0038	8.6084	L	4.9266	0.0038	600	30
506.87	4683	XING 55.9		4683	30	0.0296	8.6379	RINKER RD	4.9228	0.0296	4683	30
507.24	1945			1945	30	0.0123	8.6502	USE CR SPEEDS BAYA	4.8933	0.0123	1945	30
507.37	700		141	700	30	0.0044	8.6546	L	4.8810	0.0044	700	30
507.52	785	XING 56.5		785	30	0.0050	8.6596	SANDY SPRING	4.8766	0.0050	785	30
507.62	545		46	545	30	0.0034	8.6630	L	4.8716	0.0034	545	30
507.64	120			120	30	0.0008	8.6638	TT 30 MPH BAYARD-ALL	4.8682	0.0008	120	30
507.73	480		57	480	30	0.0030	8.6668	R	4.8674	0.0030	480	30
507.87	700			700	30	0.0044	8.6712		4.8644	0.0044	700	30
508.02	800			800	30	0.0051	8.6763		4.8600	0.0051	800	30
508.23	1100		89	1100	30	0.0019	8.6782	R	4.8549	0.0069	1100	30
508.87	3382	XING 57.9		3382	30	0.0214	8.6996	SR 712	4.8480	0.0000	3382	30
509.87	5283			5283	30	0.0334	8.7329		4.8480	0.0334	5283	30
510.05	960			960	30	0.0061	8.7390		4.8146	0.0061	960	30
510.24	1010		100	1010	30	0.0064	8.7454	R	4.8086	0.0064	1010	30
510.66	2147			2147	30	0.0136	8.7589		4.8022	0.0106	2147	30
510.85	1090		100	1090	30	0.0069	8.7658	L	4.7916	0.0069	1090	30
511.08	1200		100	1200	30	0.0076	8.7734	L	4.7848	0.0200	1200	30
512.02	4984			4984	30	0.0315	8.8048		4.7648	0.0315	4984	30
512.84	4300			4300	30	0.0271	8.8320		4.7333	0.0271	4300	30
512.95	588		89	588	30	0.0037	8.8357	R	4.7061	0.0037	588	30
513.03	398			398	30	0.0025	8.8382		4.7024	0.0025	398	30
514.01	5200			5200	30	0.0328	8.8710		4.6999	0.0328	5200	30
514.03	78		57	78	30	0.0005	8.8715	R	4.6671	0.0005	78	30
514.07	240		57	240	30	0.0015	8.8730	L REVERSE CURVE	4.6666	0.0015	240	30
514.40	1726	XING 63.0		1726	30	0.0109	8.8839	PRIVATE	4.6651	0.0109	1726	30
514.53	720	XIG 6	100	720	30	0.0045	8.8865	L PRIVATE	4.6542	0.0045	720	30
515.03	2600			2600	30	0.0164	8.9049	TIMETABLE 30 MPH	4.6496	0.0164	2600	30
515.26	1248			1248	30	0.0079	8.9128	BAYARD - ALLIANCE	4.6332	0.0136	1248	30
515.47	1100		100	1100	30	0.0069	8.9197	L	4.6196	0.0103	1100	30
516.02	2900			2900	30	0.0183	8.9380		4.6093	0.0183	2900	30
516.32	1600		100	1600	30	0.0101	8.9481	L	4.5910	0.0101	1600	30
517.03	3712	XING 66.17		3712	30	0.0225	8.9706	MAHONING AVE	4.5809	0.0234	3712	30
517.97	5000	XING 66.4		5000	30	0.0316	9.0022	SUMMIT STREET	4.5575	0.0316	5000	30
518.03	274	XING	71	274	30	0.0017	9.0039	L BROADWAY	4.5259	0.0017	274	30
518.03			50				9.0039	L JCT ALLIANCE	4.5242	0.0028	587	D 50
518.14	600		50	600	30	0.0038	9.0077	60	4.5214	0.0001	13	50
[illegible]	920	XING 127		920	30	0.0058	9.0135	L FREEDOM	4.5214	0.0014	403	D 81
518.48	880	XING 100		880	30	0.0056	9.0191	L MECHANIC AVENUE	4.5200	0.0044	1397	60
518.83	1860			1860	30	0.0117	9.0308	60 PER CR ALLI-FAIRH	4.5156	0.0058	1860	60
519.23	2090	XING 84.0		2090	30	0.0132	9.0440	LINCOLN AVE	4.5097	0.0086	2090	60
519.45	1164		100	1164	30	0.0073	9.0514	L	4.5031	0.0037	1164	60
519.84	2070	XING 84.7		2070	30	0.0131	9.0644	KEITERS CROSSING	4.4994	0.0065	2070	60
520.39	2915	XING 85.5		2016	30	0.0127	9.0772	PRIVATE	4.4929	0.0092	2915	60
520.84	2340		81	1359	A 30	0.0064	9.0836	L	4.4837	0.0074	2340	60
520.99	800		81	1292	A 50	0.0044	9.0880	L	4.4763	0.0025	800	60
521.84	4481	XING 86.5		5889	60	0.0186	9.1066	FRESHLY RD	4.4738	0.0141	4481	60
521.94	546			546	60	0.0017	9.1082		4.4596	0.0017	546	60
522.09	800		115	800	60	0.0025	9.1108	L	4.4579	0.0025	800	60
522.69	3180			3180	60	0.0100	9.1208		4.4554	0.0100	3180	60
522.84	780	XING	68	780	60	0.0025	9.1233	L PRIVATE	4.4454	0.0024	760	60
523.06	1200		68	1200	60	0.0038	9.1271	L	4.4430	0.0038	1200	60
523.42	1900	XING 88.4		1900	60	0.0060	9.1331	BEECH RD	4.4392	0.0060	1900	60
523.59	900		100	900	60	0.0028	9.1359	L	4.4332	0.0028	900	60
523.84	1300	XING 88.9		1300	60	0.0041	9.1400	PRIVATE	4.4303	0.0041	1300	60
524.07	1210		68	1210	60	0.0038	9.1438	R	4.4262	0.0038	1210	60
524.32	1342	XING	100	1342	60	0.0042	9.1481	R EASTON ST	4.4224	0.0042	1342	60
524.32		XING 90.15			60	0.0000	9.1481	PRIVATE	4.4182	0.0000		60
525.33	5311	XING 90.4		5311	60	0.0168	9.1648	PRIVATE	4.4182	0.0168	5311	60
525.34	60	XING 90.8		60	60	0.0002	9.1650	PARKS ROAD	4.4014	0.0002	60	60
525.52	940		107	940	60	0.0030	9.1680	R	4.4012	0.0030	940	60
525.66	750	XING 91.2		750	60	0.0024	9.1703	PARIS AVENUE	4.3982	0.0024	750	60
526.02	1880		70	1880	60	0.0059	9.1763	R	4.3959	0.0059	1880	60
526.25	1244			1244	60	0.0039	9.1802		4.3899	0.0039	1244	60
526.33	400		141	400	60	0.0013	9.1815	R	4.3860	0.0013	400	60
526.54	1106		141	1106	60	0.0035	9.1850	R ESHELMAN ROAD	4.3848	0.0035	1106	60
527.10	2980			2980	60	0.0094	9.1944	PRIVATE	4.3613	0.0094	2980	60
527.34	1268		83	1268	60	0.0040	9.1984	R MAPLE GROVE	4.3719	0.0040	1268	60
527.39	269		83	269	60	0.0008	9.1992	R	4.3679	0.0008	269	60
527.76	1924	XING 93.3		1924	60	0.0061	9.2053	MEEGE ROAD	4.3670	0.0061	1924	60
527.98	1180		100	1180	60	0.0037	9.2090	L	4.3609	0.0037	1180	60
528.33	1860			1860	60	0.0059	9.2149		4.3572	0.0059	1860	60
528.67	1755	XING 94.0		1755	60	0.0056	9.2204	NICKEL PLATE ROAD	4.3513	0.0081	2556	60
528.91	1280		70	1280	60	0.0040	9.2245	L	4.3433	0.0044	1292	A 50
529.33	2227	XING 94.7		2227	60	0.0070	9.2315	CHAPPEL STREET	4.3388	0.0054	1416	50
529.58	1330			1330	60	0.0042	9.2357	MAIN STREET LOUISVI	4.3335	0.0050	1330	50
529.85	1430		68	1430	60	0.0045	9.2402	R	4.3284	0.0054	1430	50
530.34	2560	XING 95.5		2560	60	0.0081	9.2483	PRIVATE	4.3230	0.0097	2560	50
530.53	989			989	60	0.0031	9.2514	FAIRHOPE	4.3133	0.0037	989	50
530.74	1120		93	2852	60	0.0090	9.2604	R	4.3096	0.0042	1120	50
531.15	2156			403	D 60	0.0014	9.2618		4.3053	0.0082	2156	50
531.24	500	JCT 6	50	500	50	0.0019	9.2637		4.2972	0.0019	500	50
531.31	376		50	376	50	0.0014	9.2651		4.2953	0.0014	376	50
531.41	500	XING 96.5		2702	A 50	0.0086	9.2737	BECK ROAD	4.2938	0.0060	2256	D 93
531.52	600	XING 97.13			68	0.0000	9.2737	BROADWAY	4.2879	0.0000	0	93
531.56	300		93		68	0.0000	9.2737	L	4.2879	0.0000	0	93
531.63	300		93		68	0.0000	9.2737	L	4.2879	0.0001	26	93
531.97	1753			1753	68	0.0049	9.2786		4.2878	0.0004	206	D 96
532.17	1100		96	98	68	0.0003	9.2789	L	4.2874	0.0047	2400	96
532.42	1300			7780	A 68	0.0181	9.2969		4.2827	0.0000		
533.42	5288				93	0.0000	9.2969		4.2827	0.0064	9168	A 66
534.42	5280	XING 100.		367	93	0.0007	9.2977	CARNAHAN AVENUE	4.2763	0.0066	2364	68
534.42		XING 100.		1388	A 93	0.0028	9.3005	8TH STREET NE	4.2697			
534.42		XIN 100.77		1465	96	0.0029	9.3034	6TH STREET NE	4.2697			
534.42		XING 100.		6719	A 96	0.0125	9.3159	5TH STREET	4.2697			
535.41	5236	XING 100.		0		0.0000	9.3159	ROWLAND AVENUE	4.2697	0.0146	5236	68
535.72	1600	XING 101.		283	107	0.0005	9.3164	2ND AVENUE NE	4.2551	0.0045	1600	68
535.72		XING 101		1228	D 10	0.0024	9.3187	TUSC AVE	4.2507			

READ DOWN WESTBOUND TRAFFIC — APPENDIX B1 STATION — READ UP EASTBOUND TRAFFIC

PAGE 11 COLUMN 2

MILES	DISTANCE, ft	HWY GRADE CROSSING	MAX MPH	DISTANCE, feet	SPEED, mph	ΔT, hours	TIME, hours	STATION	TIME, hours	ΔT, hours	DISTANCE, feet	SPEED, mph
536.72		XING	90		90	0.0167	9.3354	W&LE WANDLE 3RD S	4.2507	0.0167	RR CROSSI	
536.06	1800	XING 101.		1275	D 90	0.0031	9.3384	CANTON	4.2340	0.0050	1800	68
536.32	1400	XING	68	1400	68	0.0039	9.3423	R MCKINLEY	4.2290	0.0030	1400	68
536.44	610	XING 102.		610	68	0.0017	9.3440	CLEVELAND AVENUE	4.2251	0.0015	593	D 79
536.56	634	XING 102.		634	68	0.0018	9.3458	HIGH STREET	4.2236	0.0016	551	79
537.43	4600	XING 102.		4600	68	0.0126	9.3586	PROSPECT AVENUE	4.2220	0.0110	4600	79
538.43	5306	XING 102		5156	58	0.0144	9.3730	DUEBER AVENUE	4.2110	0.0127	5306	79
538.43		XING 103		2707	A 68	0.0069	9.3799	CLAREDON AVENUE	4.1982	0.0027	1120	79
538.43		XING 103.57					9.3799	MARYLAND AVENUE	4.1956	0.0069	2707	A 68
538.72	1500	XING	79			0.0000	9.3799	R'RAFF RD	4.1886	0.0036	1233	A 61
539.43	3780			2723	79	0.0065	9.3865		4.1850	0.0007	220	61
540.12	3600	XING 105.		3600	79	0.0086	9.3951	PERRY DRIVE	4.1843	0.0112	3600	61
540.27	800		141	800	79	0.0019	9.3970	L	4.1731	0.0103	3315	61
540.44	897			897	79	0.0022	9.3992		4.1628	0.0011	352	A 59
540.59	800			207	79	0.0005	9.3996		4.1617	0.0019	2033	59
540.59				593	D 75	0.0015	9.4012		4.1598			
540.87	1500	XING	68	1500	68	0.0042	9.4054	L GENOA STREET	4.1598	0.0053	1467	A 46
541.44	2991			2800	68	0.0078	9.4132		4.1545	0.0078	1895	46
541.48	192			52	68	0.0001	9.4133		4.1467	0.0008	192	46
541.48				331	D 68	0.0010	9.4143		4.1459			
541.74	1400		61	1400	61	0.0043	9.4186	R	4.1459	0.0014	322	A 42
542.06	1700			1700	61	0.0053	9.4239		4.1445	0.0032	704	42
542.14	400		100	400	61	0.0012	9.4251	L	4.1414	0.0018	400	42
542.44	1600			1600	61	0.0050	9.4301		4.1396	0.0072	1600	42
542.67	1187			1187	61	0.0037	9.4338		4.1323	0.0054	1187	42
542.76	500		100	500	61	0.0016	9.4353	L	4.1270	0.0023	500	42
542.89	700			612	61	0.0019	9.4372		4.1247	0.0032	700	42
542.89				88	D 61	0.0003	9.4375		4.1216			
543.20	1600	XING	59	1600	59	0.0051	9.4427	R'TTMPH 50 HIGH STF	4.1216	0.0072	1600	42
543.44	1300			1300	59	0.0042	9.4468	REED	4.1144	0.0059	1300	42
543.56	600			100	58	0.0003	9.4471		4.1085	0.0027	600	42
543.56				501	D 56	0.0018	9.4490		4.1058			
543.92	1909		46	1909	46	0.0079	9.4568	L	4.1058	0.0086	1909	42
544.01	500			371	46	0.0015	9.4583	REVERSE CURVE	4.0972	0.0023	500	42
544.01				129	D 46	0.0006	9.4589		4.0949			
544.43	2200		42	2200	42	0.0099	9.4688	R	4.0949	0.0099	2200	42
544.51	400		42	400	42	0.0018	9.4706	R	4.0850	0.0018	400	42
544.51							9.4706		4.0832	0.0033	960	D 66
544.71	1100			1100	42	0.0050	9.4756	MASSILLON	4.0799	0.0004	150	66
544.86	700		100	700	42	0.0032	9.4787	R	4.0794	0.0020	700	66
545.19	1800			1800	42	0.0081	9.4869		4.0774	0.0052	1800	66
545.39	1061		68	1061	42	0.0048	9.4916	L	4.0723	0.0030	1061	66
545.39					B&C	0.0000	9.4916	MACE TOWER	4.0692		RR CROSSI	
545.46	400			400	42	0.0018	9.4934		4.0692	0.0011	400	66
545.88	2200	XING 111.		2200	42	0.0099	9.5034	EARL ROAD	4.0681	0.0063	2200	66
546.20	1700		66	1700	42	0.0077	9.5110	L TTMPH 70/60/50	4.0618	0.0049	1700	66
546.20							9.5110		4.0569	0.0013	465	D 75
546.32	594	XING 111.		594	42	0.0027	9.5137	RC BOYDS CRN	4.0556	0.0003	129	75
546.45	700		75	700	42	0.0032	9.5169	R	4.0553	0.0018	700	75
546.62	900		75	745	42	0.0034	9.5202	R	4.0535	0.0023	900	75
546.62				2981	A 42	0.0103	9.5306		4.0513	0.0006	226	D 79
546.83	1103	XING 112.36				0.0000	9.5306	BEAUMONT AVENUE	4.0507	0.0021	877	79
547.26	2300		79	577	66	0.0017	9.5322	L	4.0486	0.0055	2300	79
547.45	1000			1000	66	0.0029	9.5350		4.0431	0.0024	1000	79
547.49	187			187	66	0.0005	9.5356		4.0407	0.0004	187	79
547.75	1400		79	1400	66	0.0040	9.5396	R	4.0403	0.0034	1400	79
548.45	3700			2816	66	0.0081	9.5477		4.0369	0.0024	1091	D 96
548.23	4100	XING 114.		1988	A 66	0.0053	9.5529	KENYON AVENUE	4.0345	0.0132	6709	96
548.42	1000		96	225	75	0.0006	9.5535	L	4.0213	0.0020	1000	96
549.44	135	XING 115.		1131	A 75	0.0028	9.5563	SR 93	4.0193	0.0006	287	D 10
550.45	5282	XING 116.		4859	79	0.0116	9.5679	CLERMONT	4.0188	0.0097	5130	100
551.20	4000	XING 116.		6451	A 79	0.0139	9.5818	ALABAMA AVE	4.0091	0.0076	4000	100
551.35	800		100	1558	96	0.0031	9.5849	L	4.0015	0.0050	2600	100
551.45	511	XING 117.		783	96	0.0015	9.5864	RAILROAD STREET	3.9965	0.0042	2157	A 96
552.45	5299			2157	A 96	0.0042	9.5906	NORTH LAWRENCE	3.9923	0.0036	1823	96
553.47	5377	XING 119.		8080	100	0.0153	9.6059	ANDERSON ROAD	3.9887	0.0106	5377	96
553.47				3162	A 10	0.0068	9.6117		3.9781			
553.47				162	105	0.0003	9.6120		3.9781			
554.19	3800	XING 119.		663	D 10	0.0013	9.6132	PRIVATE	3.9781	0.0075	3800	96
554.46	1422	XING	96	1422	96	0.0028	9.6161	L PRIVATE	3.9706	0.0028	1422	96
554.43	98	XING	96	98	96	0.0002	9.6162	L'MP 120 PRIVATE	3.9678	0.0002	98	96
555.47	5200	XING 120.		5200	96	0.0103	9.6265	SR 94	3.9676	0.0011	587	D 10
556.47	5297	XING 121.		5297	96	0.0105	9.6370	TANNERVILLE	3.9665	0.0003	181	104
557.19	3800			503	96	0.0010	9.6379		3.9662	0.0078	4128	A 97
557.19				2748	A 96	0.0053	9.6432		3.9584	0.0000		
557.19				259	101	0.0005	9.6437		3.9584	0.0028	1400	A 94
557.19				290	D 10	0.0006	9.6443		3.9556	0.0061	3000	94
557.36	900		97	900	97	0.0018	9.6460	R	3.9496	0.0172	7580	A 71
557.47	599	XING 123.		389	97	0.0008	9.6468	WALNUT	3.9323	0.0000		
557.76	1500	XING 123.		210	D 97	0.0004	9.6472	MRKET STREET	3.9323	0.0011	411	71
558.47	3789		94	3789	94	0.0076	9.6548	L	3.9313	0.0101	3789	71
558.97	2600		100	2600	94	0.0052	9.6601	L ORRVILLE	3.9211	0.0069	2600	71
558.48	2700			2700	94	0.0054	9.6655	(WLE OHIO)	3.9142	0.0072	2700	71
559.76	1500	XING 125		109	94	0.0002	9.6657	CROWN HILL	3.9070	0.0040	1500	71
559.76		XING 125		1392	D 94	0.0032	9.6689	SCHOOLHOUSE ROAI	3.9030			
560.03	1400		71	1400	71	0.0037	9.6727	R	3.9030	0.0037	1400	71
560.29	1400		108	1400	71	0.0037	9.6764	R	3.8993	0.0031	1323	D 93
560.49	1029		97	1029	71	0.0027	9.6791	R	3.8962	0.0003	148	93
560.73	1300	XING	105	1300	71	0.0035	9.6826	R MUSKRAT ROAD	3.8959	0.0000		
560.99	1365	XING	98	1365	71	0.0036	9.6862	R BACK ORRVILLE	3.8959	0.0000		
561.48	2600	XING 127.		2600	71	0.0069	9.6932	ROHRER ROAD	3.8959	0.0272	10625	A 48
562.49	5302	XING 128.		3306	71	0.0088	9.7020	PRIVATE	3.8687	0.0036	900	48
562.53	200	XING 128.		7168	A 71	0.0164	9.7184	APPLEGATE	3.8651	0.0008	200	48
563.49	5100		93	1026	93	0.0033	9.7217	L	3.8644	0.0201	5100	48
563.80	1600			1600	93	0.0033	9.7250		3.8442	0.0152	3854	46
563.97	900		95	800	93	0.0016	9.7266	L'MPH 50	3.8290	0.0011	274	A 45
564.10	700	XING 129.		0	93	0.0000	9.7266	CO RD 73	3.8279			
564.29	1000		104	0	93	0.0000	9.7266	R	3.8279	0.0003	72	45
564.40	600			74	93	0.0001	9.7267		3.8276	0.0025	600	45
564.40				2327	D 93	0.0063	9.7330		3.8251			
564.50	518		48	518	48	0.0020	9.7350	L'SMITHVILLE	3.8251	0.0181	4310	45
564.71	1123		48	1123	48	0.0044	9.7395	L	3.8069	0.0014	316	A 41
565.47	4000			4000	48	0.0158	9.7552		3.8056	0.0121	2615	41
565.51	187			187	48	0.0007	9.7560	TTMPH 35	3.7935	0.0009	187	41
566.81	1600		50	1600	48	0.0063	9.7623	R	3.7926	0.0074	1600	41
566.06	1300			1198	48	0.0047	9.7670		3.7852	0.0080	1300	41
566.05				102	D 48	0.0004	9.7674		3.7792			
566.47	2200		45	2200	45	0.0093	9.7767	L	3.7792	0.0102	2200	41
566.49	98		45	98	45	0.0004	9.7771	L	3.7691	0.0006	98	41
566.64	800			800	45	0.0034	9.7805		3.7686	0.0037	800	41
566.94	1600		49	1600	45	0.0067	9.7872	R	3.7649	0.0074	1600	41
567.06	600			474	45	0.0020	9.7892		3.7575	0.0028	600	41
567.06				126	D 45	0.0006	9.7898		3.7547			
567.46	2100		41	2100	41	0.0097	9.7995	R	3.7547	0.0097	2100	41
567.59	700		41	700	41	0.0032	9.8027	R	3.7450	0.0032	700	41
567.59							9.8027		3.7418	0.0004	94	D 44

READ DOWN WESTBOUND TRAFFIC — APPENDIX 61 STATION — READ UP EASTBOUND TRAFFIC

MILES	DISTANCE	HWY GRADE CROSSING	MAX MPH	DISTANCE	SPEED	ET HOURS	TIME HOURS	PAGE 12 COLUMN 1	TIME HOURS	ET HOURS	DISTANCE	SPEED	MILES
567.63	236			236	41	0.0011	9.8038	REVERSE CURVE	3.7414	0.0006	143	44	600.05
568.13	2600	XING	44	2600	41	0.0120	9.8158	L BEGIN 3 TR PRIVATE	3.7408	0.0112	2600	44	600.34
568.35	1200			1200	41	0.0056	9.8213		3.7296	0.0052	1200	44	600.34
568.49	700		45	700	41	0.0032	9.8246	L	3.7244	0.0030	700	44	600.55
568.62	700		45	700	41	0.0032	9.8278	L	3.7214	0.0030	700	44	600.89
568.66	218			218	41	0.0010	9.8288	REVERSE CURVE	3.7184	0.0009	218	44	601.04
568.75	500		44	500	41	0.0023	9.8311	R	3.7175	0.0022	500	44	601.24
568.75							9.8311		3.7153	0.0017	440	D 58	601.47
569.19	2300			2300	41	0.0106	9.8418		3.7136	0.0063	1860	58	601.54
569.34	800		57	800	41	0.0037	9.8454	L	3.7073	0.0027	800	56	601.89
569.45	600			600	41	0.0028	9.8482		3.7046	0.0020	600	56	601.83
569.55	500			500	41	0.0023	9.8506		3.7026	0.0017	500	56	602.38
569.78	1207	XING	56	646	41	0.0030	9.8535	L WOOSTER LIBERTY A	3.7009	0.0041	1207	56	602.45
569.78		XING	135.24				9.8535	SOUTH STREET	3.6968	0.0029	1024	D 77	602.54
569.85	1100			250	A 41	0.0011	9.8546		3.6939	0.0002	76	77	602.59
570.06	400		77	1611	44	0.0078	9.8624	R	3.6937	0.0010	400	77	602.77
570.38	1600			1600	44	0.0069	9.8693		3.6928	0.0007	282	D 82	602.77
570.48	600	XING	82	600	44	0.0026	9.8719	L PRIVATE	3.6921	0.0080	3483	82	602.77
570.49	86	XING	82	86	44	0.0003	9.8722	L PRAIRIE ROAD	3.6840	0.0019	827	A 79	603.16
571.33	4700			1827	44	0.0079	9.8800		3.6821	0.0015	705	79	603.54
571.51	700			1180	A 44	0.0044	9.8845		3.6802	0.0017	700	79	603.98
571.63	600		105	2993	56	0.0101	9.8946	R	3.6786	0.0014	600	79	604.24
571.63	35			35	56	0.0001	9.8947	REVERSE CURVE	3.6771	0.0001	36	79	604.37
571.75	600		105	600	56	0.0020	9.8968	L OK	3.6770	0.0014	600	79	604.51
572.47	3800			38	56	0.0001	9.8969		3.6756	0.0091	3800	79	604.54
573.27	4500			4138	A 56	0.0117	9.9086		3.6665	0.0108	4600	79	605.04
573.45	714		79	574	A 77	0.0014	9.9099	R KILLBUCK	3.6557	0.0017	714	78	605.47
573.45							9.9099		3.6540	0.0038	1632	D 10	605.47
574.21	4000			8264	78	0.0196	9.9287		3.6502	0.0014	507	105	606.54
574.42	1100		100	1100	78	0.0026	9.9324	R FISH	3.6488	0.0808	14701	A 70	606.54
574.45	134	XING	140.	134	79	0.0003	9.9327	MESSNER ROAD	3.6180	0.0054	1956	70	606.54
574.45						0.0000	9.9327	MP 140	3.6126	0.0042	1457	A 62	607.49
575.47	5405	XING	141.	5405	79	0.0130	9.9457	WILLOW ROAD	3.6084	0.0026	847	62	607.54
576.48	5352	XING	142.	361	79	0.0009	9.9466	FORCE ROAD	3.6058	0.0031	960	A 56	607.80
577.47	5232			6616	A 79	0.0181	9.9646		3.6028	0.0026	829	56	607.69
577.85	2000	XING	143.3			0.0000	9.9646	CEMETERY ROAD	3.5999	0.0019	549	A 51	608.54
578.08	1216		135			0.0000	9.9646	L	3.5980	0.0017	469	51	608.92
578.45	2000			5988	A 10	0.0108	9.9754		3.5963	0.0074	2000	51	609.04
578.69	1200		141			0.0000	9.9754	L	3.5888	0.0045	1200	51	609.15
578.82	700	XING	144.	173	109	0.0003	9.9757	WOOD	3.5844	0.0026	760	51	609.26
578.82		XING	144.	2586	D 10	0.0054	9.9811	ROBINSON	3.5818				609.55
579.14	1700	XING	70	1700	70	0.0046	9.9857	R MARKET STREET	3.5818	0.0063	1700	51	610.27
579.33	1000	XING	144.	1000	70	0.0027	9.9884	SHREVE MAIN STREET	3.5755	0.0037	1000	51	610.27
579.48	753		100	368	70	0.0010	9.9894	R	3.5718	0.0028	753	51	610.33
579.48				387	D 70	0.0011	9.9906		3.5690				610.57
579.72	1300		62	1300	62	0.0040	9.9945	R OK	3.5689	0.0048	1300	51	610.74
579.82	500			240	62	0.0007	9.9952	REVERSE CURVE	3.5641	0.0019	500	51	610.81
579.82				280	D 62	0.0008	9.9961		3.5623				611.05
580.08	1378		56	1182	58	0.0040	10.0001	L	3.5623	0.0051	1378	51	611.39
580.08				196	D 62	0.0007	10.0008		3.5572				611.57
580.34	1400		51	1400	51	0.0052	10.0060	L	3.5572	0.0052	1400	51	612.19
580.34							10.0060		3.5520	0.0008	238	D 57	612.57
580.44	500			500	51	0.0019	10.0078	RC	3.5511	0.0009	262	57	613.12
580.48	200		57	200	51	0.0007	10.0086	R	3.5502	0.0007	200	57	613.12
580.78	1601	XING	57	1601	51	0.0059	10.0145	R BROWNS	3.5498	0.0063	1601	57	613.12
581.48	3700	XING	147	3700	51	0.0137	10.0283	PRIVATE	3.5443	0.0042	1584	D 87	613.32
581.48							10.0283		3.5401	0.0007	306	87	613.32
581.48							10.0283		3.5394	0.0072	3136	A 77	613.44
582.31	4392	XING	148	4392	51	0.0163	10.0446	SCRUBYS	3.5321	0.0080	3277	77	613.57
582.65	1800	XING	145.	607	51	0.0023	10.0468	BIG PRAIRIE RTE 226	3.5241	0.0044	1728	A 70	613.76
582.82	867		100			0.0000	10.0468	L	3.5197	0.0020	723	70	613.91
583.46	3600			715	A 51	0.0025	10.0493		3.5177	0.0107	3972	70	614.02
583.54	277			1566	57	0.0053	10.0546		3.5069	0.0011	405	A 68	614.16
583.76	1306		100	3872	A 57	0.0111	10.0657	L	3.5058	0.0025	900	68	614.31
584.12	1800		77	2084	77	0.0066	10.0723	L	3.5033	0.0050	1800	68	614.55
584.26	700			700	77	0.0017	10.0740		3.4983	0.0019	700	68	614.54
584.48	1200		89	1200	77	0.0030	10.0770	R	3.4964	0.0028	1000	68	615.58
584.69	1100		82	1123	77	0.0026	10.0797	R	3.4936	0.0031	1100	68	616.09
584.77	400			377	D 77	0.0010	10.0807		3.4905	0.0067	2418	68	616.27
585.01	1300	XING	70	1300	70	0.0035	10.0842	R LAKEVILLE GULLS RC	3.4838	0.0064	2086	A 55	616.47
585.49	2530	XING	150.	3699	70	0.0100	10.0942	STATE ROUTE 179	3.4774	0.0045	1298	55	616.58
585.73	1300			101	D 70	0.0003	10.0945		3.4729	0.0045	1300	55	616.76
585.96	1217		68	1217	68	0.0034	10.0979	L	3.4684	0.0022	615	A 50	616.76
586.49	2800			2800	68	0.0078	10.1057		3.4662	0.0068	1799	50	616.76
587.49	5281			5401	68	0.0150	10.1207		3.4594	0.0200	5281	50	616.76
587.63	706			686	D 68	0.0018	10.1225		3.4394	0.0027	706	50	617.29
588.22	3127		56	3122	56	0.0108	10.1333		3.4367	0.0118	3122	50	618.57
588.24	92			92	56	0.0003	10.1336	REVERSE CURVE	3.4249	0.0003	92	50	619.80
588.24		XING	153.	193	D 77	0.0007	10.1343	R TR 368	3.4246	0.0000			620.61
588.49	1366		50	1366	50	0.0052	10.1395	L	3.4246	0.0052	1366	50	621.61
588.50	18			18	50	0.0001	10.1395		3.4194	0.0015	448	D 61	622.61
588.63	713			713	50	0.0027	10.1422		3.4179	0.0009	251	61	622.96
588.81	921		61	921	50	0.0035	10.1457	L	3.4170	0.0029	921	61	622.96
588.96	804		94	804	50	0.0030	10.1488	L	3.4141	0.0004	138	D 64	623.06
589.50	2850			2850	50	0.0108	10.1596		3.4137	0.0104	3517	64	623.23
590.39	4700	XING	156.	4700	50	0.0178	10.1774	TR 5	3.4033	0.0139	4700	64	623.23
590.46	554	XING	67	564	50	0.0021	10.1796	R LOUDONVILLE JEFFE	3.3894	0.0016	554	64	623.27
591.03	2848	XING	84	442	50	0.0017	10.1812	R SPRING STREET	3.3878	0.0084	2845	64	623.27
591.03				1469	A 50	0.0050	10.1862		3.3793	0.0018	872	D 77	623.38
591.51	2500			181	61	0.0006	10.1867	CP MOHICAN	3.3775	0.0045	1826	77	623.53
592.34	4400			459	A 61	0.0014	10.1881		3.3730	0.0108	4400	77	623.61
592.51	893	XING	77	7898	64	0.0234	10.2115	L PRIVATE	3.3622	0.0022	893	77	623.74
592.63	610		77	610	64	0.0018	10.2133	L	3.3600	0.0015	610	77	623.78
593.52	4700			2597	64	0.0077	10.2210		3.3585	0.0007	292	D 82	623.68
593.78	1303	XING	82	2907	A 64	0.0078	10.2287	L E PERRY PRIVATE	3.3578	0.0132	5712	82	624.61
594.52	4000			4499	77	0.0113	10.2398		3.3446	0.0019	914	D 96	624.92
594.52						0.0000	10.2398	MP 180	3.3427	0.0039	1964	98	625.03
594.71	1000	XING	160.	997	77	0.0025	10.2423	BRIDGE STREET	3.3388	0.0000			625.06
595.52	4264	XING	104	1401	A 77	0.0033	10.2456	R PRIVATE	3.3388	0.0186	8347	A 72	625.20
596.10	3100		96	96	82	0.0002	10.2458	R PERRYSVILLE	3.3202	0.0117	4447	72	625.20
596.53	2260	XING	162.	2260	82	0.0052	10.2510	PRIVATE	3.3085	0.0058	2054	A 61	625.88
597.31	4100	XING	162.	1742	82	0.0040	10.2551	TOWNSHIP ROAD 1037	3.3027	0.0031	1013	61	626.01
597.54	1197		96	5634	A 82	0.0119	10.2670	L	3.2996	0.0037	1197	61	626.18
597.60	500		96	3750	96	0.0074	10.2744	L	3.2968	0.0045	1441	61	627.01
597.82	1000			58	96	0.0001	10.2745		3.2914	0.0028	853	A 56	628.00
597.82				1478	D 96	0.0033	10.2779		3.2886				629.00
598.12	1600		72	1600	72	0.0042	10.2821	L	3.2886	0.0028	808	56	630.00
598.39	1390	XING	163.	1390	72	0.0037	10.2857	COUNTY ROAD 1075	3.2858	0.0046	1390	56	631.00
598.54	800	XING	96	800	72	0.0021	10.2878	L PRIVATE	3.2810	0.0028	800	56	631.96
598.69	800		96	800	72	0.0021	10.2899	L	3.2783	0.0028	800	56	633.00
599.20	2700			2164	72	0.0057	10.2956		3.2755	0.0093	2700	56	634.00
599.20				536	D 72	0.0015	10.2972	WB MORE TO EAST	3.2662	0.0000			634.96
599.41	1104		61	1104	61	0.0034	10.3006	R	3.2662	0.0038	1104	56	634.99
599.54	700			700	61	0.0022	10.3028		3.2624	0.0024	700	56	635.20
600.05	2700			2445	61	0.0076	10.3103		3.2600	0.0093	2700	56	635.36

READ DOWN WESTBOUND TRAFFIC — APPENDIX 61 STATION — READ UP EASTBOUND TRAFFIC

DISTANCE	HWY GRADE CROSSING	MAX MPH	DISTANCE	SPEED	ET HOURS	TIME HOURS	PAGE 12 COLUMN 2	TIME HOURS	ET HOURS	DISTANCE	SPEED
			255	D 96	0.0008	10.3112		3.2507	0.0000		
1500		56	1500	56	0.0052	10.3163	L	3.2507	0.0062	1500	56
					0.0000	10.3163		3.2455	0.0017	537	D 80
1102			1102	56	0.0036	10.3201		3.2439	0.0016	505	67
1800			1800	56	0.0062	10.3263		3.2423	0.0061	1500	67
600		69	806	55	0.0028	10.3291	L	3.2372	0.0023	800	67
1067	XING	168	1067	55	0.0037	10.3328	PRIVATE	3.2349	0.0030	1067	67
1200		69	1200	56	0.0041	10.3369	R	3.2319	0.0034	1200	67
400			400	56	0.0014	10.3383		3.2285	0.0011	400	67
800			800	56	0.0028	10.3410		3.2274	0.0023	800	67
700		72	700	56	0.0024	10.3434	R	3.2251	0.0020	700	67
2900	XING	167.	2900	56	0.0100	10.3534	PRIVATE	3.2231	0.0082	2900	67
400	XING	67	231	55	0.0008	10.3542	R LUCAS PERRYSVILL	3.2149	0.0011	400	67
457			1888	A 55	0.0058	10.3601		3.2138	0.0013	457	67
295						10.3601		3.2125	0.0008	295	67
900		67				10.3601	L	3.2117	0.0025	900	67
	XING	1168.3				10.3601	UNION STREET	3.2091	0.0006	202	D 71
	XING	168.46				10.3601	CHURCH	3.2086	0.0021	791	71
2100	XING	168.	2033	67	0.0057	10.3658	LUCAS HANLEY BRET	3.2065	0.0030	1107	71
2000		80	2000	67	0.0057	10.3715	R	3.2035	0.0053	2000	71
2200		100	2200	67	0.0062	10.3777	R	3.1982	0.0059	2200	71
1500			1500	67	0.0042	10.3819	CP LUCAS	3.1923	0.0040	1500	71
693		82	693	67	0.0020	10.3838	R	3.1883	0.0018	693	71
700		72	700	67	0.0020	10.3859	R	3.1866	0.0019	700	71
200			200	67	0.0006	10.3864		3.1846	0.0006	200	71
2600			1607	67	0.0045	10.3910		3.1841	0.0069	2600	71
2309		71	810	A 67	0.0022	10.3932	L	3.1771	0.0061	2300	71
			2483	71	0.0066	10.3998		3.1710	0.0030	1673	D 98
377	XING	171	377	71	0.0010	10.4008	PRIVATE	3.1672	0.0002	116	98
5281	XING	171	5281	71	0.0141	10.4149	SITTS ROAD	3.1670	0.0100	4884	A 87
						10.4149		3.1570	0.0038	1735	A 82
5000			5000	71	0.0133	10.4282		3.1531	0.0062	2240	82
279		113	342	71	0.0009	10.4291	L	3.1480	0.0006	279	82
300	XING	113			0.0000	10.4291	L RC ILLINOIS AVE	3.1472	0.0007	300	82
481		113	5805	A 71	0.0138	10.4428	R TT MPH 79	3.1466	0.0011	481	82
4500			1198	90	0.0025	10.4453		3.1455	0.0104	4500	82
2000			195	D 90	0.0004	10.4457		3.1351	0.0046	2000	82
800		87	880	87	0.0019	10.4476	R	3.1305	0.0014	600	82
600	XING	190	310	D 87	0.0007	10.4483	R 4TH STREET	3.1291	0.0014	600	82
600	XING	87	600	82	0.0014	10.4497	R 5TH STREET	3.1277	0.0014	600	82
1512	XING	175.	1512	82	0.0035	10.4532	MANSFIELD ORANGE	3.1263	0.0011	506	D 90
3800	XING	175.	3800	82	0.0088	10.4620	SURREY ROAD	3.1252	0.0016	729	90
		90				10.4626	B&O XING	3.1237	RR CROSSING 90 A		
300	XING	141	300	82	0.0007	10.4627	MAIN STREET	3.1237	0.0000		
1274	XING	175	1274	82	0.0029	10.4656	L W MANS LONGVIEW	3.1237	0.0000		
900			909	82	0.0021	10.4677		3.1237	0.0086	5107	A 58
400		141	406	82	0.0009	10.4686	L	3.1151	0.0000		
1268			1268	82	0.0029	10.4718		3.1151	0.0004	113	58
1850	XING	80	1546	82	0.0036	10.4751	L EMP STL MILL	3.1147	0.0059	1800	58
900	XING	177.	2503	A 82	0.0064	10.4815	PRIVATE	3.1088	0.0029	900	58
3300			1561	90	0.0033	10.4848		3.1059	0.0108	3300	58
1987		100	868	90	0.0018	10.4866	L	3.0951	0.0065	1987	58
2905			1343	A 93	0.0026	10.4894		3.0886	0.0095	2900	58
			748	93	0.0015	10.4909		3.0792	0.0000		
			1938	D 90	0.0040	10.4967		3.0792	0.0000		
1100		58	1100	58	0.0036	10.4993	R	3.0792	0.0036	1100	58
					0.0000	10.4993		3.0756	0.0031	1113	D 80
593			593	58	0.0019	10.5013	REVERSE CURVE	3.0725	0.0004	180	80
700		80	700	58	0.0023	10.5036	L	3.0721	0.0000	0	80
997		80	997	58	0.0033	10.5068	L	3.0721	0.0024	997	80
800			800	58	0.0026	10.5094		3.0697	0.0014	572	80
600		135	600	58	0.0020	10.5114	L	3.0684	0.0000		80
700	XING	179.	700	58	0.0023	10.5137	LEPPO ROAD	3.0684	0.0006	233	A 79
800		104	800	58	0.0026	10.5163	R	3.0678	0.0064	2505	A 69
1300	XING	180.	1300	58	0.0042	10.5205	SPRING MILL	3.0614	0.0024	889	69
1510		100	1510	58	0.0049	10.5255	L	3.0590	0.0041	1510	69
3500			3000	58	0.0098	10.5353		3.0549	0.0107	3600	69
2700			4610	A 58	0.0125	10.5478		3.0441	0.0074	2700	69
701		79	390	80	0.0008	10.5487	L	3.0367	0.0019	701	69
1300	XING	181.	601	D 80	0.0015	10.5502	PRIVATE	3.0348	0.0036	1300	69
600		69	600	69	0.0016	10.5519	L	3.0312	0.0016	600	69
927	XING	69	927	69	0.0025	10.5544	L ROCK ROAD	3.0296	0.0025	927	69
						10.5544		3.0271	0.0056	2611	D 10
						10.5544		3.0215	0.0007	390	102
						10.5544		3.0208	0.0400	18723	A 69
4400	XING	183.	4400	69	0.0121	10.5665	PRIVATE	2.9808	0.0026	818	60
4912	XING	184.	4912	69	0.0135	10.5800	SHELBY ROAD	2.9782	0.0028	836	A 54
5710			1688	69	0.0046	10.5846		2.9755	0.0018	527	54
9295	XING	186.	18076	A 69	0.0369	10.6215	HORNING ROAD	2.9736	0.0011	873	A 51
5308					0.0000	10.6215		2.9725	0.0031	844	51
5302	XING	188.07			0.0000	10.6215	COUNTY LINE	2.9694	0.0197	5302	51
1800	XING	188.	534	110	0.0009	10.6225	BUCYRUS STREET	2.9497	0.0067	1800	51
			3117	D 11	0.0068	10.6294		2.9430			
554	XING	60	554	60	0.0017	10.6311	R WASHINGTON	2.9430	0.0021	554	51
900			800	60	0.0026	10.6337	CRESTLINE	2.9409	0.0033	900	51
		90	RR XING	D6	0.1333	10.7670	CP CREST B4 XING	2.9376	0.1333	RR CROSSI	
205			49	60	0.0002	10.7672		2.8043	0.0007	205	51
			251	D 60	0.0008	10.7680		2.8036			
600		54	485	54	0.0017	10.7697	R	2.8036	0.0022	600	51
800			116	D 54	0.0004	10.7701		2.8013	0.0030	800	51
400		51	400	51	0.0015	10.7716	R	2.7983	0.0015	400	51
700		51	700	51	0.0026	10.7742	R	2.7968	0.0026	700	51
204			204	51	0.0008	10.7750	REVERSE CURVE	2.7942	0.0006	156	D 55
400		56	400	51	0.0015	10.7764	L	2.7937	0.0015	449	55
4006			4000	51	0.0149	10.7913		2.7921	0.0138	4006	55
1600			1600	51	0.0059	10.7972		2.7783	0.0055	1600	55
600		57	600	51	0.0022	10.7995	L	2.7728	0.0021	600	55
78			78	51	0.0003	10.7998	REV CUR TT40MPH	2.7708	0.0003	78	55
500		55	800	51	0.0030	10.8027	R	2.7705	0.0028	800	55
					0.0000	10.8027		2.7676	0.0090	4171	D 12
3600			3318	51	0.0123	10.8150		2.7587	0.0002	128	120
700	XING	133	468	A 51	0.0017	10.8167	R SR 598	2.7585	0.0000	0	120
900	XING	133	2214	56	0.0076	10.8243	R CRESTLINE ROAD	2.7585	0.0014	900	120
4330	XING	192.	4350	55	0.0151	10.8394	BIDDLE ROAD	2.7571	0.0068	4330	120
5182	XING	193.	620	55	0.0021	10.8415	BECK ROAD	2.7502	0.0028	1775	120
5288	XING	194.62			0.0000	10.8416	MAIN STREET	2.7474	0.0000		120
5284	XING	194	31026	A 56	0.0614	10.9029	WESTERN AVENUE	2.7474	0.0000		120
5289	XING	195.49				10.9029	OLENTANGY ROAD	2.7474	0.0000		
5092	XING	196.87			0.0000	10.9029	PRIVATE	2.7474	0.0387	22503	A 50
5470	XING	196.75			0.0000	10.9029	PRIVATE	2.7077	0.0272	9931	A 40
5275	XING	197.	5244	120	0.0083	10.9112	LWR LEESVILLE	2.6804	0.0127	2681	40
5254	XING	197.	3550	120	0.0056	10.9168	PRIVATE	2.6677	0.0249	5254	40
	XING	197.78		120		10.9168	MP 200 PRIVATE	2.6429			
1100	XING	198.	1991	D 12	0.0033	10.9202	PRIVATE	2.6429	0.0052	1100	40
965	XING	95	76	96	0.0001	10.9203	R ALBRIGHT	2.6377	0.0046	965	40

MILES	DISTANCE, ft	HWY/GRADE CROSSING	MAX MPH	DISTANCE, feet	SPEED, mph	dT, hours	TIME, hours	APPENDIX B1 STATION PAGE 13 COLUMN 1	TIME, hours	dT, hours	DISTANCE, feet	SPEED, mph
	READ DOWN WESTBOUND TRAFFIC								READ UP EASTBOUND TRAFFIC			
635.57	1000	XING 200.1		409	096	0.0058	10.9211	WHETSTONE STREET	2.6331	0.0047	1000	40
635.57					90	0.0667	10.9878	RR XING DELAY	2.6264	0.0667		
635.57			90	2383	D 90	0.0069	10.9947	COLSAN N&W XING	2.5617	0.0000	0	40
635.57		XING 200.79	40				10.9947	LANE STREET BUCYRU	2.5617	0.0000	0	40
635.78	1100	XING 63		1100	40	0.0052	11.0000	LTT 40 MPH WALNUT S	2.5617	0.0052	1100	40
635.78							11.0000	TRACK IN STREET	2.5565			
635.78		XING 200.93					11.0000	ALLEY TRACK IN GALEI	2.5565	0.0000	0	
635.88	500	XING 200.1		500	40	0.0024	11.0023	N SANDUSKY STATE 16	2.5565	0.0024	500	40
635.90	100		110	100	40	0.0005	11.0028	L TRACK IN GALEN STF	2.5541	0.0111	4693	D 12
635.99	500	XING 201.		500	40	0.0024	11.0052	RC TT 40 SPRING STRE	2.5430	0.0003	193	120
635.99		UG 20 TRACK IN TREET WALN					11.0052	SANDUSKY RIVER	2.5427			
636.20	1100	XING 104		1100	40	0.0052	11.0104	R SEARS STREET	2.5427	0.0000		
636.98	4186	XING 201.		4186	40	0.0198	11.0302		2.5427	0.0016	1000	120
637.99	5286	XING 203.		5280	40	0.0250	11.0552	McCRACKEN	2.5411	0.0063	5266	120
638.99	5280	XING 203.		428	40	0.0020	11.0572	SIMMS CR	2.5328	0.0083	5280	120
640.01	5390	XING 204.		17214	A 40	0.0414	11.0986	PRIVATE	2.5244	0.0085	5390	120
641.01	5272	XING 204.91					11.0986	GLENVILLE CR	2.5159	0.0083	5272	120
642.02	5298	XING 206.		15220	A 15	0.0256	11.1242	PRIVATE	2.5076	0.0084	5298	120
642.89	4593	XING 206.13				0.0000	11.1242	STRIEB CR KN	2.4993	0.0072	4593	120
644.01	5928	XING 207.10				0.0000	11.1242	MARION-MELMO	2.4920	0.0094	5928	120
644.01		XING 207.70				0.0000	11.1242	PRIVATE	2.4826			
644.01		XING 208.13				0.0000	11.1242	COUNTY LINE	2.4826			
644.01		XING 208.60				0.0000	11.1242	PRIVATE	2.4826			
645.04	5440	XING 209.		4339	120	0.0068	11.1310	NEVADA AYERS STREE	2.4826	0.0086	5440	120
645.04		XING 209.10					11.1310	MORRISON STREET	2.4741			
645.04		XING 209.27					11.1310	GOODBREAD STREET	2.4741			
646.04	5270	XING 210.		5270	120	0.0083	11.1394	EDENVILLE ROAD	2.4741	0.0061	3893	120
647.04	5290	XING 211.		5290	120	0.0083	11.1477	PRIVATE	2.4679	0.0000		
648.04	5289	XING 212.		5289	120	0.0083	11.1561	DOUGLAS RD	2.4679	0.0397	22564	A 90
649.04	5269	XING 213.		5269	120	0.0083	11.1644	PRIVATE	2.4282			
650.04	5280	XING 213.8		5280	120	0.0083	11.1727	ROCK RUN ROAD	2.4282	0.0047	2172	A 84
651.04	5284	XING 214.		3800	120	0.0088	11.1782	PRIVATE	2.4235	0.0070	3113	84
651.19	787	XING 141		0	120	0.0000	11.1782	L WILLIAMS CR	2.4165	0.0018	787	84
652.04	4500	XING 215.		0	120	0.0000	11.1782	PRIVATE	2.4147	0.0101	4500	84
652.04		XING 215.70					11.1782	PRIVATE	2.4045			
652.04		XING 216.10					11.1782	PRIVATE	2.4045			
652.04		XING 216.31					11.1782	KOINS RD	2.4045			
652.04		XING 216.89					11.1782	S 5TH STREET	2.4045			
652.04		XING 216.90					11.1782	S SANDUSKY	2.4045			
652.50	2440	XING 217.		0	120	0.0000	11.1782	UPPR SNDSKY 7TH	2.4045	0.0055	2440	84
652.50		XING 217.		9240	D 12	0.0167	11.1949	EIGHTH STREET	2.3990			
652.53	160	XING 217.34					11.1949	HAZEL STREET	2.3990	0.0004	160	84
652.53		XING 217.44			RR	0.0500	11.2449	S WARPOLE	2.3987	0.0500	RR XING SC	
652.53			90	48	90	0.0001	11.2450	USAND C&O XING	2.3487	0.0000	RR CROSSI	
652.53				383	D 90	0.0008	11.2458		2.3487			
652.62	479		84	479	84	0.0011	11.2469	L	2.3487	0.0011	479	84
653.04	2200	XING 218.		2200	84	0.0050	11.2519	TWP ROAD	2.3476	0.0050	2693	D 12
654.04	5281	XING 218.		5281	84	0.0119	11.2638	WHITE CR	2.3426	0.0076	4768	120
655.02	5185	XING 219.		3040	84	0.0069	11.2706	PRIVATE	2.3350	0.0082	5185	120
655.02		XING 220.		24675	A 84	0.0444	11.3151	MP 220 PRIVATE	2.3269			
656.02	5281	XING 220.73				0.0000	11.3151	WILL CR	2.3269	0.0083	5281	120
657.02	5281	XING 220.99				0.0000	11.3151	MIGRET CR	2.3185	0.0083	5281	120
658.02	5283	XING 221.39					11.3151	PRIVATE	2.3102	0.0083	5283	120
659.02	5280	XING 222.25				0.0000	11.3151	GAMBER CR	2.3018	0.0083	5280	120
660.02	5279	XING 220.		5279	120	0.0083	11.3234	PRIVATE	2.2935	0.0083	5279	120
661.02	5282	XING 223.		5282	120	0.0083	11.3317	KRAUS CR	2.2852	0.0083	5282	120
662.02	5284	XING 223.		5284	120	0.0083	11.3401	PRIVATE	2.2768	0.0083	5284	120
663.02	5275	XING 223.		5275	120	0.0083	11.3484	PRIVATE	2.2685	0.0083	5275	120
664.02	5279	XING 224.		5279	120	0.0083	11.3567	MAIN STREET SR	2.2602	0.0083	5279	120
664.02		XING 224.30					11.3567	ALLEY STREET	2.2518			
664.02		XING 225.80					11.3567	ULES CR	2.2518			
664.02		XING 226.84					11.3567	HELLER CR	2.2518			
664.02		XING 227.20					11.3567	PRIVATE	2.2518			
664.02		XING 227.50					11.3567	PRIVATE	2.2518			
664.02		XING 227.80					11.3567	PRIVATE	2.2518			
664.02		XING 227.84					11.3567	ANGLING ROAD	2.2518			
664.02		XING 228.38					11.3567	COUNTY LINE	2.2518			
664.02		XING 228.50					11.3567	LOUISA STREET	2.2518			
665.03	5296	XING 228.		3891	120	0.0061	11.3629	FOREST MARY STREE	2.2518	0.0078	4955	120
665.03		XING 228.21					11.3629	MP 230 GORMLY STREI	2.2440			
666.02	5275	XING 229.		5275	120	0.0083	11.3712	MARTIN STREET SR	2.2440	0.0000		
667.02	5273	XING 229.		5273	120	0.0083	11.3796		2.2440	0.0000		
668.02	5278	XING 229.		5278	120	0.0083	11.3879	CAMPBELL STREET	2.2440	0.0397	22500	A 90
669.03	5289	XING 230.		5289	120	0.0082	11.3962	BERLIN CR	2.2043	0.0000		
670.02	5269	XING 231.		2500	120	0.0038	11.4002	PRIVATE	2.2043	0.0089	4222	90
671.02	5278	XING 231.		0	120	0.0000	11.4002	PRIVATE	2.1954	0.0111	5278	90
671.31	1500	XING 234.		307	120	0.0005	11.4006	GROAT ROAD	2.1843	0.0032	1500	90
671.31		XING 234.		9240	D 12	0.0167	11.4173	WYKES CR	2.1812			90
671.31		XING 235.10					11.4173	PRIVATE	2.1812			
671.31		XING 236.08					11.4173	WALNUT STREET	2.1812			
671.31		XING 236.14					11.4173	MAIN STREET	2.1812			
671.31			90		90	0.0333	11.4506	DUNKIRK NYC XING	2.1812	0.0333	RR CR	90
671.31		XING 236.59					11.4506	RT# 135	2.1478	0.0042	2310	D 12
672.02	3771	XING 237.		3771	90	0.0079	11.4586	HOPPS ROAD	2.1437	0.0023	1461	120
673.02	5278	XING 238.		5278	90	0.0111	11.4697	WAYNE STREET	2.1414	0.0083	5278	120
674.02	5280			1951	90	0.0041	11.4738		2.1330	0.0083	5280	120
675.02	5276			22803	A 90	0.0397	11.5135		2.1247	0.0083	5278	120
676.02	5280	XING 241.57				0.0000	11.5135	TRAVERSE PK	2.1164	0.0083	5280	120
677.02	5277	XING 241.89				0.0000	11.5135	PRIVATE	2.1080	0.0083	5277	120
678.02	5278	XING 242.		1926	120	0.0031	11.5166	PETERSON CR	2.0997	0.0083	5276	120
679.02	5273	XING 243.		5273	120	0.0083	11.5249	SCOTT XING	2.0914	0.0083	5273	120
680.02	5265	XING 244.		5265	120	0.0083	11.5332	1 MILE XING	2.0830	0.0083	5265	120
681.04	5394	XING 245.		5394	120	0.0085	11.5417	ADA JOHNSON STREET	2.0747	0.0085	5394	120
682.04	5281	XING 245.		5281	120	0.0083	11.5500	MAIN STREET	2.0662	0.0083	5281	120
683.04	5276	XING 245.		5276	120	0.0083	11.5584	GILBERT STREET	2.0579	0.0083	5278	120
684.05	5377	XING 246.		5377	120	0.0085	11.5669	KLINGLER RD	2.0496	0.0085	5377	120
685.06	5303	XING 247.		5303	120	0.0084	11.5752	ST PAUL RD	2.0411	0.0084	5303	120
685.06		XING 248.10					11.5752	PRIVATE	2.0327			
685.06		XING 248.85					11.5752	COUNTY LINE	2.0327			
685.06		XING 249.55					11.5752	BENTLEY ROAD	2.0327			
685.06		XING 250.13				0.0000	11.5752	MP 250PEVEE ROAD	2.0327			
686.06	5289	XING 250.		5289	120	0.0083	11.5836	PRIVATE	2.0327	0.0083	5289	120
687.06	5286	XING 250.		5286	120	0.0083	11.5919	LAFAYETTE RD	2.0244	0.0083	5286	120
688.07	5297	XING 250.		5297	120	0.0084	11.6003	SHRIDER ROAD	2.0160	0.0084	5297	120
689.07	5304	XING 251.		5304	120	0.0084	11.6087	VINT RD	2.0077	0.0010	634	120
690.09	5371	XING 251.		5371	120	0.0085	11.6171	PHILLIPS RD	2.0067			
691.07	5202	XING 252.		5202	120	0.0082	11.6253	S HIGH STREET	2.0067	0.0397	22503	A 90
692.06	5298	XING 252.		5298	120	0.0084	11.6337	CHURCH ST	1.9669			
692.53	2400	XING 252.		2400	120	0.0038	11.6375	WASHINGTON	1.9669	0.0000		
692.59	293	XING 111		293	120	0.0006	11.6380	R RUMBAUGH RD	1.9669	0.0000		
693.08	2600	XING 253.		0	120	0.0000	11.6380	FISHER ROAD	1.9669	0.0106	4640	A 76
694.09	5336	XING 254.		0	120	0.0000	11.6380	MCCLURE XING	1.9564	0.0100	4027	76
694.66	3000	XING 254.		1896	120	0.0027	11.6406	COOL ROAD	1.9463	0.0075	3000	76
694.66		XING 255.		9240	D 12	0.0167	11.6573	THAYER ROAD	1.9389			
694.66		XING 256.93					11.6573	FETTER ROAD	1.9389			

MILES	DISTANCE, ft	HWY/GRADE CROSSING	MAX MPH	DISTANCE, feet	SPEED, mph	dT, hours	TIME, hours	APPENDIX B1 STATION PAGE 13 COLUMN 2	TIME, hours	dT, hours	DISTANCE, feet	SPEED, mph
	READ DOWN WESTBOUND TRAFFIC								READ UP EASTBOUND TRAFFIC			
694.65		XING 257.56					11.8573	METZGER ROAD	1.9389			
694.65		XING 258.42					11.8573	ROUSH ROAD	1.9389			
694.66			90		90	0.0000	11.6573	SUGAR DTI XING	1.9389	0.0000		
695.11	2373	XING 259.		2373	90	0.0068	11.6823	SUGAR ST	1.9389	0.0059	2373	76
695.11						0.0000	11.6823	MP 260	1.9329			
695.31	1100	XING 260.		600	90	0.0013	11.6836	N PINE ST	1.9329	0.0027	1100	76
695.41	500	XING 260.		148	90	0.0003	11.6839	R JACKSON	1.9302	0.0012	500	76
695.41			90		90	0.0500	11.7139	LIMA BO-NW XING	1.9290	0.0500	RR XING DE	
695.41		XING 260.		852	D 90	0.0018	11.7156	MAIN ST	1.8790	0.0000		
695.50	485	XING 76		485	76	0.0012	11.7170	R N ELIZABETH	1.8790	0.0012	485	76
695.50		XING 260.86					11.7170	N WEST ST	1.8778	0.0033	1549	D 10
695.65	800	XING 260.		800	76	0.0020	11.7190	N MCDONEL ST	1.8744	0.0001	51	100
696.09	2300	XING 100		2300	76	0.0057	11.7247	R N METCALF	1.8743	0.0028	1500	100
696.20	578	XING 100		578	76	0.0014	11.7262	N BAXTER ST	1.8715	0.0011	578	100
697.09	4700	XING 261.		4700	76	0.0117	11.7379	COLE ST	1.8704	0.0028	1613	D 12
697.09		XING 262.99					11.7379	CABLE ROAD	1.8676	0.0049	3067	120
698.09	5296	XING 263.		2624	76	0.0065	11.7444	HARTZLER RD	1.8627	0.0084	5296	120
699.09	5288	XING 264.		27143	A 76	0.0503	11.7947	EASTTOWN ROAD	1.8544	0.0083	5288	120
700.11	5393	XING 264.83				0.0000	11.7947	EAST ROAD	1.8460	0.0085	5393	120
701.11	5280	XING 265.25				0.0000	11.7947	BATY ROAD	1.8375	0.0083	5280	120
702.11	5286	XING 265.80				0.0000	11.7947	PRIVATE	1.8292	0.0083	5286	120
703.12	5286	XING 266		2061	120	0.0033	11.7980	GREENLAWN AVE	1.8208	0.0083	5286	120
704.12	5282	XING 266.		5282	120	0.0083	11.8063	MAIN ROAD	1.8125	0.0014	913	120
705.12	5281	XING 267.		5281	120	0.0083	11.8146	WAPAK ROAD	1.8111	0.0000		
705.12		XING 267.40					11.8146	PRIVATE	1.8111	0.0000		
705.12		XING 267.65					11.8146	KEMP ROAD	1.8111	0.0000		
705.12		XING 267.90					11.8146	PRIVATE	1.8111	0.0000		
705.12		XING 268.10					11.8146	PRIVATE	1.8111	0.0000		
705.12		XING 268.83					11.8146	GRUBB ROAD	1.8111	0.0000		
705.12		XING 269.20					11.8146	PRIVATE	1.8111	0.0000		
706.12	5292	XING 270.		5292	120	0.0084	11.8230	STATE RD	1.8111	0.0300	17615	A 10
707.12	5283	XING 270.		5283	120	0.0083	11.8313	OLD DELPHOS	1.7811	0.0049	2810	100
708.12	5290	XING 271.		6777	120	0.0107	11.8420	DEFIANCE TRA	1.7761	0.0100	5290	100
708.71	3100	XING 271.		1613	D 12	0.0028	11.8448	BAUGH RD	1.7661	0.0069	3100	100
709.12	2172	XING 106		2172	100	0.0041	11.8489	L PELTIER RD	1.7602	0.0041	2172	100
709.31	1000	XING 106		1000	100	0.0019	11.8508	L LEHMAN RD	1.7561	0.0019	1000	100
709.31		XING 273.68					11.8508	PRIVATE	1.7542	0.0000		
709.52	1100	XING 274.		1100	100	0.0021	11.8529	DELPHOS PIERCE ST	1.7542	0.0021	1100	100
709.88	1900	XING 274.		1900	100	0.0036	11.8565	S MAIN STREET	1.7521	0.0036	1900	100
710.07	1000	XING 100		1000	100	0.0019	11.8584	R CANAL STREET	1.7485	0.0019	1000	100
710.12	264	XING 274.		284	100	0.0005	11.8589	JEFFERSON STREET	1.7466	0.0028	1613	D 12
711.12	5277	XING 274.		5277	100	0.0100	11.8689	S CLAY STREET	1.7438	0.0062	3946	120
712.12	5286	XING 274.		5286	100	0.0100	11.8789	S BREDICK STREET	1.7376	0.0083	5286	120
713.12	5265	XING 275.		151	100	0.0003	11.8792	STATE STREET	1.7293	0.0083	5265	120
714.12	5289	XING 276.		17615	A 10	0.0300	11.9092	BRICKNER RD	1.7210	0.0083	5289	120
715.12	5284	XING 277.20			120	0.0000	11.9092	PRIVATE	1.7126	0.0083	5284	120
715.12		XING 277.88				0.0000	11.9092	BOCKEY RD	1.7043	0.0000		
715.12		XING 278.98				0.0000	11.9092	CONVERSE RD	1.7043	0.0000		
715.12		XING 279.71				0.0000	11.9092	MIDDLEPOINT	1.7043	0.0000		
716.12	5281	XING 280.		3363	120	0.0053	11.9145	MAIN STREET	1.7043	0.0061	3228	120
717.12	5281	XING 280.		5281	120	0.0083	11.9228	ADAMS STREET	1.6982	0.0000		120
718.12	5285	XING 280.		5285	120	0.0083	11.9312	MASON STREET	1.6982	0.0042	2620	A 11
719.12	5282	XING 280.		5282	120	0.0083	11.9395	DOG CREEK RD	1.6950	0.0000		
720.12	5281	XING 281.		5281	120	0.0083	11.9479	RINGWALD RD	1.6950	0.0467	24745	A 75
721.13	5297	XING 282.		8338	120	0.0132	11.9610	CHENOWITH RD	1.6484	0.0034	1114	62
722.03	4800	XING 283.		1759	D 12	0.0031	11.9641	GAMBLE RD	1.6450	0.0147	4800	62
722.13	486	XING 96		486	98	0.0009	11.9650	L HOAGLIN CTR	1.6303	0.0015	486	62
722.37	1300	XING 98		500	98	0.0010	11.9660	L GILLAND RD	1.6288	0.0040	1300	62
722.37		XING 285.88					11.9660	MENDON RD	1.6248			
722.66	1500	XING 286		185	98	0.0004	11.9663	VAN WERT WAYNE ST	1.6248	0.0046	1500	62
722.66		XING 286.		2112	D 98	0.0050	11.9713	VINE STREET	1.6203	0.0000	0	62
722.83	900	XING 62		900	62	0.0027	11.9741	R FRANKLIN STREET	1.6203	0.0022	900	62
722.83							11.9741		1.6175	0.0013	439	D 71
723.00	900	XING 71		900	62	0.0027	11.9768	R CHESTNUT STREET	1.6163	0.0012	461	71
723.02	100	XING 287.		100	62	0.0003	11.9771	RACE STREET	1.6150	0.0000	0	71
723.02		XING 287.11				0.0000	11.9771	N TYLER STREET	1.6150	0.0026	1122	D 90
723.02		XING 287.17				0.0000	11.9771	HARRISON STREET	1.6124			90
723.02		XING 287.24				0.0000	11.9771	CHERRY STREET	1.6124			
723.02		XING 287.30				0.0000	11.9771	WALNUT STREET	1.6124			
723.02		XING 287.37				0.0000	11.9771	MARKET STREET	1.6124			
723.02		XING 287.49				0.0000	11.9771	WASHINGTON STREE	1.6124			
723.02		XING 287.56				0.0000	11.9771	JEFFERSON STREET	1.6124			
723.02		XING 287.67				0.0000	11.9771	N SHANNON	1.6124			
723.02		XING 287.74				0.0000	11.9771	WALL STREET	1.6124			
723.02			90			0.0000	11.9771	CR VAN WERT IN XING	1.6124	0.0000	RR CR	90
723.02		XING 288.07				0.0000	11.9771	BURT STREET	1.6124	0.0042	2310	D 12
723.14	642	XING 288.		640	62	0.0020	11.9791	FISHER AVENUE	1.6082	0.0000	0	120
724.13	5241	XING 288.		5241	62	0.0160	11.9951	JOHN BROWN RD	1.6082	0.0040	2551	120
725.13	5264	XING 288.		4117	62	0.0126	12.0077	PRIVATE	1.6042	0.0063	5284	120
725.13				1564	A 62	0.0047	12.0124	MP 290	1.5958			
726.13	5289	XING 290.		28328	A 71	0.0533	12.0657	LIBERTY UNION	1.5959	0.0083	5289	120
727.14	5287	XING 291.27					12.0657	RICHEY CR	1.5875	0.0083	5287	120
728.14	5287	XING 292.38					12.0657	MACE XING	1.5792	0.0083	5287	120
728.14	5284	XING 293.03					12.0657	ALT RT US 30	1.5708	0.0083	5284	120
729.14		XING 293.69					12.0657	BERGNER RD	1.5625			
729.14		XING 294.71					12.0657	MAIN ST CONVOY	1.5625			
729.14		XING 294.86					12.0657	TULLY ST	1.5625			
730.14	5287	XING 295.40				0.0000	12.0657	PRIVATE	1.5625	0.0083	5287	120
731.14	5288	XING 295.		2687	120	0.0046	12.0703	PAYNE CR	1.5541	0.0083	5288	120
732.14	5286	XING 296.		5286	120	0.0083	12.0787	SHANER CR	1.5458	0.0083	5286	120
733.14	5280	XING 296.		5280	120	0.0083	12.0870	LARE CR	1.5375	0.0083	5280	120
734.14	5292	XING 297.		5292	120	0.0084	12.0953	MENTZER	1.5291	0.0084	5292	120
735.14	5288	XING 297.		5288	120	0.0083	12.1037	PRIVATE	1.5208	0.0083	5288	120
735.14		XING 298.12					12.1037	MENTZER CURC	1.5124			
735.14		XING 298.58					12.1037	CLEM XING	1.5124			
735.14		XING 299.46					12.1037	SCHOOLHOUSE	1.5124			
736.15	5282	XING 300.		5282	120	0.0083	12.1120	ST LINE RD	1.5124	0.0083	5282	120
737.15	5281	XING 301.		5281	120	0.0083	12.1204	MORGAN RD	1.5041	0.0083	5281	120
738.15	5289	XING 301.		5289	120	0.0083	12.1287	PRIVATE	1.4958	0.0083	5289	120
739.15	5285	XING 302.		5285	120	0.0083	12.1371	LORTIE RD	1.4874	0.0083	5285	120
740.15	5294	XING 303.		5294	120	0.0084	12.1454	OHIO STREET	1.4791	0.0084	5294	120
741.15	5289	XING 304.		5289	120	0.0083	12.1538	MAIN STREET	1.4707	0.0083	5289	120
742.16	5297	XING 304.		5297	120	0.0084	12.1621	WASHINGTON	1.4624	0.0084	5297	120
743.15	5274	XING 306.		5274	120	0.0083	12.1704	SNYDER RD	1.4540	0.0083	5274	120
744.16	5292	XING 306.		5292	120	0.0084	12.1788	HOFFMAN RD	1.4457	0.0084	5292	120
745.16	5281	XING 307.		5281	120	0.0083	12.1871	GROTRAIN RD	1.4373	0.0083	5281	120
745.16		XING 307.71					12.1871	WILSON RD	1.4290			
745.16		XING 308.30					12.1871	FACKLER RD	1.4290			
745.16		XING 308.86					12.1871	GARADOT RD	1.4290			
745.16		XING 309.37					12.1871	HOUK RD	1.4290			
745.16		XING 309.71					12.1871	WASHINGTON	1.4290			
745.16		XING 309.94					12.1871	FRANKE RD	1.4290			
746.16	5281	XING 310.		5281	120	0.0083	12.1955	PRIVATE	1.4290	0.0076	4814	120
747.16	5281	XING 310.		5281	120	0.0083	12.2038	PRIVATE	1.4214	0.0000	0	120
748.15	5267	XING 311.		5267	120	0.0083	12.2121	TILLMAN RD	1.4214	0.0000		

			READ DOWN WESTBOUND TRAFFIC					APPENDIX B1 STATION	READ UP EASTBOUND TRAFFIC			
MILES	DISTANCE ft	HWY/GRADE CROSSING	MAX MPH	DISTANCE feet	SPEED mph	ET hours	TIME hours	PAGE 14 COLUMN 1	TIME hours	ET hours	DISTANCE feet	SPEED mph
748.15		XING 311.75					12.2121	MINNICH ROAD	1.4214			
748.15		XING 312.50					12.2121	PRIVATE	1.4214			
748.15		XING 312.80					12.2121	PRIVATE	1.4214			
748.15		XING 313.09					12.2121	PRIVATE	1.4214			
748.15		XING 313.25					12.2121	PAULDING ROAD	1.4214			
749.18	5293	XING 313.		5293	120	0.0084	12.2205	HARTZELL ROAD	1.4214	0.0481	26233	A 79
750.18	5283	XING 315.		5283	120	0.0083	12.2288	ADAMS CENTER	1.3733	0.0125	4546	A 57
751.18	5288	XING 316.		5288	120	0.0083	12.2371	MEYER ROAD	1.3608	0.0000		
752.17	5329			5329	120	0.0084	12.2456		1.3608	0.0069	1770	A 38
753.15	5192			5192	120	0.0082	12.2538		1.3539	0.0242	4851	38
754.02	4600			2982	D 12	0.0057	12.2594		1.3297	0.0229	4600	38
754.17	749		79	1250	79	0.0030	12.2625	L	1.3068	0.0037	749	38
754.17				1097	D 79	0.0031	12.2655		1.3031			
754.32	800		57	138	57	0.0005	12.2660	LTTMPH30	1.3031	0.0040	800	38
754.32			90		57	0.0833	12.3493	RC MIKE WABXING	1.2991	0.0833	RR CROSSI	
754.32				682	D 57	0.0026	12.3520	FORT WAYNE	1.2157			
754.58	1400	XING	38	1400	38	0.0070	12.3589	R BRACKENRIDGE	1.2157	0.0070	1400	38
754.81	1200		54	1200	38	0.0060	12.3649	R	1.2088	0.0014	315	D 48
755.09	1500			1500	38	0.0075	12.3724		1.2074	0.0094	2385	48
755.18	481		49	481	38	0.0024	12.3748	R	1.1980	0.0019	481	48
755.23	489		48	489	38	0.0024	12.3772	R	1.1961	0.0019	489	48
755.28							12.3772		1.1941	0.0046	1591	D 81
756.17	4700			4700	36	0.0234	12.4006		1.1895	0.0073	3139	81
756.19	100		81	100	36	0.0005	12.4011	R	1.1822	0.0002	100	81
756.26	400		81	400	36	0.0020	12.4031	R	1.1820	0.0009	400	81
756.35	452		81	452	36	0.0023	12.4054	R	1.1810	0.0011	452	81
757.20	4500			1678	36	0.0084	12.4138		1.1800	0.0054	2874	D 12
758.21	5349	XING 323.		784	A 38	0.0033	12.4171	THOMAS ROAD	1.1746	0.0110	6975	120
759.22	5322	XING 324.		2906	48	0.0115	12.4286	HADLEY ROAD	1.1636	0.0084	5322	120
760.20	5193	XING 325.		6140	A 48	0.0175	12.4461	BASS ROAD	1.1552	0.0082	5193	120
761.20	5277	XING 326.40				0.0000	12.4461	PRIVATE	1.1470	0.0083	5277	120
762.20	5278	XING 328.		25645	A 81	0.0467	12.4927	MAIN STREET	1.1386	0.0083	5278	120
763.21	5290	XING 328.36				0.0000	12.4927	ARCOLA COLUMBIA ST	1.1303	0.0083	5290	120
764.22	5363	XING 328.		4439	120	0.0070	12.4997	PRIVATE	1.1220	0.0085	5363	120
765.21	5224	XING 329.		5224	120	0.0082	12.5080	PRIVATE	1.1135	0.0082	5224	120
765.21		XING 329.30				0.0000	12.5080	PRIVATE	1.1052	0.0000		
766.21	5296	XING 329.		5296	120	0.0084	12.5163	PRIVATE	1.1052	0.0084	5296	120
767.22	5297	XING 330.		5297	120	0.0084	12.5247	COUNTY LINE	1.0969	0.0084	5297	120
768.22	5294	XING 331.		5294	120	0.0084	12.5330	BEECHING ROAD	1.0885	0.0084	5294	120
769.22	5293	XING 332.		5293	120	0.0084	12.5414	800E?	1.0802	0.0084	5293	120
770.22	5293	XING 333.		5293	120	0.0084	12.5498	MILL STREET	1.0718	0.0084	5293	120
771.23	5292	XING 334.		5292	120	0.0084	12.5581	MOWERY ROAD	1.0635	0.0084	5292	120
772.22	5269	XING 336.		5269	120	0.0083	12.5664	PAGES CR	1.0551	0.0083	5269	120
773.23	5290	XING 338.		5290	120	0.0083	12.5748	OHIO STREET	1.0468	0.0061	3851	120
774.26	5434	XING 338.		5434	120	0.0086	12.5833	SR 9	1.0407	0.0000		120
775.24	5185	XING 338.		5185	120	0.0082	12.5915	LINE STREET SR	1.0407	0.0000		120
775.24		XING 339.63				0.0000	12.5915	POOR FARM CR	1.0407	0.0175	10663	A 11
776.23	5263	XING 341.		5263	120	0.0083	12.5998	WOLF ROAD	1.0232	0.0000		
777.24	5296	XING 342.		5296	120	0.0084	12.6082	PEARL CR 300E?	1.0232	0.0300	15337	A 80
777.54	1600	XING 343.		1600	120	0.0025	12.6107	BUTLER CR 40?	0.9932	0.0003	107	80
777.82	1503	XING 110		1503	120	0.0024	12.6131	L PRIVATE	0.9929	0.0036	1503	80
778.24	2200			2200	120	0.0035	12.6166		0.9894	0.0052	2200	80
779.24	5290			4257	120	0.0067	12.6233		0.9842	0.0125	5290	80
779.60	1900			2933	D 12	0.0056	12.6288		0.9717	0.0045	1900	80
779.89	1491		80	1491	80	0.0035	12.6324	R	0.9672	0.0035	1491	80
780.25	1900			1900	80	0.0045	12.6369		0.9636	0.0056	2933	D 12
781.15	4800			4800	80	0.0114	12.6482		0.9581	0.0029	1867	120
781.25	488	XING 122		488	80	0.0012	12.6494	L SRD 5	0.9551	0.0008	488	120
781.28	184	XING 122		184	80	0.0004	12.6498	L LARWILL DEPOT ST	0.9544	0.0003	184	120
782.25	5100	XING 347.		3628	80	0.0086	12.6584	BRINKLEY ROAD	0.9541	0.0080	5100	120
783.25	5294	XING 348.		26000	A 80	0.0475	12.7059	BLACK CR	0.9460	0.0084	5294	120
784.25	5286	XING 349.30				0.0000	12.7059	PRIVATE	0.9377	0.0083	5286	120
785.25	5287	XING 349.80				0.0000	12.7059	PRIVATE	0.9293	0.0083	5287	120
786.25	5282	XING 350.38				0.0000	12.7059	MARKET STREET	0.9210	0.0083	5282	120
787.25	5283	XING 350.		1904	120	0.0030	12.7089	PIERCETON FIRST ST	0.9126	0.0083	5283	120
788.25	5279	XING 352.		5278	120	0.0083	12.7172	VAN NESS ROAD	0.9043	0.0030	1898	120
789.25	5278	XING 353.		5278	120	0.0083	12.7256	WOOSTER ROAD	0.9013			
790.25	5293	XING 354		5293	120	0.0084	12.7339	SCHADE 250E	0.9013			
791.26	5323			5323	120	0.0084	12.7423		0.9013	0.0397	22503	A 90
792.25	5233			5233	120	0.0083	12.7506		0.8616	0.0042	2004	90
792.97	3800			3800	120	0.0060	12.7566		0.8574	0.0080	3800	90
793.26	1500	XING 122		1500	120	0.0024	12.7590	L BRONSON STREET	0.8494	0.0032	1500	90
793.45	996	XING 122		996	120	0.0016	12.7605	L MAPLE STREET	0.8462	0.0021	996	90
793.96	2700	XING 358.		390	120	0.0006	12.7611	SCOTT STREET	0.8441	0.0057	2700	90
793.96				2310	D 12	0.0042	12.7653		0.8384			
793.96			90		90	0.0333	12.7986	WARSAW B&XING	0.8384	0.0333	RR CR	90
793.96		XING 358.74					12.7986	DETROIT STREET	0.8051	0.0014	697	D 10
793.96		XING					12.7986	ALLEY?	0.8037			100
793.96		XING 358.87					12.7986	INDIANA STREET	0.8037			
793.96		XING					12.7986	ALLEY?	0.8037			
794.26	1600	XING 358.		1600	90	0.0034	12.8020	S LAKE STREET	0.8037	0.0017	903	100
795.26	5282	XING		5282	90	0.0111	12.8131	ALLEY?	0.8020	0.0100	5282	100
795.26		XING 359.25					12.8131	UNION STREET	0.7920			
795.30	178	XING 360.		178	90	0.0004	12.8135	JACKSON CR	0.7920	0.0003	175	100
796.66	1900		100	1900	90	0.0040	12.8175	R BURKET	0.7917	0.0036	1900	100
796.26	3200	XING 361.		2040	90	0.0043	12.8218	PRIVATE	0.7881	0.0028	1613	D 12
797.27	5318	XING 362.		4888	A 90	0.0097	12.8315	PRIVATE	0.7853	0.0109	6905	120
798.26	5245	XING 362.		4072	100	0.0077	12.8392	THOMAS CR 10	0.7744	0.0083	5245	120
799.26	5291	XING 363.		17615	A 10	0.0300	12.8692	PRIVATE	0.7661	0.0084	5291	120
799.26		XING 363.72					12.8692	SELBY CR 525	0.7578			
800.28	5352	XING 364.50				0.0000	12.8692	PRIVATE	0.7578	0.0084	5352	120
801.27	5220	XING 365.		1011	120	0.0016	12.8708	PRAIRIE STREET	0.7493	0.0082	5220	120
802.27	5291	XING 365.		5291	120	0.0084	12.8792	ATWOOD HARRISON S	0.7411	0.0084	5291	120
803.27	5279	XING 366.		5279	120	0.0083	12.8875	ROBINSON ROAD	0.7327	0.0083	5279	120
804.26	5227	XING 367.		5227	120	0.0082	12.8958	SNYDER CR 87	0.7244	0.0082	5227	120
805.27	5337	XING 368		5337	120	0.0084	12.9042	PRIVATE	0.7161	0.0084	5337	120
805.27		XING 368.77					12.9042	COOK CR	0.7077			
805.27		XING 369.29					12.9042	WALNUT STREET SR	0.7077			
805.27		XING 369.36					12.9042	ETNA GREEN	0.7077			
806.29	5379	XING 370.		5379	120	0.0085	12.9127	S BEECH ROAD	0.7077	0.0085	5379	120
807.27	5193	XING 372.		5193	120	0.0082	12.9209	HARRIS STREET	0.6992	0.0082	5193	120
808.30	5439	XING 373.		5439	120	0.0086	12.9294	BOURBON MAIN ST	0.6910	0.0086	5439	120
808.30		XING 373.17					12.9294	BOURBON STREET	0.6824			
808.30		XING 373.26					12.9294	CENTER STREET	0.6824			
808.30		XING 374.13					12.9294	TSCHR ST SE	0.6824			
809.28	5162	XING 375.		5162	120	0.0081	12.9376	STOCKMAN ROAD	0.6824	0.0081	5162	120
810.27	5252	XING 376.		5252	120	0.0083	12.9459	WYMER CR	0.6743	0.0083	5252	120
811.27	5270	XING 377.		5270	120	0.0083	12.9542	COMPTON CR	0.6660	0.0083	5270	120
812.27	5279	XING 377.		5279	120	0.0083	12.9625	HAWTHORNE ROAD	0.6577	0.0083	5279	120
813.27	5279	XING 378.		5279	120	0.0083	12.9709	PRIVATE	0.6494	0.0041	2573	120
814.27	5283	XING 379.		5283	120	0.0083	12.9792	S IRIS ROAD	0.6453			
815.27	5277	XING 379.		5277	120	0.0083	12.9875	PRIVATE	0.6453			
816.28	5315	XING 380.		5315	120	0.0084	12.9958	MP 380 PRIVATE	0.6453	0.0397	22503	A 90-120

			READ DOWN WESTBOUND TRAFFIC					APPENDIX B1 STATION	READ UP EASTBOUND TRAFFIC			
MILES	DISTANCE ft	HWY/GRADE CROSSING	MAX MPH	DISTANCE feet	SPEED mph	ET hours	TIME hours	PAGE 14 COLUMN 2	TIME hours	ET hours	DISTANCE feet	SPEED mph
817.25	5155	XING 380.		5155	120	0.0081	13.0041	PRIVATE	0.6056	0.0026	1233	90
818.27	5307	XING 381.		5367	120	0.0085	13.0125	KING ROAD	0.6030	0.0113	5367	90
819.10	4400	XING 383.		2090	120	0.0033	13.0158	LIBERTY STREET	0.5917	0.0093	4400	90
819.10				2310	D 12	0.0042	13.0200		0.5824			
819.10			90		90	0.0000	13.0200	PLYMOUTH NKP XING	0.5824		RR CR	90
819.10							13.0200		0.5824	0.0042	2310	D 12
819.28	944	XING 384.		944	90	0.0020	13.0220	FIFTH STREET	0.5783	0.0006	350	120
820.27	5216	XING 384.		5216	90	0.0110	13.0330	OAK ROAD	0.5777	0.0066	3500	120
821.27	5250	XING 385.		4840	90	0.0102	13.0431	PRIVATE	0.5722	0.0083	5250	120
822.27	5257	XING 385.		22503	A 90	0.0397	13.0829	PRIVATE	0.5639	0.0083	5257	120
823.28	5293	XING 386.34				0.0000	13.0829	ANGLING ROAD	0.5555	0.0084	5293	120
824.27	5269	XING 386.89				0.0000	13.0829	QUINCE ROAD	0.5472	0.0083	5269	120
825.27	5281	XING 387.70				0.0000	13.0829	REDWOOD ROAD	0.5388	0.0083	5281	120
825.27		XING 388.14					13.0829	PRIVATE	0.5305			
826.27	5268	XING 390.		4336	120	0.0068	13.0897	UNION ROAD	0.5305	0.0083	5268	120
827.27	5298			5298	120	0.0084	13.0981	DONALDSON	0.5222	0.0084	5298	120
828.28	5298	XING 393.		5298	120	0.0084	13.1064	PRIVATE	0.5138	0.0084	5298	120
829.30	5411	XING 394.		5411	120	0.0085	13.1150	SR 23	0.5055	0.0085	5411	120
830.28	5164	XING 395		5164	120	0.0082	13.1231	WALSH CR	0.4969	0.0082	5164	120
831.28	5285	XING 396.		5285	120	0.0083	13.1315	750E	0.4888	0.0083	5285	120
832.28	5288	XING 398.		5288	120	0.0083	13.1398	STARKE STREET	0.4804	0.0083	5288	120
833.28	5285	XING 398.		5265	120	0.0083	13.1481	HAMLEY JEFFERSON	0.4721	0.0083	5285	120
834.28	5279	XING 399.		5279	120	0.0083	13.1565	500E	0.4637	0.0083	5279	120
835.29	5301	XING 401.		5301	120	0.0084	13.1648	300E	0.4554	0.0084	5301	120
836.29	5282	XING 401.		5282	120	0.0083	13.1732	PRIVATE	0.4470	0.0083	5282	120
837.29	5282	XING 402.		5282	120	0.0083	13.1815	PRIVATE	0.4387	0.0083	5282	120
838.29	5270	XING 403.		5270	120	0.0083	13.1898	50E	0.4304	0.0039	2493	120
839.29	5287	XING 405.		5287	120	0.0083	13.1982	100W	0.4264			
840.29	5292	XING 405.		5292	120	0.0084	13.2065	1400S	0.4264			
841.29	5280	XING 406.		5280	120	0.0083	13.2149	SR 39	0.4264	0.0397	22503	A 90
842.32	5422	XING 407.		5422	120	0.0086	13.2234	LONG LANE	0.3867	0.0033	1556	90
843.27	5045	XING 407.		5045	120	0.0080	13.2314	PRIVATE	0.3834	0.0106	5045	90
844.11	4400	XING 408.		2090	120	0.0033	13.2347	THOMPSON STREET	0.3728	0.0093	4400	90
844.11		XING 408.		2310	D 12	0.0042	13.2389	OHIO STREET	0.3636			
844.11			90		90	0.0000	13.2389	HANNA C&O XING	0.3636		RR CR	90
844.11		XING 408.83					13.2389	450W	0.3636	0.0042	2310	D 12
844.29	986	XING 409.		966	90	0.0020	13.2409	PRIVATE	0.3594	0.0007	437	120
845.29	5281	XING 410.		5281	90	0.0111	13.2520	MP 410 SCHULL 600W	0.3587	0.0056	3500	120
846.29	5286	XING 411.		4753	90	0.0100	13.2620	FARRESTER ROAD	0.3532	0.0083	5286	120
847.29	5273	XING 411.		22503	A 90	0.0397	13.3017	TILDON 750W	0.3448	0.0083	5273	120
848.46	6182	XING 412.89				0.0000	13.3017	PRIVATE	0.3365	0.0096	6182	120
849.29	4367	XING 413.42				0.0000	13.3017	KLEINS 900W	0.3268	0.0069	4367	120
850.29	5280	XING 414.75				0.0000	13.3017	N MAIN STREET	0.3199	0.0083	5280	120
851.29	5272	XING 414.		4403	120	0.0069	13.3087	WANATAH S ILLINOIS	0.3115	0.0083	5272	120
852.28	5261	XING 415		5261	120	0.0083	13.3170	S WASHINGTON	0.3032	0.0059	3712	120
853.28	5293	XING 415		5293	120	0.0084	13.3253	LINCOLN STREET	0.2974	0.0000		
854.28	5270	XING 415.		5270	120	0.0085	13.3336	1100W	0.2974	0.0000		
854.89	3200	XING 416.		3200	120	0.0051	13.3387	OSBORN CR-CO	0.2974	0.0000		
854.89		XING 417.78					13.3387	FISHER 575E	0.2974			
854.89		XING 419.08					13.3387	DOLSON 450E	0.2974			
854.99	570	XING 133		570	120	0.0009	13.3396	R BICKEL 400E	0.2974			
855.28	1500			1500	120	0.0024	13.3419		0.2974			
855.28		XING 420.00				0.0000	13.3419	MP 420 PRIVATE	0.2974	0.0300	17615	A 10
856.28	5280	XING 420.		5280	120	0.0083	13.3503	MONTDALE ROAD	0.2674	0.0096	5037	100
857.26	5263	XING 421.		5263	120	0.0083	13.3586	PRIVATE	0.2578	0.0100	5263	100
857.41	700	XING 422.		700	120	0.0011	13.3597	CEMETERY ROAD	0.2479	0.0013	700	100
858.20	4200	XING 100		4200	120	0.0066	13.3663	R GREENWICH STRE	0.2465	0.0080	4200	100
858.28	394	XING 423.		394	120	0.0006	13.3669	VALPARAISO AXE ST	0.2386	0.0007	394	100
858.81	2800	XING 423.		2800	120	0.0044	13.3714	FRANKLIN STREET	0.2378	0.0053	2800	100
858.90	486	XING 100		486	120	0.0008	13.3721	L WASHINGTON ST	0.2325	0.0029	1551	100
859.25	2000	XING 423		2000	120	0.0032	13.3753	LAFAYETTE STREET	0.2296	0.0097	4888	A 90
860.25	5279	XING 423.		5279	120	0.0083	13.3836	NAPOLEON STREET	0.2199	0.0028	1326	90
861.25	5274	XING 424.		5274	120	0.0083	13.3919	STEVENS XING	0.2171	0.0111	5274	90
862.11	4400	XING 425.		2090	120	0.0033	13.3952	150 W	0.2060	0.0093	4400	90
862.11				2310	D 12	0.0042	13.3994		0.1967	0.0000		90
862.11			90		90	0.0667	13.4661	VALPO GTW XING	0.1967	0.0667		90
862.11		XING 426.90					13.4661	250W	0.1301	0.0042	2310	D 12
862.26	892	XING 427.		892	90	0.0019	13.4680	400W	0.1259	0.0004	261	120
863.26	5299	XING 428.		5299	90	0.0112	13.4791	EAST SALT CR	0.1255	0.0032	2000	120
864.29	5340	XING 428.		4809	90	0.0101	13.4892	W SALT CR	0.1223	0.0000		
865.28	5206	XING 430.		22503	A 90	0.0397	13.5289	MP 430 PRIVATE	0.1223	0.0000		
866.28	5291	XING 430.87				0.0000	13.5289	WHEELER PARK AVE	0.1223	0.0000		
867.28	5269	XING 431.03				0.0000	13.5289	GOTT RD 7TH	0.1223	0.0000		
868.27	5217	XING 432.36				0.0000	13.5289	TREAGER RD	0.1223	0.0533	28328	A 71
869.29	5354	XING 433.		4380	120	0.0069	13.5359	COUNTY LINE	0.0690	0.0133	4984	71
870.29	5316	XING 434.		2584	120	0.0041	13.5399	HOBART ILLINOIS ST	0.0557	0.0142	5316	71
870.40	700	XING 434.87			120	0.0000	13.5399	LINDA STREET	0.0415	0.0019	700	71
870.43		XING 434.		3432	D 12	0.0068	13.5467	CLEVELAND STREET	0.0396	0.0000		
870.58	844	XING 71		844	71	0.0023	13.5490	R LAKE PARK AVE	0.0396	0.0023	844	71
871.29	3700	XING 436.		3700	71	0.0099	13.5589	WISCONSIN STREET	0.0374	0.0018	739	D 64
871.29							13.5589		0.0356	0.0006	259	64
872.29	5320			5320	71	0.0142	13.5731		0.0350	0.0089	3653	A 71
872.49	1080			1060	71	0.0028	13.5759		0.0261	0.0261	5429	A 0-
872.49		XING 437.18				0.0000	13.5759	LIVERPOOL RD	0			
872.49							13.5759	EAST END LIVERPOOL TERMINAL				
873.29	4200	XING 438.		920	71	0.0025	13.5783	INDIANA AVENUE				
873.40	600	XING 439		7995	A 71	0.0181	13.5964	VIRGINIA STREET				
873.40						0.0000	13.5964	I-65				
873.62	1122			0		0.0000	13.5964					
873.62						0.0000	13.5964	I-80				
874.30	3600			757	95	0.0015	13.5979					
874.86	1897		D 118-0		0		13.5979					
874.86					0		13.5979					
874.86					95		13.5979					
875.10	2363			4110	D 95	0.0132	13.6111					
875.10					0		13.6111	WEST END LIVERPOOL TERMINAL				

Appendix B2 - Improved Right of Way on Unmodified Alignment Aerodynamically designed TOFC equipment operating at 120 mph maximum speed. Maximum allowable speed on curves as shown for high speed equipment only. Curve speed for conventional equipment as in Appendix B0 and Appendix B1 is substantially lower.

FRA Class 7 track where speeds over 110 mph are possible.

Other tracks classified per maximum possible speed.

Unreconstructed right of way but track re built for maximum speed up to 120 mph. $E_a = 6$ inches

Wheel load 0.6 minimum per FRA 213.307d1

Assumed c.g. of moving equipment 108 inches above top of rail

Train length 11,000 feet: 200 platforms plus 20 locomotive units @ 6,000 hp each.

20% speed reduction for reverse curves with 600 feet or less between curves.

Gradients shown above 0.5% only, positive up westbound.

Highway grade crossing locations shown are approximate

Distance Jersey Meadows to Liverpool IN 860.58 miles

Westbound trip	10.6676 hours
Eastbound trip	10.6531 hours

APPENDIX B2 — READ DOWN: WESTBOUND TRAFFIC / READ UP: EASTBOUND TRAFFIC

PAGE 1 COLUMN 1

STATION
EAST END PLATFORM
PENN STATION NY
WEST END PLATFORM
TUNNEL RESTRICTION
R - ARBITRARY SPEED LIMIT 80 MPH PER TUNNEL RE
TUNNEL RESTRICTION
TUNNEL RESTRICTION
L
RTE 169 MP 3
L
E MEADOWS
R
L
A
W MEADOWS
HACKENSACK R
HUDSON
R HUDSON
R
L HARRISON
REVERSE CURVE
R DOCK
REVERSE CURVE
L NEWARK
L
R HUNTER
A
L ELIZABETH
REVERSE CURVE
125 MPH
ELMORA
2 END 4 TRACKS
R
REVERSE CURVE
L
L METROPARK
R
L
R
R S METU 32
L
L

APPENDIX B2 — PAGE 1 COLUMN 2

STATION
NEW BRUNSWICK
DEANS
SEE PHOTO
R
MONMOUTH JUNCTION
PRINCETON JUNCTION
LAWRENCE STATION
L
TRENTON
R
MORRISVILLE PA
R
BRISTOL
REVERSE CURVE
L
REVERSE CURVE
R
REVERSE CURVE
R WY
W YARD
R WY
R DUNLAP
REVERSE CURVE
L WOODBOURNE
R
MP 10
R
R
L BUSTLETON
R STREET RD
COUNTY LINE RD

Numeric table columns (MILES, GRADIENT, DISTANCE, SPEED, TIME, etc.): [illegible]

READ DOWN — WESTBOUND TRAFFIC | APPENDIX B2 — STATION | READ UP — EASTBOUND TRAFFIC

PAGE 2 COLUMN 1

MILES	GRADIENT, PERCENT	DISTANCE, ft	MAX GRADE CROSSING	MAX MPH	DISTANCE, feet	SPEED mph	dT, hours	TIME, hours	STATION	TIME, hours	dT, hours	DISTANCE, feet	SPEED mph
73.50		5750			5743	A 11	0.00944	0.75917		9.87738	0.00833	5280	120
74.45		4600			9377	120	0.0148	0.77397		9.86905	0.00728	4600	120
74.58		680		134	680	120	0.00107	0.77504	L	9.86179	0.00107	680	120
74.86	0.60%	1480		134	1480	120	0.00234	0.77738	L	9.86072	0.00234	1480	120
75.56	0.55%	3700			3700	120	0.00584	0.78322	YORK RD	9.85838	0.00584	3700	120
75.58	0.55%	100		190	100	120	0.00016	0.78337	L	9.85254	0.00016	100	120
76.26		3580		190	3580	120	0.00565	0.78903	L	9.85238	0.00565	3580	120
76.58		1700			1700	120	0.00268	0.79171		9.84673	0.00268	1700	120
76.77		980			980	120	0.00155	0.79326		9.84405	0.00155	980	120
77.37		3200		190	3200	120	0.00505	0.79831	R	9.8425	0.00505	3200	120
77.58		1100			1100	120	0.00174	0.80004		9.83745	0.00174	1100	120
78.38		4200			4200	120	0.00663	0.80667		9.83572	0.00663	4200	120
78.58		1080		134	1080	120	0.0017	0.80837	R	9.82908	0.00135	857	120
78.73		800		134	800	120	0.00126	0.80964	R CAMP HILL	9.82774	0		
79.26		2800			2800	120	0.00442	0.81406		9.82774	0		
79.58		1680		186	1680	120	0.00265	0.81671	L	9.82774	0.00944	5743	A 110
79.90		1700		155	1700	120	0.00268	0.81939	L	9.81829	0.00251	1460	110
80.58		3580			3580	120	0.00565	0.82504		9.81578	0.00616	3580	110
81.58		5280			5623	120	0.00887	0.83392		9.80961	0.00609	5280	110
81.71		680			337	D 12	0.00096	0.83487		9.80052	0.00117	680	110
82.11		2100		110	2100	110	0.00362	0.83809	R STENTON	9.79935	0.00362	2100	110
82.58		2500			2500	110	0.0043	0.84239		9.79574	0.00056	337.3	D 12
83.58		5280			5280	110	0.00909	0.85148		9.79518	0.01645	10420	120
84.58		5280	XING29 12		3220	110	0.00564	0.85703	CONSHOHOCKEN	9.77874	0.00944	5743	A 110
85.56		5280	XING 28.5		5743	A 11	0.00944	0.86647	GRAVERS	9.76929	0.00317	1840	110
85.99		2160			3777	120	0.00596	0.87243		9.76612	0.00375	2180	110
86.47		2500		147	2500	120	0.00395	0.87638	L	9.76237	0.0043	2500	110
86.58		600			4143	120	0.00654	0.88292	EARNEST	9.75807	0.00103	600	110
87.31		3880			337	D 15	0.00056	0.88347		9.75703	0.00668	3880	110
87.58		1400		110	1400	110	0.00241	0.88588	L RAMBO	9.75035	0.00241	1400	110
87.63	0.62%	280		110	280	110	0.00048	0.88636	L	9.74794	0.00048	280	110
87.90	0.56%	1400			1400	110	0.00241	0.88877		9.74746	0.00056	327.3	D 12
88.58	0.56%	3600		124	3600	110	0.0062	0.89497	R HENDERSON	9.7469	0.00736	4663	120
88.69	0.56%	580		124	580	110	0.001	0.89597	R	9.73955	0.00092	580	120
89.58	0.68%	4700			4700	110	0.00809	0.90406		9.73863	0.00742	4700	120
89.75		880			720	110	0.00124	0.9053		9.73121	0.00139	880	120
90.08		1800		134	5743	A 11	0.00944	0.91475	L	9.72982	0.00284	1800	120
90.58		2640					0	0.91475	KING of PRUSSIA	9.72698	0.0041	2600	120
91.58		5280			4097	120	0.00647	0.92121		9.72288	0.00833	5280	120
92.58		5280			5280	120	0.00833	0.92955		9.71455	0.00833	5280	120
92.69		580			580	120	0.00092	0.93046		9.70621	0.00092	580	120
92.99		1600		233	1600	120	0.00253	0.93298	R	9.7053	0.00253	1600	120
93.58		3100			3100	120	0.00489	0.93788		9.70277	0.00489	3100	120
94.58		5280			5280	120	0.00833	0.94621	HOWELLVILLE	9.69786	0.0076	4809	120
95.11		2800			2800	120	0.00442	0.95063		9.69006	0		
95.56		2400		190	2400	120	0.00381	0.95455	R	9.69006	0		
95.71		780			780	120	0.00123	0.95578		9.69006	0		
96.22		2600		190	2600	120	0.0041	0.95988	L	9.69006	0		
96.58		1900			1900	120	0.003	0.96288		9.69006	0		
97.47	0.60%	4700			4700	120	0.00742	0.9703		9.69006	0		
97.58	0.60%	580		329	580	120	0.00092	0.97121	R	9.69006	0		
97.77	0.60%	980		329	980	120	0.00155	0.97276	R	9.69006	0.04528	22060	A 56-
98.58	0.60%	4300			4300	120	0.00679	0.97955		9.64481	0.00219	649	56
98.79	0.60%	1100			1100	120	0.00174	0.98128		9.64261	0.00417	1118	A 45-
98.90	0.56%	580		233	580	120	0.00092	0.9822	L	9.63844	0.00583	1110	A 27-
99.36	0.56%	2442			2442	120	0.00385	0.98605		9.63261	0.00446	838	27
99.43	0.56%	370						0.98605		9.62814	0.0026	370	27
99.43	0.56%				1158	120	0.00183	0.98788		9.62554	0		
99.56	0.56%	788						0.98788		9.62554	0.00553	788	27
99.79	0.56%	1100		190	1100	120	0.00174	0.98962	R	9.62001	0.00772	1100	27
99.97	0.56%	963			963	120	0.00152	0.99114		9.6123	0.00676	963	27
100.30	0.56%	1737			1737	120	0.00274	0.99388		9.60554	0.01218	1737	27
100.43	0.56%	692						0.99388	L	9.59336	0.00485	692	27
100.58		788	8	147	800	120	0.00126	0.99514	L	9.5885	0.00553	788	27
100.73		802	XING 48.15			120	0	0.99514	W HENDERSON RD	9.58298	0.00563	802	27
100.78		245			75	120	0.00012	0.99526		9.57735	0.00172	245	27
100.78			OKW		1667	D 12	0.00056	0.99581	AMTRAK 74 OHD	9.57563	0		
101.03		1313		56	1313	56	0.00444	1.00026	R	9.57563	0.00921	1313	27
101.08		260			260	56	0.00088	1.00114		9.56642	0.00182	260	27
101.08			ug 0.35		587	56	0.00169	1.00612	RAVINE ROAD	9.5646	0		
101.22		750			162	D 56	0.00061	1.00673		9.5646	0.00526	750	27
101.30		453		45	384	45	0.00153	1.00826	R	9.55934	0.00318	453	27
101.32		101			190	D 45	0.001	1.00926	REVERSE CURVE	9.55616	0.00071	101	27
101.38		274		27	274	27	0.00192	1.01118	L	9.55545	0.00192	274	27
101.42		257		52	257	27	0.00378	1.01496	L GLEN LOCH 30 MPH (	9.55353	0.00139	289.7	D 52
101.42			USE LONG SWITCH LI				0	1.01496	JCT SWITCH	9.55214	0.00306	1283	D 10(
101.61	-0.56%	980			980	27	0.00687	1.02184	T	9.54908	2.2E-05	17	107
101.90	-0.55%	1,524		179	1524	27	0.01069	1.03253	L	9.54906	0		
102.14	-0.65%	1,292			1292	27	0.00878	1.04131	T	9.54906	0		
102.26	-0.55%	673			673	27	0.00472	1.04603	T	9.54906	0		
102.49	-0.55%	1,190		174	1190	27	0.00835	1.05438	L	9.54906	0		
102.71	-0.56%	1,191		169	1191	27	0.00835	1.06273	R REVERSE CURVE	9.54906	0		
103.14	-0.55%	2,226			2226	27	0.01561	1.07835	T	9.54906	0		
103.14					1707	27	0.01197	1.09032		9.54906	0.02306	11122	A 76-
104.14	-0.55%	5,280					0	1.09032		9.52601	0.00412	1654	76
104.16					1753	A 27	0.00833	1.09865	T	9.52188	0		
104.37	-0.56%	1,230	OHD 28.38		5115	A 52	0.01444	1.1131	LO GRADE - ABANDONE	9.52188	0.00307	1,230	76
105.14	-0.56%	4,060		347	2886	A 80	0.00667	1.11976	L	9.51882	0.01009	4,050	76
105.30	-0.56%	845		347	2181	A 90	0.00444	1.12421	L	9.50873	0.00211	845	76
105.91	-0.56%	3,221			389	96	0.00073	1.12494	T	9.50662	0.00803	3,221	76
105.91	-0.56%				505	D 96	0.00111	1.12605		9.49860			
106.14	-0.56%	1,214		76	1214	76	0.00303	1.12907	R	9.49859	0.00303	1214	76
106.14	-0.56%						0	1.12907	BRADFORD HILLS - EA(	9.49557	0.0006	212.5	D 85
106.32	-0.56%	970		380	970	76	0.00242	1.13149	R	9.49497	0.00216	970	85
106.51	-0.56%	1,024			1024	76	0.00255	1.13404	T	9.49281	0.00228	1024	85
106.88	-0.55%	1,940		85	1940	76	0.00483	1.13888	L	9.49053	0.00432	1940	85
106.88	-0.55%						0	1.13888		9.4862	0.00069	436.5	D 10'
107.14	-0.55%	1,347			1347	76	0.00336	1.14224	T	9.48551	0.00171	910.5	101
107.36	-0.55%	1,173			1173	76	0.00292	1.14516	T	9.48371	0.0022	1,173	101
107.75	-0.56%	2,060		101	2060	76	0.00518	1.15034	R	9.48151	0.0039	2,060	101
107.75	-0.55%		WEST END GLEN LOC				0	1.15034		9.47761	0.00106	615.9	D 12(
108.14	-0.55%	2,027			2027	76	0.00505	1.15539	T	9.47655	0.00223	1,411	120
108.36	-0.55%	1,202			439	76	0.00109	1.15649	T	9.47432	0.0019	1,202	120
108.36	-0.55%	0			2485	A 76	0.00583	1.16232	DOWNINGTOWN	9.47243	0	0	120
108.70	-0.80%	1,777		134	55	85	0.00012	1.16244	L	9.47243	0.0028	1,777	120
108.90		1,046			1046	85	0.00233	1.16477	T	9.46962	0.00165	1,046	120
109.14		1,255		144	348	85	0.00078	1.16555	R	9.46797	0.00198	1,255	120
109.35		1,109		144			0	1.16555	R	9.46599	0.00175	1,109	120
109.80		2,376			5689	A 85	0.01139	1.17694	T	9.46424	0.00375	2,376	120
110.14		1,795		144			0	1.17694	L	9.46049	0.00283	1,795	120
110.22		466		144	10084	A 10	0.01722	1.19416	L	9.45766	0.00074	466	120
110.36		725					0	1.19416	T	9.45692	0.00114	725	120
110.57		1,087		134			0	1.19416	R	9.45578	0.00172	1,087	120
110.77	0.50%	1,087		268			0	1.19416	R	9.45406	0.00172	1,087	120
110.97	0.50%	1,036					0	1.19416	T	9.45235	0.00163	1,036	120

PAGE 2 COLUMN 2

MILES	GRADIENT, PERCENT	DISTANCE, ft	MAX GRADE CROSSING	MAX MPH	DISTANCE, feet	SPEED mph	dT, hours	TIME, hours	STATION	TIME, hours	dT, hours	DISTANCE, feet	SPEED mph
111.14	0.50%	880		190			0	1.19416	L	9.45071	0.00139	880	120
111.18	0.50%	216					0	1.19416		9.44932	0.00034	216	120
111.32	0.50%	704		155			0	1.19416	L	9.44898	0.00111	704	120
112.00	0.80%	3,610			74	120	0.00012	1.19428	T	9.44779	0.0057	3,610	120
112.14	0.50%	700		152	700	120	0.0011	1.19538	L	9.4421	0.0011	700	120
112.18	0.50%	208		152	208	120	0.00033	1.19571	L	9.44098	0.00033	208	120
112.32	0.50%	784		152	784	120	0.00124	1.19695	R REVERSE CURVE	9.44066	0.00124	784	120
113.01	0.60%	3,607			3607	120	0.00569	1.20264	T	9.43942	0.00569	3,607	120
113.14		680		134	680	120	0.00107	1.20372	R	9.43373	0.00107	680	120
113.60		2,433		134	2433	120	0.00384	1.20755	R	9.43266	0.00384	2,433	120
113.71		621			621	120	0.00098	1.20854	T	9.42882	0.00098	621	120
114.08		1,915		134	1915	120	0.00302	1.21156	L	9.42784	0.00302	1,915	120
114.14		311			311	120	0.00049	1.21205	T	9.42482	0.00049	311	120
114.56		2,240			2240	120	0.00354	1.21558	T	9.42432	0.00354	2,240	120
115.14		3,040		190	3040	120	0.0048	1.22038	L	9.42079	0.0048	3,040	120
115.21		414		190	414	120	0.00065	1.22104	L	9.41598	0.00065	414	120
115.49	0.57%	1,449			1449	120	0.00229	1.22332	T	9.41534	0.00229	1,449	120
115.96	0.57%	2,485		155	2485	120	0.00392	1.22725	R	9.41305	0.00392	2,485	120
116.14	0.57%	932			932	120	0.00147	1.22872	T	9.40913	0.00147	932	120
116.88	0.57%	3,934			3934	120	0.00621	1.23492	T	9.40766	0.00621	3,934	120
117.14	0.57%	1,348		153	1348	120	0.00212	1.23705	L	9.40145	0.00212	1,348	120
117.23	0.57%	485		153	485	120	0.00077	1.23781	L	9.39932	0.00077	485	120
117.50	0.57%	1,455			1455	120	0.0023	1.24011		9.39856	0.0023	1455	120
117.89	0.57%	2,047		169	2047	120	0.00323	1.24334	R	9.39626	0.00323	2,047	120
118.14	0.57%	1,293			1293	120	0.00204	1.24538		9.39303	0.00204	1,293	120
119.14	0.57%	5280			5280	120	0.00833	1.25372		9.39099	0.00351	2222	120
119.62	0.57%	2,536			2536	120	0.004	1.25772	T	9.38748	0		
119.80	0.57%	984		217	984	120	0.00155	1.25927	L	9.38748	0		
120.14	0.57%	1,760			1760	120	0.00278	1.26205	T	9.38748	0.00194	12,332	A 95
121.14	0.57%	5280			5280	120	0.00833	1.27038	PARKESBURG	9.36554	0		
121.23	0.57%	495			495	120	0.00078	1.27116	T	9.36554	0		
121.47		1,265		190	1265	120	0.002	1.27316	R	9.36554	0.01306	5926	A 76
122.14		3,520			3520	120	0.00556	1.27872	T	9.35248	0.0011	440	76
122.90		4,015			4015	120	0.00634	1.28506	T	9.35139	0.01001	4,015	76
123.04		770		190	770	120	0.00122	1.28627	R	9.34138	0.00192	770	76
123.14		495			495	120	0.00078	1.28705	T	9.33946	0.00123	495	76
123.54		2,145			1357	120	0.00214	1.28919	T	9.33823	0.00535	2,145	76
123.54					788	D 15	0.00139	1.29058		9.33288	0		
124.14		3,135		95	2858	95	0.0053	1.29588	R a	9.33288	[illegible]	3,365	76
124.14					477	D 95	0.00106	1.29693	ATGLEN	9.33039	0.00111	440	A 74
124.29		834		76	834	76	0.00208	1.29901	R	9.32927	0.00046	174	74
124.39		500			500	76	0.00125	1.30026	REVERSE CURVE	9.32883	0.0033	1,290	74
124.80	0.53%	2,165		107	2165	76	0.0054	1.30566	L b	9.32553	0.00369	1451	A 87
124.87	0.53%	389			389	76	0.00097	1.30663	REVERSE CURVE	9.32184	0.00069	316	87
125.14	0.53%	1,350		107	1350	76	0.00348	1.31009	R c	9.32074	0.00393	1359	87
125.34	0.53%	1,056		107	1056	76	0.00263	1.31272	R CHRISTIANA	9.31682	0.00296	1,056	87
125.95	0.61%	3,224			2224	76	0.00802	1.32075	T	9.31383	0.00811	3,224	87
126.14	0.61%	1,000		107	1000	76	0.00249	1.32329	L d	9.30472	0.00203	1,000	87
126.24	0.61%	550		107	550	76	0.00137	1.32462	L	9.30189	0.00156	550	67
126.24	10.626	550			506	76	0.00126	1.32588	REVERSE CURVE	9.30034	0.00156	550	67
126.34					44	O 76	0.00011	1.32598		9.29878	0		
126.74		2,090		74	2090	74	0.00535	1.33134	R a	9.29878	0.00591	2,090	67
126.90	-0.56%	825			680	74	0.00174	1.33308	T	9.29286	0.00233	825	67
126.90					145	D 74	0.00039	1.33347		9.29054	0		
127.14	-0.56%	1,266		67	1266	67	0.00358	1.33705	L b	9.29054	0.00356	1,266	67
127.45	-0.56%	1,650		67	1650	67	0.00466	1.34171	L GAP	9.28697	0.00466	1,650	67
127.80	-0.56%	1,870		217	1870	67	0.00529	1.347	L	9.2823	0.00529	1,870	67
128.14	-0.56%	1,760			1760	67	0.00498	1.35197	T	9.27702	0.00498	1,760	67
128.20	-0.56%	330			330	67	0.00093	1.3529		9.27204	0.00093	330	67
128.59	-0.56%	2,090		67	2090	67	0.00591	1.35881	R c	9.27111	0.00591	2,090	67
128.59	-0.56%						0	1.35881		9.2652	0.00139	583	D 92
128.88	-0.56%	1,540			1540	67	0.00435	1.36317	T	9.26381	0.00197	957	92
129.14	-0.56%	1,320		92	1320	67	0.00373	1.3669	L d	9.26184	0.00272	1,320	92
129.33	-0.56%	1,045		92	1045	67	0.00295	1.36985	L	9.25912	0.00215	1,045	92
129.33							0	1.36985		9.25697	0.00156	879.9	D 12
129.81	-0.56%	2,530			2530	67	0.00715	1.377	T	9.25542	0.00262	1,659	120
130.14	-0.56%	1,705		190	1705	67	0.00482	1.38182	L	9.2528	0.00269	1,705	120
130.61	-0.56%	2,530			2530	67	0.00715	1.38898	T	9.25011	0.00399	2,530	120
130.79	-0.56%	935		190	330	67	0.00093	1.38991	L	9.24611	0.00148	935	120
131.14	-0.56%	1,815			6563	A 67	0.01556	1.40546	T	9.24464	0.00286	1,815	120
131.94	-0.56%	4,235					0	1.40546	T	9.24177	0.0051	3,229	120
132.14	-0.56%	1,045		134			0	1.40546	R	9.23666	0		
132.75	-0.56%	3,245		134	13065	A 92	0.02333	1.4288	R	9.23666	0		
133.14	-0.56%	2,035					0	1.4288	T	9.23666	0.00944	5743	A 11
133.51		1,980					0	1.4288	T	9.22723			
133.79		1,495		104			0	1.4288	L	9.22723	0.015	8023	A 9
134.14		1,815			644	119	0.02096	1.44976	T	9.21223	0.00147	715	92
134.58		2,334			302	D 11	0.0006	1.44986	T	9.21076	0.0040	2,334	92
135.06		2,501		110	2501	110	0.00431	1.45418	L a	9.20696	0.00615	2,501	92
135.14		445			445	110	0.00077	1.45490	T	9.20081	0.00092	445	92
135.64		2,640	XING 69.1		2640	110	0.00455	1.45947	IRISHTOWN ROAD	9.19989	0.00543	2,640	92
135.83		1,045		119	1045	110	0.0018	1.46127	R b	9.19446	0.00215	1,045	92
136.08		1,320			787	110	0.00136	1.46262	T	9.19231	0.00272	1,320	92
136.08					533	D 11	0.001	1.46362		9.18959	0		
136.14		275		92	275	92	0.00057	1.46419	R a	9.18958	0.00057	275	92
136.74		3,190		92	3190	92	0.00657	1.47076	R	9.18902	0.00657	3,190	92
137.09		1,870			1870	92	0.00385	1.47461	T	9.18246	0.00017	82.28	D 95
137.14		220		95	220	92	0.00045	1.47506	L b	9.18229	0.00027	135	95
137.68		2,390		95	2390	92	0.00492	1.47998	L BIRD IN HAND	9.18201	0.00476	2,390	95
137.89		1,112			1112	92	0.00229	1.48227	T	9.17725	0.00139	788.3	D 12
138.14	0.50%	1,779		134	1779	92	0.00366	1.48593	L	9.17586	0.00032	2,103	120
138.25	0.50%	606		134	606	92	0.00125	1.48718	L	9.17254	0.00095	606	120
139.05	0.50%	4,235			3024	92	0.00623	1.4934	T	9.17159	0.00669	4,235	120
139.14		440		190			0	1.4934	L	9.1649	0.00069	440	120
139.14					1234	A 92	0.0025	1.4959		9.16421	0		
139.32		1,045		190	1482	95	0.00291	1.49882	L	9.16421	0.00165	1,045	120
139.84		1,596			1596	95	0.00318	1.502	T	9.16256	0.00252	1,596	120
139.94		1,596		134	189	95	0.00038	1.50237	R	9.16004	0.00252	1,596	120
140.14	-0.54%	1,045			12532	A 95	0.02194	1.52432	T	9.15752	0.00165	1,045	120
140.54	-0.54%	4,224					0	1.52432	T	9.15588	0.00667	4,224	120
141.14	-0.54%	1,056		233			0	1.52432	R	9.14921	0.00167	1,056	120
141.64	-0.54%	2,640		233			0	1.52432	R	9.14754	0.00417	2,640	120
142.14	-0.54%	2,640			479	120	0.00076	1.52507	T	9.14338	0.00417	2,640	120
142.19	-0.54%	275			275	120	0.00043	1.52551	T	9.13921	0.00043	275	120
142.28	-0.54%	495		152	495	120	0.00078	1.52628	R	9.13877	0.00078	495	120
142.38	-0.54%	440			440	120	0.00069	1.52698	REVERSE CURVE	9.13799	0.00069	440	120
142.56	-0.54%	990		152	990	120	0.00156	1.52855	L	9.1373	0.00156	990	120
142.60		275			275	120	0.00043	1.52898	T	9.13574	0.00043	275	120
142.79		990		134	990	120	0.00156	1.53054	L	9.1353	0.00156	990	120
143.14		1,815			1815	120	0.00286	1.53341	T	9.13374	0.00286	1,815	120
143.52		2006			2006	120	0.00317	1.53657	LANCASTER STATION	9.13088	0.00317	2006	120
144.14		3274			3274	120	0.00517	1.54174		9.12771	0.00517	3274	120
144.63		2,585			2585	120	0.00408	1.54582	DILLERVILLE - COLA BF	9.12254	0.00408	2,585	120
145.14		2,695		128	2695	120	0.00425	1.55007	R	9.11846	0.00425	2,695	120
145.33		1,045		128	1045	120	0.00165	1.55172	R	9.11421	0.00165	1,045	120

APPENDIX B2 — PAGE 3, COLUMN 1

READ DOWN — WESTBOUND TRAFFIC | STATION | READ UP — EASTBOUND TRAFFIC

MILES	GRADIENT PERCENT	DISTANCE	MAX GRADE	MAX MPH	DISTANCE	SPEED	ET HOURS	TIME HOURS	STATION	TIME HOURS	ET HOURS	DISTANCE	SPEED	MILES	GRADIENT PERCENT
146.14		4,235			4235	120	0.00868	1.55841	T	9.11298	0.00960	4,235	120	180.47	
146.79		3,465			3465	120	0.00547	1.56386	T	9.10568	0.00547	3,465	120	180.52	
147.14		1,815		134	1815	120	0.00286	1.56674	L	9.10041	0.00286	1,815	120	180.53	
147.18	0.56%	220		134	220	120	0.00035	1.56709		9.09754	0.00035	220	120	180.57	
147.82	0.56%	3,410	XING 71.6		3410	120	0.00538	1.57247	E PETER ROAD	9.09718	0.00538	3,410	120	180.57	
148.14		1,650		134	1650	120	0.0026	1.57507	L	9.09181	0.0026	1,650	120	180.66	
148.28		660		134	660	120	0.00104	1.57612	L	9.08921	0.00104	660	120	180.76	
148.54		1,455			1455	120	0.00234	1.57846	T	9.08817	0.00234	1,455	120	180.89	0.56%
148.93		2,035		134	2035	120	0.00321	1.58167	R	9.08582	0.00321	2,035	120	181.02	
149.14		1,100			1100	120	0.00174	1.58341	T	9.08261	0.00174	1,100	120	181.14	
150.14		5280			5280	120	0.00833	1.59174		9.08088	0.00833	5,280	120	181.18	
150.34	0.60%	1,056			1056	120	0.00167	1.59341	T	9.07254	0.00167	1,056	120	181.21	
150.94	0.60%	2,668		126	2668	120	0.00421	1.59762	L	9.07088	0.00421	2,668	120	181.34	
150.95	0.60%	778			778	120	0.00123	1.59885		9.06668	0.00123	778	120	181.34	
151.14	0.60%	778		134	778	120	0.00123	1.60007	R	9.06544	0.00123	778	120	181.80	
151.34	0.60%	1,056		134	1056	120	0.00167	1.60174	R 76 P	9.06421	0.00167	1,056	120	181.88	
152.04		3,724	XING 75.8		3724	120	0.00588	1.60762	PRIVATE 50 F	9.06254	0.00588	3,724	120	181.68	
152.14		500		500	500	120	0.00079	1.60841	L	9.05666	0.00079	500	120	181.76	
152.64	-0.67%	2,666		500	2666	120	0.00421	1.61262	L	9.05588	0.00421	2,666	120	181.88	
153.14	-0.67%	2,612			2612	120	0.00412	1.61674	T	9.05166	0.00412	2,612	120	182.07	
153.22	-0.67%	440			440	120	0.00069	1.61743	T	9.04754	0.00069	440	120	182.10	
[illegible]	[illegible]	[illegible]	[illegible]	[illegible]	[illegible]	[illegible]	[illegible]	[illegible]	[illegible]	[illegible]	[illegible]	[illegible]	[illegible]	[illegible]	[illegible]

APPENDIX B2 — PAGE 3, COLUMN 2

READ DOWN — WESTBOUND TRAFFIC | STATION | READ UP — EASTBOUND TRAFFIC

[illegible]	[illegible]	[illegible]	[illegible]	[illegible]	[illegible]	[illegible]	[illegible]	[illegible]	[illegible]

APPENDIX B2

READ DOWN — WESTBOUND TRAFFIC — STATION — READ UP — EASTBOUND TRAFFIC

PAGE 4 COLUMN 1

206.01	1300		82	1300	82	0.008	2.26277	L FLUSH RUN
206.10	930		82	930	82	0.00215	2.26492	L
206.10						0	2.26492	
206.01	4400		130	4400	82	0.01016	2.27508	L
206.29	1313			1313	82	0.00303	2.27812	T 201 K
206.86	3800		95	3800	82	0.00878	2.28690	R
207.59	2000		87	1487	82	0.00343	2.28033	R
207.59						0	2.29033	
207.78	1024			1451	A 82	0.00333	2.29008	
207.95	900		130	2146	87	0.00467	2.29835	L
208.05	500		130	500	87	0.00109	2.29440	L
208.31	1400			1400	87	0.00305	2.30247	T NEWPORT
208.95	1600	XING	130	1600	87	0.00392	2.20639	R PRIVATE
208.64	1600			1600	87	0.00218	2.20856	
208.91	373	XING	130	373	87	0.00081	2.30938	R PRIVATE
209.23	2476		130	2476	87	0.00539	2.21477	R
209.91	2606			1525	87	0.00332	2.21809	
210.07	825			11075	A 87	0.02083	2.23892	
210.90	4400		143			0	2.23892	L

PAGE 4 COLUMN 2

READ DOWN — WESTBOUND TRAFFIC | APPENDIX B2 — STATION | READ UP — EASTBOUND TRAFFIC

PAGE 5 COLUMN 1

MILES	GRADIENT PERCENT	DISTANCE	CROSSING	MAX MPH	DISTANCE feet	SPEED	ET hours	TIME hours	STATION	TIME hours	ET hours	DISTANCE feet	SPEED	MILES
266.11		3297			3297	110	0.00968	3.07916		7.59074	0.00227	812 A	67-	292.06
266.49		2000		110	1381	110	0.00334	3.08151	L	7.56852	0.00309	1060	67	292.06
266.49			XING 191.(		839 D	11	0.00122	3.08273	PRIVATE	7.58543			67	292.39
266.76		1100		88	1100	88	0.00287	3.0851	L MT UNION	7.58543	0.00211	1100	67	293.16
266.78		477			477	88	0.00103	3.08612	REVERSE CURVE	7.58232	0.00136	477	67	293.66
267.28		2000		98	2000	88	0.0066	3.09172	R	7.58097	0.00735	2100	67	293.95
267.60		1600		94	1600	88	0.00344	3.09516	R	7.57362	0.00452	1600	67	293.76
267.91		1662			1662	88	0.00362	3.09878		7.5691	0.00476	1662	67	293.76
268.28		2000		110	1604	88	0.00345	3.10223	L	7.56434	0.00586	2000	67	293.80
268.28					396 D	98	0.00094	3.10318		7.55889			67	294.10
268.34		300		71	300	71	0.0029	3.10398	L	7.55889	0.00066	300	67	294.15
268.37		148			67	71	0.00018	3.10416	TRC	7.55784	0.00042	148	67	294.27
268.27					81 D	87	0.00022	3.10438	MAPLETON	7.56742				294.41
269.28		4800		67	4800	87	0.01357	3.11795	RMPH60	7.56742	0.01357	4800	67	294.68
269.61		1200		67	1200	67	0.00339	3.12134	R	7.54385	0.00339	1200	67	295.14
269.61								3.12134		7.64046	0.00139	553 D	92-	295.31
269.61								3.12134		7.53907	3.5E-05	17	92	295.50
269.94		2000			2000	67	0.0065	3.12784		7.53904	0.0036	1700	92	295.87
270.22		1477		92	1477	87	0.00418	3.13202	R	7.53054	0.00394	1477	92	295.73
270.28		300			300	87	0.00080	3.13286		7.5325	0.001	533.2 D	110	295.94
270.66		2600			2600	67	0.00565	3.13852		7.5315	0.00304	1267	110	295.94
271.27		3246		134	3246	87	0.00915	3.14769	L	7.52846	0.00318	1847	110	296.16
271.48		1085		134	1085	87	0.00307	3.15076	L	7.52528	0			296.30
271.78		1600			592	87	0.00167	3.15243		7.52528	0			296.40
272.04		1400		116	6983 A	87	0.01556	3.16799	L	7.52528	0			296.54
272.27		1200		120			0	3.16799	L MILL CREEK	7.52528	0.01694	8663 A	89	296.75
272.80		2790	XING	110	8023 A	92	0.015	3.18299	L CASSVILLE	7.50833	0.00411	522	89	296.90
273.27		2600					0	3.18299		7.50722	0.00671	3151	89	297.14
273.31		178					0	3.18299		7.50052	0.00972	4203 A	74-	297.30
273.67		2600		190			0	3.18299	R	7.49079	0.00306	1196	74	297.30
274.27		1600			1667	110	0.00285	3.18584		7.48773	0.0041	1800	74	297.40
274.79		2700			813 D	11	0.00117	3.18701		7.48364	0.00222	847 A	70-	297.63
275.01		1179	XING	89	1179	89	0.00222	3.18923	R JACOBS	7.48142	0.00179	980	70	297.84
275.27		1400			4200	89	0.00894	3.19817		7.47963	0.00379	1400	70	297.86
276.26		5297			2138	89	0.00455	3.20272		7.47584	0.01433	5297	70	298.14
276.28					379 D	86	0.00083	3.20355		7.46151			70	298.19
276.52		1300		74	1300	74	0.00333	3.20688	L	7.46151	0.00352	1300	70	298.36
276.57		243			243	74	0.00062	3.2075	REVERSE CURVE	7.45799	0.00066	243	70	298.56
276.57		2100		81	2046	74	0.00516	3.21266	R	7.45734	0.00568	2100	70	299.14
276.97					84 D	74	0.00022	3.21288	REVERSE CURVE	7.45165			70	299.52
277.25		1500		70	1500	70	0.00406	3.21694	L	7.45165	0.00406	1500	70	299.72
277.44		1000	XING	70	1000	70	0.00271	3.21965	L MPH 55 4TH STREET	7.4476	0.00271	1000	70	299.86
277.59		800			800	70	0.00216	3.22181	T HUNTINGDON	7.44489	0.00216	800	70	299.85
277.76		1000		182	1000	70	0.00271	3.22452	R	7.44272	0.00271	1000	70	299.99
277.91		700			700	70	0.00189	3.22641		7.44002	0.00189	700	70	300.14
278.20		1500		89	1500	70	0.00406	3.23047	R	7.43813	0.00406	1500	70	300.42
278.24		239			239	70	0.00065	3.23112		7.43407	0.00065	239	70	301.14
278.60		3197			3197	70	0.00865	3.23977		7.43342	0.00865	3197	70	302.14
279.09		1300		144	1306	70	0.00352	3.24328	R	7.42477	0.00426	1574	70	302.95
279.26		800		70	800	70	0.00216	3.24545	R	7.42051	0.00065	239	70	302.95
279.46		800		70	464	70	0.00126	3.2467	R	7.41987	0.00294	1087	70	303.01
279.45		300			300	70	0.00081	3.24751	TDEER	7.41692	0.00081	300	70	303.07
279.64		1000		178	1000	70	0.00271	3.25022	L RC MPH 55	7.41611	0.00415	1533	70	303.07
279.81		900		144	900	70	0.00244	3.25266	L	7.41196	0.00278	991 A	65-	303.53
280.17		1600			1600	70	0.00433	3.25699		7.40919	0.00284	876	65	303.53
280.24		670		135	670	70	0.00181	3.2588	R	7.40634	0.00195	670	65	304.06
280.30		277		131	277	70	0.00075	3.25955	R	7.40439	0.00081	277	65	305.06
281.24		5000			5000	70	0.01353	3.27307		7.40358	0.01467	5000	65	305.59
281.26		166			916	70	0.00248	3.27556		7.38891	0.00175	690	65	305.84
281.36		434			20 D	6	5.8E-05	3.27561		7.38727	0			306.06
281.57		1100		68	1100	68	0.00307	3.27863	L	7.38727	0.00321	1100	65	306.06
281.75	0.51%	977		82	977	68	0.00268	3.28131	L	7.38406	0.00285	977	65	307.06
282.02	0.51%	1400	XING 206.(		1321	68	0.00363	3.28494	T WARRIOR RIDGE	7.38121	0.00408	1400	65	307.70
282.02					79 D	81	0.00022	3.28516		7.37714				308.03
282.24	0.51%	1200		68	1200	66	0.0035	3.28865	R	7.37714	0.0035	1200	65	308.06
282.49		1300		65	1300	65	0.00379	3.29244	R	7.37364	0.00379	1300	65	308.08
282.49							0	3.29244		7.36985	0.00033	119.7 D	71-	308.58
282.51		87			87	66	0.00025	3.2927	REVERSE CURVE	7.36952	0		71	308.66
282.73		1200		71	1200	65	0.0035	3.29619	L	7.36952	0.00311	1187	71	309.08
282.73							0	3.29619		7.3664	0.00066	222.9 D	81-	309.56
282.96		1200		61	1200	65	0.0035	3.29969	L	7.36586	0.00228	977.1	81	310.04
282.96							0	3.29969		7.36356	0.00022	97.33 D	95-	310.86
283.24		1500			1500	65	0.00437	3.30406		7.36334	0.00313	1453	85	310.78
283.47		1190			1190	65	0.00347	3.30753		7.36022	0.00265	1190	85	311.01
283.85		2000		124	2000	65	0.00583	3.31335	L PETERSBURG	7.35756	0.00446	2000	85	311.01
284.25		2100		85	2100	65	0.00612	3.31947	L	7.35311	0.00468	2100	85	311.04
284.25							0	3.31947		7.34843	0.00056	264 D	95	311.14
284.42		900			900	65	0.00262	3.3221		7.34787	0.00024	118	95	312.04
284.77		1886		117	823	65	0.0024	3.32449	R	7.34764				312.76
285.25		2500			1199 A	65	0.00333	3.32783		7.34764	0.02333	9114 A	59-	313.03
285.25			XING 210.(		88	71	0.00028	3.32806	PRIVATE MP 210	7.32431	0			313.29
286.25		5280	XING 210.(		2346 A	71	0.00583	3.33389	PRIVATE	7.32431	0.00346	1078	59	313.60
287.10		4600			1469 A	81	0.00276	3.33667		7.32084	0.01541	4600	59	313.68
287.22		322		126	4175	86	0.0093	3.34597	R	7.30544	0.00658	2049	59	313.08
287.22					6788 A	86	0.01361	3.35960		7.29886				313.97
287.60		2000	XING	126	83	103	0.00012	3.3597	R BARREE	7.29886	0.00222	664 A	54-	314.04
287.83		1200			1046 D	1(	0.00244	3.36215		7.29664	0.00196	560	54	314.34
288.13		1600			1600	59	0.00514	3.36728		7.29467	0.00561	1600	54	314.53
288.21		415		59	415	59	0.00133	3.36861	L	7.28906	0.00146	415	54	314.74
288.36		805		59	800	59	0.00257	3.37118	L	7.28761	0.00281	800	54	314.91
288.47		600			664	59	0.00213	3.37331	L	7.2848	0.00222	664	54	315.04
288.89		2100		35	900	59	0.00289	3.3762		7.28258	0.00772	2200	54	315.10
289.04		810			236	59	0.00071	3.37691	TUNNEL - NORTH REBL	7.27486	0.00077	220	54	315.16
289.18		830			1400	59	0.00449	3.3814	R	7.27409	0.00491	1400	54	315.48
289.31		800			800	59	0.00257	3.38397	T SPRUCE CREEK	7.26918	0.00281	800	54	315.53
289.60		1000		78	1000	59	0.00321	3.38718	R	7.26637	0.00351	1000	54	315.53
289.80		1000		84	1000	59	0.00321	3.39039	R	7.26287	0.01094	3120	54	315.76
289.80		580			1568	59	0.005	3.39538	REVERSE CURVE	7.25192	0.00306	805 A	46-	316.04
289.80					83 D	56	0.00028	3.39567		7.24887				316.19
290.14		1800		54	1800	54	0.00631	3.40198	L	7.24887	0.00615	1493	46	316.36
290.18		200		143	200	54	0.0007	3.40268	L	7.24272	0.00111	260 A	43-	316.52
290.58		2100			2100	54	0.00737	3.41006		7.24161	3.5E-05	6	43	316.63
290.84		1300		92	1300	54	0.00456	3.41461	L	7.24157	0.00673	1300	43	316.74
291.03		1077			1077	54	0.00378	3.41839		7.23685	0.00474	1077	43	316.76
291.16		700		103	700	54	0.00246	3.42084	R	7.23111	0.00308	700	43	316.76
291.37		1100		55	1100	54	0.00386	3.4247	R	7.22802	0.01169	2653	43	316.93
291.54		900		85	900	54	0.00316	3.42786	R	7.21634	0.00028	62 A	42-	317.04
291.74		1062	XING 216.(		1062	54	0.00372	3.43158	PEMBERTON	7.21606	0.00156	347	42	317.21
291.96		1100		134	1100	54	0.00386	3.43544	L	7.21449	0.00498	1100	42	317.21
292.14		1000			1000	54	0.00351	3.43895		7.20953	0.00451	1000	42	317.25
292.33		1000			451	54	0.00162	3.44056		7.20502	0.00451	1000	42	317.32
292.38					160 A	54	0.00056	3.44112		7.20052	0			317.42
292.53					246	65	0.00085	3.44197		7.20052	0			317.57
292.53					133 D	9(	0.0005	3.44247	UNION FURNACE	7.20052	0			317.66
292.84		1300		46	1205	46	0.00535	3.44782	L	7.20052	0.00586	1300	42	317.76

READ DOWN — WESTBOUND TRAFFIC | APPENDIX B2 — STATION | READ UP — EASTBOUND TRAFFIC

PAGE 5 COLUMN 2

MILES	GRADIENT PERCENT	DISTANCE	CROSSING	MAX MPH	DISTANCE feet	SPEED	ET hours	TIME hours	STATION	TIME hours	ET hours	DISTANCE feet	SPEED
292.06		450			414	46	0.0017	3.44952	TRC MPH 35	7.19405	0.00204	450	42
292.06					39 D	52	0.00017	3.44969		7.19261			42
292.39		1200		43	1200	43	0.00528	3.45497	R	7.19261	0.00541	1200	42
293.16		1400			1400	43	0.00617	3.46114		7.1872	0.00631	1400	42
293.66		2300			2300	43	0.01013	3.47127		7.18089	0.01037	2300	42
293.95		300		142	300	43	0.00132	3.47259	R	7.17051	0.00135	300	42
293.76		600			588	43	0.00259	3.47518	REVERSE CURVE	7.16916	0.00271	600	42
293.76					12 D	51	5.6E-05	3.47524		7.16646			
293.80		200		42	200	42	0.0009	3.47614	L	7.16646	0.0009	200	42
294.10		1600		44	1600	42	0.00722	3.48336	R RC	7.16555	0.00722	1600	42
294.15		247			247	42	0.00111	3.48447		7.15834	0.00111	247	42
294.27		400			400	42	0.0018	3.48627		7.15723	0.0018	400	42
294.41		1000		93	1000	42	0.00451	3.49078	L	7.15542	0.00451	1000	42
294.68		1400			1400	42	0.00631	3.49709	T BIRMINGHAM	7.15091	0.00631	1400	42
295.14		2467		126	2467	42	0.01112	3.50822	L	7.1446	0.01112	2467	42
295.31		900		126	900	42	0.00406	3.51228	L	7.13347	0.00406	900	42
295.50		1000			1000	42	0.00451	3.51679		7.12942	0.00451	1000	42
295.87		900		44	900	42	0.00406	3.52084	R	7.12491	0.00406	900	42
295.73		300			300	42	0.00135	3.5222	REVERSE URVE	7.12085	0.00135	300	42
295.94		1086		42	788	42	0.00354	3.52574	L	7.1195	0.00489	1086	42
295.94								3.52574		7.1146	0.00022	51.63 D	52
296.16		1100			1100	42	0.00056	3.5263		7.11438	0.00432	1048	46
296.30		800			800	42	0.00361	3.5299		7.11006	0.00329	800	46
296.40		1000		56	1000	42	0.00451	3.53441	L	7.10877	0.00412	1000	46
296.54		300			300	42	0.00135	3.53577	REVERSE CURVE	7.10265	0.00124	300	46
296.75		1100	XING	58	1100	42	0.00496	3.54073	R PRIVATE	7.10142	0.00453	1100	46
296.90		800			800	42	0.00361	3.54433		7.09689	0.00329	800	46
297.14		1262		73	1262	42	0.00569	3.55002	L TTMPH 35	7.09359	0.0052	1262	46
297.30		860		46	580	42	0.00358	3.5533	L TYRONE	7.0884	0.00304	560	46
297.30			XING 222.18					3.5539	WILSON ROAD	7.08486	0.00117	348 D	67
297.40		500			500	42	0.00226	3.55616	RC	7.08369	0.00043	152	67
297.63		1200		78	1200	42	0.00541	3.56157	R	7.08326	0.00339	1200	67
297.84		1100		66	1100	42	0.00496	3.56653	R	7.07987	0.00311	1100	67
297.86		600			600	42	0.00271	3.56923	RC	7.07679	0.0017	600	67
298.14		1000		67	378	42	0.0017	3.57094	L	7.07508	0.00283	1000	67
298.19		273		67	322 A	42	0.00139	3.57233	L	7.07224	0.00077	273	67
298.36		900			1772	46	0.0073	3.57962		7.07147	0.00033	125.2 D	73
298.56		1700		130	1700	46	0.007	3.58662	R	7.07113	0.00645	2477	73
299.14		2400			2400	46	0.00988	3.5965		7.06471	0.00623	2400	73
299.52		2000			1026	46	0.00423	3.60074		7.05848	0.00519	2000	73
299.72	GRAZIERVILLE END 2 TR							3.60074	GRAZIERVILLE	7.05328			
299.86		1900		73	2575 A	46	0.00861	3.60935	L	7.05328	0.00495	1900	73
299.85			XING 224.(		225	66	0.00065	3.60999	PRIVATE	7.04830	5.8E-05	21.56 D	74
299.99		600	XING 225.(		195 A	66	0.00056	3.61055	TTTMPH 60 RC PRIVAT	7.04831	0.00148	578.4	74
300.14		782	XING	74	1977	67	0.00559	3.61614	R TIPTON ROAD	7.04683	0.002	782	74
300.42		1473		74	1235 A	67	0.00333	3.61947	R	7.04482	0.00377	1473	74
301.14	0.57%	3800			3320	73	0.00861	3.62808	T TIPTON	7.04105	0.00973	3800	74
302.14		5278			4345	73	0.01127	3.63936		7.03133	0.01351	5278	74
302.95		4300		74	216 A	73	0.00056	3.63991	L	7.01782	0.01101	4300	74
302.95			XING 227.(		5017	74	0.01284	3.65275	PA RTE 502	7.00681	0.00022	89.17 D	78
303.01		300			300	74	0.00077	3.65352	T FOSTORIA RC	7.00658	0.00051	210.8	78
303.07		358		76	356	74	0.00092	3.65444	RTTMPH60	7.00606	0.00087	356	78
303.07							0	3.65444		7.00521	0.00106	487.7 D	97
303.53		2400		97	2409	74	0.00617	3.6606	R	7.00415	0.00375	1921	97
303.53							0	3.6606		7.0004	0.00111	627.7 D	11(
304.06		2800			2800	74	0.00717	3.66777		6.99928	0.00075	462.3	117
305.06		5280	XING 230.(		5133	74	0.01314	3.68091	PRIVATE	6.99854	0		
305.59		2800			882 A	74	0.00222	3.68313		6.99854	0.01833	10326 A	96
305.84		1278		130			0	3.68313	L	6.98021	0.00146	742	96
306.06		1700			6178 A	78	0.025	3.70813		6.97875	0.00237	1200	96
306.06					8870 A	97	0.01194	3.72007		6.97638	0		96
307.06		5280			400	111	0.00068	3.72075		6.97638	0.01637	5258	96
307.70		3400			455 D	11	0.00083	3.72158	T ANTIS	6.96001	0.0075	3400	96
308.03		1300		96	1300	96	0.00256	3.72415	R	6.95251	0.00221	1118	96
308.06		202			202	96	0.0004	3.72455	TMPH80(??)	6.9503	0		
308.08		108			108	96	0.00021	3.72476	REVERSE CURVE	6.9503	0.01058	4540 A	81
308.58		2600		139	2600	96	0.00513	3.72989	L	6.94575			
308.66		2500			2500	96	0.00493	3.73483		6.94575	0.00152	651	81
309.08		200			200	96	0.00039	3.73522		6.94422	0.00047	200	81
309.56		2500		197	2500	96	0.00493	3.74015	R	6.94376	0.00585	2500	81
310.04		2549			2549	96	0.00503	3.74518		6.93791	0.00596	2549	81
310.86		3200			341	98	0.00067	3.74586		6.93195	0.00743	3200	81
310.78		700		233	3982 A	96	0.0075	3.75335	L	6.92447	0.00021	90	81
311.01		1200			122	105	0.00022	3.75357	TTTMPH30	6.92428	0		
311.01					656 D	1(	0.00133	3.75491		6.92426			
311.04		142		81	142	81	0.00033	3.75524	R ALTOONA	6.92426	0.01361	4959 A	55
311.14		520		81	520	81	0.00122	3.75646	R	6.91065	0.02111	6845	55
312.04	1.56%	4800			4800	81	0.01122	3.76768	T ALTO	6.88954	0.00417	1101 A	44
312.76	1.12%	3767			3767	81	0.00885	3.77653	T SLOPE	6.88537	0.00054	125	44
313.03	1.12%	1400		110	1400	81	0.00327	3.7798	L BRICKYARD	6.88483	0.00603	1400	44
313.29	1.85%	1400	XING	110	1013	81	0.00237	3.78218	L COBURN ROAD	6.87881	0.00111	251 A	41
313.60	1.85%	3100			1465	81	0.00332	3.78551	TTTMPH 35	6.87769	0.0059	1278	41
313.68					377	86	0.00083	3.78634		6.87179			
313.08					641 D	86	0.00172	3.78806		6.87179			
313.97	1.89%	1500		55	1500	55	0.00517	3.79323	R	6.87179	0.00693	1500	41
314.04	1.89%	372			372	55	0.00128	3.79451		6.86486	0.00172	372	41
314.34	1.46%	1574			1574	55	0.00542	3.79993		6.86314	0.01386	3000	41
314.53	1.46%	1000		61	1000	55	0.00344	3.80337	R	6.84929	0.00129	337 A	36
314.74	1.82%	1100			1100	55	0.00379	3.80716	TTTMPH35	6.8479	0.00177	337	36
314.91	1.82%	900		90	900	55	0.0031	3.81026	L	6.84612	0.00473	900	36
315.04	1.82%	700			700	55	0.00241	3.81267		6.84139	0.00368	700	36
315.10	1.82%	600			640	55	0.0022	3.81487		6.83771	0.00421	800	36
315.16	1.82%				160 D	55	0.00061	3.81548		6.8335	0		
315.48	1.82%	1500		44	1500	44	0.00646	3.82194	R	6.8335	0.00789	1500	36
315.53	1.82%	276			238	44	0.00103	3.82297	TTTMPH 36 RC	6.82561	0.00145	276	36
315.53	1.82%		UG 240.80		37 D	44	0.00017	3.82314	SCOTTS RUN	6.82415	0		
315.76	1.82%	1200		41	1200	41	0.00554	3.82868	L	6.82415	0.00631	1200	36
316.04	1.82%	1500			1500	41	0.00693	3.83561		6.81784	0.00789	1500	36
316.19	1.82%	800			800	41	0.0037	3.83931		6.80995	0.00421	800	36
316.36	1.82%	900		134	800	41	0.0037	3.843	L	6.80574	0.00473	900	36
316.52	1.82%	900			900	41	0.00416	3.84716		6.80101	0.00473	900	36
316.63	1.82%	600		65	600	41	0.00277	3.84993	R	6.7965	0.00316	600	36
316.74	1.82%	600		58	600	41	0.00277	3.8527	R	6.79284	0.00316	600	36
316.76	1.82%	87			31	41	0.00014	3.85284	RC	6.78948	0.00046	87	36
316.76	1.82%				56 C	41	0.00026	3.85312		6.78903	0		
316.93	1.34%	900		36	900	36	0.00473	3.85785	L	6.78903	0.00473	900	36
317.04	1.34%	600		82	600	36	0.00316	3.86101	L HORSESHOE	6.78529	0.00316	600	36
317.21	1.77%	900		36	900	36	0.00473	3.86574	L	6.78214	0.00473	900	36
317.21							0	3.86574		6.7774	0.00084	221.9 D	53
317.25	1.77%	182			182	36	0.00096	3.8667	RC	6.77646	0.00179	566	53
317.32	1.77%	400		76	400	36	0.0021	3.86881	R	6.77467	0.00275	730 A	46
317.42	1.78%	500			500	36	0.00263	3.87144		6.77189	0.00047	113	46
317.57	1.78%	800		60	800	36	0.00421	3.87565	R	6.77143	0.00361	772 A	35
317.66	1.78%	500		134	500	36	0.00263	3.87828	R	6.76781	0.00024	45	35
317.76	1.78%	600		53	600	36	0.00316	3.88143	R	6.76757	0.00025	600	35

APPENDIX B2

READ DOWN — WESTBOUND TRAFFIC — STATION — READ UP — EASTBOUND TRAFFIC

PAGE 6 COLUMN 1

MILES	GRADIENT PERCENT	DISTANCE ft		MAX MPH	DISTANCE feet	SPEED MPH	ET HOURS	TIME HOURS	STATION	TIME HOURS	ET HOURS	DISTANCE feet	SPEED MPH
317.83	1.76%	800		57	800	36	0.00421	3.88564	R	6.76422	0.00433	800	35
318.04	1.76%	600			800	36	0.00316	3.8888		6.76	0.00325	600	35
318.23	1.76%	1000			1000	36	0.00526	3.89406		6.75675	0.00541	1000	35
318.40	1.76%	900		96	900	36	0.00473	3.89879	CMG	6.75134	0.00487	900	35
318.72	1.76%	1670			1670	36	0.00879	3.90758		6.74647	0.00904	1670	35
319.04	1.76%	1700		96	1700	36	0.00894	3.91652	R	6.73743	0.0092	1700	35
319.25	1.76%	1100		63	1100	36	0.00579	3.92231	R	6.72823	0.00595	1100	35
319.36	1.76%	800			248	36	0.0013	3.92361		6.72228	0.00325	800	35
319.57	1.76%	1100		100	720 A	36	0.00361	3.92723	R	6.71903	0.00565	1100	35
319.72	1.76%	800		46	1486	46	0.00612	3.93335	R	6.71308	0.00433	800	35
319.74	1.76%	85				46	0	3.93335	REVERSE CURVE	6.70875	0.00046	85	35
319.74	1.76%				131 D	46	0.00061	3.93396		6.70829			
319.93	1.76%	1000		36	1000	36	0.00541	3.93937	L	6.70829	0.00541	1000	36
319.93	1.76%						0	3.93937		6.70288	0.00011	21.12 D	37
319.97	1.76%	200			200	36	0.00108	3.94045	RC	6.70277	0.00092	178.9	37
320.08	1.76%	500		58	500	36	0.00271	3.94316	R	6.70185	0.00256	500	37
320.18	1.76%	500		58	500	36	0.00271	3.94586	R	6.69929	0.00256	500	37
320.18	1.76%	149			149	36	0.00081	3.94667	TRC	6.69673	0.00076	149	37
320.37	1.76%	1000		36	1000	36	0.00541	3.95208	CMPH 30(TT)	6.69597	0.00512	1000	37
320.49	1.76%	600			600	36	0.00325	3.95533		6.69085	0.00307	600	37
320.62	1.76%	700		95	700	36	0.00379	3.95911	R	6.68778	0.00358	700	37
320.77	1.76%	800		65	800	36	0.00433	3.96344	R	6.6842	0.0041	800	37
320.92	1.76%	800			800	36	0.00433	3.96777		6.6801	0.0041	800	37
321.00	1.76%	400		66	400	36	0.00216	3.96994	R	6.67601	0.00205	400	37
321.04	1.76%	200		56	200	36	0.00108	3.97102	R	6.67396	0.00102	200	37
321.08	1.76%	211			211	36	0.00114	3.97216	RC	6.67294	0.00108	211	37
321.32	1.76%	1300		37	1300	36	0.00703	3.97919	L	6.67186	0.00666	1300	37
321.32							0	3.97919		6.6652	0.00063	106.8 D	52
321.42	2.27%	500			500	36	0.00271	3.9819	T BENNY	6.66457	0.00111	304.2	52
321.51	2.27%	500		160	500	36	0.00271	3.98461	CEG -2.27%	6.66326	0.00182	500	52
321.63	2.27%	600			600	36	0.00325	3.98785	T(X I)	6.66144	0.00219	600	52
322.04	2.27%	2200		107	2040	36	0.01104	3.99889	R	6.65925	0.00801	2200	52
322.16	2.27%	607		107	159 A	36	0.00083	3.99972	R	6.65124	0.00221	607	52
322.27	2.27%	600		95	1208	37	0.00618	4.00591	R	6.64903	0.00219	600	52
322.44	1.21%	900			900	37	0.00461	4.01052		6.64685	0.00328	900	52
323.05	1.21%	3200			3200	37	0.01638	4.0269	T TUNNELS	6.64357	0.01166	3200	52
323.12	1.21%	400			400	37	0.00205	4.02894	T TUNNELS	6.63191	0.00146	400	52
323.26	0.74%	700			700	37	0.00358	4.03253	GALLITZEN	6.63046	0.00255	700	52
323.48	-1.11%	1200		52	780	37	0.00406	4.03659	L@UNT@AR E END 4 T	6.62791	0.00437	1200	52
323.48							0	4.03659		6.62354	0.00044	131.4 D	60
323.72	-1.11%	1246			1189 A	37	0.005	4.04159	EAST END S TR	6.62309	0.00357	1115	60
323.93	-1.11%	1100		95	1664	52	0.0057	4.04728	R	6.61957	0.00347	1100	60
324.27	-1.11%	1800		95	1800	52	0.00656	4.05384	R	6.6161	0.00568	1800	60
324.40	-1.09%	700			750	52	0.00255	4.05639		6.61042	0.00221	700	60
324.67	-1.09%	1400		60	1400	52	0.0051	4.06149	L	6.60821	0.00442	1400	60
324.67							0	4.06149		6.60379	0.00272	1215 D	109
324.92	-1.09%	1330			1338	52	0.00487	4.06636		6.60107	0.0004	220.5	109
325.92	-1.06%	5304			3418	52	0.01244	4.0788	WEST END 5 TRACKS	6.60067	0.01139	6146 A	96
326.38	-1.06%	2400			984 A	52	0.00333	4.08212	T CRESSON	6.58926	0.0029	1463	96
326.67	-1.06%	1536		130	4040	60	0.01529	4.09741	R	6.58636	0.00306	1536	96
326.90	-1.06%	1200			472	60	0.00133	4.09874		6.58332	0.00239	1200	96
327.96	-1.06%	5311	XING 253		9027 A	60	0.02167	4.12041	RISBINS ROAD	6.58093	0.01059	5511	96
328.19	-1.06%	1500					0	4.12041		6.57034	0.00289	1500	96
328.60	-1.06%	2162		95	724	95	0.00144	4.12185	L	6.56736	0.00431	2162	95
328.86	-1.06%	1500			1500	95	0.00299	4.12484		6.56304	0.00299	1500	95
329.07	-1.06%	962			962	95	0.00192	4.12676		6.56005	0.00196	962	95
329.24	-1.06%	900		96	900	96	0.00179	4.12856	RULLY	6.55809	0.00179	900	95
329.24							0	4.12856		6.5563	0.00072	367.1 D	106
329.77	-1.06%	2800			2800	96	0.00558	4.13414		6.55556	0.0005	287.9	106
330.11	-1.06%	1794			1794	95	0.00358	4.13772		6.55507	0.01	5680 A	95
330.49	-1.06%	2000		134	2000	95	0.00399	4.1417	R	6.54507	0.00861	4318	95
330.64	-1.06%	800			800	95	0.00159	4.1433	CASSANDRA	6.53646	0.00167	877 A	93
331.49	-1.06%	4518			3606	95	0.00719	4.15049		6.5348	0	0	93
332.37	-1.06%	4631			5660 A	96	0.01058	4.16104		6.5348	0.01029	5664	93
332.52	-1.06%	800			308	106	0.00054	4.16158		6.5245	0.00163	800	93
332.52					387 D	106	0.00072	4.1623		6.52287	0		
332.62	-1.05%	500		95	500	95	0.001	4.1633	R	6.52287	0.00102	500	93
332.71	-1.05%	500			500	95	0.001	4.1643	T PORTAGE	6.52186	0.00102	500	93
332.98	-1.05%	1400		110	1400	95	0.00279	4.16709	R	6.52084	0.00285	1400	93
333.23	-1.05%	1346		120	2691	95	0.00536	4.17245	R	6.51799	0.00274	1346	93
333.50	-1.05%	1400			55 D	95	0.00011	4.17256		6.51525	0.00285	1400	93
333.76	-1.05%	1400		93	1400	93	0.00285	4.17542	R	6.51239	0.00285	1400	93
334.03	-1.05%	1400			1400	93	0.00285	4.17827		6.50954	0.00011	55.15 D	95
334.11	-1.05%	426		95	426	93	0.00087	4.17913	L	6.50943	0.00353	1771	95
334.52	-0.04%	2200		95	2200	93	0.00448	4.18361	L	6.5059	0.00409	2059	95
335.02	-0.04%	2626			2626	93	0.00535	4.18896	WILMORE	6.50184	0		
335.61	-0.04%	3100			3100	93	0.00631	4.19526		6.50184	0		
335.94	-0.04%	1774		190	1248	93	0.00254	4.19782	R	6.50184	0.02528	10007 A	51
336.58	-0.04%	3361		190	827 A	93	0.00167	4.19943	R	6.47656	0.00347	935	51
336.87	-0.04%	1500					0	4.19945		6.47309	0.00557	1500	51
336.99	-0.63%	665			11310 A	95	0.02	4.21945		6.46752	0.00247	665	51
337.48	-0.63%	2600		200			0	4.21945	L	6.46505	0.00966	2600	51
337.79	-0.63%	1600					0	4.21948	SUMMERHILL	6.45539	0.00584	1600	51
338.28	-0.63%	2600			106	118	0.00016	4.21964		6.44945	0.0162	4365	51
338.28	-0.63%		E END MP		103 D	11	0.00017	4.21981		6.43325			
338.49	-0.63%	1100		115	54	115	8.9E-05	4.2199	R	6.43325	0.00083	221 A	49
338.49	-0.63%				1558 D	11	0.00356	4.22346		6.43241			
338.68	-0.63%	1000		51	1000	51	0.00371	4.22717	R	6.43241	0.00045	118	49
338.73	-0.63%	300			300	51	0.00111	4.22828	REVERSE CURVE	6.42197	0.00118	300	49
338.78	-0.63%	245		50	248	51	0.00092	4.2292	L	6.42081	0.00098	248	49
338.89	-0.63%	600		56	800	51	0.00223	4.23143	L	6.41985	0.00347	897	49
339.05	-0.63%	800		81	800	51	0.00297	4.2344	L	6.41638	0.00222	339 A	43
339.42	-0.63%	2000			2000	51	0.00743	4.24183	TTTMPH 40	6.41416	0.00465	1964	43
339.78	-0.63%	1865		65	1865	51	0.00693	4.24876	R SOUTH FORK	6.40951	0.00821	1865	43
340.13	-0.63%	1871			1871	51	0.00696	4.25571	T+HBSG@SUM(C734..C	6.40729	0.00824	1871	43
340.28	-0.63%	800		92	800	51	0.00297	4.25868	L	6.39905	0.00357	800	43
340.44	-0.63%	800		80	800	51	0.00297	4.26165	L	6.39553	0.00352	800	43
340.59	-0.63%	800		62	800	51	0.00297	4.26462	LMPH 35	6.392	0.00352	800	43
340.74	-0.63%	800			771	51	0.00286	4.26748		6.38848	0.00352	800	43
340.74	-0.63%				29 D	51	0.00011	4.26759		6.38496	0		
340.78	-0.63%	200		46	200	46	0.00077	4.26836	R	6.38496	0.00044	200	43
340.93	-0.63%	800		49	800	49	0.00309	4.27145	R	6.38408	0.00352	800	43
341.14	-0.63%	1100		62	1018	49	0.00394	4.27539	R REV CURVE	6.38056	0.00484	1100	43
341.14	-0.63%				81 D	49	0.00033	4.27572		6.37571			
341.31	-0.63%	900		43	900	43	0.00396	4.27969	LMPH 30	6.37571	0.00396	900	43
341.31	-0.63%						0	4.27969		6.37174	0.00033	86.96 D	57
341.48	-0.63%	900		62	900	43	0.00396	4.28366	L	6.37141	0.00317	819	49
341.51	-0.63%	199			199	43	0.00088	4.28453	RC	6.36824	0.00077	199	49
341.63	-0.63%	600		53	600	43	0.00264	4.28718	R	6.36748	0.0053	1379	49
341.78	-0.63%	800		51	800	43	0.00352	4.2907	R	6.36218	0.0025	601 A	42
341.94	-0.63%	870			870	43	0.00383	4.29453		6.35968	0.00135	299	42
342.07	-0.63%	400		78	400	43	0.00176	4.29629	L	6.35832	0.0015	400	42
342.17	-0.63%	800		85	800	43	0.00352	4.29982	L	6.35683	0.00361	800	42
342.30	-0.63%	700		72	700	43	0.00308	4.3029	LMINERAL POINT	6.35292	0.00316	700	42
342.44	-0.63%	700			700	43	0.00308	4.30598		6.34976	0.00316	700	42

APPENDIX B2

READ DOWN — WESTBOUND TRAFFIC — STATION — READ UP — EASTBOUND TRAFFIC

PAGE 6 COLUMN 2

MILES	GRADIENT PERCENT	DISTANCE ft		MAX MPH	DISTANCE feet	SPEED MPH	ET HOURS	TIME HOURS	STATION	TIME HOURS	ET HOURS	DISTANCE feet	SPEED MPH
342.50	-0.63%	900		50	600	43	0.00264	4.30863	R RC	6.34661	0.00271	600	42
342.53	-0.63%	400			400	43	0.00176	4.31039	TTTMPH 35	6.3439	0.0018	400	42
342.79	-0.63%	900		64	800	43	0.00352	4.31391	L	6.3421	0.00361	800	42
342.91	-0.63%	700		64	700	43	0.00308	4.31699	L	6.33849	0.00316	700	42
343.18	-0.63%	1400		51	1400	43	0.00617	4.32316	L	6.33533	0.00631	1400	42
343.21	-0.63%	200			200	43	0.00088	4.32404	REVERSE CURVE	6.32902	0.0009	200	42
343.36	-0.63%	800		49	800	43	0.00352	4.32756	R	6.32812	0.00361	800	42
343.42	-0.63%	300			131	43	0.00058	4.32814	RC	6.32451	0.00135	300	42
343.72	-0.63%	1600		53	539 A	43	0.00222	4.33036	L	6.32316	0.00722	1600	42
343.76	-0.63%	256			1915	49	0.00586	4.33622		6.31594	0.00129	256	42
343.82	-0.63%	215			215	48	0.00083	4.33705	PARKHILL	6.31465	0.00097	215	42
343.97	-0.63%	800		68	707	48	0.00273	4.33978	R	6.31368	0.00361	800	42
343.97	-0.63%				93 D	46	0.00039	4.34017		6.31008	0		
344.12	-0.63%	800		42	800	42	0.00361	4.34378	R	6.31008	0.00361	800	42
344.20	-0.63%	400			400	42	0.0018	4.34558	RC	6.30647	0.0018	400	42
344.33	-0.63%	700		45	700	42	0.00316	4.34874	L	6.30467	0.00316	700	42
344.46	-0.63%	700		42	700	42	0.00316	4.3519	L	6.30151	0.00316	700	42
344.46	-0.63%						0	4.3519		6.29835	0.00011	25.23 D	44
344.48	-0.63%	100			100	42	0.00045	4.35235	RC	6.29824	0.00032	74.77	44
344.65	-0.63%	900		84	900	42	0.00406	4.35641	R	6.29792	0.00387	900	44
344.79	-0.63%	700		63	700	42	0.00316	4.35956	R	6.29405	0.0005	126 D	53
344.86	-0.63%	377		63	377	42	0.0017	4.36125	R	6.29355	0.00339	949	53
344.86	-0.63%						0	4.36125	AD	6.29015	0.00017	47.96 D	56
345.10	-0.63%	1300			1300	42	0.00588	4.36713		6.28999	0.00423	1252	56
345.33	-0.63%	1200		56	1200	42	0.00541	4.37254	R	6.28575	0.00406	1200	56
345.35	-0.63%						0	4.37254		6.2817	0.00017	50.6 D	59
345.58	-0.63%	1200			1200	42	0.00541	4.37795		6.28153	0.00369	1149	59
345.79	-0.63%	1200		59	1200	42	0.00541	4.38336	L	6.27784	0.00385	1200	59
345.97	-0.63%	1000		78	1000	42	0.00451	4.38787	L	6.27399	0.00321	1000	59
346.24	-0.63%	1400		59	1400	42	0.00631	4.39418	L	6.27078	0.00449	1400	59
346.28	-0.63%	200			200	42	0.0009	4.39508	RC	6.26628	0.00064	200	59
346.47	-0.63%	1000		59	1000	42	0.00451	4.39959	R	6.26564	0.00321	1000	59
346.47	-0.63%						0	4.39959		6.26243	0.00058	187.7 D	66
346.51	-0.63%	246		69	246	42	0.00111	4.4007	R	6.26185	0.00058	58.27	66
346.53	-0.63%	600		190	177	42	0.0008	4.4015	R	6.26132	0.00165	600	66
346.78	-0.63%	400			140 A	40	0.00083	4.40234		6.25967	0.0022	800	66
347.18	-0.63%	2120			810	44	0.00349	4.40582	T CONEMAUGH TOWER	6.25748	0.00072	283	66
347.27	-0.63%	900		107	484 A	44	0.00194	4.40777	R	6.25675	0		
347.29	-0.63%	100			7458	50	0.00931	4.41708	RC	6.25675	0.00806	2594 A	52
347.52	-0.63%	1200		166	1200	50	0.00455	4.42163	L	6.2487	0.00387	1063	52
347.75	-0.63%	1200			1200	50	0.00455	4.42617		6.24483	0.00437	1200	52
347.84	-0.63%	476		151	476	50	0.0018	4.42797	L	6.24046	0.00173	476	52
347.90	-0.63%	300			300	50	0.00114	4.42911	RC	6.23872	0.00109	300	52
348.03	-0.63%	700		137	700	50	0.00265	4.43176	R	6.23763	0.00255	700	52
348.20	-0.63%	900		65	900	50	0.00341	4.43517	R	6.23508	0.00328	900	52
348.41	-0.63%	1100		72	658	50	0.00249	4.43766	R	6.2318	0.00401	1100	52
348.60	-0.63%	1600			1121 A	50	0.00389	4.44155		6.2278	0.00364	1000	52
348.78	-0.63%	1600		89	78	59	0.00025	4.44181	R	6.22416	0.00364	1000	52
348.99	-0.63%	1061		75	1687 A	59	0.005	4.44681	R	6.22051	0.00386	1061	52
349.14	-0.63%	800			1400	69	0.00386	4.45067	T JOHNSTOWN	6.21665	0.00291	800	52
349.20		800		204	600	69	0.00165	4.45231	R	6.21373	0.00219	600	52
349.40		800			496	69	0.00137	4.45368	W END 3 TR	6.21155	0.00158	434	52
349.46					302 D	66	0.00084	4.45452	E END 1 TR SANG HOLL	6.20997			
349.44		200		52	200	52	0.00073	4.45535	R	6.20997	0.00472	1183 A	36
349.63		1000			1000	52	0.00364	4.459		6.20525	0.00216	433	36
349.82		1000		86	1000	52	0.00364	4.46264	L	6.20309	0.00498	1000	36
349.90		438		181	438	52	0.0016	4.46423	L	6.1981	0.00218	438	36
350.28		2000			2000	52	0.00728	4.47152		6.19592	0.00997	2000	36
350.43		800		156	800	52	0.00291	4.47443	R	6.18595	0.00399	800	36
350.81		2000		86	2000	52	0.00728	4.48172	R	6.18197	0.00997	2000	36
351.10		1529		54	1529	52	0.00557	4.48728	R	6.172	0.00762	1529	36
351.25		800			800	52	0.00291	4.4902		6.16438	0.00399	800	36
351.48		1200		67	1200	52	0.00437	4.49457	R	6.16038	0.00598	1200	36
351.63		800			733	52	0.00085	4.49542	DORNOCK	6.15441	0.00399	800	36
351.63					733 A	52	0.00083	4.49625		6.15042			
351.63					118	54	0.00041	4.49667		6.15042			
351.63					216 D	54	0.00089	4.49755		6.15042			
351.82		1000		38	1000	38	0.00499	4.50254	L	6.15042	0.00498	1000	36
351.99		539		36	539	36	0.00443	4.50697	L	6.14544	0.00443	848	36
351.99							0	4.50697		6.14101	0.00022	46.53 D	42
352.03		200			200	38	0.001	4.50797	RC	6.14076	0.00069	153.1	42
352.22		1000		42	1000	38	0.00498	4.51296	R	6.14009	0.00151	1000	42
352.22							0	4.51296		6.13858	0.00128	300.9 D	65
352.39		900		130	900	38	0.00449	4.51744	R	6.13431	0.00157	839.1	65
352.56		1000		65	1000	38	0.00498	4.52242	R	6.13274	0.00291	1000	65
352.56							0	4.52242		6.12982	0.00022	78.61 D	69
352.61		200			200	38	0.001	4.52342	REVERSE CURVE	6.1296	0.00033	121.4	69
352.77		800		75	800	38	0.00399	4.5274	L	6.12927	0.0022	800	69
352.78		100			100	38	0.0005	4.5279		6.12707	0.00027	100	69
353.09		1600			1600	38	0.00797	4.53585		6.1268	0.00439	1600	69
353.39		1600		105	1600	38	0.00797	4.54385	R	6.12241	0.00439	1600	69
353.51		500			800	38	0.00399	4.54784		6.11801	0.0022	800	69
353.66		800		181	600	38	0.00299	4.55083	R	6.11582	0.00165	600	69
353.73		400		74	400	38	0.00199	4.55282	R	6.11417	0.0011	400	69
353.79		282			282	38	0.00141	4.55423	RC	6.11307	0.00077	282	69
353.82		700		68	700	38	0.00349	4.55772	L	6.1123	0.00192	700	69
353.92							0	4.55772		6.11038	0.00161	710.3 D	96
354.00		4000			815	38	0.00406	4.56178	R2	6.10877	0.00085	440.7	96
354.78		566		190	275 A	38	0.00111	4.56289	L	6.10791	0.01029	4947 A	84
354.83		257		190	257	42	0.00116	4.56405	L	6.09764	0.00272	2962 A	72
355.78		5000			708	42	0.00319	4.56726		6.09041	0.00106	745	72
355.93		800			2705 A	42	0.00044	4.5677		6.08846	0.0021	800	72
356.63		4000		84	787 A	66	0.00222	4.56992	L a	6.08538	0.01052	4000	72
356.73		250			5490	69	0.01507	4.58399	RC SANG HOLLOW	6.07583	0.00063	250	72
356.77		237		86	3604 A	69	0.00889	4.59288	R b	6.07531	0.00067	237	72
356.77		20		66	475	64	0.00107	4.59395	R	6.07468	5.3E-05	20	72
357.53		4000			4000	64	0.00902	4.61297		6.07463	0.01052	4000	72
357.72		1000		64	725	64	0.00164	4.61461	L	6.06411	0.00263	1000	72
357.72					275 D	67	0.00067	4.61528		6.06148			
358.29		2996		72	2996	72	0.00788	4.62316	L SEWARD	6.06148	0.00788	2996	72
358.29							0	4.62316	E END 2 TR SANG HOL	6.0536	0.00287	1392 D	120
358.67		2600			2000	72	0.00526	4.62842	SO-SEWARD	6.05063	0.00102	646.3	120
358.71		214			214	72	0.00056	4.62898	T	6.04961	0.00034	214	120
359.66		5000		190	5000	72	0.01315	4.64213	R	6.04927	0.00789	5000	120
360.08		2265		190	2266	72	0.00596	4.64809	R	6.04168	0.00048	303	120
360.60		3000			1921	72	0.004	4.65209	W END 2 TR SANG HOL	6.0412	0		
361.00		2138			13584 A	72	0.02778	4.67987	T NEW FLORENCE	6.0412	0		
361.63		3000		120			0	4.67987	R	6.0412	0		
362.00		3000		120			0	4.67987	R	6.0412	0		
362.38		3000					0	4.67987		6.0412	0		
362.54		626		196			0	4.67987	L	6.0412	0		
362.68		504		130			0	4.67987	L	6.0412	0		
363.45		4000					0	4.67987		6.0412	0.02194	12582 A	95
363.45					5743 A	11	0.00644	4.68631	MP 286	6.01926	0		
363.65		1056				120	0	4.68631	[illegible]	6.01926	0.01306	5920 A	76

APPENDIX B2

READ DOWN — WESTBOUND TRAFFIC | STATION | READ UP — EASTBOUND TRAFFIC

PAGE 7 COLUMN 1

MILES	GRADIENT PERCENT	DISTANCE, ft	[illegible]	MAX MPH	DISTANCE, ft	[illegible]	[illegible]	TIME, hours	STATION
363.71		300	X963	286.6		120	0	4.66831	PRIVATE
363.79		400		152	400	120	0.00062	4.66894	R REVERSE CURVE
363.86		400		152	400	120	0.00063	4.69058	CMPH 50
364.36		2730			223	120	0.00036	4.69093	
364.38					786	C 15	0.00139	4.69232	
365.28		4637		95	4399	95	0.00060	4.70191	R
365.28					477	D 95	0.00108	4.70207	
365.68		3000		76	2936	76	0.00732	4.70939	R
365.91		276				76	0	4.70639	RC
365.61					317	D 76	0.00084	4.71083	INDIA
366.20		1600		59	1590	59	0.00482	4.71616	L
366.57		2000		59	2000	59	0.00647	4.72167	L LOCKPORT
366.90		1700			1700	58	0.00048	4.72702	ENGLISHS
367.20		1615		59	1615	59	0.0052	4.73222	L
367.58		1900		59	1900	59	0.0061	4.73832	CBOLWAR
367.71		300			300	59	0.00257	4.74086	
367.86		300		75	300	59	0.00289	4.74375	R
[illegible]									

PAGE 7 COLUMN 2

[illegible]
[illegible]

APPENDIX B2

PAGE 8 COLUMN 1

READ DOWN WESTBOUND TRAFFIC | STATION | READ UP EASTBOUND TRAFFIC

MILES	GRADIENT, PERCENT	DISTANCE, ft	[illegible]	MAX MPH	DISTANCE, feet	SPEED, mph	ET, hours	TIME, hours	STATION	TIME, hours	ET, hours	DISTANCE, feet	SPEED, mph	MILES
419.76		415			415	67	0.00117	5.4442		5.23304	0.00117	415	67	427.75
419.90		632			632	67	0.00179	5.44599		5.23277	0.00179	632	67	427.79
419.97		368		112	368	67	0.00104	5.44703	R b	5.23098	0.00104	368	67	427.96
420.08		553		112	553	67	0.00156	5.44859	R	5.22994	0.00156	553	67	428.10
420.12		233		134	233	67	0.00066	5.44925	R	5.22838	0.00066	233	67	428.17
420.15		135		102	135	67	0.00038	5.44963	R c	5.22772	0.00038	135	67	428.21
420.16		79		102	79	67	0.00022	5.44985	R	5.22734	0.00022	79	67	428.26
420.26	0.50%	499		102	499	67	0.00141	5.45126	R	5.22712	0.00141	499	67	428.26
420.31	0.50%	305		90	305	67	0.00086	5.45213	R	5.22571	0.00086	305	67	428.36
420.35	0.50%	196			196	67	0.00055	5.45268		5.22484	0.00055	196	67	428.47
420.40		266			266	67	0.00075	5.45343		5.22429	0.00075	266	67	428.47
420.50		539		67	539	67	0.00152	5.45496	R d	5.22354	0.00152	539	67	428.56
420.54		195			195	67	0.00055	5.45551		5.22201	0.00055	195	67	428.74
420.72		940			940	67	0.00266	5.45816		5.22146	0.0012	422.8	67	428.83
420.73		60		117	60	67	0.00017	5.45833	R e	5.22027				428.12
420.92		1000		117	1000	67	0.00283	5.46116	R	5.22027				429.20
421.08		840		117	840	67	0.00237	5.46353	R	5.22027				429.31
421.11		160			160	67	0.00045	5.46398		5.22027	0.00778	2418 A	50-	429.42
421.16		260			260	67	0.00073	5.46472		5.21249	0.00072	119	50	428.50
421.30		734		185	734	67	0.00207	5.4668	R	5.21177	0.00278	734	50	429.52
421.32		97		185	97	67	0.00027	5.46707	R	5.20899	0.00037	97	50	429.56
421.45		706			706	67	0.002	5.46907	BECKS RUN	5.20863	0.00267	706	50	429.68
421.49		197			197	67	0.00056	5.46962		5.20596	0.00075	197	50	429.89
421.57		421			421	67	0.00119	5.47081		5.2052	0.00159	421	50	429.80
421.68		579		95	579	67	0.00164	5.47245	R b	5.20361	0.00219	579	50	430.07
421.69		62		95	62	67	0.00018	5.47263	R	5.20142	0.00023	62	50	430.11
421.81		652			652	67	0.00184	5.47447		5.20118	0.00247	652	50	430.21
421.87		288		120	288	67	0.00081	5.47528	L	5.19871	0.00108	288	50	430.29
421.91	-0.50%	239		170	239	67	0.00068	5.47595	L	5.19763	0.00091	239	50	430.38
422.02	-0.50%	560		134	560	67	0.00155	5.47751	L a	5.19672	0.00208	560	50	430.45
422.06	-0.50%	211		95	211	67	0.0006	5.4781	WILLIAMSBURG	5.19464	0.0008	211	50	430.62
422.06		37		95	37	67	0.0001	5.47821	L	5.19384	0.00014	37	50	430.82
422.15		439			439	67	0.00124	5.47945		5.1937	0.00166	439	50	430.94
422.24		524			524	67	0.00148	5.48093		5.19204	0.00198	524	50	430.96
422.36		634			634	67	0.00179	5.48272		5.19006	0.0024	634	50	430.99
422.42		269		78	269	67	0.00076	5.48348	L b	5.18766	0.00102	269	50	431.01
422.43		97		134	97	67	0.00027	5.48376	L	5.18663	0.00037	97	50	431.09
422.44		44		134	44	67	0.00012	5.48388		5.18627	0.00017	44	50	431.20
422.49		273		95	273	67	0.00077	5.48465	L c	5.1861	0.00103	273	50	431.38
422.62		883			494	67	0.0014	5.48605		5.18506	0.00194	484 A	44-	431.44
422.62					1570 A	67	0.00444	5.49049		5.18312	0.00160	240	44	431.58
422.66		217						5.49049		5.18209	0.00082	217	44	431.56
422.78		619		95			0	5.49049	L a	5.18115	0.00066	619	44	431.77
422.81		164					0	5.49049		5.17849	0.00071	164	44	431.98
422.93		630					0	5.49049		5.17778	0.00271	630	44	432.11
423.00		370		95		78	0	5.49049	L b	5.17507	0.00159	370	44	432.15
423.04		175				78	0	5.49049		5.17348	0.00075	175	44	432.21
423.16		622				78	0	5.49049		5.17273	0.00268	622	44	432.33
423.19		203					0	5.49049	J&L TUNNEL	5.17006	0.00087	203	44	432.34
423.19					83	76	0.00021	5.4907		5.16917	0			432.42
423.20		36			458 D	78	0.00139	5.49209		5.16917	0.00015	36	44	432.46
423.25		297		50	297	50	0.00113	5.49322	L c	5.16902	0.00128	297	44	432.50
423.33		385		105	385	50	0.00146	5.49468	L	5.16774	0.00166	385	44	432.52
423.38		282			282	50	0.00107	5.49574	REVERSE CURVE	5.16608	0.00121	282	44	432.72
423.40		111			111	50	0.00042	5.49616		5.16487	0.00048	111	44	432.85
423.51		556		58	556	50	0.00211	5.49827	R a	5.16439	0.00239	556	44	432.90
423.57		333			333	50	0.00126	5.49953	REVERSE	5.162	0.00143	333	44	432.91
423.59		111			111	50	0.00042	5.49995	STATION 0+00 FMcKY	5.16057	0.00048	111	44	432.96
423.73		720		54	720	50	0.00273	5.50268	L e	5.16009	0.0031	720	44	433.10
423.76		194			194	50	0.00073	5.50341		5.15699	0.00084	194	44	433.20
423.94		899			899	50	0.00341	5.50682		5.15615	0.00387	899	44	433.29
423.96		101		186	101	50	0.00038	5.5072	L	5.15228	0.00043	101	44	433.46
423.97		103		186	103	50	0.00039	5.50759		5.15185	0.00044	103	44	433.56
424.08		535			535	50	0.00203	5.50962	REVERSE CURVE	5.15141	0.0023	535	44	433.82
424.12		262		186	262	50	0.00099	5.51061	R	5.1491	0.00113	262	44	433.85
424.14		100			100	50	0.00038	5.51099		5.14797	0.00043	100	44	434.04
424.15		45			45	50	0.00017	5.51116	REVERSE CURVE	5.14754	0.00019	45	44	434.18
424.20		268			268	50	0.00102	5.51218	OLIVER YARD	5.14735	0.00115	268	44	434.20
424.21		189		152	189	50	0.00072	5.51289	L	5.1462	0.00081	189	44	434.23
424.33		498			498	50	0.00189	5.51478		5.14538	0.00214	498	44	434.42
424.37		185			185	50	0.0007	5.51548		5.14324	0.0008	185	44	434.61
424.49		651		86	651	50	0.00247	5.51794	R a	5.14244	0.0028	651	44	434.60
424.52		164			164	50	0.00062	5.51857	REVERSE CURVE	5.13964	0.00071	164	44	434.82
424.55		147			147	50	0.00056	5.51912		5.13894	0.00063	147	44	434.92
424.60		285		54	505	50	0.00191	5.52104	L b	5.1383	0.00123	285	44	434.99
424.66		302			83 D	50	0.00033	5.52137	REVERSE CURVE	5.13706	0.0013	302	44	435.01
424.69		156		44	156	44	0.00067	5.52204	R c	5.13577	0.00067	156	44	435.18
424.71		109		72	109	44	0.00047	5.52251	R	5.1351	0.00106	298.7 D	83-	435.36
424.73		122		72	122	44	0.00053	5.52304	R	5.13404	0.00037	122	63	435.37
424.74		11		72	11	44	4.7E-05	5.52308	LIBERTY BRIDGE	5.13368	3.3E-05	11	63	435.41
424.79		270		63	270	44	0.00116	5.52425	R	5.13364	0.00061	270	63	435.50
424.90		600			600	44	0.00258	5.52683		5.13263	0.00026	98.07 D	68-	435.58
424.97		350		134	350	44	0.00151	5.52833	R	5.13250	0.00208	853.9	68	435.62
425.14		914			914	44	0.00393	5.53227		5.13018	0.00255	914	68	435.66
425.16		86			86	44	0.00037	5.53264	SMITHFIELD STREET	5.12763	0.00024	86	68	435.75
425.20		220		76	220	44	0.00095	5.53359	R	5.12738	0.00061	220	68	435.76
425.28		450			450	44	0.00194	5.53552	MP 0 P&LE STATION	5.12678	0.00125	450	68	435.81
425.32		200		68	200	44	0.00086	5.53638	L REVERSE CURVE	5.12553	0.00056	200	68	435.89
425.32						44	0	5.53638		5.12497	0.00056	214.1 D	79-	435.91
425.33		50			50	44	0.00022	5.5366		5.12441	0.00177	728.9	78	435.94
425.33							0	5.5366		5.12264	0.00115	486	78	436.04
425.52		1000			1000	44	0.0043	5.5409		5.12153	0	0	78	436.09
425.66		750			750	44	0.00323	5.54413		5.12153	0.00111	452 A	76-	436.13
425.71		250		134	250	44	0.00108	5.54521	R FORT PITT BRIDGE	5.12042	0.00062	250	76	436.20
425.79		400		134	400	44	0.00172	5.54693	R	5.1198	0.001	400	76	436.32
425.90		600			600	44	0.00258	5.54951		5.1188	0.0015	600	76	436.51
426.09		1000			1000	44	0.0043	5.55382	10802	5.11731	0.00145	581	76	436.64
426.13		200			200	44	0.00086	5.55468		5.11586	0.00222	899 A	72-	436.70
426.20		400		78	400	44	0.00172	5.5564		5.11364	0.00033	150	72	436.88
426.26		300		78	300	44	0.00129	5.55769	L a	5.11331	0.00079	300	72	436.95
426.28		100			100	44	0.00043	5.55812	L	5.11245	0.00026	100	72	437.07
426.40		650			650	44	0.0028	5.56092		5.11219	0.00171	650	72	437.20
426.47		350		110	350	44	0.00151	5.56243	R a	5.11048	0.00092	350	72	437.20
426.49		100		110	100	44	0.00043	5.56286	R	5.10956	0.00026	100	72	437.26
426.66		900		138	900	44	0.00387	5.56673	R WEST END BRIDGE	5.1093	0.00237	900	72	437.38
426.85		1000		138	615	44	0.00266	5.56939	R	5.10693	0.00263	1000	72	437.40
426.87		100		138				5.56939	R	5.1043	0.00026	100	72	437.82
427.04		900			2233 A	44	0.00778	5.57717		5.10403	0.00237	900	72	437.94
427.17		700			868 A	63	0.0025	5.57967		5.10167	0.00184	700	72	437.83
427.19		130		214			0	5.57967	R	5.09983	0.00034	130	72	437.96
427.23		170		214	231	68	0.00064	5.58031	R	5.09948	0.00045	170	72	438.02
427.41		1000		214			0	5.58031	R	5.09904	0.00263	1000	72	438.06
427.56		800		174	1692 A	68	0.00444	5.58476	R	5.09641	0.00224	850	72	438.20
427.60		150					0	5.58476	REVERSE CURVE	5.09417	0.00039	150	72	438.21
427.62		100					0	5.58476		5.09378	0.00026	100	72	438.32

APPENDIX B2

PAGE 8 COLUMN 2

READ DOWN WESTBOUND TRAFFIC | STATION | READ UP EASTBOUND TRAFFIC

GRADIENT, PERCENT	DISTANCE, ft	[illegible]	MAX MPH	DISTANCE, feet	SPEED, mph	ET, hours	TIME, hours	STATION	TIME, hours	ET, hours	DISTANCE, feet	SPEED, mph
	450		78			0	5.58476	L a	5.09351	0.00118	450	72
	450			358	76	0.00089	5.58565		5.09233	0.00118	450	72
	1000			1000	76	0.00249	5.58814		5.09115	0.00263	1000	72
	600			600	76	0.0015	5.58963		5.08852	0.00158	600	72
	400		68	400	76	0.001	5.59063	L b	5.08694	0.00105	400	72
	190			190	76	0.00047	5.59111	REVERSE CURVE	5.08588	0.0005	190	72
	260			173	76	0.00043	5.59154	CORLISS	5.08538	0.00068	260	72
				87 D	76	0.00022	5.59176		5.0847	0		
	550		72	550	72	0.00145	5.59321	R a	5.0847	0.00145	550	72
	550		72	550	72	0.00145	5.59466	R	5.08325	0.00145	550	72
						0	5.59466		5.08181	0.00022	86.03 D	70
	450			450	72	0.00118	5.59584		5.08159	0.00091	383.2	76
	1000	XING 3.4		1000	72	0.00263	5.59847	RIVER AVENUE	5.08068	0.00249	1000	76
	1000			1000	72	0.00263	5.6011		5.07819	0.00249	1000	76
	1000			1000	72	0.00263	5.60373		5.0757	0.00249	1000	76
	450			450	72	0.00118	5.60491		5.0732	0.00112	450	76
	550			550	72	0.00145	5.60636		5.07208	0.00137	550	76
	600			600	72	0.00158	5.60794		5.07071	0.0015	600	76
	400		107	400	72	0.00105	5.60899	R x	5.06922	0.001	400	76
	100		107	100	72	0.00026	5.60925	R	5.06822	0.00025	100	76
	250			250	72	0.00066	5.60991	REVERSE CURVE	5.06797	0.00062	250	76
	600		107	600	72	0.00158	5.61149	L a	5.06735	0.0015	600	76
	50			50	72	0.00013	5.61162		5.06585	0.00012	50	76
	1000			1000	72	0.00263	5.61425		5.06573	0.00249	1000	76
	1000			1000	72	0.00263	5.61688		5.06324	0.00249	1000	76
	220			220	72	0.00058	5.61746		5.06074	0.00055	220	76
	550			550	72	0.00145	5.61891		5.0602	0.00137	550	76
	230		78	230	72	0.00061	5.61951	L DAVIS ISLAND c	5.05883	0.00057	230	76
	800		78	660	72	0.00171	5.62122	L	5.05825	0.00162	660	76
	350			300	72	0.00079	5.62201		5.05663	0.00106	478.5 D	95
	1000			600	72	0.00158	5.62359		5.05558	0.00174	873.5	95
	1000			989 A	72	0.00222	5.62581	NEVILLE ISLAND	5.05384	0.00199	1000	95
	800			1161	76	0.00294	5.62875		5.05184	0.0016	800	95
	300		95	300	76	0.00075	5.6295	R a	5.05085	0.0006	300	95
						0	5.6295		5.05005	0.00133	783.3 D	11
	100			100	76	0.00025	5.62975		5.04871	0.00067	416.7	110
	400			400	76	0.001	5.63075		5.04805	0		
	600			600	76	0.0015	5.63224		5.04805	0		
	1000			1000	76	0.00249	5.63473		5.04805	0		
	240			240	76	0.0006	5.63533		5.04805	0		
	760		131	760	76	0.00189	5.63723	L	5.04805	0		
	50		131	50	76	0.00012	5.63735	L	5.04805	0		
	950			950	76	0.00237	5.63972		5.04805	0		
	1000			1000	76	0.00249	5.64221		5.04805	0		
	800			800	76	0.00199	5.6442		5.04805	0		
	200		190	200	76	0.0005	5.6447	L	5.04805	0		
	300		190	300	76	0.00075	5.64545	L	5.04805	0		
	640		152	640	76	0.00159	5.64704	L	5.04805	0		
	60			60	76	0.00015	5.64719	EMSWORTH DAM	5.04805	0		
	450			450	76	0.00112	5.64832	REVERSE CURVE	5.04805	0		
	360		152	200	76	0.0005	5.64881	R	5.04805	0		
	210					0	5.64881		5.04805	0		
	40		152			0	5.64881	L	5.04805	0		
	1000		152	5026 A	76	0.01306	5.66187	L	5.04805	0.01361	6028 A	50
	700		152			0	5.66187	L	5.03444	0.00109	600	104
	250					0	5.66187	REVERSE CURVE	5.03334	0.00046	250	104
	50		152	6149 A	95	0.01139	5.67326	R	5.03289	9.1E-06	50	104
	300		152			0	5.67326	R	5.0328	0.00055	300	104
	700					0	5.67326		5.03225	0.00127	700	104
	550					0	5.67326		5.03098	0.001	550	104
	450					0	5.67326		5.02997	0.00082	450	104
	1000					0	5.67326		5.02916	0.00182	1000	104
	1000					0	5.67326		5.02733	0.00182	1000	104
	800					0	5.67326		5.02551	0.00146	800	104
	200		134			0	5.67326	R	5.02406	0.00036	200	104
	1000		134			0	5.67326	R	5.02369	0.00182	1000	104
	700		134			0	5.67326	R	5.02187	0.00127	700	104
	120	XING 9.95				0	5.67326	PRIVATE?	5.0206	0.00022	120	104
	180					0	5.67326		5.02038	0.00033	180	104
	1000					0	5.67326		5.02005	0.00182	1000	104
	1000	XING 9.4				0	5.67326	LENORA AVENUE	5.01823	0.00182	1000	104
	1000			227	109	0.00039	5.67364		5.01641	0.00182	1000	104
	100			156 D	15	0.00028	5.67392		5.01459	0.00018	100	104
	550		104	550	104	0.001	5.67492	L a	5.0144	0.001	550	104
	350			350	104	0.00064	5.67556		5.0134	0.00011	61.6 D	13
	90			90	104	0.00016	5.67572		5.01329	0.00062	346.4	106
	910			910	104	0.00166	5.67738		5.01267			
	950	XING 10.3		950	104	0.00173	5.67911	BROADWAY STREET	5.01267			
	50		152	50	104	9.1E-05	5.6792	L CORAOPOLIS	5.01267			
	200	XING	152	200	104	0.00036	5.67957	L MULBERRY STREET	5.01267			
	500	XING 10.6		500	104	0.00091	5.68048	RC MILL STREET	5.01267			
	300		152	300	104	0.00055	5.68102	R	5.01267			
	300			300	104	0.00055	5.68157		5.01267			
	250		152	250	104	0.00046	5.68203	R	5.01267			
	450			450	104	0.00082	5.68284	REVERSE CURVE	5.01267			
	50			50	104	9.1E-05	5.68294		5.01267			
	300		152	300	104	0.00055	5.68348	L	5.01267	0		
	400			400	104	0.00073	5.68421	REVERSE CURVE	5.01267	0		
	100	XING	152	100	104	0.00018	5.68439	R MAIN STREET	5.01267	0		
	150			150	104	0.00027	5.68467		5.01267	0.00972	5142 A	94
	550			550	104	0.001	5.68567		5.00296	0.00071	350	94
	250		190	250	104	0.00046	5.68612	R	5.00225	0.0005	250	94
	200	XING 10.9		200	104	0.00036	5.68649	WATT STREET	5.00174	0.0004	200	94
	400			400	104	0.00073	5.68722		5.00134	0.00081	400	94
	900			600	104	0.00109	5.68831		5.00053	0.00121	600	94
	1000			1000	104	0.00182	5.69013		4.99933	0.00201	1000	94
	700	XING 11.4		700	104	0.00127	5.6914	PRIVATE?	4.99731	0.00141	700	94
	300		131	300	104	0.00055	5.69195	R	4.9959	0.0006	300	94
	1000		131	1000	104	0.00182	5.69377	R	4.9953	0.00201	1000	94
	250	XING	131	250	104	0.00046	5.69423	R	4.99329	0.0005	250	94
	750	XING 11.5		400	104	0.00073	5.69496	PRIVATE?	4.99279	0.00151	750	94
	650					0	5.69496		4.99127	0.00131	650	94
				4468 A	10	0.00778	5.70273	12	4.98996	0		
	350					0	5.70273		4.98996	0.00071	350	94
	650					0	5.70273		4.98925	0.00131	650	94
	350		164			0	5.70273	R	4.98794	0.00071	350	94
	900		164			0	5.70273	R	4.98724	0.00181	900	94
	100		121			0	5.70273	R	4.98542	0.0002	100	94
	1000		121			0	5.70273	R	4.97515	0.00201	1000	94
	600		121			0	5.70273	R STOOPS FERRY	4.97313	0.00121	600	94
	400			515	113	0.00086	5.7036		4.97192	0.00081	400	94
	200			577 D	11	0.00126	5.70486	REVERSE CURVE	4.97112	0.0004	200	94
	740		94	740	94	0.00149	5.70614	L	4.97071	0.00149	740	94
	60		94	60	94	0.00012	5.70626	L a	4.96922	0.00012	60	94
	600		94	600	94	0.00121	5.70747	L	4.9691	0.00121	600	94

APPENDIX B2

READ DOWN — WESTBOUND TRAFFIC | STATION | READ UP — EASTBOUND TRAFFIC

PAGE 9 COLUMN 1

MILES	GRADIENT PERCENT	DISTANCE ft	HWY GRADE CROSSING	MAX MPH	DISTANCE ft	SPEED MPH	ET HOURS	TIME HOURS	STATION		TIME HOURS	ET HOURS	DISTANCE ft	SPEED MPH	MILES	GRADIENT PERCENT
436.50		400			400	94	0.00061	5.70628			4.96799	0.00073	414.8 D	108	449.32	
436.58		950			950	94	0.00181	5.71019			4.96711	0.00017	90.23	108	449.36	
436.59		50		119	50	94	0.0001	5.71028	L b		4.96694	8.8E-06	50	108	449.37	
436.73		750		119	750	94	0.00151	5.7118	L		4.96686	0.00132	750	108	449.70	
438.78		200			200	94	0.0005	5.71231			4.96554	0.00044	250	108	449.75	
438.97		1000			1000	94	0.00201	5.71432			4.9651	0.00175	1000	108	449.95	
439.16		1000			1000	94	0.00201	5.71634			4.96335	0.00175	1000	108	450.07	
439.19		180			180	94	0.00036	5.7167			4.96159	0.00032	180	108	450.07	
439.19							0	5.7167		11	4.96128	0.0025	1408 A	105	450.14	
439.35		820			820	94	0.00165	5.71835			4.95875	0.00045	250	105	450.16	
439.54		1000			1000	94	0.00201	5.72037			4.95835	0.0016	1000	105	450.19	
439.72		950	XING 14.5					5.72037	PRIVATE?		4.95652	0.00171	950	105	450.30	
439.73		50		106	50	94	0.0001	5.72047	R GLENWILLARD		4.95481	9E-05	50	105	450.35	
439.91		1000		106	1000	94	0.00201	5.72248	R		4.95472	0.0018	1000	105	450.41	
439.93		100		106	100	94	0.0002	5.72268	R		4.95291	0.00018	100	105	450.41	
440.10		900			900	94	0.00181	5.7245			4.95273	0.00157	900	105	450.52	
440.20		500			500	94	0.00101	5.72551			4.95115	0.0009	500	105	450.56	
440.29		482			480	94	0.00093	5.72643			4.95021	0.00083	460	105	450.65	
440.29		40		106	40	94	8.1E-05	5.72651	R		4.94936	7.2E-05	40	106	450.71	
440.48		1000		106	1000	94	0.00201	5.72853	R		4.94935	0.0018	1000	106	450.79	
440.57		450		130	450	94	0.00091	5.72943	R		4.9475	0.00081	450	106	450.80	
440.97		100			100	94	0.0002	5.72964			4.94669	0.00099	500	106	450.94	
440.88		1000	XING 10.6				0	5.72964	PRIVATE		4.9457	0.0018	1000	105	451.01	
441.06		1000			4673 A	94	0.00884	5.73852			4.9439	0.0018	1000	105	451.09	
441.19		795					0	5.73852	COUNTY LINE ALLE-BEA		4.94209	0.00135	750	106	451.20	
441.24		290					0	5.73852			4.94074	0.00045	250	106	451.26	
441.30		780			471	106	0.00085	5.73937	SOUTH HEIGHTS		4.94029	0.00135	750	106	451.38	
441.35		250		106	250	106	0.00045	5.73982	R		4.93893	0.00045	250	106	451.32	
441.57		750		106	750	106	0.00135	5.74118	R		4.93848	0.00135	750	106	451.32	
441.62		250			250	106	0.00045	5.74163			4.93713	0.00083	460 D	121	451.49	
441.81		1000			1000	105	0.0018	5.74343			4.9363	0.00119	755	120	451.51	1.82%
442.00		1000			1000	105	0.0018	5.74524			4.93511	0.00158	1000	120	451.51	MAX
442.18		1000			1000	106	0.0018	5.74704			4.93353	0.00158	1000	120	451.51	WEST
442.21		120			120	106	0.00022	5.74726			4.93196	0.00019	120	120	451.62	BOUND
442.38		880			880	106	0.00159	5.74884			4.93178	0.00136	860	120	451.62	
442.51		700			700	106	0.00126	5.75011			4.93042	0.0011	700	120	451.62	
442.57		300		190	300	106	0.00054	5.75065	R		4.92932	0.00047	300	120	451.63	
442.76		1000		190	1000	106	0.0018	5.75245	R		4.92879	0.00158	1000	120	451.54	
442.93		940		190	940	106	0.0017	5.75415	R		4.92721	0.00148	940	120	451.64	
442.95		60			60	105	0.00011	5.75425			4.92573	9.5E-05	60	120	451.58	
443.13		1000			1000	106	0.0018	5.75606			4.92564	0.00158	1000	120	451.58	MAX
443.21		380			380	106	0.00069	5.75674			4.92406	0.0006	380	120	451.66	EAST
443.21		20			20	106	3.8E-05	5.75678			4.92346	3.2E-05	20	120	451.56	BOUND
443.27		300		152	300	106	0.00054	5.75732	L		4.92342	0.00047	300	120	451.60	-1.90%
443.33		300			300	106	0.00054	5.75786	REVERSE CURVE		4.92295	0.00047	300	120	451.66	
443.33		40			40	106	7.2E-05	5.75793	WEST ECONOMY		4.92248	6.3E-05	40	120	451.86	
443.37		200		152	200	106	0.00036	5.7583	R		4.92242	0.00032	200	120	451.82	
443.51		780			780	105	0.00137	5.75967			4.9221	0.0006	375	120	451.93	
443.65		720			720	105	0.0013	5.76096			4.9216				451.83	
443.66		200		152	30	105	5.4E-05	5.76102	L		4.9216				451.84	
443.70		30					0	5.76102	REVERSE CURVE		4.9215				451.84	
443.73		150					0	5.76102			4.9215				451.65	
443.79		300		152			0	5.76102	R		4.9215				451.66	
443.89		540					0	5.76102			4.9215				451.67	
444.08		1000					0	5.76102			4.9215				451.66	
444.19		590					0	5.76102			4.9215				451.68	
444.19					8260 A	10	0.01389	5.77491		16	4.9215				451.68	
444.27		390					0	5.77491			4.9215				451.69	
444.27		30		160			0	5.77491	L		4.9215				451.70	
444.46		1000		160			0	5.77491	L ALIQUIPPA		4.9215				451.71	
444.53		360		160			0	5.77491	L		4.9215				451.72	
444.65		660					0	5.77491			4.9215				451.73	
444.82		900					0	5.77491			4.9215				451.74	
444.84		100		134			0	5.77491	R		4.9215				451.75	
444.95		600		134			0	5.77491	R		4.9215				451.79	
445.03		400					0	5.77491			4.9215				451.75	
445.20		900					0	5.77491			4.9215				451.77	
445.22		100					0	5.77491			4.9215				451.77	
445.27		250			231	120	0.00036	5.77527			4.9215				451.77	
445.38		600		134	600	120	0.00095	5.77622	L		4.9215				451.78	
445.41		150			150	120	0.00024	5.77646	WEST ALIQUIPPA		4.9215				451.78	
445.60		1000			1000	120	0.00156	5.77802			4.9215	0.02194	12532 A	93	451.79	
445.79		1000			1000	120	0.00156	5.77961			4.89896				451.80	
445.98		1000			1000	120	0.00158	5.78119			4.89058	0.00168	900	96	451.80	
446.13		800			800	120	0.00126	5.78246			4.88767	0.00164	923	96	451.80	
446.16		200		134	200	120	0.00032	5.78277	L		4.88622	0.00187	827 A	93	451.80	
446.20		210			210	120	0.00033	5.7831			4.88456	0.00187	16	93	451.81	
446.32		590		134	590	120	0.00093	5.78403	L		4.88289	0.00027	134	93	451.81	
446.35		200		104	200	120	0.00032	5.78435	L		4.88262	0.00041	200	93	451.81	
446.47		600		104	600	120	0.00095	5.78529	L		4.88221	0.00122	600	93	451.91	
446.54		400			400	120	0.00063	5.78593			4.88099	0.00081	400	93	451.91	
446.73		1000			1000	120	0.00156	5.7875	HOPEWELL TOWNSHIP		4.88017	0.00204	1000	93	451.68	
446.75		100			100	120	0.00016	5.78766			4.88814	0.0002	100	93	452.01	
446.92		900		186	900	120	0.00142	5.78908	L		4.88793	0.00183	900	93	452.45	
446.96		300		186	300	120	0.00047	5.78956	L		4.8861	0.00081	400	93	452.82	
447.07		500			500	120	0.00079	5.79035	REVERSE CURVE		4.88549	0.00102	500	93	452.73	
447.11		200		131	200	120	0.00032	5.79066	R		4.88447	0.00041	200	95	452.86	
447.20		480		131	480	120	0.00076	5.79142	R		4.88407	0.00072	350	95	452.90	
447.30		520		131	520	120	0.00082	5.79224	R		4.88335	0			453.48	
447.32		80		131	80	120	0.00013	5.79237	R		4.88335	0			453.52	
447.49		920			920	120	0.00145	5.79382			4.88335	0			453.70	
447.68		1000			1000	120	0.00158	5.7954			4.88335				453.85	
447.77		480			200	120	0.00032	5.79571			4.88335				453.99	
447.83		650		134	134	120	0.00021	5.79592	R		4.88335				454.13	
447.90		150		134	78	120	4.4E-05	5.79597	R		4.88335				454.27	
447.90					786 D	15	0.00138	5.79736			4.88335				454.30	
447.96		300		96	300	95	0.0006	5.79796	R x		4.88335				455.85	
447.97		100			100	95	0.0002	5.79816	S COLONA		4.88335	0.01028	4647 A	76	456.71	
448.01		200		90	200	95	0.0004	5.79856	R X		4.87307				456.93	
448.06		250			250	95	0.0005	5.79906			4.87307	0	0	76	456.28	
448.08		100			100	95	0.0002	5.79926			4.87307	0.00019	77	76	456.48	
448.09		60		134	60	95	0.00016	5.79941	R		4.87288	0			456.68	
448.20		540			540	95	0.00106	5.80049			4.87288	0			456.88	
448.25		280			280	95	0.00056	5.80104			4.87288	0			457.03	
448.37		650			595	95	0.00119	5.80223			4.87288	0			457.37	
448.37					55 D	95	0.00011	5.80234			4.87288	0			457.58	
448.44		350		93	350	93	0.00071	5.80305	L b		4.87288	0			457.92	
448.63		1000		93	1000	95	0.00204	5.80509	L COLONA		4.87288	0			457.66	
448.80		900		93	900	95	0.00183	5.80692	L		4.87288	0			458.06	
448.82		100			100	95	0.0002	5.80713			4.87288	0			458.28	
449.01		1000			1000	95	0.00204	5.80918			4.87288	0			458.78	
449.07		350			350	93	0.00071	5.80988			4.87288	0			458.49	
449.20		650		107	650	93	0.00132	5.8112	L b		4.87288	0			458.49	
449.21		80		107	80	93	0.00016	5.81136	L b'		4.87288	0			458.96	
449.21							0	5.81136		24	4.87288	0.02028	8125 A	30	450.06	

APPENDIX B2

READ DOWN — WESTBOUND TRAFFIC | STATION | READ UP — EASTBOUND TRAFFIC

PAGE 9 COLUMN 2

GRADIENT PERCENT	DISTANCE ft	HWY GRADE CROSSING	MAX MPH	DISTANCE ft	SPEED MPH	ET HOURS	TIME HOURS	STATION		TIME HOURS	ET HOURS	DISTANCE ft	SPEED MPH
	70		107	70	93	0.00014	5.8115	L b'		4.86926	0.00061	1081	30
	850			800	93	0.00173	5.81324	MONACA S		4.8461	0.00328	44 A	29
	1000			1000	93	0.00204	5.81527			4.84562	0.00474	720.2	28
	1000	XING	110	1000	93	0.00204	5.81731	L 4TH STREET		4.84106	0.00093	1000	29
	100		110	100	93	0.0002	5.81751	L MONACA		4.83436	0.00006	100	29
	900			900	93	0.00183	5.81935			4.83399	0.00049	900	29
	600			600	93	0.00122	5.82057			4.82802	0.00087	600	29
						0	5.82057	OHIO RIVER BRANCH 1		4.8241	0		29
	400			24	93	4.8E-05	5.82062			4.8241	0.00061	400	28
				376 D	92	0.00083	5.82145			4.82149			28
	250		76	250	76	0.00061	5.82206	R		4.82149	0.00163	250	28
	750		76	750	76	0.00182	5.82388	R		4.81985	0.0049	750	28
	90		76	80	76	0.00019	5.82407	R		4.81496	0.00092	60	29
	320			320	76	0.00078	5.82485			4.81444	0.00209	320	29
						0	5.82485	OHIO RIVER S BANK		4.81236	0		29
	600			600	76	0.00148	5.82633			4.81236	0.00392	600	29
	700			700	76	0.0017	5.826			4.80843	0.00457	700	29
						0	5.826	OHIO RIVER N BANK		4.80386	0		29
	300			300	76	0.00073	5.82873			4.80386	0.00196	300	29
	400		50	400	76	0.00097	5.8297	R		4.8019	0.00261	400	29
	600		50	600	76	0.00146	5.83116	R		4.79928	0.00392	600	29
	200		50	200	76	0.00049	5.83165	R BEAVER		4.79537	0.00131	200	29
	400			400	78	0.00097	5.83262			4.79406	0.00261	400	29
	400		150	400	78	0.00097	5.83359	R beaver station		4.79145	0.00261	400	29
	570			420	78	0.00102	5.83461			4.78884	0.00372	570	29
	330					0	5.83461		0	4.78511	0.00216	330	29
	100		190			0	5.83461	L		4.78296	0.00065	100	29
	200		190	20	75	4.8E-05	5.83466	L BEAVER NEWSPAPER		4.7823	0.00131	200	29
				760 D	74	0.00587	5.83702			4.781			29
	906		30	906	30	0.00571	5.84205	L		4.781	0.00591	906	29
	100			100	30	0.00083	5.84287			4.77509	0.00065	100	29
	20		35	20	30	0.00013	5.84379	L		4.77444	0.00013	20	29
	20			20	30	0.00013	5.84392			4.77431	0.00013	20	29
	20		34	21	30	0.00013	5.84405	L		4.77418	0.00013	20	29
	10			9 D	30	5.0E-05	5.84411	REVERSE CURVE		4.77404	6.5E-05	10	29
	25		29	25	29	0.00016	5.84427	R		4.77398	0.00016	25	29
	5		34	5	29	3.3E-05	5.84431	R		4.77382	0.00028	48.2 D	36
	60		34	60	29	0.00039	5.84469	R		4.77354	4.9E-05	8.8	34
	50			50	29	0.00033	5.84498	SHARON ROAD		4.77349	0.00028	50	34
	65			65	29	0.00042	5.84538	REVERSE CURVE		4.77321	0.00066	85	34
	14		78	14	29	9.1E-05	5.84547	L		4.77285	7.3E-05	14	34
	5			6	29	3.9E-05	5.84551	REVERSE CURVE		4.77277	3.3E-05	6	34
	10		88	10	29	6.5E-05	5.84558	R		4.77274	5.5E-05	10	34
	5	XX		5	29	3.3E-05	5.84561	TRACK IN 5TH AVENUE		4.77268	2.9E-05	5	34
	100	XX		100	29	0.00065	5.84627	SHARON ROAD		4.77265	0.00058	100	34
	100	XX		100	29	0.00065	5.84692	TO BUFFALO STREET		4.7721	0.00058	100	34
	100	XX		100	29	0.00065	5.84757	REVERSE CURVE		4.77154	0.00058	100	34
	25	XX		25	29	0.00016	5.84773	BUFFALO STREET		4.77096	0.00014	25	34
	35		34	35	29	0.00023	5.84796	L		4.77084	0.00019	35	34
	40			40	29	0.00026	5.84822	REVERSE CURVE		4.77065	0.00022	40	34
	15	XING		15	29	0.0001	5.84833	TUSCARAWAS ROAD		4.77043	8.8E-05	15	34
	24		34	24	29	0.00016	5.84848	R		4.77034	0.00013	24	34
	60			60	28	0.00039	5.84888			4.7702	0.00035	60	34
	60			60	28	0.00039	5.84927			4.76987	0.00035	60	34
	30		34	30	28	0.0002	5.84947	R		4.76953	0.00017	30	34
	10	XING		10	28	6.5E-05	5.84953	WALNUT LANE		4.76937	0.00061	20.63 D	36
	35			35	29	0.00023	5.84976	REVERSE CURVE		4.76876	0.00012	24.47	36
	55		62	55	29	0.00036	5.85012	L		4.76864	0.00029	55	36
	10		96	10	29	6.5E-05	5.85018	L		4.76834	5.2E-05	10	36
	55		96	55	29	0.00036	5.85054	L		4.76829	0.00029	56	36
	45	XING		45	28	0.00029	5.85084	SPRING LANE		4.768	0.00024	45	36
	100			100	28	0.00065	5.85148			4.76826	0.00053	100	36
	25			25	28	0.00016	5.85165			4.76773	0.00013	25	36
	35	XING	60	35	29	0.00023	5.85188	L DIVISION LANE		4.7676	0.00067	35	36
	10		67	10	29	6.5E-05	5.85195	L		4.75742	5.3E-05	10	36
	30			30	29	0.0002	5.85214			4.75737	0.00016	30	36
	65			65	29	0.00042	5.85257			4.75721	0.00034	65	36
	35		36	35	29	0.00023	5.8528	L		4.75687	0.00018	35	36
	15	XING	38	15	29	9.8E-05	5.8529	L DIVISION LANE		4.76669	7.8E-05	15	38
	30	XING		30	29	0.0002	5.8531	RC PA HWY 68		4.7666	0.00017	33 D	39
	53		39	56	29	0.00036	5.85346	R		4.76644	0.00026	53	39
	20		39	26	29	0.00017	5.85363	R		4.76618	9.7E-05	26	39
	16		50	16	29	0.0001	5.85368	R track removed - pike lt		4.76609	0.00022	49.11 D	43
	10		91	10	29	6.5E-05	5.85375	R		4.76588	1.7E-05	3.863	43
	11		43	11	29	7.2E-05	5.85382	R		4.76586	0		43
	15			15	29	9.8E-05	5.85392	REVERSE CURVE		4.76586	0		43
	15		43	15	29	9.8E-05	5.85402	L		4.76586	8.8E-05	15	43
	3	XING		3	29	2E-05	5.85404	BEAVER INDUSTRIAL		4.76578	1.3E-05	3	43
	10		43	10	29	6.5E-05	5.8541	R TRACKS WERE re ST		4.76577	4.4E-05	10	43
						0	5.8541	TRACKS HAVE BEEN RE		4.76572	0.00323	1395 D	10
						0	5.8541	VANPORT		4.76250	0.00047	96.2	105
	375			375	28	0.00245	5.85655	W BEAV 143		4.76222	0	0	
	700	XING 2.04		700	28	0.00457	5.86112	PRIVATE		4.76222	0	0	103
	2000		103	2000	29	0.01302	5.87414	L		4.76222	0.00368	2000	103
	379			379	29	0.00674	5.88188			4.75855	0.00062	379	103
	600		102	5241	29	0.03423	5.91611	L MP1180(T)		4.75802	0.0011	600	103
	600		102	232 A	29	0.00135	5.9176	SPRC		4.75663	0.00147	800	103
	600			583	34	0.00314	5.92064			4.75406	0.0011	600	103
	900		130	102 A	34	0.00056	5.92119	R		4.75326	0.00166	900	103
	1900			423	38	0.00223	5.92342			4.7516	0.00027	145	103
	900		106	220 A	38	0.00111	5.92453	R		4.75133	0		
	978			740 A	29	0.00111	5.92564			4.75133	0		
	400			11128 A	43	0.02889	5.95453			4.75133	0		
	700		100			0	5.95453	R		4.75133	0		
	781		106			0	5.95453	R		4.75133	0.0175	6479 A	79
	3200					0	5.95453			4.73385	0.00108	438	79
	3200	XING 6.19				0	5.95453	PRIVATE		4.73276	0.00767	3200	78
	1175		134			0	5.95453	RE IND		4.72511	0.00761	3173	78
	900			0		0	5.95453			4.7175	0.00139	572 A	77
	2100	XING 7.60		429	96	0.00164	5.95616	WABASH STREET		4.71611	0.00105	426	77
	1089		126	2614 A	96	0.005	5.96116	LW IND		4.71506	0.00269	1089	77
	2100	XING 8.06		318	102	0.00013	5.96129	PRIVATE		4.71236	0.00517	2100	77
				641 D	80	0.00128	5.96257			4.70722	0		
	850		79	800	79	0.00192	5.96449	L		4.70722	0.00197	800	77
	1800			1800	79	0.00432	5.96881			4.70526	0.00443	1800	77
	1100		200	1100	79	0.00264	5.97144	L		4.70082	0.00271	1100	77
	1283			1283	79	0.00308	5.97452			4.69812	0.00316	1283	77
	300		94	300	79	0.00072	5.97524			4.69496	0.00074	300	77
	900		94	900	79	0.00216	5.9774	R		4.69422	0.00221	900	77
	1200			1154	79	0.00277	5.98016			4.69201	0.00295	1200	77
				46 D	75	0.00011	5.98027			4.68906			
	1100		77	1100	77	0.00271	5.98298	R		4.68906	0.00511	2076	77
						0	5.98298			4.68395	0.00167	685 A	74
	900			900	77	0.00221	5.98519			4.68228	0.00222	900	74
	1175		119	1175	77	0.00289	5.98808	R		4.68006	0.00301	1175	74

APPENDIX B2 — PAGE 10 — COLUMN 1

READ DOWN — WESTBOUND TRAFFIC | STATION | READ UP — EASTBOUND TRAFFIC

MILES	GRADIENT PERCENT	DISTANCE ft	RR GRADE CROSSING	MAX MPH	DISTANCE ft	SPEED MPH	ET hours	TIME hours	STATION	TIME hours (EB)
459.96		5276		100	5274	77	0.01298	6.00167	RMIDLAND	4.67706
459.98		200		150	200	77	0.00071	6.00178	R	4.68354
460.67		2900			2957	77	0.00876	6.01564		4.66132
460.67					1401 A	77	0.00392	6.01283		4.95518
460.84		1500		99	58	82	0.00015	6.0128	L	4.66818
460.86					182 D	82	0.00044	6.01385		4.65334
461.06		900		74	900	74	0.0023	6.01626	L MPH45	4.65334
461.26		1100			1100	74	0.00282	6.01967		4.65023
461.3?		800		134	400	74	0.00154	6.0066	L	4.64884
461.53		800			800	74	0.00300	6.02280	MKX. YD LEAD	4.64468
461.64		600		101	600	74	0.00184	6.02418	L	4.64207
461.76		586			589	74	0.00146	6.00984		4.64011
461.86		700		134	700	74	0.00179	6.02746	L	4.63825
462.43		2900			2616	74	0.00721	6.03464		4.63687
462.43					84 D	74	0.00022	6.03466		4.6268
462.53		531		78	800	78	0.00216	6.03702	L'OKWB	4.6265
462.59		219		78	206	78	0.00073	6.03775		4.62470
462.70		900	XING 12.9		900	78	0.00244	6.04019	CHURCH STREET	4.62368
462.86		876		107	878	78	0.00156	6.04175	R MPH45	4.62394
462.9?		200		107		78	0	6.04175	R	4.61673
462.96		200			6	78	1.9E-05	6.04177	MPH 45	4.61608
462.96					225 D	77	0.00067	6.04242	REVERSE CURVE	4.61747
463.07		621		58	621	58	0.00306	6.0446	URC	4.61747
463.07							0	6.0446	LITTLE BEAVER	4.61536
463.13		300			300	58	0.00098	6.04547		4.61248
463.25		800	XING 152		800	58	0.00196	6.04743	R PRIVATE	4.61274
463.36		600		152	600	58	0.00196	6.04529	L REVERSE CURVE	4.61274
463.87		2700			2700	58	0.00887	6.05621	CHWB CHICOF-K LINE	4.61274
464.58		8332	XING 15.1		3387	58	0.01741	6.07562	PUTNAM STREET	4.59022
464.96		400	XING 15.2		400	58	0.00131	6.07693	MYRTLE STREET	4.5926
465.00		208	XING 152		208	58	0.00084	6.07776	L WALTERS STREET	4.59124
465.10		600		152	500	58	0.00163	6.0794	R REVERSE CURVE	4.59638
465.31		1100	XING 16.5		312	58	0.00102	6.08041	MPH 35(TT) VIRGINIA ST	4.58886
465.50		1000		129			0	6.08041	L	4.58697
465.71		1100		142			0	6.08041	L	4.58108
465.88		900			3618 A	58	0.01167	6.09208		4.57708
466.37		2800			3607 A	73	0.01083	6.10291	US 30	4.57462
466.56		1000		115			0	6.10291	L	4.56603
466.63		400			62	90	0.00077	6.10368		4.56205
466.63					406 D	90	0.00094	6.10462	REVERSE CURVE	4.56129
466.73		500		73	500	73	0.0013	6.10592	R	4.56129
466.92		500		110	500	73	0.0013	6.10662	R MPH 35(TT)	4.5666
466.58		320				73	0	6.10663		4.56791
466.94		300			298	73	0.00077	6.1074		4.55563
466.94					332 D	73	0.00094	6.10834		4.55681
466.96		200	XING 58		200	56	0.00066	6.10902	R PRIVATE	4.55581
466.96							0	6.10902		4.55614
467.11		700			700	56	0.00237	6.11138		4.55491
467.18		282	XING 60		282	56	0.00085	6.11224	R BROADWAY	4.55292
467.18							0	6.11224		4.55213
467.23		400	XING 13.4		400	56	0.00135	6.11359	TT MPH 35 WASHINGTC	4.55198
467.35		600	XING 63		600	56	0.00203	6.11562	R MARKET STREET	4.55196
467.35			XING 18.96				0	6.11562	JACKSON STREET	4.54811
467.46		600			600	56	0.00203	6.11765	E LIVERPOOL	4.54661
467.57		600		92	600	58	0.00203	6.11968	R	4.5478
467.69		600			600	56	0.00203	6.12171	MPH 35	4.54603
467.80		600		95	600	56	0.00203	6.12374	R	4.54430
467.88		400			400	56	0.00135	6.12509		4.54277
467.93		300			300	56	0.00101	6.1261		4.54172
468.03		500		182	500	56	0.00169	6.1278	L	4.54065
468.20		900			900	56	0.00304	6.13084		4.53961
468.29		500		72	500	56	0.00169	6.13253	L	4.53725
468.29							0	6.13253		4.53368
468.41		600		56	600	56	0.00203	6.13456	L	4.53492
468.62		1100			1100	56	0.00372	6.13828		4.53437
468.79		900		94	900	56	0.00304	6.14133	L	4.53226
468.83		215			215	56	0.00073	6.14205		4.53106
468.40		3000			1233	56	0.00417	6.14622		4.52906
469.52		1200		95	500 A	56	0.00167	6.14789	R	4.52252
469.98		1877			443	60	0.0014	6.14929	MPH 50(TT)	4.52181
470.14		1400			941 A	60	0.00167	6.15096	BELT LINE	4.51868
470.48		1800		95	400	83	0.00130	6.15233	L	4.51118
470.65		900		90	1891 A	63	0.00472	6.15705	L	4.50878
470.87		1131		101	1306	72	0.0087	6.16575	L'OKWB	4.50824
471.02		600			4895 A	72	0.01139	6.17715		4.50304
471.19		900		252			0	6.17715	R	4.50076
471.32		700			88	90	0.00018	6.17733		4.49824
471.44		600		110	600	90	0.00126	6.1786	L	4.49636
471.59		800			361	90	0.00116	6.17976	MPH 50(TT)	4.49496
471.59					249 D	90	0.00050	6.18031		4.49281
471.82		1200		80	1295	80	0.00287	6.18328	R	4.49281
471.88		300			300	80	0.00071	6.18399		4.49281
471.99		500			500	80	0.00118	6.18518		4.48281
472.15		900		90	900	80	0.00213	6.18731	R	4.48327
472.41		1300	XING 22.4		1100	80	0.0026	6.18981	18TH STREET	4.48211
472.41					280 D	80	0.00072	6.19063	WELLSVILLE	4.4763
472.58		900		67	900	67	0.00254	6.19316	L	4.4783
472.71		700			700	67	0.00198	6.19515		4.47252
472.84		700		120	700	67	0.00198	6.19713	L	4.46867
472.86		200			200	67	0.00057	6.1977		4.46862
473.03		800			800	67	0.00228	6.19996		4.46578
473.20		900		81	900	67	0.00254	6.20251	L	4.46241
473.47		1400			1400	67	0.00396	6.20646		4.45663
473.66		1000		67	1000	67	0.00283	6.20926	L	4.45273
473.89		1222			1222	67	0.00345	6.21274		4.44863
473.98		500			500	67	0.00141	6.21416		4.44338
474.10		600		107	600	67	0.0017	6.21586	L'WOKWB	4.43128
474.17		400			39	67	0.00011	6.21596	RC	4.43878
474.17					361 D	67	0.00122	6.21718		4.43707
474.31		700		45	700	45	0.00295	6.22013	R/L OKWB	4.43707
474.31							0	6.22013		4.43412
474.44		700	XING 25.7		700	45	0.00295	6.22308	TT MPH 35 PRIVATE	4.53389
474.59		800		50	800	45	0.00337	6.22644	L YEL OR OK WB	4.43148
474.70		600			600	45	0.00253	6.22897	MPH 35	4.42843
474.88		902		54	902	45	0.00401	6.23298	R'WOKWB	4.42871
475.15		1400		54	1400	45	0.00589	6.23887	R'WOKWB	4.42521
475.36		1100			1100	45	0.00463	6.2435		4.4183
475.45		495		95	490	45	0.00206	6.24556	L	4.41796
475.72		1400		95	1400	45	0.00589	6.25147	L'OW POST	4.41038
475.85		700			700	45	0.00295	6.25442		4.40358
476.04		1000			1000	45	0.00421	6.25863		4.40195
476.27		1200		78	1200	45	0.00505	6.26368	L	4.39706
476.49		1200			962	45	0.00275	6.26643		4.39126
476.49			XING 27.8		419 A	45	0.00167	6.26809	GANDOS CROSSING	4.38544
476.91		2196		72	1581	50	0.00409	6.27219	R	4.38544

APPENDIX B2 — PAGE 10 — COLUMN 2

READ DOWN — WESTBOUND TRAFFIC | STATION | READ UP — EASTBOUND TRAFFIC

MILES	GRADIENT PERCENT	DISTANCE ft	RR GRADE CROSSING	MAX MPH	DISTANCE ft	SPEED MPH	ET hours	TIME hours	STATION	TIME hours (EB)
477.06		600		92	497 A	58	0.00167	6.27386	R	4.27476
477.21		800		93	2385	54	0.00606	6.28027	R	4.27980
477.36		600		60	110	54	0.00039	6.28291	HOV	4.26701
477.67		1800	XING 28.9		1292 A	54	0.00417	6.26807	HAMMONDSVILLE WAY	4.36312
477.80		700		101	796	83	0.0021	6.28986	L	4.30535
477.91		679		83	306 D	83	0.00183	6.29021	L	4.30195
478.09		900		39	900	39	0.00437	6.29458	L	4.34916
478.12		225			225	39	0.00109	6.29567	REVC	4.34479
478.35		1100		41	1100	39	0.00534	6.30192	R	4.34468
478.50		900			900	39	0.00437	6.30028		4.3367
478.76		1400		57	1400	39	0.0068	6.31219	L'OKWB	4.3392
478.97		800	XING 30.5		800	39	0.00389	6.31607	EAST AVENUE	4.3326
479.11		1000		92	1000	39	0.00486	6.32093	R	4.33087
479.31		1100			1100	39	0.00534	6.32627	IRONDALE	4.32708
479.58		1300		64	1300	38	0.00631	6.33258	L	4.32262
479.81	0.71%	1300		67	1300	38	0.00631	6.33889	L	4.31739
479.92	0.71%	500			500	39	0.00287	6.34176		4.31307
479.97	0.71%	300			300	39	0.00146	6.34322		4.30257
480.37	0.71%	2100		56	985	39	0.00479	6.348	R	4.30211
480.40		149			117 A	39	0.00056	6.34856	REVC	4.30172
480.70		1600		50	1266	41	0.00558	6.35414	L'OKWB	4.30114
480.95		1300		62	738 A	41	0.00306	6.35719	L	4.29433
481.35		1800			4814	50	0.01828	6.37543	NEW SALISBURY	4.2901
481.40		900		91	900	50	0.00341	6.37884	R	4.28378
481.54		700	XING 32.9		706	50	0.00269	6.38149	PRIVATE	4.28032
481.79		1200		74	1200	50	0.00455	6.38603	L'OKWB	4.27747
481.90		656			596	50	0.00224	6.38827		4.27276
482.16		900			900	50	0.00341	6.39255		4.28936
482.27		900		65	900	50	0.00341	6.39609	L'OKWB	4.26951
482.44		900			871	50	0.0033	6.39839		4.25726
482.44					29 D	50	0.00011	6.3985		4.25821
482.68		1300		46	1300	46	0.00613	6.40463	R	4.25871
482.68							0	6.40463		4.25358
482.68							0	6.40463		4.25241
482.72		190			190	48	0.00075	6.40538	REVC	4.25228
482.85		700		68	700	48	0.00278	6.40816	L'OKWB	4.24589
482.91		300			300	48	0.00118	6.40933		4.24892
483.44		2800			2806	48	0.01105	6.42037		4.24132
483.65		1100		87	1100	48	0.00434	6.42471	R	4.23799
483.93	0.67%	1471			1471	48	0.0058	6.43052		4.23343
484.02	0.67%	500			500	48	0.00197	6.43249		4.23094
484.19	0.67%	900	XING 73		900	48	0.00355	6.43604	L McGARY	4.22896
484.66	0.67%	2500	XING 35.7		2500	48	0.00986	6.44591	NOLAN	4.22541
484.90	0.67%	1267		153	539	48	0.00213	6.44803	R	4.21556
484.92	0.67%	100			845 A	48	0.00306	6.45109		4.21056
485.25		1700	XING 36.1		1886	58	0.0057	6.45679	PRIVATE	4.21019
485.45		1100		58	1100	58	0.00372	6.46051	R	4.20746
485.51		300	XING 36.5		207	58	0.0007	6.46121	REV C KAYTT	4.19911
485.57					93 D	56	0.00033	6.46154		4.18792
485.67		1900		56	1900	56	0.0072	6.46874	L'ROGERS	4.19792
485.89		300			71	56	0.00027	6.46901	RC	4.19043
485.89	0.92%				29 D	56	0.00011	6.46912		4.19087
485.91	0.92%	100		48	100	48	0.00035	6.46951	RYOK	4.19063
486.18	0.92%	1431		46	1431	48	0.00565	6.47516	RYOK	4.18584
486.18	0.92%		XING 37.33				0	6.47516	MAIN STREET	4.18389
486.27	0.92%	500			500	48	0.00197	6.47713	RC SALINEVILLE	4.18306
486.43	0.92%	800		95	800	48	0.00316	6.48029	L	4.18229
486.60	0.92%	900			900	46	0.00356	6.48384		4.17948
486.77	0.92%	900		95	900	48	0.00356	6.48739	L	4.17632
486.90	0.92%	300			600	48	0.00316	6.49055		4.17397
487.20	0.92%	1500			1500	48	0.00592	6.49647		4.17038
487.33	0.92%	700	XING 69		700	48	0.00276	6.49923	R FOUNDRY HILL ROAD	4.1861
487.38	0.92%	300			300	48	0.00118	6.50041	REVC	4.16285
487.50	0.92%	600		76	600	46	0.00237	6.50278	L	4.16159
487.58	0.92%	450			400	48	0.00156	6.50436	REVC	4.15949
487.77	0.92%	1000		54	1000	48	0.00395	6.5083	R	4.15609
487.79	0.92%	116			116	48	0.00046	6.50878	REV C	4.15458
487.91	0.92%	600		62	600	48	0.00237	6.51115	L	4.15414
488.12	0.82%	1100		92	1106	48	0.00434	6.51547	L TT MPH 35	4.1523
488.13	0.82%		XING 38.38				0	6.51547	DOBSON	4.14894
488.16	0.92%	208			208	46	0.00082	6.51629	REV C	4.14839
488.28	0.92%	700		92	575	46	0.00227	6.51856	R	4.14030
488.34	0.92%	500		58			0	6.51856	R	4.14652
488.61	0.92%	900			781 A	46	0.00222	6.52078		4.14289
488.91	0.92%	1600		152	3683	64	0.00934	6.53012	L MP	4.14188
489.01	0.82%	500		152	500	54	0.00176	6.53188	L40	4.14168
489.14	0.82%	700			700	54	0.00246	6.53434		4.14160
489.26	0.82%	800	XING 134		800	54	0.00281	6.53714	R HULL ROAD	4.13263
489.40	0.92%	600			600	54	0.0021	6.53924		4.13047
489.90	0.94%	2613		85	2276	54	0.00883	6.54756	L MPH 35(TT)	4.12615
490.04	0.94%	768		95	1105 A	54	0.00361	6.55119	L SUMMITVILLE DIV PO	4.1163
490.20	0.98%	800			707	62	0.00216	6.55335		4.11538
490.20	0.94%				1674 A	62	0.00506	6.55842		4.11206
490.32	0.94%	1700		72			0	6.55842	R	4.11206
490.53	0.94%	600			415	72	0.00109	6.55973		4.10692
490.74	0.94%	600		109	600	72	0.00158	6.58120	R	4.10864
490.90	0.94%	500			500	72	0.0021	6.5634		4.10137
490.95	-0.86%	350			258	72	0.00068	6.56405		4.10834
490.95	-0.86%				40 D	72	0.00011	6.56416		4.09721
491.30	-0.86%	1300		70	1300	78	0.00362	6.56778	L'OKWB@54	4.09721
491.27	-0.86%	400			45	79	0.00013	6.56791	MPH 39	4.09228
491.27	-0.86%				382 D	70	0.00141	6.56932		4.09077
491.43	-0.65%	600		50	600	50	0.00303	6.57195	RYOK	4.09077
491.43	-0.64%						0	6.57195	REVERSE CURVE	4.08774
491.56	-0.84%	700		54	700	50	0.00265	6.57460	L'OKWB	4.08791
491.56	-0.84%						0	6.57460		4.08527
491.67	-0.84%	600			600	50	0.00227	6.5769	SUMMITVILLE	4.08299
491.73	-0.84%	310		86	316	50	0.00117	6.57806	LW END 2 TRACKS	4.08297
491.86	-0.71%	500			500	50	0.00303	6.58111	MPH 40(TT)	4.08087
491.89	-0.71%	32			32	50	0.00017	6.58123		4.08132
491.96	-0.71%	362		68	302	50	0.00189	6.58312	R	4.08121
492.44	-0.71%	2400	XING 45.4		2400	50	0.00908	6.59221	PRIVATE	4.07105
492.89	0.77%	2400			2400	50	0.00908	6.6013		4.07196
492.89							0	6.6013		4.06148
493.89	0.72%	5200			3256	50	0.01234	6.61364		4.06081
494.81	0.70%	3500			457 A	50	0.00167	6.61531		4.05981
494.82	0.70%	1100		95	263	54	0.00085	6.61616	L	4.04954
494.91	0.70%	460			3779 A	54	0.02417	6.64033		4.04939
495.58	0.79%	5200	XING 46.3		3647	95	0.00608	6.64641		4.04827
495.88	0.77%	1600			56 D	95	0.00011	6.64652		4.03986
496.22		900		93	800	93	0.00163	6.64815	L	4.03322
496.89	-0.77%	3400			3104	93	0.00632	6.65447		4.02128
496.93					329 D	93	0.00063	6.6563		4.02283
497.17	-0.90%	1200		76	1200	78	0.00291	6.65822	L KENSINGTON	4.02283
497.12							0	6.66622		4.01991

APPENDIX B2

READ DOWN — WESTBOUND TRAFFIC — STATION — READ UP — EASTBOUND TRAFFIC

PAGE 11 COLUMN 1

MILES	GRADIENT PERCENT	DISTANCE ft	CROSSING	MAX MPH	DISTANCE feet	SPEED mph	ET hours	TIME hours	STATION	TIME hours	ET hours	DISTANCE feet	SPEED mph	MILES
497.31	-0.70%	1000			1000	75	0.00243	6.66034	MPH 40(TT)	4.01097	0.00113	508.7	95	530.43
497.45	-0.70%	900		95	900	75	0.00219	6.66283	R	4.01783	0.00179	900	95	530.43
497.75	-0.79%	1600			1600	75	0.00389	6.66672	WEST KENSINGTON	4.01604	0.00319	1600	95	530.64
497.98	-0.79%	524		95	524	75	0.00127	6.66799	R	4.01285	0.00104	524	95	530.77
497.92		185		95	185	75	0.00045	6.66844	R	4.01181	0.00037	185	95	531.11
498.48		3039			3039	75	0.00738	6.67582		4.01144	0.00139	786.3 D	120	531.31
498.49			WEST END SMITHS FERRY KEY					6.67582		4.01005	0.00368	2251	120	531.31
498.88		2061			2061	75	0.005	6.68082	MP	4.0065	0.00325	2061	120	531.39
499.74		4500			1691	75	0.00411	6.68493		4.00326	0.0071	4500	120	531.60
499.79		300		164	5474 A	78	0.01194	6.69687		3.99614	0.00047	300	120	532.13
499.86		476	XING 51.11		17532 A	95	0.02164	6.71851	BROGAN ROAD	3.99567	0.00075	476	120	532.13
500.99		5340	XING 51.53				0	6.71851	L LIPPINCOT ROAD	3.99492	0.00843	5342	120	532.24
501.05		800	XING 51.86				0	6.71851	PRIVATE	3.98648	0.00126	800	120	532.37
501.14		500		269			0	6.71851	R	3.98522	0.00079	500	120	532.54
501.85		3752	XING 52.28				0	6.71851	PRIVATE	3.98443	0.00592	3752	120	532.73
502.02		900					0	6.71851	MPH 40	3.97851	0.00142	900	120	532.75
502.37		1600	XING	134			0	6.71851	R US HWY 30	3.97709	0.00253	1600	120	532.75
502.48		800					0	6.71851	E ROCHESTER	3.97457	0.00126	800	120	533.19
502.67	-0.77%	2073		134	1346	120	0.00212	6.72064		3.9733	0.00327	2073	120	533.37
502.93		300		134	300	120	0.00047	6.72141	L	3.97003	0.00047	300	120	533.41
503.06		714			714	120	0.00113	6.72254		3.96956	0.00113	714	120	533.68

PAGE 11 COLUMN 2

APPENDIX B2

READ DOWN — WESTBOUND TRAFFIC | STATION | READ UP — EASTBOUND TRAFFIC

PAGE 12 COLUMN 1

MILES	GRADIENT, PERCENT	DISTANCE, ft	MIN GRADE CROSSING	MAX MPH	DISTANCE, feet	SPEED, MPH	ST. hours	TIME, hours	STATION	TIME, hours	ST. hours	DISTANCE, feet	SPEED, MPH	MILES	GRADIENT, PERCENT
564.38	0.59%	2000			2000	120	0.00316	7.40487		3.28573	0.00549	2000	69	594.47	
564.61	0.59%	1200		190	1200	120	0.00189	7.40676	L	3.28024	0.00329	1200	69	594.85	0.66%
564.74	0.59%	700	XING 144.		2302	120	0.00363	7.4104	WOOD	3.27695	0.00192	700	69	594.96	
564.74			XING 144.		816 D	11	0.00144	7.41184	ROBINSON	3.27503				595.07	
565.07	0.59%	1700	XING	94	1700	94	0.00343	7.41527	R a MARKET STREET	3.27503	0.00467	1700	69	595.19	
565.26	0.59%	1000	XING 144.		1000	94	0.00201	7.41728	SHREVE MAIN STREET	3.27036	0.00274	1000	69	595.47	
565.40	0.59%	753		134	487	94	0.00094	7.41822	R	3.26762	0.00207	753	69	596.19	
565.40					296 D	94	0.00061	7.41883		3.26556				596.19	
565.64	0.59%	1300		83	1300	83	0.00287	7.4218	R b	3.26556	0.00357	1300	69	596.19	
565.74		500			337	83	0.00077	7.42257	REVERSE CURVE	3.26198	0.00137	500	69	596.26	
565.74					163 D	83	0.00039	7.42296		3.26061				596.49	
566.00		1378		76	1229	78	0.00306	7.42602	L c	3.26061	0.00378	1378	69	596.64	
566.00					149 D	78	0.00039	7.42641		3.25683				596.74	
566.27		1400		69	1400	69	0.00384	7.43025	L	3.25683	0.00384	1400	69	596.98	
566.27							0	7.43025		3.25298	0.00059	148.9 D	76	597.32	
566.36		500			500	69	0.00137	7.43163	RC	3.2526	0.00068	301.1	76	597.49	
566.40		200		76	200	69	0.00055	7.43217	R d	3.25172	0.0005	200	76	598.11	
566.70		1601	XING	76	1601	69	0.00439	7.43657	R BROWNS	3.25122	0.00399	1601	76	598.40	
566.70							0	7.43657		3.24723	0.00106	476.5 D	90	599.04	
566.70							0	7.43657		3.24618	0.00064	321.5	95	599.06	
567.40		3700	XING 147.		3700	69	0.01016	7.44673	PRIVATE	3.24554	0.00861	4058 A	83	599.04	
568.23		4382	XING 148.		4382	69	0.01206	7.45878	SCRUBYS	3.23695	0.00738	3236	83	599.20	
568.58		1600	XING 148.		607	69	0.00167	7.46045	BIG PRAIRIE RTE 226	3.22954	0.00411	1800	83	599.25	
568.74		887		134	3359 A	69	0.00833	7.46878	L	3.22543	0.00202	887	83	599.36	
569.41		3600			3231 A	83	0.00694	7.47572		3.22341	0.00799	3600	83	599.49	
569.46		277			309	93	0.00063	7.47635		3.21542	0.00063	277	83	599.68	
569.70		1300		134	268 D	92	0.00058	7.47691	L	3.21479	0.00297	1300	83	599.83	
570.16		2400		83	2400	83	0.00548	7.48239	L s	3.21182	0.00548	2400	83	599.95	
570.18							0	7.48239		3.20635	0.00044	204.2 D	91	600.08	
570.22		300			300	83	0.00068	7.48307	REVERSE CURVE	3.2059	0.0002	35.84	91	600.23	
570.41		1000		95	1000	83	0.00226	7.48535	R b	3.2057	0.00208	1000	91	600.48	
570.61		1100		110	1100	83	0.00251	7.48786	R	3.20362	0.00153	734	91	600.78	
570.68		400	XING 150.		400	83	0.00091	7.48877	LAKEVILLE GULLS RO	3.2021	0			601.50	
570.94		1300	XING	93	1300	83	0.00297	7.49174	Re STATE ROUTE 179	3.2021	0.01111	4864 A	74	602.01	
571.41		2500			2500	83	0.0057	7.49745		3.19098	0.00091	240	74	602.14	
571.66		1300			1300	83	0.00297	7.50041		3.19007	0.00333	1300	74	602.38	
571.89		1217		91	1217	83	0.00278	7.50319	L c	3.18704	0.00444	1646 A	86	602.60	
572.42		2800			1883	83	0.0043	7.50749		3.1826	0.00528	1633	86	602.68	
572.42					2906 A	83	0.00556	7.51304		3.17734	0			602.68	
573.42		5281			3931	91	0.00818	7.52122		3.17734	0.01515	5281	58	602.66	
573.55		700			411 D	91	0.00094	7.52217		3.16218	0.00201	700	66	603.51	
574.10		2900	XING	74	2900	74	0.00742	7.52959	R TR 386 a	3.16018	0.00832	2900	66	604.44	
574.15		288			122	74	0.00031	7.5299	REVERSE CURVE	3.15185	0.00082	288	66	605.53	
574.15					164 D	74	0.00044	7.53034		3.15103				606.52	
574.42		1400		66	1400	66	0.00402	7.53436	L b	3.15103	0.00402	1400	86	607.53	
574.55		700			700	66	0.00201	7.53637		3.14702	0.00089	347.3 D	82	608.54	
574.55							0	7.53637		3.14613	0.00081	282.7	82	608.58	
574.72		900		82	900	66	0.00258	7.53895	L c	3.14531	0.00208	900	82	608.88	
574.72							0	7.53895		3.14323	0.00027	98.08 D	10	608.96	
574.87		804		128	804	66	0.00231	7.54126	L	3.14301	0.00155	705.4	86	609.16	
575.42		2900	XING 155.		2900	86	0.00632	7.54958	TR 5	3.14146	0.00629	2900	86	609.15	
576.31		4700			4700	66	0.01349	7.56307		3.13507	0.01035	4700	86	609.15	OH SR 61 - MOVE TO AVOID 3 LEVEL CROSSINK
576.42		554	XING	89	554	68	0.00163	7.56466	R d LOUDONVILLE JEFF	3.12472	0.00122	554	86	609.19	
576.96		2846	XING	86	442	68	0.00127	7.56593	R SPRING STREET	3.1235	0.00627	2840	86	609.19	
576.96					3717 A	66	0.00944	7.57537		3.11725	0.001	901.6 D	10	609.31	
577.43		2800			1110 A	62	0.0025	7.57787	CP MOHICAN	3.11623	0.00364	1998	104	609.46	
578.26		4400			4484	88	0.00987	7.58776		3.11259	0.00861	4400	104	609.55	
578.43		898	XING	104	898	86	0.00197	7.58971	L PRIVATE	3.10468	0.00163	893	104	609.67	
578.55		610		104	610	68	0.00134	7.59106	L mon	3.10296	0.00111	610	104	609.70	
579.44		4700			2587	68	0.00572	7.59678		3.10184	0.00033	188.3 D	110	609.78	
579.68		1303	XING	110	6993 A	66	0.01389	7.61067	L/E PERRY PRIVATE	3.10151	0.01001	5815	110	610.54	
580.44		4000			413	104	0.00075	7.61142		3.08149	0.00039	233.1 D	117	610.64	
580.44							0	7.61142	MP 580	3.08111	0.00044	272.9	117	610.86	
580.63		1000	XING 160.		987	104	0.00162	7.61323	BRIDGE STREET	3.09066	0.00635	3831 A	110	610.97	
581.44	0.65%	4284	XING	140	2986 A	10	0.00528	7.61861	R PRIVATE	3.08427	0.01111	8084 A	97	611.12	
582.03		3100		129	3017	110	0.00519	7.62271	R PERRYSVILLE	3.07316	4.9E-05	25	97	611.12	
582.45		2260	XING 162.		5743 A	11	0.00944	7.63316	PRIVATE	3.07311	0.00441	2260	97	611.60	
583.23		4100	XING 162.		1981	120	0.00313	7.63628	TOWNSHIP ROAD 1007	3.0687	0.01064	5005 A	82	611.93	
583.46		1197		129	1197	120	0.00189	7.63817	L	3.05815	0.00217	938	82	612.11	
583.55		500		129	500	120	0.00079	7.63895	L	3.05598	0.00115	500	82	612.30	
583.74		1000			298	120	0.00042	7.63938	0	3.05483	0.00231	1000	82	613.92	
583.74					732 D	12	0.00128	7.64066		3.05252	0.005	7066 A	74	614.92	
584.04		1600		97	1600	97	0.00312	7.64378	L PERRYSV	3.04752	0.00208	806	74	614.92	
584.31		1390	XING 163.		1390	97	0.00271	7.64649	COUNTY ROAD 1075	3.04548	0.00358	1390	74	615.92	
584.46		800	XING	129	800	97	0.00156	7.64806	L PRIVATE	3.0419	0.00205	800	74	615.92	
584.61		800		129	800	97	0.00156	7.64962	L	3.03985	0.00206	800	74	616.92	
585.12		2700			2306	97	0.0045	7.65412	x	3.0378	0.00691	2700	74	617.38	
585.12					394 D	97	0.00085	7.65496		3.03089				617.68	
585.33	0.51%	1104		82	1104	82	0.00255	7.6575	R a	3.03089	0.00283	1104	74	618.92	
585.46	0.51%	700			700	82	0.00162	7.65912		3.02806	0.00179	700	74	619.92	
585.98	0.51%	2700			2317	82	0.00581	7.66493		3.02627	0.00691	2700	74	620.92	
585.98					183 D	82	0.00044	7.66538		3.01936				621.17	
586.26	0.51%	1500		74	1500	74	0.00384	7.66922	L b	3.01936	0.00384	1500	74	621.31	
586.26							0	7.66922		3.01552	0.00083	358.6 D	89	621.50	
586.47	0.51%	1102			1102	74	0.00282	7.67204		3.01469	0.00158	743.4	89	621.50	
586.81	0.51%	1800			1800	74	0.00461	7.67684		3.01311	0.00383	1800	89	621.50	
586.96	0.51%	800		93	800	74	0.00205	7.67869	L c	3.00928	0.0017	800	89	621.50	
587.16	0.51%	1067	XING 166.		1067	74	0.00273	7.68142	PRIVATE	3.00758	0.00227	1067	89	621.71	
587.39	0.51%	1200		93	1200	74	0.00307	7.68449	R d	3.0053	0.00285	1200	89	621.71	
587.47	0.51%	400			400	74	0.00102	7.68552		3.00276	0.00085	400	89	621.90	
587.62	0.51%	800			800	74	0.00205	7.68757	COULTER	3.0019	0.0017	800	89	621.92	
587.75	0.51%	700		97	700	74	0.00179	7.68936	R a	3.0002	0.00149	700	89	621.91	
588.30	0.80%	2900	XING 167.		2900	74	0.00742	7.69678	PRIVATE	2.99871	0.00617	2900	89	621.91	
588.38	0.60%	400	XING	89	231	74	0.00059	7.69737	R LUCAS f	2.99254	0.00086	400	89	622.12	
588.46	0.60%	457			4203 A	74	0.00972	7.70709	0	2.99169	0.00097	457	89	622.12	
588.52	0.60%	295						7.70709		2.99071	0.00063	295	89	623.91	
588.69	0.80%	900		68				7.70709	L g	2.99008	0.00192	900	89	623.92	
588.69			XING 168.3					7.70709	UNION STREET	2.98817	0.00053	161.9 D	95	624.82	
588.69			XING 168.46					7.70709	CHURCH	2.98764	0.00166	381.1	95	624.82	
589.09	0.60%	2100	XING 168.89				0	7.70709	LUCAS HARLEY BRETZ	2.98615	0.00221	1107	95	624.92	
589.46	0.60%	2000		107	1716	69	0.00566	7.71075	R h	2.98397	0.00369	2000	95	625.94	
589.88	0.60%	2200		134	2200	69	0.00468	7.71543	R	2.97999	0.00439	2200	95	626.94	
590.17	0.60%	1500			1500	89	0.00319	7.71862	CP LUCAS	2.9756	0.00289	1500	95	627.94	
590.30	0.60%	693		110	693	89	0.00147	7.7201	R i	2.97261	0.00128	693	95	627.94	
590.43	0.60%	700		96	700	89	0.00149	7.72159	R	2.97123	0.0014	700	95	628.81	
590.47		200			200	89	0.00043	7.72201		2.96983	0.0004	200	95	628.81	
590.96		2600			1607	89	0.00342	7.72543		2.96943	0.00518	2600	95	629.93	
591.39		2300		95	2163 A	89	0.00444	7.72988	L j	2.96425	0.00459	2300	95	629.93	
591.39							0	7.72988		2.95966	0.00117	649.9 D	116	629.93	
591.47		377	XING 171.		1807	95	0.005	7.73288	PRIVATE	2.9985	0.00273	1673	116	630.96	
592.47		5281	XING 171.		5281	95	0.01053	7.74341	SITTS ROAD	2.95577	0			630.96	
593.41		5000			5000	95	0.00997	7.75338		2.95577	0.00586	3318 A	110	631.96	
593.47		279		152	279	95	0.00056	7.75393	L	2.95021	0.01639	9689 A	96	632.96	
593.52		300	XING	152	63	95	0.00013	7.75406	L/RC ILLINOIS AVENUE	2.93382	0			633.96	
593.61		481		152			0	7.75406	RTT MPH 70	2.93382	0			633.96	

PAGE 12 COLUMN 2

MILES	GRADIENT, PERCENT	DISTANCE, ft	MIN GRADE CROSSING	MAX MPH	DISTANCE, feet	SPEED, MPH	ST. hours	TIME, hours	STATION	TIME, hours	ST. hours	DISTANCE, feet	SPEED, MPH
		4800			8281 A	96	0.015	7.76806		2.93382	0.00397	1888	90
		2000					0	7.76806		2.92985	0.00421	2000	90
		600		116	59	113	9.9E-05	7.76816	R a	2.92564	0.00126	600	90
		600	XING	134	98 D	11	0.00017	7.76832	R 4TH STREET	2.92438	0.00126	600	90
		600	XING	110	600	110	0.00163	7.77036	R 5TH STREET b	2.92311	0.00126	600	90
		1512	XING 175.		1512	110	0.0026	7.77296	MANSFIELD ORANGE S	2.92185	0.00318	1512	90
		3800	XING 175.		3213	110	0.00553	7.77849	SURREY ROAD	2.91867	0.008	3800	90
					587 D	11	0.00111	7.7796		2.91067	0		
				90		90	0	7.7796	B&O XING c	2.91067	0	RR CR	90
		300	XING	190	300	90	0.00063	7.78024	L MAIN STREET	2.91067	0.00096	455	90
		1274	XING 176.		1274	90	0.00268	7.78292	W MANS LONGVIEW AV	2.90972	0		
		900			900	90	0.00169	7.78461		2.90972	0.00806	3574 A	76
		400		190	400	90	0.00084	7.78560	L	2.90166	0		
		1268			1268	90	0.00267	7.78832		2.90166	0.00027	113	78
		1800	XING	174	1800	90	0.00379	7.79211	L EMP STL MILL	2.90139	0.00437	1800	78
		900	XING 177.		900	90	0.00189	7.794	PRIVATE	2.89702	0.00219	900	78
		3300			3300	90	0.00694	7.80095		2.89483	0.00801	3300	78
		1987		124	858	90	0.00181	7.80276	L	2.88682	0.00482	1987	78
		2900			5826 A	90	0.01276	7.81553		2.88199	0.00704	2900	78
					288	106	0.00044	7.81597		2.87495	0		
					756 D	10	0.00156	7.81753		2.87495	0		
		1100		78	1100	78	0.00267	7.8202	R a	2.87495	0.00267	1100	76
							0	7.8202		2.87228	0.00144	694 D	10
		568			568	78	0.00144	7.82164	REVERSE CURVE	2.87084	0.0005	277	104
		700		107	700	78	0.0017	7.82334	L a	2.87033	0		
		997		107	997	78	0.00242	7.82576	L	2.87033	0		
		800			800	78	0.00194	7.8277		2.87033	0		
		600		179	600	78	0.00146	7.82916	L	2.87033	0		
		700	XING 179.		700	78	0.0017	7.83086	LEPPO ROAD	2.87035	0		
		800		139	800	78	0.00194	7.8328	R	2.87035	0.00889	4830 A	93
		1300	XING 180.		1300	78	0.00316	7.83596	SPRING MILL	2.86144	0.00181	889	93
		1510		134	1510	78	0.00367	7.83962	L	2.85963	0.00306	1510	93
		3800			3000	78	0.00728	7.84691		2.85656	0.00794	3800	93
		2706			4647 A	78	0.01028	7.85719		2.84861	0.0056	2700	93
		701		107			0	7.85719	L f	2.84312	0.00143	701	93
		1300	XING 181.94				0	7.85719	PRIVATE	2.84169	0.00265	1300	93
		600		93		93	0	7.85719	L g	2.83904	0.00122	600	93
		927	XING	93		93	0	7.85719	L ROCK ROAD	2.83782	0.00189	927	93
								7.85719		2.83593	0.0015	843.5 D	12
							0	7.85719	TOLEDO JCT	2.83443	0.0054	3424	120
		4400	XING 183.		4081	93	0.00831	7.8655	PRIVATE	2.82903	0.00194	17191 A	81
		4912	XING 184.		4912	93	0.01	7.8755	SHELBY ROAD	2.78708	0.00472	1923 A	73
		5710			1628	93	0.00344	7.87894		2.78236	0.00147	565	72
		5296	XING 186.		13359 A	93	0.02361	7.90255	HORNING ROAD	2.7809	0.00022	835 A	69
		5308			200	120	0.00032	7.90287	begin raise 188.59	2.78867	0.00032	844	69
		5302	XING 188.		5302	120	0.00837	7.91125	COUNTY LINE	2.78836	0.01455	5302	66
		1600	XING 188.		1718	120	0.00271	7.91396	BUCYRUS STREET	2.7738	0.00454	1600	66
					1150 D	17	0.00217	7.91611		2.78686			
		554	XING	81	554	81	0.0013	7.91741	R a WASHINGTON	2.76686	0.00152	554	69
		900			900	81	0.0021	7.91951	CRESTLINE	2.76534	0.00247	900	69
				90			0.08333	8.00284	CP CREST S B4 XING	2.76287	0.08333	RR CROSS	
								8.00284		2.67954			
		200			15	61	4.6E-05	8.00288		2.67954	0.00065	200	69
					181 D	61	0.00044	8.00333		2.67889	0		
		600		73	1317	73	0.00342	8.00675	R c	2.67889	0.00165	600	69
		800			63 D	73	0.00022	8.00697		2.67734	0.0022	800	69
		400		69	400	69	0.0011	8.00807	R d	2.67510	0.0011	400	69
		700		69	700	69	0.00192	8.00999	R	2.67405	0.00182	700	69
		704			704	69	0.00056	8.01055	REVERSE CURVE	2.67213	0.00028	104.9 D	74
		400		74	400	69	0.0011	8.01165	L e	2.67138	0.00128	488.1	74
		4000			4000	69	0.01098	8.02263	end raise 189.60	2.67057	0.01024	4000	74
		1600			1600	69	0.00439	8.02702		2.66033	0.0041	1600	74
		600		76	600	69	0.00165	8.02867	L CREST WEST	2.65624	0.00154	600	74
		78			78	69	0.00021	8.02888	REV CUR TT40MPH	2.6547	0.0002	78	74
		800		74	800	69	0.0022	8.03108	R CREST WEST WEST	2.6545	0.00205	800	74
			XING 191.02				0	8.03108	R SR 598	2.65246	0.00256	1309 D	12
		3600	XING 191.		3318	69	0.00911	8.04018	R CRESTLINE ROAD	2.6499	0.00382	2291	120
		700	XING	176	1051 A	69	0.00278	8.04296	BIDDLE ROAD	2.64628	0.0011	700	120
		900	XING	176	631	74	0.00213	8.04509	BECK ROAD	2.64518	0.00147	900	120
		4380	XING 194.		4380	74	0.01121	8.0563	MAIN STREET	2.64376	0.00691	4380	120
		5192	XING 194.		1420	74	0.00363	8.05993	WESTERN AVENUE	2.63686	0.00819	5192	120
		5288	XING 195.43				0	8.05993	OLENTANGY ROAD	2.62885	0.00835	5288	120
			XING 195.		19895 A	74	0.03611	8.09604	PRIVATE	2.62031	0.00374	2368	120
		5284	XING 196.25			120	0	8.09604	PRIVATE	2.61657	0		
			XING 197.05			120	0	8.09604	LWR LEESVILLE	2.61657	0.02583	14432 A	00
		5289	XING 197.		735	120	0.00116	8.0972	PRIVATE	2.59074	0.00526	9129 A	40
		5092	XING 197.		5092	120	0.00804	8.10524	PRIVATE	2.58546	0.00228	481	40
			XING 198.20					8.10524	PRIVATE	2.58318			
		5470	XING 198.		5470	120	0.00863	8.11387	R ALBRIGHT	2.58318			
		5275			5275	120	0.00833	8.1222	east end	2.58318			
		5254			5254	120	0.00829	8.13049		2.58318	0.07466	5254	40
		1100			1100	120	0.00174	8.13223		2.5383	0.00921	1100	40
		965		129		120	0	8.13223		2.5331	0.00457	965	40
		1000	XING 200.		88	120	0.00014	8.13237	WHETSTONE STREET	2.52853	0.00473	1000	40
					924 D	12	0.00167	8.13403		2.52379			
				90		90	0.06667	8.2007	COLSAN N&W XING	2.52379	0.06667	RR CR	40
			XING 200.		963 D	90	0.00278	8.20348	LANE STREET BUCYRU	2.45712	0		40
		1100	XING	85	1100	40	0.00521	8.20868	LTT 40 MPH WALNUT S	2.45712	0.00521	1100	40
			XING 200.93				0	8.20868	ALLEY TRACK IN GALEI	2.45192	0	0	40
		500	XING 200.		500	40	0.00237	8.21105	N SANDUSKY STATE H	2.45192	0.00237	500	40
		100		118	100	40	0.00047	8.21153	L TRACK IN GALEN STR	2.44955	0.00047	100	40
		600	XING 201.		500	40	0.00237	8.21389	RC TT 40 SPRING STRE	2.44908	0.00237	500	40
			XING 201.20					8.21389	SANDUSKY RIVER - TR	2.44671			
		1100	XING	112	7558 A	40	0.02194	8.23584	R SEARS STREET	2.44671	0.00444	1677 D	12
			XING 201.		2312	86	0.00515	8.24098	STREET AT GRADE NO	2.44226	0.00558	3409	120
		4166						8.24098	end raise 202.98	2.43668			
		5286	XING 203.16				0	8.24098	McCRACKEN	2.43668	0.00834	5286	120
		5280	XING 203.66				0	8.24098	SIMMS CR	2.42834	0.00833	5280	120
			XING 204		15973 A	35	0.02917	8.27016	PRIVATE	2.42071	0.00851	5360	120
			XING 204.		0	120	0	8.27016	GLENVILLE CR	2.4117	0.00832	5272	120
		5390	XING 205.60			120	0	8.27016	PRIVATE	2.40338	0.00854	5399	120
		5272	XING 206.		641	120	0.00101	8.27117	STRIED CR KN	2.39502	0.00725	4593	120
		5296	XING 207.		5928	120	0.00936	8.28052	MARION MELMO	2.38777	0.00936	5928	120
			XING 207.70				0	8.28052	PRIVATE	2.37841			
		4593	XING 208		4593	120	0.00725	8.28777	COUNTY LINE	2.37841			
			XING 208.60				0	8.28777	PRIVATE	2.37841			
		5928	XING 209.		5928	120	0.00936	8.29713	NEVADA AYERS STREE	2.37841	0	0	120
			XING 209.10				0	8.29713	MORRISON STREET	2.37841			
			XING 209.27			120	0	8.29713	GOODBREAD STREET	2.37841			
		5440			5440	120	0.00859	8.30571	MP 210	2.37841	0.00859	5440	120
			XING 210.61			120	0	8.30571	EDENVILLE ROAD	2.36983	0	0	120
		5270	XING 211.		5270	120	0.00832	8.31403	PRIVATE	2.36983	0.00832	5270	120
		5290	XING 212.		5290	120	0.00835	8.32238	DOUGLAS RD	2.36151	798		
		5269	XING 213.		5269	120	0.00835	8.33073	PRIVATE	2.36151	0		
			XING 213.8		0	120	0	8.33073	ROCK RUN ROAD	2.36151	0.05972	22501 A	90

APPENDIX B2

READ DOWN — WESTBOUND TRAFFIC | STATION | READ UP — EASTBOUND TRAFFIC

PAGE 13 COLUMN 1

MILES	STATION	NAME
634.96	KING 214	PRIVATE
635.96	KING 100	L WILLIAMS CR
635.96	KING 215.50	PRIVATE
636.96	KING 215.70	PRIVATE
636.96	KING 236	PRIVATE
637.11		east end
637.11	KING 216.31	KOHNS RD
637.11	KING 216.70	S 5TH STREET
637.11	KING 216.90	S SANDUSKY
637.96		UPPER SANDUSKY
638.46	KING 217	7TH STREET
638.46	KING 217	EIGHTH STREET
638.46	KING 217.34	HAZEL STREET
638.46	KING 217.44	S WARPOLE
638.46		USAKE-CAC KING
638.46		
638.56		
638.96	KING 218	TWP ROAD
638.96		end nowe 219.07
638.96	KING 218.19	WHITE CR
639.96	KING 219	PRIVATE
640.95	KING 220	MP 220 PRIVATE
640.95	KING 220	WILL CR
640.95	KING 220.90	MIGREY CR
641.95	KING 221	PRIVATE
642.95	KING 222	GAMBER CR
642.95	KING 222.50	PRIVATE
643.95	KING 223	KRAUS CR
643.95	KING 223	PRIVATE
643.95	KING 223	PRIVATE
644.95	KING 224	MAIN STREET BR
644.95	KING 224.30	ALLEY STREET
645.95	KING 225	UKLER CR
646.95	KING 226	KELLER CR
647.95	KING 227	PRIVATE
647.95	KING 227.20	PRIVATE
647.95	KING 227.80	PRIVATE
647.95	KING 227.94	ANGLING ROAD
648.95	KING 228	COUNTY LINE
648.95	KING 228.50	LOUISA STREET
648.95	KING 228	FOREST MARY STREE
649.95	KING 229	MP 229 DORMEY STREE
649.95	KING 229	MARTIN STREET CR
649.95	KING 229	UNNAMED FOREST
649.95	KING 229	CAMPBELL STREET
650.95	KING 230	PERLIN CR
651.95	KING 231	PRIVATE
652.95	KING 231	PRIVATE
653.95	KING 234	GREAT ROAD
654.95		end
654.95	KING 234	WYKES CR
655.95	KING 235	PRIVATE
656.95	KING 236	WALNUT STREET
657.23	KING 236	MAIN STREET
657.23		DUNKIRK NTC KING
657.23	KING 236.69	RT# 120
657.94	KING 237	HOPPE ROAD
658.94	KING 238	WAYNE STREET
659.94		end nowe 239.00
660.94	KING 241	TRAVERSE PK
661.94	KING 241.50	PRIVATE
661.94	KING 242.81	PETERSON CR
662.94	KING 243	SCOTT KING
663.94	KING 244	1 MILE KING
664.94	KING 245	ADA JOHNSON STREET
665.94	KING 245	MAIN STREET
665.94	KING 245	GILBERT STREET
665.94	KING 246	KLINGLER RD
666.98	KING 247	ST PAUL RD
667.98	KING 248	PRIVATE
668.98	KING 248	COUNTY LINE
669.98	KING 249	BENTLEY ROAD
670.98	KING 250	MP 250 PEVEE ROAD
670.98	KING 250	PRIVATE
670.98	KING 250	LAFAYETTE RD
670.98	KING 250	SHRIDER ROAD
671.98	KING 251	VINT RD
671.98	KING 251	PHILLIPS RD
672.98	KING 252	S HIGH STREET
673.98	KING 252	CHURCH ST
674.99	KING 252	WASHINGTON
676.01	KING 253	FURMBAUGH RD
677.99	KING 253	FISHER ROAD
678.00	KING 254	MCCLURS KING
678.45	KING 254	COOL ROAD
678.51	KING 149	THAYER ROAD
679.00	KING 256	FETTER ROAD
680.01	KING 257	METZGER ROAD
680.58	KING 258	R ROUSH ROAD
680.58		
680.58		SUGAR ST KING
680.93	KING 260	SUGAR ST
681.28	KING 260	N PINE ST
681.33	KING 260	N JACKSON
681.33		LIMA RD 505 KING
681.33	KING 260.52	MAIN ST
681.42	KING 101	N ELIZABETH
681.42	KING 260.66	N WEST ST
681.58	KING 260	N MCDONEL ST
682.01	KING 134	N METCALF
682.12	KING 134	N BAXTER ST
683.01	KING 261	COLE ST
683.01	KING 262	CABLE ROAD
683.01	KING 263	HANTZLER RD
684.01	KING 264	EASTTOWN ROAD
685.01	KING 264	EAST ROAD
685.01	KING 266.25	SATY ROAD
686.04	KING 266	PRIVATE
686.04	KING 266.41	GREENE AVE
687.04	KING 266	MAIN ROAD
687.04	KING 287.25	MAPPAS ROAD
688.04	KING 287	PRIVATE
688.04	KING 287.60	KEMP ROAD
688.04	KING 287.80	PRIVATE
689.04	KING 288	PRIVATE

PAGE 13 COLUMN 2

STATION	NAME
KING 269	GRUBB ROAD
KING 269	PRIVATE
	MP 270
KING 270	STATE RD
KING 270.65	OLD DELPHOS
KING 271	DEFIANCE TRA
KING 271.77	BAUGH RD
KING 272	PELTIER RD
KING 272	LEHMAN RD
KING 134	L PRIVATE
KING 134	L DELPHOS PIERCE ST
KING 274	S MAIN STREET
KING 274	S CANAL STREET
KING 134	JEFFERSON STREET
KING 274	S CLAY STREET
KING 274	S BREDICK STREET
KING 275	STATE STREET
KING 276	BRICKNER RD
KING 277	PRIVATE
KING 277	BOCKEY RD
KING 278	CONVERSE RD
KING 279	MIDDLE POINT
	MP 280
KING 280	MAIN STREET
KING 280	ADAMS STREET
KING 280	MASON STREET
KING 280	DUCK CREEK RD
KING 281	KINGWALD RD
KING 282	CHENOWETH RD
KING 283	GAMBLE RD
KING 131	L HOAGLIN CTR
KING 131	L GILLILAND RD
KING 286.46	MENDON RD
KING 286	VAN WERT WAYNE STR
KING 286	VINE STREET
KING 94	N FRANKLIN STREET
KING 94	N CHESTNUT STREET
KING 287	RACE STREET
KING 287.11	N TYLER STREET
KING 287.17	HARRISON STREET
KING 287.24	CHERRY STREET
KING 287.30	WALNUT STREET
KING 287.37	MARKET STREET
KING 287.46	WASHINGTON STREET
KING 287.56	JEFFERSON STREET
KING 287.67	N SHANNON
KING 287.74	WALL STREET
90	CR VAN WERT IN KING
KING 288.97	BURT STREET
KING 288	FISHER AVENUE
KING 289	JOHN BROWN RD
KING 289	PRIVATE
	MP 290
KING 290	LIBERTY UNION
KING 291	RICKEY CR
KING 292	MACE KING
KING 293	ALT RT US 30
KING 293.56	BURDHAUS RD
KING 294.71	MAIN ST CONVOY
KING 294.86	TULLY ST
KING 295	PRIVATE
KING 295	WAYNE CR
KING 296	SHANER CR
KING 296	SARR CR
KING 297	MENTZER
KING 297	PRIVATE
KING 298.12	MENTZER CURC
KING 298.88	OLEM KING
KING 298.46	SCHOOL HOUSE
	MP 300
KING 300	ST LINE RD
KING 301	MORGAN RD
KING 301	PRIVATE
KING 302	LORTIE RD
KING 303	OHIO STREET
KING 304	MAIN STREET
KING 304	WASHINGTON
KING 306	SNYDER RD
KING 306	HOFFMAN RD
KING 307	GROTRAIN RD
KING 307.71	WILSON RD
KING 308.30	PACKLER RD
KING 308.65	GAZAGOT RD
KING 309.37	HOOK RD
KING 309.77	WASHINGTON
KING 309.64	FRANKE RD
	MP 310
KING 310	PRIVATE
KING 310	PRIVATE
KING 311	TILLMAN RD
KING 311.70	MINNICH ROAD
KING 312.50	PRIVATE
KING 312.80	PRIVATE
KING 312.99	PRIVATE
KING 312.26	PAULDING ROAD
KING 313	CITY-YOUNGS HARTZL
KING 315	WEST-CH ADAMS CEN
KING 316	MEYER ROAD
	4REF
	begin table 319.01
106	L
76	L TEMPL
90	NEVOUR MILE WASON
	FORT WAYNE
KING 51	S BRACKENRIDGE
72	N
95	R
64	R
	end none 120.25
106	R
106	R

APPENDIX B2

PAGE 14 COLUMN 1

READ DOWN WESTBOUND TRAFFIC — STATION — READ UP EASTBOUND TRAFFIC

MILES	GRADIENT, PERCENT	DISTANCE, ft	WAYSIDE CROSSING	MAX MPH	DISTANCE, feet	SPEED, mph	ET, hours	TIME, hours	STATION	TIME, hours	ET, hours	DISTANCE, feet	SPEED, mph	MILES	GRADIENT, PERCENT
747.27		452		108	452	51	0.00166	9.51602	R	1.13970	0.00078	452	108	828.21	
743.12		4600	XING 323.4		1678	51	0.00822	9.52286	THOMAS ROAD	1.13897	0.00087	401.3	D 12	827.21	
744.14		5349	XING 324.1		1894	A 51	0.00956	9.52841	HADLEY ROAD	1.1383	0.01491	9440	120	828.24	
745.14		5322	XING 325.1		1878	84	0.00985	9.53428	BABB ROAD	1.12339	0.0084	5322	120	829.20	
746.13		5183	XING 326.		14062	A 84	0.03028	9.56454	PRIVATE	1.11499	0.0062	5193	120	829.20	
747.13		5277	XING 326.		4663	A 10	0.01139	9.57593	MAIN STREET	1.10879	0.00838	5277	120	830.03	
748.13	1.60%	5278	XING 328.		4846	120	0.00733	9.58326	ARCOLA COLUMBIA STR	1.09846	0.00833	5278	120	830.03	
749.13		5290	XING 328.1		5290	120	0.00835	9.59161	PRIVATE	1.09013	0.00835	5290	120	830.03	
750.14		5363	XING 329.1		5363	120	0.00846	9.60007	PRIVATE	1.08178	0.00846	5363	120	830.03	
751.12		5224	XING 329.		5224	120	0.00824	9.60832	PRIVATE	1.07332	0.00824	5224	120	830.21	
751.13							0	9.60832		1.06508	0			831.21	-0.83%
752.14		5296	XING 329.		5296	120	0.00836	9.61668	PRIVATE	1.06508	0.00836	5296	120	831.21	
753.14		5297	XING 330.		5297	120	0.00836	9.62504	COUNTY LINE	1.05672	0.00836	5297	120	832.21	
754.14		5294	XING 331.		5294	120	0.00836	9.63339	BEECHING ROAD	1.04836	0.00836	5294	120	833.21	
755.14		5293	XING 332.		5293	120	0.00835	9.64174	800E?	1.04	0.00835	5293	120	834.38	
756.15		5293	XING 333.		5293	120	0.00835	9.6501	MILL STREET	1.03165	0.00835	5293	120	835.21	
757.15	-0.74%	5292	XING 334.		5292	120	0.00835	9.65845	MOWERY ROAD	1.02329	0.00835	5292	120	836.21	
758.15		5269	XING 336.		5269	120	0.00832	9.66677	PAGES CR	1.01494	0.00832	5269	120	837.21	
759.15		5290	XING 336.		5290	120	0.00835	9.67512	SR 9	1.00663	0.00835	5290	120	838.20	
760.16		5434	XING 338.		5434	120	0.00856	9.68369	LINE STREET SR	0.99828	0.00858	5434	120	839.20	
761.16		5185	XING 338.		5185	120	0.00818	9.69188	POOR FARM CR	0.9897	0.00819	5185	120	840.20	
762.16		5263	XING 341.		5263	120	0.00831	9.70018	WOLF ROAD	0.98152	0.00744	4716	120	840.81	
763.16	0.61%	5286	XING 342.1		5286	120	0.00836	9.70854	PEARL CR 300E?	0.97407	0.01722	7336	A 107	840.92	
763.46	0.61%	1600	XING 343.		1600	120	0.00253	9.71107	BUTLER CR 407	0.96165	0.00019	107	107	841.20	
763.75		1503	XING	147	1503	120	0.00237	9.71344	L PRIVATE	0.96160	0.00286	1503	107	842.20	
764.16		2200			2200	120	0.00347	9.71691		0.959	0.00380	2200	107	843.20	0.51%
765.17		5290			5290	120	0.00835	9.72526		0.95511	0.00336	5290	107	843.33	0.51%
765.53		1900			433	D 12	0.00072	9.72598		0.94574	0.00336	1900	107	844.13	
765.81		1491		107	1491	107	0.00284	9.72882	R	0.94238	0.00284	1491	107	844.20	
766.17		1900			1900	107	0.00336	9.73198		0.93874	0.00072	432.6	D 12	844.73	
767.08		4800			4800	107	0.0085	9.74048		0.93902	0.00889	6287	120	844.82	
767.17		488	XING	164	486	107	0.00086	9.74134	L SRD 57	0.92913	0.00077	488	120	845.20	
767.20		184	XING	164	3812	107	0.00675	9.74809	L LARWILL DEPOT STR	0.92836	0.00028	184	120	846.20	0.54%
768.17		5100	XING 347.		7386	A 10	0.01222	9.76031	BRINKLEY ROAD	0.92807	0.00809	5100	120	847.20	0.57%
769.17		5294	XING 348.		897	120	0.00142	9.76173	BLACK CR	0.92002	0.00835	5294	120	848.05	
770.17		5286	XING 349.		5286	120	0.00834	9.77057	PRIVATE	0.91166	0.00834	5286	120	848.03	
771.18		5287	XING 349.		5287	120	0.00834	9.77642	PRIVATE	0.90332	0.00834	5287	120	848.03	
772.18		5282	XING 350.		5282	120	0.00834	9.78675	MARKET STREET	0.89498	0.00834	5282	120	848.03	
773.18		5283	XING 350.		5283	120	0.00834	9.79509	PIERCETON FIRST STR	0.88664	0.00834	5283	120	849.20	
774.18		5279	XING 362.		5279	120	0.00833	9.80342	VAN NESS ROAD	0.8783	0.003	1899	120	849.21	0.60%
775.18		5278	XING 353.		5276	120	0.00833	9.81175	WOOSTER ROAD	0.8753				850.22	
776.18	-1.64%	5293	XING 354		5293	120	0.00835	9.82011	SCHADE 250E	0.8753				851.20	
777.19		5323			5323	120	0.0084	9.82851		0.8753	0.03872	22503	A 90	852.21	
778.18		5233			5233	120	0.00826	9.83677		0.83556	0.00422	2004	90	853.21	
778.90		3800			3800	120	0.006	9.84278	Begin raise 357.96	0.83136	0.006	3800	90	854.19	
779.18		1500	XING	164	1500	120	0.00237	9.84512	L BRONSON STREET	0.82537	0.00316	1500	90	855.21	
779.37	-0.73%	996	XING	164	996	120	0.00157	9.8467	L MAPLE STREET	0.82021	0.0021	996	90	856.22	0.56%
779.85	-0.73%	2700	XING 358.		1778	120	0.0028	9.84951	SCOTT STREET	0.81812	0.00566	2700	90	856.35	0.55%
779.88					924	D 12	0.00167	9.85117		0.81245				856.35	
779.88				90		90	0.03333	9.88451	WARSAW B&XING	0.81245	0.03333	RR CR	90	856.51	
779.88			XING 358.74				0	9.88451	DETROIT STREET	0.7791	0.00167	924	D 12	857.21	
779.88			XING				0	9.88451	ALLEY?	0.77742				858.22	
779.88			XING 358.87				0	9.88451	INDIANA STREET	0.77743				858.42	0.57%
779.88			XING				0	9.88451	ALLEY?	0.77743				858.42	
780.18	-0.73%	1600	XING 359.		1600	90	0.00337	9.88787	S LAKE STREET	0.77743	0.00107	676	120	858.42	
781.18		5282	XING		5282	90	0.01112	9.89899	ALLEY?	0.77637	0.00834	5282	120	858.21	
781.16			XING 359.25				0	9.89899	UNION STREET	0.76803				859.33	
781.27		176	XING 360.		176	90	0.00037	9.89936	JACKSON CR	0.76803	0.00028	176	120	859.33	
781.58		1900		134	1900	90	0.004	9.90336	R BURKET	0.76775	0.003	1900	120	859.54	
782.18		3200	XING 361.		2040	90	0.00429	9.90765	PRIVATE	0.78475	0.00505	3200	120	859.54	
783.19		5318	XING 362.		14432	A 90	0.02583	9.93349	PRIVATE	0.7697	0.00839	5318	120	860.22	
784.19		5245	XING 362.		0	0	0	9.93349	THOMAS CR 10	0.75131	0.00828	5245	120	860.56	
785.19		5291	XING 363.		2592	120	0.00408	9.93756	PRIVATE	0.74303	0.00835	5291	120	860.56	
785.19			XING 363.72				0	9.93756	SELBY CR 525	0.73468					
786.20		5352	XING 364.		5352	120	0.00845	9.94601	PRIVATE	0.73468	0.00845	5352	120		
787.19		5220	XING 365.		5220	120	0.00824	9.95425	PRAIRIE STREET	0.72623	0.00824	5220	120		
788.19		5291	XING 365.		5291	120	0.00835	9.9626	ATWOOD HARRISON ST	0.71799	0.00835	5291	120		
789.19		5278	XING 366.		5279	120	0.00833	9.97092	ROBINSON ROAD	0.70964	0.00833	5279	120		
790.18		5227	XING 367.		5227	120	0.00825	9.97918	SNYDER CR 67	0.70131	0.00825	5227	120		
791.19		5337	XING 368.		5337	120	0.00842	9.9876	PRIVATE	0.69306	0.00842	5337	120		
791.19			XING 368.77				0	9.9876	COOK CR	0.68464					
791.19			XING 368.29				0	9.9876	WALNUT STREET SR	0.68464					
791.19			XING 368.36				0	9.9876	ETNA GREEN	0.68464					
792.21		5378	XING 370.		5379	120	0.00849	9.99609	S BEECH ROAD	0.68464	0.00849	5378	120		
793.19		5193	XING 372.		5193	120	0.0082	10.0043	HARRIS STREET	0.67615	0.0082	5193	120		
794.22		5438	XING 373.		5438	120	0.00858	10.0129	BOURBON MAIN STREE	0.66795	0.00858	5438	120		
794.22			XING 373.17				0	10.0129	BOURBON STREET	0.65937					
794.22			XING 373.26				0	10.0129	CENTER STREET	0.65937					
794.22			XING 374.13				0	10.0129	TSCHR ST 68	0.65937					
795.20		5182	XING 376.		5182	120	0.00815	10.021	STOCKMAN ROAD	0.65937	0.00815	5182	120		
796.20		5252	XING 376.		5252	120	0.00829	10.0293	WYMER CR	0.65122	0.00829	5252	120		
797.19		5270	XING 377.		5270	120	0.00832	10.0376	COMPTON CR	0.64298	0.00832	5270	120		
798.19		5279	XING 377.		5279	120	0.00833	10.046	HAWTHORNE ROAD	0.63481	0.00833	5279	120		
799.19		5279	XING 378.		5279	120	0.00833	10.0543	PRIVATE	0.62628	0.00406	2573	120		
800.20		5283	XING 379.		5283	120	0.00834	10.0626	S IRIS ROAD	0.62222					
801.19		5277	XING 379.		5277	120	0.00833	10.071	PRIVATE	0.62222					
801.19			XING 380.20				0	10.071	MP 380 PRIVATE	0.62222					
802.20		5315	XING 380.		5315	120	0.00839	10.0793	PRIVATE	0.62222	0.03872	22503	a 90-120		
803.18		5155	XING 381.		5155	120	0.00814	10.0875	KING ROAD	0.5826	0.00256	1203	90		
804.19		5367			5367	120	0.00847	10.096		0.5799	0.01129	5367	90		
805.03		4400	XING 383.		3478	120	0.00549	10.1014	LIBERTY STREET	0.56661	0.00926	4400	90		
805.03					924	D 12	0.00167	10.1031		0.55935					
805.03				90		90	0	10.1031	PLYM NKP XING	0.55935		RR CR	90		
805.03			XING 384.07				0	10.1031	FIFTH STREET	0.55935	0.00167	924	D 120-90		
805.21		944	XING 384.		944	90	0.00199	10.1051	OAK ROAD	0.55768	3.2E-05	20	120		
806.19		5216	XING 385.		5216	90	0.01098	10.1161	PRIVATE	0.55765	0.00822	5216	120		
807.19		5286	XING 385.		4840	90	0.01019	10.1263	PRIVATE	0.54942	0.00833	5280	120		
808.20		5287	XING 386.		14432	A 95	0.02583	10.1521	ANGLING ROAD	0.54109	0.00834	5287	120		
809.20		5283	XING 386.99				0	10.1521	QUINCE ROAD	0.53274	0.00836	5293	120		
810.20		5288	XING 387.		1857	120	0.00293	10.155	REDWOOD ROAD	0.52439	0.00832	5288	120		
811.20		5281	XING 388.		5281	120	0.00833	10.1634	PRIVATE	0.51607	0.00833	5281	120		
812.19		5265	XING 390.		5265	120	0.00831	10.1717	UNION ROAD	0.50774	0.00831	5265	120		
813.20		5296			5298	120	0.00836	10.18	DONALDSON	0.49942	0.00836	5296	120		
814.20		5296	XING 393.		5298	120	0.00836	10.1884	PRIVATE	0.49106	0.00836	5296	120		
815.23		5411	XING 394.		5411	120	0.00854	10.1969	SR 23	0.4827	0.00854	5411	120		
816.20		5164	XING 395.		5164	120	0.00815	10.2051	WALSH CR	0.47416	0.00815	5164	120		
817.20		5285	XING 396.		5285	120	0.00834	10.2134	700E	0.46601	0.00834	5285	120		
818.21		5286	XING 398.		5288	120	0.00835	10.2218	STRIKE STREET	0.45767	0.00835	5286	120		
819.21		5285	XING 398.		5285	120	0.00834	10.2301	HAMLET JEFFERSON S	0.44932	0.00834	5285	120		
820.21		5279	XING 399.		5279	120	0.00833	10.2384	500E	0.44098	0.00833	5279	120		
821.21		5301	XING 401.		5301	120	0.00837	10.2468	300E	0.43265	0.00837	5301	120		
822.21		5282	XING 401.		5282	120	0.00834	10.2551	PRIVATE	0.42428	0.00834	5282	120		
823.21		5282	XING 402.		5282	120	0.00834	10.2635	PRIVATE	0.41595	0.00834	5282	120		
824.21		5270	XING 403.		5270	120	0.00832	10.2718	50E	0.40761	0.00393	2493	120		
825.21		5297	XING 405.		5297	120	0.00834	10.2801	100W	0.40367					

APPENDIX B2

PAGE 14 COLUMN 2

READ DOWN WESTBOUND TRAFFIC — STATION — READ UP EASTBOUND TRAFFIC

DISTANCE, ft	WAYSIDE CROSSING	MAX MPH	DISTANCE, feet	SPEED, mph	ET, hours	TIME, hours	STATION	TIME, hours	ET, hours	DISTANCE, feet	SPEED, mph
5292	XING 405.		5292	120	0.00835	10.2885	140S	0.40367			
5280	XING 406.		5280	120	0.00833	10.2968	SR 39	0.40367	0.03872	22503	A 90
5422	XING 407.		5422	120	0.00856	10.3054	LONG LANE	0.36395	0.00327	1556	90
5045	XING 407.		5045	120	0.00796	10.3133	PRIVATE	0.36068	0.01062	5045	90
	XING 408.61				0	10.3133	THOMPSON STREET	0.35006			
4400	XING 408.		3478	120	0.00548	10.3188	OHIO STREET	0.35006	0.00926	4400	90
			924	D 12	0.00167	10.3205	HANNA C&O XING	0.3408			
	XING	90		90	0	10.3205	450W	0.3408		RR CR	90
	XING 409.36				0	10.3205	PRIVATE	0.3408	0.00167	924	D 12
968	XING 410.		968	90	0.00203	10.3225	MP 410 SCHULL 600W	0.33914	6.6E-05	42	120
5281	XING 411.		5281	90	0.01111	10.3336	FARRESTER ROAD	0.33907	0.00833	5281	120
	XING 411.59				0	10.3336	TILDON 750W	0.33074	0		120
5289	XING 412.		4755	90	0.01	10.3437	PRIVATE	0.33074	0.00834	5285	120
5273	XING 413.		14432	A 90	0.02583	10.3695	KLEINS 900W	0.32238	0.00832	5273	120
6182	XING 414.75				0	10.3695	N MAIN STREET	0.31407	0.00976	6182	120
4367	XING 414.		1602	120	0.00303	10.3725	WANATAH S ILLINOIS	0.30432	0.00689	4367	120
5280	XING 415.		5280	120	0.00833	10.3809	S WASHINGTON	0.29742	0.00833	5280	120
5272	XING 416.		5272	120	0.00832	10.3892	LINCOLN STREET	0.28909	0.00832	5272	120
5251	XING 415.		5251	120	0.00829	10.3975	1100W	0.28077	0.00829	5251	120
5293	XING 416.		5293	120	0.00835	10.4058	OSBRN CR CO	0.27246	0.00836	5293	120
5270	XING 417.		5270	120	0.00832	10.4141	FISHER 575E	0.26413	0.00832	5270	120
3200	XING 419.		3200	120	0.00505	10.4192	DOLSON 400E	0.25581	0.00832	5270	120
570	XING	178	570	120	0.0009	10.4201	R BICKEL 400E	0.24749			
1500	XING 420.		1500	120	0.00237	10.4224	PRIVATE	0.24749			
5280	XING 420.		5280	120	0.00833	10.4308	MUNTDALE ROAD	0.24749	0.00833	5280	120
5263	XING 421.		5263	120	0.00831	10.4391	PRIVATE	0.23916	0.00831	5263	120
700	XING 422.		700	120	0.0011	10.4402	CEMETERY ROAD	0.23085	0.00016	101	120
4200	XING	134	4200	120	0.00663	10.4468	R GREENWICH STREET	0.23069	0		
394	XING 422.		394	120	0.00062	10.4474	AXE STREET	0.23069			
2800	XING 423.		2800	120	0.00442	10.4519	FRANKLIN STREET	0.23069	0		
486	XING	134	486	120	0.00077	10.4526	L VALPARAISO WASHIN	0.23069	0		
2000	XING 423.		2000	120	0.00316	10.4558	LAFAYETTE STREET	0.23069	0.02583	14432	A 90
5279	XING 423.		5279	120	0.00833	10.4641	NAPOLEON STREET	0.20486	0.00279	1326	90
5274	XING 424.		5274	120	0.00832	10.4724	STEVENS XING	0.20207	0.0111	5274	90
4400	XING 425		3476	120	0.00549	10.4779	150W	0.19097	0.00926	4400	90
			924	D 12	0.00167	10.4796	VALPO GTW XING	0.18171	0		90
	XING	90		90	0.06667	10.5463	250W	0.18171	0.06667		90
	XING 427.80				0	10.5463	400W	0.11505	0.00167	924	D 12
892	XING 428.		892	90	0.00188	10.5481	EAST SALT CR	0.11338	1.1E-05	7	120
5299	XING 428.		5299	90	0.01115	10.5593	W SALT CR	0.11337	0.0083	5260	120
5340			4808	90	0.01012	10.5694	end raise 429.28	0.10507	0.00843	5340	120
5206	XING 430.		14432	A 90	0.02583	10.5952	PRIVATE	0.09664	0.00822	5206	120
5291	XING 430.87				0	10.5952	WHEELER PARK AVE	0.08842	0.00573	3630	120
5289	XING 431.		1585	120	0.00258	10.5982	GOTT RD 7TH	0.08269	0.02194	12582	A 95
5217	XING 432.		5212	120	0.00823	10.6064	TREAGER RD	0.06075	0		
5354	XING 433.		5364	120	0.00845	10.6149	COUNTY LINE	0.06075	0.00984	4984	95
5316	XING 434.		5278	120	0.00825	10.6231	HOBART ILLINOIS STRE	0.05081	0.00954	4557	95
700	XING 434.87			120	0	10.6231	LINDA STREET	0.04167	0		
	XING 434.		788	D 12	0.00139	10.6245	CLEVELAND STREET	0.04167			
844	XING	95	844	95	0.00166	10.6262	R LAKE PARK AVE	0.04167	0		95
3700	XING 435.		3700	95	0.00738	10.6336	WISCONSIN STREET	0.04167	0		
5320			5320	95	0.01061	10.6442		0.04167	0		
1080			1080	95	0.00211	10.6463		0.04167	0.04167	12383	A 0-1
	XING 437.16				0	10.6463	LIVERPOOL ROAD				
						10.646	EAST END LIVERPOOL TERMINAL				
4200	XING 438.		920	95	0.00183	10.6481	INDIANA AVENUE				
800	XING 439.		5261	A 95	0.015	10.6631	VIRGINIA STREET				
					0	10.6631	I-65 ONO				
1122				95	0	10.6631					
					0	10.6631	I-80 ONO				
3600			385	113	0.00061	10.6635					
1897			1873	D 15	0.00028	10.67					
				0		10.67	WEST END LIVERPOOL TERMINAL				

Appendix B3 - Improved right of way on unmodified alignment. Streamlined high-speed passenger train operating at 200 mph maximum speed.
Maximum allowable speed on curves as shown for high speed equipment only.
FRA Class 9 track where speeds over 180 mph are possible.
Other tracks classified per maximum possible speed.
Unreconstructed right of way but track re built for maximum speed up to 200 mph. $E_a = 6.635$ inches, $E_u = 4$ inches
Wheel load 0.6 minimum per FRA 213.307d1
Assumed c.g. of moving equipment 48 inches above top of rail
Train length 1,360 feet: 10 non-articulated streamlined passenger cars, 8 locomotive units @ 10,000 hp each. Maximum train braking rate 5 mph/second.
20% speed reduction for reverse curves with 600 feet or less between curves.
Gradients shown in Appendix B2.
Highway grade crossing locations shown are approximate
Distance New York Penn Station to Chicago Union Station 891.61 miles

Westbound trip	7.3670 hours
Eastbound trip	7.4227 hours

APPENDIX B3

PAGE 1 COLUMN 1

MILES	DISTANCE, ft	NEW GRADE CROSSING	MAX MPH	DISTANCE, ft (READ DOWN WESTBOUND TRAFFIC)	SPEED, mph	dT, hours	TIME, hours	STATION	TIME, hours (READ UP EASTBOUND TRAFFIC)	dT, hours	DISTANCE, ft	SPEED, mph
								EAST END PLATFORM	7.4224	0.0008	33	D 15
-0.258	1860		15					PENN STATION NEW YORK	7.4216	0.0168	1327	15
0.00				69	A 5	0.0017	0.0017	WEST END PLATFORM	7.4048	0.0025	495	D 60
0.23	1200		15	1131	15	0.0143	0.0160		7.4023	0.0022	755	60
0.86	3310		60	3017	A 15	0.0051	0.0212		7.4001	0.0104	3310	60
1.00	778		226	3971	60	0.0097	0.0309	R	7.3896	0.0025	778	60
2.00	5288		60	5288	60	0.0167	0.0475		7.3872	0.0167	5288	60
2.92	4860		60	4860	60	0.0153	0.0629		7.3706	0.0153	4860	60
2.92							0.0629		7.3552	0.0032	1489	D 117
3.00	427		117	427	60	0.0013	0.0642	L RTE 1&9	7.3520	0.0007	458	117
3.38	2006		117	938	60	0.0029	0.0672	L	7.3513	0.0008	920	117
3.38					60	0.0000	0.0672	E MEADOWS	7.3506	0.0022	1807	D 157
4.00	3274			4192	A 60	0.0087	0.0759		7.3483	0.0023	268	157
5.00	5281			7870	A 11	0.0108	0.0866		7.3460	0.0000		
5.45	2377						0.0865	FEET	7.3480	0.0000		
5.73	1473		226				0.0865	R	7.3460	0.0000		
5.85	589						0.0866		7.3460	0.0182	11129	A 60
6.01	838						0.0865	A	7.3288	0.0028	834	60
6.01				101	181	0.0001	0.0866	W MEADOWS	7.3272	0.0000		60
6.18	526			3274	D 18	0.0056	0.0922		7.3272	0.0017	529	60
6.17	330		60	330	60	0.0010	0.0933	HACKENSACK R	7.3256	0.0010	330	60
6.17							0.0933		7.3249	0.0054	3128	D 158
6.17				1360	60	0.0043	0.0975		7.3191	0.0004	352	158
7.00	4409			11389	A 60	0.0185	0.1160		7.3187	0.0000		
7.39	2022					0.0000	0.1160	HUDSON	7.3187	0.0000		
8.04	3443					0.0000	0.1160	R HUDSON	7.3187	0.0000		
8.28	1273		233			0.0000	0.1160	R	7.3187	0.0000		
8.37	487					0.0000	0.1160		7.3167	0.0149	10624	A 73
9.15	4095					0.0000	0.1160		7.3016	0.0038	1360	73
9.13				59	158	0.0001	0.1161		7.2983	0.0000		
9.13				2880	D 12	0.0047	0.1208		7.2983			
9.36	1205		73	1205	73	0.0031	0.1239	L HARRISON	7.2983	0.0031	1205	73
9.36							0.1239		7.2952	0.0019	928	D 108
9.43	371			371	73	0.0010	0.1249	REVERSE CURVE	7.2933	0.0001	47	108
9.61	955		108	955	73	0.0025	0.1274	R	7.2932	0.0006	350	108
9.66	265			54	73	0.0001	0.1275	REVERSE CURVE	7.2926	0.0005	272	108
9.78	531		162	2196	A 73	0.0047	0.1321	L	7.2921	0.0000		
9.85	424					0.0000	0.1321		7.2921	0.0019	1054	A 97
10.07	1168		253	217	104	0.0004	0.1325	L NEWARK	7.2902	0.0021	1096	97
10.12	265			205	D 16	0.0004	0.1329		7.2881	0.0005	265	97
10.24	637		97	637	97	0.0012	0.1342	L	7.2876	0.0013	637	97
10.24							0.1342		7.2864	0.0021	1289	D 135
10.35	584			584	97	0.0011	0.1353	dx	7.2843	0.0001	58	130
10.48	686			686	97	0.0013	0.1366		7.2842	0.0000	9	130
10.59	581		135	80	97	0.0002	0.1368	R HUNTER	7.2842	0.0007	509	135
11.35	4013			4080	A 97	0.0076	0.1444		7.2835	0.0017	1320	D 165
11.55	1054			5040	A 13	0.0079	0.1523	A	7.2816	0.0043	3747	165
11.88	1832		165			0.0000	0.1523	L	7.2776	0.0021	1832	165
12.38	2553					0.0000	0.1523		7.2754	0.0010	919	D 200
13.37	5262			11149	A 15	0.0115	0.1638		7.2744	0.0001	64	183
13.56	1002					0.0000	0.1638		7.2743	0.0000		
13.94	2004		277			0.0000	0.1638	R	7.2743	0.0000		
14.37	2287						0.1638		7.2743	0.0000		
14.39	106					0.0000	0.1638		7.2743	0.0216	16393	A 94
14.53	742		277			0.0000	0.1638	L	7.2527	0.0008	436	94
15.37	4389	4389				0.0000	0.1638		7.2518	0.0013	594	A 86
15.45	434	4823		717	200	0.0007	0.1645		7.2505	0.0007	330	86
15.45		4833		4571	D 20	0.0059	0.1704		7.2498			
15.61	867	5700	94	854	94	0.0017	0.1721	L ELIZABETH	7.2496	0.0016	867	86
15.64	163	5863			94	0.0000	0.1721	REVERSE CURVE	7.2479	0.0004	163	86
15.64		5863		211	D 94	0.0004	0.1726	TT 55 MPH	7.2475			
16.09	2380	8103	86	2330	86	0.0051	0.1777	R	7.2475	0.0051	2330	86
16.09							0.1777	ELMORA	7.2424	0.0063	4782	D 200
16.37	1517			1380	86	0.0030	0.1807	E END 5 TRACKS	7.2361	0.0054	5670	200
17.37	5283				86	0.0000	0.1807		7.2307	0.0000		
18.35	5269			16319	A 86	0.0228	0.2035		7.2307	0.0063	6309	A 182
19.32	5258					0.0000	0.2035		7.2244	0.0006	586	182
19.52	774			806	182	0.0004	0.2040		7.2238	0.0008	774	182
19.74	1187		187	1187	182	0.0012	0.2052	R	7.2230	0.0012	1187	182
20.12	2014			1360	182	0.0014	0.2066	TO	7.2218	0.0010	1008	D 200
20.38	1342		330	6309	A 18	0.0063	0.2129	L	7.2206	0.0024	2495	200
20.51	695					0.0000	0.2129	HERE	7.2184	0.0000		
20.65	809		272			0.0000	0.2129	R	7.2184	0.0000		
20.87	1016					0.0000	0.2129	RAHWAY	7.2184	0.0000		
20.90	160					0.0000	0.2129	L	7.2184	0.0000		
20.95	214					0.0000	0.2129		7.2184	0.0000		
21.14	1016		330			0.0000	0.2129	L	7.2184	0.0046	5027	A 186
21.37	1230			134	200	0.0001	0.2130		7.2138	0.0001	65	186
21.62	1295			790	D 26	0.0006	0.2136		7.2134	0.0010	1018	186
21.95	1781		186	1781	186	0.0018	0.2156	R	7.2124	0.0000		
22.02	378			378	186	0.0004	0.2160	REVERSE CURVE	7.2124	0.0013	1297	A 167
22.13	540		218	540	186	0.0006	0.2166	L	7.2111	0.0000		
22.37	1295			442	186	0.0006	0.2170		7.2111	0.0000		
22.96	3108			4780	A 18	0.0048	0.2218	COLONIA	7.2111	0.0000		
22.94				131	186	0.0001	0.2219		7.2111	0.0000		
22.96				340	D 15	0.0003	0.2222		7.2111	0.0000		
23.08	621		190	541	190	0.0005	0.2229	R	7.2111	0.0000		
23.08				265	190	0.0003	0.2230		7.2111	0.0000		
23.20	621			436	D 15	0.0004	0.2236		7.2111	0.0000		
23.35	826		182	826	182	0.0009	0.2243	L	7.2111	0.0000		
23.38	158		162	158	182	0.0002	0.2245	L	7.2111	0.0105	9123	A 146
23.79	2156			425	182	0.0004	0.2249		7.2006	0.0018	1360	146
23.79				1730	D 17	0.0020	0.2269		7.1988	0.0000		
24.06	1421		146	1421	146	0.0018	0.2288	R	7.1988	0.0018	1421	146
24.15	474			474	146	0.0005	0.2294	REVERSE CURVE	7.1970	0.0006	474	146
24.38	1105		146	1105	146	0.0014	0.2308	ISELIN	7.1964	0.0008	719	146
24.68	1602		146	1602	146	0.0021	0.2329	L	7.1956	0.0000		
24.80	701		146	248	146	0.0003	0.2332	L A	7.1956	0.0027	2010	A 136
24.92	681			453	D 14	0.0005	0.2338	METROPARK	7.1928	0.0031	2178	135
25.18	1361		135	1361	135	0.0019	0.2357	R	7.1897	0.0011	778	A 136
25.18				1360	135	0.0019	0.2376		7.1886			
25.18				500	A 13	0.0007	0.2383		7.1886			
25.37	985			109	138	0.0001	0.2385	MENLO PARK	7.1886	0.0011	762	130
25.48	598			314	D 13	0.0004	0.2389		7.1875	0.0006	588	130
25.92	2284		130	2284	130	0.0033	0.2420	L	7.1868	0.0033	2284	130
26.10	979			979	130	0.0014	0.2437		7.1835	0.0006	891	D 147
26.39	1522		147	385	130	0.0006	0.2442	R	7.1824	0.0005	408	147
26.84	2398		147	3008	A 15	0.0042	0.2484	R	7.1819	0.0075	5199	A 114
27.37	2755			1986	147	0.0024	0.2508	METUCHEN	7.1744	0.0023	1360	114
27.37				588	A 14	0.0008	0.2516		7.1721			
27.53	860			282	150	0.0003	0.2519		7.1721	0.0014	860	114
27.53				1364	D 14	0.0020	0.2538		7.1707			
27.79	1397		114	1397	114	0.0023	0.2563	R SOUTH METUCHEN	7.1707	0.0023	1397	114
27.79				1360	114	0.0023	0.2585	0.0250	7.1684	0.0022	2432	D 172
28.12	2095			11121	A 11	0.0145	0.2730		7.1662	0.0001	75	172
28.36	813		233			0.0000	0.2730	L	7.1661	0.0006	500	172
28.47	573		233			0.0000	0.2730	L	7.1645	0.0006	573	172
28.65	937					0.0000	0.2730		7.1639	0.0010	937	172
29.00	1874		172			0.0000	0.2730	L	7.1629	0.0021	1874	172
29.36	1874					0.0000	0.2730		7.1608	0.0018	1528	D 200
30.14	4135					0.0000	0.2730		7.1587	0.0042	4481	200
30.35	1074		233	9296	A 17	0.0094	0.2825	R	7.1560	0.0010	1074	200
30.37	107					0.0000	0.2825		7.1540	0.0001	107	200
31.36	5273					0.0000	0.2825		7.1539	0.0050	5273	200

APPENDIX B3

PAGE 1 COLUMN 2

MILES	DISTANCE, ft	NEW GRADE CROSSING	MAX MPH	DISTANCE, ft (READ DOWN WESTBOUND TRAFFIC)	SPEED, mph	dT, hours	TIME, hours	STATION	TIME, hours (READ UP EASTBOUND TRAFFIC)	dT, hours	DISTANCE, ft	SPEED, mph
31.53	875					0.0000	0.2825		7.1489	0.0006	875	200
32.04	2676		229	830	200	0.0006	0.2831	L	7.1481	0.0025	2676	200
32.36	1686			1698	200	0.0016	0.2847		7.1456	0.0016	1696	200
32.44	420			420	200	0.0004	0.2851		7.1440	0.0004	420	200
32.65	1103		228	1103	200	0.0010	0.2861	L	7.1436	0.0010	1103	200
33.35	3727			3727	200	0.0035	0.2897	NEW BRUNSWICK	7.1426	0.0035	3727	200
34.36	5290			5290	200	0.0050	0.2947		7.1391	0.0050	5290	200
34.93	3029			3028	200	0.0029	0.2976		7.1341	0.0029	3029	200
35.35	2208		204	2208	200	0.0021	0.2996	L	7.1312	0.0021	2208	200
36.37	5404			5404	200	0.0051	0.3047		7.1291	0.0051	5404	200
37.36	5202			5202	200	0.0049	0.3097		7.1240	0.0049	5202	200
38.36	5303			5303	200	0.0050	0.3147		7.1191	0.0050	5303	200
39.36	5258			5258	200	0.0050	0.3197		7.1141	0.0049	5138	200
40.36	5285			5285	200	0.0050	0.3247		7.1092	0.0000		
40.44	430			430	200	0.0004	0.3251	DEANS	7.1092	0.0063	6309	A 18
40.66	1188		273	175	200	0.0002	0.3252	L	7.1029	0.0007	713	182
40.78	645			1008	D 18	0.0010	0.3262		7.1022	0.0007	645	182
41.36	3013		182	3013	182	0.0031	0.3294	R	7.1015	0.0031	3013	182
42.35	5253			1360	182	0.0014	0.3308	MONMOUTH JCT	7.0984	0.0010	1008	D 20
43.36	5314			6909	A 18	0.0067	0.3384		7.0974	0.0081	9559	200
44.36	5282			6825	200	0.0064	0.3478		7.0893	0.0050	5282	200
45.38	5286			5286	200	0.0050	0.3526		7.0839	0.0050	5286	200
46.36	5284			5284	200	0.0050	0.3576		7.0783	0.0050	5284	200
47.36	5276			5276	200	0.0050	0.3626		7.0739	0.0050	5276	200
48.36	5274			5274	200	0.0050	0.3675		7.0683	0.0050	5274	200
48.42	317			317	200	0.0003	0.3681	PRINCETON JUNCTION	7.0633	0.0003	317	200
49.36	4968			4968	200	0.0047	0.3726		7.0630	0.0047	4968	200
50.36	5268			5268	200	0.0050	0.3776		7.0583	0.0050	5268	200
51.38	5295			5295	200	0.0050	0.3826		7.0533	0.0050	5295	200
51.68	1671			1671	200	0.0016	0.3844		7.0483	0.0016	1671	200
51.90	1148		281	1148	200	0.0011	0.3855	R LAWRENCE STATION	7.0467	0.0011	1148	200
52.36	2453			2453	200	0.0023	0.3878		7.0456	0.0023	2453	200
53.36	5291			5291	200	0.0050	0.3928		7.0433	0.0050	5291	200
54.36	5277			5277	200	0.0050	0.3976		7.0383	0.0073	7744	200
55.36	5280			5280	200	0.0050	0.4026		7.0310	0.0063	6306	A 18
56.36	5276			5276	200	0.0050	0.4076		7.0247	0.0016	1534	A 17
57.17	4273			4273	200	0.0040	0.4116		7.0231	0.0000		
57.35	990		268	3682	200	0.0035	0.4153	L	7.0231	0.0000		
57.46	532		286		200	0.0000	0.4153	L	7.0231	0.0000		
58.10	3402			1272	D 20	0.0013	0.4166	TRENTON	7.0231	0.0000		
58.23	691		182	691	177	0.0007	0.4173	R	7.0231	0.0000		
58.35	638			638	177	0.0007	0.4180	A	7.0231	0.0000		
58.96	3236		177	2600	177	0.0026	0.4206	L	7.0231	0.0000		
59.27	1606			0	177	0.0000	0.4206	MORRISVILLE PA	7.0231	0.0226	15868	A 73
59.36	500		233		177	0.0000	0.4206	R	7.0005	0.0007	260	73
59.57	1100			23	177	0.0000	0.4206		6.9998	0.0029	1100	73
59.87				3813	D 17	0.0068	0.4266		6.9969			
59.72	800		73	800	73	0.0021	0.4287	[illegible]	6.9969	0.0021	800	73
59.78	300			300	73	0.0006	0.4296	REVERSE CURVE	6.9948	0.0002	98	D 76
59.96	586		76	960	73	0.0025	0.4320	L	6.9946	0.0015	653	76
60.02	300		76	80	73	0.0002	0.4322	L	6.9931	0.0007	300	76
60.04	100			164	A 73	0.0004	0.4326	REVERSE CURVE	6.9924	0.0020	721	A 59
60.19	800		76	956	76	0.0024	0.4350	R	6.9904	0.0025	760	59
60.30	580			243	76	0.0006	0.4356		6.9879	0.0019	580	59
60.30				337	D 76	0.0009	0.4366		6.9860			
60.43	700		59	705	59	0.0022	0.4388	R	6.9860	0.0022	700	59
60.43							0.4388		6.9838	0.0013	600	D 83
60.53	500			560	59	0.0016	0.4404	REVERSE CURVE	6.9825	0.0011	500	83
60.72	1000		83	860	59	0.0028	0.4432	L	6.9814	0.0011	500	83
60.78	300			1101	A 59	0.0029	0.4461	REVERSE CURVE	6.9803	0.0008	407	D 15
60.96	1000		186	339	83	0.0008	0.4469	R	6.9794	0.0001	59	95
61.17	1100			69	83	0.0001	0.4470		6.9793	0.0008	425	A 94
61.17				780	A 83	0.0017	0.4487		6.9785			
61.17				290	94	0.0006	0.4492		6.9785	0.0038	1629	A 96
61.40	1200		94	542	94	0.0011	0.4503	R	6.9747	0.0030	1066	68
61.48	300			667	D 96	0.0016	0.4519	REVERSE CURVE	6.9717	0.0008	300	66
61.61	600		66	800	66	0.0023	0.4541	L	6.9709	0.0023	800	66
61.63				1360	66	0.0039	0.4580		6.9686	0.0028	1360	D 11
61.86	1880			4308	A 66	0.0085	0.4666		6.9657	0.0001	52	117
61.96							0.4666		6.9656	0.0047	2670	A 91
62.79	4380			103	120	0.0002	0.4667		6.9609	0.0028	1360	91
62.79				788	D 12	0.0014	0.4681		6.9581			
62.95	800		95	800	95	0.0016	0.4697	R	6.9581	0.0017	800	91
62.96	100			91	95	0.0002	0.4699	W YARD	6.9564	0.0021	1005	91
62.98	100			109	D 95	0.0002	0.4701		6.9543	0.0002	100	91
63.08	500		91	500	91	0.0010	0.4712	R	6.9541	0.0010	500	91
63.08							0.4712		6.9531	0.0034	2219	D 15
63.08	0			1360	91	0.0028	0.4740		6.9497	0.0002	177	153
63.08	0			4674	A 91	0.0074	0.4814		6.9495	0.0000		
63.98	4980					0.0000	0.4814		6.9495	0.0052	3924	A 13
64.53	3000					0.0000	0.4814		6.9443	0.0015	1024	132
64.96	2280		132	5846	132	0.0064	0.4877	R	6.9428	0.0000		
65.08	685		132		132	0.0000	0.4877	R	6.9428	0.0074	4436	A 94
65.57	2600			1360	D 11	0.0021	0.4898	REVERSE CURVE	6.9354	0.0027	1360	94
65.98	2100		94	2100	94	0.0042	0.4941	L WOODBOURNE	6.9327	0.0042	2100	94
65.98				1360	94	0.0027	0.4968		6.9286	0.0063	3843	D 16
66.96	5280			12508	A 94	0.0174	0.5163		6.9223	0.0004	371	139
67.38	2200					0.0000	0.5163		6.9228	0.0000		
67.62	1280		233			0.0000	0.5163	R	6.9228	0.0000		
67.72	500					0.0000	0.5163		6.9228	0.0000		
67.96	1300		286			0.0000	0.5163	R	6.9228	0.0000		
68.11	760		266	528	170	0.0006	0.5168	R	6.9228	0.0076	7153	A 16
68.74	3300			248	D 17	0.0003	0.5171		6.9152	0.0006		
68.96	1200		198	1200	165	0.0014	0.5185	R	6.9152	0.0000		
69.09	680		165	680	165	0.0008	0.5193	R	6.9152	0.0079	6298	A 13
69.56	2600			1280	165	0.0015	0.5207		6.9073	0.0019	1360	135
69.59				1320	D 16	0.0017	0.5224		6.9054			
69.83	1300		130	1300	135	0.0018	0.5242	L	6.9054	0.0018	1300	135
69.83				1360	135	0.0018	0.5261	HOLLAND	6.9036	0.0000		
69.96	700			1427	A 13	0.0019	0.5281		6.9036	0.0010	705	135
70.06	500					0.0000	0.5281		6.9026	0.0007	500	135
70.19	750		165	367	143	0.0005	0.5286	R	6.9019	0.0010	750	135
70.49	1580			326	D 14	0.0004	0.5290		6.9009	0.0022	1580	135
70.70	1100		135	1100	135	0.0015	0.5306	L	6.8987	0.0015	1100	135
70.96	1400			1360	135	0.0018	0.5325	1916.885071	6.8972	0.0019	1272	D 16
70.98	80			8052	A 13	0.0076	0.5401	2406.140007	6.8956	0.0001	74	184
71.38	2100		190		0	0.0000	0.5401	R	6.8956	0.0000		
71.70	1700				0	0.0000	0.5401		6.8956	0.0000		
71.87	900	XING 12.91			0	0.0000	0.5401	L BUSTLETON	6.8956	0.0000		
71.96	500			75	184	0.0001	0.5402		6.8956	0.0076	6033	A 13
72.36	2060			1272	D 18	0.0016	0.5418		6.8879	0.0019	1360	135
72.89	2800		135	2800	135	0.0039	0.5457	R STREET RD	6.8860	0.0039	2800	135
72.96	400			405	135	0.0008	0.5463		6.8821	0.0036	3184	D 20
73.66	3686			980	130	0.0013	0.5476		6.8785	0.0008	886	200
73.91	1300		202	17442	A 13	0.0194	0.5671	L	6.8777	0.0012	1300	200
73.96	300	XING 10.95				0.0000	0.5671	COUNTY LINE	6.8765	0.0003	300	200
74.96	5280					0.0000	0.5671		6.8762	0.0050	5280	200
75.96	5280					0.0000	0.5671		6.8712	0.0025	2651	200
76.96	5280			5444	200	0.0052	0.5722		6.8687	0.0115	11149	A 16
77.84	4800			1874	D 20	0.0019	0.5742		6.8572	0.0016	1380	165
77.96	680		165	680	165	0.0008	0.5750	L	6.8556	0.0004	380	165
78.24	1480		165	1480	165	0.0017	0.5767	L	6.8548	0.0012	1480	165
78.95	3700			1360	165	0.0015	0.5782	YORK RD	6.8531	0.0019	1874	D 20
78.96	100		231	11149	A 16	0.0130	0.5912	L	6.8512	0.0006	587	200

APPENDIX B3

PAGE 2 COLUMN 1

MILES	DISTANCE FT	HWY GRADE CROSSING	MAX MPH	READ DOWN WESTBOUND TRAFFIC DISTANCE FT	SPEED MPH	ET HOURS	TIME HOURS	STATION	READ UP EASTBOUND TRAFFIC TIME HOURS	ET HOURS	DISTANCE FT	SPEED MPH
79.64	3580		233			0.0000	0.5612	L	6.8506	0.0034	3580	200
79.96	1700					0.0000	0.5612		6.8472	0.0000		
80.15	980			0	200	0.0000	0.5612		6.8472	0.0000		
80.76	3200		233	751	200	0.0007	0.5919	R	6.8472	0.0000		
80.96	1100			3426	200	0.0114	0.6033		6.8472	0.0115	11149	A 165
81.76	4200			1874	D 20	0.0019	0.6053		6.8357	0.0016	1360	165
81.96	1080		165	1080	165	0.0012	0.6065	R	6.8341	0.0012	1080	165
82.12	800		165	800	165	0.0009	0.6074	R CAMP HILL	6.8329	0.0009	800	165
82.65	2800			1380	165	0.0016	0.6090		6.8320	0.0013	1191	D 168
82.96	1680		190	7501	A 18	0.0080	0.6170	L	6.8307	0.0001	64	168
83.29	1700		190	3646	A 18	0.0035	0.6205	L	6.8306	0.0000		
83.96	3580					0.0000	0.6205		6.8306	0.0152	13101	A 135
84.96	5280			17	200	0.0000	0.6205		6.8154	0.0010	680	135
85.09	680			3194	D 20	0.0036	0.6241		6.8144	0.0010	680	135
85.49	2100		135	2100	135	0.0029	0.6271	R STENTON	6.8134	0.0029	2100	135
85.96	2500			1386	135	0.0019	0.6290		6.8105	0.0036	3194	D 200
86.96	5280			17442	A 13	0.0194	0.6484		6.8069	0.0052	5504	200
87.96	5280	XING 29.18				0.0000	0.6484	CONSHOHOCKEN	6.8017	0.0000		
88.96	5280	XING 29.82				0.0000	0.6484	GRAVERS	6.8017	0.0000		
88.96	0			696	200	0.0006	0.6491	MP 30	6.8017	0.0056	9642	A 181
89.38	2180			1062	D 20	0.0011	0.6501		6.7961	0.0000		
89.85	2500		181	2500	181	0.0026	0.6527	L	6.7961	0.0000		
89.96	600			800	181	0.0008	0.6534	EARNEST	6.7961	0.0126	10800	A 135
90.70	3880			1748	181	0.0018	0.6552		6.7823	0.0019	1380	135
90.70				2132	D 18	0.0026	0.6577		6.7804	0.0000		
90.96	1400		135	1400	135	0.0020	0.6597	L RAMBO	6.7804	0.0020	1400	135
91.02	280		135	280	135	0.0004	0.6601	L	6.7784	0.0004	280	135
91.28	1400			1386	135	0.0019	0.6620		6.7780	0.0010	780	D 153
91.96	3600		153	3435	A 13	0.0045	0.6665	R HENDERSON	6.7770	0.0052	4240	153
92.07	580		153	785	153	0.0010	0.6675	R	6.7718	0.0007	580	153
92.96	4700			1360	153	0.0017	0.6692		6.7711	0.0007	560	D 165
93.15	880			2858	A 15	0.0034	0.6726		6.7704	0.0058	5070	165
93.47	1800		165	3162	165	0.0036	0.6762	L	6.7646	0.0021	1800	165
93.96	2600			1386	165	0.0015	0.6778	KING of PRUSSIA	6.7625	0.0019	1874	D 64
94.96	5280			11149	A 16	0.0115	0.6893		6.7606	0.0057	5006	200
95.96	5280					0.0000	0.6893		6.7549	0.0050	5280	200
96.07	580					0.0000	0.6893		6.7499	0.0005	580	200
96.38	1600		206	2831	200	0.0027	0.6920	R	6.7494	0.0015	1600	200
96.96	3100			3100	200	0.0029	0.6949		6.7479	0.0029	3100	200
97.96	5280			5280	200	0.0050	0.6999	HOWELLVILLE	6.7450	0.0050	5280	200
98.49	2800			2800	200	0.0027	0.7026		6.7400	0.0027	2800	200
98.96	2480		233	2460	200	0.0023	0.7049	R	6.7373	0.0023	2460	200
99.11	780			780	200	0.0007	0.7056		6.7350	0.0007	780	200
99.60	2600		233	2600	200	0.0025	0.7081	L	6.7343	0.0014	1434	200
99.96	1900			1900	200	0.0018	0.7099		6.7329			
100.85	4700			4700	200	0.0045	0.7144		6.7329	0.0066	9642	A 181
100.96	580		404	580	200	0.0005	0.7148	R	6.7263	0.0000		
101.15	980		404	980	200	0.0008	0.7158	R	6.7263	0.0000		
101.96	4300			4300	200	0.0041	0.7199		6.7263	0.0000		
102.17	1100			1100	200	0.0010	0.7210		6.7263	0.0000		
102.28	580		296	580	200	0.0006	0.7215	L	6.7263	0.0000		
102.75	2442			2442	200	0.0023	0.7238		6.7263	0.0000		
102.82	370						0.7238		6.7263	0.0000		
102.82			233	1158	200	0.0011	0.7249	R	6.7263	0.0000		
102.96	786			962	200	0.0009	0.7258		6.7263	0.0000		
103.35	2065						0.7258		6.7263	0.0000		
103.36		OH		156	200	0.0001	0.7260	RTE 352 OHO 1.46	6.7263	0.0000		
103.68	1737			1062	D 20	0.0011	0.7270		6.7263	0.0244	17231	A 69-1
103.82	692						0.7270	L	6.7019	0.0000		
103.96	786		181		161	0.0000	0.7270	L	6.7019	0.0049	1771	69
104.12	802				161		0.7270		6.6970	0.0016	522	A 554
104.12		XING 45.18			181		0.7270	W HENDERSON 45.18	6.6954			
104.16	245			46	181	0.0000	0.7271		6.6954	0.0006	179	56
104.16		OKW		4107	D 18	0.0062	0.7333	AMTRAK 14 OHO 0.65	6.6948			
104.41	1313		69	1313	69	0.0036	0.7369	R	6.6948	0.0016	405	A 41-5
104.46	280			280	69	0.0007	0.7376		6.6932	0.0001	26	41
104.46		UG		495	69	0.0014	0.7390	RAVINE ROAD 0.35	6.6931			
104.60	750			255	D 66	0.0008	0.7397		6.6931	0.0035	750	41
104.69	463		56	357	56	0.0012	0.7410	R	6.6896	0.0022	463	41
104.71	101			187	D 56	0.0008	0.7418		6.6874	0.0005	101	41
104.76	274		41	274	41	0.0013	0.7430	L	6.6869	0.0013	274	41
104.76							0.7430		6.6856	0.0013	354	D 64
104.81	257		64	257	41	0.0012	0.7442	L	6.6843	0.0001	22	64
104.81					0		0.7442		6.6842	0.0068	4473	D 187
104.81							0.7442	GLEN LOCH 30 MPH TT ON	6.6774	0.0003	300	186
104.99	980			980	41	0.0045	0.7487	T	6.6771	0.0000		
105.28	1524		219	380	41	0.0018	0.7505	L	6.6771	0.0000		
105.52	1252			732	A 41	0.0026	0.7531	T	6.6771	0.0000		
105.65	675			17759	A 64	0.0253	0.7784	T	6.6771	0.0000		
105.87	1190		213	0		0.0000	0.7784	L	6.6771	0.0000		
106.10	1191		201	0		0.0000	0.7784	R REVERSE CURVE	6.6771	0.0000		
106.52	2226			0		0.0000	0.7784	T	6.6771	0.0000		
107.52	5280					0.0000	0.7784		6.6771	0.0000		
107.75	1230	OHD 26.36				0.0000	0.7784	LO-GRADE - ABANDONED	6.6771	0.0000		
108.52	4050		426			0.0000	0.7784	L	6.6771	0.0000		
108.68	845		426			0.0000	0.7784	L	6.6771	0.0259	17406	A 93-1
109.29	3221			221	182	0.0002	0.7786	T	6.6502	0.0028	1360	93
109.29				3590	D 12	0.0049	0.7836		6.6474	0.0000		
109.52	1214		93	1214	93	0.0025	0.7861	R	6.6474	0.0025	1214	93
109.52						0.0000	0.7861	BRADFORD HILLS - EAST CALN T	6.6449	0.0006	318	D 104
109.70	970		463	970	93	0.0020	0.7880	R	6.6443	0.0012	652	104
109.90	1074			390	93	0.0008	0.7888	T	6.6431	0.0019	1024	104
110.26	1940		104	941	A 93	0.0018	0.7906	L	6.6412	0.0035	1940	104
110.26				2060	104	0.0055	0.7961		6.6377	0.0012	705	D 125
110.52	1347			3467	A 10	0.0042	0.8002	T	6.6365	0.0010	642	125
110.74	1173					0.0000	0.8002	T	6.6355	0.0018	1173	125
111.14	2080		125			0.0000	0.8002	R	6.6337	0.0032	2080	125
111.14		WEST END		1133	125	0.0017	0.8020		6.6305	0.0022	1701	D 165
111.52	2027			7606	A 12	0.0101	0.8121	T	6.6283	0.0004	326	165
111.75	1202					0.0000	0.8121	DOWNINGTOWN	6.6279	0.0014	1202	165
112.08	1777		165			0.0000	0.8121	L	6.6265	0.0020	1777	165
112.28	1046					0.0000	0.8121	T	6.6245	0.0007	602	D 177
112.52	1255		177			0.0000	0.8121	R	6.6238	0.0018	1699	177
112.73	1109		177	3326	A 18	0.0037	0.8158	R	6.6220	0.0019	1795	177
113.18	2376					0.0000	0.8158	T	6.6201	0.0037	3526	A 165
113.52	1795		177	0	177	0.0000	0.8158	L	6.6164	0.0002	149	165
113.61	466		177	670	177	0.0007	0.8165	L	6.6162	0.0006	466	165
113.75	725			802	D 17	0.0007	0.8172	T	6.6157	0.0008	725	165
113.95	1087		165	1087	165	0.0012	0.8184	R	6.6149	0.0012	1087	165
114.16	1087		330	1087	165	0.0012	0.8187	R	6.6137	0.0014	1302	D 190
114.35	1035			273	165	0.0003	0.8209	T	6.6123	0.0008	820	190
114.52	880		233	6122	A 16	0.0066	0.8256	L	6.6115	0.0009	880	190
114.56	216					0.0000	0.8256	T	6.6106	0.0002	216	190
114.70	754		190		186	0.0006	0.8266	L	6.6104	0.0008	803	190
115.39	3610			100	186	0.0001	0.8267	T	6.6096	0.0014	1379	A 186
115.52	700		186	700	186	0.0007	0.8274	L	6.6082	0.0000		
115.56	209		186	209	186	0.0002	0.8276	L	6.6082	0.0000		
115.71	784		186	784	186	0.0008	0.8284	R REVERSE CURVE	6.6082	0.0066	6123	A 165
116.39	3607			1380	186	0.0014	0.8298	T	6.6016	0.0016	1380	165

APPENDIX B3

PAGE 2 COLUMN 2

MILES	DISTANCE FT	HWY GRADE CROSSING	MAX MPH	READ DOWN WESTBOUND TRAFFIC DISTANCE FT	SPEED MPH	ET HOURS	TIME HOURS	STATION	READ UP EASTBOUND TRAFFIC TIME HOURS	ET HOURS	DISTANCE FT	SPEED MPH
116.39				686	A 16	0.0007	0.8305		6.6000			
116.39				370	168	0.0004	0.8309		6.6000			
116.39				1191	D 16	0.0013	0.8321		6.6000			
116.52	660		186	660	186	0.0009	0.8329	R	6.6000	0.0008	660	165
116.98	2438		165	2433	165	0.0028	0.8357	R	6.5992	0.0028	2433	165
117.10	621			621	165	0.0007	0.8364	T	6.5964	0.0007	621	165
117.46	1915		165	1915	165	0.0022	0.8386	L	6.5957	0.0022	1915	165
117.52	351			351	165	0.0004	0.8390	T	6.5935	0.0019	1874	D 19
117.52				1049	165	0.0012	0.8402	COATESVILLE	6.5916	0.0005	572	200
117.94	2240			7502	A 16	0.0074	0.8475	T	6.5911	0.0000		
118.52	3040		233			0.0000	0.8475	L	6.5911	0.0000		
118.60	414		233			0.0000	0.8475	L	6.5911	0.0036	3646	A 19
118.87	1449					0.0000	0.8475	T	6.5876	0.0014	1360	190
119.34	2485		190	1078	190	0.0011	0.8486	R	6.5862	0.0025	2485	190
119.34				1360	190	0.0014	0.8500		6.5837	0.0003	340	D 19
119.34				1054	A 19	0.0010	0.8510		6.5834	0.0003	349	196
119.52	952			2173	198	0.0021	0.8531	T	6.5831	0.0007	2817	A 18
120.26	3934			279	D 15	0.0003	0.8534	T	6.5824	0.0014	1360	188
120.52	1346		168	1346	188	0.0014	0.8548	L	6.5810	0.0014	1346	188
120.61	485		186	485	188	0.0005	0.8553	L	6.5796	0.0005	485	188
120.89	1455			1455	188	0.0015	0.8567	T	6.5791	0.0015	1455	188
121.27	2047		188	2047	188	0.0021	0.8586	R	6.5776	0.0021	2047	188
121.32	1298			1360	188	0.0014	0.8602	T	6.5755	0.0007	603	D 20
122.52	5280			4341	A 18	0.0042	0.8644		6.5748	0.0056	5890	200
123.00	2586			3408	200	0.0032	0.8676	T	6.5692	0.0018	1870	200
123.19	984		206	984	200	0.0009	0.8686	L	6.5674	0.0000	0	200
123.52	1760			1760	200	0.0017	0.8702	T	6.5674	0.0000	0	200
124.52	5280			5280	200	0.0050	0.8752	PARKESBURG	6.5674	0.0000	0	200
124.61	495			495	200	0.0005	0.8757	T	6.5674	0.0000		
124.85	1265		230	1265	200	0.0012	0.8769	R	6.5674	0.0000	0	200
125.52	3520			3520	200	0.0033	0.8802	T	6.5674	0.0000		
126.28	4015			3506	200	0.0034	0.8836	T	6.5674	0.0000		
126.43	770		233	0	200	0.0000	0.8836	R	6.5674	0.0233	20035	A 11
126.52	495			0	200	0.0000	0.8836	T	6.5441	0.0013	805	117
126.93	2145			0	200	0.0000	0.8836	T	6.5428	0.0042	2332	A 94
126.93				3858	D 20	0.0046	0.8882		6.5386	0.0000		
127.52	3135		117	2423	117	0.0039	0.8921	R	6.5386	0.0025	1360	94
127.52				712	D 11	0.0013	0.8934	ATGLEN	6.5361	0.0000		
127.68	834		94	834	94	0.0017	0.8951	R	6.5361	0.0017	834	94
127.77	500			500	94	0.0010	0.8961	REVERSE CURVE	6.5344	0.0021	1260	D 13
128.18	2164		132	860	94	0.0017	0.8978	L	6.5323	0.0020	1408	132
128.26	389			4406	A 94	0.0071	0.9049	REVERSE CURVE	6.5303	0.0006	389	132
128.52	1380		132			0.0000	0.9049	R	6.5297	0.0020	1389	132
128.72	1066		132		132	0.0000	0.9049	R CHRISTIANA	6.5277	0.0004	280	132
128.72				1066	132	0.0015	0.9064		6.5273	0.0000		
129.33	3224			1956	A 13	0.0026	0.9091	T	6.5273	0.0080	4740	A 90
129.52	1000		165	237	143	0.0003	0.9094	L	6.5193	0.0005	260	90
129.52				444	D 14	0.0006	0.9100	0	6.5188	0.0000		
129.62	550		132	0	132	0.0000	0.9100	L	6.5188	0.0040	2119	90
129.73	550			0	132	0.0000	0.9100	REVERSE CURVE	6.5143	0.0012	536	A 82
129.75				1368	D 13	0.0023	0.9123		6.5131	0.0000		
130.12	2090		90	2090	90	0.0044	0.9167	R	6.5131	0.0012	535	82
130.28	825			623	90	0.0013	0.9180	T	6.5119	0.0019	825	82
130.28				202	D 90	0.0004	0.9185		6.5100	0.0000		
130.52	1265		82	365	82	0.0008	0.9193	L	6.5100	0.0028	1265	82
130.52				900	82	0.0021	0.9214	GAP	6.5071	0.0000		
130.83	1650		82	1650	82	0.0038	0.9252	L	6.5071	0.0038	1650	82
131.19	1870		266	1360	82	0.0031	0.9284	L	6.5033	0.0043	1870	82
131.19				1904	A 82	0.0003	0.9286		6.4990	0.0000		
131.52	1760			34	106	0.0001	0.9287	T	6.4990	0.0041	1760	82
131.56	330				106	0.0000	0.9287		6.4949	0.0008	330	82
131.58				862	D 10	0.0013	0.9300		6.4941	0.0000		
131.98	2090		82	2090	82	0.0048	0.9348	R	6.4941	0.0046	2090	82
131.98				1360	82	0.0031	0.9380		6.4893	0.0017	887	D 11
132.27	1540			2666	A 82	0.0051	0.9431	T	6.4876	0.0011	653	113
132.52	1320		113	0		0.0000	0.9431	L	6.4865	0.0022	1320	113
132.72	1045		113		113	0.0000	0.9431	L	6.4843	0.0018	1045	113
132.72				1239	113	0.0021	0.9452		6.4825	0.0044	3580	D 19
133.20	2530			9592	A 11	0.0126	0.9579	T	6.4781	0.0002	245	193
133.52	1705		233	0	165	0.0000	0.9579	L	6.4779	0.0000		
134.00	2530			0	165	0.0000	0.9579	T	6.4779	0.0000		
134.18	935		233	2425	A 16	0.0027	0.9606	L	6.4779	0.0000		
134.52	1815			120	174	0.0001	0.9607	T	6.4779	0.0090	8555	A 16
135.32	4235			447	D 17	0.0005	0.9612	T	6.4689	0.0016	1380	165
135.52	1045		165	1045	165	0.0012	0.9624	R	6.4673	0.0012	1045	165
136.13	3245		165	3245	165	0.0037	0.9661	R	6.4661	0.0037	3245	165
136.52	2035			1390	165	0.0016	0.9677	T	6.4624	0.0003	298	D 17
136.89	1980			5782	A 16	0.0063	0.9739	T	6.4621	0.0001	98	171
137.18	1485		202			0.0000	0.9739	L	6.4620			
137.52	1815			160	185	0.0002	0.9741	T	6.4620	0.0097	7895	A 13
137.96	2334			2347	D 15	0.0028	0.9769	T	6.4523	0.0019	1360	135
138.44	2501		135	2501	135	0.0036	0.9804	L	6.4504	0.0035	2501	135
138.52	445			445	135	0.0006	0.9810	T	6.4469	0.0014	991	135
138.52				915	135	0.0013	0.9823		6.4455	0.0000		
139.02	2640	XING 59.18		2010	A 13	0.0027	0.9850	BUSHTOWN ROAD	6.4458	0.0047	3087	A 11
139.22	1045		146	820	146	0.0011	0.9861	R	6.4408	0.0001	40	113
139.47	1320			6	146	0.0000	0.9861	T	6.4407	0.0022	1320	113
139.47				1254	D 14	0.0018	0.9879		6.4385			
139.52	275		113	275	113	0.0005	0.9884	R	6.4385	0.0005	275	113
140.12	3190		113	3190	113	0.0053	0.9937	R	6.4380	0.0053	3190	113
140.12				1360	113	0.0023	0.9960		6.4327			113
140.48	1870			506	A 11	0.0008	0.9968	T	6.4327	0.0002	135	D 11
140.52	220		117	224	117	0.0004	0.9972	L BIRD IN HAND	6.4325	0.0032	1955	117
140.97	2390		117	2390	117	0.0039	1.0011	L	6.4293	0.0039	2390	117
140.97				1360	117	0.0022	1.0033		6.4254			
141.10	1112			6806	A 11	0.0118	1.0151	T	6.4254	0.0027	1986	D 16
141.52	1779		165		165	0.0000	1.0151	L	6.4227	0.0010	906	165
141.63	605		165		165	0.0000	1.0151	L	6.4217	0.0007	605	165
142.44	4235				165	0.0000	1.0151	T	6.4210	0.0009	865	D 16
142.44						0.0000	1.0151		6.4201	0.0003	250	182
142.52	440		233			0.0000	1.0151	L	6.4198	0.0000		
142.72	1045		233			0.0000	1.0151	L	6.4198	0.0053	4840	A 16
143.02	1595			565	165	0.0006	1.0157	T	6.4145	0.0016	1260	165
143.32	1595		165	1595	165	0.0018	1.0175	R	6.4129	0.0016	1595	165
143.32							1.0175		6.4111	0.0019	1815	D 19
143.52	1045			1045	165	0.0012	1.0187	T	6.4092	0.0001	122	199
143.52				315	165	0.0004	1.0191		6.4091	0.0000		
144.32	4224					0.0000	1.0191	T	6.4091	0.0000		
144.52	1056		286	6122	A 19	0.0066	1.0257	R	6.4091	0.0000		
144.52						0.0000	1.0257		6.4091	0.0046	4661	A 18
145.02	2640		286			0.0000	1.0257	R	6.4045	0.0000		
145.52	2640			3603	A 18	0.0035	1.0292	T	6.4045	0.0000		
145.52				335	186	0.0003	1.0295		6.4045	0.0000		
145.57	275			560	D 15	0.0006	1.0301	T	6.4045	0.0000		
145.67	495		186	495	186	0.0005	1.0306	R	6.4045	0.0066	6122	A 18
145.75	440			624	186	0.0006	1.0312	REVERSE CURVE	6.3979	0.0001	95	165
145.94	990		166	0	186	0.0000	1.0312	T	6.3978	0.0011	990	165
145.99	275			1081	D 16	0.0012	1.0324	T	6.3967	0.0003	275	165
146.18	990		165	990	165	0.0011	1.0335	L	6.3964	0.0011	990	165

APPENDIX B3

PAGE 3 COLUMN 1

MILES	DISTANCE, FT	HWY GRADE CROSSING	MAX SPEED	DISTANCE, FT	SPEED, mph	ET, HOURS	TIME, HOURS	STATION	TIME, HOURS	ET, HOURS	DISTANCE, FT	SPEED, mph
			READ DOWN WESTBOUND TRAFFIC					STATION	READ UP EASTBOUND TRAFFIC			
146.52	1815			1360	165	0.0016	1.0351 T		6.3953	0.0010	919	D 183
146.52							1.0351		6.3943	0.0003	277	183
146.90	2006			6464	A 16	0.0069	1.0420	LANCASTER STATION	6.3940	0.0000		
147.52	3274						1.0420		6.3940	0.0079	7524	A 157
148.01	2585			342	187	0.0003	1.0424 T		6.3861	0.0016	1380	157
148.01				1514	D 16	0.0017	1.0440	DILLERVILLE - COLA BRANCH	6.3845	0.0000		
148.52	2695		157	2695	157	0.0033	1.0473 R		6.3845	0.0033	2695	157
148.72	1045		157	1045	157	0.0013	1.0485 R		6.3812	0.0013	1045	157
148.72							1.0485		6.3799	0.0014	1243	D 182
149.52	4235			1360	157	0.0016	1.0502 T		6.3785	0.0003	257	182
149.52				3535	A 1	0.0063	1.0564		6.3782	0.0063	4840	A 165
150.18	3465			191	176	0.0002	1.0566 T		6.3729	0.0016	1360	165
150.18				654	D 17	0.0007	1.0573		6.3713			
150.52	1815		165	1815	165	0.0021	1.0594 L		6.3713	0.0021	1815	165
150.56	220		165	220	165	0.0003	1.0596 L		6.3692	0.0003	220	165
150.56				1360	165	0.0016	1.0612		6.3689	0.0003	296	D 166
150.56				1602	A 16	0.0018	1.0630		6.3686	0.0002	152	171
151.21	3410	XING 71.88		152	171	0.0002	1.0632	E PETER ROAD	6.3684	0.0018	1602	A 165
151.21				298	D 17	0.0003	1.0635		6.3666	0.0016	1360	165
151.52	1650		165	1650	165	0.0019	1.0654 L		6.3650	0.0019	1650	165
151.64	660		165	660	165	0.0008	1.0662 L		6.3631	0.0008	660	165
151.93	1485			1485	165	0.0017	1.0679 T		6.3623	0.0017	1485	165
152.31	2015		165	2015	165	0.0023	1.0702 R		6.3606	0.0065	5705	165
152.52	1100			1100	165	0.0013	1.0715 T		6.3541	0.0028	2408	A 156
153.52	5280			280	165	0.0003	1.0718		6.3513	0.0004	304	155
153.52				4507	A 16	0.0049	1.0767		6.3509	0.0000		
153.72	1056			288	161	0.0003	1.0770 T		6.3509	0.0013	1056	155
153.72				1281	D 15	0.0014	1.0784		6.3496			
154.23	2668		155	2668	155	0.0033	1.0817 L		6.3496	0.0033	2668	155
154.37	778			778	155	0.0010	1.0826 T		6.3463	0.0006	469	D 166
154.37				582	155	0.0007	1.0834		6.3457	0.0004	309	165
154.37				2406	A 15	0.0028	1.0862		6.3453			
154.52	778		165			0.0000	1.0862 R		6.3453	0.0009	778	165
154.72	1056		165	206	165	0.0002	1.0864 R		6.3444	0.0012	1056	165
155.43	3724	XING 75.82		6806	A 16	0.0073	1.0937	PRIVATE	6.3432	0.0012	1061	D 186
155.52	500		369			0.0000	1.0937 L		6.3420	0.0001	118	186
156.03	2668		369			0.0000	1.0937 L		6.3419	0.0000		
156.52	2613					0.0000	1.0937 T		6.3419	0.0066	6122	A 165
156.52				585	188	0.0006	1.0943		6.3353			
156.60	440			1191	D 18	0.0013	1.0956 T		6.3353	0.0000		
156.83	1210		165	1210	165	0.0014	1.0970 L		6.3353	0.0049	4068	A 147
157.13	1395			771	165	0.0009	1.0979 T		6.3304	0.0016	1360	147
157.13				824	D 16	0.0010	1.0989		6.3286	0.0000		
157.33	1045		147	1045	147	0.0013	1.1002 L		6.3286	0.0013	1045	147
157.33							1.1002		6.3273	0.0010	824	D 161
157.52	990			990	147	0.0013	1.1015 T		6.3263	0.0000	18	165
157.52							1.1015		6.3263	0.0049	4068	A 147
158.52	5280			370	147	0.0006	1.1020		6.3214	0.0016	1360	147
158.52				4025	A 14	0.0047	1.1066		6.3198			
158.67	778	XING 79.13		288	164	0.0003	1.1070	MOORES MILL ROAD	6.3198	0.0010	778	147
158.67				775	D 16	0.0009	1.1079		6.3188			
158.87	1056		147	1056	147	0.0014	1.1093 R		6.3188	0.0014	1056	147
159.28	2168			1360	147	0.0018	1.1110 T		6.3172	0.0024	2181	D 191
159.28				9243	A 14	0.0106	1.1216		6.3148	0.0003	281	191
159.52	1278		202	0		0.0000	1.1216 R		6.3145	0.0000		
159.61	500		202			0.0000	1.1216 R		6.3145	0.0000		
160.38	4057					0.0000	1.1216 T		6.3145	0.0000		
160.52	723		196			0.0000	1.1216 R		6.3145			
160.68	834	XING 0	195	150	183	0.0002	1.1217	K MARKET ST MT JOY	6.3145	0.0083	7850	A 165
161.08	2112			819	D 18	0.0010	1.1227 T		6.3062	0.0016	1360	165
161.52	2334		165	2334	165	0.0027	1.1254 R		6.3046	0.0027	2334	165
161.55	165		165	165	165	0.0002	1.1256 R		6.3019	0.0002	165	165
162.30	3960			1360	165	0.0016	1.1272 T		6.3017	0.0009	812	D 181
162.52	1155		233	6122	A 16	0.0066	1.1338 L		6.3008	0.0001	116	181
162.75	1210		233	0		0.0000	1.1338 L		6.3007	0.0000		
163.23	2530			0		0.0000	1.1338 T		6.3007	0.0000		
163.52	1540		186	2804	186	0.0029	1.1366 L		6.3007	0.0000		
163.52				109	D 18	0.0001	1.1367		6.3007			
163.76	1278		184	984	184	0.0010	1.1377 L		6.3007	0.0000		
164.27	2668			0	184	0.0000	1.1377 T		6.3007	0.0167	13990	A 117
164.27					A 18	0.0000	1.1377		6.2840			
164.27				0	184	0.0000	1.1377		6.2840			
164.27				2958	D 18	0.0037	1.1415		6.2840			
164.52	1334		117	1334	117	0.0022	1.1436 R	ELIZABETHTOWN	6.2840	0.0022	1334	117
164.86	1779		117	1779	117	0.0029	1.1465 R		6.2818	0.0029	1779	117
164.99	723			723	117	0.0012	1.1477 T		6.2788	0.0012	723	117
164.99				617	117	0.0010	1.1487		6.2777			
165.29	1556		165	1066	A 11	0.0017	1.1504 L		6.2777	0.0025	1556	117
165.40	611			180	125	0.0003	1.1506 T		6.2752	0.0010	611	117
165.40				284	D 12	0.0004	1.1511		6.2747			
165.52	611		117	611	117	0.0010	1.1521 L		6.2742	0.0010	611	117
165.71	1000		117	1000	117	0.0016	1.1537 L		6.2732	0.0016	1000	117
165.71				1360	117	0.0022	1.1559		6.2716	0.0046	3859	D 200
166.52	4280					0.0000	1.1559 T		6.2670	0.0004	421	200
167.24	3779			20033	A 11	0.0233	1.1792 T		6.2666	0.0036	3779	200
167.49	1334		233			0.0000	1.1792		6.2630	0.0013	1334	200
167.52	167					0.0000	1.1792 T		6.2617	0.0002	167	200
168.52	5280					0.0000	1.1792		6.2615	0.0011	1117	200
169.52	5280					0.0000	1.1792		6.2604	0.0127	12165	A 181
169.52				742	200	0.0007	1.1799		6.2477			
170.29	4083			2085	D 20	0.0022	1.1821 T		6.2477	0.0016	1360	181
170.52	1198		181	1198	181	0.0014	1.1835 L		6.2461	0.0002	195	181
170.68	825		181	825	181	0.0010	1.1845 L		6.2459	0.0000		
171.17	2585			1360	181	0.0016	1.1861 T		6.2459	0.0000		
171.52	1870		165	1016	A 18	0.0012	1.1873 R		6.2459	0.0106	7876	A 117
171.52				2079	165	0.0024	1.1898		6.2353			
171.81	1556		165	1556	165	0.0018	1.1914 R		6.2353	0.0042	2332	A 94
172.28	2445			460	165	0.0005	1.1920 T		6.2311	0.0002	42	94
172.28				1365	D 16	0.0027	1.1946		6.2309	0.0000		
172.52	1278		117	566	117	0.0009	1.1955 R		6.2309	0.0026	1176	94
172.52				712	D 11	0.0013	1.1968		6.2283	0.0000		
172.87	1870		94	1870	94	0.0038	1.2006 R		6.2283	0.0038	1870	94
172.87				1362	94	0.0027	1.2033	REVERSE CURVE	6.2245	0.0000		94
172.87				1299	A 94	0.0024	1.2056		6.2245	0.0008	415	D 108
173.46	3080		108	421	108	0.0007	1.2065 L		6.2237	0.0047	2665	108
173.52	330			330	108	0.0006	1.2071 T		6.2190	0.0027	1858	D 156
173.52				1030	108	0.0018	1.2089	ROYALTON - ROYALTON BRANCH	6.2163	0.0003	231	156
174.12	3168			5636	A 10	0.0083	1.2172 T		6.2160	0.0026	2105	A 146
174.26	723		251			0.0000	1.2172 L		6.2134	0.0000	26	146
174.33	389					0.0000	1.2172 T		6.2134	0.0005	389	146
174.49	834		571				1.2172 L		6.2129	0.0011	834	146
174.51	111					0.0000	1.2172 T		6.2118	0.0001	111	146
174.52	16		146			0.0000	1.2172 L	MIDDLETOWN	6.2117	0.0001	56	146
174.81	1556		146	171	146	0.0002	1.2174 L		6.2116	0.0020	1556	146
174.89	389			389	146	0.0005	1.2179	REVERSE CURVE	6.2096	0.0016	1521	D 178
175.09	1058		178	971	146	0.0013	1.2192 R		6.2078	0.0000	5	178
175.29	1056			6453	A 14	0.0076	1.2268 T		6.2078	0.0008	789	178
175.49	1064		190			0.0000	1.2268 L		6.2070	0.0000		

APPENDIX B3

PAGE 3 COLUMN 2

MILES	DISTANCE, FT	HWY GRADE CROSSING	MAX SPEED	DISTANCE, FT	SPEED, mph	ET, HOURS	TIME, HOURS	STATION	TIME, HOURS	ET, HOURS	DISTANCE, FT	SPEED, mph
			READ DOWN WESTBOUND TRAFFIC					STATION	READ UP EASTBOUND TRAFFIC			
175.52	187					0.0000	1.2268 T		6.2070	0.0000		
175.81	1556					0.0000	1.2268 T		6.2070	0.0000		
175.96	778		571	55	178	0.0001	1.2269 L		6.2070	0.0054	4837	A 16
176.43	2445			635	D 17	0.0007	1.2276 T	FLYOVER	6.2016	0.0016	1360	160
176.52	500		160	500	160	0.0006	1.2282 R		6.2000	0.0006	500	160
176.81	1556		160	1556	160	0.0018	1.2301 R		6.1994	0.0018	1556	160
177.41	3168			1360	160	0.0016	1.2317 T		6.1976	0.0022	2112	D 20
177.52	556		369			0.0000	1.2317 L		6.1954	0.0015	1812	200
177.72	1045		369			0.0000	1.2317 L		6.1939	0.0010	1045	200
178.46	3905			12400	A 16	0.0130	1.2447 T		6.1929	0.0037	3905	200
178.52	330		219			0.0000	1.2447 R		6.1892	0.0003	330	200
178.97	2395		219			0.0000	1.2447 R		6.1889	0.0006	601	200
179.52	2885					0.0000	1.2447 T		6.1883	0.0000		
180.52	5280			3930	200	0.0037	1.2484	STEELTON	6.1883			
181.14	3300					0.0000	1.2484 T		6.1883	0.0115	11149	A 16
181.49	1815		272			0.0000	1.2484 R		6.1768	0.0000		
181.52	165				200	0.0000	1.2484 T		6.1768	0.0000		
181.96	2354			1874	D 20	0.0019	1.2503 T		6.1768	0.0000		
182.20	1278		165	1278	165	0.0015	1.2518 R		6.1768	0.0000		
182.42	1167			1167	165	0.0013	1.2531 T		6.1768	0.0000		
182.50	389		202	189	165	0.0002	1.2533 L		6.1768	0.0000		
182.52	111			6122	A 16	0.0066	1.2599 T		6.1768	0.0132	9719	A 11
182.85	1760			266	166	0.0003	1.2602 T		6.1636	0.0015	1360	110
182.85				3289	D 11	0.0042	1.2644		6.1621	0.0000		
182.98	660		110	660	110	0.0011	1.2656 R		6.1621	0.0012	723	110
183.52	2860			1360	110	0.0023	1.2679 T		6.1609	0.0087	3753	A 48
183.52				1313	A 11	0.0026	1.2705		6.1522	0.0000		
183.86	1778			111	121	0.0005	1.2710 T		6.1522	0.0032	822	48
183.86				1954	D 12	0.0035	1.2745		6.1490			
183.91	269		56	269	58	0.0009	1.2753 L	HARRISBURG	6.1490	0.0011	269	48
183.92	54		58	54	58	0.0002	1.2756 L		6.1479	0.0002	54	48
183.96	215			60	58	0.0002	1.2757 T		6.1477	0.0006	215	48
183.96				156	D 58	0.0006	1.2763	REVERSE CURVE	6.1469			
184.03	377		48	377	48	0.0015	1.2778 R		6.1469	0.0015	377	48
184.14	560			590	46	0.0023	1.2801	REVERSE CURVE	6.1454	0.0006	173	D 59
184.14						0.0000	1.2801		6.1448	0.0012	420	59
184.27	700		59	700	48	0.0028	1.2829 L		6.1436	0.0022	700	59
184.42	754			67	48	0.0003	1.2831 T		6.1413	0.0008	293	D 74
184.52	535			984	A 48	0.0031	1.2862 T		6.1405	0.0035	1372	74
184.56	218			460	74	0.0012	1.2874 T		6.1370	0.0000		74
184.59	163		74	163	74	0.0004	1.2878 R		6.1370	0.0000		
184.73	707			707	74	0.0018	1.2896 T		6.1370	0.0021	717	A 56
184.98	1361			1018	74	0.0026	1.2922 T		6.1349	0.0048	1360	56
184.98				343	D 74	0.0010	1.2932		6.1303			
185.07	435		56	435	56	0.0015	1.2947 L		6.1303	0.0015	435	56
185.07	0			6		0.0000	1.2947	REVERSE CURVE	6.1288	0.0022	864	D 95
185.15	435		130	435	56	0.0015	1.2961 R		6.1266	0.0000	24	95
185.27	653			653	56	0.0022	1.2983 T		6.1266	0.0025	200	95
185.46	986		110	272	56	0.0009	1.2993 R		6.1240	0.0020	980	95
185.52	327			2056	A 56	0.0077	1.3070 T		6.1220	0.0007	327	95
185.65	701					0.0000	1.3070 T		6.1213	0.0014	701	95
185.68	162		95		95	0.0000	1.3070 L		6.1198	0.0003	162	95
186.52	4439			1202	95	0.0024	1.3094 T		6.1196	0.0036	2431	D 18
186.52				10133	A 95	0.0148	1.3242		6.1160	0.0003	249	160
187.52	5281					0.0000	1.3242 T		6.1157	0.0000		
187.52				86	165	0.0001	1.3243		6.1157	0.0154	10402	A 68
188.42	4736			2857	D 16	0.0043	1.3286 T		6.1003	0.0029	1360	88
188.45	163		68	163	68	0.0004	1.3290 R		6.0974	0.0004	165	88
188.48	162			163	68	0.0004	1.3293	REVERSE CURVE	6.0970	0.0004	163	88
188.52	219		68	218	68	0.0005	1.3298 L		6.0966	0.0005	218	88
188.52						0.0000	1.3298	LINGLESTOWN	6.0961	0.0006	180	D 94
188.60	437			437	88	0.0009	1.3307 T		6.0958	0.0037	1614	94
188.84	1257		94	1360	88	0.0029	1.3336 L		6.0921	0.0026	1257	94
188.84				1734	A 88	0.0033	1.3370		6.0896	0.0000		
189.39	2887			8	108	0.0000	1.3370 T		6.0896	0.0038	1360	67
189.39		EAST END		1052	D 10	0.0023	1.3393		6.0858	0.0000		
189.43	219		67	219	67	0.0006	1.3399 L	ROCKVILLE	6.0858	0.0006	219	67
189.53	547		67	547	67	0.0015	1.3414 L		6.0852	0.0015	547	67
189.58	218		67	218	67	0.0006	1.3420 L		6.0837	0.0006	218	67
189.58				1360	67	0.0038	1.3459		6.0831	0.0038	2015	D 13
190.29	3756					0.0000	1.3459 T		6.0793	0.0024	1741	135
190.29				6505	A 67	0.0117	1.3576	SUSQUEHANNA RIVER	6.0769	0.0000		
190.46	925		135			0.0000	1.3576 R		6.0769	0.0013	925	135
190.53	381					0.0000	1.3576	MARYSVILLE	6.0756	0.0005	381	135
190.71	916					0.0000	1.3576 T		6.0751	0.0013	916	135
191.26	2909		135	1022	135	0.0014	1.3591 R		6.0738	0.0023	1672	135
191.53	1465			1360	135	0.0019	1.3610 T		6.0715	0.0085	4129	A 10
191.53				1427	A 13	0.0019	1.3628		6.0650			
191.53				22	143	0.0000	1.3629		6.0650	0.0058	3139	103
192.06	2797			1443	D 14	0.0022	1.3652 T		6.0592	0.0014	727	A 95
192.55	2581		103	2581	103	0.0047	1.3699 L		6.0578	0.0001	75	95
192.62	335		103	1053	103	0.0019	1.3718 L		6.0577	0.0007	335	95
192.80	960			232	D 10	0.0004	1.3723 T		6.0570	0.0013	854	95
192.87	392		95	391	95	0.0008	1.3731 R		6.0557	0.0001	49	A 94
193.00	671			671	95	0.0013	1.3744 T		6.0556	0.0013	660	94
193.01	70			43	95	0.0001	1.3745		6.0543	0.0014	700	94
193.01		XING 114.2		28	D 9	0.0001	1.3745	PRIVATE	6.0529			
193.24	1200		94	1200	94	0.0024	1.3770 L		6.0529	0.0024	1200	94
193.24							1.3770		6.0505	0.0032	2046	D 15
193.50	1370			1370	94	0.0028	1.3797		6.0473	0.0001	85	151
194.05	2900			2900	94	0.0058	1.3856		6.0472	0.0000		
194.27	1176		179	5425	A 94	0.0088	1.3943 R		6.0472	0.0000		
194.50	1200	XING 115.3		66	138	0.0001	1.3944	PRIVATE	6.0472	0.0117	7653	A 98
195.35	4500	XING 115.9		1385	D 1	0.0022	1.3966	PRIVATE	6.0355	0.0026	1360	98
195.50	791		96	791	98	0.0015	1.3982 R		6.0329	0.0015	791	98
195.70	1084		98	1084	98	0.0021	1.4003 R		6.0314	0.0021	1084	98
195.70									6.0293	0.0024	1507	D 14
196.27	3000			1360	98	0.0026	1.4029		6.0269	0.0003	191	141
196.50	1200		147	5211	A 98	0.0084	1.4113 R		6.0266	0.0073	4507	A 10
197.07	3000		147	39	139	0.0001	1.4113 R		6.0193	0.0002	137	107
197.19	665			1155	D 13	0.0018	1.4131		6.0191	0.0000		
197.35	800		107	800	107	0.0014	1.4145 R		6.0191	0.0000		
197.50	800		138	800	107	0.0014	1.4160 R		6.0191	0.0078	3363	A 50
197.71	1100		110	587	107	0.0010	1.4170 R		6.0113	0.0021	580	50
197.86	800			0	107	0.0000	1.4170		6.0092	0.0030	800	50
197.86				1313	D 10	0.0032	1.4202	-1	6.0062	0.0000		
198.06	1100		50	1100	50	0.0042	1.4243 L		6.0062	0.0042	1100	50
198.06							1.4243		6.0020	0.0002	61	D 54
198.12	270			270	50	0.0010	1.4253	TRC TT MP106	6.0016	0.0007	209	54
198.27	800		54	800	50	0.0030	1.4284 R		6.0011	0.0028	800	54
198.27				290	50	0.0011	1.4295		5.9983	0.0038	1604	D 12
198.48	1100			114	A 50	0.0004	1.4299		5.9944	0.0002	134	125
198.49	100			100	54	0.0004	1.4302		5.9942	0.0000	0	125
199.27	4096		125	1260	54	0.0044	1.4347 R	DUNCANNON	5.9942	0.0060	3300	125
199.27						0.0000	1.4347		5.9882	0.0009	624	D 14
199.37	500			5466	A 54	0.0110	1.4457		5.9880	0.0012	618	141
199.82	2402			2590	A 12	0.0037	1.4494		5.9871	0.0018	1380	141
200.39	3000		141	1380	141	0.0018	1.4512 L		5.9853	0.0040	3000	141

READ DOWN WESTBOUND TRAFFIC — APPENDIX B3 STATION — READ UP EASTBOUND TRAFFIC

PAGE 4 COLUMN 1 — PAGE 4 COLUMN 2

MILES	DISTANCE FT	HWY GRADE CROSSING	MAX MPH	DISTANCE FT	SPEED MPH	ET HOURS	TIME HOURS	STATION	TIME HOURS	ET HOURS	DISTANCE FT	SPEED MPH	MILES	DISTANCE FT	HWY GRADE CROSSING	MAX MPH	DISTANCE FT	SPEED MPH	ET HOURS	TIME HOURS	STATION	TIME HOURS	ET HOURS	DISTANCE FT	SPEED MPH
200.64	1300		141	1300	141	0.0017	1.4530	L	5.9815	0.0017	1300	141	233.36	1300		12.	198	A 81	0.0013	1.7251	L MP148	5.7039	0.0023	1074	80
200.99	1680			1386	141	0.0018	1.4546		5.9798	0.0013	1077	D 166	233.45	500					0.0000	1.7251	TRC	5.7016	0.0011	500	80
201.24	1300		165	0		0.0000	1.4548	R	5.9785	0.0024	2100	166	233.77	1680		60	1524	90	0.0030	1.7283	R	5.7006	0.0035	1680	90
201.39	800			5230	A 14	0.0046	1.4613		5.9759	0.0005	296	D 171	233.77							1.7283		5.6970	0.0016	820	D 117
201.54	800			1602	A 16	0.0018	1.4631		5.9756	0.0014	1304	171	233.90	700			700	90	0.0015	1.7297		5.6966	0.0006	20	117
201.80	1886	XING 1	171			0.0000	1.4631	L AQUEDUCT	5.9742	0.0021	1886	171	234.11	1100		117	660	90	0.0014	1.7311	L	5.6955	0.0006	400	117
201.90							1.4631		5.9721	0.0034	358	D 176	234.75	3285		117	2695	A 80	0.0046	1.7358	L	5.6949	0.0053	3295	117
202.39	2600	XING 123.3		1084	171	0.0017	1.4643	PRIVATE	5.9717	0.0000	46	176	234.96	1200			2298	117	0.0037	1.7397		5.6896	0.0019	1200	117
203.38	5285	XING 129.6		8023	A 17	0.0083	1.4725	PRIVATE	5.9717	0.0160	12800	A 118	235.11	800		117	800	117	0.0013	1.7409	R	5.6877	0.0022	1360	117
204.39	5279					0.0000	1.4725		5.9557	0.0003	160	118	235.26	800		117	800	117	0.0013	1.7422	R	5.6855	0.0013	800	117
204.62	1200			80	198	0.0001	1.4728		5.9554	0.0020	1200	116	235.47	1100			1100	117	0.0018	1.7440		5.6842	0.0018	1100	117
204.62				3661	D 15	0.0044	1.4771		5.9534	0.0000			235.61	700		156	700	117	0.0011	1.7452	R	5.6824	0.0017	1072	117
205.01	2080	XING 1	115	2083	116	0.0034	1.4805	L PRIVATE	5.9534	0.0034	2083	116	236.12	2685			2685	117	0.0044	1.7495		5.6807	0.0053	2863	A 87
205.36	1800			1800	116	0.0028	1.4834		5.9500	0.0029	1800	116	236.46	1900			1002	117	0.0016	1.7511	T DENHOLM	5.6754	0.0030	1380	87
205.39	182			182	116	0.0003	1.4837		5.9471	0.0003	160	116	236.48				888	D 11	0.0017	1.7529		5.6724			
205.96	3000		116	3000	116	0.0049	1.4886	L	5.9469	0.0067	4114	116	236.84	1900		87	1900	87	0.0041	1.7569	L	5.6724	0.0041	1900	87
206.19	1200			1200	116	0.0020	1.4906		5.9401	0.0006	378	A 113	236.84							1.7569		5.6683	0.0038	2459	D 15
206.36	900		127	1951	116	0.0025	1.4931	R	5.9396	0.0010	608	113	236.97	700			700	87	0.0015	1.7585		5.6645	0.0027	2243	156
206.50	762		160	101	D 116	0.0002	1.4933	R	5.9385	0.0011	762	113	237.12	781		156	680	87	0.0014	1.7599	L	5.6618	0.0000		
206.71	1100		113	1100	113	0.0018	1.4951	R	5.9372	0.0018	1100	113	237.23	581		156	8671	A 87	0.0145	1.7744	L	5.6618	0.0000		
206.71		XING 127.30					1.4951	PRIVATE	5.9354	0.0031	183	D 132	237.98	4000					0.0000	1.7744		5.6616	0.0017	1360	156
206.93	1200		155	1300	113	0.0023	1.4974	R	5.9343	0.0002	172	132	238.12	700		156			0.0000	1.7744	R	5.6601	0.0008	700	156
207.16	1200	XING 1	120	2616	A 15	0.0040	1.5014	R PRIVATE	5.9341	0.0038	2485	A 114	238.16	238		156			0.0000	1.7744	R	5.6593	0.0003	285	156
207.39	1200			81	132	0.0001	1.5015		5.9305	0.0024	1544	114	238.52	1800					0.0000	1.7744		5.6590	0.0002	185	D 16
207.60	1100			649	D 12	0.0010	1.5025		5.9277	0.0024	1349	A 101	238.75	1200		180			0.0000	1.7744	L	5.6588	0.0035	2815	160
207.90	1600	XING 1	114	1600	114	0.0027	1.5052	L PRIVATE	5.9253	0.0001	73	101	239.11	1800			1886	156	0.0013	1.7757	CP HAWSTONE	5.6552	0.0005	434	D 16
208.14	1287			877	114	0.0015	1.5066		5.9252	0.0024	1287	101	239.11						0.0000	1.7757	160	5.6546	0.0000	26	160
208.14				410	D 11	0.0007	1.5074		5.9228	0.0000			239.26	800			800	156	0.0010	1.7767	T 291.6	5.6548	0.0026	2292	A 16
208.39	1300		101	1300	101	0.0024	1.5098	L	5.9228	0.0024	1300	101	239.45	1014		109	1014	156	0.0012	1.7779	L	5.6522	0.0011	927	A 15
208.57	930		101	930	101	0.0017	1.5115	L	5.9204	0.0017	930	101	240.11	3600			3600	156	0.0042	1.7822		5.6511	0.0000		75
208.57				1080	101	0.0026	1.5141		5.9187	0.0020	1257	D 137	240.27	800			800	156	0.0010	1.7831		5.6511	0.0000		75
208.57				3683	A 10	0.0060	1.5201		5.9187	0.0002	154	137	240.54	1471		190	1471	156	0.0018	1.7849	L	5.6511	0.0000		75
209.40	4400		160			0.0000	1.5201	L	5.9168	0.0044	2840	A 117	240.67	1700			1700	156	0.0021	1.7870		5.6511	0.0000		75
209.40				181	131	0.0002	1.5203	MP 130	5.9121	0.0041	2582	117	241.13	1400		156		156	0.0000	1.7870	R	5.6511	0.0140	8344	A 90
209.65	1313			508	D 11	0.0008	1.5211	T 201.6	5.9080	0.0022	1264	A 106	241.32	1000		156		156	0.0000	1.7870	R	5.6371	0.0051	1473	90
210.37	3800		117	3440	117	0.0056	1.5266	R	5.9058	0.0024	1360	106	241.47	800			819	156	0.0010	1.7880		5.6340	0.0016	713	A 79
210.37				360	D 11	0.0006	1.5273		5.9034	0.0000			241.47				2381	D 15	0.0027	1.7918		5.6324			
210.97	3250		106	3250	106	0.0057	1.5330	R	5.9034	0.0027	3200	106	241.77	1586		96	1086	90	0.0033	1.7950	L	5.6324	0.0018	780	79
210.97						0.0000	1.5330		5.8977	0.0030	2102	D 160	241.89	800			327	90	0.0007	1.7957	REVERSE CURVE	5.6306	0.0014	400	79
211.17	1024			1024	106	0.0018	1.5348		5.8947	0.0000	17	160	241.89				273	D 90	0.0006	1.7963		5.6292			
211.34	900		160	336	106	0.0006	1.5354	L	5.8947	0.0000	6	160	242.08	1050		79	1000	79	0.0024	1.7987	R	5.6292	0.0024	1000	79
211.43	500		160			0.0000	1.5354	L	5.8947	0.0004	300	160	242.23	800		79	800	79	0.0019	1.8005	R	5.6268	0.0019	800	79
211.70	1400			3903	A 10	0.0125	1.5479	T NEWPORT	5.8943	0.0017	1400	160	242.63	2114			1390	79	0.0033	1.8038	SHAWNEE	5.6249	0.0011	522	D 89
212.04	1800	XING 1	160			0.0000	1.5479	R PRIVATE	5.8926	0.0021	1800	160	243.06	2300		96	1442	A 79	0.0031	1.8069	R	5.6238	0.0060	4338	99
212.23	1000					0.0000	1.5479		5.8905	0.0017	1000	160	243.20	700		99	2512	99	0.0044	1.8113	R	5.6156	0.0010	494	A 93
212.30	373	XING 1	160			0.0000	1.5479	R PRIVATE	5.8885	0.0004	373	160	243.20				1431	99	0.0027	1.8141		5.6145			
212.77	2478		160	572	160	0.0007	1.5486	R	5.8889	0.0029	2478	160	243.50	1400			169	D 96	0.0003	1.8144		5.6145	0.0028	1360	93
213.30	2800					0.0000	1.5486		5.8860	0.0008	737	D 150	243.74	1791		93	1291	93	0.0026	1.8170	L	5.6117	0.0036	1732	93
213.45	825			3686	A 15	0.0045	1.5531		5.8852	0.0011	1057	175	243.81	900			900	93	0.0018	1.8188		5.6082	0.0000		
214.29	4400		175	2889	175	0.0029	1.5560	L	5.8841	0.0000			244.07	800		96	460	93	0.0009	1.8196	R	5.6082	0.0051	1806	A 54
214.89	3175			1380	175	0.0015	1.5575		5.8841	0.0000			244.22	800		98	787	A 93	0.0015	1.8213	R	5.6031	0.0002	60	54
214.89				1431	A 17	0.0015	1.5590		5.8841	0.0126	10147	A 128	244.39	900		102	590	102	0.0011	1.8224	R	5.6029	0.0032	830	54
215.29	2100		384	136	180	0.0001	1.5591	L	5.8715	0.0028	1921	128	244.46	400				102		1.8224	TRC MP[illegible]	5.5997	0.0014	400	54
215.29				2349	D 18	0.0029	1.5620		5.8687	0.0004	279	A 126	244.46				1098	D 10	0.0027	1.8250		5.5983			
215.63	1800		128	1800	128	0.0027	1.5647	L	5.8683	0.0014	986	126	244.67	1100		54	1100	54	0.0039	1.8289	L	5.5983	0.0039	1100	54
215.70	400			400	128	0.0006	1.5653	T RC MP(ISETT)	5.8669	0.0008	400	126	244.67							1.8289		5.5944	0.0014	486	D 75
215.84	719		128	719	128	0.0011	1.5663	R	5.8663	0.0016	1037	126	244.87	1054			1064	54	0.0037	1.8326		5.5930	0.0014	536	79
216.28	2300			1380	128	0.0020	1.5683		5.8647	0.0000		91	245.06	1000		74	309	54	0.0011	1.8337	L	5.5916	0.0024	1000	79
216.41	3000			4986	A 17	0.0067	1.5750		5.8647	0.0000		91	245.34	1500		79	1051	A 54	0.0027	1.8364	L LEWISTOWN	5.5892	0.0036	1500	79
216.94	713		155			0.0000	1.5750	R	5.8647	0.0000		91	245.34				1143	79	0.0027	1.8391		5.5856	0.0036	2126	D 14
217.28	1800			1257	155	0.0015	1.5765	T MILLERSTOWN	5.8647	0.0057	3363	A 96.4	245.95	3200			1360	79	0.0033	1.8424		5.5820	0.0014	1574	144
217.74	3414		283	242	155	0.0003	1.5768	L	5.8590	0.0027	1382	96	246.06	567		144	7580	A 79	0.0125	1.8549	R	5.5806	0.0007	567	144
217.74				2172	D 15	0.0033	1.5801		5.8563				246.99	4607		144			0.0000	1.8549	R	5.5799	0.0065	4607	144
218.27	2800		96	2800	96	0.0055	1.5856	L	5.8580	0.0154	7780	96	247.08	900		144	228	144	0.0028	[illegible]	R	5.5734	0.0007	500	144
218.63	1900			1900	96	0.0027	1.5883		5.8408	0.0030	1946	D 150	247.52	2300			1380	144	0.0018	1.8570		5.5727	0.0013	1049	D 16
218.80	902	XING 1	156	6288	A 96	0.0099	1.5983	R PRIVATE	5.8379	0.0001	77	180	247.68	872		283	4414	A 14	0.0054	1.8624	R	5.5714	0.0002	200	167
219.05	1300					0.0000	1.5983		5.8378	0.0000			247.99	1800			168	164	0.0001	1.8625		5.5712	0.0000		
219.27	1200		157			0.0000	1.5983	R	5.8378	0.0000			248.23	1300	XING 168.4		190	D 16	0.0002	1.8627	GRANVILLE	5.5712	0.0000		
219.29	100		156	18	144	0.0000	1.5983	[illegible]	5.8378	0.0076	5457	A 117	248.39	812		160	810	160	0.0010	1.8637	L	5.5712	0.0000	1736	A 16
219.94	3400			1034	D 14	0.0018	1.5986		5.8302	0.0022	1386	117	248.74	1868		160	1868	160	0.0022	1.8659		5.5692			
220.24	1620		117	1620	117	0.0028	1.6074	L DURWARD CURVE	5.8080	0.0035	2160	117	248.99	1300			1300	160	0.0015	1.8674		5.5692	0.0150	10205	A 91
220.29	246		117	246	117	0.0004	1.6028	L	5.8285	0.0042	2332	A 94.1	248.99				60	160	0.0001	1.8675		5.5542			
220.86	3000			1360	117	0.0022	1.6050		5.8203	0.0022	1017	A 79.6	248.99				1617	A 16	0.0003	1.8678		5.5542			
220.89	700		165	1025	A 11	0.0025	1.6075	R	5.8161	0.0001	60	79	249.64	4500	XING 170.6		726	161	0.0003	1.8681	PRIVATE	5.5542	0.0028	1380	91
220.99				70	129	0.0001	1.6076		5.8180				249.84		XING 170.8		2687	D 16	0.0038	1.8719	ANDERSON	5.5514			
221.03	500			1145	D 12	0.0019	1.6096		5.8180	0.0012	500	79	249.95	771		91	721	91	0.0016	1.8735	L	5.5514	0.0016	771	91
221.24	600		94	415	94	0.0006	1.6104	R	5.8168	0.0019	800	79	250.34	1800		91	1900	91	0.0040	1.8775	L	5.5498	0.0040	1900	91
221.24				381	D 94	0.0008	1.6112	REVERSE CURVE	5.8149				250.34				560	91	0.0014	1.8789		5.5458	0.0007	377	D 10
221.45	1100		79	1100	79	0.0026	1.6138	L	5.8149	0.0026	1100	79	250.45	700			700	91	0.0015	1.8803		5.5451	0.0006	326	104
221.45							1.6138		5.8123	0.0002	71	D 82	250.72	1300		104	1110	A 86	0.0022	1.8825	R	5.5445	0.0024	1300	104
221.56	800			800	79	0.0014	1.6153	REVERSE CURVE	5.8121	0.0012	529	82	250.72				593	104	0.0016	1.8841		5.5421	0.0013	779	D 12
221.75	1000		101	760	79	0.0018	1.6171	R THOMPSONTOWN	5.8108	0.0023	1000	82	250.85	1200		177	2804	A 10	0.0046	1.8887	R	5.5408	0.0008	421	127
221.80	282			177	A 79	0.0004	1.6175	T RC MP 155	5.8086	0.0007	282	82	250.98	179	XING 172.20				0.0000	1.8887	PRIVATE	5.5402	0.0002	118	D 13
222.01	1100		82	1445	82	0.0035	1.6209	L	5.8078	0.0025	1100	82	251.51	2274	XING 173.3		1500	A 12	0.0023	1.8910	PRIVATE	5.5400	0.0032	2166	130
222.01							1.6209		5.8054	0.0046	2950	D 164	251.86	2500		156	146	137	0.0002	1.8912	R	5.5368	0.0028	1515	A 11
222.24	1200			1206	82	0.0028	1.6237		5.8008	0.0002	186	164	251.98				745	D 13	0.0011	1.8923		5.5340	0.0051	3141	117
222.27	184			160	82	0.0004	1.6240		5.8006	0.0000			252.20	1174		117	1174	117	0.0019	1.8942	R	5.5289	0.0014	830	A 11
222.48	1100		160	8810	A 82	0.0138	1.6378	R	5.8006	0.0000			252.20				1360	117	0.0022	1.8964		5.5275			
223.24	4000			187	161	0.0002	1.6382		5.8006	0.0086	6606	A 131	252.20				828	A 11	0.0016	1.8976		5.5275			
224.13	4700			827	D 15	0.0011	1.6393		5.7921	0.0020	1360	131	252.73	2406			31	124	0.0000	1.8979		5.5275	0.0023	1360	110
224.24	596	XING 1	131	596	131	0.0006	1.6401	R PRIVATE	5.7901	0.0009	586	131	252.73				480	D 11	0.0008	1.8987		5.5252			
224.34	496		131	496	131	0.0007	1.6408	R	5.7892	0.0007	496	131	252.96	1300		116	1300	110	0.0022	1.9009	R	5.5252	0.0022	1300	110
225.24	4800			1360	131	0.0025	1.6425		5.7885	0.0017	1330	D 162	252.98		XING 174.6		1360	110	0.0023	1.9032	HORNINGFORD	5.5230	0.0019	1267	D 14
225.66	2340			5529	A 13	0.0106	1.6534		5.7868	0.0003	292	102	253.39	2166		286	5064	A 11	0.0076	1.9108	R	5.5211	0.0004	263	144
226.26	2000		199	0		0.0000	1.6534	L	5.7866	0.0000			253.98	3100	XING 175.3		56	144	0.0001	1.9109	PRIVATE	5.5207	0.0069	4732	A 11
226.27	100		199	0		0.0000	1.6534	L	5.7865	0.0138	9756	A 102	254.43	2579			1168	D 14	0.0017	1.9126		5.5138	0.0023	1360	113
226.75	2500			96	171	0.0001	1.6535		5.7727	0.0025	1380	102	254.98	2900	XING 1	113	2900	113	0.0048	1.9174	L SWIGARES	5.5115	0.0048	2900	113
226.75				2760	D 17	0.0038	1.6574		5.7702				255.98	5286	XING 177.1		1360	113	0.0023	1.9197	PRIVATE	5.5098	0.0019	1297	D 14
227.22	2500		102	2500	102	0.0046	1.6620	R TUSCARORA	5.7702	0.0046	2500	102	256.02	161	XING 177.6		6408	A 11	0.0076	1.9276	STATE HWY 103 McVEY TO	5.5047	0.0003	200	147
227.31	466		102	466	102	0.0009	1.6629	R	5.7656	0.0009	466	102	256.53	2700		165	0	148	0.0000	1.9276	L	5.5044	0.0077	5524	A 11
227.31		XING 148.6					1.6629	PRIVATE	5.7647	0.0009	501	D 119	256.53				1340	D 14	0.0019	1.9295		5.4967	0.0023	1360	113
227.31				1380	102	0.0025	1.6654		5.7638	0.0001	49	119	256.96	2300		113	2300	113	0.0038	1.9334	L MATAWANNA	5.4944	0.0063	8735	113
227.80	2600			1909	A 10	0.0033	1.6687		5.7637	0.0032	2000	119	256.98	100			150	113	0.0002	1.9336		5.4881	0.0003	260	A 11
228.22	2200		119	1531	119	0.0024	1.6711	L	5.7605	0.0035	2200	119	256.98				1360	113	0.0021	1.9357		5.4878			
228.62	2100		119	2100	119	0.0033	1.6744	L	5.7570	0.0033	2100	119	256.98				1100	A 11	0.0018	1.9375		5.4878			
228.87	1350			1350	119	0.0021	1.6766		5.7537	0.0007	479	D 164	257.53	2900			144	122	0.0002	1.9377		5.4878	0.0023	1380	111
229.14	1400		132	10	119	0.0000	1.6766	R	5.7530	0.0001	58	132	257.53				375	D 12	0.0006	1.9383		5.4855			
229.42	1500			3896	A 11	0.0049	1.6815	180	5.7529	0.0014	1000	132	257.85	1700		111	1700	111	0.0029	1.9412	R	5.4855	0.0029	1700	111
229.42	6		132	157	141	0.0002	1.6817	R PORT ROYAL	5.7515	0.0051	3265	A 107	257.98	680			680	111	0.0012	1.9424		5.4826	0.0019	1319	D 14
229.78	1900			1237	D 14	0.0019	1.6836		5.7464	0.0024	1360	107	258.95	5100			680	111	0.0012	1.9436		5.4807	0.0006	17	146
230.12	1767		107	1767	107	0.0031	1.6868	R	5.7440	0.0125	6772	107	258.95				3824	A 11	0.0056	1.9492		5.4807			
230.17	270		107	270	107	0.0005	1.6872	R	5.7320	0.0022	1181	A 94.4	258.95				78	138	0.0001	1.9494		5.4807			
231.11	5090			1360	107	0.0024	1.6896		5.7298	0.0030	1224	A 73.4	258.95				508	D 12	0.0008	1.9502		5.4807			
231.11				4807	A 10	0.0073	1.6969		5.7286				258.98	148	XING 1	124	149	124	0.0002	1.9504	R PRIVATE	5.4807	0.0076	5232	A 11
231.64	2800			13	141	0.0000	1.6970		5.7256	0.0015	500	73	258.98					124	0.0000	1.9504	RYDE MP 180	5.4731			
231.64		XING 152.8		1820	D 14	0.0026	1.6996	PRIVATE	5.7253				259.13	800		124	800	124	0.0012	1.9516	R	5.4731	0.0003	180	112
231.80	800		94	286	94	0.0006	1.7001	L RC	5.7253	0.0021	800	73	259.20	400			400	124	0.0006	1.9523		5.4728	0.0007	400	112
231.80				514	D 94	0.0012	1.7013		5.7252				259.36	800		156	395	124	0.0006	1.9528	R	5.4721	0.0014	800	112
232.12	1731		73	1731	73	0.0045	1.7058	R T1 MP 159	5.7232	0.0042	1814	73	259.36				410	D 12	0.0007	1.9535		5.4707			
232.20	400			500	73	0.0010	1.7068		5.7190	0.0016	557	A 56.1	259.49	711		112	711	112	0.0012	1.9547	R	5.4707	0.0012	711	112
232.46	1400		73	1129	73	0.0029	1.7098	R RC	5.7174	0.0044	1360	56	259.49									5.4695	0.0017	1159	D 14
232.46				271	D 73	0.0008	1.7105		5.7130				259.98	2500			1360	112	0.0023	1.9570		5.4678	0.0003	227	143
232.58	600		59	600	59	0.0019	1.7125	L MP 160	5.7130	0.0019	600	59	260.97	5222			8434	A 11	0.0092	1.9662	T VINEYARD	5.4675	0.0147	8376	A 58
232.58		XING 153.9		1360	59	0.0044	1.7168	JUNIATA STREET	5.7111	0.0012	462	D 76	261.58	3200			353	152	0.0004	1.9666	TRC & CURVE	5.4528	0.0044	1380	58
233.00	2210			901	A 56	0.0028	1.7196	T MIFFLIN	5.7099	0.0041	1758	61	261.58				2885	D 15	0.0052	1.9718		5.4484			
233.11	600		31	489	31	0.0011	1.7206	L	5.7058	0.0014	600	61	261.70	1137		58	1137	58	0.0037	1.9755	R RC	5.4484	0.0037	1137	56
233.11				1380	31	0.0032	1.7238		5.7044	0.0005	226	D 304	261.79				1380	56	0.0044	1.9800		5.4447			56

APPENDIX B3

PAGE 5 COLUMN 1

MILES	DISTANCE, ft	HWY GRADE CROSSING	MAX MPH	Δ DISTANCE, ft	SPEED, mph	ΔT, hours	TIME, hours	STATION	TIME, hours	ΔT, hours	Δ DISTANCE, ft	SPEED, mph
				READ DOWN WESTBOUND TRAFFIC					READ UP EASTBOUND TRAFFIC			
261.79				43	A 56	0.0001	1.9801		5.4447	0.0001	17	D 59
261.96	900		59	0	59	0.0000	1.9801	L	5.4446	0.0028	863	59
262.14	917		59	414	59	0.0013	1.9814	L	5.4418	0.0029	917	59
262.14						0.0000	1.9814		5.4389	0.0032	1497	D 117
262.50	1900		165	1360	59	0.0044	1.9858	L	5.4357	0.0000	3	117
262.69	1000			4213	A 59	0.0068	1.9846		5.4357	0.0023	1400	117
262.82	700		117			0.0000	1.9846	R	5.4334	0.0011	700	117
262.82							1.9946		5.4323	0.0010	665	D 136
262.97	800		135	2580	A 11	0.0039	1.9984	R	5.4313	0.0002	135	136
263.07	500		165	1380	135	0.0019	2.0004	R	5.4311	0.0031	2678	D 191
263.84	4065			10800	A 13	0.0128	2.0132		5.4280	0.0002	158	191
263.97	705		286			0.0000	2.0132	R	5.4278	0.0000		
264.52	2874		286			0.0000	2.0132	R	5.4278	0.0000		
264.97	2400		385			0.0000	2.0132	R	5.4278	0.0000		
265.40	2260		385			0.0000	2.0132	R	5.4278	0.0000		
265.97	3000			81	161	0.0001	2.0133		5.4278	0.0163	14143	A 135
266.41	2300			2132	D 18	0.0026	2.0158		5.4115	0.0019	1360	135
266.75	1834		135	1834	135	0.0026	2.0184	L'NEWTON HAMILTON	5.4096	0.0023	1634	135
266.94	1000			1000	135	0.0014	2.0198		5.4073	0.0007	466	D 147
267.02	402			380	135	0.0005	2.0203		5.4066	0.0012	906	147
267.87	4500		147	2225	A 13	0.0030	2.0233	R	5.4054	0.0056	4500	147
268.38	2700		147	5017	147	0.0065	2.0298	R	5.3998	0.0018	1377	147
268.87	2591			1380	147	0.0018	2.0315		5.3978	0.0036	2225	A 135
268.87				3289	A 14	0.0040	2.0356		5.3948			
268.87				93	167	0.0001	2.0356		5.3948	0.0056	3626	A 106
269.50	3297			1176	D 18	0.0015	2.0371		5.3892	0.0024	1360	108
269.88	2000		135	1098	135	0.0015	2.0386	L	5.3868	0.0035	2000	108
269.88		XING 191.6		962	D 12	0.0015	2.0401	PRIVATE	5.3833			
270.09	1100		108	1100	108	0.0019	2.0420	L'MT UNION	5.3833	0.0019	1100	108
270.18	477			477	108	0.0008	2.0428	REVERSE CURVE	5.3814	0.0008	472	D 122
270.18				885	108	0.0015	2.0444		5.3806	0.0000	5	122
270.67	2600		122	1647	A 10	0.0027	2.0471	R	5.3806	0.0040	2600	122
270.97	1600		122	1670	122	0.0026	2.0497	R	5.3766	0.0009	584	122
271.29	1682			1360	122	0.0021	2.0518		5.3757	0.0069	3786	A 82
271.29				870	A 12	0.0013	2.0531		5.3688			
271.67	2000		135	185	128	0.0003	2.0534	L	5.3688	0.0021	912	82
271.67				1297	D 12	0.0022	2.0556		5.3667			
271.73	300		86	288	88	0.0006	2.0563	L	5.3667	0.0007	300	82
271.75	148				88	0.0002	2.0563	T'RC	5.3660	0.0003	148	82
271.75				190	D 58	0.0003	2.0566	MAPLETON	5.3657			
272.66	4800		82	4800	82	0.0111	2.0677	R'MPH60	5.3657	0.0111	4800	82
272.89	1200		82	1200	82	0.0028	2.0705	R	5.3546	0.0028	1200	82
272.89							2.0705		5.3518	0.0022	1197	D 122
272.89							2.0705		5.3496	0.0002	100	122
273.33	2300			1360	82	0.0031	2.0736		5.3494	0.0018	1120	A 113
273.61	1477		113	2696	A 82	0.0051	2.0787	R	5.3476	0.0023	1360	113
273.66	300			51	113	0.0001	2.0788		5.3453	0.0031	2267	D 168
274.04	2000			1060	113	0.0018	2.0806		5.3422	0.0001	117	168
274.66	3246		165	6311	A 11	0.0090	2.0896	L	5.3421	0.0000		
274.86	1086		165	171	162	0.0002	2.0898	L	5.3421	0.0069	5660	A 143
275.16	1600			389	D 15	0.0005	2.0903		5.3352	0.0019	1427	A 135
275.43	1400		143	1074	143	0.0014	2.0917	L	5.3333	0.0002	160	135
275.66	1200		147	326	D 14	0.0004	2.0921	L'MILL CREEK	5.3331	0.0017	1200	135
276.18	2790	XING 1	135	2790	135	0.0039	2.0960	L CASSVILLE	5.3314	0.0039	2790	135
276.18							2.0960		5.3275	0.0016	1224	D 140
276.18							2.0960		5.3259	0.0000	16	163
276.66	2500			1360	135	0.0019	2.0980		5.3259	0.0074	5812	A 135
276.69	178			7583	A 13	0.0094	2.1073		5.3185	0.0189	1527	A 125
277.35	3500		233			0.0000	2.1073	R	5.2996	0.0000		86
277.66	1600			271	170	0.0003	2.1076		5.2996	0.0000		86
278.17	2700			2464	D 17	0.0033	2.1110		5.2996	0.0031	1896	A 110
278.39	1179	XING 1	110	1179	110	0.0020	2.1130	R JACOBS	5.2965	0.0081	4713	110
278.66	1400			1360	110	0.0023	2.1153		5.2884	0.0034	1803	A 90
278.66				3575	A 11	0.0056	2.1206	200	5.2850			
279.66	5297			237	136	0.0003	2.1212		5.2850	0.0029	1360	90
279.66				1526	D 13	0.0026	2.1237		5.2821			
279.91	1300		90	1300	90	0.0027	2.1264	L	5.2821	0.0042	1985	90
279.95	243			243	90	0.0005	2.1270	REVERSE CURVE	5.2779	0.0005	290	A 86
280.35	2100		100	1997	90	0.0042	2.1312	R	5.2775	0.0030	1360	86
280.35				105	D 90	0.0002	2.1314	REVERSE CURVE	5.2743			
280.64	1500		86	1500	86	0.0035	2.1347	L	5.2743	0.0035	1500	86
280.82	1000	XING 2	86	1000	86	0.0022	2.1369	L'MPH 55 4TH STREET	5.2710	0.0022	1000	86
280.98	800			800	86	0.0018	2.1386	T'HUNTINGDON	5.2688	0.0013	690	D 110
281.17	1000		199	560	86	0.0012	2.1399	R	5.2675	0.0019	1110	110
281.30	700			2083	A 86	0.0040	2.1439		5.2656	0.0012	700	110
281.58	1500		110	547	110	0.0008	2.1449	R	5.2644	0.0048	2783	110
281.63	739			236	110	0.0004	2.1453		5.2596	0.0040	2093	A 86
282.23	3197			1121	110	0.0019	2.1472		5.2556	0.0001	60	86
282.23				2038	A 11	0.0033	2.1505		5.2555			
282.48	1300		177	94	126	0.0001	2.1506	R	5.2555	0.0029	1300	86
282.48				1244	D 12	0.0022	2.1528		5.2526			
282.63	800		86	800	86	0.0018	2.1546	R	5.2526	0.0018	800	86
282.78	800		86	800	86	0.0018	2.1563	R	5.2508	0.0018	800	86
282.84	300			300	86	0.0007	2.1570	T'DEER	5.2490	0.0030	1790	D 140
283.03	1000		218	1000	86	0.0022	2.1592	L'RC MPH 55	5.2460	0.0002	158	140
283.20	900		177	60	86	0.0001	2.1593	L	5.2458	0.0000		
283.50	1600			6948	A 86	0.0113	2.1706		5.2458	0.0000		
283.63	670		161			0.0000	2.1706	R	5.2458	0.0000		
283.68	277		161			0.0000	2.1706	R	5.2458	0.0106	6439	A 86
284.63	5000					0.0000	2.1706		5.2352	0.0030	1360	86
284.74	600			101	143	0.0001	2.1707		5.2322	0.0013	600	86
284.74				1940	D 14	0.0032	2.1739		5.2309	0.0000		
284.95	1100		85	1100	86	0.0025	2.1764	L	5.2309	0.0039	1757	86
285.14	977		113	977	86	0.0022	2.1786	L	5.2270	0.0008	360	A 79
285.40	1400	XING 208.6		1256	86	0.0028	2.1814	T'WARRIOR RIDGE	5.2262	0.0033	1360	79
285.40				144	D 86	0.0003	2.1817		5.2229	0.0000		
285.63	1200		79	1200	79	0.0029	2.1846	R	5.2229	0.0029	1200	79
285.87	1300		79	1300	79	0.0031	2.1877	R	5.2200	0.0031	1300	79
285.87							2.1877		5.2169	0.0005	220	D 88
285.89	87			87	79	0.0002	2.1879	REVERSE CURVE	5.2164	0.0000	1	88
286.12	1200		86	1200	79	0.0029	2.1908	L	5.2164	0.0023	1068	88
286.12				73	79	0.0002	2.1910		5.2141	0.0006	302	D 94
286.34	1200		99	582	A 79	0.0013	2.1923	L	5.2135	0.0017	898	99
286.34				545	88	0.0012	2.1935		5.2118	0.0014	818	D 99
286.63	1500			180	88	0.0003	2.1938		5.2104	0.0002	123	124
286.85	1190			860	A 88	0.0017	2.1955		5.2102	0.0040	2369	A 104
286.85				340	98	0.0007	2.1962		5.2062	0.0000		
287.23	2000		153	2655	A 98	0.0045	2.2007	L PETERSBURG	5.2062	0.0025	1360	104
287.23				42	123	0.0001	2.2008		5.2037	0.0000		
287.23				680	D 12	0.0011	2.2018		5.2037	0.0000		
287.63	2100		104	2100	104	0.0036	2.2056	L	5.2037	0.0038	2100	104
287.63							2.2056		5.1999	0.0022	1456	D 144
287.80	900			900	104	0.0016	2.2073		5.1977	0.0000	30	144
288.16	1885		144	460	104	0.0008	2.2081	R	5.1977	0.0017	1300	144
288.63	2500			5881	A 10	0.0086	2.2167		5.1960	0.0019	1806	D 176
288.63		XING 210.35					2.2167	PRIVATE 210	5.1941	0.0002	173	176
289.63	5288	XING 210.5		8592	A 14	0.0103	2.2270	PRIVATE	5.1939	0.0068	5892	A 155
290.54	4800					0.0000	2.2270		5.1871	0.0017	1380	155
290.80	322		155			0.0000	2.2270	R	5.1854	0.0004	322	155
290.94	1800	XING 2	155	3204	155	0.0039	2.2309	R'BARREE	5.1850	0.0103	7313	A 110
291.33	3100			1748	D 15	0.0025	2.2334		5.1747	0.0018	1045	110
291.59	315						2.2334	L	5.1729	0.0005	315	110
291.74	805		110	1300	110	0.0022	2.2357	L	5.1724	0.0014	805	110
291.74						0.0000	2.2357		5.1710			
291.86	600			1360	110	0.0023	2.2380	L	5.1710	0.0010	600	110
292.27	2190			1181	A 11	0.0019	2.2399	T'MPH 60WB	5.1700	0.0030	1745	110

APPENDIX B3

PAGE 5 COLUMN 2

MILES	DISTANCE, ft	HWY GRADE CROSSING	MAX MPH	Δ DISTANCE, ft	SPEED, mph	ΔT, hours	TIME, hours	STATION	TIME, hours	ΔT, hours	Δ DISTANCE, ft	SPEED, mph
				READ DOWN WESTBOUND TRAFFIC					READ UP EASTBOUND TRAFFIC			
292.43	810			542	120	0.0009	2.2408	T TUNNEL	5.1670	0.0024	1325	A 96
292.43				337	D 1	0.0006	2.2414		5.1646	0.0000		
292.54	630		110	630	110	0.0011	2.2424	R'MPH40EB	5.1646	0.0011	590	96
292.54				220	110	0.0004	2.2428		5.1635	0.0000		
292.70	800			157	110	0.0003	2.2431	T'SPRUCE CREEK	5.1635	0.0004	227	96
292.70				423	D 11	0.0008	2.2439		5.1631	0.0000		
292.89	1000		96	1000	96	0.0020	2.2458	R	5.1631	0.0042	1803	A 66
293.06	1000		103	800	96	0.0016	2.2474	R	5.1589	0.0022	770	66
293.19	690			77	96	0.0002	2.2476	REVERSE CURVE	5.1567	0.0017	880	66
293.19				713	D 96	0.0017	2.2492		5.1550			
293.53	1800		66	1800	66	0.0052	2.2544	L	5.1550	0.0052	1800	66
293.53							2.2544		5.1498	0.0006	206	D 76
293.57	200		175	200	66	0.0006	2.2550	L	5.1492	0.0018	732	76
293.96	2100			1160	66	0.0033	2.2583		5.1474	0.0067	2706	76
294.21	1300		76	469	A 66	0.0013	2.2596	L	5.1407	0.0010	371	A 68
294.41	1077			3379	76	0.0084	2.2680		5.1397	0.0018	660	68
294.55	700		126	169	D 76	0.0004	2.2684	R	5.1379	0.0019	700	68
294.75	1100		68	1100	68	0.0031	2.2715	R	5.1360	0.0031	1100	68
294.92	900		104	900	68	0.0025	2.2740	R	5.1329	0.0018	678	D 10
295.13	1062	XING 216.6		460	68	0.0013	2.2753	PEMBERTON	5.1310	0.0002	104	102
295.33	1100		165	2403	A 68	0.0052	2.2805	L	5.1308	0.0063	2720	A 58
295.52	1000			0		0.0000	2.2805		5.1245	0.0052	1560	58
295.71	1000					0.0000	2.2805		5.1193	0.0006	163	A 53
295.71				206	104	0.0004	2.2809		5.1187			
295.71				1098	D 10	0.0026	2.2834	UNION FURNACE	5.1187			
295.96	1300		56	1300	58	0.0042	2.2877	L	5.1187	0.0032	907	53
296.05	453			372	58	0.0012	2.2889	T'RC MPH 35	5.1155	0.0016	453	53
296.05				81	D 58	0.0003	2.2891		5.1139			
296.27	1200		53	1200	53	0.0043	2.2934	R	5.1139	0.0165	4340	53
296.54	1400			1360	53	0.0049	2.2983		5.0984	0.0003	95	A 50
296.97	2300			2156	A 53	0.0054	2.3037		5.0981	0.0017	460	50
297.03	300		210			0.0000	2.3037	R	5.0964	0.0011	300	50
297.14	800			127	96	0.0003	2.3040		5.0953	0.0023	600	50
297.14				967	D 96	0.0025	2.3065		5.0930			
297.18	200		50	200	50	0.0008	2.3072	L	5.0930	0.0008	200	50
297.18							2.3072		5.0922	0.0042	1962	D 12
297.48	1600		54	1360	50	0.0052	2.3124	R RC	5.0880	0.0001	87	128
297.53	247			114	A 50	0.0004	2.3128		5.0879	0.0000		
297.61	400			773	54	0.0027	2.3155		5.0879	0.0000		
297.80	1000		115	713	54	0.0025	2.3180	L	5.0879	0.0000		
298.06	1400			4145	A 54	0.0090	2.3270	T'BIRMINGHAM	5.0879	0.0113	5805	A 54
298.53	2487		155			0.0000	2.3270	L	5.0766	0.0041	1181	54
298.70	900		156	397	115	0.0007	2.3276	L	5.0725	0.0001	19	A 53
298.89	1000			1512	D 11	0.0034	2.3310		5.0724	0.0006	160	53
299.06	900		54	1184	54	0.0042	2.3352	R	5.0718	0.0032	900	53
299.12	300			16	D 54	0.0001	2.3352	REVERSE URVE	5.0686	0.0011	300	53
299.32	1085		53	1085	53	0.0039	2.3391	L	5.0675	0.0039	1085	53
299.32							2.3391		5.0636	0.0010	327	D 71
299.53	1100			1100	53	0.0039	2.3430		5.0626	0.0021	773	71
299.68	800			260	53	0.0008	2.3440		5.0605	0.0021	800	71
299.87	1000		71	884	A 53	0.0021	2.3460	L	5.0584	0.0027	1000	71
299.93	300			1156	71	0.0031	2.3491	REVERSE CURVE	5.0557	0.0040	1518	71
300.14	1100	XING 2	72	1060	71	0.0028	2.3519	R PRIVATE	5.0517	0.0017	584	A 56
300.29	800			76	A 71	0.0002	2.3522		5.0500	0.0003	98	56
300.53	1282		90	1724	72	0.0045	2.3567	L'TTMPH 35	5.0497	0.0043	1282	56
300.53				300	D 72	0.0009	2.3576		5.0454	0.0000		
300.69	860		56	860	56	0.0029	2.3605	L'TYRONE	5.0454	0.0029	860	56
300.69		XING 222.18					2.3605	WILSON ROAD	5.0425	0.0014	502	D 81
300.78	500			500	56	0.0017	2.3622	RC	5.0411	0.0001	40	81
301.01	1200		95	860	56	0.0028	2.3651	R	5.0410	0.0027	1150	81
301.22	1100		81	1087	A 56	0.0030	2.3681	R	5.0383	0.0026	1100	81
301.22				1713	81	0.0040	2.3721		5.0357	0.0001	24	D 82
301.33	800			80	A 81	0.0001	2.3722	RC	5.0356	0.0013	575	82
301.52	1000		82	180	82	0.0004	2.3726	L	5.0343	0.0023	1000	82
301.52							2.3726		5.0320	0.0012	670	D 10
301.57	273		103	273	82	0.0006	2.3733	L	5.0308	0.0000	0	105
301.74	900			900	82	0.0021	2.3753		5.0308	0.0014	885	D 12
302.07	1700		180	187	82	0.0004	2.3758	R	5.0294	0.0003	198	129
302.52	2400			4427	A 82	0.0099	2.3856		5.0291	0.0073	4201	A 90
302.90	2000			1	135	0.0000	2.3856		5.0218	0.0029	1360	90
302.90				1466	D 13	0.0025	2.3881	GRAZIERVILLE E END 3 TR	5.0189	0.0000		
303.26	1900	XING 2	90	1900	90	0.0040	2.3921	L PRIVATE	5.0189	0.0040	1900	90
303.37	600	XING 225.06		600	90	0.0013	2.3934	RC'TTMPH 80 PRIVATE	5.0149	0.0013	600	90
303.52	782	XING 2	90	782	90	0.0016	2.3950	R TIPTON ROAD	5.0136	0.0016	782	90
303.80	1473		90	1473	90	0.0031	2.3981	R	5.0120	0.0031	1473	90
304.52	3800			1360	90	0.0029	2.4010	T'TIPTON	5.0089	0.0027	1646	D 13
304.52				5832	A 90	0.0097	2.4107		5.0062	0.0002	140	139
304.52				140	139	0.0002	2.4108		5.0060	0.0067	5832	A 90
305.52	5278			1646	D 13	0.0027	2.4135		4.9993	0.0029	1360	90
306.33	4300		96	4300	90	0.0090	2.4226	L	4.9934	0.0090	4300	90
306.33		XING 227.93					2.4226	PA RTE 802	4.9844	0.0003	138	D 95
306.39	300			300	90	0.0006	2.4233	T'FOSTORIA RC	4.9841	0.0003	164	95
306.46	358		95	358	90	0.0006	2.4240	R'TTMPH 60	4.9838	0.0007	358	95
306.46				702	90	0.0015	2.4255		4.9831	0.0000		
306.92	2408		95	373	A 90	0.0006	2.4262	R	4.9831	0.0040	2408	95
306.92				1334	95	0.0027	2.4289		4.9789	0.0046	3271	D 17
307.45	2800			1360	95	0.0027	2.4316		4.9737	0.0003	245	177
308.45	5283	XING 230.7		9898	A 95	0.0144	2.4460	PRIVATE	4.9734	0.0048	4577	A 16
308.98	2800					0.0000	2.4460		4.9686	0.0033	2800	160
309.22	1276		160	913	160	0.0011	2.4471	L	4.9653	0.0016	1276	160
309.45	1200			1200	160	0.0014	2.4485		4.9638	0.0016	1360	160
310.44	5258			160	160	0.0002	2.4487		4.9622	0.0101	7505	A 11
310.44				5758	A 16	0.0068	2.4555	E END 4 TR	4.9521			
311.16	3800			377	181	0.0004	2.4559	T'ANTIS	4.9521	0.0022	1360	118
311.16				2762	D 18	0.0036	2.4594	EAST END POPLAR RUN	4.9499			
311.41	1300		116	1300	118	0.0021	2.4615	R	4.9499	0.0021	1300	118
311.45	202			202	118	0.0003	2.4618	T'MPH60(TT)	4.9478	0.0027	1999	D 15
311.47	108			108	118	0.0002	2.4620	RC	4.9451	0.0002	184	166
311.96	2600		166	1050	118	0.0017	2.4637	L	4.9449	0.0014	1200	166
312.43	2500			8999	A 11	0.0119	2.4755		4.9436	0.0000		
312.47	200					0.0000	2.4755		4.9436	0.0000		
312.95	2500		207			0.0000	2.4755	R	4.9436	0.0000		
313.43	2549					0.0000	2.4755		4.9436	0.0000		
314.03	3200			2182	A 16	0.0024	2.4780		4.9436	0.0154	11036	A 99
314.17	700		286			0.0000	2.4780	L	4.9281	0.0003	160	99
314.39	1200			215	174	0.0002	2.4782	T'TTMPH30	4.9278	0.0023	1200	99
314.39				3003	D 17	0.0042	2.4824	ALTOONA	4.9255			
314.42	143		99	142	99	0.0003	2.4826	R	4.9255	0.0003	142	99
314.52	520		99	520	99	0.0010	2.4836	R	4.9252	0.0010	600	99
315.43	4800			1360	99	0.0026	2.4862	T'ALTO	4.9242	0.0030	1996	D 15
316.15	3787			6725	A 99	0.0102	2.4964	T'SLOPE	4.9212	0.0002	191	153
316.15				6	147	0.0000	2.4964	E END 3 TR	4.9210			
316.15				498	D 14	0.0007	2.4971		4.9210			
316.41	1400		135	1400	135	0.0020	2.4991	L BRICKYARD	4.9210	0.0045	3435	A 13
316.68	1400	XING 2	135	100	135	0.0001	2.4992	L COBURN ROAD	4.9165	0.0117	6505	A 67
317.07	2100			0	135	0.0000	2.4992	T'TTMPH 35	4.9048	0.0038	1360	67
317.07				1385	135	0.0019	2.5012		4.9010			
317.07				2015	D 13	0.0038	2.5049		4.9010			
317.36	1500		67	1500	67	0.0042	2.5092	R	4.9010	0.0042	1500	67
317.43	372			372	67	0.0011	2.5102		4.8968	0.0004	145	D 74
317.73	1574			998	67	0.0028	2.5130		4.8964	0.0046	1801	74
317.92	1000		99	2005	A 67	0.0046	2.5175	R	4.8918	0.0017	1000	74
318.12	1100			47	98	0.0001	2.5176	T'TTMPH35	4.8901	0.0034	1342	74
318.12				634	D 98	0.0014	2.5190		4.8867			
318.29	900		74	900	74	0.0023	2.5213	L	4.8867	0.0021	798	A 54
318.43	700			700	74	0.0018	2.5231		4.8846	0.0058	1862	54

APPENDIX B3

PAGE 6 COLUMN 1 / PAGE 6 COLUMN 2

READ DOWN WESTBOUND TRAFFIC — STATION — READ UP EASTBOUND TRAFFIC

MILES	DISTANCE	HWY GRADE CROSSING	MAX MPH	DISTANCE	SPEED MPH	ET HOURS	TIME HOURS	STATION	TIME HOURS	ET HOURS	DISTANCE	SPEED MPH	MILES	DISTANCE	HWY GRADE CROSSING	MAX MPH	DISTANCE	SPEED MPH	ET HOURS	TIME HOURS	STATION	TIME HOURS	ET HOURS	DISTANCE	SPEED MPH
318.58	600			425	74	0.0011	2.5242		4.8786	0.0004	514	A 50<	343.82	800		98	514	98	0.0010	2.8028	L	4.5945	0.0016	955	A 61
318.58				375	D 74	0.0011	2.5253		4.8784				343.82				961	D 96	0.0012	2.8040		4.5927			
318.86	1500		54	1500	54	0.0053	2.5306	R	4.8764	0.0041	1084	50	343.97	600		76	800	76	0.0020	2.8060	L MPH 35	4.5927	0.0017	500	61
318.91	276			215	54	0.0008	2.5313	TTTMPH 38 RC	4.8743	0.0010	276	50	344.12	600			498	78	0.0012	2.8072		4.5910	0.0034	1268	61
318.91		UG 240.69		81	D 54	0.0002	2.5315	SCOTTS RUN	4.8733				344.12				961	D 74	0.0005	2.8081		4.5871			
319.14	1200		50	1200	50	0.0045	2.5361	L	4.8733	0.0045	1200	50	344.16	200		61	200	61	0.0006	2.8087	R	4.5871	0.0008	272	A 53
319.43	1500			1360	50	0.0052	2.5412		4.8688	0.0027	1071	D 98<	344.31	800		61	800	55	0.0025	2.8112	R	4.5862	0.0009	260	53
319.58	800			1944	A 50	0.0051	2.5464		4.8661	0.0001	76	98	344.52	1100		77	968	61	0.0030	2.8142	R REV CURVE	4.5803	0.0039	1100	53
319.58							2.5464		4.8660	0.0042	1880	A 70<	344.52				194	D 61	0.0004	2.8148		4.5814			
319.73	800		165			0.0000	2.5464	L	4.8616	0.0030	900	A 44	344.69	900		55	900	53	0.0030	2.8178	L MPH 30	4.5814	0.0032	900	53
319.90	900			81	91	0.0002	2.5465		4.8568	0.0000	73	44	344.69							2.8178		4.5782	0.0013	458	D 57
320.01	600		98	1215	D 98	0.0051	2.5516	R	4.8565	0.0026	600	44	344.86	900		77	900	53	0.0032	2.8219	L	4.5769	0.0015	442	77
320.13	600		70			0.0000	2.5516	R	4.8556	0.0026	600	44	344.90	189			189	53	0.0007	2.8217	RC	4.5750	0.0007	189	57
320.14	57			252	70	0.0007	2.5523	RC	4.8538	0.0004	57	44	345.01	600		68	261	53	0.0008	2.8227	R	4.5751	0.0017	600	68
320.14				435	D 70	0.0014	2.5537		4.8029				345.16	920		98	458	A 53	0.0015	2.8241	R	4.5734	0.0012	497	D 53
320.31	900		44	900	44	0.0039	2.5576	L	4.8529	0.0039	900	44	345.33	870			1241	96	0.0036	2.8277		4.5722	0.0025	1173	86
320.43	600		54	600	44	0.0026	2.5602	L HORSESHOE	4.8490	0.0026	600	44	345.40	400		86	1391	A 86	0.0033	2.8310	L	4.5697	0.0009	400	68
320.60	900		44	900	44	0.0039	2.5641	L	4.8464	0.0036	900	44	345.56	600		104	40	91	0.0001	2.8311	L	4.5680	0.0026	1200	86
320.60							2.5641		4.8425	0.0012	336	D 66	345.56				79	D 91	0.0002	2.8313		4.5662			
320.63	182			182	44	0.0008	2.5648	RC	4.8413	0.0001	46	65	345.68	700		66	700	86	0.0015	2.8028	L MINERAL POINT	4.5662	0.0015	640	A 76
320.71	400		94	402	44	0.0017	2.5666	R	4.8412	0.0006	200	65	345.82	700			128	88	0.0003	2.8331		4.5647	0.0008	360	76
320.80	500			500	44	0.0022	2.5687		4.8406	0.0016	500	65	345.82				572	D 86	0.0014	2.8345		4.5638			
320.93	800		74	278	44	0.0012	2.5699	R	4.8391	0.0025	800	65	345.93	600		62	660	62	0.0018	2.8083	R RC	4.5638	0.0015	600	78
321.05	500		165	707	A 44	0.0024	2.5723	R	4.8386	0.0015	500	65	346.01	400			400	62	0.0017	2.8376	TTTMPH 35	4.5620	0.0016	400	76
321.16	600		65	920	65	0.0027	2.5750	R	4.8353	0.0017	600	65	346.16	900		75	800	92	0.0024	2.8400	L	4.5613	0.0027	965	76
321.31	800		70	800	65	0.0025	2.5774	R	4.8338	0.0023	800	65	346.29	700		78	160	62	0.0005	2.8405	L	4.5560	0.0022	785	A 90
321.43	650			581	65	0.0016	2.5790		4.8313	0.0027	1388	D 114	346.29				141	A 62	0.0004	2.8409		4.5568			
321.62	1000			198	A 65	0.0006	2.5796		4.8286	0.0003	196	114	346.29				342	98	0.0010	2.8418		4.5568			
321.79	900		117	602	70	0.0016	2.5812	LMG	4.8283	0.0000			346.29				57	D 98	0.0002	2.8421		4.5568			
322.10	1870			5401	A 70	0.0066	2.5880		4.8283	0.0039	3026	A 78	346.56	1400		63	1546	63	0.0046	2.8467	L	4.5568	0.0037	1160	80
322.43	1700		117	90	114	0.0001	2.5882	R	4.8224	0.0056	2575	78	346.60	200			54	D 62	0.0002	2.8469	REVERSE CURVE	4.5531	0.0009	200	60
322.43				1014	D 11	0.0020	2.5902		4.8168				346.75	800		60	800	60	0.0025	2.8494	R	4.5525	0.0025	800	60
322.63	1100		76	1000	76	0.0024	2.5926	R	4.8166	0.0026	912	A 50	346.81	300			300	60	0.0009	2.8503	RC	4.5500	0.0049	1037	80
322.75	600			1368	78	0.0033	2.5959		4.8140	0.0016	400	A 43	347.11	1900		86	1000	60	0.0033	2.8537	L	4.5461	0.0010	304	A 51
322.96	1100		283	430	D 78	0.0012	2.5973	R	4.8125	0.0021	475	43	347.16	288			267	A 60	0.0006	2.8543		4.5441	0.0018	345	51
323.11	800		56	811	56	0.0021	2.5992	R	4.8104	0.0036	800	43	347.20	215			634	65	0.0024	2.8567	RC	4.5429	0.0008	215	51
323.12	85				56	0.0000	2.5992	REVERSE CURVE	4.8068	0.0004	85	43	347.36	808		65	582	65	0.0018	2.8584	R	4.5420	0.0030	800	51
323.31	1000		43	188	D 54	0.0007	2.6000	L	4.8065	0.0044	1000	43	347.36				238	D 65	0.0008	2.8592		4.5390			
323.31				1000	43	0.0044	2.6044		4.8021	0.0002	39	D 46<	347.51	800		51	500	51	0.0030	2.8621	R	4.5390	0.0030	800	51
323.35	200			200	43	0.0009	2.6052	RC	4.8019	0.0007	181	46	347.51							2.8621		4.5360	0.0001	31	D 53
323.45	500		70	500	43	0.0022	2.6074	R	4.8012	0.0021	500	46	347.56	400			400	51	0.0015	2.8636	RC	4.5359	0.0013	369	53
323.54	500		70	500	43	0.0022	2.6096	R	4.7991	0.0021	500	46	347.72	700		55	706	51	0.0026	2.8662	L	4.5346	0.0025	700	53
323.57	149			148	43	0.0007	2.6103	TRC	4.7970	0.0006	149	46	347.85	700		58	260	51	0.0010	2.8672	L	4.5321	0.0026	700	58
323.76	1000		46	11	43	0.0000	2.6104	L MPH 30(TT)	4.7964	0.0041	1000	46	347.85				39	A 51	0.0003	2.8675		4.5296	0.0001	16	D 54
323.87	600			81	A 43	0.0003	2.6107		4.7923	0.0025	600	46	347.87	100			401	53	0.0014	2.8688	RC	4.5295	0.0003	84	54
324.00	700		94	2293	46	0.0094	2.6201	R	4.7886	0.0029	700	46	348.04	900		54	900	53	0.0032	2.8721	R	4.5292	0.0032	900	54
324.16	800		67	65	46	0.0002	2.6204	R	4.7869	0.0033	800	46	348.17	700		65	480	53	0.0018	2.8738	R	4.5290	0.0006	192	D 65
324.31	800			663	A 46	0.0026	2.6230		4.7856	0.0033	600	46	348.24	377		65	19	A 53	0.0001	2.8738	R	4.5254	0.0026	865	65
324.38	400		104		65	0.0000	2.6230	R	4.7803	0.0016	400	46	348.24							2.8738	AO	4.5228	0.0002	78	D 88
324.42	200		98	25	65	0.0001	2.6230	R	4.7787	0.0008	200	46	348.49	1300			961	54	0.0034	2.8772		4.5226	0.0034	1221	88
324.46	211			1364	65	0.0040	2.6270	RC	4.7776	0.0006	211	46	348.71	1200		88	416	A 54	0.0013	2.8788	R	4.5192	0.0033	1200	88
324.46				309	D 67	0.0011	2.6281		4.7776				348.71				862	65	0.0019	2.8805		4.5159	0.0002	82	D 72
324.71	1300		46	1300	46	0.0054	2.6334	L	4.7770	0.0054	1300	46	348.71							2.8805		4.5157	0.0041	1540	72
324.71							2.6334		4.7719	0.0043	1945	D 124	348.94	1200			276	A 65	0.0008	2.8813		4.5116	0.0013	438	A 61
324.80	500			500	46	0.0021	2.6355	PRENNY	4.7673	0.0003	173	124	349.17	1200		72	2224	72	0.0059	2.8871	L	4.5103	0.0011	380	61
324.90	800		186	500	46	0.0021	2.6375	LEB -2.27%	4.7670	0.0000			349.17				1306	A 72	0.0030	2.8901		4.5092			
325.01	600			386	46	0.0015	2.6390	T(R 1) RC	4.7670	0.0015	629	A 117	349.36	1005		95	8	93	0.0000	2.8901	L	4.5092	0.0031	1000	61
325.43	2200		132	4638	A 46	0.0087	2.6467	R	4.7655	0.0012	753	117	349.36				723	D 92	0.0018	2.8919		4.5061			
325.54	807		195			0.0000	2.6467	R	4.7643	0.0010	607	117	349.62	1400		61	1400	61	0.0045	2.8963	L	4.5061	0.0043	1400	61
325.66	600		117			0.0000	2.6467	R	4.7633	0.0006	381	117	349.66	200			200	61	0.0006	2.8969	RC	4.5018	0.0006	215	D 72
325.83	900			1815	A 31	0.0040	2.6527		4.7627	0.0083	4079	A 65	349.85	1000		72	1000	61	0.0031	2.9000	R	4.5012	0.0026	985	72
326.43	3200					0.0000	2.6527	TTUNNELS	4.7544	0.0068	700	63	349.85				180	61	0.0005	2.9005		4.4986	0.0007	269	D 85
326.51	400			497	130	0.0007	2.6534	TTUNNELS	4.7536	0.0012	400	63	349.90	246		85	436	A 61	0.0013	2.9017	R	4.4979	0.0000	0	85
326.64	700			1867	D 12	0.0037	2.6571	GALLITZEN	4.7524	0.0021	700	63	350.01	900		237	248	72	0.0007	2.9024	R	4.4979	0.0026	1496	D 13
326.87	1200		63	1200	63	0.0036	2.6607	L GUN T@AR E END 4 TR	4.7503	0.0036	1200	63	350.16	800			514	72	0.0014	2.9037		4.4953	0.0008	611	132
326.87							2.6607		4.7467	0.0023	1005	D 105	350.56	2129			720	A 72	0.0017	2.9056	T'CONEMAUGH TOWER	4.4944	0.0007	505	132
327.10	1246			1246	65	0.0037	2.6645	EAST END 6 TR	4.7444	0.0004	194	105	350.66	900		132	5063	A 85	0.0086	2.9142	R	4.4937	0.0000		
327.31	1100		117	114	65	0.0003	2.6648	R	4.7440	0.0047	2257	A 74	350.68	100					0.0000	2.9142	RC	4.4937	0.0000		
327.65	1800		117	1576	A 63	0.0036	2.6687	R	4.7393	0.0017	660	74	350.91	1200		204			0.0000	2.9142	L	4.4937	0.0000		
327.79	700			1472	90	0.0030	2.6717		4.7376	0.0018	700	74	351.13	1200					0.0000	2.9142		4.4937	0.0000		
327.79				436	D 16	0.0010	2.6727		4.7358	0.0000			351.22	476		162			0.0000	2.9142	L	4.4937	0.0057	3584	A 10
328.05	1400		74	1400	74	0.0036	2.6763	L	4.7358	0.0036	1400	74	351.28	300					0.0000	2.9142	RC	4.4880	0.0026	1357	A 88
328.05							2.6763		4.7322	0.0059	3949	D 180	351.41	750		168	100	132	0.0001	2.9144	R	4.4854	0.0039	1819	88
328.30	1338			1336	74	0.0034	2.6797		4.7263	0.0002	218	180	351.41				968	D 13	0.0016	2.9159		4.4815			
328.30				22	74	0.0001	2.6797	260	4.7261	0.0000			351.58	900		104	449	104	0.0008	2.9167	R	4.4815	0.0008	264	A 84
329.31	5304			3802	A 74	0.0070	2.6868	WEST END 6 TRACKS	4.7261	0.0119	3982	A 111	351.58				451	D 10	0.0009	2.9176	E END 4 TR	4.4808			
329.76	2400			11056	A 11	0.0142	2.7010	TORESSON	4.7142	0.0000			351.79	1190		88	1986	88	0.0043	2.9219	R	4.4808	0.0005	360	84
330.05	1536		233			0.0000	2.7010	R	4.7142	0.0000			351.98	1000			101	D 86	0.0002	2.9222	W END 4 TR	4.4803	0.0023	1000	64
330.28	1200					0.0000	2.7010		4.7142	0.0000			352.17	1000		84	1000	84	0.0023	2.9244	R	4.4778	0.0023	1000	64
331.29	5311	XING 260.3		189	173	0.0002	2.7012	RISBINS ROAD	4.7142	0.0127	9680	A 117	352.17							2.9244		4.4756	0.0004	207	D 82
331.57	1500			2382	D 17	0.0031	2.7043		4.7015	0.0022	1350	117	352.37	1061		92	1061	84	0.0024	2.9268	R	4.4751	0.0005	164	82
331.98	2162		117	2159	117	0.0036	2.7078	L	4.6993	0.0035	2152	117	352.52	800			566	84	0.0007	2.9275	TJOHNSTOWN	4.4748	0.0037	1530	A 64
332.26	1500			1500	117	0.0024	2.7102		4.6958	0.0024	1500	117	352.64	900		251	1036	A 84	0.0022	2.9296	R	4.4711	0.0057	560	64
332.45	982			982	117	0.0016	2.7118		4.6934	0.0016	962	117	352.79	800			57	98	0.0001	2.9297	W END 3 TR	4.4694	0.0024	800	64
332.62	900		117	900	117	0.0015	2.7133	REILLY	4.6918	0.0016	900	117	352.79				898	D 96	0.0019	2.9316	E END 1 TR SANS HOLLOW	4.4670			
332.62							2.7133		4.6903	0.0029	2181	D 169	352.83	200		64	200	64	0.0006	2.9322	R	4.4670	0.0006	209	64
333.15	2800			1386	117	0.0022	2.7156		4.6874	0.0000	12	169	353.03	1500			1000	64	0.0030	2.9350		4.4664	0.0019	837	D 89
333.49	1794			8868	A 11	0.0118	2.7273		4.6874	0.0012	1041	A 185	353.20	1000		99	360	64	0.0011	2.9362	L	4.4645	0.0022	1190	99
333.87	2000		180			0.0000	2.7273	R	4.6862	0.0035	3064	165	353.29	456		222	2128	A 84	0.0049	2.9411	L	4.4623	0.0016	578	D 56
334.02	800					0.0000	2.7273	CASSANDRA	4.6827	0.0000			353.87	2500					0.0000	2.9411		4.4613	0.0001	40	117
334.85	4519			4178	A 16	0.0046	2.7319		4.6827	0.0118	8886	A 117	353.82	800		180	839	A 99	0.0012	2.9423	R	4.4612	0.0202	1266	A 10
335.73	4651					0.0000	2.7319		4.6709	0.0022	1380	117	354.20	2000		106	3113	106	0.0056	2.9478	R	4.4586	0.0210	560	106
335.81	500			178	180	0.0002	2.7321		4.6687	0.0013	800	117	354.20				1380	106	0.0024	2.9503		4.4560	0.0033	1865	106
335.91				2744	D 10	0.0036	2.7356		4.6674				354.20				117	A 10	0.0002	2.9505		4.4547	0.0098	1504	A 82
336.00	500		117	606	117	0.0008	2.7364	R	4.6674	0.0008	500	117	354.49	1509		115	136	107	0.0003	2.9608	R	4.4509	0.0012	534	87
336.10	500			500	117	0.0008	2.7372	TPORTAGE	4.6666	0.0003	587	D 176	354.84	800			683	D 10	0.0014	2.9622		4.4497	0.0043	1468	A 46
336.36	1400		135	860	117	0.0014	2.7386	R	4.6657	0.0001	66	133	354.86	1200		82	1200	82	0.0026	2.9649	R	4.4454	0.0025	586	46
336.36				2445	A 11	0.0037	2.7423		4.6655				355.02	500			124	82	0.0003	2.9652	DORNOCK	4.4431	0.0033	800	46
336.62	1346		147	114	134	0.0002	2.7424	R	4.6656	0.0046	3630	A 114	355.02				676	D 82	0.0020	2.9672		4.4358			
336.65	1400			727	D 12	0.0011	2.7435		4.6616	0.0029	1960	114	355.21	1050		46	1006	46	0.0041	2.9713	L	4.4358	0.0041	1006	46
337.15	1400		114	1400	114	0.0023	2.7458	R	4.6593	0.0023	1400	114	355.37	689		46	889	46	0.0037	2.9650	L	4.4357	0.0037	889	46
337.41	1400			1380	114	0.0023	2.7481		4.6570	0.0002	192	D 117	355.37							2.9650		4.4320	0.0002	56	D 50
337.49	426		117	381	A 11	0.0006	2.7487	L	4.6568	0.0028	1724	117	355.41	200			200	46	0.0008	2.9658	RC	4.4318	0.0006	144	50
337.91	2290		117	1386	117	0.0022	2.7508	L	4.6540	0.0036	2230	117	355.60	1800		52	1030	46	0.0041	2.9699	R	4.4313	0.0036	1000	52
338.41	2626					0.0000	2.7508	WILMORE	4.6504	0.0004	3639	D 176	355.60							2.9699		4.4277	0.0016	542	D 60
338.99	3100			3856	A 11	0.0063	2.7563	WEST END POP RUN	4.6470	0.0004	375	176	355.77	900		235	160	46	0.0007	2.9708	R	4.4261	0.0006	356	80
339.33	1774		233	10674	A 14	0.0124	2.7686	R	4.6465	0.0000			355.96	1000		80	180	46	0.0067	2.9713	R	4.4253	0.0024	1060	50
339.97	3361		233			0.0000	2.7686	R	4.6466	0.0000			355.96				1020	52	0.0037	2.9750		4.4229	0.0023	1269	D 12
340.25	1500					0.0000	2.7686		4.6466	0.0104	8818	A 141	356.00	200			1167	A 52	0.0033	2.9783	REVERSE CURVE	4.4206	0.0000	15	121
340.38	660			2406	A 10	0.0024	2.7711		4.6362	0.0000			356.15	600		93	373	60	0.0009	2.9792	L	4.4206	0.0000		
340.87	2600		255			0.0000	2.7711	L	4.6362	0.0000			356.17	100			109	80	0.0002	2.9795		4.4206	0.0056	3116	A 80
341.17	1600			347	191	0.0003	2.7714	SUMMERHILL	4.6362	0.0137	7736	A 69	356.47	1600			289	80	0.0006	2.9801		4.4150	0.0007	328	90
341.66	2600			2436	D 16	0.0028	2.7742		4.6225	0.0008	290	68	356.78	1600		128	850	A 80	0.0019	2.9820	R	4.4143	0.0066	368	A 85
341.87	1100		141	6	141	0.0000	2.7742	R	4.6217	0.0033	1930	63	356.78				110	93	0.0002	2.9822		4.4136			
341.87				2234	D 14	0.0043	2.7785	E END MIN POINT	4.6184				356.78				2401	A 93	0.0043	2.9865		4.4136			
342.06	1000		63	1000	63	0.0030	2.7815	R	4.6184	0.0030	1000	63	356.98	600			119	117	0.0002	2.9867		4.4136	0.0002	78	85
342.12	300			300	63	0.0009	2.7824	REVERSE CURVE	4.6154	0.0002	78	D 67	357.04	600		272	820	D 11	0.0015	2.9882	R	4.4133	0.0013	800	85
342.16	248		67	246	63	0.0007	2.7832	L	4.6152	0.0013	472	67	357.12	400		90	400	90	0.0008	2.9891	R	4.4120	0.0009	400	85
342.28	600		87	600	63	0.0018	2.7850	L	4.6139	0.0017	600	67	357.17	282			154	90	0.0003	2.9894	RC	4.4111	0.0006	262	85
342.43	800		99	212	63	0.0005	2.7856	L	4.6122	0.0015	938	D 94	357.17				126	D 90	0.0003	2.9897	280	4.4105			
342.81	2000			167	A 63	0.0005	2.7861	TTTMPH 40	4.6107	0.0002	85	94	357.30	700		65	700	65	0.0015	2.9912	L	4.4106	0.0016	700	85
342.81				981	67	0.0028	2.7889		4.6105	0.0015	719	A 84	357.30				1980	65	0.0030	2.9943	WEST END JOHNSTOWN	4.4089	0.0036	2240	D 15
342.81				861	A 67	0.0022	2.7910		4.6090				358.06	4000					0.0000	2.9943	SO	4.4053	0.0001	70	150
342.81				579	84	0.0013	2.7923		4.6080	0.0031	1359	84	358.17	568		203	7582	A 85	0.0122	3.0064	L	4.4052	0.0000		
343.16	1885		84	1885	84	0.0042	2.7966	R SOUTH FORK	4.6063	0.0080	3536	64	358.22	257		232			0.0000	3.0064	L	4.4052	0.0103	6960	A 10
343.52	1871			1360	84	0.0031	2.7996	T-MISO-@SUM(C734 .C850	4.5979	0.0010	441	A 70	359.16	5000			101	146	0.0001	3.0066		4.3849	0.0610	580	103
343.67	600		112	1006	A 84	0.0022	2.8018	L	4.5969	0.0024	645	76	359.31	800			1570	D 14	0.0024	3.0066		4.3839	0.0015	800	103

READ DOWN WESTBOUND TRAFFIC — APPENDIX B3 STATION — READ UP EASTBOUND TRAFFIC

PAGE 7 COLUMN 1

MILES	DISTANCE ft	HWY GRADE CROSSING	MAX MPH	DISTANCE ft	SPEED mph	ET HOURS	TIME HOURS	STATION	TIME HOURS	ET HOURS	DISTANCE ft	SPEED mph
360.07	4000		103	4000	103	0.0074	3.0163	L	4.3824	0.0074	4000	103
360.07									4.3850	0.0002	92	D 106
360.11	200			100	105	0.0002	3.0165	RC SAND HOLLOW	4.3848	0.0045	2506	106
360.15	237		106	237	108	0.0006	3.0169	R	4.3803	0.0005	268	A 103
360.16	20		106	20	103	0.0000	3.0170	R	4.3796	0.0000	6	
360.92	4500			1003	103	0.0018	3.0188		4.3798	0.0000	0	
360.92				2094	A 10	0.0038	3.0226	SQ-SEWARD	4.3798		0	
360.92				150	123	0.0002	3.0228		4.3786	0.0024	1231	A 98-1
360.92				683	D 12	0.0011	3.0240		4.3774	0.0008	360	98
361.11	1000		103	504	100	0.0009	3.0249	L	4.3766	0.0022	1000	86
361.11				496	D 10	0.0010	3.0256		4.3744			
361.67	2906		88	2906	88	0.0067	3.0326	L SEWARD	4.3744	0.0064	2946	88
361.67							3.0326	E END 2 TR SAND HOLLOW	4.3680	0.0055	3776	D 168
362.05	2000	OK		1390	88	0.0029	3.0366		4.3627	0.0002	239	168
362.09	214			12140	A 85	0.0174	3.0529	T	4.3626	0.0000		
363.04	5000		233			0.0000	3.0529	R	4.3625	0.0000		
363.47	2285		233			0.0000	3.0529	R	4.3625	0.0000		
364.04	3000			137	167	0.0002	3.0531	W END 2 TR SAND HOLLOW	4.3625	0.0106	9245	A 147
364.44	2139			921	D 16	0.0011	3.0543	T NEW FLORENCE	4.3519	0.0016	1366	147
365.01	3000		147	3000	147	0.0039	3.0581	R	4.3501	0.0039	3000	147
365.39	2000		147	2000	147	0.0026	3.0607	R	4.3462	0.0026	2000	147
365.77	2000			2000	147	0.0026	3.0633		4.3436	0.0026	2000	147
365.92	820		233	820	147	0.0011	3.0643	L	4.3410	0.0011	820	147
366.08	824		147	824	147	0.0011	3.0654	L	4.3399	0.0011	824	147
366.84	4000			1380	147	0.0018	3.0671		4.3388	0.0006	448	D 157
366.84						0.0000	3.0671	290	4.3382	0.0002	135	157
367.04	1056					0.0000	3.0671		4.3380			
367.04				5109	A 14	0.0061	3.0733	T CONNPT JCT 3 TR	4.3380			
367.09	300	XING 290.8				0.0000	3.0733	PRIVATE	4.3380	0.0095	8587	A 117
367.17	400		186			0.0000	3.0733	R REVERSE CURVE	4.3285	0.0041	2606	117
367.24	400		186			0.0000	3.0733	L MPH 56	4.3244	0.0042	2332	A 94-1
367.76	2720			225	169	0.0003	3.0735		4.3202	0.0039	1951	94
367.76				2181	D 10	0.0028	3.0764		4.3163			
368.66	4837		117	4125	117	0.0067	3.0831	R	4.3163	0.0030	1524	A 73-6
368.68				712	D 11	0.0013	3.0843		4.3133			
369.24	3000		94	2781	94	0.0056	3.0899	R	4.3133	0.0028	1085	73
369.30	275				94	0.0000	3.0899	RC	4.3105	0.0007	275	73
369.30				514	D 94	0.0012	3.0911	INDIA	4.3098			
369.58	1500		73	1500	73	0.0039	3.0950	L	4.3098	0.0039	1500	73
369.96	2000		73	2000	73	0.0052	3.1002	L LOCKPORT	4.3058	0.0052	2000	73
370.28	1700			1350	73	0.0035	3.1037	ENDTRNS	4.3007	0.0028	1474	D 117
370.59	1619		117	3652	A 73	0.0072	3.1108	L	4.2979	0.0001	47	104
370.95	1900		117	318	117	0.0006	3.1114	L BOLIVAR	4.2978	0.0064	4890	A 73-1
371.10	800			684	D 11	0.0012	3.1126		4.2882	0.0006	326	A 80-1
371.27	905		95	900	95	0.0018	3.1144	R	4.2884	0.0066	2297	98
371.49	1161		117	456	96	0.0010	3.1154	R	4.2818	0.0002	72	A 64-4
371.49				388	D 95	0.0016	3.1170		4.2816			
371.83	1800		98	1800	98	0.0052	3.1221	R	4.2818	0.0023	791	64
371.94	588			591	95	0.0015	3.1237	RC	4.2793	0.0017	589	64
371.94				36	D 95	0.0001	3.1236		4.2776			
372.49	2900		64	2900	64	0.0086	3.1324	L	4.2776	0.0060	2900	64
372.57	400		64	400	64	0.0012	3.1336	L	4.2696	0.0012	409	64
372.57							3.1335		4.2678	0.0028	1407	D 117
372.69	601			631	64	0.0018	3.1354		4.2649	0.0000	24	117
372.83	1300		117	729	64	0.0022	3.1376	R	4.2649	0.0008	500	117
373.17	1806			4033	A 64	0.0082	3.1458		4.2641	0.0026	1602	117
373.42	1800		165	828	A 11	0.0016	3.1472	R	4.2615	0.0000		
373.47	264		185	186	124	0.0003	3.1475	R	4.2615	0.0049	2702	A 86-1
374.00	2800			1066	D 12	0.0019	3.1494		4.2566	0.0036	2061	88
374.33	1700		89	1700	89	0.0038	3.1531	R	4.2515	0.0005	349	A 64-4
374.42	500		101	373	89	0.0006	3.1536	R	4.2502	0.0013	540	48
374.57	800			127	D 86	0.0003	3.1541		4.2489	0.0098	4382	84
375.12	2800		84	2806	84	0.0066	3.1607	R	4.2391	0.0010	441	A 76-1
375.33	1119			1119	84	0.0025	3.1632		4.2381	0.0000	10	76
375.43	500		100	241	84	0.0005	3.1637	R	4.2381	0.0012	500	76
375.46	244		106	1217	A 84	0.0025	3.1662	R TORRANCE	4.2368	0.0008	244	76
375.59	600			68	160	0.0001	3.1664	E END 3 TR (SIDING)	4.2363	0.0015	600	76
375.59				626	D 16	0.0013	3.1677	REVERSE CURVE	4.2346			
375.97	2000		76	2000	76	0.0050	3.1727	L	4.2348	0.0050	2000	76
376.19	1200		117	1360	76	0.0034	3.1761	L	4.2298	0.0007	289	D 171
376.33	706		110	725	A 76	0.0017	3.1777	L	4.2291	0.0035	1611	88
376.42	500		98	500	88	0.0011	3.1788	L	4.2256	0.0061	2691	86
376.50	430		88	435	88	0.0009	3.1797	L	4.2195	0.0017	725	A 76
376.50				1950	88	0.0029	3.1827		4.2178			
376.50				1616	A 88	0.0033	3.1860		4.2178			
377.26	4000			7	107	0.0000	3.1860	GRAYS	4.2178	0.0034	1360	76
377.26				832	d 10	0.0017	3.1877		4.2144			
377.39	700		76	700	76	0.0017	3.1895	R	4.2144	0.0017	700	76
377.51	600		76	600	76	0.0016	3.1910	R	4.2127	0.0015	600	76
377.51							3.1910		4.2112	0.0010	449	D 94-1
377.54	177			177	76	0.0004	3.1914	RC	4.2102	0.0001	36	94
378.07	2600		54	1193	76	0.0029	3.1944	L	4.2101	0.0050	2600	94
378.07		XING 301.8					3.1944	T 880 PA	4.2051	0.0015	712	D 117
378.38	1600			3492	A 76	0.0067	3.2011		4.2036	0.0014	898	117
378.39	100	XING 3	117			0.0000	3.2011	R PRIVATE	4.2024	0.0002	100	117
378.64	1288	XING 3	117	1114	117	0.0018	3.2029	R PRIVATE	4.2022	0.0021	1280	117
379.40	4000	XING 302.6		1360	117	0.0022	3.2051	T 880 PA	4.2001	0.0030	2784	D 171
380.40	5278			10433	A 11	0.0130	3.2181		4.1971	0.0000	11	171
380.96	3000			105	169	0.0002	3.2183		4.1971	0.0000		
380.99	167		202		169	0.0000	3.2183	L	4.1971	0.0136	10488	A 117
381.32	1700			2181	D 16	0.0029	3.2212		4.1835	0.0022	1360	117
381.39	400		117	400	117	0.0006	3.2218	R	4.1813	0.0006	400	117
381.58	1000		117	1000	117	0.0016	3.2234	R MILLWOOD	4.1807	0.0016	1000	117
381.58							3.2234		4.1791	0.0027	1966	D 165
381.77	1000			1050	117	0.0018	3.2251		4.1764	0.0000	7	165
381.90	677		165	386	117	0.0006	3.2256	L	4.1764	0.0004	305	165
382.03	700		233	0		0.0000	3.2256	L	4.1760			
382.33	1600			0		0.0000	3.2256		4.1760	0.0022	1900	165
382.46	483			6		0.0000	3.2256		4.1738	0.0008	483	160
382.62	700		165	0	0	0.0000	3.2256	R	4.1732	0.0008	700	165
383.39	4100			8686	A 11	0.0116	3.2374	2 TR (W OF SIDING)	4.1724	0.0026	2241	165
383.54	2587					0.0000	3.2374	T DERRY	4.1698	0.0036	3887	165
384.33	2400	XING 3	165			0.0000	3.2374	R VALLEY STREET	4.1685	0.0000		
384.45	287		165	4586	165	0.0056	3.2431	R	4.1665	0.0118	9886	A 117
385.39	5000			1360	165	0.0016	3.2446		4.1547	0.0011	680	117
385.39				1864	A 16	0.0021	3.2467		4.1536			
385.53	706			155	172	0.0002	3.2469		4.1536	0.0011	700	117
385.53				2331	D 17	0.0031	3.2499		4.1525			
385.97	2300		117	2306	117	0.0038	3.2537	R BRADENVILLE	4.1525	0.0038	2306	117
385.97							3.2537		4.1487	0.0041	3287	D 190
386.28	1600			1360	117	0.0022	3.2559		4.1446	0.0000	7	190
387.09	4300			8669	A 11	0.0113	3.2672		4.1438	0.0060	7561	A 165
387.28	989		165			0.0000	3.2672	L	4.1366	0.0000		
388.28	5268		165	1909	165	0.0022	3.2699	L	4.1366	0.0016	1360	160
388.39	600		165	600	165	0.0007	3.2706	L	4.1350	0.0007	600	165
389.15	4000			1360	165	0.0016	3.2722	T LATROBE	4.1343	0.0014	1302	D 180
389.28	700		330	7153	A 16	0.0076	3.2798	R	4.1329	0.0000	6	180
390.25	5112		330	359	189	0.0004	3.2802	R	4.1329	0.0059	5647	A 172
390.25				900	D 18	0.0009	3.2812		4.1270			
390.66	2200		172	1030	172	0.0011	3.2823	R BEATTY	4.1270	0.0070	5822	A 147
390.66				1170	D 17	0.0014	3.2837		4.1200			
391.09	2225		147	2225	147	0.0029	3.2866	R	4.1200	0.0018	1360	147
391.12	177		147	177	147	0.0002	3.2868	R	4.1182	0.0002	177	147
392.09	5100			1380	147	0.0018	3.2885		4.1180	0.0026	2207	D 154
393.09	5278			14083	A 14	0.0153	3.3038		4.1156	0.0004	360	182
393.35	1400					0.0000	3.3038		4.1151	0.0136	12270	A 142
393.74	2074		344			0.0000	3.3038	L	4.1015	0.0000		

READ DOWN WESTBOUND TRAFFIC — APPENDIX B3 STATION — READ UP EASTBOUND TRAFFIC

PAGE 7 COLUMN 2

MILES	DISTANCE ft	HWY GRADE CROSSING	MAX MPH	DISTANCE ft	SPEED mph	ET HOURS	TIME HOURS	STATION	TIME HOURS	ET HOURS	DISTANCE ft	SPEED mph
394.08	1800			20	187	0.0000	3.3038	DONOHOE	4.1015	0.0036	2225	A 13
394.61	2800			3016	D 10	0.0034	3.3074		4.0985	0.0019	1380	136
394.87	1389		136	1339	136	0.0018	3.3092	L	4.0966	0.0019	1339	136
395.30	2300		136	2300	136	0.0032	3.3128	L	4.0947	0.0032	2300	136
395.43	682			682	136	0.0010	3.3134		4.0915	0.0010	682	136
395.77	1800		136	1800	136	0.0025	3.3159	R	4.0905	0.0025	1800	136
395.85	400			400	136	0.0008	3.3165		4.0880	0.0015	1417	D 16
395.85						0.0000	3.3165	320	4.0862	0.0000	226	167
395.96	600			600	135	0.0006	3.3173		4.0860			167
396.25	1500		272	360	136	0.0005	3.3178	L	4.0860	0.0000		
396.51	1397		404			0.0000	3.3178	L	4.0859	0.0065	4765	A 13
396.85	1800			8483	A 13	0.0103	3.3280		4.0774	0.0000		
397.74	4700					0.0000	3.3282	T GREENSBURG	4.0774	0.0000		
397.85	579		233			0.0000	3.3282	R	4.0774	0.0039	2593	A 11
398.12	1400		233	171	173	0.0002	3.3284	R	4.0735	0.0022	1390	117
398.12				2382	D 17	0.0031	3.3315		4.0713			
398.42	1594		117	1594	117	0.0026	3.3341	R	4.0713	0.0026	1594	117
398.86	2320			1380	117	0.0022	3.3363		4.0687	0.0023	1654	D 15
399.18	1700		233	7401	A 11	0.0101	3.3463	L	4.0664	0.0001	63	158
399.78	3200					0.0000	3.3463		4.0663	0.0096	7166	A 11
399.85	335		330			0.0000	3.3463	L	4.0566	0.0000		
399.91	311		330			0.0000	3.3463	L	4.0565	0.0006	361	A 11
400.44	2800			83	158	0.0001	3.3464		4.0559	0.0023	1390	114
400.44				1602	D 11	0.0025	3.3489		4.0536			
400.81	900		114	900	114	0.0015	3.3504	R	4.0536	0.0015	900	114
400.63	100			100	114	0.0002	3.3506		4.0521	0.0028	2087	D 16
401.63	5306			1285	114	0.0021	3.3527	T RADEBAUGH	4.0493	0.0003	267	165
401.95	1700			9297	A 11	0.0124	3.3651		4.0490	0.0000		
402.83	3568		165			0.0000	3.3651	L [illegible]	4.0490			
403.12	2606		186			0.0000	3.3651	L	4.0490	0.0158	11149	A 90
403.43	1592			1550	168	0.0018	3.3669		4.0332	0.0027	1360	90
403.43				77	166	0.0000	3.3670		4.0305			
403.43				2689	D 16	0.0039	3.3708		4.0305			
403.63	1100		95	1100	95	0.0022	3.3730	R	4.0305	0.0060	2980	95
403.74	582		95	582	95	0.0011	3.3742	R	4.0245	0.0019	905	A 82
404.42	3600	XING 326.8		1360	96	0.0027	3.3769	WRIGLEYS	4.0226	0.0042	1640	82
404.42				1310	A 96	0.0024	3.3793		4.0184			
404.42				174	100	0.0003	3.3796		4.0184			
404.42				258	D 10	0.0015	3.3811		4.0184	0.0001	90	A 81
404.61	1000		82	1000	82	0.0023	3.3834	L	4.0183	0.0011	460	81
404.63	100			100	82	0.0002	3.3836		4.0172	0.0026	1601	81
404.78	800			24	D 82	0.0001	3.3837		4.0133	0.0001	58	A 80
404.93	800		81	800	81	0.0021	3.3858	R	4.0132	0.0001	90	80
405.20	1300			1276	81	0.0030	3.3868	T MANOR	4.0131	0.0031	1300	80
405.20				74	D 81	0.0001	3.3868		4.0100			
405.41	1105		80	1105	80	0.0026	3.3915	L	4.0100	0.0028	1105	80
405.41							3.3915		4.0074	0.0071	1060	D 11
405.63	1200			1200	80	0.0028	3.3943		4.0003	0.0009	531	117
405.63				160	90	0.0004	3.3947	330	4.0044			
405.90	1400			3291	A 80	0.0060	3.4006	T SHAFTON	4.0044	0.0016	1002	117
406.62	3610		117	2536	117	0.0041	3.4050	R	4.0028	0.0062	3810	117
406.62				1360	117	0.0022	3.4072		3.9966	0.0037	2858	D 18
407.52	4720		222	8686	A 11	0.0118	3.4190	R IRWIN	3.9929	0.0002	174	164
407.59	378		222			0.0000	3.4190	R	3.9927	0.0000		
408.44	4500					0.0000	3.4190		3.9927	0.0059	5444	A 16
408.49	385					0.0000	3.4190		3.9868	0.0000		
408.76	1400		165	1036	165	0.0012	3.4202	R LARIMER	3.9868	0.0065	5230	A 14
408.76				1360	165	0.0016	3.4218		3.9803			
408.76				1041	A 16	0.0012	3.4230		3.9803			
409.44	3600			211	169	0.0002	3.4232		3.9803	0.0014	1075	141
409.49	285			1273	D 16	0.0016	3.4248	JAN152014	3.9789	0.0004	285	141
409.66	900		141	900	141	0.0012	3.4260	L ARDARA	3.9785	0.0012	900	141
409.78	602		185	600	141	0.0008	3.4268	L ARA	3.9773	0.0016	1325	D 17
410.44	3500			790	141	0.0010	3.4278	MIDPOINT	3.9757	0.0004	290	170
410.82	2000			5230	A 14	0.0065	3.4342		3.9754	0.0000		
411.34	2749		166	2250	166	0.0026	3.4368	R	3.9754	0.0000		
411.42	403		185	1798	185	0.0021	3.4389	RCR TRAFF	3.9754	0.0150	11206	A 10
411.42				1041	A 18	0.0012	3.4401	TRAFFORD	3.9604			
412.35	4900			21	168	0.0000	3.4401		3.9604	0.0024	1360	166
412.35				2478	D 16	0.0034	3.4435		3.9580			
412.86	2700		108	2700	108	0.0047	3.4482	L	3.9580	0.0047	2700	106
413.38	2743		122	1360	108	0.0024	3.4506	L	3.9533	0.0008	472	D 12
413.63	1300		122	1547	A 10	0.0027	3.4533	L	3.9525	0.0048	2867	122
413.63				1360	127	0.0021	3.4554		3.9479	0.0010	614	A 11
413.63				1340	A 12	0.0020	3.4574		3.9469			
414.06	2306			122	131	0.0002	3.4576		3.9468	0.0015	640	117
414.06				539	D 13	0.0008	3.4584	PITCAIRN YARD	3.9454	0.0011	649	117
414.35	1500		117	1500	117	0.0024	3.4608	R WALL	3.9443	0.0000		117
414.60	1301			1301	117	0.0022	3.4630		3.9443	0.0070	3602	A 74
414.87	1450			195	117	0.0003	3.4633		3.9373	0.0035	1360	74
414.87				1205	D 11	0.0024	3.4657		3.9348			
415.19	1700		74	1700	74	0.0044	3.4701	R	3.9338	0.0044	1700	74
415.40	1083			1083	74	0.0028	3.4729	WILMERDING	3.9294	0.0009	411	D 91
415.46						0.0000	3.4729	CP WING F PERRY BR	3.9285	0.0008	188	91
415.54	748		99	257	74	0.0007	3.4736	L	3.9282	0.0038	1632	A 62
415.60	306			1275	A 74	0.0026	3.4764		3.9244	0.0007	60	62
415.64	1300			79	94	0.0002	3.4766		3.9242	0.0040	1300	62
415.84				722	D 94	0.0018	3.4783		3.9202	0.0001		
416.05	1066		62	1066	62	0.0033	3.4816	L	3.9202	0.0086	2684	62
416.12	400		62	400	62	0.0012	3.4829	L	3.9114	0.0012	400	62
416.54	2208			1380	62	0.0042	3.4870		3.9102	0.0010	298	A 54
416.61	400		164	562	A 62	0.0026	3.4896	L	3.9092	0.0017	408	54
416.69	406			65	83	0.0001	3.4897		3.9075	0.0036	682	54
416.69				583	D 83	0.0016	3.4913		3.9046			
416.77	400		54	400	54	0.0014	3.4927	L WESTINGHOUSE BR	3.9046	0.0002	58	A 52
416.82	300			300	54	0.0011	3.4938	RC	3.9043	0.0007	180	52
416.96	700		54	700	54	0.0025	3.4963	R	3.9036	0.0026	700	52
417.05	480		62	460	54	0.0017	3.4979	R	3.9011	0.0017	480	52
417.05				31	D 54	0.0001	3.4980		3.8994			
417.16	600		52	600	52	0.0022	3.5002	R	3.8994	0.0022	600	52
417.25	500			500	52	0.0018	3.5021	RC	3.8972	0.0004	114	D 59
417.35	480		58	480	52	0.0017	3.5038	L	3.8968	0.0028	868	59
417.48	705			389	52	0.0014	3.5052		3.8940	0.0056	1605	58
417.57	500		74	245	A 52	0.0008	3.5058	L	3.8882	0.0020	537	A 42
418.05	2500			735	58	0.0024	3.5083		3.8860	0.0061	1360	42
418.05				811	A 58	0.0017	3.5100	PT PERRY TUNNEL	3.8801			
418.05				1153	74	0.0030	3.5130	MONONGAHELA BR MP 10	3.8801			
418.05				544	D 74	0.0018	3.5148	JCT PORT PERRY BR	3.8801			
418.15	524		42	524	42	0.0024	3.5171	R	3.8801	0.0024	524	42
418.15						0.0000	3.5171	mp 18 (OLD MP 10)	3.8777	0.0002	50	D 46
418.18	205			205	42	0.0009	3.5180	REVERSE CURVE	3.8775	0.0006	148	48
418.26	400		46	400	42	0.0018	3.5198	L	3.8769	0.0016	400	46
418.26							3.5196		3.8753	0.0029	1127	D 96
418.47	1106			700	42	0.0034	3.5232		3.8724	0.0000	23	98
418.62	800		165	117	A 42	0.0005	3.5238	R	3.8724	0.0014	750	98
418.75	700		99	488	46	0.0020	3.5256	R	3.8710	0.0013	700	98
418.91	856		147	2731	A 46	0.0069	3.5327	R	3.8697	0.0018	1118	D 13
419.12	1100			4646	A 98	0.0077	3.5404		3.8679	0.0001	45	132
419.14	105		141			0.0000	3.5404	R	3.8678	0.0000		99
419.27	700		141			0.0000	3.5404	R	3.8678	0.0000		89
419.43	800					0.0000	3.5404		3.8678	0.0000		98
419.46	180		165			0.0000	3.5404	L	3.8678	0.0000		89
419.65	1000					0.0000	3.5404		3.8678	0.0000		89
419.91	1400		165			0.0000	3.5404	L	3.8678	0.0094	5395	A 85
420.14	1200					0.0000	3.5404		3.8584	0.0018	780	80
420.14				4	186	0.0000	3.5404	L	3.8566	0.0007	0	80
420.25	580			1774	D 13	0.0031	3.5436	CONNECTION ALIGNMENT	3.8568	0.0014	580	80

APPENDIX B3

READ DOWN WESTBOUND TRAFFIC — STATION — READ UP EASTBOUND TRAFFIC

PAGE 8 COLUMN 1

MILES	DISTANCE, ft	HWY GRADE CROSSING	MAX MPH	DISTANCE, ft	SPEED, mph	ET, hours	TIME, hours	STATION	TIME, hours	ET, hours	DISTANCE, ft	SPEED, mph
420.32	385		80	501	80	0.0012	3.5447	L	3.8562	0.0009	386	80
420.56	1264			1380	80	0.0032	3.5479	PRIOR CONNECTION REM	3.8543	0.0021	1069	D 117
420.63	364		165	3281	A 80	0.0063	3.5542	R	3.8522	0.0000	559	117
420.90	1402					0.0000	3.5542		3.8513	0.0014	890	117
421.14	1254		117		117	0.0000	3.5542	L	3.8499	0.0000	0	80
421.14						0.0000	3.5542		3.8496	0.0000	0	80
421.14						0.0000	3.5542		3.8499	0.0060	3172	A 82-1
421.66	2746		117	1200	117	0.0019	3.5561	L	3.8439	0.0000		
421.66		PRR		91	117	0.0001	3.5562	AMITY STREET	3.8439	0.0031	1380	82
421.66		P&LE		1022	D 74	0.0019	3.5582	AMITY STREET	3.8408	0.0000		
421.67	46		82	46	82	0.0001	3.5583	L	3.8408	0.0001	46	82
421.81	758			758	82	0.0018	3.5600		3.8407	0.0020	1058	D 118
421.84	152		119	152	82	0.0003	3.5603	L EASTWARD TRACK	3.8387	0.0000	22	118
421.84	19		119	19	82	0.0000	3.5604	REVERSE CURVE	3.8387	0.0000		
421.87	149			149	82	0.0003	3.5607		3.8387	0.0000		
421.90	129		116	129	82	0.0003	3.5610	R EASTWARD TRACK	3.8387	0.0002	129	118
422.03	703			160	82	0.0004	3.5614		3.8385	0.0011	703	118
422.22	1000			3302	A 82	0.0062	3.5675		3.8374	0.0003	187	118
422.41	1000					0.0000	3.5675		3.8371	0.0000		
422.53	660					0.0000	3.5675		3.8371	0.0000		
422.60	340					0.0000	3.5675		3.8371	0.0055	2990	A 87-1
422.79	1000			618	A 11	0.0010	3.5685		3.8316	0.0018	829	87
422.79				61	128	0.0001	3.5686		3.8296	0.0000		
422.89	540			1109	D 12	0.0020	3.5706	GLENWOOD BRIDGE	3.8298	0.0012	540	87
422.98	490		87	490	87	0.0010	3.5716	R	3.8268	0.0010	490	87
423.09	596		87	596	87	0.0018	3.5729	R	3.8276	0.0013	596	87
423.17	415			415	87	0.0006	3.5738		3.8263	0.0009	476	D 104
423.17							3.5738		3.8254	0.0002	133	104
423.29	632			632	87	0.0014	3.5752		3.8252	0.0000		
423.36	368		136	313	87	0.0007	3.5759	R	3.8252	0.0000		
423.46	553		138	1585	A 87	0.0031	3.5790	R	3.8252	0.0000		
423.51	233		165			0.0000	3.5790	R	3.8252	0.0035	1712	A 82-1
423.53	135		126			0.0000	3.5790	R	3.8217	0.0000	15	82
423.55	79		126			0.0000	3.5790	R	3.8217	0.0002	79	82
423.64	499		126			0.0000	3.5790	R	3.8215	0.0012	499	82
423.70	305		110			0.0000	3.5790	R	3.8203	0.0007	305	82
423.74	196					0.0000	3.5790		3.8196	0.0005	196	82
423.74				64	108	0.0001	3.5791		3.8191	0.0000		
423.79	266			662	D 10	0.0013	3.5804		3.8191	0.0006	266	82
423.89	539		82	539	82	0.0012	3.5817	R	3.8185	0.0012	539	82
423.93	196			196	82	0.0005	3.5821		3.8173	0.0029	1647	D 134
423.93						0.0000	3.5821		3.8144	0.0002	158	134
424.10	940			940	82	0.0022	3.5843		3.8142	0.0000		
424.12	80		144	80	82	0.0001	3.5844	R	3.8142	0.0000		
424.30	1000		144	166	82	0.0004	3.5848	R	3.8142	0.0000		
424.46	840		144	3920	A 82	0.0072	3.5920	R	3.8142	0.0000		
424.49	160					0.0000	3.5920		3.8142	0.0000		
424.54	290					0.0000	3.5920		3.8142	0.0000		
424.68	734		227			0.0000	3.5920	R	3.8142	0.0037	2445	A 117-
424.70	97		227			0.0000	3.5920	R	3.8105	0.0001	36	117
424.83	706					0.0000	3.5920	BECKS RUN	3.8104	0.0011	706	117
424.87	197					0.0000	3.5920		3.8093	0.0003	197	117
424.87				119	123	0.0002	3.5922		3.8090	0.0000		
424.95	421			211	D 12	0.0003	3.5925		3.8090	0.0007	421	117
425.06	579		117	579	117	0.0009	3.5934	R	3.8083	0.0009	579	117
425.07	62		117	62	117	0.0001	3.5935	R	3.8074	0.0000	11	117
425.20	652			652	117	0.0011	3.5946		3.8074	0.0000		
425.25	286		147	286	117	0.0005	3.5950	L	3.8074	0.0000		
425.30	239		147	239	117	0.0004	3.5954	L	3.8074	0.0000		
425.40	550		165	550	117	0.0009	3.5963	L	3.8074	0.0000		
425.44	211		117	211	117	0.0003	3.5967	L	3.8074	0.0000		
425.45	37		117	37	117	0.0001	3.5967	L	3.8074	0.0040	2263	A 95-1
425.53	439			439	117	0.0007	3.5974		3.8034	0.0004	202	95
425.63	524			524	117	0.0008	3.5983		3.8030	0.0010	524	95
425.75	634			397	117	0.0006	3.5988		3.8020	0.0013	634	95
425.80	289		95	348	A 11	0.0006	3.5995	L	3.8007	0.0005	289	95
425.82	97		165			0.0000	3.5995	L	3.8002	0.0008	419	D 110
425.82				195	120	0.0003	3.5998		3.7994	0.0002	135	109
425.83	44		165	104	D 12	0.0002	3.6000		3.7992	0.0000		
425.88	273		117	273	117	0.0004	3.6004	L	3.7992	0.0000		
426.01	683			663	117	0.0011	3.6015		3.7992	0.0000		
426.05	217						3.6015		3.7992	0.0000		
426.17	619		117			0.0000	3.6015	L	3.7992	0.0000		
426.20	164					0.0000	3.6015		3.7992	0.0000		
426.32	630			630	117	0.0010	3.6025		3.7992	0.0044	2219	A 82-1
426.39	370		117	370	117	0.0006	3.6031	L	3.7948	0.0007	324	82
426.39						0.0000	3.6031		3.7941	0.0003	136	82
426.42	175			175	117	0.0003	3.6034		3.7938	0.0015	615	A 70-4
426.54	622			622	117	0.0010	3.6044		3.7923	0.0001	46	70
426.58	203			205	117	0.0003	3.6047		3.7922	0.0005	203	70
426.58				34	117	0.0001	3.6048		3.7917	0.0000		
426.58	36			1022	D 11	0.0019	3.6067		3.7917	0.0001	38	70
426.64	297		82	297	82	0.0007	3.6074	L	3.7916	0.0015	541	70
426.71	385		131	385	82	0.0009	3.6083	L	3.7901	0.0005	174	A 66-1
426.77	262			125	82	0.0003	3.6086	REVERSE CURVE	3.7896	0.0007	249	66
426.79	111			268	D 11	0.0007	3.6092		3.7889	0.0003	111	66
426.89	556		70	556	70	0.0015	3.6107	R	3.7886	0.0016	558	66
426.95	333			333	70	0.0009	3.6116	REVERSE	3.7870	0.0010	333	66
426.95				31	70	0.0001	3.6117	STATION 29+75 P&LE	3.7860	0.0000		
426.98	111			80	D 70	0.0002	3.6119	CURVE	3.7860	0.0003	111	66
427.11	720		66	720	66	0.0021	3.6140	L	3.7857	0.0021	720	66
427.15	194			194	66	0.0006	3.6146		3.7836	0.0005	194	66
427.32	899			899	66	0.0026	3.6172		3.7830	0.0026	899	66
427.34	101		229	101	66	0.0003	3.6174	L	3.7804	0.0003	101	66
427.36	103		229	103	66	0.0003	3.6177	L	3.7801	0.0003	103	66
427.46	535			63	66	0.0002	3.6179	REVERSE CURVE	3.7798	0.0015	535	66
427.51	262		229	2693	A 66	0.0058	3.6237	R	3.7783	0.0008	262	66
427.53	100					0.0000	3.6237		3.7775	0.0003	100	66
427.54	46					0.0000	3.6237	REVERSE CURVE	3.7772	0.0001	45	66
427.59	268					0.0000	3.6237		3.7771	0.0008	268	66
427.62	189		166			0.0000	3.6237	L	3.7763	0.0005	189	66
427.72	466					0.0000	3.6237		3.7758	0.0017	434	66
427.75	185					0.0000	3.6237		3.7746	0.0014	439	A 54-1
427.80	651		106			0.0000	3.6237	R	3.7732	0.0015	481	54
427.91	164			164	106	0.0003	3.6240	REVERSE CURVE	3.7718	0.0008	164	54
427.93	147			147	106	0.0003	3.6243		3.7710	0.0005	147	54
427.99	285		66	354	66	0.0010	3.6253	L	3.7706	0.0010	285	54
428.05	303			211	D 66	0.0007	3.6260	REVERSE CURVE	3.7695	0.0011	303	54
428.08	156		54	156	54	0.0005	3.6265	R	3.7684	0.0005	156	54
428.10	109		88	109	54	0.0004	3.6269	R	3.7679	0.0013	466	D 78-4
428.10						0.0000	3.6269		3.7666	0.0001	47	78
428.12	122		88	122	54	0.0004	3.6273	LIBERTY BRIDGE	3.7665			78
428.12	11		88	11	54	0.0000	3.6274		3.7665			78
428.17	270		76	276	54	0.0009	3.6283	R	3.7665			78
428.29	600			600	54	0.0021	3.6304		3.7665	0.0003	118	D 83-1
428.35	356		165	248	54	0.0009	3.6313	R	3.7667	0.0040	1748	83
428.53	914			985	A 54	0.0028	3.6342	SMITHFIELD STREET	3.7622			
428.54	58			546	A 78	0.0015	3.6356		3.7622	0.0002	96	83
428.58	220		94			0.0000	3.6356	R	3.7620	0.0005	220	83
428.58				14	88	0.0000	3.6356	MP 0 P&LE STATION	3.7615	0.0000		
428.67	480			125	D 88	0.0003	3.6359	REVERSE CURVE	3.7615	0.0010	450	83
428.71	200		83	200	83	0.0005	3.6364	L	3.7605	0.0005	200	83
428.71						0.0000	3.6364		3.7600	0.0017	896	D 116
428.72	50			50	83	0.0001	3.6365		3.7583	0.0002	112	114
428.91	1000			1000	83	0.0023	3.6388		3.7581	0.0000		
429.05	750			310	83	0.0007	3.6395		3.7581	0.0000		
429.09	250		165	2279	A 83	0.0044	3.6439	R FORT PITT BRIDGE	3.7581	0.0000		

PAGE 8 COLUMN 2

MILES	DISTANCE, ft	HWY GRADE CROSSING	MAX MPH	DISTANCE, ft	SPEED, mph	ET, hours	TIME, hours	STATION	TIME, hours	ET, hours	DISTANCE, ft	SPEED, mph
429.17	400		165			0.0000	3.6439	R	3.7581	0.0034	1882	A 95
429.28	600					0.0000	3.6439		3.7547	0.0003	160	95
429.47	1000			160	110	0.0003	3.6442		3.7544	0.0020	1500	95
429.51	200			481	D 11	0.0008	3.6450		3.7524	0.0004	200	95
429.59	400		95	400	95	0.0008	3.6458	L	3.7520	0.0008	400	95
429.64	300		95	300	95	0.0006	3.6464	L	3.7512	0.0006	300	95
429.66	100			100	95	0.0002	3.6466		3.7506	0.0021	1232	D 13
429.79	650			650	95	0.0013	3.6479		3.7485	0.0002	172	132
429.85	350		135	350	95	0.0007	3.6486	R	3.7483	0.0000		
429.87	100		135	100	95	0.0002	3.6488	R	3.7483	0.0000		
430.04	900		169	160	95	0.0003	3.6491	R WEST END BRIDGE	3.7483	0.0000		
430.23	1000		169			0.0000	3.6491	R	3.7483	0.0000		
430.25	100		169				3.6491	R	3.7483	0.0000		
430.42	900			4512	A 95	0.0074	3.6566		3.7483	0.0000		
430.55	700						3.6566		3.7483	0.0000		
430.58	130		260				3.6566	R	3.7483	0.0000		
430.61	170		263				3.6566	R	3.7483	0.0081	4436	A 94
430.80	1000		293				3.6566	R	3.7402	0.0005	260	94
430.96	850		152				3.6566	R	3.7397	0.0017	850	94
430.99	150			30	133	0.0000	3.6566	REVERSE	3.7380	0.0003	150	94
431.01	100			1298	D 13	0.0022	3.6588	CURVE	3.7377	0.0002	100	94
431.09	450		94	450	94	0.0009	3.6597	L	3.7375	0.0009	450	94
431.18	450			450	94	0.0009	3.6606		3.7366	0.0024	1170	94
431.37	1000			910	94	0.0018	3.6624		3.7342	0.0008	370	A 89
431.48	500			987	A 94	0.0019	3.6643		3.7334	0.0011	510	89
431.56	400		100			0.0000	3.6643	L	3.7323	0.0009	400	89
431.59	190					0.0000	3.6643	REVERSE CURVE	3.7314	0.0004	190	89
431.64	260			98	105	0.0002	3.6645	CORLISS	3.7310	0.0006	260	89
431.64				455	D 10	0.0008	3.6654		3.7304			
431.75	580		89	550	89	0.0012	3.6666	R	3.7304	0.0012	550	89
431.85	560		89	580	89	0.0012	3.6677	R	3.7292	0.0012	550	89
431.85							3.6677		3.7280	0.0024	1394	D 13
431.94	450			450	89	0.0010	3.6687		3.7256	0.0001	56	132
432.13	1000	XING 3.4		910	89	0.0019	3.6706	RIVER AVENUE	3.7255	0.0000		
432.31	1000			4186	A 89	0.0072	3.6778		3.7255	0.0014	1000	132
432.50	1000					0.0000	3.6778		3.7241	0.0014	674	132
432.59	450					0.0000	3.6778		3.7227	0.0000		
432.69	550					0.0000	3.6778		3.7227	0.0000		
432.81	800					0.0000	3.6778		3.7227	0.0000		
432.88	400		132			0.0000	3.6778	R	3.7227	0.0000		
432.90	100		132			0.0000	3.6778	R	3.7227	0.0000		
432.95	250					0.0000	3.6778	REVERSE CURVE	3.7227	0.0000		
433.06	600		132			0.0000	3.6778	L	3.7227	0.0000		
433.07	50			1986	A 12	0.0028	3.6806		3.7227	0.0000		
433.26	1050					0.0000	3.6806		3.7227	0.0074	4436	A 94
433.45	1000					0.0000	3.6806		3.7153	0.0012	590	94
433.49	220					0.0000	3.6806		3.7141	0.0004	220	94
433.60	550			129	140	0.0002	3.6808		3.7137	0.0011	550	94
433.60				1579	D 14	0.0026	3.6833	5	3.7126	0.0000		
433.64	230		94	230	94	0.0005	3.6838	L DAVIS ISLAND	3.7126	0.0005	230	94
433.76	650		94	650	94	0.0013	3.6851	L	3.7121	0.0013	850	94
433.76							3.6851		3.7108	0.0013	712	D 11
433.83	350			350	94	0.0007	3.6858		3.7095	0.0001	38	117
434.02	1000			1000	94	0.0020	3.6878		3.7094	0.0010	600	117
434.21	1000			10	94	0.0000	3.6878	NEVILLE ISLAND	3.7084	0.0016	1000	117
434.32	600			2332	A 94	0.0042	3.6920		3.7068	0.0010	600	117
434.38	300		117			0.0000	3.6920	R	3.7058	0.0005	300	117
434.38							3.6920		3.7053	0.0024	1794	D 16
434.40	100					0.0000	3.6920		3.7029	0.0000	6	161
434.47	400					0.0000	3.6920		3.7029	0.0000		161
434.59	600			658	117	0.0011	3.6931		3.7029	0.0000		161
434.78	1000			260	117	0.0004	3.6935		3.7029	0.0004	300	161
434.82	240			7870	A 11	0.0106	3.7041		3.7025	0.0003	240	161
434.97	760		161			0.0000	3.7041	L	3.7022	0.0009	760	161
434.98	50		161			0.0000	3.7041	L	3.7013	0.0001	50	161
435.16	950					0.0000	3.7041		3.7012	0.0014	1272	D 16
435.34	1000					0.0000	3.7041		3.6998	0.0007	576	186
435.50	800					0.0000	3.7041		3.6991	0.0002	216	186
435.53	200		186			0.0000	3.7041	L	3.6989	0.0000		
435.59	300		186			0.0000	3.7041	L	3.6989	0.0000		
435.71	640		186			0.0000	3.7041	L	3.6989	0.0000		
435.72	60					0.0000	3.7041	REVERSE CURVE	3.6989	0.0000		
435.81	450					0.0000	3.7041		3.6989	0.0000		
435.87	300		186			0.0000	3.7041	R	3.6989	0.0000		
435.91	210					0.0000	3.7041	REVERSE CURVE	3.6989	0.0000		
435.91	40		186			0.0000	3.7041	L	3.6989	0.0000		
436.10	1000		186			0.0000	3.7041	L	3.6989	0.0000		
436.23	700		186			0.0000	3.7041	L	3.6989	0.0000		
436.28	250			4929	A 16	0.0055	3.7096	REVERSE CURVE	3.6989	0.0000		
436.29	50		186			0.0000	3.7096	R	3.6989	0.0000		
436.35	300		186			0.0000	3.7096	R	3.6989	0.0000		
436.48	700					0.0000	3.7096		3.6989	0.0000		
436.59	650					0.0000	3.7096		3.6989	0.0066	6122	A 16
436.67	450					0.0000	3.7096		3.6923	0.0000		
436.86	1000					0.0000	3.7096		3.6923	0.0000		
437.05	1000			35	179	0.0000	3.7096		3.6923	0.0000		
437.20	800			706	D 17	0.0008	3.7104		3.6923	0.0000		
437.24	200		165	200	165	0.0002	3.7106	R	3.6923	0.0000		
437.43	1000		165	1000	165	0.0011	3.7118	R	3.6923	0.0000		
437.56	700		165	700	165	0.0008	3.7126	R	3.6923	0.0000		
437.58	120	XING 8.95		1360	165	0.0016	3.7142	PRIVATE?	3.6923	0.0000		
437.58				243	A 16	0.0003	3.7144	9	3.6923	0.0000		
437.62	180					0.0000	3.7144		3.6923	0.0000		
437.81	1000					0.0000	3.7144		3.6923	0.0095	7402	A 12
438.00	1000	XING 9.4				0.0000	3.7144	LENORA AVENUE	3.6828	0.0004	260	128
438.19	1000			156	196	0.0002	3.7146	2313620.1	3.6824	0.0015	1000	128
438.20	100			1639	D 18	0.0021	3.7167		3.6809	0.0001	100	128
438.31	500		128	550	128	0.0008	3.7175	L	3.6808	0.0008	550	128
438.38	350			350	128	0.0006	3.7181		3.6800	0.0028	2296	D 17
438.39	90			90	128	0.0001	3.7182		3.6772	0.0001	95	179
438.56	910			810	128	0.0013	3.7195		3.6771	0.0000		
438.74	950	XING 10.3		10	128	0.0000	3.7196	BROADWAY STREET	3.6771	0.0000		
438.75	50		186				3.7196	L CORAOPOLIS	3.6771	0.0000		
438.79	200	XING 1	186				3.7196	L MULBERRY STREET	3.6771	0.0000		
438.89	500	XING 10.5					3.7196	MILL STREET REV CURVE	3.6771	0.0000		
438.94	300		186				3.7196	R	3.6771	0.0000		
439.00	300			5346	A 12	0.0083	3.7279		3.6771	0.0000		
439.05	250		186				3.7279	R	3.6771	0.0000		
439.13	460						3.7279	REVCURVE	3.6771	0.0000		
439.14	50						3.7279		3.6771	0.0000		
439.20	300		186				3.7279	L	3.6771	0.0000		
439.27	400						3.7279	REVERSE CURVE	3.6771	0.0000		
439.29	100	XING 1	186				3.7279	R MAIN STREET	3.6771	0.0000		
439.32	150						3.7279		3.6771	0.0000		
439.43	650						3.7279		3.6771	0.0000		
439.47	250		293				3.7279	R	3.6771	0.0000		
439.51	200	XING 10.9					3.7279	WATT STREET	3.6771	0.0000		
439.59	400						3.7279		3.6771	0.0000		
439.70	600						3.7279		3.6771	0.0055	4929	A 16
439.89	1000			0	160	0.0000	3.7279		3.6716	0.0000		
440.02	700	XING 11.4		0		0.0000	3.7279	PRIVATE?	3.6716	0.0000		
440.08	300		161	1604	161	0.0019	3.7298	R	3.6716	0.0000		
440.27	1000		161	1000	161	0.0012	3.7310	R	3.6716	0.0035	2990	A 14
440.32	250	XING 1	161	250	161	0.0003	3.7312	R PRIVATE?	3.6681	0.0000		
440.46	750	XING 11.6		750	161	0.0009	3.7321	PRIVATE?	3.6681	0.0000		
440.58	650			910	161	0.0007	3.7328		3.6681	0.0000		
440.85	350			350	161	0.0005	3.7326		3.6681	0.0000		

APPENDIX B3

PAGE 9 COLUMN 1

MILES	DISTANCE, ft	RWY GRADE CROSSING	MAX MPH	DISTANCE, ft (Westbound, read down)	SPEED, mph	ET, HOURS	TIME, HOURS	STATION	TIME, HOURS (Eastbound, read up)	ET, HOURS	DISTANCE, ft	SPEED, mph
440.77	650			650	161	0.0000	3.7328		3.6681	0.0000		
440.84	350		200	701	161	0.0008	3.7337	R	3.6681	0.0000		
441.01	900		200	589	D 16	0.0007	3.7344	R	3.6681	0.0000		
441.03	100		148	90	148	0.0001	3.7345	R	3.6681	0.0071	4960	A 117
441.22	1000		148			0.0000	3.7345	R	3.6610	0.0003	183	117
441.33	600		148			0.0000	3.7345	R STOOPS FERRY	3.6607	0.0010	660	117
441.41	400			9	170	0.0000	3.7345		3.6597	0.0006	400	117
441.44	200			2231	D 17	0.0029	3.7374	REVERSE CURVE	3.6591	0.0003	200	117
441.58	740		117	740	117	0.0012	3.7386	L	3.6588	0.0012	740	117
441.59	60		117	60	117	0.0001	3.7387	L	3.6576	0.0001	60	117
441.71	600		117	600	117	0.0010	3.7397	L	3.6575	0.0010	600	117
441.78	400			400	117	0.0006	3.7403		3.6565	0.0016	1119	D 146
441.96	960			960	117	0.0015	3.7419		3.6549	0.0003	231	146
441.97	50		146	10	117	0.0000	3.7419	L	3.6546	0.0001	50	146
442.12	750		146	4603	A 11	0.0066	3.7485	L	3.6545	0.0029	2296	146
442.16	250					0.0000	3.7485		3.6516	0.0000		
442.35	1000					0.0000	3.7485		3.6516	0.0000		
442.54	1000					0.0000	3.7485		3.6516	0.0000		
442.58	180					0.0000	3.7485		3.6516	0.0000		
442.73	820					0.0000	3.7485		3.6516	0.0032	2354	A 133
442.92	1000			855	146	0.0011	3.7496		3.6484	0.0006	410	133
443.10	950	XING 14.9		532	D 14	0.0007	3.7503	PRIVATE?	3.6478	0.0014	950	133
443.11	50		133	50	133	0.0001	3.7504	R GLENWILLARD	3.6464	0.0001	50	133
443.30	1000		133	1000	133	0.0014	3.7518	R	3.6463	0.0014	1000	133
443.32	100		133	100	133	0.0001	3.7520	R	3.6449	0.0001	100	133
443.49	900			900	133	0.0013	3.7532		3.6448	0.0013	975	D 164
443.58	500			460	133	0.0007	3.7539		3.6436	0.0003	222	156
443.67	460					0.0000	3.7539		3.6432	0.0000		
443.68	40		233			0.0000	3.7539	R	3.6432	0.0000		
443.87	1000		233			0.0000	3.7539	R	3.6432	0.0000		
443.95	450		233	4919	A 13	0.0064	3.7603	R	3.6432	0.0000		
444.06	560					0.0000	3.7603		3.6432	0.0000		
444.25	1000	XING 15.6				0.0000	3.7603	PRIVATE	3.6432	0.0000		
444.44	1000					0.0000	3.7603		3.6432	0.0067	5083	A 129
444.58	750					0.0000	3.7603	COUNTY LINE ALLE BEAVE	3.6365	0.0005	360	129
444.63	250			150	158	0.0002	3.7605		3.6360	0.0004	250	129
444.77	750			1221	D 15	0.0016	3.7621	SOUTH HEIGHTS	3.6356	0.0011	750	129
444.81	250		129	250	129	0.0004	3.7624	R	3.6345	0.0004	250	129
444.96	750		129	750	129	0.0011	3.7635	R	3.6341	0.0011	750	129
445.00	250			250	129	0.0004	3.7639		3.6330	0.0038	3251	D 197
445.19	1000			1000	129	0.0015	3.7654		3.6292	0.0002	176	197
445.38	1000			110	129	0.0002	3.7655		3.6290	0.0000		
445.57	1000			13383	A 12	0.0151	3.7806		3.6290	0.0000		
445.59	120					0.0000	3.7806		3.6290	0.0000		
445.76	880					0.0000	3.7806		3.6290	0.0000		
445.89	700					0.0000	3.7806		3.6290	0.0000		
445.95	300		233			0.0000	3.7806	R	3.6290	0.0038	3893	A 186
446.14	1000		233			0.0000	3.7806	R	3.6252	0.0000		
446.32	940		233			0.0000	3.7806	R	3.6252	0.0000		
446.33	90					0.0000	3.7806		3.6252	0.0000		
446.52	1000					0.0000	3.7806		3.6252	0.0010	960	186
446.59	380					0.0000	3.7806		3.6247	0.0004	380	186
446.59	20					0.0000	3.7806		3.6238	0.0000	20	186
446.65	300		186			0.0000	3.7806	L	3.6238	0.0003	300	186
446.71	300					0.0000	3.7806	REVERSE CURVE	3.6235	0.0003	300	186
446.72	40					0.0000	3.7806	WEST ECONOMY	3.6232	0.0000	40	186
446.75	200		186			0.0000	3.7806	R	3.6232	0.0008	818	186
446.90	760					0.0000	3.7806		3.6224			
447.03	720					0.0000	3.7806		3.6224			
447.07	200		186			0.0000	3.7806	L	3.6224			
447.09	60					0.0000	3.7806	REVERSE CURVE	3.6224			
447.12	180					0.0000	3.7806		3.6224			
447.17	300		186			0.0000	3.7806	R	3.6224			
447.28	580					0.0000	3.7806		3.6224			
447.47	1000					0.0000	3.7806		3.6224			
447.58	590					0.0000	3.7806		3.6224			
447.65	380					0.0000	3.7806		3.6224	0.0008	6122	A 165
447.66	30		233	1031	A 18	0.0010	3.7817	L	3.6158			
447.84	1000		233			0.0000	3.7817	L ALIQUIPPA	3.6158			
447.91	350		233			0.0000	3.7817	L	3.6158			
448.03	650			130	189	0.0001	3.7818		3.6158	0.0005	460	165
448.20	900			1246	D 16	0.0013	3.7831		3.6153	0.0010	900	165
448.22	100		165	100	165	0.0001	3.7832	R	3.6143	0.0001	100	165
448.34	600		165	600	165	0.0007	3.7839	R	3.6142	0.0007	600	165
448.41	400			400	165	0.0005	3.7844		3.6135	0.0005	400	165
448.58	900			900	165	0.0010	3.7854		3.6130	0.0010	900	165
448.60	100			100	165	0.0001	3.7855		3.6120	0.0001	100	165
448.65	250			250	165	0.0003	3.7858		3.6119	0.0003	250	165
448.76	600		165	600	165	0.0007	3.7865	L	3.6116	0.0007	600	165
448.79	150			150	165	0.0002	3.7867	WEST ALIQUIPPA	3.6109	0.0002	150	165
448.98	1000			1000	165	0.0011	3.7878		3.6107	0.0011	1000	165
449.17	1000			1000	165	0.0011	3.7890		3.6096	0.0011	1000	165
449.36	1000			1000	165	0.0011	3.7901		3.6085	0.0005	464	165
449.51	800			800	165	0.0009	3.7910		3.6080			
449.55	200		165	200	165	0.0002	3.7913	L	3.6080			
449.59	210			210	165	0.0002	3.7915		3.6080			
449.70	580		165	580	165	0.0007	3.7922	L	3.6080	0.0009	779	A 162
449.74	200		208	200	165	0.0002	3.7924	L	3.6071			
449.85	600		208	600	165	0.0007	3.7931	L	3.6071			
449.93	400			400	165	0.0005	3.7936		3.6071			
450.12	1000			180	165	0.0002	3.7937	HOPEWELL TOWNSHIP	3.6071			
450.14	100			1854	A 16	0.0024	3.7961		3.6071	0.0000		
450.31	900		229			0.0000	3.7961	L	3.6071			
450.36	300		229	298	172	0.0003	3.7964	L	3.6071			
450.46	800			480	D 17	0.0006	3.7970	REVERSE CURVE	3.6071			
450.50	200		162	200	162	0.0002	3.7972	R	3.6071			
450.59	480		162	480	162	0.0006	3.7978	R	3.6071			
450.69	520		162	400	162	0.0005	3.7983	R	3.6071			
450.70	80		162	0	162	0.0000	3.7983	R	3.6071			162
450.88	920			0	162	0.0000	3.7983		3.6071	0.0109	8107	A 117
451.06	1000					0.0000	3.7983		3.5962	0.0003	210	117
451.15	450					0.0000	3.7983		3.5959	0.0007	480	117
451.25	550		165			0.0000	3.7983	R	3.5952	0.0009	550	117
451.28	150		165	1428	162	0.0017	3.7999	R	3.5943	0.0002	150	117
451.28				1841	D 16	0.0025	3.8024		3.5941			
451.34	300		117	300	117	0.0005	3.8029	R	3.5941	0.0005	300	117
451.36	100			100	117	0.0002	3.8031		3.5936	0.0002	100	117
451.40	200		117	200	117	0.0003	3.8034	R	3.5934	0.0003	200	117
451.44	250			250	117	0.0004	3.8038		3.5931	0.0005	321	D 128
451.46	100			100	117	0.0002	3.8040		3.5926	0.0002	100	128
451.48	80		165	80	117	0.0001	3.8041	R	3.5924			
451.58	540			540	117	0.0008	3.8050		3.5924			
451.63	280			280	117	0.0005	3.8054		3.5924			
451.76	650			582	117	0.0009	3.8064		3.5924			
451.76				68	D 11	0.0001	3.8065		3.5924	0.0023	1460	A 115
451.82	350		115	350	115	0.0006	3.8070	L	3.5901	0.0006	350	115
452.01	1000		115	1000	115	0.0016	3.8087	L COLONA	3.5895	0.0016	1000	115
452.18	900		115	900	115	0.0015	3.8102	L	3.5879	0.0015	900	115
452.20	100			100	115	0.0002	3.8103		3.5864	0.0006	577	D 131
452.39	1000			1000	115	0.0016	3.8120		3.5858	0.0008	523	131
452.46	350			260	115	0.0004	3.8124		3.5847	0.0005	350	131
452.58	650		131	2214	A 11	0.0033	3.8157	L	3.5842	0.0001	58	131
452.59	80		131			0.0000	3.8157	L	3.5841	0.0000		
452.61	70		131			0.0000	3.8157	L	3.5841	0.0000		
452.77	850					0.0000	3.8157		3.5841	0.0000		
452.96	1000			36	131	0.0001	3.8157		3.5841	0.0000		
453.15	1000	XING 2	135	634	A 13	0.0007	3.8164	L 4TH STREET	3.5641	0.0000		

APPENDIX B3

PAGE 9 COLUMN 2

MILES	DISTANCE, ft	RWY GRADE CROSSING	MAX MPH	DISTANCE, ft (Westbound, read down)	SPEED, mph	ET, HOURS	TIME, HOURS	STATION	TIME, HOURS (Eastbound, read up)	ET, HOURS	DISTANCE, ft	SPEED, mph
453.17	100		135	956	135	0.0013	3.8176	L MONACA	3.5841	0.0070	4222	A 95
453.34	900			561	135	0.0008	3.8185		3.5771	0.0007	360	95
453.45	600					0.0000	3.8185	OHIO RIVER BRANCH TO K	3.5764	0.0012	600	95
453.59	400				135	0.0000	3.8185		3.5752	0.0008	400	95
453.53				1349	D 12	0.0022	3.8207		3.5744	0.0000		
453.57	250		95	250	95	0.0005	3.8212	R	3.5744	0.0005	250	95
453.72	750		95	750	95	0.0015	3.8227	R	3.5739	0.0015	750	95
453.73	80		95	80	95	0.0002	3.8229	R	3.5724	0.0002	80	95
453.79	320			320	95	0.0006	3.8235		3.5722	0.0006	320	95
453.79						0.0000	3.8235	OHIO RIVER S BANK	3.5716	0.0000		
453.91	600			600	95	0.0012	3.8247		3.5716	0.0017	838	95
454.04	700			440	95	0.0009	3.8256		3.5699	0.0000		
454.04				248	A 95	0.0005	3.8261	OHIO RIVER N BANK	3.5699	0.0000		
454.09	300				98	0.0000	3.8261		3.5699	0.0000		
454.17	400		98		98	0.0000	3.8261	R	3.5699	0.0000		
454.26	600		98		98	0.0000	3.8261	R	3.5699	0.0000		
454.32	200		98	200	98	0.0004	3.8265	R BEAVER	3.5699	0.0000		95
454.40	400			400	98	0.0008	3.8273		3.5699	0.0073	2602	A 37
454.47	400		233	400	98	0.0006	3.8280	R	3.5626	0.0008	180	37
454.58	570			560	98	0.0011	3.8291		3.5618	0.0009	570	37
454.58						0.0000	3.8291	26	3.5609	0.0016	305	37
454.64	330					0.0000	3.8291		3.5573	0.0002	40	A 35
454.66	100		233	585	A 98	0.0010	3.8302	L	3.5571	0.0006	85	35
454.70	200		233	11	104	0.0000	3.8302	L BEAVER NEWSPAPERS B	3.5565	0.0011	200	35
454.70				1386	D 98	0.0037	3.8339	REVISE TO 180 DEGREE C	3.5555	0.0000		
454.87	905		37	905	37	0.0046	3.8386	L	3.5555	0.0049	905	35
454.89	100			100	37	0.0005	3.8390		3.5506	0.0005	100	35
454.89	20		117	20	37	0.0001	3.8391	L	3.5501	0.0001	20	35
454.90	20			20	37	0.0001	3.8392		3.5500	0.0001	20	35
454.90	20		42	9	37	0.0000	3.8393	L	3.5499	0.0001	20	35
454.90	10			21	D 37	0.0001	3.8394	REVERSE CURVE	3.5498	0.0001	10	35
454.91	25		95	25	35	0.0001	3.8395	R	3.5497	0.0001	25	35
454.91	5		42	5	35	0.0000	3.8396	R	3.5496	0.0001	21	D 37
454.92	50		42	50	35	0.0003	3.8398	R	3.5495	0.0002	34	37
454.93	50	XX		50	35	0.0003	3.8401	SHARON ROAD	3.5493	0.0003	50	37
454.94	65			65	35	0.0004	3.8405	REVERSE CURVE	3.5490	0.0003	65	37
454.94	14		94	14	35	0.0001	3.8405	L	3.5487	0.0001	14	37
454.95	6			6	35	0.0000	3.8406	REVERSE CURVE	3.5486	0.0000	6	37
454.95	10		108	10	35	0.0001	3.8406	R	3.5486	0.0001	10	37
454.95	5			5	35	0.0000	3.8406	TRACK IN 5TH AVENUE	3.5485	0.0000	5	37
454.97	100			100	35	0.0006	3.8412	SHARON ROAD	3.5485	0.0005	100	37
454.99	100			100	35	0.0005	3.8417	TO BUFFALO STREET	3.5480	0.0005	100	37
455.00	100			100	35	0.0006	3.8423	REVERSE CURVE	3.5475	0.0005	100	37
455.01	25	XX		25	35	0.0001	3.8424	BUFFALO STREET	3.5470	0.0001	25	37
455.02	35		42	35	35	0.0002	3.8426	L	3.5469	0.0002	35	37
455.02	40			40	35	0.0002	3.8428	REVERSE CURVE	3.5467	0.0002	40	37
455.03	16	XING		16	35	0.0001	3.8429	TUSCARAWAS ROAD	3.5465	0.0001	16	37
455.03	24		42	24	35	0.0001	3.8430	R	3.5464	0.0001	24	37
455.04	60			60	35	0.0003	3.8434		3.5463	0.0003	60	37
455.05	60			60	35	0.0003	3.8437		3.5460	0.0003	60	37
455.06	30		42	30	35	0.0002	3.8438	R	3.5457	0.0002	30	37
455.06	10	XING		10	35	0.0001	3.8439	WALNUT LANE	3.5456	0.0001	10	37
455.07	35			35	35	0.0002	3.8441	REVERSE CURVE	3.5454	0.0002	35	37
455.08	55		78	55	35	0.0003	3.8444	L	3.5452	0.0003	55	37
455.08	10		95	10	35	0.0001	3.8444	L	3.5449	0.0001	10	37
455.09	56		95	56	35	0.0003	3.8447	L	3.5448	0.0003	56	37
455.10	45	XING		45	35	0.0002	3.8450	SPRING LANE	3.5445	0.0002	45	37
455.12	100			100	35	0.0006	3.8456		3.5443	0.0005	100	37
455.12	25			25	35	0.0001	3.8457		3.5438	0.0001	25	37
455.13	35	XING	74	35	35	0.0002	3.8458	L DIVISION LANE	3.5437	0.0002	35	37
455.13	10		82	10	35	0.0001	3.8459	L	3.5435	0.0001	10	37
455.14	30			30	35	0.0002	3.8461		3.5434	0.0002	30	37
455.15	65			65	35	0.0004	3.8464		3.5432	0.0003	65	37
455.16	35		44	35	35	0.0002	3.8466	L	3.5429	0.0002	35	37
455.16	15	XING	44	15	35	0.0001	3.8467	L DIVISION LANE	3.5427	0.0001	15	37
455.16	30	XING		30	35	0.0002	3.8468	RC PA HWY 68	3.5426	0.0002	30	37
455.18	55		48	10	35	0.0001	3.8469	R	3.5424	0.0003	55	37
455.18	20		48	40	A 35	0.0002	3.8471	R	3.5421	0.0001	20	37
455.18	16		57	41	37	0.0002	3.8473	R track removed - p&le to C&	3.5420	0.0001	16	37
455.18	10		38	10	37	0.0001	3.8474	R	3.5419	0.0001	11	D 35
455.19	11		54	11	37	0.0001	3.8474	R	3.5418	0.0000		38
455.19	15			15	37	0.0001	3.8475	REVERSE CURVE	3.5418	0.0006	209	D 53
455.19	15		53	15	37	0.0001	3.8476	L	3.5410	0.0000		53
455.19	3	XING		3	37	0.0000	3.8476	RC BEAVER INDUSTRIAL	3.5410	0.0000		
455.19	10		53	10	37	0.0001	3.8476	R TRACKS WERE IN 5TH S'	3.5410	0.0000		53
455.19						0.0000	3.8476	TRACKS HAVE BEEN REMC	3.5410	0.0052	2757	D 14
455.19						0.0000	3.8476	RC VANPORT	3.5358	0.0001	113	147
455.27	375			375	37	0.0019	3.8496	W BEAV 143	3.5357	0.0000		
455.40	700	XING 3.64		700	37	0.0038	3.8531	PRIVATE	3.5357	0.0000	0	147
455.83	2300		147	221	37	0.0011	3.8543	L	3.5357	0.0000	5	147
456.00	879			27	A 37	0.0001	3.8544		3.5357			
456.11	600		186	419	A 38	0.0017	3.8562	L MPH50(TT)	3.5357	0.0000		
456.11				7012	A 53	0.0135	3.8695	REVERSE CURVE	3.5357			
456.26	800		186			0.0000	3.8695	R	3.5357	0.0025	1878	A 13
456.38	600			0		0.0000	3.8695		3.5332	0.0023	1673	137
456.55	900		137	106	135	0.0001	3.8696	R	3.5309	0.0016	1127	A 13
456.91	1900			154	D 13	0.0000	3.8698		3.5293	0.0027	1868	130
457.08	900		130	900	130	0.0015	3.8712	R	3.5266	0.0006	615	A 12
457.26	978			1428	130	0.0021	3.8733		3.5257	0.0011	780	126
457.38	600			150	D 13	0.0002	3.8735		3.5248	0.0009	600	126
457.51	700		126	700	126	0.0011	3.8748	R	3.5237	0.0011	700	126
457.66	761		164	580	120	0.0010	3.8756	R	3.5225	0.0022	1713	D 16
458.26	3200			7681	A 12	0.0099	3.8856		3.5204	0.0003	254	166
458.87	3200	XING 6.19		1700	165	0.0020	3.8875	PRIVATE	3.5201	0.0000		
459.09	1173		165	753	165	0.0008	3.8883	R E INDUSTRY	3.5201	0.0000		
459.26	900				165	0.0000	3.8883		3.5201	0.0037	3101	A 15
459.66	2100	XING 7.85		580	D 16	0.0007	3.8890	WABASH STREET	3.5164	0.0000		
459.87	1089		153	1089	153	0.0013	3.8903	L W INDUSTRY	3.5164	0.0121	8115	A 97
460.27	2100	XING 8.06		47	153	0.0001	3.8904	PRIVATE	3.5043	0.0027	1360	97
460.27				2063	D 15	0.0031	3.8935		3.5016			
460.42	800		97	800	97	0.0016	3.8951	L	3.5016	0.0016	800	97
460.76	1800			1360	97	0.0027	3.8977		3.5000	0.0010	560	D 11
460.97	1100		295	1832	A 97	0.0033	3.9010	L	3.4990	0.0039	2340	115
461.21	1283			981	115	0.0016	3.9026		3.4951	0.0005	315	115
461.27	300		115	300	115	0.0005	3.9031		3.4946	0.0036	2008	A 95
461.44	900		115	120	115	0.0002	3.9033	R	3.4910	0.0003	180	95
461.66	1200			1364	115	0.0022	3.9056		3.4907	0.0024	1200	95
461.66				616	D 11	0.0011	3.9067		3.4883			
461.87	1100		95	1100	95	0.0022	3.9089	R	3.4883	0.0022	1100	95
461.87						0.0000	3.9089		3.4861	0.0016	895	D 12
462.04	900			900	95	0.0018	3.9107		3.4845	0.0000	5	123
462.26	1175		148	460	95	0.0009	3.9116	R	3.4845	0.0109	7068	123
463.26	5278		123	3011	A 95	0.0052	3.9168	R MIDLAND	3.4736	0.0002	134	A 12
463.32	290		188	4342	123	0.0067	3.9235	R	3.4734	0.0058	3184	A 91
463.32				2184	A 12	0.0032	3.9267		3.4676			
463.32				66	137	0.0001	3.9267		3.4678			
464.06	3900			570	D 13	0.0008	3.9276		3.4678	0.0005	260	91
464.27	1100		122	132	122	0.0002	3.9278	L	3.4673	0.0023	1100	91
464.27				968	D 12	0.0017	3.9295		3.4650			
464.44	900		91	900	91	0.0019	3.9314	L MPH45	3.4650	0.0019	900	91
464.65	1100			1100	91	0.0023	3.9337		3.4631	0.0019	1077	D 12
464.65							3.9337		3.4612	0.0001	83	125
464.76	800		195	260	91	0.0005	3.9342	L	3.4611	0.0012	777	125
464.91	800			3318	A 91	0.0058	3.9400	MIDL YD LEAD	3.4599	0.0000		
465.02	805		125			0.0000	3.9400	L	3.4599	0.0000		
465.13	568					0.0000	3.9400		3.4599	0.0000		
465.26	700		165			0.0000	3.9400	L	3.4599	0.0071	3892	A 67

READ DOWN — WESTBOUND TRAFFIC — APPENDIX B3 — STATION — READ UP — EASTBOUND TRAFFIC

PAGE 10 COLUMN 1

MILES	DISTANCE	XING / STATION	MAX MPH	DISTANCE	SPEED	GRADE	TIME	STATION	TIME	GRADE	DISTANCE	SPEED
465.81	2806			1457	128	0.0002	3.9427		3.4628	0.0042	1917	86
465.87		WEST END		1134	D 12	0.0021	3.9443		3.4486			
465.91	631		96	531	68	0.0012	3.9455	L OKWS	3.4498	0.0031	681	A 70
465.96	260	XING 19.86		268	86	0.0006	3.9461	CHURCH STREET	3.4465	0.0006	350	70
466.14	900		152	900	86	0.0026	3.9481	RMP/45	3.4457	0.0018	678	70
466.26	678			678	86	0.0015	3.9486		3.4426			
466.30	200		152	0	86	0.0000	3.9486	R	3.4438	0.0005	206	70
466.34	200			34	86	0.0001	3.9487	MPH 45	3.4434	0.0005	200	70
466.34				366	D 86	0.0006	3.9606	REVERSE CURVE	3.4429			
466.46	631		70	631	70	0.0017	3.9623	L	3.4429	0.0017	631	70
466.46							3.9623	LITTLE BEAVER	3.4412	0.0066	3670	D 121
466.52	300			300	70	0.0006	3.9631	REVERSE CURVE	3.4356	0.0008	734	121
466.63	600	XING 1	186	600	70	0.0016	3.9647	R PRIVATE	3.4348	0.0000		
466.74	400		186	460	70	0.0012	3.9659	L CURVE	3.4348	0.0000		
467.25	2700			10556	A 70	0.0173	3.9732	OKWS OHIO/PA LINE	3.4348	0.0000		
468.26	5332	XING 15.10				0.0000	3.9732	PUTNAM STREET	3.4348	0.0038	3200	A 156
468.34	450					0.0000	3.9732		3.4310	0.0000		
468.39	255	XING 1	156			0.0000	3.9732	L RC MYRTLE STREET	3.4310	0.0000		
468.48	300	XING 1	169			0.0000	3.9732	R WALTERS STREET	3.4310	0.0000		
468.69	1100	XING 16.11				0.0000	3.9732	M-H(S)(TT) VIRGINIA STREET	3.4310	0.0042	3357	A 142
468.88	1000		158	473	158	0.0008	3.9738	L	3.4268	0.0000		
469.09	1100		174	1100	158	0.0013	3.9751	L	3.4268	0.0000		
469.26	900			900	158	0.0003	3.9754		3.4268	0.0000		
469.75	2600			704	D 15	0.0008	3.9763	LW 30	3.4266	0.0103	8447	A 80
469.94	1000		142	2197	142	0.0020	3.9782	LTC	3.4165	0.0000	416	80
470.02	400				142	0.0000	3.9792		3.4156	0.0009	1305	A 69
470.02				1768	D 14	0.0029	3.9821		3.4128			
470.11	500		96	500	96	0.0011	3.9831	R OKWS	3.4126	0.0007	240	69
470.21	500		156	500	96	0.0011	3.9842	RMP/50(TT)	3.4121	0.0014	500	69
470.27	320			130	96	0.0009	3.9843		3.4107	0.0004	320	69
470.32	300				96	0.0000	3.9846		3.4096	0.0008	300	69
470.32				400	D 96	0.0012	3.9858		3.4096			
470.36	200	XING 1	99	100	69	0.0005	3.9862	R PRIVATE	3.4090	0.0005	200	69
470.38						0.0008	3.9867		3.4085	0.0083	105	D 74
470.48	700			700	69	0.0016	3.9881		3.4067	0.0005	590	74
470.54	252	XING 1	74	212	66	0.0007	3.9888	R BROADWAY	3.4067	0.0000	250	74
470.54						0.0000	3.9888		3.4061	0.0007	66	D 74
470.62	400	XING 16.40		400	69	0.0011	3.9899	TT MPH 39 WASHINGTON ST	3.4058	0.0006	211	74
470.73	600	XING 1	74		69	0.0000	3.9898	R MARKET STREET	3.4051	0.0015	500	74
470.78		XING 16.50				0.0000	3.9898	JACKSON STREET	3.4036	0.0019	947	D 112
470.85	800			296	A 69	0.0008	3.9906	L LIVERPOOL	3.4017	0.0001	55	112
470.96	500		113	718	74	0.0016	3.9924	R	3.4016	0.0000		
471.07	600			195	A 74	0.0005	3.9928	MPH/35	3.4016	0.0000		
471.19	600		115	805	78	0.0029	3.9958	R	3.4016	0.0000		
471.28	400			2454	A 78	0.0049	3.9967		3.4016	0.0000		
471.32	300					0.0000	3.9967		3.4016	0.0040	2136	A 68
471.41	500		199	78	109	0.0005	3.9986	L	3.3976	0.0019	460	68
471.58	900			607	D 16	0.0012	4.0010		3.3966	0.0019	101	88
471.65	300		88	020	88	0.0011	4.0021	L	3.3647	0.0011	500	88
471.65							4.0021		3.3636	0.0016	630	D 116
471.75	500		117	500	88	0.0011	4.0034	L	3.3620	0.0001	82	116
472.00	1100			780	88	0.0018	4.0051		3.3619	0.0015	800	116
472.17	900		116	2838	A 88	0.0048	4.0099	L	3.3606	0.0009	900	116
472.23	215					0.0000	4.0099		3.3601	0.0001	34	D 117
472.28	3000			128	A 11	0.0002	4.0101		3.3600	0.0000	3161	117
473.01	1200		117	2866	117	0.0047	4.0148	R	3.3629	0.0019	1200	117
473.25	1300			1337	117	0.0022	4.0170	MPH 55(TT)	3.3620	0.0030	2144	117
473.53	1400			2867	117	0.0046	4.0216		3.3783	0.0014	638	A 110
473.87	1800		117	1233	D 11	0.0004	4.0222	L	3.3768	0.0023	1380	110
474.04	1000		110	900	110	0.0015	4.0237	L	3.3745	0.0015	900	110
474.25	1131		126	1131	110	0.0019	4.0256	L OKWS	3.3730	0.0008	460	D 124
474.40	800			222	110	0.0004	4.0260		3.3722	0.0000	4	124
474.57	900		310	2458	A 11	0.0037	4.0297	R	3.3722	0.0000		
474.71	700					0.0000	4.0297		3.3722	0.0053	3037	A 95
474.80	600		105			0.0000	4.0297	L	3.3669	0.0011	560	95
474.97	600			24	128	0.0000	4.0298	MPH 50(TT)	3.3658	0.0016	800	95
474.97				1088	D 17	0.0018	4.0316		3.3642			
475.21	1260		96	1258	96	0.0025	4.0341	R	3.3642	0.0033	1258	96
475.27	300			300	96	0.0006	4.0347		3.3617	0.0003	174	D 102
475.56	500			500	96	0.0009	4.0356		3.3614	0.0000	23	102
475.56				560	96	0.0011	4.0368		3.3614			
475.53	800		117	1067	A 96	0.0016	4.0384	R	3.3614	0.0031	1523	A 82
475.79	1300	XING 25.41		25	107	0.0000	4.0387	18TH STREET	3.3583	0.0001	1586	82
475.79				683	D 10	0.0014	4.0400	WELLSVILLE	3.3582			
475.96	700		82	800	82	0.0021	4.0421	L	3.3582	0.0021	940	82
476.10	700			700	82	0.0016	4.0437		3.3531	0.0016	851	D 99
476.23	700		147	560	82	0.0015	4.0451	L	3.3523	0.0010	549	99
476.27	200			1286	A 82	0.0026	4.0477		3.3504	0.0025	395	99
476.42	800					0.0000	4.0479		3.3500	0.0029	475	99
476.59	900		99	1628	99	0.0051	4.0510	L	3.3481	0.0024	1265	A 82
476.85	1400			451	D 16	0.0009	4.0520		3.3463	0.0021	1360	82
477.04	1000		82	1000	82	0.0020	4.0540	L	3.3434	0.0028	1114	63
477.27	1200			750	82	0.0016	4.0556		3.3408	0.0024	1226	A 54
477.37	500				82	0.0000	4.0556		3.3314	0.0028	560	54
477.48	600		152	1482	82	0.0034	4.0563	L OKWS	3.3361	0.0021	690	54
477.56	400				82	0.0000	4.0593	RC	3.3346	0.0014	400	54
477.56				468	D 82	0.0019	4.0626		3.3326			
477.63	700		54	700	64	0.0025	4.0633	P/C OKWS	3.3326	0.0025	700	54
477.69						0.0000	4.0633		3.3301	0.0004	136	D 76
477.82	700	XING 29.73		700	54	0.0025	4.0657	TT MPH 39 PRIVATE	3.3297	0.0017	564	62
477.87	800		62	900	54	0.0023	4.0680	L	3.3280	0.0024	800	62
478.03	600			266	A 64	0.0010	4.0690	REVERSE CURVE TTMPH	3.3264	0.0010	25	D 66
478.27	950		78	1202	62	0.0037	4.0727	R	3.3254	0.0026	1247	66
478.53	1400			143	A 62	0.0005	4.0732	R	3.3219	0.0040	1400	98
478.74	1100			2811	98	0.0081	4.0813		3.3178	0.0028	1335	D 75
478.84	495		117	2826	A 98	0.0065	4.0875	L	3.3150	0.0002	120	116
479.10	1400		117			0.0000	4.0875	L YELLOW CREEK	3.3148	0.0038	2135	A 96
479.33	700					0.0000	4.0875		3.3110	0.0038	1856	96
479.42	1000			25	108	0.0000	4.0875		3.3072	0.0010	300	A 84
479.42				347	D 10	0.0007	4.0882		3.3062			
479.65	1200		85	1200	85	0.0024	4.0906	L	3.3062	0.0007	2036	86
479.88	1000			1012	85	0.0020	4.0927		3.3053	0.0016	697	A 77
479.88		XING 37.69		189	D 90	0.0004	4.0930	DANDOS CROSSING	3.2997	0.0000		
480.29	2100		68	2728	98	0.0059	4.0989	R	3.2997	0.0016	662	77
480.44	800		113	269	D 89	0.0006	4.0995	R	3.2981	0.0000	130	A 74
480.60	800		77	714	77	0.0018	4.1013	R	3.2976	0.0050	2691	74
480.60				66	D 77	0.0002	4.1015		3.2926			
480.73	800		74	500	74	0.0020	4.1035	RC	3.2926	0.0031	484	A 48
481.05	1600	XING 38.66		1000	74	0.0026	4.1070	HAMMONDSVILLE WASHN	3.2878	0.0030	86	48
481.18	700		120	195	A 74	0.0010	4.1075	L	3.2875	0.0036	700	48
481.29	575		76	796	76	0.0019	4.1094	L	3.2847	0.0029	670	48
481.29				564	D 76	0.0017	4.1110		3.2824			
481.46	900		46	900	46	0.0036	4.1146	L	3.2824	0.0036	900	46
481.50	723			125	46	0.0008	4.1156	REVC	3.2788	0.0001	24	D 50
481.50						0.0000	4.1156		3.2787	0.0003	116	50
481.71	1100		50	1100	46	0.0048	4.1198	R	3.2780	0.0042	1100	50
481.68	900			35	46	0.0001	4.1200		3.2738	0.0011	352	D 70
482.15	1400		70	72	A 46	0.0003	4.1202	L	3.2727	0.0000	1046	70
482.30		XING 30.04		800	50	0.0030	4.1233	EAST AVENUE	3.2774	0.0000	128	D 76
482.30						0.0000	4.1233		3.2871	0.0017	672	76
482.46	1000		76	483	50	0.0017	4.1250	R	3.2854	0.0025	1000	76
482.70	1100			727	A 50	0.0023	4.1273	ROUNDALE	3.2829	0.0003	13	D 80
482.94	1300		116	1573	76	0.0045	4.1318	L	3.2826	0.0025	1141	80
483.15	1100		83	285	A 76	0.0008	4.1323	L	3.2803	0.0016	420	80
483.30	790			1506	76	0.0034	4.1360		3.3243	0.0018	800	80
483.30				225	A 76	0.0008	4.1368		3.2025			
483.38	300			2470	62	0.0057	4.1425		3.2026	0.0016	315	A 81

READ DOWN — WESTBOUND TRAFFIC — APPENDIX B3 — STATION — READ UP — EASTBOUND TRAFFIC

PAGE 10 COLUMN 2

MILES	DISTANCE	XING / STATION	MAX MPH	DISTANCE	SPEED	GRADE	TIME	STATION	TIME	GRADE	DISTANCE	SPEED
483.38				440	D 82	0.0012	4.1437		3.2499			
483.76	2100		61	2100	61	0.0005	4.1502	R	3.2499	0.0038	1215	61
483.78	145			145	61	0.0005	4.1507	REVC	3.2461	0.0005	145	61
484.09	1600		61	1600	61	0.0050	4.1556	L OKWS	3.2456	0.0050	1600	61
484.31	1200		101	1200	61	0.0037	4.1593	L	3.2406	0.0017	566	D 61
484.82	1800			150	61	0.0005	4.1598	NEW GALILEE	3.2389	0.0049	2371	91
484.79	900		117	1627	A 61	0.0030	4.1632	R	3.2340	0.0014	660	91
484.92	700	XING 32.56		1463	61	0.0032	4.1666	PRIVATE	3.2326	0.0015	700	91
485.15	1200		91	1206	91	0.0035	4.1680	L OKWS	3.2311	0.0025	1200	91
485.31	865			805	91	0.0018	4.1711		3.2286	0.0011	552	91
485.49	900			505	91	0.0011	4.1721		3.2275	0.0041	1647	A 69
485.65	900		104	327	A 91	0.0008	4.1727	L OKWS	3.2234	0.0016	469	69
485.82	900			1070	95	0.0021	4.1749		3.2219	0.0028	900	69
485.82				813	D 80	0.0020	4.1769		3.2190			
486.07	1300		69	1300	69	0.0042	4.1810	R	3.2190	0.0042	1300	69
486.07						0.0000	4.1810		3.2148	0.0013	300	D 83
486.10						0.0000	4.1810		3.2135	0.0005	110	83
486.10	150			150	69	0.0007	4.1817	REVC	3.2132	0.0004	0	83
486.24	700		83	700	69	0.0022	4.1839	L OKWS	3.2132	0.0006	280	83
486.29	300			300	69	0.0010	4.1848		3.2126	0.0024	1040	D 12
486.82	2800			170	69	0.0005	4.1854		3.2103	0.0000	1	124
487.03	1100		120	1101	A 69	0.0029	4.1883	R	3.2103	0.0063	3668	A 90
487.31	1471			476	A 80	0.0019	4.1902		3.2040	0.0018	995	90
487.40	500			2135	A 90	0.0051	4.1949		3.2022	0.0011	500	90
487.40				49	D 120	0.0001	4.1950		3.2011			
487.40				858	D 12	0.0017	4.1967		3.2011			
487.58	900	XING 3	90	900	90	0.0019	4.1986	L McGARY	3.2011	0.0082	3605	96
488.05	2500	XING 35.71		1360	90	0.0029	4.2015	NIXON	3.1929	0.0026	1202	A 69
488.29	1267		102	2786	A 90	0.0062	4.2067	R	3.1901	0.0027	931	69
488.31	100			67	118	0.0000	4.2117		3.1874	0.0003	100	69
488.53	1100	XING 36.11		1344	D 11	0.0027	4.2144	PRIVATE	3.1871	0.0009	305	A 61
488.64	1100		85	1747	88	0.0034	4.2178	R	3.1862	0.0030	1034	61
488.69	300	XING 36.50			88	0.0000	4.2178	REV C McKAY	3.1832	0.0002	38	A 59
488.83				153	D 63	0.0004	4.2183		3.1830			
489.15	1800		61	1800	61	0.0059	4.2242	ROGERS	3.1830	0.0040	1180	59
489.27	100			68	61	0.0002	4.2244	RC	3.1780	0.0003	100	59
489.27				35	D 61	0.0001	4.2245		3.1787			
489.29	100		59	100	59	0.0003	4.2248	ROK	3.1787	0.0030	100	59
489.56	1431		59	1431	59	0.0046	4.2294	ROK	3.1754	0.0046	1431	59
489.66		XING 37.17				0.0000	4.2294	MAIN STREET	3.1708	0.0004	148	D 67
489.66	500			500	59	0.0016	4.2310	R/C SALINEVILLE	3.1704	0.0010	352	67
489.81	800		67	800	59	0.0026	4.2336	L	3.1694	0.0023	800	67
489.89	300			66	59	0.0002	4.2338		3.1671	0.0010	461	D 65
490.13	900		117	301	A 59	0.0008	4.2347	L	3.1661	0.0031	1386	65
490.20	300			566	67	0.0028	4.2375		3.1630	0.0044	1966	65
490.39	1000			723	A 67	0.0023	4.2398		3.1586	0.0024	672	A 65
490.72	700	XING 3	85	2517	65	0.0049	4.2464	R FOUNDRY HILL ROAD	3.1562	0.0002	90	66
490.78	300			306	65	0.0007	4.2461	REVC	3.1560	0.0020	300	66
490.89	600		94	179	65	0.0004	4.2465	L	3.1551	0.0017	600	66
490.89	400			421	D 86	0.0011	4.2475	REVC	3.1534	0.0011	400	66
491.15	1000		84	1000	66	0.0028	4.2504	R	3.1523	0.0029	1000	66
491.18	116			116	66	0.0003	4.2507	REVC	3.1494	0.0006	200	D 76
491.28	600		76	600	66	0.0017	4.2524	L	3.1488	0.0010	500	76
491.50	1160		76	644	66	0.0018	4.2543	L/MPH 35	3.1478	0.0007	100	76
491.50		XING 39.53		459	A 66	0.0013	4.2555	DOGOON	3.1448	0.0009	421	D 93
491.54	208			556	76	0.0018	4.2570	REVC	3.1439	0.0001	37	93
491.67	700		101	1152	76	0.0029	4.2599	R	3.1438	0.0029	420	93
491.82	800		92	1091	A 76	0.0024	4.2623	R	3.1420	0.0016	200	92
491.89	900			157	93	0.0001	4.2624		3.1413	0.0032	1326	D 13
492.12	1000		93	1200	93	0.0024	4.2648	L MP	3.1381	0.0033	219	134
492.39	500		156	2401	A 93	0.0043	4.2691	L 46	3.1348	0.0034	2249	A 11
492.52	700			348	A 11	0.0008	4.2727		3.1314	0.0005		
492.67	800	XING 4	105	144	106	0.0002	4.2729	R HULL ROAD	3.1309	0.0011	709	117
492.79	600			164	D 10	0.0002	4.2731		3.1345	0.0051	604	117
493.28	2613		117	2616	117	0.0042	4.2773	MPH 35(TT)	3.1332	0.0051	2157	A 88
493.43	866		117	866	117	0.0011	4.2784	SUMMITVILLE DIV POINT	3.1281	0.0017	550	88
493.59	800			5	117	0.0000	4.2808		3.1268	0.0037	800	88
493.59				672	D 11	0.0018	4.2826		3.1232			
493.90	1700		88	1700	88	0.0037	4.2863	R	3.1232	0.0004	1700	88
494.02	600			600	88	0.0013	4.2876		3.1218	0.0026	1071	A 89
494.18	800		129	625	88	0.0013	4.2888	R	3.1192	0.0003	280	69
494.28	800			692	A 88	0.0014	4.2902		3.1180	0.0051	1131	69
494.34					88	0.0000	4.2902		3.1154	0.0006	300	A 61
494.34				439	D 82	0.0011	4.2913		3.1145			
494.58	1300		99	1300	69	0.0035	4.2948	L OKWS@54	3.1145	0.0036	960	61
494.60	100			247	69	0.0007	4.2955	MPH 35	3.1110	0.0012	400	61
494.60				150	D 69	0.0004	4.2959	REVERSE CURVE	3.1103			
494.61	900		61	900	61	0.0025	4.2984	ROK	3.1103	0.0025	900	61
494.81						0.0000	4.2984	REVERSE CURVE	3.1078	0.0002	93	D 86
494.84	700		98	700	61	0.0022	4.3006	L CURVE OKWS	3.1075	0.0004	140	86
494.91						0.0000	4.3006		3.1071	0.0029	1068	D 11
495.06	800			800	61	0.0019	4.3025	SUMMITVILLE	3.1043	0.0006		117
495.12	310		117	90	61	0.0002	4.3027	LW END 2 TRACKS	3.1043	0.0000		117
495.37	1300			186	A 61	0.0005	4.3032	MPH 40(TT)	3.1043	0.0013	800	117
495.27	32			514	96	0.0015	4.3047		3.1030	0.0001	32	117
495.37	500		117	2601	A 96	0.0080	4.3127	R	3.1029	0.0008	500	117
495.37		XING 43.40				0.0000	4.3127	PRIVATE	3.1021	0.0039	2301	D 17
496.37						0.0000	4.3127		3.0982	0.0001	153	170
496.25	4800			3645	A 11	0.0116	4.3243		3.0981	0.0000		
496.28	1600			219	164	0.0003	4.3246		3.0981	0.0149	10176	A 11
496.86	3600	XING 48.50		1857	D 16	0.0024	4.3271		3.0842	0.0022	1300	117
496.70	1100		117	1100	117	0.0018	4.3289	L	3.0820	0.0018	1100	117
498.23	400			400	117	0.0007	4.3296		3.0802	0.0016	1119	D 14
498.28	2200			900	117	0.0015	4.3311		3.0786	0.0003	236	146
498.75				1020	A 11	0.0065	4.3376		3.0783	0.0070	4808	A 11
498.29				212	143	0.0003	4.3372		3.0713			
499.47	1000			1056	D 14	0.0015	4.3387		3.0713	0.0022	1360	113
499.62	800		115	900	116	0.0023	4.3409	L	3.0691	0.0086	2008	A 93
500.25	3400			1300	115	0.0024	4.3433		3.0605	0.0007	1360	93
500.28				1003	A 11	0.0016	4.3439		3.0628			
500.28				702	128	0.0003	4.3442		3.0628			
500.28				386	D 11	0.0016	4.3458		3.0628			
500.28	1200		96	1200	96	0.0024	4.3482	L KENSINGTON	3.0628	0.0024	1200	96
500.50						0.0000	4.3482		3.0604	0.0012	584	D 11
500.69	1000			1000	96	0.0020	4.3502	MPH 40(TT)	3.0592	0.0006	416	117
500.68	900			916	95	0.0007	4.3509	R	3.0567	0.0015	800	117
501.37	1800		117	2283	A 95	0.0040	4.3549	R	3.0572	0.0026	1800	117
501.27	524		117	491	117	0.0008	4.3557	R	3.0546	0.0008	524	117
501.30	185		117	185	117	0.0003	4.3568	R	3.0536	0.0003	185	117
501.30						0.0000	4.3568		3.0535	0.0046	3808	D 20
502.37	5100			1393	117	0.0022	4.3581	MP	3.0489	0.0012	1241	206
503.32	4800	XING 51.18		3888	A 11	0.0105	4.3686	BROWN ROAD 50	3.0477	0.0038	4061	200
503.18	300	XING 5				0.0000	4.3686	L LIPPINCOT ROAD	3.0439	0.0000	0	0
505.27	476	XING 54.05				0.0000	4.3686	PRIVATE	3.0439			
504.28	5342	XING 55.28		5432	166	0.0063	4.3702	PRIVATE	3.0439	0.0000	0	
504.43	800					0.0000	4.3762		3.0438	0.0000	0	
504.52	500		130			0.0000	4.3762	R	3.0438	0.0114	13149	A 16
505.24	2750			5858	A 16	0.0072	4.3834		3.0324	0.0005	450	186
505.41	800					0.0000	4.3834	MPH 40	3.0319	0.0010	900	165
505.71	1600	XING 5	185	383	188	0.0004	4.3838	R US HWY 30	3.0309	0.0018	1600	166
505.86	800			1191	D 15	0.0013	4.3851	E ROCHESTER	3.0291	0.0009	830	168
506.23	2075		185	2075	165	0.0024	4.3875	L	3.0282	0.0024	2075	166
506.31	500		185	385	165	0.0004	4.3879	L	3.0258	0.0003	300	166
506.44	714			714	166	0.0004	4.3884		3.0255	0.0019	1074	D 11
506.44		JCT 54.8		165		0.0004	4.3888	JCT MAHONING BEGINNIN	3.0236	0.0000		200
506.49	213			213	182	0.0002	4.3890	BAYARD	3.0236	0.0000		200
506.56	400	XING 5	200	400	165	0.0006	4.3896	R BAYARD STREET	3.0236	0.0000		200

APPENDIX B3

PAGE 11 COLUMN 1

MILES	DISTANCE, ft	HWY GRADE CROSSING	MAX MPH	DISTANCE, ft	SPEED, mph	dT, hours	TIME, hours	STATION	TIME, hours	dT, hours	DISTANCE, ft	SPEED, mph
				READ DOWN WESTBOUND TRAFFIC					READ UP EASTBOUND TRAFFIC			
506.64	416			33	165	0.0000	4.3894		3.0236	0.0000	9	200
506.71	370				165	0.0000	4.3894		3.0236	0.0002	200	200
506.71				11149	A 16	0.0115	4.4009		3.0234			
507.64	4900	XING 56.86		0			4.4009	PRIVATE	3.0234	0.0046	4900	200
508.64	5276	XING 57.57		380	200	0.0004	4.4012	PRIVATE	3.0188	0.0040	4219	200
508.75	600			0	0	0.0000	4.4012		3.0148	0.0000	0	
509.03	1472		117	1472	200	0.0014	4.4026	R	3.0148	0.0000	0	
509.64	3200			3200	200	0.0030	4.4057		3.0148	0.0063	6300	A 182
510.64	5280			4306	200	0.0041	4.4098		3.0085	0.0053	4840	A 185
510.81	900			1674	D 20	0.0019	4.4117	HUGLE RUN	3.0032	0.0016	1360	165
511.01	1074	XING 5	165	1074	165	0.0012	4.4129	L PRIVATE	3.0016	0.0012	1074	165
511.63	3300			1360	165	0.0016	4.4145		3.0004	0.0007	654	D 178
511.69	300			3586	A 16	0.0040	4.4184		2.9997	0.0003	283	178
511.82	700		196			0.0000	4.4184	R/REVC	2.9994	0.0040	3686	A 165
511.96	700		186	283	178	0.0003	4.4187	L'REVC	2.9954	0.0005	477	165
512.12	883			654	D 17	0.0007	4.4195		2.9949	0.0010	883	165
512.46	1800		165	1800	165	0.0021	4.4215	L	2.9939	0.0021	1800	165
512.63	900			900	165	0.0010	4.4226		2.9918	0.0011	973	D 184
513.63	5236			460	165	0.0005	4.4231	PARISONDARD	2.9907	0.0002	159	184
513.63				5444	A 16	0.0059	4.4290		2.9905	0.0059	5444	A 165
513.97	1800			159	184	0.0002	4.4292		2.9846	0.0016	1360	165
513.97				973	D 18	0.0011	4.4302		2.9830			
514.14	900		165	900	165	0.0010	4.4313	R/PARKS RD	2.9830	0.0010	900	165
514.64	2560			1360	165	0.0016	4.4328		2.9820	0.0017	1641	D 196
514.79	800			11149	A 18	0.0115	4.4443		2.9803	0.0018	1819	196
515.40	3229		233			0.0000	4.4443	R	2.9785	0.0031	3229	196
515.65	1300	XING 64.86				0.0000	4.4443	PRIVATE	2.9754	0.0016	1684	196
516.61	5086	XING 65.08		3600	200	0.0034	4.4476	PRIVATE	2.9738	0.0101	9625	A 185
517.63	5383	XING 65.71		1375	200	0.0013	4.4481	PRIVATE	2.9637	0.0005	460	165
517.80	900			1674	D 20	0.0019	4.4510	NKP RD	2.9632	0.0010	909	165
518.08	1473		165	1473	165	0.0017	4.4527	R	2.9622	0.0004	309	165
518.63	2900			2900	165	0.0033	4.4560		2.9618	0.0049	4068	A 147
518.89	1384			540	165	0.0008	4.4568	LOUISVILLE	2.9569	0.0018	1360	147
518.89				824	D 16	0.0010	4.4576		2.9551			
519.27	2000		82	2000	147	0.0026	4.4602	L'COTTAGE STS NIMISHILL	2.9551	0.0026	2000	147
519.27						0.0000	4.4602		2.9525	0.0012	973	D 168
519.27						0.0000	4.4602	ABUTS MAIN	2.9513	0.0025	2242	168
519.69	2200			1360	147	0.0018	4.4620		2.9488	0.0000		
520.05	1900			4068	A 14	0.0049	4.4669		2.9488	0.0043	3652	A 153
520.26	1124		165			0.0000	4.4669	R	2.9445	0.0000		
520.62	1900			1696	165	0.0019	4.4688	FAIRHOPE-JCT MAIN	2.9445	0.0003	289	153
520.71	500	XING 96.58		500	165	0.0006	4.4694	BEK ROAD	2.9442	0.0006	500	153
520.71				40	165	0.0000	4.4695		2.9436			
520.83	600	XING 97.13		680	D 20	0.0007	4.4701	BROADWAY	2.9436	0.0007	600	153
520.88	300		153	300	153	0.0004	4.4705	L	2.9429	0.0004	300	153
520.94	300		153	300	153	0.0004	4.4709	L	2.9425	0.0004	300	153
521.27	1753			1753	153	0.0022	4.4730		2.9421	0.0003	228	D 156
521.48	1100		158			0.0000	4.4730	L	2.9418	0.0031	2626	158
521.73	1300			1140	A 15	0.0014	4.4744		2.9387	0.0013	1144	D 181
521.73						0.0000	4.4744		2.9374	0.0002	171	181
522.73	5268	XING 100.2		1320	159	0.0016	4.4760	CARNAHAN AVENUE	2.9372	0.0000		
523.73	8260	XING 100.47				0.0000	4.4760	8TH STREET NE	2.9372	0.0000		
523.73		XIN 100.77		12501	A 15	0.0126	4.4887	6TH STREET NE	2.9372	0.0215	16029	A 90-1
524.72	5236	XING 100.8		0		0.0000	4.4887	5TH STREET	2.9157	0.0000		
525.02	1600	XING 100.96				0.0000	4.4887	ROWLAND AVENUE	2.9157	0.0029	1360	90
525.02		XING 101.0		223	198	0.0002	4.4889	2ND AVENUE NE	2.9128			
525.02		XING 101.1		4620	D 15	0.0061	4.4949	TUBC AVE	2.9128			
525.02		XING 1	90		90	0.0167	4.5116	WILLE WANDLE 3RD STREE	2.9128	0.0167	RR CROSSINK	
525.36	1800	XING 101.8		1360	90	0.0029	4.5144	CANTON	2.8961	0.0038	1800	90
525.63	1400	XING 1	112	2007	A 90	0.0042	4.5186	R MCKINLEY	2.8923	0.0029	1400	90
525.75	610			425	112	0.0007	4.5193		2.8894	0.0013	610	90
525.87	634	XING 102.0		652	D 11	0.0012	4.5206	CLEVELAND AVENUE	2.8881	0.0013	634	90
525.87					90	0.0000	4.5206		2.8865	0.0039	2567	D 160
526.74	4600	XING 102.1		1360	90	0.0029	4.5234	HIGH STREET	2.8826	0.0004	303	160
527.74	5306	XING 102.6		4595	A 90	0.0078	4.5312	PROSPECT AVENUE	2.8825	0.0000		
527.74		XING 102.7		3021	A 12	0.0050	4.5362	DUEBER AVENUE	2.8825			
527.74		XING 103.3		234	148	0.0003	4.5365	CLAREDON AVENUE	2.8825	0.0074	5678	A 131
527.74		XING 103.5		696	D 14	0.0009	4.5374	MARYLAND AVENUE	2.8751	0.0020	1360	131
528.03	1500	XING 1	131	1500	131	0.0022	4.5396	R/RAFF RD	2.8731	0.0022	1500	131
528.74	3780			1360	131	0.0020	4.5416		2.8709	0.0012	918	D 153
529.42	3600	XING 105.5		6446	A 13	0.0083	4.5498	PERRY DRIVE	2.8697	0.0012	944	153
529.57	800		233			0.0000	4.5498	L	2.8685	0.0094	6857	A 112
529.74	897					0.0000	4.5498		2.8591	0.0047	2768	112
529.90	800			14	169	0.0000	4.5499		2.8544	0.0014	800	112
529.90				2057	D 18	0.0028	4.5527		2.8530			
530.18	1500	XING 1	112	1500	112	0.0025	4.5552	L GENOA STREET	2.8530	0.0019	1086	A 101
530.75	2991			2830	112	0.0048	4.5600		2.8511	0.0022	1168	101
530.78	192			0	112	0.0000	4.5600		2.8489	0.0004	192	101
530.78				344	D 11	0.0006	4.5606	PERRY HEIGHTS	2.8485			
531.05	1400		101	1400	101	0.0026	4.5633	R	2.8485	0.0026	1400	101
531.05							4.5633		2.8459	0.0016	945	D 128
531.37	1700			1360	101	0.0026	4.5658		2.8443	0.0001	70	128
531.45	400		165	3502	A 98	0.0056	4.5714	L	2.8442	0.0055	3349	A 101-
531.75	1600					0.0000	4.5714		2.8387	0.0007	363	A97-1
531.97	1187					0.0000	4.5714		2.8380	0.0003	160	97
532.07	500		165			0.0000	4.5714	L	2.8377	0.0010	500	97
532.20	700			128	127	0.0002	4.5716		2.8367	0.0014	700	97
532.20				1097	D 12	0.0019	4.5735		2.8353			
532.50	1900	XING 1	97	1600	93	0.0033	4.5768	R/TTMPH 50 HIGH STREET	2.8353	0.0013	680	97
532.75	1300			1300	93	0.0026	4.5794	REED	2.8340	0.0032	1460	A 75-4
532.86	600			113	93	0.0002	4.5797		2.8306	0.0055	2169	75
532.86				487	D 92	0.0011	4.5808		2.8253	0.0006	240	A 70-
533.23	1909		75	1909	73	0.0050	4.5857	L	2.8247	0.0023	860	70
533.32	500			377	73	0.0010	4.5867	REVERSE CURVE	2.8224	0.0014	500	70
533.32				123	D 73	0.0003	4.5870		2.8210			
533.74	2200		70	2200	67	0.0062	4.5932	R	2.8210	0.0060	2200	70
533.81	400		70	400	67	0.0011	4.5944	R	2.8150	0.0011	400	70
533.81						0.0000	4.5944		2.8139	0.0024	1164	D 113
534.02	1100			1100	67	0.0031	4.5975	MASSILLON	2.8115	0.0018	1099	113
534.15	700		166	260	67	0.0007	4.5982	R	2.8097	0.0012	700	113
534.49	1900			1276	A 87	0.0031	4.6013		2.8085	0.0023	1360	113
534.70	1061		113			0.0000	4.6013	L	2.8062	0.0018	1061	113
534.70		NO DIAMOND				0.0000	4.6013	MACE TOWER RR XING	2.8044			
534.70				2136	A 90	0.0040	4.6052		2.8044	0.0000		
534.77	400			2333	113	0.0039	4.6091		2.8044	0.0009	527	A 108
535.19	2200	XING 111.36		162	D 11	0.0003	4.6094	EARL ROAD	2.8035	0.0024	1360	108
535.51	1700		108	1700	108	0.0030	4.6124	L TTMPH 70/60/50	2.8011	0.0030	1700	108
535.51						0.0000	4.6124		2.7981	0.0008	544	D 124
535.62	584	XING 111.92		584	108	0.0010	4.6134	RC BOYDS CRN	2.7972	0.0002	150	124
535.75	700		124	700	108	0.0012	4.6147	R	2.7970	0.0009	600	124
535.93	900		124	86	108	0.0001	4.6148	R	2.7961	0.0014	900	124
535.93				1381	A 10	0.0023	4.6171		2.7947	0.0004	282	D 131
536.13	1100	XING 117.36		813	120	0.0013	4.6184	BEAUMONT AVENUE	2.7943	0.0012	841	131
536.57	2300		131	1611	A 12	0.0024	4.6206	L	2.7931	0.0053	2300	131
536.76	1000			1432	131	0.0021	4.6229		2.7898	0.0014	1000	131
536.79	187			187	131	0.0003	4.6231		2.7884	0.0003	187	131
537.06	1400		131	1400	131	0.0020	4.6252	R	2.7881	0.0020	1400	131
537.76	3700			1360	131	0.0020	4.6271		2.7861	0.0015	1144	D 158
538.54	4100	XING 114.05		5209	A 13	0.0068	4.6339	KENYON AVENUE	2.7846	0.0080	6656	158
538.73	1000		158	2231	158	0.0027	4.6366	L	2.7766	0.0012	1000	158
538.75	135			135	158	0.0002	4.6368		2.7754	0.0015	1358	D 185
539.75	5282	XING 115.1		1225	158	0.0015	4.6382	SR 93	2.7739	0.0008	917	185
540.51	4000	XING 116.1		1718	A 15	0.0020	4.6403	CLERMONT	2.7730	0.0063	5782	A 165
540.51	4000	XING 116.1		1718	A 15	0.0020	4.6403	CLERMONT	2.7730	0.0063	5782	A 165
540.51				9175	A 16	0.0056	4.6459		2.7667	0.0016	1360	165
540.51		XING 116.77		245	183	0.0008	4.6461	ALABAMA AVE	2.7661			
540.51				918	D 18	0.0010	4.6471		2.7651			

APPENDIX B3

PAGE 11 COLUMN 2

MILES	DISTANCE, ft	HWY GRADE CROSSING	MAX MPH	DISTANCE, ft	SPEED, mph	dT, hours	TIME, hours	STATION	TIME, hours	dT, hours	DISTANCE, ft	SPEED, mph
				READ DOWN WESTBOUND TRAFFIC					READ UP EASTBOUND TRAFFIC			
540.66	800		165	800	165	0.0009	4.6480	L	2.7651	0.0009	800	165
540.76	511	XING 117.0		511	165	0.0042	4.6522	RAILROAD STREET N LAW	2.7642	0.0018	1699	D 19
541.76	5299			849	165	0.0010	4.6532	NORTH LAWRENCE	2.7624	0.0002	225	197
542.78	5377	XING 119.2		11149	A 16	0.0115	4.6647	ANDERSON ROAD	2.7622	0.0124	11703	A 16
543.50	3800	XING 119.64		3476	200	0.0023	4.6671	PRIVATE	2.7498	0.0016	1360	158
543.50				2205	D 20	0.0023	4.6694		2.7482			
543.77	1422	XING 1	158	1422	158	0.0017	4.6711	L PRIVATE	2.7482	0.0017	1422	158
543.79	96	XING 1	156	96	158	0.0001	4.6712	L'MP 120 PRIVATE	2.7465	0.0001	96	158
544.77	5200	XING 120.9		1360	158	0.0016	4.6729	SR 94	2.7464	0.0021	1973	D 19
545.78	5297	XING 121.8		467	A 15	0.0006	4.6734	TANNERVILLE	2.7443	0.0001	68	198
545.78				8404	A 19	0.0088	4.6822		2.7442	0.0115	10876	A 19
546.50	3800			376	189	0.0004	4.6825		2.7327	0.0016	1360	160
546.50				1464	D 1	0.0016	4.6841		2.7311	0.0000		
546.67	900		160	900	160	0.0011	4.6852	R	2.7311	0.0006	484	160
546.78	599	XING 123.8		368	160	0.0004	4.6856	WALNUT	2.7305	0.0014	1156	A 1
547.06	1500	XING 123.8		231	D 16	0.0003	4.6859	MARKET STREET	2.7291	0.0017	1360	155
547.78	3789		155	3789	155	0.0046	4.6905	L ORRVILLE	2.7274	0.0034	2749	155
548.27	2600		165	1360	155	0.0017	4.6922	L	2.7240	0.0000		
548.78	2700			4280	A 15	0.0049	4.6971	(W&LE OHIO)	2.7240	0.0090	6480	A 11
549.07	1900	XING 125.11		349	172	0.0004	4.6975	CROWN HILL	2.7150	0.0022	1360	117
549.07		XING 125.38		2331	D 17	0.0031	4.7006	SCHOOLHOUSE ROAD	2.7128	0.0000		
549.33	1400		117	1400	117	0.0023	4.7029	R	2.7128	0.0023	1400	117
549.60	1400		179	1360	117	0.0022	4.7051	R	2.7105	0.0024	1747	D 16
549.79	1029		180	7635	A 11	0.0087	4.7137	R	2.7081	0.0008	882	160
550.04	1300	XING 1	174			0.0000	4.7137	R MUSKRAT ROAD	2.7073	0.0001	94	D 16
550.30	1365	XING 1	162			0.0000	4.7137	R BACK ORRVILLE	2.7072	0.0030	2571	162
550.79	2600	XING 127.6		472	A 15	0.0006	4.7143	ROHRER ROAD	2.7042	0.0007	843	D 17
550.79				1579	A 16	0.0030	4.7173		2.7035	0.0003	296	175
551.80	5302	XING 126.0		1450	168	0.0016	4.7189	PRIVATE	2.7032	0.0065	5803	A 15
551.83	200	XING 128.7		706	D 16	0.0008	4.7197	APPLEGATE	2.6967	0.0009	723	153
552.80	5100		155	5100	153	0.0063	4.7261	L	2.6958	0.0000		153
553.10	1600			1360	153	0.0017	4.7277		2.6958	0.0000		
553.27	900		156			0.0000	4.7277	L'MPH 60	2.6958	0.0000		
553.41	700	XING 129.57				0.0000	4.7277	CO RD 73	2.6958	0.0148	9377	A 79
553.59	1000		172			0.0000	4.7277	R	2.6810	0.0018	760	79
553.71	600			922	153	0.0011	4.7288		2.6792	0.0014	600	79
553.71				2518	D 1	0.0041	4.7330		2.6778			
553.81	516		79	518	79	0.0012	4.7342	L'SMITHVILLE	2.6778	0.0012	518	79
554.02	1123		79	1123	79	0.0027	4.7369	L	2.6766	0.0033	1123	79
554.78	4000			1360	79	0.0033	4.7402		2.6733	0.0016	784	D 10
554.81	187			177	A 79	0.0004	4.7406	TTMPH 35	2.6717	0.0001	42	107
554.81				1904	A 82	0.0038	4.7444		2.6716	0.0040	2021	A 82
554.81				84	109	0.0002	4.7446		2.6676	0.0057	2470	82
554.81				662	D 10	0.0013	4.7459		2.6619	0.0010	430	A 74
555.12	1600		82	1600	82	0.0037	4.7496	R	2.6609	0.0002	90	74
555.36	1300			1117	82	0.0026	4.7522		2.6607	0.0033	1300	74
555.36				183	D 82	0.0004	4.7526		2.6574			
555.78	2200		74	2200	74	0.0056	4.7583	L	2.6574	0.0056	2200	74
555.80	96		74	96	74	0.0003	4.7585	L	2.6518	0.0037	1428	74
555.95	800			800	74	0.0020	4.7606		2.6481	0.0008	310	A 67
556.25	1600		80	560	74	0.0014	4.7620	R	2.6473	0.0021	780	67
556.25				311	A 74	0.0008	4.7628		2.6452			
556.36	600			1049	80	0.0025	4.7652		2.6452	0.0017	600	67
556.36				260	D 80	0.0007	4.7660		2.6436			
556.76	2100		67	2100	67	0.0059	4.7719	R	2.6435	0.0059	2100	67
556.90	700		67	700	67	0.0020	4.7739	R	2.6376	0.0020	700	67
556.90						0.0000	4.7739		2.6356	0.0003	123	D 73
556.94	236			236	67	0.0007	4.7745	REVERSE CURVE	2.6353	0.0003	113	73
557.43	2600	XING 1	73	1124	67	0.0032	4.7777	L E END 3 TR PRIVATE	2.6350	0.0104	4005	73
557.66	1200			256	A 67	0.0007	4.7784		2.6246	0.0001	53	A 72
557.79	700		73	2500	73	0.0067	4.7851	L	2.6245	0.0012	442	72
557.79				307	A 70	0.0008	4.7859		2.6233			
557.92	700		79	996	79	0.0024	4.7883	L	2.6233	0.0018	700	72
557.97	218			185	D 7	0.0004	4.7886	REVERSE CURVE	2.6215	0.0008	218	72
558.06	500		72	500	72	0.0013	4.7900	R	2.6208	0.0013	500	72
558.06				1300	72	0.0036	4.7936		2.6190	0.0012	536	D 11
558.50	2300			1377	A 72	0.0031	4.7967		2.6184	0.0045	2235	94
558.50							4.7967		2.6139	0.0001	69	A 93
558.65	800		94	363	94	0.0007	4.7974	L	2.6138	0.0005	287	93
558.76	600			573	94	0.0012	4.7985		2.6133	0.0012	600	93
558.86	500			27	D 94	0.0001	4.7986		2.6121	0.0010	500	93
559.08	1207	XING 1	93	1207	93	0.0025	4.8011	L'WOOSTER LIBERTY AVE	2.6111	0.0025	1207	93
559.08		XING 135.24				0.0000	4.8011	SOUTH STREET	2.6086	0.0019	1134	D 12
559.29	1100			1100	93	0.0022	4.8033		2.6067	0.0000	16	128
559.37	400		128	260	93	0.0005	4.8038	R	2.6067	0.0005	350	128
559.67	1800			3885	A 93	0.0066	4.8104		2.6062	0.0004	270	D 13
559.79	600	XING 1	135	1109	A 12	0.0016	4.8120	L PRIVATE	2.6058	0.0027	1930	135
559.80	68	XING 1	135			0.0000	4.8120	L PRAIRIE ROAD W END 3	2.6031	0.0001	68	135
560.69	4700			6780	A 13	0.0085	4.8205		2.6030	0.0021	1717	D 17
560.82	700					0.0000	4.8205		2.6008	0.0028	2617	173
560.93	600		173			0.0000	4.8205	R	2.5980	0.0000		120
560.94	35					0.0000	4.8205	REVERSE CURVE	2.5980	0.0000		120
561.05	600		173			0.0000	4.8205	L'OK	2.5980	0.0000		120
561.05				1683	A 16	0.0019	4.8224		2.5980			
561.77	3800					0.0000	4.8224		2.5980	0.0000		120
561.77				832	A 17	0.0009	4.8233		2.5980			
561.77				1488	176	0.0016	4.8249		2.5980	0.0115	9241	A 13
562.63	4500			2064	D 16	0.0026	4.8274		2.5865	0.0020	1360	130
562.76	714		130	714	130	0.0010	4.8285	R/KILLBUCK	2.5845	0.0010	714	130
563.52	4000			1360	130	0.0020	4.8305		2.5835	0.0039	3368	D 18
563.73	1100		166	7071	A 13	0.0090	4.8395	R/FISH	2.5796	0.0016	1712	200
563.75	134	XING 140.10				0.0000	4.8395	MESSNER ROAD	2.5780	0.0016	1716	200
564.78	5405	XING 141.8		11149	A 18	0.0115	4.8510	WILLOW ROAD	2.5764	0.0000		
565.79	5362	XING 142.94				0.0000	4.8510	FORCE ROAD	2.5764	0.0000		
566.78	5232			1643	200	0.0018	4.8526		2.5764	0.0000		
567.16	2000	XING 143.3		200	200	0.0002	4.8528	CEMETERY ROAD	2.5764	0.0000		
567.39	1216		222	1216	200	0.0012	4.8539	L	2.5764	0.0000		
567.77	2000					0.0000	4.8539		2.5764	0.0236	20169	A 11
568.00	1200		233			0.0000	4.8539	L	2.5528	0.0010	620	116
568.13	700	XING 144.3		1807	200	0.0017	4.8556	WOOD	2.5519	0.0026	1821	A 10
568.13		XING 144.4		3893	D 20	0.0047	4.8603	ROBINSON	2.5493			
568.45	1700	XING 1	116	1700	116	0.0028	4.8631	R MARKET STREET	2.5493	0.0025	1360	102
568.64	1000	XING 144.8		1000	116	0.0016	4.8647	SHREVE MAIN STREET	2.5468	0.0019	1000	102
568.78	753		166	305	116	0.0005	4.8652	R	2.5449	0.0015	752	A 93
568.78				448	D 11	0.0008	4.8660		2.5434			
569.03	1300		102	1300	107	0.0024	4.8684	R/CK	2.5434	0.0018	860	93
569.12	500			243	102	0.0005	4.8689	REVERSE CURVE	2.5416	0.0010	500	93
569.12				257	D 10	0.0005	4.8693		2.5406			
569.38	1378		93	1378	93	0.0028	4.8721	L	2.5406	0.0028	1378	93
569.38						0.0000	4.8721		2.5378	0.0001	27	D 94
569.65	1400		100	1360	93	0.0028	4.8749	L	2.5377	0.0028	1373	94
569.74	500			69	A 93	0.0001	4.8750	RC	2.5349	0.0010	500	94
569.78	200		94	200	94	0.0004	4.8756	R	2.5339	0.0004	200	94
570.09	1601	XING 1	94	1081	94	0.0022	4.8776	R BROWNS	2.5335	0.0032	1601	94
570.79	3700	XING 147.1		1360	94	0.0027	4.8804	PRIVATE	2.5303	0.0042	2893	D 18
571.62	4392	XING 148.1		11218	A 9	0.0160	4.8964	SCRUBYS	2.5261	0.0000	27	189
571.96	1800	XING 148.56				0.0000	4.8964	BIG PRAIRIE RTE 226	2.5261	0.0000		
572.13	887		165			0.0000	4.8964	L	2.5261	0.0000		
572.79	3500			1750	165	0.0020	4.8984		2.5261	0.0158	11578	A 10
572.84	277					0.0000	4.8984		2.5103	0.0001	90	102
573.08	1300		165	42	165	0.0000	4.8984	L	2.5102	0.0024	1300	102
573.09				2467	D 15	0.0035	4.9019		2.5078			
573.54	2400		102	2400	102	0.0045	4.9064	L'RC	2.5078	0.0045	2400	102
573.54						0.0000	4.9064		2.5033	0.0008	482	D 11
573.60	300			300	102	0.0006	4.9069		2.5025	0.0007	403	117
573.79	1000		117	1000	102	0.0019	4.9088	R REVERSE CURVE	2.5018	0.0005	300	117
574.90	1100		135	60	102	0.0001	4.9089	R	2.5013	0.0004	295	A 11

APPENDIX B3

PAGE 12 COLUMN 1

MILES	READ DOWN WESTBOUND TRAFFIC	STATION	READ UP EASTBOUND TRAFFIC	MILES
574.07	400 1384 A 10 0.0024 4.9113		2.5009 0.0078 4728 115	603.22
574.32	1300 XING 1 115 115 0.0000 4.9113	PLAKEVILLE GULLS ROAD	2.4931 0.0006 374 A 112	603.33
574.79	2500 XING 150.9 1380 115 0.0022 4.9136	STATE ROUTE 179	2.4925 0.0001 60 112	603.46
574.79	1728 A 14 0.0027 4.9163		2.4924	603.81
574.79	1484 128 0.0022 4.9186		2.4924	603.86
575.04	1300 583 D 13 0.0006 4.9194		2.4924 0.0022 1300 117	604.13
575.27	1217 112 1217 112 0.0021 4.9214	L	2.4902 0.0021 1217 112	604.88
575.60	2800 1300 112 0.0023 4.9237		2.4881 0.0016 1078 D 141	605.40
575.80	0.0000 4.9237		2.4865 0.0002 118 141	605.53
575.80	6282 A 13 0.0076 4.9314		2.4863 0.0100 6226 A 91-1	605.78
576.80	6281 277 146 0.0004 4.9317		2.4763 0.0054 2564 91	605.89
576.93	700 1312 D 14 0.0031 4.9346		2.4709 0.0013 502 A 824	606.06
577.48	2900 XING 1 91 2900 91 0.0060 4.9406	R TR 365	2.4696 0.0026 1074 82	606.06
577.54	288 58 91 0.0001 4.9409	REVERSE CURVE	2.4671 0.0007 286 82	606.90
577.54	229 D 91 0.0005 4.9414		2.4664	607.83
577.80	1400 82 1405 82 0.0032 4.9447	L	2.4664 0.0032 1400 82	608.91
577.83	700 700 82 0.0016 4.9463		2.4632 0.0011 510 D 101	609.31
578.10	900 101 900 82 0.0015 4.9478	L	2.4621 0.0020 1060 101	610.92
578.10	0.0000 4.9476		2.4601 0.0025 1630 D 146	611.92
578.26	804 136 1446 A 12 0.0030 4.9506	L	2.4576 0.0001 97 146	612.26
578.26	4.9506		2.4575 0.0050 5436 A 110	612.26
578.81	2900 XING 155.2 152 101 0.0003 4.9511	YR 5	2.4495 0.0004 435 A 106	612.37
579.70	4700 5636 A 10 0.0066 4.9586		2.4467 0.0016 886 106	612.54
579.70	326 142 0.0004 4.9601		2.4473	612.54
579.70	1183 D 14 0.0018 4.9618		2.4473	612.58
579.80	564 XING 1 119 1936 116 0.0033 4.9662	RLOUDONVILLE JEFFERSON	2.4473 0.0010 564 108	612.58
579.80	127 D 11 0.0002 4.9664		2.4463	612.69
580.34	2848 XING 1 109 2848 108 0.0051 4.9705	R SPRING STREET	2.4463 0.0051 2848 106	612.69
580.34	0.0000 4.9705		2.4412 0.0024 1608 D 48	612.84
580.34	4.9705		2.4388 0.0003 210 146	612.92
580.81	2500 1360 106 0.0024 4.9729	CP MOHICAN	2.4385 0.0051 3714 A 126	613.05
580.85	4400 2752 A 10 0.0046 4.9775		2.4334 0.0020 1360 128	613.09
581.62	893 XING 1 128 2172 128 0.0012 4.9806	L PRIVATE	2.4314 0.0013 893 128	613.16
581.93	810 128 810 128 0.0009 4.9815	L	2.4301 0.0009 810 128	613.16
582.62	4700 790 128 0.0011 4.9826		2.4292 0.0004 270 D 135	613.16
582.82	3534 A 12 0.0046 4.9872		2.4288 0.0062 4430 135	613.16
582.82	120 147 0.0002 4.9873		2.4226	613.92
582.82	496 D 14 0.0007 4.9880		2.4226	613.92
583.07	1303 XING 1 135 1303 135 0.0018 4.9898	LE FERRY PRIVATE	2.4226 0.0018 1303 135	614.22
583.83	4000 1380 135 0.0019 4.9917		2.4208 0.0013 983 D 158	614.34
583.83	7334 A 13 0.0091 5.0008		2.4196 0.0000 41 158	614.35
584.02	1000 XING 169.9 42 169 0.0000 5.0009	BRIDGE STREET	2.4195 0.0012 1000 156	614.50
584.52	4264 XING 1 172 528 D 16 0.0006 5.0015	R PRIVATE	2.4183 0.0051 4264 158	614.50
585.41	3100 158 3100 158 0.0037 5.0052	R PERRYSVILLE	2.4132 0.0087 3100 159	615.19
585.84	2260 XING 160.3 2260 158 0.0027 5.0079	PRIVATE	2.4055 0.0027 2260 164	615.82
586.62	4100 XING 162.8 3500 158 0.0042 5.0121	TOWNSHIP ROAD 1037	2.4028 0.0018 1500 158	615.49
586.84	1167 158 158 0.0000 5.0121	L	2.4020 0.0094 6605 A 118	616.32
586.94	500 156 158 0.0000 5.0121	L	2.3956 0.0006 360 119	617.30
587.13	1000 1575 158 0.0019 5.0140		2.3950 0.0015 1000 119	618.30
587.13	1722 D 12 0.0024 5.0164		2.3934	619.30
587.43	1600 119 1600 119 0.0026 5.0190	L	2.3934 0.0026 1600 119	620.31
587.83	1380 XING 163.8 1360 115 0.0022 5.0212	COUNTY ROAD 1075	2.3608 0.0007 440 D 131	621.27
587.64	800 XING 1 155 2987 A 118 0.0046 5.0258	L PRIVATE	2.3902 0.0003 207 131	622.31
588.00	800 156 0.0000 5.0258	L	2.3865 0.0060 8683 A 101	623.31
588.51	2700 118 136 0.0002 5.0259		2.3836 0.0028 1300 101	624.30
588.51	1217 D 13 0.0019 5.0279		2.3813	624.30
588.72	1104 101 1104 101 0.0021 5.0299	R	2.3813 0.0042 2230 101	624.51
588.85	700 700 101 0.0013 5.0312		2.3771 0.0018 912 A 30-1	624.63
589.36	2700 860 101 0.0012 5.0325		2.3753 0.0029 1360 96	624.88
589.36	1218 A 16 0.0022 5.0346		2.3724	624.88
589.36	137 116 0.0002 5.0349		2.3724	624.88
589.36	688 D 11 0.0013 5.0361		2.3724	624.88
589.64	1500 96 1500 96 0.0032 5.0393	L	2.3724 0.0032 1500 96	625.08
589.64	0.0000 5.0393		2.3692 0.0013 718 D 104	625.09
589.85	1102 1102 96 0.0023 5.0416		2.3679 0.0006 384 114	625.18
590.16	1600 258 96 0.0005 5.0422		2.3673 0.0030 1600 114	625.30
590.35	800 114 2255 A 96 0.0042 5.0463	L	2.3643 0.0013 800 114	625.30
590.55	1067 XING 168.6 1154 114 0.0017 5.0482	PRIVATE	2.3630 0.0018 1067 114	625.30
590.77	1200 114 1200 114 0.0020 5.0502	R	2.3612 0.0029 1200 114	625.51
590.85	400 400 114 0.0007 5.0509		2.3582 0.0003 171 D 132	626.30
591.00	600 600 114 0.0013 5.0522		2.3588 0.0036 2176 119	627.30
591.00	180 114 0.0003 5.0525		2.3564	628.30
591.00	541 A 11 0.0010 5.0535		2.3634	629.33
591.13	700 119 3497 119 0.0040 5.0575	R	2.3554 0.0018 1080 A 119	630.32
591.68	2800 XING 187.6 302 D 11 0.0003 5.0580	PRIVATE	2.3536 0.0038 1360 110	631.32
591.76	400 XING 1 110 400 110 0.0007 5.0587	R LUCAS PERRYSVILLE	2.3513 0.0007 400 110	632.19
591.85	467 467 110 0.0008 5.0595		2.3506 0.0008 467 110	633.32
591.90	295 295 110 0.0005 5.0600		2.3498 0.0005 295 110	633.32
592.07	900 110 900 110 0.0015 5.0615	L	2.3493 0.0015 900 110	633.37
592.07	XING 168.3 110 0.0000 5.0615	UNION STREET	2.3478 0.0012 781 D 132	633.82
592.07	XING 168.46 110 0.0000 5.0615	CHURCH	2.3466 0.0006 19 132	634.35
592.47	2100 XING 168.6 1360 110 0.0023 5.0639	LUCAS HANLEY BRETZ	2.3466 0.0019 1300 132	634.35
592.85	2000 132 2837 A 11 0.0046 5.0685	R	2.3447 0.0029 2000 132	634.35
593.27	3200 156 1188 132 0.0017 5.0701	R	2.3416 0.0015 1009 132	634.35
593.35	1500 498 A 13 0.0007 5.0708	CP LUCAS	2.3400 0.0022 1074 A 118	635.34
593.68	683 135 1613 135 0.0027 5.0735	R	2.3394 0.0022 1360 118	636.35
593.68	631 D 17 0.0009 5.0745		2.3514	637.35
593.81	700 118 708 118 0.0011 5.0756	R	2.3344 0.0003 2010 118	638.35
593.85	200 206 118 0.0003 5.0759		2.3312 0.0032 130 A 117	639.35
593.85	2506 118 0.0041 5.0800	MP 170	2.3310 0.0020	640.35
594.34	2600 34 D 11 0.0001 5.0801		2.3310 0.0022 1360 117	640.50
594.78	2300 117 2300 117 0.0037 5.0838	L	2.3288 0.0037 2300 117	641.35
594.78	0.0000 5.0838		2.3251 0.0042 3288 D 197	641.35
594.84	377 XING 171.3 377 117 0.0006 5.0844	PRIVATE	2.3209 0.0002 164 197	641.35
595.85	5281 XING 171.8 985 117 0.0016 5.0860	BITTS ROAD	2.3207 0.0021 2080 A 189	641.35
596.80	5000 14330 A 11 0.0181 5.1041		2.3186 0.0000	641.35
596.85	279 186 0.0000 5.1041	L	2.3186 0.0000	641.35
596.91	300 XING 1 186 0.0000 5.1041	ERC ILLINOIS AVENUE	2.3186 0.0000 186	641.84
597.00	481 186 0.0000 5.1041	RTT MPH 70	2.3186 0.0127 10986 A 143	641.84
597.85	4800 62 184 0.0001 5.1041		2.3060 0.0019 1427 A 136	641.84
598.23	2000 1986 D 18 0.0023 5.1064		2.3041 0.0002 150 135	641.84
598.34	600 143 1373 143 0.0018 5.1082	R	2.3039 0.0007 548 135	641.84
598.46	600 XING 1 135 328 D 14 0.0005 5.1087	R 4TH STREET	2.3032 0.0000 135	641.84
598.57	600 XING 1 135 600 135 0.0008 5.1095	R 5TH STREET	2.3032 0.0000 135	641.93
598.86	1512 XING 175.0 1360 135 0.0019 5.1114	MANSFIELD ORANGE STREET	2.3032 0.0087 5228 A 50-1	642.35
598.86	2010 A 13 0.0027 5.1141		2.2945	643.33
599.58	3800 XING 175.3 4 146 0.0000 5.1141	SURREY ROAD	2.2945 0.0029 1360 90	644.33
599.58	1908 D 14 0.0031 5.1172		2.2916 0.0000	645.33
599.58	90 0.0000 5.1172	BIG XING	2.2918 0.0000 RR CRC 90	646.33
599.58	XING 175.57 0.0000 5.1172	MAIN STREET	2.2918 0.0035 2248 D 153	647.33
599.63	300 XING 1 233 300 90 0.0006 5.1178	LW MANB LOGVIEW AVE	2.2881 0.0003 229 153	648.33
599.87	1274 1080 90 0.0022 5.1201		2.2878 0.0000	649.33
600.05	800 8884 A 90 0.0132 5.1333		2.2878 0.0000	650.33
600.12	400 270 0.0000 5.1333	L	2.2878 0.0005 400 153	651.33
600.36	1268 0.0000 5.1333		2.2873 0.0016 1268 153	652.33
600.70	1800 XING 1 153 0.0000 5.1333	LEMP STL MILL	2.2857 0.0017 1341 153	653.33
600.87	900 XING 177.2 0.0000 5.1333	PRIVATE	2.2840 0.0000	653.33
601.30	3300 2377 A 10 0.0028 5.1361		2.2840 0.0000 0	653.33
601.87	1997 146 0.0000 5.1361	L	2.2840 0.0122 8186 A 90-1	653.33
602.42	2600 0.0000 5.1361		2.2718 0.0027 1360 90	653.33
602.42	63 163 0.0001 5.1362		2.2691 0.0000	653.33
602.42	2048 D 16 0.0037 5.1400		2.2691 0.0000	653.33
602.63	1100 96 1100 96 0.0022 5.1421	R	2.2691 0.0022 1100 96	653.33
602.63	0.0000 5.1421		2.2669 0.0020 1264 D 130	653.33
602.74	595 585 96 0.0012 5.1433	REVERSE CURVE	2.2649 0.0001 98 132	653.33
602.88	700 132 700 96 0.0014 5.1447	L	2.2648 0.0000 0 132	653.33
603.04	687 132 87 96 0.0001 5.1448	L	2.2648 0.0014 997 132	653.33

APPENDIX B3

PAGE 12 COLUMN 2

MILES	READ DOWN WESTBOUND TRAFFIC	STATION	READ UP EASTBOUND TRAFFIC
800	219 4242 A 96 0.0070 5.1518		2.2104 0.0021 1633 D 10
900	219 0.0000 5.1518	L	2.2613 0.0002 166 169
700	XING 179.57 0.0000 5.1518	LEPPO ROAD	2.2611 0.0000
800	171 0.0000 5.1518	R	2.2611 0.0018 1041 A 16
1300	XING 160.6 9203 A 13 0.0081 5.1592	SPRING MILL	2.2593 0.0010 444 185
1510	165 0.0000 5.1592	L	2.2583 0.0000 185
3900	1045 162 0.0019 5.1610		2.2583 0.0048 6977 A 13
2700	1532 D 14 0.0017 5.1628		2.2496 0.0036 2240 A 11
701	131 90 131 0.0001 5.1637	L	2.2459 0.0001 60 114
1300	XING 161.9 611 D 13 0.0009 5.1646	PRIVATE	2.2458 0.0022 1300 114
800	114 800 114 0.0010 5.1656	L	2.2436 0.0010 600 114
927	XING 1 114 927 114 0.0010 5.1671	L ROCK ROAD	2.2426 0.0015 927 114
	1360 114 0.0023 5.1694		2.2411 0.0048 3991 D 20
6405	XING 183.88 0.0000 5.1694	PRIVATE	2.2363 0.0022 5484 200
4912	XING 164.2 20416 A 11 0.0240 5.1934	SHELBY ROAD	2.2341 0.0000
5710	0.0000 5.1934		2.2341 0.0287 31942 A 98
5295	XING 156.4 1500 200 0.0015 5.1949	HORNING ROAD	2.2044 0.0042 725 98
5309	0.0000 5.1949		2.2002 0.0015 725 A 90
5302	XING 186.6 5302 200 0.0050 5.1998	COUNTY LINE	2.1987 0.0003 134 90
1800	XING 188.1 920 200 0.0009 5.2008	BUCYRUS STREET	2.1984 0.0001 96 A 89
	4429 D 20 0.0058 5.2064		2.1983
594	XING 1 99 305 99 0.0006 5.2070	R WASHINGTON	2.1983 0.0006 260 89
900	249 D 96 0.0005 5.2075	CRESTLINE	2.1977 0.0019 900 89
	90 90 0.0053 5.2806	CP CREST 84 XING	2.1958 0.0833 RR CROSS
200	174 90 0.0004 5.2812		2.1120 0.0004 200 89
	26 D 90 0.0001 5.2812		2.1121
900	69 926 69 0.0015 5.2926	R	2.1121 0.0013 600 88
	1568 69 0.0034 5.2959		2.1106 0.0002 80 D 68
800	102 D 98 0.0002 5.2961		2.1106 0.0015 720 97
400	85 400 85 0.0003 5.2978	R	2.1091 0.0009 400 88
700	85 750 65 0.0018 5.2986	R stop for 3 sec dwell wb	2.1082 0.0014 700 82
204	204 85 0.0005 5.2980	REVERSE CURVE	2.1068 0.0024 204 92
400	92 400 85 0.0009 5.2989	L	2.1084 0.0006 400 82
	756 85 0.0017 5.3016		2.1056 0.0001 56 D 13
	5.3016		2.1050 0.0001 38 130
	5.3016		2.1054 0.0069 4147 A 94
4000	486 A 90 0.0010 5.3025		2.0985 0.0008 440 94
	118 92 0.0002 5.3028	MP 190	2.0976 0.0005 230 A 91
	5.3028		2.0971 0.0014 682 91
1900	171 A 92 0.0003 5.3032		2.0957 0.0012 600 91
800	94 4684 94 0.0094 5.3126	L	2.0945 0.0002 78 91
78	91 D 94 0.0002 5.3128	REV CUR 774MPH	2.0945 0.0017 800 91
300	91 506 91 0.0017 5.3145	R	2.0928 0.0081 4987 D 20
	0.0000 5.3145		2.0846 0.0000 46 200
3900	1360 91 0.0028 5.3173		2.0846 0.0006 0 200
700	XING 1 219 22866 A 91 0.0280 5.3453	R SR 598	2.0845 0.0006 500 200
900	XING 1 219 0.0000 5.3453	R CRESTLINE ROAD	2.0880 0.0041 4380 200
4380	XING 192.60 0.0000 5.3453	BIGGLE ROAD	2.0818 0.0049 5182 200
5182	XING 193.80 0.0000 5.3453	BECK ROAD	2.0770 0.0050 5288 200
5088	XING 194.07 0.0000 5.3453	MAIN STREET	2.0720 0.0050 5294 200
5294	XING 194.75 1379 200 0.0013 5.3466	WESTERN AVENUE	2.0670 0.0046 4215 200
5289	XING 195.7 5289 200 0.0050 5.3516	OLENTANGY ROAD	2.0620
4380	XING 196.7 5602 200 0.0048 5.3564	PRIVATE	2.0630 0.0000
6000	XING 196.6 5602 200 0.0048 5.3584	PRIVATE	2.0630 0.0135 11887 A 15
5476	XING 196.3 5476 200 0.0050 5.3616	PRIVATE	2.0495 0.0146 3804 A 90
5275	XING 197.6 5275 200 0.0050 5.3666	LWR LOSSVILLE	2.0349
5284	XING 197.6 3049 200 0.0029 5.3695	PRIVATE	2.0349 0.0082 2139 A 40
	XING 197.79 200 0.0000 5.3695	MP 200 PRIVATE	2.0287 0.0017 369 46
1140	XING 1 168.2 2205 D 2 0.0023 5.3718	PRIVATE	2.0270 0.0047 1000 40
985	XING 1 138 992 158 0.0007 5.3725	R ALBRIGHT	2.0223
1000	XING 200.9 0 158 0.0000 5.3725	WHETSTONE STREET	2.0223 0.0067 RR CROSS
	2473 D 12 0.0038 5.3763		1.9548
	XING 200.70 30 30 0.0667 5.4430	COLSAN NEW XING	1.9556 0.0062 1100 40
	0.0000 5.4430	LANE STREET BUCYRUS	1.9554 0.0000 0 40
1100	XING 2 101 1100 30 0.0023 5.4453	LTT 40 MPH WALNUT STREET	1.9554 0.0024 500 40
	XING 202.98 0.0000 5.4453	ALLEY TRACK IN GALEN ST	1.9530 0.0006 100 40
500	XING 203.97 260 30 0.0005 5.4458	N SANDUSKY STATE HWY	1.9480 0.0024 500 40
100	145 1175 A 90 0.0023 5.4481	L TRACK IN GALEN STREET	1.9456 0.0024 40
500	XING 201.10 0.0000 5.4481	RD TT 40 SPRING STREET	1.9451
	XING 201.20	SANDUSKY RIVER TRACK	1.9451 0.0089 5002 D 20
1100	XING 2 130 4593 A 10 0.0071 5.4552	R SEARS STREET	1.9362 0.0000 40 200
4186	XING 201.45 0.0000 5.4552	STREET AT GRADE NO NAME	1.9362 0.0046 4600 200
5286	XING 203.16 0.0000 5.4552	McCRACKEN	1.9316 0.0050 5286 200
5260	XING 203.66 0.0000 5.4552	SIMMS CR - IN STREET	1.9266 0.0051 5390 200
5286	XING 204.75 16942 A 13 0.0186 5.4738	PRIVATE	1.9215 0.0050 5272 200
5277	XING 204.9 4623 200 0.0044 5.4784	GLENVILLE CR	1.9165 0.0050 5286 200
5208	XING 205.9 5208 200 0.0050 5.4834	PRIVATE	1.9115 0.0043 4518 200
4503	XING 206.1 4503 200 0.0043 5.4877	STREB CR KN	1.9072 0.0056 5826 200
5928	XING 207.1 5928 200 0.0056 5.4934	MARION-MELMO	1.9016
	XING 207.70 0.0000 5.4934	PRIVATE	1.9016
	XING 208.13 0.0000 5.4934	COUNTY LINE	1.9016
	XING 208.60 0.0000 5.4934	PRIVATE	1.9016 0.0052 5440 200
5440	XING 208.0 5440 200 0.0052 5.4986	NEVADA AYERS STREET	1.8964
	XING 209.10 0.0000 5.4986	MORRISON STREET	1.8964
	XING 209.27 0.0000 5.4986	GOODBREAD STREET	1.8964
	0.0000 5.4986	MP 210	1.8964 0.0050 5276 200
5270	XING 210.61 5270 200 0.0050 5.5036	EDENVILLE ROAD	1.8914 0.0050 5260 200
5260	XING 211.6 5260 200 0.0050 5.5086	PRIVATE	1.8864 0.0049 5146 200
5269	XING 212.15 5269 200 0.0050 5.5136	DOUGLAS RD	1.8815
5286	XING 213.2 5286 200 0.0050 5.5186	PRIVATE	1.8815 0.0097 22803 A 90
5280	XING 213.80 5280 200 0.0050 5.5236	ROCK RUN ROAD	1.8418 0.0000
8264	XING 214.3 5284 200 0.0050 5.5286	PRIVATE	1.8418 0.0000
787	XING 2 233 787 200 0.0007 5.5293	L WILLIAMS CR	1.8418 0.0000
4600	XING 215.5 2421 200 0.0023 5.5316	PRIVATE	1.8418
	XING 215.70 0.0000 5.5316	PRIVATE	1.8418
	XING 216.10 0.0000 5.5316	PRIVATE	1.8418
	XING 216.31 0.0000 5.5316	KOINS RD	1.8418
	XING 216.88 0.0000 5.5316	S 5TH STREET	1.8418
	XING 216.90 0.0000 5.5316	S SANDUSKY	1.8418
2600	XING 217.06 206 0.0000 5.5316	UPPER SANDUSKY YTHS	1.8418 0.0029 1360 90
	XING 217.2 4879 D 10 0.0061 5.5377	EIGHTH STREET	1.8389
	XING 217.34 98 0.0000 5.5377	HAZEL STREET	1.8389
	XING 217.44 98 0.0000 5.5377	S WARPOLE	1.8389
	90 90 0.0500 5.5877	L&AND C&O XING	1.8388 0.0500 RR CRC 90
	0.0000 5.5877		1.7888 0.0081 4879 D 20
479	XING 2 138 479 90 0.0010 5.5887	L TWP ROAD	1.7826 0.0000 31 200
2200	XING 219.1 581 90 0.0019 5.5906	WHITE CR	1.7826 0.0003 0 200
5281	XING 218.6 5729 A 90 0.0064 5.5968	PRIVATE	1.7797 0.0049 3280 200
5189	XING 220.2 0.0000 5.5968	MP 220 PRIVATE	1.7748 0.0050 5186 200
5281	XING 220.7 18642 A 12 0.0186 5.6187	MILL CR	1.7698 0.0050 5281 200
5281	XING 220.58 0.0000 5.6187	MOREY CR	1.7648 0.0050 5283 200
5283	XING 221.3 4889 200 0.0047 5.6233	PRIVATE	1.7598 0.0050 5280 200
5280	XING 222.3 5280 200 0.0050 5.6283	GAMBER CR	1.7548 0.0050 5279 200
5279	XING 222.8 5279 200 0.0050 5.6333	PRIVATE	1.7498 0.0050 5282 200
5282	XING 223.2 5282 200 0.0050 5.6383	KRAUS CR	1.7448 0.0050 5284 200
5284	XING 223.4 5284 200 0.0050 5.6434	PRIVATE	1.7398 0.0050 5275 200
5275	XING 223.9 5275 200 0.0050 5.6483	PRIVATE	1.7348 0.0050 5279 200
5279	XING 224.2 5279 200 0.0050 5.6533	MAIN STREET CR	1.7298
	XING 224.35 0.0000 5.6533	ALLEY STREET	1.7298
	XING 226.80 0.0000 5.6533	LILES CR	1.7298
	XING 226.84 0.0000 5.6533	HELLER CR	1.7298
	XING 227.30 0.0000 5.6533	PRIVATE	1.7298
	XING 227.50 0.0000 5.6533	PRIVATE	1.7298
	XING 227.86 0.0000 5.6533	ANSLING ROAD	1.7298
	XING 227.94 0.0000 5.6533	COUNTY LINE	1.7298
	XING 228.38 0.0000 5.6533	LOUISA STREET	1.7298
	XING 228.60 0.0000 5.6533		1.7298
5290	XING 228.91 5290 200 0.0050 5.6584	PORENT MARY STREET	1.7298 0.0050 5290 200
5275	XING 229.21 5275 200 0.0050 5.6634	GORMLY STREET	1.7248 0.0050 5275 200

READ DOWN WESTBOUND TRAFFIC — APPENDIX B3 STATION — READ UP EASTBOUND TRAFFIC

PAGE 13 COLUMN 1

MILES	DISTANCE FT	HWY GRADE CROSSING	MAX MPH	DISTANCE FT	SPEED MPH	ST. HOURS	TIME HOURS	STATION	TIME HOURS	ST. HOURS	DISTANCE FT	SPEED MPH	MILES
656.33	5273	XING 229.3		5273	200	0.0050	5.8664	MARTIN STREET RR	1.7198	0.0037	3858	200	718.44
657.33	5278	XING 226.7		5278	200	0.0050	5.8735	CAMPBELL STREET	1.7161	0.0061	22871	A 90	718.44
658.33	5289	XING 230.6		5289	200	0.0050	5.6764	BERLIN CR	1.6880	0.0000			718.44
659.33	5269	XING 221.8		5269	200	0.0050	5.8663	PRIVATE	1.6880	0.0000		90	718.44
660.33	5278	XING 231.7		2694	200	0.0020	5.6863	PRIVATE	1.6880	0.0000		90	719.45
660.81	1500	XING 234.1			200	0.0000	5.6853	GROVE ROAD	1.6860	0.0029	1380	90	720.45
660.61		XING 234.9		4676	D 20	0.0061	5.6914	WYKES CR	1.6651				721.45
660.61		XING 235.10			90	0.0000	5.6914	PRIVATE	1.6651				722.45
660.61		XING 236.06				0.0000	5.6914	WALNUT STREET	1.6651				723.45
660.61		XING 236.14				0.0000	5.6914	MAIN STREET	1.6651				724.45
660.61			90		90	0.0333	5.7248	DUNKIRK NYC XING	1.6651	0.0333	RR CRC	90	724.45
660.91		XING 236.49				0.0000	5.7248	RTE 125	1.6518	0.0261	4678	D 200	724.45
661.33	5771	XING 237.81		1960	90	0.0029	5.7276	HOPPS ROAD	1.6457	0.0000	20	200	724.45
662.33	5278	XING 239.7		22871	A 90	0.0081	5.7358	WAYNE STREET	1.6457	0.0041	4350	200	724.45
663.33	5290					0.0000	5.7358		1.6416	0.0050	5290	200	725.45
664.33	5278					0.0000	5.7358		1.6366	0.0050	5278	200	726.45
665.33	5280	XING 241.6		836	200	0.0008	5.7366	TRAVERSE CH	1.6019	0.0050	5289	200	727.45
666.33	5277	XING 241.6		5277	200	0.0050	5.7816	PRIVATE	1.9266	0.0050	5277	200	728.46
667.33	5279	XING 242.8		5078	200	0.0050	5.7466	PETERSON CR	1.6016	0.0050	5276	200	729.46
668.33	5273	XING 240.6		5273	200	0.0050	5.7716	SCOTT XING	1.8186	0.0050	5273	200	730.46
669.32	5265	XING 244.8		5266	200	0.0050	5.7766	1 MILE XING	1.6116	0.0050	5266	200	731.46
670.34	5354	XING 245.58		5394	200	0.0061	5.7817	ADA JOHNSON STREET	1.8066	0.0061	6964	200	732.46
671.34	5281	XING 245.6		5281	200	0.0050	5.7867	MAIN STREET	1.6015	0.0050	5281	200	733.46
672.34	5276	XING 246.7		5276	200	0.0050	5.7817	GILBERT STREET	1.5965	0.0050	5276	200	734.46
673.36	5377	XING 246.8		5377	200	0.0051	5.7967	KLINGLER RD	1.5915	0.0051	5377	200	734.46
674.37	5300	XING 247.6		5300	200	0.0050	5.8018	ST PAUL RD	1.5864	0.0050	5300	200	734.46
674.37		XING 248.10			200	0.0000	5.8018	PRIVATE	1.5814				734.46
674.37		XING 248.65			200	0.0000	5.8018	COUNTY LINE	1.5814				734.46
674.37		XING 249.65			200	0.0000	5.8018	BENTLEY ROAD	1.5814				734.46
674.37		XING 250.15			200	0.0000	5.8018	MP 250/PEVEE ROAD	1.5814				734.46
675.37	5269	XING 250.4		5269	200	0.0050	5.8068	PRIVATE	1.5614	0.0050	5269	200	735.46
676.37	5286	XING 250.7		5066	200	0.0050	5.8118	LAFAYETTE RD	1.5764	0.0050	5296	200	736.46
677.37	5287	XING 250.8		5287	200	0.0050	5.8168	SHREDER ROAD	1.5714	0.0050	5287	200	737.46
678.38	5304	XING 251.3		5304	200	0.0050	5.8218	VINT RD	1.5664	0.0050	5304	200	737.46
679.39	5371	XING 251.6		5371	200	0.0051	5.8269	PHILLIPS RD	1.5614	0.0051	5371	200	737.96
680.36	5202	XING 252.2		5202	200	0.0049	5.8316	S HIGH STREET	1.5563	0.0004	96	200	738.46
681.32	5298	XING 252.3		8696	200	0.0060	5.8367	CHURCH ST	1.5562	0.0060	6908	A 160	739.46
681.84	2406	XING 252.4		1066	D 20	0.0010	5.8382	WASHINGTON	1.5499	0.0000			740.47
681.69	283	XING 2	182	280	182	0.0003	5.8385	R RUMBAUGH RD	1.5499	0.0000			741.48
682.58	2500	XING 253.6		1360	187	0.0014	5.8401	FISHER ROAD	1.5499	0.0000			742.49
683.40	5536	XING 254.1		4285	A 18	0.0046	5.8457	MCCLURE XING	1.5499	0.0215	16067	A 90	743.33
683.96	3000	XING 254.86				0.0000	5.8457	COOL ROAD	1.5060	0.0028	1380	90	743.47
683.96		XING 255.64				0.0000	5.8457	THAYER ROAD	1.5251				743.47
683.96		XING 256.92				0.0000	5.8457	FETTER ROAD	1.5251				748.62
683.96		XING 257.58		346	186	0.0002	5.8460	METZGER ROAD	1.5251				743.62
683.96		XING 258.42		4446	D 19	0.0026	5.8516	ROUSH ROAD	1.5251				748.62
683.96			90		90	0.0000	5.8518	SUGAR CRT XING	1.5251	0.0000		90	743.83
683.96							5.8518		1.5251	0.0012	616	D 111	744.12
684.41	2373	XING 259.6		1380	90	0.0029	5.8546	SUGAR ST	1.5239	0.0001	69	111	744.40
684.41				1925	A 100	0.0036	5.8582	MP 260	1.5238	0.0026	1925	A 90	744.49
684.62	1100	XING 260.1		69	111	0.0001	5.8585	N PINE ST	1.5202	0.0018	890	90	744.49
684.72	560	XING 260.2		616	D 11	0.0012	5.8597	R JACKSON	1.5184	0.0013	560	90	744.49
684.72			90		90	0.0000	5.8597	LIMA RD NW XING	1.5171	0.0030		90	744.58
684.72		XING 260.50			90	0.0000	5.8597	MAIN ST	1.4673	0.0042	2805	D 165	744.56
684.81	485	XING 2	125	485	90	0.0010	5.8107	R N ELIZABETH	1.4631	0.0008	780	165	745.47
684.81		XING 260.66				0.0000	5.8107	N WEST ST	1.4622	0.0000			745.49
684.96	800	XING 260.7		800	90	0.0017	5.8124	N MCDONEL ST	1.4622	0.0030		165	746.57
685.40	2300	XING 2	165	75	90	0.0002	5.8126	R N METCALF	1.4602	0.0000		165	745.65
685.50	576	XING 2	165	3116	A 90	0.0026	5.8181	N BAXTER ST	1.4822	0.0007	576	165	746.51
686.39	4700	XING 261.8		6406	A 17	0.0110	5.8291	COLE ST	1.4615	0.0019	1874	D 200	747.52
686.39		XING 262.36				0.0000	5.8261	CABLE ROAD	1.4566	0.0027	2606	200	748.53
687.40	5266	XING 262.5		11149	A 16	0.0115	5.8406	HARTZLER RD	1.4588	0.0050	5098	200	749.51
688.40	5288	XING 264.05				0.0000	5.8406	EASTTOWN ROAD	1.4319	0.0000	5286	200	750.51
689.42	5393	XING 264.9		807	200	0.0008	5.8414	EAST ROAD	1.4468	0.0051	5808	200	751.51
690.42	5280	XING 265.2		5280	200	0.0050	5.8464	BATY ROAD	1.4418	0.0050	5280	200	752.51
691.42	5288	XING 266.8		5286	200	0.0050	5.8514	PRIVATE	1.4383	0.0050	5098	200	753.53
692.42	5085	XING 266.4		5285	200	0.0050	5.8564	GREENLAWN AVE	1.4318	0.0050	5285	200	754.52
693.42	5292	XING 266.9		5262	200	0.0050	5.8614	MAIN ROAD	1.4268	0.0050	5030	200	754.52
694.42	5081	XING 267.2		5281	200	0.0050	5.8664	WAPAK ROAD	1.4218	0.0050	5281	200	755.52
694.42		XING 267.40				0.0000	5.8664	PRIVATE	1.4163				756.52
694.42		XING 267.65				0.0000	5.8664	KEMP ROAD	1.4168				757.52
694.42		XING 267.90				0.0000	5.8664	PRIVATE	1.4168				758.53
694.42		XING 268.10				0.0000	5.8664	PRIVATE	1.4168				759.53
694.42		XING 268.83				0.0000	5.8664	GRUBB ROAD	1.4168				760.53
695.43	5282	XING 269.2		5292	200	0.0050	5.8714	PRIVATE	1.4168	0.0050	5292	200	761.53
696.43	5283	XING 270.4		5283	200	0.0050	5.8764	STATE RD	1.4118	0.0061	1164	200	762.53
697.43	5280	XING 270.6		8676	200	0.0082	5.8828	OLD DELPHOS	1.4107	0.0115	11149	A 165	763.56
698.01	3100	XING 271.2		1874	D 20	0.0018	5.8846	DEFIANCE TRAIL	1.3992	0.0016	1390	165	764.54
698.43	2172	XING 2	165	2172	165	0.0025	5.8870	BAUGH RD	1.3976	0.0025	2172	165	765.54
698.62	1000	XING 2	165	1000	165	0.0011	5.8882	L PELTIER RD	1.3951	0.0011	1000	165	766.54
698.62		XING 272.60				0.0000	5.8882	L LEHMAN RD	1.3940	0.0003	286	D 170	766.85
698.82		XING 273.06					5.8882	PRIVATE	1.3937	0.0001	104	170	767.13
698.82	1100	XING 274.3		1100	165	0.0013	5.8895	DELPHOS PIERCE STREET	1.3936	0.0015	1230	A 165	767.55
699.18	1900	XING 274.3		1900	165	0.0022	5.8916	S MAIN STREET	1.3921	0.0018	1380	165	767.55
699.37	1000	XING 2	165	1000	165	0.0011	5.8928	R CANAL STREET	1.3905	0.0011	1000	165	768.56
699.43	284	XING 274.7		284	165	0.0003	5.8931	JEFFERSON STREET	1.3894	0.0018	1671	D 200	769.31
700.43	5277	XING 274.7		1076	165	0.0012	5.8943	S CLAY STREET	1.3875	0.0047	5015	200	769.19
701.43	5280	XING 274.9		11149	A 16	0.0130	5.9073	S BREDICK STREET	1.3828	0.0038	3600	200	769.55
702.43	5295	XING 275.0		3805	200	0.0034	6.0107	STATE STREET	1.3790	0.0050	5285	200	770.46
703.43	5289	XING 276.6		5289	200	0.0050	6.0158	BRICKNER RD	1.3740	0.0050	5286	200	770.55
704.43	5264	XING 277.2		5284	200	0.0050	6.0206	PRIVATE	1.3690	0.0050	5284	200	770.59
704.43		XING 277.68				0.0000	6.0206	BOOKEY RD	1.3640				771.55
704.43		XING 278.93				0.0000	6.0206	CONVERSE RD	1.3640				772.56
704.43		XING 279.71				0.0000	6.0206	MIDDLEPOINT	1.3640				773.56
705.43	5281	XING 280.0		5281	200	0.0050	6.0256	MAIN STREET	1.3640	0.0050	5281	200	774.56
706.43	5281	XING 280.2		5281	200	0.0050	6.0306	ADAMS STREET	1.3590	0.0050	5281	200	775.56
707.43	5285	XING 280.3		5285	200	0.0050	6.0356	MASON STREET	1.3540	0.0033	5285	200	776.56
708.43	5282	XING 280.7		5282	200	0.0050	6.0406	DOG CREEK RD	1.3207	0.0010	1015	200	777.56
709.43	5281	XING 281.0		5281	200	0.0050	6.0456	RINGWALD RD	1.3197	0.0000			778.56
710.43	5267	XING 282.3		8032	200	0.0076	6.0534	CHENOWETH RD	1.3197	0.0000			779.56
711.34	4800	XING 283.3		2085	D 20	0.0022	6.0556	GAMBLE RD	1.3197	0.0127	12165	A 161	780.57
711.43	480	XING 2	161		161	0.0000	6.0556	L HOAGLIN CTR	1.3070	0.0153	9405	A 103	781.56
711.88	1300	XING 2	161		161	0.0000	6.0556	L GILLAND RD	1.2637	0.0016	800	103	782.28
711.68		XING 286.38			161	0.0000	6.0656	MENDON RD	1.2922				782.57
711.96	1500	XING 286.7		922	161	0.0011	6.0666	VAN WERT WAYNE STREET	1.2022	0.0022	1100	A 96	782.75
711.96		XING 286.86		2384	D 16	0.0034	6.0601	VINE STREET	1.2850				783.27
712.13	900	XING 2	105	250	99	0.0005	6.0606	R FRANKLIN STREET	1.2905	0.0008	360	90	783.27
712.31	900	XING 2	117	1401	99	0.0027	6.0632	R CHESTNUT STREET	1.2892	0.0019	900	90	783.27
712.32	105	XING 287.04			99	0.0000	6.0632	RACE STREET	1.2873	0.0002	100	90	783.27
712.32		XING 287.11		349	D 96	0.0006	6.0637	N TYLER STREET	1.2871				783.27
712.32		XING 287.17			90	0.0000	6.0637	HARRISON STREET	1.2871				783.27
712.32		XING 287.24				0.0000	6.0637	CHERRY STREET	1.2871				783.27
712.32		XING 287.30				0.0000	6.0637	WALNUT STREET	1.2871				783.57
712.32		XING 287.37				0.0000	6.0637	MARKET STREET	1.2871				784.57
712.32		XING 287.49				0.0000	6.0637	WASHINGTON STREET	1.2871				784.57
712.32		XING 287.55				0.0000	6.0637	JEFFERSON STREET	1.2871				784.57
712.32		XING 287.67				0.0000	6.0637	N SHANNON	1.2871				784.60
712.32		XING 287.74				0.0000	6.0637	WALL STREET	1.2871				784.96
712.32			90		90	0.0000	6.0637	CR VAN WERT IN XING	1.2871	0.5000	RR CRC	90	785.57
712.32		XING 288.57				0.0000	6.0637	BURT STREET	1.2871	0.0081	4679	D 120	786.58
712.45	642	XING 286.3		642	90	0.0014	6.0651	FISHER AVENUE	1.2810	0.0060	0	200	787.57
713.44	5241	XING 288.8		736	35	0.0015	6.0666	JOHN BROWN RD	1.2810	0.0011	1204	200	788.57
714.44	5284	XING 286.59				0.0000	6.0666	PRIVATE	1.2799	0.0050	5284	200	788.57
714.44						0.0000	6.0666	MP 290	1.2749				789.59
715.44	5288	XING 290.1		22871	A 90	0.0281	6.0947	LIBERTY UNION	1.2749	0.0050	5209	200	790.57
716.44	5287	XING 291.2				0.0000	6.0947	RICHEY CR	1.2699	0.0050	5287	200	791.58
717.44	5287	XING 292.3		2046	200	0.0076	6.0976	MACE XING	1.2649	0.0050	5287	200	792.58

READ DOWN WESTBOUND TRAFFIC — APPENDIX B3 STATION — READ UP EASTBOUND TRAFFIC

PAGE 13 COLUMN 2

MILES	DISTANCE FT	HWY GRADE CROSSING	MAX MPH	DISTANCE FT	SPEED MPH	ST. HOURS	TIME HOURS	STATION	TIME HOURS	ST. HOURS	DISTANCE FT	SPEED MPH
718.44	5294	XING 293.0		5294	200	0.0050	6.1026	ALT RT US 30	1.2598	0.0050	5294	200
718.44		XING 293.59				0.0000	6.1026	BERGNER RD	1.2548			
718.44		XING 294.71				0.0000	6.1026	MAIN ST CONVOY	1.2548			
718.44		XING 294.98				0.0000	6.1026	TULLY ST	1.2548			
719.45	5287	XING 295.4		5287	200	0.0050	6.1076	PRIVATE	1.2548	0.0050	5287	200
720.45	5288	XING 296.8		5288	200	0.0050	6.1126	PAYNE CR	1.2498	0.0050	5288	200
721.45	5268	XING 296.3		5286	200	0.0050	6.1176	BRANER CR	1.2448	0.0050	5268	200
722.45	5290	XING 296.9		5286	200	0.0050	6.1226	LARE CR	1.2398	0.0050	5290	200
723.45	5292	XING 297.1		5290	200	0.0050	6.1276	MENTZER	1.2348	0.0050	5292	200
724.45	5288	XING 297.9		5288	200	0.0050	6.1326	PRIVATE	1.2299	0.0050	5288	200
724.45		XING 298.12				0.0000	6.1326	MENTZER CURC	1.2249			
724.45		XING 298.86				0.0000	6.1326	CLEM XING	1.2249			
724.45		XING 299.46				0.0000	6.1326	SCHOOLHOUSE	1.2249			
724.45						0.0000	6.1326	MP 300	1.2249	0.0000		
725.45	5282	XING 300.2		5282	200	0.0050	6.1376	ST LINE RD	1.2249	0.0050	5282	200
726.45	5281	XING 301.3		5281	200	0.0050	6.1426	MORGAN RD	1.2199	0.0050	5281	200
727.45	5289	XING 301.6		5289	200	0.0050	6.1476	PRIVATE	1.2149	0.0050	5289	200
728.46	5285	XING 302.9		5285	200	0.0050	6.1526	LORTIE RD	1.2099	0.0050	5285	200
729.46	5294	XING 303.9		5294	200	0.0050	6.1576	OHIO STREET	1.2049	0.0050	5294	200
730.46	5289	XING 303.9		5289	200	0.0050	6.1626	MAIN STREET	1.1999	0.0050	5289	200
731.46	5297	XING 304.1		5297	200	0.0050	6.1677	WASHINGTON	1.1949	0.0050	5297	200
732.46	5274	XING 306.3		5274	200	0.0050	6.1726	SNYDER RD	1.1899	0.0050	5274	200
733.46	5282	XING 306.9		5292	200	0.0050	6.1777	HOFFMAN RD	1.1849	0.0050	5282	200
734.46	5281	XING 307.1		5281	200	0.0050	6.1827	GROTRIAN RD	1.1799	0.0050	5281	200
734.46		XING 307.71				0.0000	6.1827	WILSON RD	1.1749			
734.46		XING 308.30				0.0000	6.1827	FACKLER RD	1.1749			
734.46		XING 308.55				0.0000	6.1827	GARARDOT RD	1.1749			
734.46		XING 309.37				0.0000	6.1827	HOUK RD	1.1749			
734.46		XING 308.71				0.0000	6.1827	WASHINGTON	1.1749			
734.46		XING 309.94				0.0000	6.1827	FRANKE RD	1.1749			
735.46	5261	XING 310.6		5261	200	0.0050	6.1877	PRIVATE	1.1749	0.0050	5261	200
736.46	5281	XING 310.9		5281	200	0.0050	6.1927	PRIVATE	1.1699	0.0050	5281	200
737.46	5267	XING 311.6		5267	200	0.0050	6.1977	TILLMAN RD	1.1649	0.0085	9871	200
737.48		XING 311.75				0.0000	6.1977	MINNICH ROAD	1.1584			
737.96		XING 312.50				0.0000	6.1977	PRIVATE	1.1584			
738.46	5293	XING 312.8		5293	200	0.0050	6.2027	PRIVATE	1.1584	0.0050	5293	200
739.46	5283	XING 313.9		5283	200	0.0050	6.2077	PRIVATE	1.1534	0.0206	18220	A 15
740.47	5288	XING 313.2		5298	200	0.0050	6.2127	PAULDING ROAD	1.1328	0.0000	0	
741.48	5328	XING 313.44		5329	200	0.0050	6.2177	HARTZELL ROAD	1.1328	0.0005	5802	A 76
742.49	5192	XING 315.5		5192	200	0.0049	6.2226	ADAMS CENTER	1.1232	0.0005	213	76
743.33	4800	XING 318.2		329	200	0.0003	6.2229	MEYER ROAD	1.1228	0.0016	587	A 63
743.47	749		170		200	0.0000	6.2229	L	1.1212	0.0017	366	63
743.47				5620	D 20	0.0068	6.2298		1.1195			
748.62	909		76	536	76	0.0015	6.2312	UTTAPPHOO	1.1196	0.0024	300	63
743.62			96		76	0.0830	6.3145	REVOLUR MIKE WABXING	1.1171	0.0831		
748.62				265	D 76	0.0007	6.3152	FORT WAYNE	1.0338			
743.83	1400	XING 3	63	1400	63	0.0042	6.3194	R BRACKENRIDGE	1.0338	0.0042	1400	63
744.12	1200		63	1200	63	0.0036	6.3230	R	1.0296	0.0006	310	D 78
744.40	1500			160	63	0.0005	6.3236		1.0289	0.0058	2880	76
744.49	431		40	685	A 63	0.0019	6.3054	R	1.0230	0.0012	461	76
744.49				2450	60	0.0058	6.3112		1.0218			
744.49				46	D 96	0.0001	6.3313	MP 320	1.0218	0.0000		
744.58	468		78	489	78	0.0012	6.3305	R	1.0218	0.0012	468	78
744.56				1360	78	0.0035	6.3358		1.0206	0.0081	1700	D 13
745.47	4700			5117	A 78	0.0090	6.3450		1.0175	0.0043	2998	133
745.49	100		133			0.0000	6.3450	R	1.0132	0.0001	100	133
746.57	400		133			0.0000	6.3450	R	1.0131	0.0006	400	133
745.65	452		133			0.0000	6.3450	R	1.0125	0.0006	452	133
746.51	4000					0.0000	6.3466		1.0119	0.0057	3272	D 20
747.52	5343	XING 323.0		17786	A 13	0.0208	6.3669	THOMAS ROAD	1.0062	0.0062	6577	200
748.53	5322	XING 324.93				0.0000	6.3659	HADLEY ROAD	1.0020	0.0050	5322	200
749.51	5158	XING 325.5		395	200	0.0004	6.3662	BASS ROAD	0.9970	0.0049	5180	200
750.51	5277	XING 326.8		5277	200	0.0050	6.3712	PRIVATE	0.9971	0.0050	5277	200
751.51	5278	XING 328.1		5278	200	0.0050	6.3762	MAIN STREET	0.9871	0.0050	5278	200
752.51	5285	XING 328.3		5290	200	0.0050	6.3812	ARCOLA COLUMBIA STREET	0.9821	0.0050	5290	200
753.53	5363	XING 328.9		5363	200	0.0051	6.3863	PRIVATE	0.9771	0.0051	5363	200
754.52	5224	XING 329.0		5224	200	0.0049	6.3913	PRIVATE	0.9720	0.0049	5224	200
754.52		XING 328.30				0.0000	6.3913	PRIVATE	0.9671	0.0000		200
755.52	5298	XING 329.9		5298	200	0.0050	6.3963	PRIVATE	0.9671	0.0050	5298	200
756.52	5297	XING 330.5		5297	200	0.0050	6.4013	COUNTY LINE	0.9621	0.0050	5297	200
757.53	5294	XING 331.5		5294	200	0.0050	6.4063	BEECHING ROAD	0.9571	0.0050	5294	200
758.53	5298	XING 332.6		5280	200	0.0050	6.4113	BOGEY	0.9521	0.0050	5280	200
759.53	5293	XING 333.8		5293	200	0.0050	6.4163	MILL STREET	0.9471	0.0050	5293	200
760.53	5292	XING 334.7		5292	200	0.0050	6.4214	MOWERY ROAD	0.9421	0.0050	5292	200
761.53	5269	XING 336.5		5269	200	0.0050	6.4263	PASSIE CR	0.9371	0.0050	5269	200
762.53	5250	XING 338.4		5230	200	0.0050	6.4314	OHIO STREET	0.9321	0.0050	5250	200
763.56	5434	XING 338.61		5434	200	0.0051	6.4365	SR 9	0.9271	0.0051	5434	200
764.54	5185	XING 338.9		5185	200	0.0048	6.4414	LINE STREET SR	0.9220	0.0049	5185	200
765.54	5263	XING 339.9		5263	200	0.0050	6.4464	POOR FARM CR	0.9171	0.0039	4157	200
766.54	5298	XING 341.3		4234	200	0.0040	6.4504	WOLF ROAD	0.9132	0.0066	6642	A 13
766.85	1600	XING 342.0		1082	D 20	0.0011	6.4515	PEARL CR 300E?	0.9066	0.0011	1080	181
767.13	1505	XING 3	181	1503	181	0.0016	6.4530	BUTLER CR 400	0.9065	0.0000	0	181
767.55	2200	XING 343.8		1360	181	0.0014	6.4546	L PRIVATE	0.9065	0.0000		
767.55				4276	A 16	0.0063	6.4607		0.9056	0.0000		
768.56	5290			101	189	0.0001	6.4608		0.9055	0.0000		
769.31	1500			3293	D 13	0.0037	6.4645		0.9056	0.0136	11209	A 1
769.19	1495		132	1421	132	0.0021	6.4667	R	0.8920	0.0025	1380	132
769.55	1500			1390	132	0.0028	6.4695		0.8900	0.0036	2311	D 20
770.46	4800			18048	A 13	0.0193	6.4889		0.8862	0.0032	3369	200
770.55	484	XING 3	200			0.0000	6.4889	L BIRD ??	0.8830	0.0005	484	200
770.59	164	XING 3	202			0.0000	6.4889	L LASWELL DEPOT STREET	0.8825	0.0002	184	200
771.55	5106	XING 347.30				0.0000	6.4889	BRINKLEY ROAD	0.8823	0.0048	5100	200
772.56	5294	XING 348.61				0.0000	6.4889	BLACK CR	0.8775	0.0050	5294	200
773.56	5286	XING 349.3		1863	A 15	0.0016	6.4905	PRIVATE	0.8725	0.0050	5298	200
774.56	5287	XING 349.8		3048	200	0.0068	6.4973	PRIVATE	0.8675	0.0050	5287	200
775.56	5282	XING 350.3		5282	200	0.0050	6.5023	MARKET STREET	0.8625	0.0050	5282	200
776.56	5283	XING 350.6		5280	200	0.0050	6.5073	PIERCETON FIRST STREET	0.8575	0.0050	5283	200
777.56	5279	XING 352.5		5279	200	0.0050	6.5123	VAN NESS ROAD	0.8525	0.0050	5279	200
778.56	5278	XING 353.4		5278	200	0.0050	6.5173	WOOSTER ROAD	0.8475	0.0050	5278	200
779.56	5283	XING 354.7		5293	200	0.0050	6.5223	SCHADE 350E	0.8426	0.0008	814	200
780.57	5323			5323	200	0.0050	6.5274		0.8417	0.0000		
781.56	5233			5233	200	0.0050	6.5323		0.8417	0.0000		
782.28	3860			2900	200	0.0027	6.5351		0.8417	0.0000		
782.57	1506	XING 3	202	1417	200	0.0013	6.5364	L BRONSON STREET	0.8417	0.0000		
782.75	986	XING 3	202		200	0.0000	6.5364	L MAPLE STREET	0.8417	0.0061	22871	A 90
783.27	2700	XING 358.24			200	0.0000	6.5364	SCOTT STREET	0.8136	0.0029	1380	90
783.27				4679	D 20	0.0061	6.5425		0.8107	0.0000		
783.27			90		90	0.0333	6.5759	WARSAW DIXING	0.8107	0.0333	RR CRC	90
783.27		XING 358.74				0.0000	6.5759	DETROIT STREET	0.7774	0.0047	3292	D 17
783.27		XING				0.0000	6.5759	ALLEY?	0.7727	0.0000	25	174
783.27		XING 358.87				0.0000	6.5759	INDIANA STREET	0.7727	0.0000		
783.27		XING				0.0000	6.5759	ALLEY?	0.7727	0.0000		
783.57	1600	XING 359.0		1380	90	0.0029	6.5787	S LAKE STREET	0.7727	0.0027	2426	A 16
784.57	8050	XING		11922	A 90	0.0165	6.5953	ALLEY?	0.7700	0.0014	1182	185
784.57		XING 359.25				0.0000	6.5953	UNION STREET	0.7686	0.0000		
784.57						0.0000	6.5953	MP 360	0.7686	0.0000		
784.60	176	XING 360.50				0.0000	6.5953	JACKSON CR	0.7686	0.0002	178	185
784.96	1900		185			0.0000	6.5953	R BUFFET	0.7684	0.0027	1900	185
785.57	3200	XING 361.3				0.0000	6.5953	PRIVATE	0.7662	0.0019	1874	D 20
786.58	5318	XING 362.5		11149	A 16	0.0115	6.6068	PRIVATE	0.7643	0.0013	8144	200
787.57	5245	XING 362.26				0.0000	6.6068	THOMAS CR 50	0.7586	0.0050	8745	200
788.57	5291	XING 363.2		3683	200	0.0038	6.6106	PRIVATE	0.7530	0.0050	5291	200
789.57		XING 363.72				0.0000	6.6106	SELBY CR 525	0.7480	0.0000		200
789.59	5352	XING 364.8		5352	200	0.0051	6.6157	PRIVATE	0.7480	0.0051	5352	200
790.57	5220	XING 365.3		5220	200	0.0049	6.6206	PRAIRIE STREET	0.7429	0.0049	5220	200
791.58	5291	XING 365.45		5291	200	0.0050	6.6256	ATWOOD HARRISON STREET	0.7380	0.0050	5291	200
792.58	5278	XING 366.8		5278	200	0.0050	6.6306	ROBINSON ROAD	0.7330	0.0050	5279	200

APPENDIX B3

PAGE 14 COLUMN 1

READ DOWN WESTBOUND TRAFFIC — STATION — READ UP EASTBOUND TRAFFIC

MILES	DISTANCE ft	HWY GRADE CROSSING	MAX MPH	DISTANCE ft	SPEED mph	ET, HOURS	TIME, HOURS	STATION	TIME, HOURS	ET, HOURS	DISTANCE ft	SPEED mph
793.57	5227	XING 367.4		5227	200	0.0049	6.6356	SNYDER CR 87	0.7280	0.0049	5227	200
794.58	5337	XING 368.1		5337	200	0.0051	6.6406	PRIVATE	0.7231	0.0051	5337	200
794.58		XING 368.77				0.0000	6.6406	COOK CR	0.7180	0.0000		
794.58		XING 369.29				0.0000	6.6406	WALNUT STREET SR	0.7180	0.0000		
794.58		XING 369.36				0.0000	6.6406	ETNA GREEN	0.7180	0.0000		
794.58						0.0000	6.6406	MP 370	0.7180	0.0000		
795.60	5379	XING 370.9		5379	200	0.0051	6.6457	S BEECH ROAD	0.7180	0.0051	5379	200
796.58	5193	XING 372.9		5193	200	0.0049	6.6507	HARRIS STREET	0.7129	0.0049	5193	200
797.61	5439	XING 373.0		5439	200	0.0052	6.6558	BOURBON MAIN STREET	0.7080	0.0052	5439	200
797.61		XING 373.17				0.0000	6.6558	BOURBON STREET	0.7028	0.0000		
797.61		XING 373.29				0.0000	6.6558	CENTER STREET	0.7028	0.0000		
797.61		XING 374.13				0.0000	6.6558	TSCHR ST 8E	0.7028	0.0000		
798.59	5162	XING 375.1		5162	200	0.0049	6.6907	STOCKMAN ROAD	0.7028	0.0049	5162	200
799.58	5252	XING 376.2		5252	200	0.0050	6.6657	WYMER CR	0.6979	0.0050	5252	200
800.58	5270	XING 377.0		5270	200	0.0050	6.6707	COMPTON CR	0.6929	0.0050	5270	200
801.58	5279	XING 377.8		5279	200	0.0050	6.6757	HAWTHORNE ROAD	0.6879	0.0050	5279	200
802.58	5279	XING 378.8		5279	200	0.0050	6.6807	PRIVATE	0.6829	0.0050	5279	200
803.58	5283	XING 379.1		5283	200	0.0050	6.6857	S IRIS ROAD	0.6779	0.0050	5283	200
804.58	5277	XING 379.6		5277	200	0.0050	6.6907	PRIVATE	0.6729	0.0014	1483	200
805.59	5315	XING 380.2		5315	200	0.0050	6.6957	PRIVATE	0.6715	0.0000		
806.56	5155	XING 380.5		5155	200	0.0049	6.7006	PRIVATE	0.6715	0.0000		
807.58	5367	XING 381.0		5000	200	0.0047	6.7053	KING ROAD	0.6715	0.0281	22671	A 90
808.41	4400	XING 383.2		86	200	0.0001	6.7054	LIBERTY STREET	0.6434	0.0029	1360	90
808.41				4679	D 20	0.0061	6.7115		0.6405			
808.41			90		90	0.0000	6.7115	PLYM NKP XING	0.6405		RR CRC	90
808.41						0.0000	6.7115		0 0.6405	0.0061	4679	D 200
808.59	944	XING 384.0		944	90	0.0020	6.7135	FIFTH STREET	0.6344	0.0048	5061	200
809.58	5216	XING 384.6		416	90	0.0009	6.7144	OAK ROAD	0.6296	0.0000		
810.58	5280	XING 385.7		22671	A 90	0.0281	6.7425	PRIVATE	0.6296	0.0016	1700	200
811.58	5287	XING 385.90				0.0000	6.7425	PRIVATE	0.6280	0.0050	5287	200
812.58	5293	XING 386.34				0.0000	6.7425	ANGLING ROAD	0.6230	0.0050	5293	200
813.58	5269	XING 386.6		3058	200	0.0031	6.7456	QUINCE ROAD	0.6180	0.0050	5269	200
814.58	5281	XING 387.7		5281	200	0.0050	6.7506	REDWOOD ROAD	0.6130	0.0050	5281	200
814.58		XING 388.14				0.0000	6.7506	PRIVATE	0.6080			200
815.56	5268	XING 390.8		5268	200	0.0050	6.7556	UNION ROAD	0.6080	0.0050	5268	200
816.58	5298			5298	200	0.0050	6.7606	DONALDSON	0.6030	0.0050	5298	200
817.58	5298	XING 393.5		5298	200	0.0050	6.7656	PRIVATE	0.5980	0.0050	5298	200
818.61	5411	XING 394.1		5411	200	0.0051	6.7707	SR 23	0.5930	0.0051	5411	200
819.59	5164	XING 395.5		5164	200	0.0049	6.7756	WALSH CR	0.5879	0.0049	5164	200
820.59	5285	XING 396.5		5285	200	0.0050	6.7806	750E	0.5830	0.0050	5285	200
821.59	5288			5288	200	0.0050	6.7806		0.5780	0.0050	5288	200
822.59	5285	XING 398.0		5285	200	0.0050	6.7906	STARKE STREET	0.5730	0.0050	5285	200
823.59	5279	XING 398.5		5279	200	0.0050	6.7956	HAMLET JEFFERSON STRE	0.5680	0.0050	5279	200
824.59	5301	XING 399.11		5301	200	0.0050	6.8007	500E	0.5630	0.0050	5301	200
825.60	5282	XING 401.1		5282	200	0.0050	6.8057	300E	0.5580	0.0050	5282	200
826.60	5282	XING 401.8		5282	200	0.0050	6.8107	PRIVATE	0.5530	0.0050	5282	200
827.59	5270	XING 402.1		5270	200	0.0050	6.8156	PRIVATE	0.5480	0.0050	5270	200
828.59	5287	XING 403.3		5287	200	0.0050	6.8207	50E	0.5430	0.0019	2018	200
829.60	5292	XING 405.2		5282	200	0.0050	6.8257	100W	0.5411			
830.60	5280	XING 405.7		5280	200	0.0050	6.8307	1400S	0.5411	0.0000		
831.62	5422	XING 406.2		5422	200	0.0051	6.8358	SR 39	0.5411	0.0000		
832.58	6046	XING 407.2		4766	200	0.0045	6.8403	LONG LANE	0.5411	0.0281	22671	A 90
832.58		XING 4307 70				0.0000	6.8403	PRIVATE	0.5130			
833.41	4400	XING 408.61			200	0.0000	6.8403	THOMPSON STREET	0.5130	0.0029	1360	90
833.41		XING 408.7		4679	D 20	0.0061	6.8464	OHIO STREET	0.5101			
833.41			90		90	0.0000	6.8464	HANNA C&O XING	0.5101		RR CRC	90
833.41		XING 408.83				0.0000	6.8464	450W	0.5101	0.0061	4679	D 200
833.60	966	XING 409.2		968	90	0.0020	6.8485	PRIVATE	0.5040	0.0002	177	200
834.60	5281			394	90	0.0008	6.8493		0.5038	0.0038	4000	200
834.60		XING 410.35				0.0000	6.8493	MP 410 SCHULL 600W	0.5000	0.0000		200
835.60	5286	XING 411.3		22671	A 90	0.0281	6.8774	FARRESTER ROAD	0.5000	0.0050	5286	200
836.60	5273	XING 411.89				0.0000	6.8774	TILDON 750W	0.4950	0.0050	5273	200
837.77	6182	XING 412.89				0.0000	6.8774	PRIVATE	0.4900	0.0059	6182	200
838.59	4367	XING 413.4		3323	200	0.0031	6.8806	KLEINS 500W	0.4841	0.0041	4367	200
839.59	5280	XING 414.7		5280	200	0.0050	6.8856	N MAIN STREET	0.4800	0.0050	5280	200
840.59	5272	XING 414.8		5272	200	0.0050	6.8905	WANATAH S ILLINOIS	0.4750	0.0050	5272	200
841.59	5251	XING 415.1		5251	200	0.0050	6.8955	S WASHINGTON	0.4700	0.0050	5251	200
842.59	5293	XING 415.1		5293	200	0.0050	6.9005	LINCOLN STREET	0.4650	0.0050	5293	200
843.59	5270	XING 415.4		5270	200	0.0050	6.9055	1100W	0.4600	0.0050	5270	200
844.19	3200	XING 416.5		3200	200	0.0030	6.9086	OSBRN CR-CO	0.4550	0.0050	5270	200
844.30	570	XING 4 219		570	200	0.0005	6.9091	FISHER 575E	0.4500	0.0004	472	200
844.59	1500	XING 419.0		1500	200	0.0014	6.9105	DOLSON 450E	0.4496	0.0014	1500	200
844.59		XING 419.76				0.0000	6.9105	R BICKEL 400E	0.4482			
844.59		XING 420.00				0.0000	6.9105	MP 420 PRIVATE	0.4482			
845.59	5280	XING 420.2		5280	200	0.0050	6.9155	MONTDALE ROAD	0.4482	0.0050	5280	200
846.58	5263	XING 421.0		4089	200	0.0039	6.9194	PRIVATE	0.4432	0.0050	5263	200
846.71	700	XING 422.2		1674	D 20	0.0019	6.9213	CEMETERY ROAD	0.4382	0.0015	1600	200
847.51	4200	XING 4 165		4200	165	0.0046	6.9262	R GREENWICH STREET	0.4367	0.0115	11149	A 165
847.58	394	XING 423.4		394	165	0.0005	6.9266	AXE STREET	0.4252	0.0000		
848.11	2850	XING 423.8		966	165	0.0011	6.9277	FRANKLIN STREET	0.4252	0.0000		
848.11				1290	A 16	0.0015	6.9292		0.4252			
848.11				298	170	0.0003	6.9295		0.4252			
848.11				246	D 17	0.0003	6.9298		0.4252			
848.21	486	XING 4 165		486	165	0.0006	6.9303	L VALPARAISO WASHINGTO	0.4252	0.0000		
848.59	2000	XING 423.8		1360	165	0.0016	6.9319	LAFAYETTE STREET	0.4252	0.0000		
849.59	5278	XING 423.8		5275	A 16	0.0141	6.9460	NAPOLEON STREET	0.4252	0.0000		
850.58	5274	XING 424.8		5274	200	0.0050	6.9510	STEVENS XING	0.4252	0.0168	11522	A 90-1
851.42	4400	XING 425.7		361	200	0.0003	6.9513	150 W	0.4096	0.0029	1360	90
851.42				4679	D 20	0.0061	6.9574		0.4057	0.0000		
851.42			90		90	0.0667	7.0241	VALPO GTW XING	0.4057	0.0667		90
851.42		XING 426.93				0.0000	7.0241	250W	0.3390	0.0061	4679	D 200
851.59	892	XING 427.5		892	90	0.0019	7.0260	400W	0.3329	0.0019	1980	200
852.59	5299	XING 428.3		468	90	0.0010	7.0270	EAST SALT CR	0.3310	0.0050	5299	200
853.60	5340	XING 428.68				0.0000	7.0270	W SALT CR	0.3260	0.0051	5340	200
854.59	5209	XING 430.2		22671	A 90	0.0281	7.0551	MP 430 PRIVATE	0.3209	0.0049	5209	200
855.59	5291	XING 430.67				0.0000	7.0551	WHEELER PARK AVE	0.3160	0.0000		
856.59	5289	XING 431.0		3286	200	0.0031	7.0582	GOTT RD 7TH	0.3160	0.0233	20035	A 117
857.58	5212	XING 432.3		5212	200	0.0049	7.0631	TREAGER RD	0.2927	0.0000		
858.59	5354	XING 433.0		5354	200	0.0051	7.0682	COUNTY LINE	0.2927			
859.60	5316	XING 434.5		700	200	0.0007	7.0689	HOBART ILLINOIS STREET	0.2927	0.0011	690	117
859.73	700	XING 434.8		1457	200	0.0014	7.0703	LINDA STREET	0.2916	0.0011	700	117
859.73		XING 434.80		3856	D 20	0.0046	7.0749	CLEVELAND STREET	0.2905	0.0000		
859.88	844	XING 4 117		844	117	0.0014	7.0762	R LAKE PARK AVE	0.2905	0.0014	844	117
860.59	3700	XING 435.8		1360	117	0.0022	7.0784	WISCONSIN STREET	0.2891	0.0046	3858	D 200
861.60	5320			5417	A 11	0.0076	7.0861		0.2846	0.0012	1235	200
861.60	1060	XING 437.16				0.0000	7.0861	LIVERPOOL RD E END LIV	0.2833	0.0000		
862.60	4290					0.0000	7.0861		0.2833	0.0000		
862.71	600					0.0000	7.0861		0.2833	0.0000		
862.71		OHD				0.0000	7.0861	I-65	0.2833	0.0000		
862.92	1122					0.0000	7.0861		0.2833	0.0000		
862.92		OHD				0.0000	7.0861	I-80	0.2833	0.0000		
863.60	3600	XING 438.95				0.0000	7.0861	INDIANA AVENUE	0.2833	0.0000		
863.96	1897			0		0.0000	7.0861		0.2833	0.0000		
863.96				14616	A 15	0.0157	7.1018	WEST END LIVERPOOL TE	0.2833	0.0000		
863.96		XING 439.73					7.1018	VIRGINIA STREET	0.2833			
864.41	2363	XING 440.28				0.0000	7.1018	21ST STREET	0.2833	0.0000		
864.41		XING 440.36				0.0000	7.1018	BROADWAY	0.2833	0.0281	22671	A 90
865.41	5263	XING 440.4		2500	200	0.0024	7.1041	WASHINGTON S	0.2552	0.0029	1360	90
865.41		XING 440.66				0.0000	7.1041	16TH STREET	0.2523	0.0000		90
865.41		XING 440.70				0.0000	7.1041	MADISON STREET	0.2523	0.0000		60
865.41		XING 440.84				0.0000	7.1041	JACKSON AVE	0.2523	0.0000		90
865.41		XING 441.00				0.0000	7.1041	HARRISON STREET	0.2523	0.0000		90
865.41		XING 441.14				0.0000	7.1041	15TH AVENUE	0.2523	0.0000		60
865.41		XING 441.23				0.0000	7.1041	13TH AVENUE	0.2523	0.0000		80
865.41		XING 441.6		561	200	0.0005	7.1047	GRANT STREET	0.2523	0.0000		90
865.41		XING 441.7		4679	D 20	0.0061	7.1108	10TH AVENUE	0.2523	0.0000		60

APPENDIX B3

PAGE 14 COLUMN 2

READ DOWN WESTBOUND TRAFFIC — STATION — READ UP EASTBOUND TRAFFIC

MILES	DISTANCE ft	HWY GRADE CROSSING	MAX MPH	DISTANCE ft	SPEED mph	ET, HOURS	TIME, HOURS	STATION	TIME, HOURS	ET, HOURS	DISTANCE ft	SPEED mph
865.41			90		90	0.0000	7.1106	TOLLESTON CSX CROSSIN	0.2525	0.0000		90
865.41		XING 441.9		1360	90	0.0029	7.1136	9TH AVENUE	0.2525	0.0066	4107	D 19
866.41	5298	OHD 442.02		17386	A 90	0.0263	7.1400	IHB RR	0.2467	0.0003	255	190
867.41	5293	XING 442.25				0.0000	7.1400	TAFT STREET	0.2464	0.0000		
867.61	1050	XING 442.76				0.0000	7.1400	5TH AVENUE	0.2464	0.0000		
867.90	1500	XING 4 202				0.0000	7.1400	R CSSB OVERHEAD	0.2464	0.0000		
868.41	2700	XING 444.10				0.0000	7.1400	CLARKE ROAD	0.2464	0.0000		
868.42	5305					0.0000	7.1400	EJ&E RR	0.2464	0.0000		
869.82	2115	OHD 445.40				0.0000	7.1400	EJ&E OHD	0.2464	0.0239	19023	A 90
869.97	793					0.0000	7.1400	0	0.2225	0.0015	699	90
869.97		XING 445.5		1452	187	0.0015	7.1414	PRIVATE	0.2210	0.0000		
870.09	661			3941	D 18	0.0054	7.1468		0.2210	0.0014	661	90
870.09			90	1360	90	0.0029	7.1497	CSX CLARKE JCT	0.2196	0.0000		90
870.42	1718			1100	A 90	0.0022	7.1518	JCT PRR - PC/CR - NYC	0.2196	0.0003	135	D 95
870.61	1000				104	0.0000	7.1518	PC/CR CONNECTION TO CI	0.2193	0.0002	101	90
870.76	807		104	7	104	0.0006	7.1518	R	0.2191	0.0047	1929	A 60
870.76				1056	D 10	0.0024	7.1543	CP 501 EX-NYC TO WHITIN	0.2144	0.0043	1360	60
870.78	60		60	80	60	0.0003	7.1545	JCT NYC - ARBITRARY SPE	0.2101	0.0003	80	60
871.08	1600		110	1360	60	0.0043	7.1588	R	0.2098	0.0051	1600	60
871.44	1900		251	3356	A 60	0.0073	7.1661	R	0.2047	0.0026	1247	D 11
871.76	1700			7541	A 11	0.0106	7.1768	VIA NYC	0.2019	0.0000	0	110
872.40	3380					0.0000	7.1768		0.2019	0.0036	2666	D 17
872.72	1700		215			0.0000	7.1768	L	0.1983	0.0002	202	174
872.76	209			136	156	0.0002	7.1769		0.1981	0.0064	5506	A 15
873.16	2100			181	D 16	0.0002	7.1771		0.1917	0.0017	1360	152
873.18	115		152	115	152	0.0001	7.1773	L	0.1900	0.0001	115	152
873.22	200			200	152	0.0002	7.1775	REVERSE CURVE	0.1899	0.0002	200	152
873.24	125		152	125	152	0.0002	7.1777	R	0.1897	0.0002	125	152
873.32	400			400	152	0.0005	7.1782		0.1895	0.0002	135	D 15
873.34	125		166	125	152	0.0002	7.1783	R	0.1893	0.0001	116	155
873.40	300			300	152	0.0004	7.1787	REVERSE CURVE	0.1892	0.0000		
873.42	115		166	115	152	0.0001	7.1789	L	0.1892	0.0000		
873.76	1600			36	152	0.0000	7.1789		0.1892	0.0000		
873.94	930			4371	A 15	0.0076	7.1865		0.1892	0.0165	10267	A 70
874.18	1250		176			0.0000	7.1865	R	0.1727	0.0000		
874.76	3100			84	170	0.0001	7.1866		0.1727	0.0012	430	70
874.90	740			0		0.0000	7.1866		0.1715	0.0020	740	70
874.94	190		173	0	170	0.0000	7.1866	L	0.1695	0.0005	190	70
874.94				3520	D 17	0.0056	7.1922	CP 506-CP 507	0.1690	0.0000		
874.97	150		70	150	70	0.0004	7.1926	INDIANA HARBOR CANAL	0.1690	0.0004	150	70
875.76	4200			1360	70	0.0037	7.1963	RETURN TO PRR	0.1686	0.0034	1272	70
875.76				2083	A 70	0.0045	7.2008		0.1652			
875.76				88	101	0.0002	7.2010		0.1652	0.0000		
875.78	86			777	D 13	0.0017	7.2027		0.1652	0.0002	86	70
875.81	177		70	172	70	0.0005	7.2031	L	0.1650	0.0005	172	70
875.86	244		70	244	70	0.0007	7.2038	R REVERSE CURVE	0.1645	0.0007	244	70
875.86				1360	70	0.0037	7.2075	JCT NYC - PC/CR - PRR	0.1638	0.0028	1429	D 12
875.86		XING 451.7				0.0000	7.2075	LAKE AVENUE	0.1610	0.0000	30	121
875.86				3532	A 70	0.0071	7.2146	AMTRAK HAMMOND WHITI	0.1610	0.0000		
876.58	3831					0.0000	7.2146	PRR IN PLACE TO WESTER	0.1610	0.0048	2812	A 94
876.58		XING 452.1		95	115	0.0002	7.2147	CALUMET AVENUE	0.1561	0.0057	2831	94
876.92	1850			644	D 11	0.0012	7.2159		0.1504	0.0001	69	A 93
877.15	1200		94	1200	94	0.0024	7.2183	R REVERSE CURVE	0.1503	0.0017	815	93
877.38	1200		94	1673	94	0.0034	7.2217	L	0.1486	0.0005	235	A 90
877.47	500			27	D 94	0.0001	7.2217		0.1481	0.0007	310	90
877.62	760		93	780	93	0.0016	7.2233	L	0.1474	0.0016	760	90
877.63	70		93	169	93	0.0004	7.2237	L	0.1458	0.0001	70	90
877.67	200			81	D 93	0.0002	7.2239	REVERSE CURVE	0.1457	0.0004	200	90
877.88	1100		90	1100	90	0.0023	7.2262	R	0.1453	0.0023	1100	90
878.47	3100			1360	90	0.0029	7.2291		0.1430	0.0018	966	D 12
878.60	711		141	3116	A 90	0.0056	7.2346	L	0.1412	0.0005	330	122
878.95	1850		132		132	0.0000	7.2346	L	0.1407	0.0085	4401	A 70
878.99	210			126	121	0.0002	7.2348	REVERSE CURVE	0.1322	0.0006	235	70
879.02	180		132	1429	D 12	0.0028	7.2376	R	0.1316	0.0004	160	70
879.22	1065						7.2376		0.1312	0.0026	965	70
879.30	400	UG 454	70	2615	70	0.0076	7.2453	CALUMET RIVER UG	0.1286	0.0011	400	70
879.60	1600						7.2453		0.1275	0.0043	1600	70
879.76	860			860	70	0.0023	7.2476	AT BRIDGE MAX SPEED PE	0.1232	0.0034	1637	D 13
879.99	1200		132	500	70	0.0014	7.2489	L REVERSE CURVE	0.1198	0.0000	0	132
880.09	500		132	5891	A 70	0.0107	7.2596	R	0.1198	0.0000	0	132
880.22	700			5064	A 13	0.0066	7.2662		0.1198	0.0014	1106	D 15
880.36	734		156			0.0000	7.2662	R	0.1184	0.0000	0	156
880.59	1200					0.0000	7.2662		0.1184	0.0025	2147	D 19
881.61	5421			10628	A 15	0.0114	7.2776	53400	0.1161	0.002	1953	198
882.61	5250					0.0000	7.2776	902	0.1142	0.0000		
883.59	5184					0.0000	7.2776		0.1142	0.0000		
884.59	5275			1709	194	0.0017	7.2793	28718	0.1142	0.0000		
884.89	1600			3612	D 15	0.0043	7.2836		0.1142	0.0000		
884.95	303		117	303	117	0.0005	7.2841	L ENGLEWOOD	0.1142	0.0230	19699	A 11
885.37	2222			1360	117	0.0022	7.2863		0.0912	0.0000		117
885.37						0.0000	7.2863	METRA (CRI&P)	0.0912	0.007	3709	A 72
885.59	1178			947	117	0.0015	7.2878		0.0839	0.003	1206	72
885.67	154			1247	D 11	0.003	7.2903		0.0807	4E-04	154	72
885.95	1200		72	1200	72	0.0032	7.2936	R	0.0803	0.003	1200	72
885.85				1360	72	0.0036	7.2971		0.0772	0.002	785.6	D 10
885.85				1694	A 72	0.004	7.3006		0.0756	1E-04	61	102
885.85				270	98	0.0005	7.3013		0.0754	0.0040	1583	A 61
886.56	3770			446	D 98	0.0009	7.3022		0.0711	0.0032	1360	61
886.57	30		61	30	61	0.0001	7.3023	R	0.0679	0.0001	30	61
886.63	310		81	310	61	0.0007	7.3030	R	0.0678	0.0000		
886.93	1580			1380	81	0.0032	7.3062		0.0678	0.0044	1873	61
887.00	400		82	50	A 61	0.0001	7.3063	L	0.0634	0.0000		
887.20	1066			1620	82	0.0037	7.3100		0.0634	0.0003	117	A 79
887.20				320	82	0.0007	7.3107		0.0631	0.0031	1276	79
887.30	516		89	183	A 82	0.0004	7.3111	L	0.0600	0.0013	514	A 68
887.52	1154			1167	85	0.0026	7.3137		0.0587	0.0021	760	68
887.57	281		85	171	85	0.0004	7.3141	R	0.0566	0.0017	800	68
887.60	120		95	120	85	0.0003	7.3144	R	0.0549	0.0005	171	68
888.63	5456			1360	85	0.0030	7.3174		0.0544	0.0003	120	68
889.67	5506			10624	A 85	0.0159	7.3333		0.0541	0.0052	3124	D 16
890.28	3250			254	180	0.0003	7.3336		0.0489	0.0001	118	161
890.28				2044	D 10	0.0029	7.3365		0.0488	0.0124	8903	A 10
890.35	380		108	380	108	0.0007	7.3372	L REVERSE CURVE	0.0364	0.0041	2079	A 83
890.41	340		108	340	108	0.0006	7.3378	R	0.0323	0.0007	320	83
890.55	700			700	D 10	0.0014	7.3392		0.0316	0.0008	340	83
890.57	146		83	146	83	0.0003	7.3395	L	0.0308	0.0016	700	83
890.62	230		83	230	83	0.0005	7.3400	L	0.0292	0.0003	146	83
890.68	320		83	320	83	0.0007	7.3407	R REVERSE CURVE	0.0289	0.0005	230	83
890.68			90				7.3407	IC(CN) RR CROSSING	0.0284	0.0007	320	83
890.74	310		70	310	83	0.0007	7.3414	S BRANCH CHICAGO RIVE	0.0276	0.0016	829	D 11
890.79	300		132	300	83	0.0007	7.3421	L	0.0260	0.0002	115	112
890.97	950	XING 466.3		750	83	0.0017	7.3438	LUMBER STREET	0.0258	0.0000		
891.04	320		132	2857	A 83	0.0054	7.3492	R	0.0258	0.0000		
891.13	500					0.0000	7.3492	REVERSE CURVE	0.0258	0.0000		
891.21	420		132			0.0000	7.3492	L	0.0258	0.0000		
891.34	677					0.0000	7.3492		0.0258	0.0068	3283	A 67
891.45	600		165			0.0000	7.3492	L	0.0190	0.0000		67
891.50	240			88	115	0.0001	7.3493		0.0190	0.0016	574	67
891.74	1250			1281	D 11	0.0027	7.3520		0.0174	0.0033	926	A 38
891.84	550		67	1053	67	0.0030	7.3550	R	0.0141	0.0000	410	38
892.02	950			447	D 67	0.0016	7.3566		0.0129	0.0047	950	38
892.13	550		38	550	38	0.0027	7.3593	L REVERSE CURVE	0.0073	0.0029	583	38
892.21	475			1360	38	0.0068	7.3661	CHICAGO UNION STATION	0.0044	0.0044	440	A 0
892.23	90						7.3661	EAST END PLATFORM	0.0000			
892.32	490		186	201	38	0.0010	7.3671	R				
892.46	750			212	D 38	0.0021	7.3692	WEST END PLATFORM				

Appendix B4 - Improved Right of Way on Partially rebuilt Alignment; Sufficient for High-Speed TOFC Operations Aerodynamically designed TOFC equipment operating at 120 mph maximum speed. Maximum allowable speed on curves as shown for high speed equipment only.
Minimum of FRA Class 7 track where speeds over 110 mph are possible
Other tracks classified per maximum possible speed.
E_a = 6 inches; wheel load 0.6 minimum per FRA 213.307d1
Assumed c.g. of moving equipment 108 inches above top of rail
Train length 11,000 feet: 200 platforms plus 20 locomotive units @ 6,000 hp each.
20% speed reduction for reverse curves with 600 feet or less between curves.
Railroad grade crossings eliminated except where they do not affect allowable speed. Other highly effective improvements made to right of way.
Class 7 track where speeds over 110 mph are possible
Other tracks classified per maximum possible speed.
Highway grade crossing locations shown are approximate
Travel Distance Jersey Meadows to Liverpool IN 859.06 miles

Westbound trip	9.8499 hours
Eastbound trip	9.7956 hours

APPENDIX B4

READ DOWN — WESTBOUND TRAFFIC | STATION | READ UP — EASTBOUND TRAFFIC

PAGE 1 COLUMN 1

MILES	DISTANCE, ft	HWY GRADE CROSSING	MAX MPH	DISTANCE, ft	SPEED, MPH	ET, HOURS	TIME, HOURS	STATION	TIME, HOURS	ET, HOURS	DISTANCE, ft	SPEED, MPH
0.00								E MEADOWS	9.7990	0.0000		
0.62	3274								9.7960	0.0260	525	D 89
0.62									9.7780	0.0087	2748	60
1.62	5281								9.7673	0.0167	5281	60
2.07	2377								9.7507	0.0078	2377	60
2.35	1479		160					R	9.7432	0.0047	1479	60
2.46	589								9.7380	0.0019	589	90
2.62	836								9.7368	0.0026	836	60
2.62				0		0.0000	0.0000	W MEADOWS	9.7341	0.0000		60
2.72	526			5328	A 64	0.0000	0.0200		9.7340	0.0017	528	60
2.78	330		60		60	0.0000	0.0200	HACKENSACK R	9.7323	0.0010	330	60
2.78						0.0000	0.0200		9.7313	0.0011	411	D 89
2.78						0.0000	0.0200		9.7302	0.0001	40	80
3.62	4409			1839	60	0.0081	0.0281		9.7301	0.0114	4218	A 59
4.00	2022			2027	60	0.0084	0.0326	HUDSON	9.7187	0.0052	1762	60
4.65	3443		160	3443	60	0.0108	0.0434	R	9.7130	0.0111	3443	59
4.89	1275		190	1128	60	0.0036	0.0468	R	9.7020	0.0041	1272	58
4.99	547			4131	A 60	0.0111	0.0580		9.6979	0.0016	487	59
5.75	4035					0.0000	0.0580		9.6963	0.0100	4015	59
5.75				110	60	0.0003	0.0583		9.6854	0.0000		
5.75				428	D 60	0.0012	0.0595		9.6854			
5.98	1205		59	1205	59	0.0038	0.0633	L HARRISON	9.6834	0.0039	1205	59
5.98						0.0000	0.0633		9.6795	0.0011	405	D 79
6.05	371			371	59	0.0012	0.0645	REVERSE CURVE	9.6784	0.0030		79
6.23	955		66	565	59	0.0031	0.0676	R DOCK	9.6784	0.0032	921	79
6.28	265			265	59	0.0029	0.0684	REVERSE CURVE	9.6782	0.0006	265	79
6.38	531		131	531	59	0.0017	0.0701	L	9.6786	0.0015	531	79
6.46	424			424	59	0.0014	0.0715		9.6743	0.0010	424	79
6.68	1164		160	1186	59	0.0037	0.0752	L NEWARK	9.6733	0.0028	1148	79
6.73	285			285	59	0.0009	0.0761		9.6720	0.0008	265	79
6.85	637		79	637	59	0.0020	0.0781	L	9.6698	0.0016	637	79
6.85						0.0000	0.0781		9.6683	0.0017	859	D 115
6.96	584			584	59	0.0019	0.0800		9.6666	0.0007	411	150
7.09	686			686	59	0.0022	0.0822		9.6659	0.0000		
7.20	581		110	581	59	0.0019	0.0841	R HUNTER	9.6659	0.0019	1130	110
7.96	4013			4013	59	0.0125	0.0970		9.6638	0.0000		
8.16	1054			520	59	0.0017	0.0986		9.6638	0.0006		
8.51	1832		134	3685	A 59	0.0108	0.1095	L	9.6630	0.0256	19882	A 78
8.99	2553			5250	79	0.0054	0.1148		9.6384	0.0000		
9.99	5262			11918	A 78	0.0238	0.1386		9.6384	0.0033	1287	A 70
10.18	1009					0.0000	0.1385		9.6361	0.0000	18	70
10.56	2004		222			0.0000	0.1385	R	9.6363	0.0004	2004	70
10.99	2287			1472	A 11	0.0025	0.1410		9.6296	0.0081	2287	70
11.01	106					0.0000	0.1410		9.6238	0.0000	106	70
11.15	742		222			0.0000	0.1410	L	9.6232	0.0020	742	70
11.98	4399					0.0000	0.1410		9.6212	0.0119	4399	70
12.06	434			434	113	0.0008	0.1418		9.6093	0.0012	434	70
12.06				1026	D 11	0.0021	0.1439		9.6081			
12.23	867		76	932	76	0.0022	0.1461	L ELIZABETH	9.6081	0.0023	867	70
12.26	163				76	0.0000	0.1461	REVERSE CURVE	9.6057	0.0004	163	70
12.26				128	D 70	0.0003	0.1464	TT 55 MPH	9.6053			
12.70	2330		70	2330	70	0.0063	0.1527	R	9.6053	0.0063	2330	70
12.70						0.0000	0.1527	ELMORA	9.5990	0.0008	1360	D 120
12.99	1517			1517	70	0.0041	0.1568	R END 6 TRACKS	9.5982	0.0002	124	120
13.99	5285			5285	70	0.0143	0.1751		9.5960	0.0083	5285	120
14.99	5289			4250	70	0.0114	0.1865		9.5877	0.0083	5289	120
15.99	5258			19746	A 70	0.0383	0.2208		9.5793	0.0083	5258	120
16.13	774					0.0000	0.2208		9.5710	0.0012	774	120
16.36	1187		149			0.0000	0.2208	R	9.5698	0.0019	1187	120
16.74	2014					0.0000	0.2208		9.5679	0.0032	2014	120
16.99	1342		266			0.0000	0.2208	L	9.5648	0.0021	1342	120
17.13	695					0.0000	0.2208		9.5626	0.0011	695	120
17.30	909		101			0.0000	0.2208	R	9.5615	0.0014	909	120
17.49	1015					0.0000	0.2208		9.5601	0.0016	1016	120
17.52	160					0.0000	0.2208		9.5585	0.0003	160	120
17.56	214					0.0000	0.2208		9.5583	0.0003	214	120
17.75	1016		208			0.0000	0.2208	L	9.5579	0.0016	1016	120
17.99	1280					0.0000	0.2208		9.5563	0.0019	1280	120
18.23	1285					0.0000	0.2208		9.5544	0.0020	1285	120
18.57	1785		110	236	120	0.0004	0.2210	R	9.5523	0.0012	729	120
18.64	378			376	120	0.0006	0.2216	REVERSE CURVE	9.5512	0.0000		
18.74	540		178	540	120	0.0009	0.2225	L	9.5512	0.0000		
18.99	1268			1268	120	0.0020	0.2247		9.5512	0.0000		
19.58	3108			3108	120	0.0049	0.2296		9.5512	0.0094	5743	A 110
19.69	621		150	621	120	0.0010	0.2306	R	9.5417	0.0015	802	110
19.83	621			821	120	0.0010	0.2316		9.5402	0.0000		
19.97	828		149	828	120	0.0013	0.2329	L	9.5402	0.0000		
20.00	156		149	156	120	0.0002	0.2331	L	9.5402	0.0036	2062	A 106
20.43	2158			2158	120	0.0034	0.2365		9.5366	0.0037	2062	106
20.68	1421		149	1421	120	0.0022	0.2388	A	9.5329	0.0025	1421	106
20.77	474			474	120	0.0007	0.2395	REVERSE CURVE	9.5304	0.0008	474	106
20.98	1105		149	1105	120	0.0017	0.2413	L	9.5296	0.0020	1105	106
21.28	1602		149	1602	120	0.0025	0.2438	L ISELIN	9.5275	0.0003	162	106
21.41	703		149	1045	120	0.0016	0.2454	L	9.5272	0.0000		106
21.54	681			327	D 12	0.0008	0.2462	METROPARK	9.5272	0.0000		106
21.80	1361		110	1361	110	0.0023	0.2485	R	9.5272	0.0000		106
21.99	985			1486	110	0.0026	0.2509		9.5272	0.0106	5654	A 95
22.10	598			127	D 11	0.0002	0.2511		9.5167	0.0004	207	93
22.53	2284		106	2284	106	0.0041	0.2552	L	9.5162	0.0047	2284	93
22.72	979			979	106	0.0017	0.2569		9.5116	0.0020	979	93
23.01	1522		120	1522	106	0.0027	0.2596	R	9.5096	0.0031	1522	93
23.46	2398		120	2398	106	0.0040	0.2636	R	9.5065	0.0049	2398	93
23.98	2755			2755	106	0.0048	0.2684		9.5016	0.0056	2755	93
24.15	860			861	106	0.0009	0.2697		9.4960	0.0018	860	93
24.15				379	D 10	0.0007	0.2704		9.4943			
24.41	1397		93	1397	93	0.0028	0.2730	R S METUCHEN	9.4943	0.0028	1397	93
24.41						0.0000	0.2730		9.4914	0.0015	843	D 120
24.81	2096			2095	93	0.0043	0.2775		9.4899	0.0020	1252	120
24.98	913		100	913	93	0.0019	0.2794	L	9.4879	0.0014	913	120
25.09	573		160	573	93	0.0012	0.2806	L	9.4865	0.0008	573	120
25.27	937			937	93	0.0019	0.2825		9.4858	0.0015	937	120
25.62	1874		140	1874	93	0.0038	0.2863	L	9.4841	0.0030	1874	120
25.98	1874			1874	93	0.0038	0.2901		9.4812	0.0030	1874	120

PAGE 1 COLUMN 2

MILES	DISTANCE, ft	HWY GRADE CROSSING	MAX MPH	DISTANCE, ft	SPEED, MPH	ET, HOURS	TIME, HOURS	STATION	TIME, HOURS	ET, HOURS	DISTANCE, ft	SPEED, MPH
26.76	4135			2734	80	0.0056	0.2957		9.4780	0.0065	4135	120
26.96	1074		190	13307	A 90	0.0236	0.3193	R	9.4717	0.0012	1074	120
26.98	107					0.0000	0.3193		9.4705	0.0002	107	120
27.98	5275					0.0000	0.3193		9.4696	0.0083	5275	120
28.15	875					0.0000	0.3193		9.4615	0.0014	875	120
28.65	2676		167			0.0000	0.3193	L	9.4601	0.0042	2676	120
28.97	1696					0.0000	0.3193		9.4559	0.0027	1696	120
29.05	430			165	120	0.0003	0.3196		9.4532	0.0007	426	120
29.26	1107		164	1103	120	0.0017	0.3213	L	9.4525	0.0017	1103	120
29.97	3727			3727	120	0.0059	0.3272	NEW BRUNSWICK	9.4508	0.0059	3727	120
30.97	5290			5290	120	0.0083	0.3355		9.4449	0.0083	5290	120
31.54	3029			3029	120	0.0048	0.3403		9.4366	0.0048	3029	120
31.96	2206		167	2206	120	0.0035	0.3438	L	9.4318	0.0035	2206	120
32.99	5404			5404	120	0.0085	0.3523		9.4283	0.0085	5404	120
33.97	5262			5262	120	0.0082	0.3605		9.4198	0.0082	5262	120
34.98	5303			5303	120	0.0084	0.3689		9.4116	0.0084	5303	120
35.67	5259			5268	120	0.0083	0.3772		9.4032	0.0083	5259	120
36.97	5286			5286	120	0.0083	0.3855		9.3949	0.0083	5285	120
37.05	430			430	120	0.0007	0.3862	DEANS	9.3865	0.0007	430	120
37.28	1184		222	1183	120	0.0019	0.3881	L	9.3858	0.0019	1183	120
37.40	646			646	120	0.0010	0.3891	SEE PHOTO	9.3840	0.0010	645	120
37.97	3013		149	3013	120	0.0048	0.3938	R	9.3830	0.0048	3013	120
38.97	5253			5253	120	0.0083	0.4021	MONMOUTH JUNCTIO	9.3782	0.0083	5253	120
39.97	5314			5314	120	0.0084	0.4105		9.3698	0.0084	5314	120
40.97	5282			5282	120	0.0083	0.4188		9.3616	0.0083	5282	120
41.97	5286			5286	120	0.0083	0.4272		9.3532	0.0083	5286	120
42.97	5284			5284	120	0.0083	0.4355		9.3449	0.0083	5284	120
43.97	5276			5276	120	0.0083	0.4438		9.3365	0.0083	5276	120
44.97	5274			5274	120	0.0083	0.4522		9.3282	0.0083	5274	120
45.03	317			317	120	0.0005	0.4527		9.3199	0.0005	317	120
45.97	4968			4968	120	0.0078	0.4605	PRINCETON JUNCTIO	9.3194	0.0078	4968	120
46.97	5266			5266	120	0.0083	0.4688		9.3115	0.0083	5266	120
47.97	5295			5295	120	0.0084	0.4772		9.3032	0.0084	5295	120
48.29	1671			1671	120	0.0026	0.4798		9.2948	0.0026	1671	120
48.51	1145		217	1148	120	0.0018	0.4816	R LAWRENCE STATIO	9.2922	0.0018	1145	120
48.97	2463			2463	120	0.0039	0.4855		9.2904	0.0038	2463	120
49.98	5291			5291	120	0.0084	0.4939		9.2865	0.0084	5291	120
50.97	5277			5277	120	0.0083	0.5022		9.2782	0.0002	140	120
51.97	5290			5290	120	0.0083	0.5105		9.2780	0.0000		
52.97	5276			5276	120	0.0083	0.5188		9.2780	0.0000		
53.78	4273			4273	120	0.0067	0.5256		9.2780	0.0000		
53.97	990		235	990	120	0.0016	0.5272	L	9.2780	0.0439	21546	A 58
54.07	582		230	582	120	0.0009	0.5280	L	9.2341	0.0083	2595	59
54.72	3402			3402	120	0.0054	0.5334	TRENTON	9.2257	0.0044	1095	A 48
54.85	691		140	691	120	0.0011	0.5345	R	9.2213	0.0003	66	48
54.97	638			638	120	0.0010	0.5355		9.2210	0.0025	638	48
55.58	3236			3236	120	0.0051	0.5406		9.2185	0.0128	3236	48
55.88	1600			1560	120	0.0025	0.5431	MORRISVILLE PA	9.2058	0.0063	1600	48
55.98	500		130	0	100	0.0000	0.5431	R	9.1984	0.0020	500	48
56.19	1100				100	0.0000	0.5431		9.1975	0.0043	1100	48
56.19				1601	D 10	0.0054	0.5485		9.1931			
56.34	800		65	800	58	0.0026	0.5483	R STA x	9.1931	0.0032	800	48
56.39	300			300	58	0.0010	0.5501	REVERSE CURVE	9.1900	0.0012	300	48
56.58	960		62	960	58	0.0031	0.5532	L D	9.1885	0.0020	960	48
56.64	300		62	300	58	0.0010	0.5541	L	9.1849	0.0012	300	48
56.66	100			100	58	0.0003	0.5545	REVERSE CURVE	9.1837	0.0004	100	48
56.81	800		62	800	58	0.0026	0.5570	R D	9.1833	0.0032	800	48
56.92	560			467	58	0.0013	0.5583		9.1870	0.0023	560	48
56.92				173	D 56	0.0006	0.5589		9.1779			
57.05	700		48	700	48	0.0028	0.5617	R D	9.1778	0.0028	700	48
57.05							0.5617		9.1751	0.0003	90	D 54
57.14	500			500	48	0.0020	0.5637	REVERSE CURVE	9.1748	0.0014	410	54
57.33	1000		60	1000	48	0.0039	0.5676	Ld	9.1734	0.0035	1000	54
57.39	300			300	48	0.0012	0.5688	REVERSE CURVE	9.1699	0.0011	300	54
57.58	1000		110	1000	48	0.0039	0.5727	R	9.1688	0.0035	1000	54
57.79	1100			1100	48	0.0043	0.5771		9.1653	0.0039	1100	54
58.02	1200		76	1200	48	0.0047	0.5816	Ra	9.1614	0.0042	1200	54
58.07	300			300	48	0.0012	0.5830	REVERSE CURVE	9.1572	0.0011	300	54
58.22	800		54	800	48	0.0032	0.5862	Lt	9.1562	0.0028	800	54
58.22						0.0000	0.5862		9.1534	0.0011	375	D 74
58.58	1890			1890	48	0.0074	0.5935		9.1523	0.0029	1500	74
59.41	4380			2920	48	0.0115	0.6051		9.1494	0.0112	4380	74
59.56	800		76			0.0000	0.6051	RaW	9.1372	0.0020	500	74
59.56	0			500	A 48	0.0022	0.6073	W YARD	9.1362	0.0000		
59.58	100			1269	54	0.0060	0.6133		9.1352	0.0003	100	74
59.60	100			100	54	0.0004	0.6138		9.1349	0.0003	100	74
59.69	500		74	500	54	0.0018	0.6156	RaW	9.1346	0.0013	500	74
59.69							0.6156		9.1334	0.0008	207	D 83
59.69	0						0.6156		9.1326	0.0001	65	85
59.69	0						0.6156		9.1327	0.0036	1808	A 75
60.58	4680			3246	54	0.0114	0.6270		9.1291	0.0063	2546	75
61.15	3000			3412	A 54	0.0100	0.6370		9.1228	0.0075	3000	75
61.58	2280		107	3398	74	0.0085	0.6454	R DUNLAP	9.1153	0.0057	2260	76
61.71	680		107	680	74	0.0017	0.6472	R	9.1096	0.0017	680	76
62.18	2500			366	74	0.0008	0.6481	REVERSE CURVE	9.1079	0.0052	2500	76
62.58	2100		76	440	A 74	0.0011	0.6492	L WOODBOURNE	9.1017	0.0052	2100	76
62.58				3600	78	0.0095	0.6587		9.0965	0.0024	1065	D 12
63.58	5280			5280	76	0.0130	0.6716		9.0940	0.0016	412	100
64.00	2200			2200	76	0.0055	0.6770		9.0923	0.0131	5743	A 110
64.24	1260		196	1280	76	0.0032	0.6805	R	9.0792	0.0011	640	110
64.33	500			500	76	0.0012	0.6817		9.0781	0.0009	500	110
64.58	1300		233	1300	76	0.0032	0.6850	R	9.0772	0.0022	1300	110
64.73	780		233	440	76	0.0011	0.6861	R	9.0750	0.0013	780	110
65.35	3000					0.0000	0.6861		9.0737	0.0057	3000	110
65.58	1200		134			0.0000	0.6861	R	9.0680	0.0021	1200	110
65.58	0			10215	A 76	0.0255	0.7116		9.0659	0.0000		110
65.71	660		134			0.0000	0.7116	R	9.0659	0.0012	660	110
66.20	2600					0.0000	0.7116		9.0647	0.0046	2600	110
66.20						0.0000	0.7116		9.0603	0.0000		110
66.45	1300		110			0.0000	0.7116	L	9.0603	0.0022	1300	110
66.45						0.0000	0.7116	HOLLAND	9.0580	0.0000		110
66.58	700					0.0000	0.7116		9.0580	0.0020	1143	110
66.67	500					0.0000	0.7116		9.0561	0.0009	500	110

APPENDIX B4

PAGE 2 COLUMN 1

				READ DOWN WESTBOUND TRAFFIC				STATION	READ UP EASTBOUND TRAFFIC			
MILES	DISTANCE, ft	HWY GRADE CROSSING	MAX MPH	ΔDISTANCE, ft	SPEED, mph	ΔT, hours	TIME, hours		TIME, hours	ΔT, hours	ΔDISTANCE, ft	SPEED, mph
66.81	700		134			0.0000	0.7118	R	9.0552	0.0012	700	110
67.11	1580			186	110	0.0003	0.7120		9.0540	0.0020	1140	110
67.31	1100		110	1100	110	0.0019	0.7139	L	9.0520	0.0019	1100	110
67.58	1400			1400	110	0.0024	0.7163		9.0501	0.0024	1400	110
67.60	80			80	110	0.0001	0.7164		9.0477	0.0001	80	110
67.99	2100		155	2100	110	0.0036	0.7200	R	9.0476	0.0036	2100	110
68.31	1700			1700	110	0.0029	0.7229		9.0440	0.0029	1700	110
68.49	900	XING 1	134	900	110	0.0015	0.7245	L BUSTLETON	9.0410	0.0015	900	110
68.58	500			500	110	0.0009	0.7254		9.0396	0.0009	500	110
68.97	2080			2080	110	0.0036	0.7289		9.0386	0.0036	2080	110
69.50	2800	UG	110	2800	110	0.0048	0.7336	R STREET RD	9.0351	0.0048	2800	110
69.58	400			400	110	0.0007	0.7344		9.0302	0.0006	337	D 12
70.28	3680			3680	110	0.0063	0.7406		9.0297	0.0066	3740	120
70.52	1300		164	1300	110	0.0022	0.7430	L	9.0236	0.0021	1300	120
70.58	300	XING 15.95		300	110	0.0005	0.7435	COUNTY LINE RD	9.0217	0.0005	300	120
71.58	5280			5280	110	0.0091	0.7526		9.0212	0.0083	5280	120
72.58	5280			40	110	0.0001	0.7527		9.0129	0.0083	5280	120
73.58	5280			5743	A 11	0.0094	0.7621		9.0046	0.0083	5280	120
74.45	4600			9377	120	0.0148	0.7769		8.9962	0.0073	4600	120
74.58	680		134	680	120	0.0011	0.7780	L	8.9890	0.0011	680	120
74.86	1480		134	1480	120	0.0023	0.7803	L	8.9879	0.0023	1480	120
75.56	3700			3700	120	0.0058	0.7862	YORK RD	8.9856	0.0058	3700	120
75.58	100		190	100	120	0.0002	0.7863	L	8.9797	0.0002	100	120
75.58	0						0.7863		8.9796	0.0000	0	120
76.26	3580		190	3580	120	0.0057	0.7920	L	8.9796	0.0057	3580	120
76.58	1700			1700	120	0.0027	0.7947		8.9739	0.0027	1700	120
76.77	980			980	120	0.0015	0.7962		8.9712	0.0015	980	120
77.37	3200		190	3200	120	0.0051	0.8013	R	8.9697	0.0051	3200	120
77.58	1100			1100	120	0.0017	0.8030		8.9646	0.0017	1100	120
78.38	4200			4200	120	0.0066	0.8096		8.9629	0.0066	4200	120
78.58	1060		134	1060	120	0.0017	0.8113	R	8.9563	0.0014	857	120
78.73	800		134	800	120	0.0013	0.8126	R CAMP HILL	8.9549	0.0000		
79.26	2800			2800	120	0.0044	0.8170		8.9549	0.0000		
79.58	1680		155	1680	120	0.0027	0.8197	L	8.9549	0.0094	5743	A 11C
79.90	1700		155	1700	120	0.0027	0.8224	L	8.9455	0.0025	1460	110
80.58	3580			3580	120	0.0057	0.8280		8.9430	0.0062	3580	110
81.58	5280			5623	120	0.0089	0.8369		8.9368	0.0091	5280	110
81.71	680			337	D 12	0.0006	0.8374		8.9277	0.0012	680	110
82.11	2100		110	2100	110	0.0036	0.8411	R STENTON	8.9265	0.0036	2100	110
82.58	2500			2500	110	0.0043	0.8454		8.9229	0.0006	337	D 12
83.58	5280			5280	110	0.0091	0.8546		8.9224	0.0164	10420	120
84.58	5280	XING 29.18		3220	110	0.0056	0.8602	CONSHOHOCKEN	8.9059	0.0094	5743	A 11C
85.58	5280	XING 29.52		5743	A 11	0.0094	0.8694	GRAVERS	8.8965	0.0032	1840	110
85.99	2180			3777	120	0.0060	0.8754		8.8933	0.0038	2180	110
86.47	2500		147	2500	120	0.0039	0.8793	L	8.8896	0.0043	2500	110
86.58	600			4143	120	0.0065	0.8859	EARNEST	8.8853	0.0010	600	110
87.31	3880			337	D 12	0.0006	0.8864		8.8842	0.0067	3880	110
87.58	1400		110	1400	110	0.0024	0.8889	L RAMBO	8.8775	0.0024	1400	110
87.63	280		110	280	110	0.0005	0.8893	L	8.8751	0.0005	280	110
87.90	1400			1400	110	0.0024	0.8917		8.8747	0.0006	337	D 12
88.58	3600		124	3600	110	0.0062	0.8979	R HENDERSON	8.8741	0.0074	4683	120
88.69	580		124	580	110	0.0010	0.8989	R	8.8667	0.0009	580	120
89.58	4700			4700	110	0.0081	0.9070		8.8658	0.0074	4700	120
89.75	880			720	110	0.0012	0.9083		8.8584	0.0014	880	120
90.09	1800		134	5743	A 11	0.0094	0.9177	L	8.8570	0.0028	1800	120
90.58	2600					0.0000	0.9177	KING of PRUSSIA	8.8542	0.0041	2600	120
91.58	5280			4097	120	0.0065	0.9242		8.8501	0.0080	5280	120
92.58	5280			5280	120	0.0083	0.9325		8.8417	0.0083	5280	120
92.69	580			580	120	0.0009	0.9334		8.8334	0.0009	580	120
92.99	1600		233	1600	120	0.0025	0.9360	R	8.8325	0.0025	1600	120
93.58	3100			3100	120	0.0049	0.9409		8.8300	0.0049	3100	120
94.58	5280			5280	120	0.0083	0.9492	HOWELLVILLE	8.8251	0.0083	5280	120
95.11	2800			2800	120	0.0044	0.9536		8.8167	0.0044	2800	120
95.58	2480		190	2480	120	0.0039	0.9575	R	8.8123	0.0039	2480	120
95.73	780			780	120	0.0012	0.9588		8.8084	0.0012	780	120
96.22	2600		190	2600	120	0.0041	0.9629	L	8.8072	0.0041	2600	120
96.58	1900			1900	120	0.0030	0.9659		8.8031	0.0030	1900	120
97.47	4700			4700	120	0.0074	0.9733		8.8001	0.0074	4700	120
97.58	580		329	580	120	0.0009	0.9742	R	8.7927	0.0009	580	120
97.77	980		329	980	120	0.0015	0.9757	R	8.7917	0.0015	980	120
98.58	4300			4300	120	0.0068	0.9825		8.7902	0.0068	4300	120
98.79	1100			1100	120	0.0017	0.9843		8.7834	0.0017	1100	120
98.90	580		233	580	120	0.0009	0.9852	L	8.7817	0.0009	580	120
99.36	2442			2442	120	0.0039	0.9890		8.7808	0.0039	2442	120
99.36		UG 2.07					0.9890	STREAM	8.7769	0.0000		120
100.88	3000						0.9890		8.7769	0.0000		120
100.88		OHD 1.46				0.0000	0.9890	RTE 352	8.7769	0.0000		120
102.39	4000		163	0	120	0.0000	0.9890	L	8.7769	0.0000		120
102.39	3000	XING 45.18				0.0000	0.9890	W HENDERSON	8.7769	0.0000		120
102.39	1500	UG 0.35		0		0.0000	0.9890	RAVINE ROAD	8.7769	0.0000		120
103.91	2000		163	0		0.0000	0.9890	R	8.7769	0.0000		120
103.91	1000					0.0000	0.9890	GLEN LOCH	8.7769	0.0000		120
104.44	2797						0.9890	R	8.7769	0.0000		120
104.44	2000			42797	120	0.068	1.0586		8.7769	0.068	42797	120
105.95	2000			0		0.0000	1.0586	T	8.7094	0.0000		120
107.47	1000		174	0		0.0000	1.0586	L	8.7094	0.0000		120
	1000		163					R REVERSE CURVE				
	6000											
	3500		347					L				
	4000											
	1500		163					R				
	2500											
	2080		163					L				
107.85	2027			2027	120	0.0032	1.0598		8.7094	0.0032	2027	120
108.08	1202			1202	120	0.0019	1.0617	T	8.7062	0.0019	1202	120
108.08	0			0		0.0000	1.0617	DOWNINGTOWN	8.7043	0.0000	0	120
108.41	1777		134	1777	120	0.0028	1.0645	L	8.7043	0.0028	1777	120
108.61	1046			1046	120	0.0017	1.0661	T	8.7015	0.0017	1046	120
108.85	1255		144	1255	120	0.0020	1.0681	R	8.6998	0.0020	1255	120
109.06	1109		144	1109	120	0.0018	1.0699	R	8.6978	0.0018	1109	120
109.51	2376			2376	120	0.0036	1.0736	T	8.6961	0.0038	2376	120
109.85	1795		144	1795	120	0.0028	1.0764	L	8.6923	0.0028	1795	120

APPENDIX B4

PAGE 2 COLUMN 2

				READ DOWN WESTBOUND TRAFFIC				STATION	READ UP EASTBOUND TRAFFIC			
MILES	DISTANCE, ft	HWY GRADE CROSSING	MAX MPH	ΔDISTANCE, ft	SPEED, mph	ΔT, hours	TIME, hours		TIME, hours	ΔT, hours	ΔDISTANCE, ft	SPEED, mph
109.94	466		144	466	120	0.0007	1.0772	L	8.6895	0.0007	466	120
110.08	725			725	120	0.0011	1.0783	T	8.6888	0.0011	725	120
110.28	1087		134	1087	120	0.0017	1.0800	R	8.6876	0.0017	1087	120
110.49	1087		208	1087	120	0.0017	1.0817	R	8.6859	0.0017	1087	120
110.68	1035			1035	120	0.0016	1.0834	T	8.6842	0.0016	1035	120
110.85	880		190	880	120	0.0014	1.0848	L	8.6825	0.0014	880	120
110.89	216			216	120	0.0003	1.0851		8.6812	0.0003	216	120
111.03	754		155	754	120	0.0012	1.0863	L	8.6808	0.0012	754	120
111.72	3610			3610	120	0.0057	1.0920	T	8.6796	0.0057	3610	120
111.85	700		102	700	120	0.0011	1.0931	L	8.6739	0.0011	700	120
111.89	209		152	209	120	0.0003	1.0934	L	8.6728	0.0003	209	120
112.04	784		152	784	120	0.0012	1.0947	R REVERSE CURVE	8.6725	0.0012	784	120
112.72	3607			3607	120	0.0057	1.1004	T	8.6713	0.0057	3607	120
112.85	680		134	680	120	0.0011	1.1014	R	8.6656	0.0011	680	120
113.31	2433		134	2433	120	0.0038	1.1053	R	8.6645	0.0038	2433	120
113.43	621			621	120	0.0010	1.1063	T	8.6606	0.0010	621	120
113.79	1915		134	1915	120	0.0030	1.1093	L	8.6597	0.0030	1915	120
113.85	311			311	120	0.0005	1.1098	COATESVILLE	8.6566	0.0005	311	120
114.27	2240			2240	120	0.0035	1.1133	T	8.6562	0.0035	2240	120
114.85	3040		190	3040	120	0.0048	1.1181	L	8.6526	0.0048	3040	120
114.93	414		190	414	120	0.0007	1.1188	L	8.6478	0.0007	414	120
115.20	1449			1449	120	0.0023	1.1210	T	8.6472	0.0023	1449	120
115.67	2485		155	2485	120	0.0039	1.1250	R	8.6449	0.0039	2485	120
115.85	932			932	120	0.0015	1.1264	T	8.6410	0.0015	932	120
116.60	3934			3934	120	0.0062	1.1326	T	8.6395	0.0062	3934	120
116.85	1346		153	1346	120	0.0021	1.1348	L	8.6333	0.0021	1346	120
116.94	485		153	485	120	0.0008	1.1355	L	8.6312	0.0008	485	120
117.22	1455			1455	120	0.0023	1.1378		8.6304	0.0023	1455	120
117.61	2047		159	2047	120	0.0032	1.1411	R	8.6281	0.0032	2047	120
117.85	1293			1293	120	0.0020	1.1431	T	8.6249	0.0020	1293	120
118.85	5280			5280	120	0.0083	1.1514		8.6228	0.0035	2222	120
119.33	2536			2536	120	0.0040	1.1554	T	8.6193	0.0000		
119.52	984		217	984	120	0.0016	1.1570	L	8.6193	0.0000		
119.85	1760			1760	120	0.0028	1.1598	T	8.6193	0.0219	12532	A 95
120.85	5280			5280	120	0.0083	1.1681	PARKESBURG	8.5974	0.0000		
120.94	495			495	120	0.0008	1.1689	T	8.5974	0.0000		
121.18	1265		190	1265	120	0.0020	1.1709	R	8.5974	0.0131	5926	A 76
121.85	3520			3520	120	0.0056	1.1764	T	8.5843	0.0011	440	76
122.61	4015			4015	120	0.0063	1.1828	T	8.5832	0.0100	4015	76
122.76	770		190	770	120	0.0012	1.1840	R	8.5732	0.0019	770	76
122.85	495			495	120	0.0008	1.1848	T	8.5713	0.0012	495	76
123.26	2143			1357	120	0.0021	1.1869	T	8.5701	0.0053	2146	76
123.26				786	D 12	0.0014	1.1883		8.5647	0.0000		
123.85	3135		95	2658	95	0.0053	1.1936	R a	8.5647	0.0064	3356	76
123.85				477	D 95	0.0011	1.1947	ATGLEN	8.5584	0.0011	440	A 74
124.01	834		76	834	76	0.0021	1.1967	R	8.5552	0.0004	174	74
124.10	500			500	76	0.0012	1.1980	REVERSE CURVE	8.5548	0.0033	1290	74
124.51	2168		107	2168	76	0.0054	1.2034	L b	8.5515	0.0039	1451	A 67
124.59	389			389	76	0.0010	1.2044	REVERSE CURVE	8.5476	0.0009	316	67
124.85	1389		107	1389	76	0.0035	1.2078	R c	8.5467	0.0039	1389	67
125.05	1056		107	1056	76	0.0026	1.2104	R CHRISTIANA	8.5428	0.0030	1056	67
125.66	3224			3224	76	0.0080	1.2185	T	8.5398	0.0091	3224	67
125.85	1000		107	1000	76	0.0025	1.2210	L d	8.5307	0.0028	1000	67
125.95	550		107	550	76	0.0014	1.2223	L	8.5279	0.0016	550	67
126.06	550			506	76	0.0013	1.2236	REVERSE CURVE	8.5263	0.0015	550	67
126.06				44	D 76	0.0001	1.2237		8.5248	0.0000		
126.45	2090		74	2090	74	0.0053	1.2291	R a	8.5248	0.0069	2090	67
126.61	825			680	74	0.0017	1.2308	T	8.5168	0.0023	825	67
126.61				145	D 74	0.0004	1.2312		8.5166	0.0000		
126.85	1265		67	1265	67	0.0036	1.2348	L b	8.5165	0.0036	1265	67
127.16	1650		67	1650	67	0.0047	1.2394	L GAP	8.5129	0.0047	1650	67
127.52	1870		217	1870	67	0.0053	1.2447	L	8.5083	0.0053	1870	67
127.85	1760			1760	67	0.0050	1.2497	T	8.5030	0.0050	1760	67
127.91	330			330	67	0.0009	1.2506		8.4980	0.0009	330	67
128.31	2090		67	2090	67	0.0059	1.2565	R c	8.4971	0.0059	2090	67
128.31						0.0000	1.2565		8.4912	0.0014	583	D 92
128.60	1540			1540	67	0.0044	1.2609	T	8.4898	0.0020	957	92
128.85	1320		92	1320	67	0.0037	1.2646	L d	8.4878	0.0027	1320	92
129.05	1045		92	1045	67	0.0030	1.2676	L	8.4851	0.0022	1045	92
129.05						0.0000	1.2676		8.4829	0.0016	871	D 12
129.53	2530			2530	67	0.0072	1.2747	T	8.4814	0.0020	1659	120
129.85	1705		190	1705	67	0.0048	1.2795	L	8.4788	0.0027	1705	120
130.33	2530			2530	67	0.0072	1.2867	T	8.4761	0.0040	2530	120
130.51	935		190	330	67	0.0009	1.2876	L	8.4721	0.0015	935	120
130.85	1815			6583	A 67	0.0156	1.3032	T	8.4706	0.0029	1815	120
131.65	4235					0.0000	1.3032	T	8.4677	0.0051	3229	120
131.85	1045		134			0.0000	1.3032	R	8.4625	0.0000		
132.46	3245		134	13055	A 92	0.0233	1.3265	R	8.4626	0.0000		
132.85	2035						1.3265	T	8.4626	0.0094	5743	A 11
133.23	1980						1.3265	T	8.4532			
133.51	1485		164				1.3265	L	8.4532	0.0150	8023	A 92
133.85	1815			644	119	0.0010	1.3275	T	8.4382	0.0015	715	92
134.29	2334			302	D 11	0.0005	1.3280	T	8.4367	0.0048	2334	92
134.77	2501		110	2501	110	0.0043	1.3323	L a	8.4319	0.0051	2501	92
134.85	445			445	110	0.0006	1.3331	T	8.4268	0.0009	445	92
135.35	2640	XING 59.18		2640	110	0.0045	1.3377	IRISHTOWN ROAD	8.4259	0.0054	2640	92
135.55	1045		119	1045	110	0.0016	1.3395	R b	8.4204	0.0022	1045	92
135.80	1320			787	110	0.0014	1.3408	T	8.4183	0.0027	1320	92
135.80				533	D 11	0.0010	1.3418		8.4156	0.0000		
135.85	275		92	275	92	0.0006	1.3424	R a	8.4156	0.0006	275	92
136.45	3190		92	3190	92	0.0066	1.3489	R	8.4150	0.0066	3190	92
136.81	1870			1870	92	0.0038	1.3528	T	8.4084	0.0002	82	D 95
136.85	220		95	220	92	0.0005	1.3533	L b	8.4083	0.0003	138	95
137.30	2390		95	2390	92	0.0049	1.3582	L BIRD IN HAND	8.4080	0.0048	2390	95
137.51	1112			1112	92	0.0023	1.3605	T	8.4032	0.0014	788	D 12
137.85	1779		134	1779	92	0.0037	1.3641	L	8.4018	0.0033	2103	120
137.85						0.0000	1.3641		8.3985	0.0000		
137.96	605		134	605	92	0.0012	1.3654	L	8.3985	0.0010	605	120
138.77	4235			3024	92	0.0062	1.3716	T	8.3976	0.0067	4235	120
138.85	440		190			0.0000	1.3716	L	8.3909	0.0007	440	120
138.85				1234	A 92	0.0025	1.3741		8.3902	0.0000		
139.05	1045		190	1462	95	0.0029	1.3770	L	8.3902	0.0016	1045	120

APPENDIX B4

PAGE 3 COLUMN 1

MILES	DISTANCE ft	HWY GRADE CROSSING	MAX MPH	DISTANCE ft	SPEED mph	ST. hours	TIME hours	STATION	TIME hours	ST. hours	DISTANCE ft	SPEED mph
				READ DOWN WESTBOUND TRAFFIC				STATION	READ UP EASTBOUND TRAFFIC			
139.35	1595			1598	96	0.0002	1.3802	T	8.3886	0.0025	1595	120
139.65	1595		134	589	96	0.0004	1.3806	R	8.3860	0.0025	1595	120
139.85	1046			12632 A	96	0.0019	1.4006	T	8.3835	0.0016	1046	120
139.85						0.0000	1.4026		8.3818	0.0000		120
140.65	4224					0.0000	1.4026	T	8.3818	0.0067	4224	120
140.85	1056		233			0.0000	1.4026	R	8.3752	0.0017	1056	120
140.85						0.0000	1.4025		8.3735	0.0000		120
141.35	2640		233			0.0000	1.4025	R	8.3735	0.0042	2640	120
141.85	2640			479	120	0.0006	1.4033	T	8.3693	0.0042	2640	120
141.85						0.0000	1.4033		8.3652	0.0000		120
141.90	275			275	120	0.0004	1.4037	T	8.3652	0.0004	275	120
142.00	495		152	495	120	0.0006	1.4045	R	8.3647	0.0008	495	120
142.08	440			440	120	0.0007	1.4052	REVERSE CURVE	8.3640	0.0007	440	120
142.27	990		152	990	120	0.0016	1.4067	L	8.3633	0.0016	990	120
142.32	275			275	120	0.0004	1.4072	T	8.3617	0.0004	275	120
142.51	990		134	990	120	0.0016	1.4087	L	8.3613	0.0016	990	120
142.85	1815			1815	120	0.0029	1.4116	T	8.3597	0.0029	1815	120
143.23	2006			2006	120	0.0032	1.4148	LANCASTER STATION	8.3568	0.0032	2006	120
143.85	3274			3274	120	0.0052	1.4199		8.3537	0.0052	3274	120
144.34	2585			2585	120	0.0041	1.4240	DILLERVILLE COL BR	8.3485	0.0041	2585	120
144.85	2695		128	2695	120	0.0043	1.4283	R	8.3444	0.0043	2695	120
145.05	1045		128	1045	120	0.0018	1.4298	R	8.3402	0.0016	1045	120
145.85	4235			4235	120	0.0067	1.4366	T	8.3385	0.0067	4235	120
146.51	3465			3465	120	0.0056	1.4421	T	8.3318	0.0055	3465	120
146.85	1815		134	1815	120	0.0029	1.4449	L	8.3264	0.0029	1815	120
146.89	220		134	220	120	0.0003	1.4453		8.3235	0.0003	220	120
147.54	3410	XING 71.88		3410	120	0.0054	1.4507	E PETER ROAD	8.3232	0.0054	3410	120
147.85	1650		134	1650	120	0.0026	1.4533	L	8.3178	0.0026	1650	120
147.98	660		134	660	120	0.0010	1.4543	L	8.3152	0.0010	660	120
148.26	1485			1485	120	0.0023	1.4566	T	8.3141	0.0023	1485	120
148.64	2035		134	2035	120	0.0032	1.4598	R	8.3118	0.0032	2035	120
148.85	1100			1100	120	0.0017	1.4616	T	8.3086	0.0017	1100	120
149.85	5280			5280	120	0.0083	1.4699		8.3068	0.0083	5280	120
149.85						0.0000	1.4699		8.2986	0.0000		120
150.05	1056			1056	120	0.0017	1.4716	T	8.2986	0.0017	1056	120
150.56	2668		126	2668	120	0.0042	1.4758	L	8.2968	0.0042	2668	120
150.70	778			778	120	0.0012	1.4770		8.2926	0.0012	778	120
150.85	778		134	778	120	0.0012	1.4783	R	8.2914	0.0012	778	120
151.05	1056		134	1056	120	0.0017	1.4799	R 75 P	8.2902	0.0017	1056	120
151.76	3724	XING 75.82		3724	120	0.0059	1.4858	PRIVATE 50 F	8.2885	0.0059	3724	120
151.85	500		300	500	120	0.0008	1.4866	L	8.2826	0.0008	500	120
152.36	2668		300	2668	120	0.0042	1.4908	L	8.2818	0.0042	2668	120
152.85	2612			2612	120	0.0041	1.4949	T	8.2776	0.0041	2612	120
152.93	440			440	120	0.0007	1.4956	T	8.2735	0.0007	440	120
153.16	1210		134	1210	120	0.0019	1.4975	L	8.2728	0.0019	1210	120
153.46	1595			1595	120	0.0025	1.5001		8.2709	0.0025	1595	120
153.66	1045		120	1045	120	0.0016	1.5017	L	8.2684	0.0016	1045	120
153.85	990			990	120	0.0016	1.5033	T	8.2667	0.0016	990	120
154.85	5280			5280	120	0.0083	1.5116		8.2652	0.0083	5280	120
155.00	778	XING 79.13		778	120	0.0012	1.5128	MOORES MILL ROAD	8.2568	0.0012	778	120
155.20	1056		120	1056	120	0.0017	1.5145	R	8.2556	0.0017	1056	120
155.61	2168			2168	120	0.0034	1.5179	T	8.2540	0.0034	2168	120
155.85	1278		164	1278	120	0.0020	1.5199	R	8.2505	0.0020	1278	120
155.95	500		164	500	120	0.0008	1.5207	R	8.2485	0.0008	500	120
156.73	4057			4057	120	0.0064	1.5271		8.2477	0.0016	1034	120
156.85	723		168	723	120	0.0011	1.5283	R	8.2461			
157.01	834	XING 8	159	834	120	0.0013	1.5296	R MARKET ST MT JOY	8.2461			
157.41	2112			2112	120	0.0033	1.5329	T	8.2461			
157.85	2334		134	2334	120	0.0037	1.5366	R	8.2461	0.0218	12630 A	96
157.88	165		134	165	120	0.0003	1.5369	R	8.2241	0.0000		
158.63	3960			3960	120	0.0063	1.5431		8.2241	0.0012	619	95
158.85	1155		190	1155	120	0.0018	1.5449	L	8.2229	0.0023	1156	95
159.08	1210		190	1210	120	0.0019	1.5468	L	8.2206	0.0024	1210	95
159.56	2530			2530	120	0.0040	1.5508	T	8.2182	0.0050	2530	95
159.85	1540		152	1540	120	0.0024	1.5533	L	8.2132	0.0031	1540	95
160.09	1278		150	1278	120	0.0020	1.5553	L	8.2101	0.0026	1278	95
160.60	2666			1580	120	0.0030	1.5582	T	8.2075	0.0053	2666	95
160.60				788 D	12	0.0014	1.5596		8.2022	0.0000		
160.85	1334		95	1334	95	0.0027	1.5623	R a	8.2022	0.0027	1334	95
161.19	1779		96	1779	95	0.0036	1.5658	R ELIZABETHTOWN	8.1996	0.0036	1779	95
161.32	723			723	95	0.0014	1.5673	T	8.1960	0.0014	723	95
161.62	1556		134	1556	95	0.0031	1.5704	L	8.1946	0.0031	1556	95
161.73	611			611	95	0.0012	1.5716	T	8.1915	0.0012	611	95
161.85	611		95	611	95	0.0012	1.5728	L b	8.1903	0.0012	611	95
162.04	1000		95	1000	95	0.0020	1.5748	L	8.1890	0.0020	1000	95
162.04						0.0000	1.5748		8.1870	0.0014	783 D	125
162.85	4280			4280	95	0.0085	1.5833	T	8.1856	0.0010	958	120
163.57	3779			3779	95	0.0075	1.5908	T	8.1846	0.0060	3779	120
163.82	1334		190	1334	95	0.0027	1.5935		8.1787	0.0000		
163.85	167			167	95	0.0003	1.5939	T	8.1787	0.0000		
164.85	5280			1440	95	0.0029	1.5967		8.1787	0.0000		
165.85	5280			12532 A	95	0.0219	1.6187		8.1787	0.0000		
165.85						0.0000	1.6187	T	8.1787	0.0350	16458 A	78
165.85						0.0000	1.6187		8.1437	0.0000		
166.67	4082			670	120	0.0011	1.6197	T	8.1437	0.0013	921	76
166.85	1198		131	1198	120	0.0019	1.6216	L	8.1424	0.0030	1198	76
166.85						0.0000	1.6216		8.1394	0.0000		
167.01	825		131	825	120	0.0013	1.6229	L	8.1394	0.0021	825	76
167.50	2585			2585	120	0.0041	1.6270	T	8.1373	0.0064	2585	76
167.85	1870		134	1870	120	0.0030	1.6300	R	8.1308	0.0047	1870	76
167.85						0.0000	1.6300		8.1267	0.0000		76
168.15	1556		134	1556	120	0.0025	1.6324	R	8.1262	0.0039	1556	76
168.61	2445			1180	120	0.0019	1.6343	T	8.1223	0.0061	2445	76
168.61				1305 D	15	0.0024	1.6367		8.1162	0.0000		
168.85	1278		76	1278	76	0.0032	1.6399	R	8.1162	0.0032	1278	76
168.85						0.0000	1.6399		8.1131	0.0000		
169.20	1870		76	1800	76	0.0047	1.6446	R	8.1131	0.0047	1870	76
169.20						0.0000	1.6446	REVERSE CURVE	8.1084	0.0007	286 D	88
169.20						0.0000	1.6446		8.1077	0.0060	2791	88
169.79	3080		88	3080	76	0.0077	1.6523	L	8.1017	0.0000	0	88
169.85	330			330	76	0.0008	1.6531	T	8.1017	0.0017	941 D	116
169.85						0.0000	1.6531	ROYALTON BRANCH	8.1000	0.0000		110

APPENDIX B4

PAGE 3 COLUMN 2

MILES	DISTANCE ft	HWY GRADE CROSSING	MAX MPH	DISTANCE ft	SPEED mph	ST. hours	TIME hours	STATION	TIME hours	ST. hours	DISTANCE ft	SPEED mph
				READ DOWN WESTBOUND TRAFFIC				STATION	READ UP EASTBOUND TRAFFIC			
170.45	3168			3168	76	0.0079	1.6810	T	8.1000	0.0041	2652	119
170.59	725		204	723	76	0.0018	1.6828	L	8.0959	0.0012	723	119
170.66	389			389	76	0.0010	1.6837	T	8.0946	0.0006	389	119
170.82	834		480	834	76	0.0021	1.6858	L	8.0942	0.0013	834	119
170.84	111			111	76	0.0003	1.6861	T	8.0928	0.0002	111	119
170.85	56		119	56	76	0.0001	1.6862	L MIDDLETOWN	8.0927	0.0001	56	119
171.15	1550		119	1558	76	0.0039	1.6701	L	8.0926	0.0025	1556	119
171.22	349			389	76	0.0010	1.6711	REVERSE CURVE	8.0901	0.0001	70 D	12
171.42	1056		145	364	76	0.0009	1.6720	R	8.0900	0.0022	1410	120
171.62	1056			3373 A	76	0.0076	1.6796	T	8.0878	0.0017	1056	120
171.82	1056		150			0.0000	1.6798	L	8.0861	0.0017	1056	120
171.85	167			14364 A	88	0.0261	1.7069	T	8.0845	0.0003	167	120
172.15	1556					0.0000	1.7069	T	8.0842	0.0025	1556	120
172.25	778		486			0.0000	1.7059	L	8.0818	0.0012	778	120
172.76	2445					0.0000	1.7058	T FLYOVER	8.0805	0.0039	2445	120
172.85	500		130			0.0000	1.7059	R HIGHSPIRE	8.0767	0.0008	500	120
173.14	1556		130			0.0000	1.7059	R	8.0759	0.0025	1556	120
173.74	3168					0.0000	1.7059		8.0734	0.0050	3168	120
173.85	556		300			0.0000	1.7059	L	8.0684	0.0009	556	120
174.05	1045		300			0.0000	1.7059	L	8.0675	0.0016	1045	120
174.79	3905			701 A	111	0.0011	1.7070	T	8.0659	0.0062	3905	120
174.85	330		178	302	120	0.0006	1.7078	R	8.0597	0.0006	330	120
175.30	2395		178	2395	120	0.0038	1.7113	R	8.0590	0.0036	2395	120
175.85	2885			2885	120	0.0046	1.7159	T	8.0554	0.0046	2885	120
176.85	5280			5280	120	0.0083	1.7242	STEELTON	8.0508	0.0083	5280	120
177.48	3300			3300	120	0.0052	1.7294	T	8.0425	0.0052	3300	120
177.82	1815		222	1815	120	0.0029	1.7323	R	8.0373	0.0029	1815	120
177.85	165			165	120	0.0003	1.7325	T	8.0345	0.0003	165	120
178.29	2334			2334	120	0.0037	1.7362	T	8.0342	0.0037	2334	120
178.53	1278		134	1278	120	0.0020	1.7382	R	8.0305	0.0020	1278	120
178.76	1167			1167	120	0.0018	1.7401	T	8.0285	0.0018	1167	120
178.83	389		154	389	120	0.0006	1.7407	L	8.0267	0.0006	389	120
178.85	111			111	120	0.0002	1.7409	T	8.0261	0.0002	111	120
179.18	1760			1760	120	0.0028	1.7437	T	8.0259	0.0009	600	120
179.31	660		153	660	120	0.0010	1.7447	R	8.0249			
179.85	2860			2866	120	0.0046	1.7492	T	8.0249			
180.19	1778			1778	120	0.0028	1.7520	T	8.0249			
180.24	269			269	120	0.0004	1.7524		8.0249			
180.24		STATE TOWER 104.38				0.0000	1.7524	HARRISBURG	8.0249			
180.25	54			54	120	0.0001	1.7525		8.0249			
180.29	215			215	120	0.0003	1.7529		8.0249			
180.36	377			377	120	0.0006	1.7535		8.0249			
180.47	593			593	120	0.0009	1.7544		8.0249			
180.60	700		163	700	120	0.0011	1.7555	L	8.0249			
180.75	754			754	120	0.0012	1.7567	T	8.0249			
180.85	539			539	120	0.0008	1.7575	T	8.0249			
180.89	218			218	120	0.0003	1.7579	T	8.0249			
180.92	163		163	163	120	0.0003	1.7581	R	8.0249			
181.06	707			707	120	0.0011	1.7593	T	8.0249			
181.06	0	HERR STREET 105.20				0.0000	1.7560	WEST END AMTRAK -	8.0249			
181.31	1365			1361	120	0.0021	1.7614	T	8.0249			
181.40	435			435	120	0.0007	1.7621		8.0249			
181.48	439			435	120	0.0007	1.7628		8.0249			
181.60	653			653	120	0.0010	1.7638	T	8.0249			
181.79	980		183	980	120	0.0016	1.7654	R	8.0249			
181.85	327			327	120	0.0005	1.7658	T	8.0246			
181.98	701			701	120	0.0011	1.7670	T	8.0249	0.0339	18006 A	76
182.01	162		165	162	120	0.0008	1.7872	L	7.9910	0.0039	1531 A	71
182.85	4419			4419	120	0.0070	1.7742	T	7.9872	0.0028	891	71
182.85				0		0.0000	1.7740		7.9845	0.0078	4144 A	56
183.85	5281			5281	120	0.0083	1.7825	T	7.9767	0.0039	1129	56
183.85				3383	120	0.0053	1.7879		7.9728			
184.75	4736			1373 D	12	0.0027	1.7906	T	7.9726	0.0163	4736	56
184.78	163		71	163	71	0.0004	1.7910	R	7.9565	0.0006	163	55
184.81	160			163	71	0.0004	1.7914	REVERSE CURVE	7.9560	0.0006	163	55
184.85	218		71	218	71	0.0006	1.7920	L	7.9554	0.0008	218	55
184.65						0.0000	1.7920	LINGLESTOWN	7.9547	0.0000		
184.93	437			437	71	0.0012	1.7932	T	7.9547	0.0015	437	55
185.17	1257		77	1000	71	0.0027	1.7956	L	7.9532	0.0043	1257	50
185.72	2897			2852	71	0.0076	1.8035	T	7.9488	0.0100	2897	55
185.77				288 D	71	0.0009	1.8044		7.9388			
185.72							1.8044	EAST END ROCKVILLE	7.9389			
185.76	219		55	219	55	0.0008	1.8051	L	7.9389	0.0008	219	56
185.76						0.0000	1.8051	ROCKVILLE	7.9381	0.0000		
185.86	547		56	547	56	0.0019	1.8070	L	7.9381	0.0019	547	56
185.86						0.0000	1.8070		7.9362	0.0000		
185.91	218		56	218	55	0.0008	1.8078	L	7.9362	0.0008	218	55
185.91						0.0000	1.8078		7.9355	0.0018	591 D	64
186.62	3756			3756	55	0.0129	1.8207	T	7.9336	0.0071	3165	84
186.62						0.0000	1.8207	SUSQUEHANNA RIVE	7.9267	0.0000		
186.79	925		110	925	55	0.0032	1.8239	R	7.9267	0.0021	925	84
186.86	361			361	55	0.0013	1.8252	T	7.9246	0.0011	505	84
186.86						0.0000	1.8252	MARYSVILLE	7.9235	0.0039	1961 A	78
187.04	916			916	55	0.0032	1.8284	T	7.9196	0.0008	727 A	77
187.50	2909		110	2909	55	0.0100	1.8384	R	7.9191	0.0036	1120	77
187.86	1495			1456	55	0.0050	1.8434	T	7.9153	0.0036	1455	77
188.39	2797			658	55	0.0023	1.8456	T	7.9127	0.0068	2797	77
188.88	2581		84	4950 A	86	0.0196	1.8600	L	7.9056	0.0053	2581	77
188.95	335		84	975	81	0.0023	1.8616	L	7.8995	0.0008	335	77
189.13	960			70 D	81	0.0002	1.8617	T	7.8987	0.0025	960	77
189.20	392		78	391	78	0.0009	1.8627	R	7.8963	0.0010	391	77
189.33	671			671	78	0.0016	1.8643	T	7.8954	0.0017	671	77
189.33		BANKS		DIV POST			1.8643	BANKS	7.8937		BANKS TOW	
189.34	76			48	78	0.0001	1.8644		7.8937	0.0017	700	77
189.34		XING 114.25		23 D	78	0.0001	1.8645	PRIVATE	7.8920			
189.56	1163		77	1163	77	0.0029	1.8673	L	7.8920	0.0030	1230	77
189.83	1407			1407	77	0.0036	1.8708		7.8890	0.0002	89 D	80
189.83					77	0.0000	1.8708		7.8886	0.0032	1338	80
190.38	2900			2900	77	0.0071	1.8779		7.8856	0.0096	4076	80
190.60	1176		145	1176	77	0.0029	1.8808	R	7.8760	0.0000		80
190.83	1200	XING 115.32		1200	77	0.0030	1.8838	PRIVATE	7.8760	0.0028	1200	80
191.68	4500	XING 115.90		5724	77	0.0141	1.8978	PRIVATE	7.8732	0.0058	690	80

APPENDIX B4 — PAGE 4 COLUMN 1

READ DOWN: WESTBOUND TRAFFIC — STATION — READ UP: EASTBOUND TRAFFIC

MILES	DISTANCE, ft	HWY GRADE CROSSING	MAX MPH	DISTANCE, ft	SPEED, mph	ET, HOURS	TIME, HOURS	STATION	TIME, HOURS	ET, HOURS	DISTANCE, ft	SPEED, mph
191.83	791		80	803	A 77	0.0019	1.8898	R	7.8716	0.0000		
192.03	1084		85	0	80	0.0000	1.8898	R	7.8716	0.0183	8074	A 41-
192.60	3000			2846	80	0.0067	1.9005		7.8532	0.0127	2836	41
192.83	1200		120	1200	80	0.0028	1.9094	R	7.8441	0.0056	1200	41
193.40	3000		120	3000	80	0.0071	1.9160	R	7.8385	0.0139	3000	41
193.52	686			665	80	0.0016	1.9180		7.8217	0.0031	665	41
193.68	800		67	800	80	0.0019	1.9198	R	7.8186	0.0037	800	41
193.83	800		110	800	80	0.0019	1.9218	R	7.8149	0.0037	800	41
194.04	1100		89	1100	80	0.0026	1.9244	R	7.8112	0.0051	1100	41
194.19	800			108	80	0.0003	1.9247		7.8061	0.0037	800	41
194.19				692	D 80	0.0022	1.9268		7.8024			
194.40	1100		41	1100	41	0.0051	1.9318	L	7.8024	0.0051	1100	41
194.40							1.9319		7.7973	0.0002	50	D 45-
194.45	270			270	41	0.0012	1.9332	TRC MPH30	7.7971	0.0008	220	45
194.60	800		45	800	41	0.0037	1.9369	R	7.7962	0.0034	800	45
194.60							1.9369		7.7928	0.0031	1198	D 101
194.81	1100			1100	41	0.0051	1.9420		7.7897	0.0000		101
194.83	100			100	41	0.0005	1.9424		7.7897	0.0000		
195.60	4096		101	4098	41	0.0189	1.9613	R DUNCANNON	7.7897	0.0077	4096	101
195.60							1.9613		7.7820	0.0006	444	D 115
195.70	500			500	41	0.0023	1.9637		7.7813	0.0001	56	115
196.15	2402			2402	41	0.0111	1.9747		7.7812	0.0040	2402	115
196.72	3000		115	1730	41	0.0080	1.9827	L	7.7772	0.0049	3000	115
196.97	1300		115	316	A 41	0.0014	1.9841	L	7.7723	0.0021	1300	115
197.32	1880			754	46	0.0030	1.9871		7.7701	0.0010	610	115
197.57	1300		134	0		0.0000	1.9871	R	7.7691			
197.57							1.9871	WEST END ROCKVILL	7.7691			
197.72	800			13102	A 45	0.0322	2.0193		7.7691			
197.87	800						2.0193		7.7691	0.0044	9490	A 95
198.23	1896	XING 1 139				0.0006	2.0199	L AQUEDUCT	7.7647	0.0056	2831	95
198.72	2600	XING 123.3		8962	A 10	0.0122	2.0316	PRIVATE	7.7590	0.0025	1234	A 92-
199.72	5285	XING 123.80				0.0000	2.0316	PRIVATE	7.7565	0.0009	456	92
200.15	2270						2.0316		7.7556			
200.15							2.0316	EAST END NEWPORT	7.7556			
200.72	3000			1190	A 11	0.0019	2.0335		7.7556	0.0109	5229	92
200.95	1200			683	117	0.0009	2.0344		7.7447	0.0025	1200	92
200.95				684	D 11	0.0012	2.0357		7.7423			
201.34	2083	XING 1	95	2083	95	0.0042	2.0398	L PRIVATE	7.7423	0.0043	2083	92
201.69	1800			1800	95	0.0038	2.0434		7.7380	0.0037	1800	92
201.72	180			100	95	0.0002	2.0436		7.7343	0.0013	605	92
201.72		XING 127.3		82	D 11	0.0002	2.0438	PRIVATE	7.7329			
202.29	3000		92	3000	92	0.0062	2.0499	L	7.7329	0.0067	3006	A 82-
202.52	1200			1200	92	0.0025	2.0524		7.7283	0.0016	801	82
202.69	900		103	900	92	0.0019	2.0543	R	7.7247	0.0021	900	82
202.83	752		138	752	92	0.0015	2.0558	R	7.7226	0.0017	752	82
203.04	1100		82	1100	92	0.0023	2.0581	R	7.7209	0.0025	1100	82
203.26	1200		126	1200	92	0.0025	2.0606	R	7.7184	0.0028	1200	82
203.49	1200	XING 1	98	1200	92	0.0025	2.0630	R PRIVATE	7.7156	0.0028	1200	82
203.72	1200			1200	92	0.0025	2.0655		7.7128	0.0028	1200	82
203.93	1100			1100	92	0.0023	2.0677		7.7101	0.0025	1100	82
204.23	1600	XING 1	93	1600	92	0.0033	2.0710	L PRIVATE	7.7075	0.0037	1600	82
204.47	1287			1032	92	0.0021	2.0730		7.7038	0.0030	1287	82
204.47				285	D 92	0.0006	2.0737		7.7008			
204.72	1300		82	1300	82	0.0030	2.0767	L LOSH RUN	7.7008	0.0030	1300	82
204.90	930		82	930	82	0.0021	2.0789	L	7.6978	0.0021	930	82
204.90						0.0000	2.0789		7.6957	0.0003	124	D 87-
205.73	4400		130	4400	82	0.0102	2.0890	L	7.6954	0.0095	4276	87
205.98	1313			1313	82	0.0030	2.0921		7.6861	0.0029	1313	87
206.70	3800		95	3800	82	0.0088	2.1008	R	7.6832	0.0083	3800	87
207.30	3200		87	1487	82	0.0034	2.1043	R	7.6750	0.0070	3200	87
207.30						0.0000	2.1040		7.6680	0.0016	1002	D 120
207.50	1024			1481	A 82	0.0033	2.1076		7.6662	0.0000	22	120
207.67	900		130	2146	87	0.0047	2.1123	L	7.6661	0.0014	900	120
207.76	500		130	500	87	0.0011	2.1134	L	7.6647	0.0008	500	120
208.03	1400			1400	87	0.0030	2.1164	T NEWPORT	7.6639	0.0016	823	120
208.37	1800	XING 1	130	1800	87	0.0039	2.1203	R PRIVATE	7.6629	0.0000		
208.56	1000			1000	87	0.0022	2.1225		7.6629	0.0000		
208.63	373	XING 1	130	373	87	0.0008	2.1233	R PRIVATE	7.6629	0.0000		
209.10	2475		130	2478	87	0.0054	2.1287	R	7.6629	0.0147	8729	A 104
209.63	2800			1625	87	0.0033	2.1320		7.6482	0.0008	489	104
209.79	825			11070	A 87	0.0208	2.1529		7.6473	0.0015	805	104
210.62	4400		140			0.0000	2.1529	L	7.6458	0.0008	420	104
211.20	3000						2.1529		7.6451			
211.20							2.1529	WEST END NEWPORT	7.6451			
211.22	116			420	113	0.0007	2.1536		7.6451	0.0009	498	A 107
211.62	2100		288	288	D 11	0.0005	2.1541	L	7.6442	0.0035	1925	105
211.96	1800		104	2170	104	0.0040	2.1580	L	7.6407	0.0183	8821	A 78-
212.03	400			30	D 11	0.0001	2.1581	T TRC MPH78(TT)	7.6223	0.0006	254	78
212.17	719		103	719	103	0.0013	2.1594	R	7.6217	0.0017	719	78
212.61	2300			2300	103	0.0042	2.1636		7.6200	0.0056	2300	78
213.14	2800			2800	103	0.0051	2.1688		7.6144	0.0068	2800	78
213.27	713		126	713	103	0.0013	2.1701	R	7.6076	0.0017	713	78
213.61	1800			1800	103	0.0033	2.1734	T MILLERSTOWN	7.6059	0.0044	1800	78
214.07	2414		239	1750	103	0.0032	2.1766	L	7.6015	0.0059	2414	78
214.07				884	D 10	0.0014	2.1780		7.5956			
214.60	2800		78	2800	78	0.0068	2.1848	L	7.5956	0.0068	2800	78
214.96	1900			1900	78	0.0046	2.1894		7.5888	0.0003	140	D 84
215.13	902	XING 1	127	902	78	0.0022	2.1916	R PRIVATE	7.5885	0.0003	137	84
215.38	1300			1300	78	0.0032	2.1948		7.5882	0.0111	4981	A 85-
215.60	1200		128	1200	78	0.0029	2.1977	R	7.5771	0.0018	631	85
215.62	100		129	100	78	0.0002	2.1979	R	7.5752	0.0002	100	85
216.27	3400			3400	78	0.0083	2.2062		7.5750	0.0058	3400	85
216.57	1620		95	1620	78	0.0039	2.2101	L DURWARD	7.5696	0.0047	1620	85
216.62	246		95	249	78	0.0006	2.2107	L	7.5603	0.0007	246	85
217.19	3000			329	78	0.0008	2.2115		7.5596	0.0087	3000	85
217.32	700		134	3872	A 78	0.0089	2.2204	R	7.5509	0.0020	700	85
217.42	500					0.0000	2.2204		7.5488	0.0015	500	85
217.57	800		95	104	91	0.0002	2.2206	R	7.5474	0.0023	800	95
217.57				595	D 10	0.0014	2.2221		7.5450			
217.78	1100		95	1100	65	0.0032	2.2253	L	7.5450	0.0032	1100	65
217.78							2.2253		7.5418	0.0001	19	D 66-
217.89	800			500	65	0.0017	2.2270	REVERSE CURVE	7.5418	0.0017	581	66

APPENDIX B4 — PAGE 4 COLUMN 2

READ DOWN: WESTBOUND TRAFFIC — STATION — READ UP: EASTBOUND TRAFFIC

MILES	DISTANCE, ft	HWY GRADE CROSSING	MAX MPH	DISTANCE, ft	SPEED, mph	ET, HOURS	TIME, HOURS	STATION	TIME, HOURS	ET, HOURS	DISTANCE, ft	SPEED, mph
218.08	1000		82	1000	65	0.0029	2.2268	R THOMPSONTOWN	7.5401	0.0026	1000	66
218.13	282			282	65	0.0006	2.2308	TRC TT MPH66	7.5372	0.0005	282	66
218.34	1100		96	1100	65	0.0032	2.2340	L	7.5364	0.0032	1100	96
218.34							2.2340		7.5333	0.0023	1045	D 10
218.57	1200			1200	65	0.0035	2.2375		7.5310	0.0066	3722	107
218.60	184			184	65	0.0005	2.2380		7.5244	0.0000		
218.81	1100		120	1100	65	0.0032	2.2412	R	7.5244	0.0000		
219.57	4000			4003	65	0.0117	2.2529		7.5244	0.0000		
220.46	4700			1534	65	0.0045	2.2573		7.5244	0.0000		
220.57	598	XING 1	107	182	A 68	0.0008	2.2576	R PRIVATE	7.5244	0.0000		
220.67	496		107	3796	68	0.0080	2.2659	R	7.5244	0.0226	9254	A 83
221.58	4800			13208	A 65	0.0283	2.2942		7.5018	0.0070	3060	83
222.21	3340						2.2942		7.4948	0.0078	3346	83
222.59	2000		160	0		0.0000	2.2942	L	7.4870	0.0046	2000	63
222.61	100		182	0		0.0000	2.2942	L	7.4825	0.0002	100	83
223.08	2500			139	107	0.0002	2.2945		7.4822	0.0057	2500	83
223.08				569	D 10	0.0013	2.2958		7.4765			83
223.55	2500		83	2500	83	0.0057	2.3015	R TUSCARORA	7.4765	0.0057	2500	83
223.64	485		83	485	83	0.0011	2.3026	R MEXICO	7.4708	0.0011	485	83
223.64		XING 148.6				0.0000	2.3026	PRIVATE	7.4698	0.0005	201	D 92
223.64						0.0000	2.3026		7.4693	0.0005	244	92
223.64							2.3026		7.4688	0.0037	1575	A 87
224.13	2600			2600	83	0.0059	2.3085		7.4654	0.0012	500	87
224.55	2200		97	2200	83	0.0050	2.3135	L a	7.4642	0.0048	2200	87
224.95	2100		97	2100	83	0.0048	2.3183	L	7.4594	0.0034	1582	87
225.20	1350			1350	83	0.0031	2.3214		7.4560	0.0000		
225.47	1400		107	1400	83	0.0032	2.3246	R b	7.4560	0.0000		
225.75	1500			1350	83	0.0031	2.3277		7.4560	0.0072	3117	A 76
225.75	0						2.3277	R PORT ROYAL	7.4488	0.0086	3188	A 59
226.11	1900			1242	A 83	0.0064	2.3341		7.4306	0.0012	360	59
226.45	1767		67	2586	87	0.0056	2.3398	R c	7.4367	0.0061	1903	59
226.50	270		87	270	87	0.0006	2.3402	R	7.4326	0.0044	1263	A 46
227.44	5000			5000	87	0.0109	2.3511		7.4282	0.0153	3868	46
227.98	2800			2857	87	0.0065	2.3566		7.4129	0.0111	2800	46
227.98		XING 152.6		285	D 87	0.0006	2.3573	PRIVATE	7.4018	0.0000		
228.13	800		76	463	76	0.0012	2.3584	L TRC & MT PLEASANT	7.4018	0.0020	800	46
228.13				337	D 76	0.0009	2.3594		7.3997	0.0000		
228.45	1731		59	1731	59	0.0056	2.3649	R MPH 50	7.3997	0.0068	1731	46
228.53	400			400	59	0.0013	2.3662		7.3919	0.0016	400	46
228.80	1400		59	1227	59	0.0039	2.3701	R RC	7.3903	0.0055	1400	46
228.80				173	D 59	0.0006	2.3707		7.3848			
228.91	600		48	600	48	0.0024	2.3731	L MPH40	7.3848	0.0024	600	46
228.91		XING 153.90				0.0000	2.3731	JUNIATA STREET	7.3824	0.0010	301	D 66
229.33	2210			2210	48	0.0087	2.3818	T MIFFLIN	7.3814	0.0056	1909	66
229.44	600		68	600	48	0.0024	2.3842	L e	7.3759	0.0017	600	66
229.44						0.0000	2.3842		7.3742	0.0004	164	D 74
229.69	1300		100	1300	48	0.0051	2.3893	L MPH 45	7.3737	0.0029	1130	74
229.78	500			500	48	0.0020	2.3913	TRC	7.3708	0.0013	500	74
230.10	1680		74	1680	48	0.0066	2.3979	R b	7.3696	0.0043	1680	74
230.10							2.3979		7.3653	0.0005	207	D 83
230.23	700			700	48	0.0026	2.4007		7.3648	0.0003	138	93
230.44	1100		95	1100	48	0.0043	2.4050	L c	7.3645	0.0017	2947	A 71
231.07	3335		95	2910	48	0.0115	2.4165	L	7.3628	0.0048	1605	71
231.29	1200			2367	A 48	0.0078	2.4243		7.3580	0.0032	1200	71
231.44	800		95	443	66	0.0013	2.4256	R d	7.3548	0.0021	800	71
231.60	800		95			0.0000	2.4256	R	7.3520	0.0021	800	71
231.80	1100			1046	A 66	0.0044	2.4300		7.3505	0.0029	1100	71
231.94	700		127	529	74	0.0014	2.4314	R	7.3476	0.0019	700	71
232.45	2885			1306	74	0.0033	2.4347		7.3457	0.0072	2885	71
232.81	1900			2925	A 74	0.0069	2.4416	EDENHOLM	7.3385	0.0051	1900	71
232.81				46	85	0.0001	2.4417		7.3334			
232.81				320	D 85	0.0008	2.4425		7.3334			
233.17	1900		71	1900	71	0.0051	2.4476	L s	7.3334	0.0051	1900	71
233.17						0.0000	2.4476		7.3284	0.0021	971	D 10
233.30	700			700	71	0.0019	2.4495		7.3263	0.0006	470	108
233.45	781		127	781	71	0.0021	2.4516	L	7.3256	0.0000	0	
233.56	581		127	581	71	0.0015	2.4531	L	7.3255	0.0000	0	
234.31	4000			4000	71	0.0107	2.4638		7.3256	0.0000	0	
234.45	700		127	700	71	0.0019	2.4656	R	7.3256	0.0000	0	
234.49	235		127	235	71	0.0006	2.4663	R	7.3256	0.0000	0	
234.85	1900			1900	71	0.0051	2.4713		7.3256	0.0000	0	
235.08	1200		130	1200	71	0.0032	2.4745	L	7.3255	0.0247	12635	A 74
235.44	1901			703	71	0.0024	2.4769	CP HAWSTONE	7.3008	0.0000	0	
235.44				3011	A 71	0.0197	2.4967		7.3008	0.0000	0	
235.58	800					0.0000	2.4967		7.3006	0.0009	346	74
235.78	1014		160			0.0000	2.4967	L	7.2998	0.0050	1836	A 65
236.45	3800					0.0000	2.4967		7.2949	0.0046	1643	65
236.60	800			3198	A 58	0.0058	2.5026		7.2901	0.0023	800	65
236.88	1471		168			0.0000	2.5026	L	7.2878	0.0043	1471	65
237.20	1700					0.0000	2.5026		7.2835	0.0050	1700	65
237.46	1400		127			0.0000	2.5026	R	7.2785	0.0041	1400	65
237.65	1000		127			0.0000	2.5026	R	7.2744	0.0029	1000	65
237.80	800			405	107	0.0007	2.5032		7.2715	0.0023	800	65
237.80				875	D 10	0.0018	2.5050		7.2692			
238.10	1586		74	1586	74	0.0041	2.5091	L a	7.2692	0.0046	1586	65
238.22	600			417	74	0.0011	2.5102	REVERSE CURVE	7.2646	0.0017	600	65
238.22				183	D 74	0.0005	2.5107		7.2628			
238.43	1000		65	1000	65	0.0029	2.5136	R b	7.2628	0.0005	150	65
238.56	800		65	800	65	0.0023	2.5159	R	7.2624	0.0065	2430	A 45
238.66	512			512	65	0.0015	2.5174		7.2540			
238.96	1602			2114	65	0.0082	2.5256	SHAWNEE	7.2540	0.0056	1300	45
239.39	2300		85	2300	65	0.0067	2.5303	R c	7.2460	0.0087	2300	45
239.53	700		87	700	65	0.0020	2.5323	R	7.2388	0.0029	700	45
239.83	1600			1600	65	0.0047	2.5370		7.2359	0.0067	1600	45
240.07	1291		78	1291	65	0.0038	2.5407	L d	7.2292	0.0054	1291	45
240.24	900			900	65	0.0026	2.5433		7.2237	0.0038	900	45
240.40	800		78	900	65	0.0029	2.5457	R e	7.2199	0.0034	800	45
240.55	800		78	800	65	0.0023	2.5480	R	7.2166	0.0034	800	45
240.72	900		83	400	65	0.0012	2.5492	R	7.2132	0.0038	900	45
240.79	400			65	65	0.0014	2.5506	TRC MPH 65	7.2094	0.0017	400	45
240.79				323	D 65	0.0011	2.5517		7.2077			
241.00	1100		45	1100	45	0.0046	2.5563	L f	7.2077	0.0046	1100	45

APPENDIX B4

PAGE 5 COLUMN 1

MILES	DISTANCE ft	HWY GRADE CROSSING	MAX MPH	DISTANCE ft	SPEED mph	ET hours	TIME hours	STATION	TIME hours	ET hours	DISTANCE ft	SPEED mph
			READ DOWN WESTBOUND TRAFFIC						READ UP EASTBOUND TRAFFIC			
241.00						0.0000	2.5568		7.2031	0.0011	204	D 64
241.20	1054			1054	45	0.0044	2.5607		7.2021	0.0002	780	64
241.20							2.5607	EAST END LEWISTOWN	7.1998			
241.39	1000		64	1000	45	0.0042	2.5650	L	7.1998	0.0030	1006	64
241.68	1500		64	1500	45	0.0063	2.5713	L'LEWISTOWN	7.1960	0.0044	1500	64
241.68						0.0000	2.5713		7.1924	0.0024	1078	D 10
242.28	3200			3200	45	0.0135	2.5847		7.1900	0.0001	81	107
242.39	562		117	562	45	0.0024	2.5871	R	7.1899			
243.32	4907		117	3684	45	0.0156	2.6028	R	7.1899			
243.41	500		117	2261	A 45	0.0076	2.6104	R	7.1898			
243.85	2300			1260	64	0.0036	2.6142		7.1899			
244.01	672		180	0		0.0000	2.6142	R	7.1889	0.0225	11062	A 74
244.32	1600			11740	A 64	0.0261	2.6403		7.1674	0.0031	1220	74
244.56	1300	XING 169.40				0.0000	2.6403	GRANVILLE	7.1643	0.0033	1300	74
244.72	812		130		103	0.0000	2.6403	L	7.1610	0.0021	812	74
244.72							2.6403	WEST END LEWISTOWN	7.1589			
245.07	1868		130		103	0.0000	2.6403		7.1589	0.0048	1868	74
245.32	1300					0.0000	2.6403		7.1541	0.0033	1300	74
246.17	4500	XING 170.65		226	103	0.0004	2.6407	PRIVATE	7.1508	0.0115	4500	74
246.17		XING 170.84		753	D 10	0.0016	2.6423	ANDERSON	7.1393			
246.31	771		74	771	74	0.0020	2.6443	L a	7.1393	0.0020	771	74
246.67	1900		74	1900	74	0.0049	2.6492	L	7.1373	0.0049	1900	74
246.67						0.0000	2.6492		7.1324	0.0006	297	D 86
246.81	700			700	74	0.0016	2.6510		7.1318	0.0010	443	86
247.05	1300		86	1300	74	0.0035	2.6543	R b	7.1308	0.0029	1300	86
247.05						0.0000	2.6543		7.1279	0.0002	102	D 89
247.28	1200		103	1200	74	0.0031	2.6574	R	7.1277	0.0023	1098	89
247.31	179	XING 172.26		179	74	0.0005	2.6578	PRIVATE	7.1254	0.0004	179	89
247.84	2774	XING 173.35		2774	74	0.0071	2.6649	PRIVATE	7.1250	0.0059	2774	89
248.31	2500		127	2500	74	0.0064	2.6713	R	7.1191	0.0053	2500	89
248.54	1174		90	1174	74	0.0030	2.6743	R a	7.1138	0.0025	1174	89
249.07	2800			1173	74	0.0030	2.6773		7.1113	0.0060	2800	89
249.31	1300		89			0.0000	2.6773	R b	7.1053	0.0028	1300	89
249.31		XING 174.84				0.0000	2.6773	HORNINGFORD	7.1025	0.0002	80	D 92
249.72	2168		271			0.0000	2.6773	R	7.1024	0.0043	2068	92
250.31	3100	XING 175.31		2026	A 74	0.0069	2.6843	PRIVATE	7.0981	0.0048	2313	92
250.76	2379			1276	A 85	0.0026	2.6871		7.0933	0.0000		
251.31	2900	XING 1	92	9271	89	0.0197	2.7068	L SWIGARTS	7.0933	0.0006	258	A 91
252.31	5295	XING 177.16		458	89	0.0010	2.7078	PRIVATE	7.0926	0.0066	2809	91
252.35	191	XING 177.61		929	A 89	0.0019	2.7097	STATE HWY 103 MOVE	7.0860	0.0004	191	91
252.86	2700		134	6604	92	0.0140	2.7237	L	7.0865	0.0050	2700	91
253.30	2300		92	2300	92	0.0047	2.7284	L'MATAWANNA a	7.0806	0.0048	2300	91
253.31	100			100	92	0.0002	2.7286	C	7.0761	0.0002	100	91
253.86	2900			27	D 92	0.0001	2.7287		7.0759	0.0060	2900	91
254.19	1700		91	1700	91	0.0036	2.7322	R b	7.0699	0.0035	1700	91
254.31	680			680	91	0.0014	2.7337		7.0663	0.0014	680	91
255.28	5100			5100	91	0.0106	2.7443		7.0640	0.0106	5100	91
255.31	146	XING 1	101	146	91	0.0000	2.7446	R PRIVATE a	7.0543	0.0003	148	91
255.46	800		101	800	91	0.0017	2.7462	R RYDE	7.0540	0.0017	800	91
255.54	400			400	91	0.0008	2.7471		7.0523	0.0008	400	91
255.69	800		127	800	91	0.0017	2.7487	R	7.0515	0.0017	800	91
255.82	711		91	711	91	0.0015	2.7502	R b	7.0496	0.0239	9017	A 47
256.31	2800			2800	91	0.0054	2.7556		7.0256	0.0104	2578	47
256.31							2.7556	EAST END VINEYARD	7.0155			
257.30	5222			5222	91	0.0109	2.7665	T'VINEYARD	7.0155	0.0210	5222	47
257.91	3200			5183	91	0.0108	2.7773	T'PIG 8 CURVE	6.9945	0.0129	3200	47
257.91				861	D 91	0.0004	2.7797		6.9816			
258.12	1137		47	1137	47	0.0046	2.7843	R'RC a	6.9816	0.0046	1137	47
258.12						0.0000	2.7843		6.9770	0.0001	14	D 48
258.29	900		48	900	47	0.0036	2.7878	L	6.9770	0.0036	866	48
258.47	917		48	917	47	0.0037	2.7916	L	6.9736	0.0036	917	48
258.47						0.0000	2.7916		6.9699	0.0026	986	D 95
258.83	1900		134	1900	47	0.0077	2.7993	L	6.9672	0.0018	914	95
259.02	1000			1000	47	0.0040	2.8033		6.9654	0.0020	1000	95
259.15	700		95	700	47	0.0028	2.8061	R b	6.9634	0.0014	700	95
259.15						0.0000	2.8061		6.9620	0.0008	451	D 110
259.30	800		110	800	47	0.0032	2.8094	R	6.9612	0.0006	349	110
259.30							2.8094	WEST END VINEYARD	6.9606			
259.40	500		134	500	47	0.0020	2.8114	R	6.9606	0.0009	500	110
260.17	4085			4085	47	0.0165	2.8278		6.9597	0.0070	4085	110
260.30	700		233	198	47	0.0008	2.8286	R	6.9527	0.0012	700	110
260.85	2874		233	140	A 47	0.0006	2.8292	R	6.9515	0.0049	2874	110
261.30	2400		314	1677	48	0.0065	2.8356	R	6.9466	0.0041	2400	110
261.73	2280		314			0.0000	2.8358	R	6.9424	0.0030	2280	110
262.30	3000			10079	A 48	0.0264	2.8622		6.9385	0.0062	3000	110
262.30							2.8622	EAST END NEWTON H	6.9330			
262.74	2300					0.0000	2.8622		6.9333	0.0006	282	110
262.74	10		110				2.8622		6.9328			
263.08	1824		110	6789	A 95	0.0125	2.8747	L'NEWTON HAMILTON	6.9328	0.0000		
263.27	1000					0.0000	2.8747		6.9328	0.0000		
263.35	402					0.0000	2.8747		6.9328	0.0178	3342	A 88
264.20	4500		120			0.0000	2.8747	R	6.9151	0.0000		
264.71	2700		120	4807	110	0.0083	2.8830	R	6.9151	0.0067	3112	88
265.20	2591			2591	110	0.0045	2.8874		6.9084	0.0033	1501	88
265.20						0.0000	2.8874		6.9051	0.0106	4492	A 71
265.83	3297			3297	110	0.0057	2.8931		6.8946	0.0022	812	A 67
266.21	2000		110	1381	110	0.0023	2.8955	L	6.8923	0.0031	1093	67
266.21		XING 191.6		638	D 11	0.0012	2.8967	PRIVATE	6.8892			67
266.42	1100		58	1100	88	0.0024	2.8990	L'MT UNION	6.8892	0.0031	1100	67
266.51	477			477	88	0.0010	2.9001	REVERSE CURVE	6.8861	0.0013	477	67
267.00	2900		98	2600	88	0.0066	2.9067	R	6.8847	0.0073	2900	67
267.30	1800		98	1600	88	0.0034	2.9091	R	6.8774	0.0045	1600	67
267.62	1682			1682	88	0.0038	2.9127		6.8729	0.0048	1682	67
268.00	2000		110	1604	88	0.0035	2.9162	L	6.8681	0.0057	2000	67
268.00				396	D 88	0.0009	2.9171		6.8625			67
268.06	300		71	300	71	0.0008	2.9179	L	6.8625	0.0008	300	67
268.08	148			67	71	0.0002	2.9181	TRC	6.8616	0.0004	148	67
268.08				81	D 82	0.0002	2.9183	MAPLETON	6.8612			
268.99	4800		67	4800	67	0.0136	2.9319	R TTMPH60	6.8612	0.0136	4800	67
269.22	1200		67	1200	67	0.0034	2.9353	R	6.8478	0.0034	1200	67
269.22							2.9353		6.8442	0.0014	583	D 92
269.22							2.9353		6.8428	0.0000	17	92

APPENDIX B4

PAGE 5 COLUMN 2

MILES	DISTANCE ft	HWY GRADE CROSSING	MAX MPH	DISTANCE ft	SPEED mph	ET hours	TIME hours	STATION	TIME hours	ET hours	DISTANCE ft	SPEED mph
			READ DOWN WESTBOUND TRAFFIC						READ UP EASTBOUND TRAFFIC			
269.66	2300			2300	67	0.0065	2.9418		6.8428	0.0035	1700	92
269.94	1477		92	1477	67	0.0042	2.9460	R	6.8393	0.0030	1477	92
269.99	300			300	67	0.0008	2.9468		6.8363	0.0010	533	D 11
270.37	2000			2000	67	0.0057	2.9525		6.8353	0.0030	1767	110
270.99	3246		134	3246	67	0.0092	2.9616	L	6.8322	0.0032	1847	110
271.19	1085		134	1085	67	0.0031	2.9647	L	6.8290	0.0000		
271.49	1600			592	67	0.0017	2.9664		6.8290	0.0000		
271.76	1400		116	6583	A 67	0.0156	2.9819	L	6.8290	0.0000		
271.99	1200		120			0.0000	2.9819	L'MILL CREEK	6.8290	0.0169	8952	A 89
272.52	2796	XING 1	110	8023	A 92	0.0150	2.9969	L CASSVILLE	6.8121	0.0011	522	89
272.99	2500					0.0000	2.9969		6.8110	0.0067	3151	89
273.02	178					0.0000	2.9969		6.8043	0.0097	4203	A 74
273.02							2.9969	WEST END NEWTON HAMILTON	6.7946			
273.69	3500		190			0.0000	2.9969	R	6.7946	0.0031	1198	74
273.99	1800			1657	110	0.0029	2.9998		6.7915	0.0041	1600	74
274.50	2700			613	D 11	0.0012	3.0010		6.7874	0.0022	847	A 70
274.72	1179	XING 1	69	1179	69	0.0022	3.0032	R JACOBS	6.7852	0.0018	660	70
274.99	1400			4200	69	0.0089	3.0121		6.7834	0.0038	1400	70
275.99	5287			2138	69	0.0046	3.0167		6.7796	0.0143	5287	70
275.99				359	D 82	0.0008	3.0175		6.7653			70
276.24	1300		74	1300	74	0.0033	3.0208	L a	6.7653	0.0035	1300	70
276.28	243			243	74	0.0006	3.0215	REVERSE CURVE	6.7618	0.0007	243	70
276.68	2100		81	2016	74	0.0052	3.0266	R	6.7611	0.0057	2100	70
276.68				84	D 74	0.0002	3.0268	REVERSE CURVE	6.7554			70
276.97	1500		70	1500	70	0.0041	3.0308	L	6.7554	0.0041	1500	70
277.15	1000	XING 2	70	1000	70	0.0027	3.0336	L 4TH STREET	6.7514	0.0027	1000	70
277.31	800			800	70	0.0022	3.0358	T'HUNTINGDON	6.7487	0.0022	800	70
277.50	1000		162	1000	70	0.0027	3.0385	R	6.7465	0.0027	1000	70
277.63	700			700	70	0.0019	3.0404		6.7438	0.0019	700	70
277.91	1500		89	1500	70	0.0041	3.0444	R c	6.7419	0.0041	1500	70
277.96	238			239	70	0.0006	3.0451		6.7378	0.0006	239	70
278.56	3197			3197	70	0.0086	3.0537		6.7372	0.0086	3197	70
278.81	1300		144	1300	70	0.0035	3.0572	R	6.7285	0.0043	1574	70
278.96	800		70	800	70	0.0022	3.0594	R a	6.7243	0.0006	239	70
279.11	800		70	464	70	0.0013	3.0607	R	6.7236	0.0029	1087	70
279.17	300			300	70	0.0008	3.0615	T'DEER	6.7207	0.0008	300	70
279.36	1000		176	1000	70	0.0027	3.0642	L'RC MPH 55	6.7199	0.0041	1533	70
279.53	900		144	900	70	0.0024	3.0666	L	6.7157	0.0028	991	A 65
279.83	1600			1600	70	0.0043	3.0709		6.7130	0.0026	976	65
279.96	670		131	670	70	0.0018	3.0727	R	6.7101	0.0020	670	65
280.01	277		131	277	70	0.0007	3.0735	R	6.7082	0.0008	277	65
280.96	5000			5000	70	0.0135	3.0870		6.7074	0.0146	5000	65
280.99	166			910	70	0.0025	3.0895		6.6925	0.0017	600	65
281.07	434			20	D 60	0.0001	3.0896		6.6910	0.0000		
281.07							3.0896	EAST END WARRIOR R	6.6910			
281.28	1100		69	1100	69	0.0030	3.0926	L	6.6910	0.0032	1100	65
281.47	977		92	977	69	0.0027	3.0953	L	6.6878	0.0028	977	65
281.73	1400	XING 206.6		1321	69	0.0036	3.0986	T'WARRIOR RIDGE	6.6850	0.0041	1400	65
281.73				79	D 81	0.0002	3.0991		6.6809			
281.96	1200		65	1200	65	0.0035	3.1026	R	6.6808	0.0035	1200	65
281.96							3.1026		6.6774			
282.20	1300		65	1300	65	0.0038	3.1064	R	6.6774	0.0038	1300	65
282.20						0.0000	3.1064		6.6736	0.0003	120	D 71
282.22	87			87	65	0.0003	3.1066	REVERSE CURVE	6.6733	0.0000		71
282.45	1200		71	1200	65	0.0035	3.1101	L	6.6733	0.0031	1167	71
282.45						0.0000	3.1101		6.6702	0.0006	223	D 81
282.68	1200		81	1200	65	0.0035	3.1136	L	6.6696	0.0023	977	81
282.68						0.0000	3.1136		6.6673	0.0002	97	D 85
282.96	1500			1500	65	0.0044	3.1180		6.6671	0.0031	1403	86
282.96							3.1180		6.6640			
283.18	1190			1190	65	0.0035	3.1215		6.6640	0.0027	1190	85
283.56	2000		124	2000	65	0.0058	3.1273	L PETERSBURG	6.6613	0.0045	2000	85
283.96	2100		85	2100	65	0.0061	3.1334	L	6.6569	0.0047	2100	85
283.96						0.0000	3.1334		6.6522	0.0006	284	D 95
284.13	900			900	65	0.0026	3.1360		6.6516	0.0002	118	95
284.49	1885		117	823	65	0.0024	3.1384	R	6.6514			
284.96	2500			1139	A 65	0.0030	3.1416		6.6514	0.0233	9114	A 59
284.96		XING 210.38		86	71	0.0002	3.1420	PRIVATE	6.6281	0.0000		
285.96	5089	XING 210.56		2346	A 71	0.0056	3.1476	PRIVATE	6.6281	0.0035	1078	59
286.87	4800			1469	A 81	0.0028	3.1506		6.6246	0.0154	4800	59
286.93	322		126	4175	85	0.0088	3.1599	R	6.6092	0.0066	2049	59
286.93				6786	A 80	0.0136	3.1735		6.6026			
287.31	2000	XING 2	128	63	103	0.0001	3.1737	R'BARREE	6.6026	0.0022	664	A 54
287.54	1200			1045	D 10	0.0024	3.1761		6.6004	0.0020	580	54
287.84	1600			1600	59	0.0051	3.1812		6.5984	0.0056	1600	54
287.92	415		59	415	59	0.0013	3.1825	L SOUTH TUNNEL	6.5928	0.0015	415	54
288.07	805		59	800	59	0.0026	3.1851	L SOUTH TUNNEL	6.5914	0.0028	800	54
288.19	600			664	59	0.0021	3.1873		6.5886	0.0023	664	54
288.60	2190			900	59	0.0029	3.1902		6.5862	0.0077	2230	54
288.76	810			220	59	0.0007	3.1909	TUNNEL	6.5785	0.0008	220	54
288.88	630			1400	59	0.0045	3.1954	R	6.5778	0.0049	1400	54
289.03	800			800	59	0.0026	3.1979	T'SPRUCE CREEK	6.5728	0.0028	800	54
289.22	1000		78	1000	59	0.0032	3.2011	R	6.5700	0.0035	1000	54
289.41	1000		84	1000	59	0.0032	3.2043	R	6.5665	0.0109	2120	54
289.52	590			1568	59	0.0050	3.2093	REVERSE CURVE	6.5556	0.0031	808	A 46
289.52				83	D 56	0.0003	3.2096		6.5525			
289.86	1800		54	1800	54	0.0063	3.2159	L	6.5525	0.0061	1480	46
289.90	200		143	200	54	0.0007	3.2166	L	6.5464	0.0011	250	A 43
290.29	2100			2100	54	0.0074	3.2240		6.5453	0.0000	8	43
290.54	1300		82	1300	54	0.0046	3.2286	L	6.5452	0.0057	1300	43
290.74	1077			1077	54	0.0038	3.2323		6.5395	0.0047	1077	43
290.88	700		103	700	54	0.0025	3.2348	R	6.5348	0.0031	700	43
291.08	1100		95	1100	54	0.0039	3.2386	R	6.5317	0.0117	2663	43
291.26	900		95	900	54	0.0032	3.2418	R	6.5200	0.0003	62	A 42
291.46	1062	XING 216.6		1062	54	0.0037	3.2455	PEMBERTON	6.5197	0.0016	347	42
291.66	1100		134	1100	54	0.0039	3.2494	L	6.5182	0.0050	1100	42
291.85	1000			1000	54	0.0035	3.2529		6.5132	0.0045	1000	42
292.04	1000			461	54	0.0016	3.2545		6.5087	0.0045	1000	42
292.04				160	A 54	0.0006	3.2551		6.5042	0.0000		
292.04				240	55	0.0008	3.2559		6.5042	0.0000		
292.04				133	D 55	0.0005	3.2564	UNION FURNACE	6.5042	0.0000		
292.29	1300		46	1300	46	0.0054	3.2618	L	6.5042	0.0059	1300	42

READ DOWN — WESTBOUND TRAFFIC | APPENDIX B4 — STATION | READ UP — EASTBOUND TRAFFIC

PAGE 6 COLUMN 1

MILES	DISTANCE ft	HWY GRADE CROSSING	MAX MPH	V DISTANCE ft	SPEED mph	dT, hours	TIME, hours	STATION	TIME, hours	dT, hours	DISTANCE ft	SPEED mph
292.38	455			414	46	0.0017	3.2635	TRC MPH 35	6.4060	0.0020	453	42
292.38				39	D 55	0.0002	3.2636		6.4053			42
292.60	1200		43	1200	43	0.0053	3.2689	R	6.4863	0.0054	1200	42
292.87	1400			1400	43	0.0062	3.2751		6.4609	0.0063	1400	42
293.30	2300			2300	45	0.0101	3.2852		6.4846	0.0104	2300	42
293.36	300		142	300	43	0.0013	3.2865	R	6.4742	0.0014	300	42
293.47	600			588	43	0.0026	3.2891	REVERSE CURVE	6.4728	0.0027	600	42
293.47				12	D 51	0.0001	3.2892		6.4701			
293.51	200		42	200	42	0.0009	3.2901	L	6.4701	0.0009	200	42
293.51						0.0000	3.2901		6.4692	0.0030		
293.81	1600		44	1600	42	0.0072	3.2973	R RC	6.4682	0.0072	1600	42
293.86	247			247	42	0.0011	3.2984		6.4620	0.0011	247	42
293.94	400			400	42	0.0016	3.3002		6.4609	0.0016	400	42
294.13	1000		93	1000	42	0.0045	3.3047	L	6.4591	0.0045	1000	42
294.39	1400			1400	42	0.0063	3.3110	T BIRMINGHAM	6.4546	0.0063	1400	42
294.86	2467		126	2467	42	0.0111	3.3222	L	6.4483	0.0111	2467	42
295.03	900		126	900	42	0.0041	3.3262	L	6.4371	0.0041	900	42
295.22	1000			1000	42	0.0045	3.3307		6.4331	0.0045	1000	42
295.39	900		44	900	42	0.0041	3.3348	R	6.4286	0.0041	900	42
295.45	300			300	42	0.0014	3.3361	REVERSE URVE	6.4240	0.0014	300	42
295.65	1086		42	798	42	0.0036	3.3397	L	6.4232	0.0049	1086	42
295.65							3.3397		6.4183	0.0002	52	D 48-
295.86	1100			1100	42	0.0006	3.3402		6.4180	0.0043	1048	46
296.01	800			800	42	0.0036	3.3436		6.4137	0.0033	800	46
296.20	1000		58	1000	42	0.0045	3.3484	L	6.4104	0.0041	1000	46
296.26	300			300	42	0.0014	3.3497	REVERSE CURVE	6.4063	0.0012	300	46
296.47	1100	XING32	58	1100	42	0.0050	3.3547	R PRIVATE	6.4051	0.0045	1100	46
296.62	800			800	42	0.0036	3.3583		6.4006	0.0033	800	46
296.86	1262		73	1262	42	0.0057	3.3640	L TT MPH 35	6.3973	0.0052	1262	46
297.02	860		46	860	42	0.0036	3.3679	L TYRONE	6.3921	0.0036	860	46
297.02		XING 222.16					3.3679	WILSON ROAD	6.3885	0.0012	348	D 67-
297.11	500			500	42	0.0023	3.3701	RC	6.3874	0.0004	152	67
297.34	1200		78	1200	42	0.0054	3.3755	R	6.3869	0.0034	1200	67
297.55	1100		66	1100	42	0.0050	3.3805	R	6.3835	0.0031	1100	67
297.66	600			600	42	0.0027	3.3832	RC	6.3804	0.0017	600	67
297.85	1000		67	378	42	0.0017	3.3848	L	6.3787	0.0028	1000	67
297.90	273		67	322	A 42	0.0014	3.3863	L	6.3759	0.0006	273	67
298.08	900			1772	46	0.0073	3.3936		6.3751	0.0003	123	D 73-
298.40	1700		130	1706	46	0.0070	3.4006	R	6.3748	0.0064	2477	73
298.85	2400			2400	46	0.0099	3.4105		6.3684	0.0062	2400	73
299.23	2000			1028	46	0.0042	3.4147	E END 3 TR	6.3621	0.0050	2000	73
299.59	1900		73	2575	A 46	0.0086	3.4233	L	6.3570	0.0049	1900	73
299.59		XING 224.6(		225	58	0.0006	3.4239	PRIVATE	6.3620	0.0001	22	D 74-
299.70	600	XING 225.0(		186	A 58	0.0006	3.4245	RC PRIVATE	6.3620	0.0015	578	74
299.85	782	XING 2	74	1977	57	0.0066	3.4301	R TIPTON ROAD	6.3505	0.0020	782	74
300.13	1473		74	1235	A 67	0.0033	3.4334	R	6.3485	0.0038	1473	74
300.85	3800			3820	73	0.0086	3.4420	T TIPTON	6.3447	0.0097	3800	74
301.85	5278			4345	73	0.0113	3.4533		6.3350	0.0135	5276	74
302.66	4300		74	216	A 73	0.0006	3.4539	L	6.3215	0.0110	4300	74
302.66		XING 227.9(		5017	74	0.0128	3.4667	PA RTE 502	6.3105	0.0002	69	D 78-
302.72	300			300	74	0.0008	3.4675	T FOSTORIA RC	6.3103	0.0005	211	78
302.79	358		78	358	74	0.0009	3.4684	RTT MPH 60	6.3097	0.0008	356	78
302.79						0.0000	3.4684		6.3089	0.0015	488	D 97-
303.25	2409		97	2409	74	0.0062	3.4746	R	6.3078	0.0038	1921	97
303.25							3.4746	WEST END WARRIOR	6.3041			
303.25						0.0000	3.4746		6.3041	0.0011	626	D 11!
303.78	2800			2800	74	0.0072	3.4817		6.3030	0.0007	462	117
304.78	5293	XING 230.7(		5133	74	0.0131	3.4949	PRIVATE	6.3022	0.0000		
305.31	2800			892	A 74	0.0022	3.4971		6.3022	0.0183	10336	A 95-
305.55	1276		130			0.0000	3.4971	L	6.2839	0.0015	742	96
305.78	1200			6179	A 78	0.0250	3.5221		6.2824	0.0024	1200	96
305.78				6670	A 97	0.0119	3.5340		6.2800	0.0000		96
306.77	5256			400	111	0.0007	3.5347		6.2800	0.0104	5258	96
307.49	3800			468	D 11	0.0008	3.5355	T ANTIS	6.2697	0.0075	3800	96
307.49							3.5355	EAST END POPLAR R	6.2622			
307.74	1300		96	1300	96	0.0026	3.5381	R	6.2622	0.0022	1119	96
307.78	202			202	96	0.0004	3.5385	T MPH 80(TT)	6.2600	0.0000		
307.80	108			108	96	0.0002	3.5387	REVERSE CURVE	6.2600	0.0106	4940	A 81-
308.29	2600		135	2600	96	0.0051	3.5438	L	6.2494			
308.76	2500			2500	96	0.0049	3.5488		6.2494	0.0015	661	81
308.80	200			200	96	0.0004	3.5492		6.2479	0.0005	200	81
309.28	2500		197	2500	96	0.0049	3.5541	R	6.2474	0.0056	2500	81
309.76	2540			2540	96	0.0050	3.5591		6.2418	0.0060	2540	81
310.36	3200			341	96	0.0007	3.5598		6.2358	0.0075	3200	81
310.50	700		230	3982	A 96	0.0075	3.5673	L	6.2281	0.0002	30	81
310.72	1200			122	105	0.0002	3.5675	T TT MPH 30	6.2278	0.0000		
310.72				668	D 1(	0.0013	3.5689		6.2279			
310.75	142		81	142	81	0.0000	3.5692	R ALTOONA	6.2279	0.0136	4958	A 55-
310.85	520		81	520	81	0.0012	3.5704	R	6.2143	0.0236	6840	55
311.76	4800			4800	81	0.0112	3.5816	T ALTO	6.1907	0.0042	1101	A 44-
312.48	3787			3787	81	0.0089	3.5905	T SLOPE	6.1866	0.0006	125	44
312.74	1400		110	1400	81	0.0030	3.5936	L BRICKYARD	6.1860	0.0060	1400	44
313.01	1400	XING 2	110	1013	81	0.0024	3.5961	L COBURN ROAD	6.1800	0.0011	251	A 41-
313.40	2100			1468	A 81	0.0033	3.5995	T TT MPH 35	6.1789	0.0059	1278	41
313.40				377	56	0.0008	3.6003		6.1730			
313.40				641	D 6(	0.0017	3.6020		6.1730			
313.69	1600		55	1500	56	0.0062	3.6072	R	6.1730	0.0069	1500	41
313.76	372			372	56	0.0013	3.6085		6.1661	0.0017	372	41
314.06	1574			1574	56	0.0054	3.6139		6.1643	0.0139	3000	41
314.25	1000		61	1000	56	0.0034	3.6173	R	6.1505	0.0014	337	A 38-
314.45	1100			1100	56	0.0038	3.6211	T TT MPH 35	6.1491	0.0018	337	36
314.63	900		60	900	56	0.0031	3.6242	L	6.1473	0.0047	900	38
314.76	700			700	56	0.0024	3.6266		6.1426	0.0037	700	38
314.91	800			640	56	0.0022	3.6288		6.1389	0.0042	800	36
314.91				160	D 55	0.0006	3.6294		6.1347	0.0000		
315.19	1500		44	1500	44	0.0065	3.6359	R	6.1347	0.0079	1500	38
315.25	276			239	44	0.0010	3.6369	T TT MPH 35 RC	6.1268	0.0015	276	36
315.25		UG 240.69		37	D 44	0.0002	3.6371	SCOTTS RUN	6.1254	0.0000		
315.47	1200		41	1200	41	0.0055	3.6426	L	6.1254	0.0063	1200	36
315.76	1500			1500	41	0.0069	3.6495		6.1190	0.0079	1500	36
315.91	800			800	41	0.0037	3.6533		6.1112	0.0042	800	36
316.06	800		134	800	41	0.0037	3.6569	L	6.1069	0.0042	800	36

READ DOWN — WESTBOUND TRAFFIC | APPENDIX B4 — STATION | READ UP — EASTBOUND TRAFFIC

PAGE 6 COLUMN 2

MILES	DISTANCE ft	HWY GRADE CROSSING	MAX MPH	V DISTANCE ft	SPEED mph	dT, hours	TIME, hours	STATION	TIME, hours	dT, hours	DISTANCE ft	SPEED mph
316.73	900			900	41	0.0042	3.6611		6.1027	0.0047	900	36
316.34	800		65	600	41	0.0026	3.6639	R	6.0980	0.0033	600	36
316.46	600		58	600	41	0.0028	3.6666	R	6.0946	0.0032	600	36
316.47	87			31	41	0.0001	3.6668	RC	6.0917	0.0005	87	36
316.47				56	D 41	0.0003	3.6671		6.0912	0.0000		
316.64	900		36	900	36	0.0047	3.6718	L	6.0912	0.0047	900	36
316.76	600		52	600	36	0.0030	3.6750	L HORSESHOE	6.0865	0.0032	600	36
316.93	900		36	900	36	0.0047	3.6797	L	6.0833	0.0047	900	36
316.93						0.0000	3.6797		6.0786	0.0006	222	D 53
316.96	182			182	36	0.0010	3.6807	RC	6.0777	0.0016	500	53
317.04	400		70	400	36	0.0021	3.6828	R	6.0759	0.0028	730	A 46
317.13	500			500	36	0.0026	3.6854		6.0731	0.0005	113	46
317.29	800		60	800	35	0.0042	3.6896	R	6.0726	0.0036	772	A 35
317.38	500		134	500	36	0.0026	3.6922	R	6.0690	0.0002	40	35
317.49	600		53	600	36	0.0032	3.6954	R	6.0688	0.0032	600	35
317.64	800		67	800	35	0.0042	3.6996	R	6.0656	0.0043	800	35
317.76	600			600	36	0.0032	3.7027		6.0612	0.0032	600	35
317.95	1000			1000	38	0.0053	3.7080		6.0580	0.0054	1000	35
318.12	900		95	900	38	0.0047	3.7127	L MG	6.0525	0.0049	900	35
318.43	1670			1670	36	0.0088	3.7215		6.0477	0.0090	1670	35
318.76	1700		96	1700	36	0.0089	3.7305	R	6.0386	0.0092	1700	35
318.96	1100		63	1100	35	0.0058	3.7363	R	6.0294	0.0060	1100	35
319.08	600			248	35	0.0013	3.7376		6.0235	0.0032	600	35
319.29	1100		100	720	A 36	0.0036	3.7412	R	6.0202	0.0060	1100	35
319.44	800		46	1486	48	0.0061	3.7473	R	6.0143	0.0043	800	35
319.45	85				46	0.0000	3.7473	REVERSE CURVE	6.0100	0.0005	85	36
319.45				131	D 46	0.0006	3.7479		6.0095			
319.64	1000		36	1000	35	0.0054	3.7533	L	6.0095	0.0054	1000	35
319.64						0.0000	3.7533		6.0041	0.0001	21	D 37
319.68	200			200	35	0.0011	3.7544	RC	6.0040	0.0009	179	37
319.78	500		58	500	35	0.0027	3.7571	R	6.0031	0.0026	500	37
319.87	500		58	500	35	0.0027	3.7598	R	6.0005	0.0026	500	37
319.90	140			140	35	0.0008	3.7606	T RC	5.9979	0.0008	140	37
320.09	1000		38	1000	35	0.0054	3.7660	L MPH 30(TT)	5.9972	0.0051	1000	37
320.20	600			600	35	0.0032	3.7693		5.9921	0.0031	600	37
320.34	700		96	700	35	0.0038	3.7731	R	5.9890	0.0036	700	37
320.49	800		56	800	35	0.0043	3.7774	R	5.9854	0.0041	800	37
320.64	800			800	35	0.0043	3.7817		5.9813	0.0041	800	37
320.71	400		66	400	35	0.0022	3.7839	R	5.9772	0.0020	400	37
320.75	200		55	200	35	0.0011	3.7850	R	5.9752	0.0010	200	37
320.79	211			211	35	0.0011	3.7861	RC	5.9741	0.0011	211	37
321.04	1300		37	1300	35	0.0070	3.7931	L	5.9731	0.0067	1300	37
321.04						0.0000	3.7931		5.9664	0.0008	196	D 52
321.13	500			500	35	0.0027	3.7958	T BENNY	5.9656	0.0011	304	52
321.23	500		190	500	35	0.0027	3.7986	L EB -2.27%	5.9646	0.0016	500	52
321.34	600			600	35	0.0032	3.8018	T(# 1)	5.9606	0.0022	600	52
321.76	2200		107	2040	35	0.0110	3.8128	R	5.9605	0.0060	2200	52
321.87	607		107	158	A 36	0.0008	3.8137	R	5.9624	0.0022	607	52
321.99	600		96	1208	37	0.0062	3.8199	R	5.9602	0.0022	600	52
322.16	900			900	37	0.0046	3.8245		5.9481	0.0033	900	52
322.76	3200			3200	37	0.0164	3.8408	T TUNNELS	5.9448	0.0117	3200	52
322.84	400			400	37	0.0020	3.8429	T TUNNELS	5.9331	0.0015	400	52
322.97	700			700	37	0.0036	3.8466	GALLITZEN	5.9317	0.0026	700	52
323.20	1200		50	793	37	0.0041	3.8505	L@UNT@AP E END 4	5.9291	0.0044	1200	52
323.20						0.0000	3.8505		5.9247	0.0004	131	D 60
323.43	1246			1169	A 37	0.0060	3.8566	EAST END 5 TR	5.9243	0.0035	1115	60
323.64	1100		96	1564	52	0.0057	3.8617	R	5.9208	0.0035	1100	60
323.98	1800		96	1800	52	0.0058	3.8675	R	5.9173	0.0057	1800	60
324.12	700			700	52	0.0025	3.8703		5.9116	0.0022	700	60
324.38	1400		60	1400	52	0.0051	3.8754	L	5.9094	0.0044	1400	60
324.38						0.0000	3.8754		5.9050	0.0027	1215	D 1!
324.63	1338			1338	52	0.0049	3.8803		5.9023	0.0004	228	109
325.64	5304			3416	52	0.0124	3.8928	WEST END 5 TRACKS	5.9019	0.0114	6146	A 95
326.09	2400			984	A 52	0.0033	3.8961	T CRESSON	5.8906	0.0029	1463	95
326.38	1536		190	4840	60	0.0153	3.9114	R	5.8876	0.0031	1536	95
326.61	1200			422	60	0.0013	3.9127		5.8845	0.0004	1200	95
327.62	5311	XING 253.3		9027	A 60	0.0217	3.9344	RISBINS ROAD	5.8821	0.0106	5311	95
327.90	1500					0.0000	3.9344		5.8715	0.0050	1500	95
328.31	2162		90	724	95	0.0014	3.9358	L	5.8665	0.0043	2162	95
328.60	1500			1500	95	0.0030	3.9388		5.8642	0.0030	1500	95
328.78	982			982	95	0.0019	3.9407		5.8613	0.0020	982	95
328.95	900		95	900	95	0.0018	3.9425	R ULLY	5.8593	0.0018	900	95
328.95						0.0000	3.9425		5.8575	0.0007	387	D 10
329.48	2800			2800	95	0.0056	3.9481		5.8568	0.0005	288	108
329.82	1794			1794	95	0.0036	3.9517		5.8563	0.0105	5069	A 95
330.20	2000		134	2000	95	0.0040	3.9557	R	5.8457	0.0086	4319	95
330.35	800			800	95	0.0016	3.9573	CASSANDRA	5.8371	0.0017	827	A 93
332.21	4619			3606	95	0.0072	3.9644		5.8354	0.0000	0	93
332.09	4631			5608	A 95	0.0106	3.9750		5.8354	0.0102	5054	93
332.24	800			308	106	0.0006	3.9755		5.8252	0.0018	800	93
332.24				387	D 16	0.0007	3.9763		5.8230	0.0000		
332.33	500		95	500	95	0.0010	3.9773	R	5.8236	0.0010	500	93
332.43	500			500	95	0.0010	3.9782	T PORTAGE	5.8226	0.0010	500	93
332.69	1400		110	1400	95	0.0028	3.9810	R	5.8215	0.0029	1400	93
332.95	1346		120	2691	95	0.0054	3.9864	R	5.8166	0.0027	1346	93
333.21	1400			65	D 95	0.0001	3.9865		5.8159	0.0028	1400	93
333.48	1400		93	1400	93	0.0029	3.9894	R	5.8130	0.0028	1400	93
333.74	1400			1400	93	0.0029	3.9922		5.8102	0.0001	55	D 95
333.82	426		96	426	93	0.0009	3.9931	L	5.8101	0.0035	1771	95
334.24	2200		95	2200	93	0.0045	3.9976	L	5.8066	0.0041	2039	95
334.74	2626			2625	93	0.0053	4.0029	WILMORE	5.8025	0.0000		
334.74							4.0029	WEST END POPLAR R	5.8025			
335.37	3100			3100	93	0.0063	4.0092		5.8025	0.0000		
335.66	1774		190	1246	93	0.0026	4.0118	R	5.8025	0.0253	10067	A 51
336.30	3361		190	827	A 93	0.0017	4.0134	R	5.7772	0.0036	936	51
336.58	1500					0.0000	4.0134		5.7737	0.0056	1500	51
336.71	686			11310	A 95	0.0200	4.0334		5.7682	0.0025	685	51
337.20	2600		208			0.0000	4.0334	L	5.7657	0.0097	2600	51
337.50	1600					0.0000	4.0334	SUMMERHILL	5.7560	0.0060	1600	51
337.99	2600			180	158	0.0002	4.0336		5.7501	0.0162	4363	51
337.99				103	D 11	0.0002	4.0338	E END MINERAL POIN	5.7339			
336.20	1100		116	54	115	0.0001	4.0338	R	5.7339	0.0006	221	A 49

APPENDIX B4

PAGE 7 COLUMN 1

READ DOWN — WESTBOUND TRAFFIC | STATION | READ UP — EASTBOUND TRAFFIC

MILES	DISTANCE ft	HWY GRADE CROSSING	MAX MPH	DISTANCE ft	SPEED mph	ET HOURS	TIME HOURS	STATION	TIME HOURS	ET HOURS	DISTANCE ft	SPEED mph
338.20				1558	C-11	0.0036	4.0374		5.7331			
338.39	1000		51	1000	51	0.0037	4.0411	R	5.7331	0.0004	116	49
338.45	300			300	51	0.0011	4.0422	REVERSE CURVE	5.7326	0.0012	300	46
338.50	248		56	248	51	0.0009	4.0432	L	5.7315	0.0010	248	49
338.61	600		58	600	51	0.0022	4.0454	L	5.7306	0.0036	887	46
338.76	800		61	800	51	0.0030	4.0484	L	5.7270	0.0022	536	A 43
339.14	2000			2000	51	0.0074	4.0558	TTTMPH 40	5.7248	0.0087	1964	43
339.49	1865		58	1865	51	0.0069	4.0627	R SOUTH FORK	5.7162	0.0087	1865	40
339.85	1871			1871	51	0.0069	4.0697		5.7076	0.0082	1871	43
340.00	800		92	800	51	0.0030	4.0726	L	5.6997	0.0035	800	43
340.15	800		90	800	51	0.0030	4.0756	L	5.6962	0.0035	800	43
340.30	800		82	800	51	0.0030	4.0786	L MPH 35	5.6927	0.0035	800	43
340.45	800			771	51	0.0029	4.0814		5.6891	0.0035	800	43
340.45				29	D 51	0.0001	4.0815		5.6856	0.0000		
340.49	200		40	200	42	0.0008	4.0823	R	5.6856	0.0008	200	43
340.64	800		48	800	48	0.0031	4.0854	R	5.6847	0.0035	800	40
340.85	1100		42	1018	49	0.0039	4.0893	R REV CURVE	5.6812	0.0046	1100	43
340.85				81	D 46	0.0003	4.0897		5.6764			
341.02	900		43	900	43	0.0040	4.0936	L MPH 30	5.6764	0.0046	900	43
341.02						0.0000	4.0936		5.6724	0.0063	85	D 46
341.19	900		82	900	43	0.0040	4.0976	L	5.6721	0.0032	819	49
341.23	199			199	43	0.0009	4.0985	RC	5.6689	0.0006	196	49
341.34	600		50	600	43	0.0026	4.1011	R	5.6681	0.0050	1570	49
341.49	800		81	800	43	0.0036	4.1048	R	5.6632	0.0005	101	A 42
341.66	870			870	43	0.0038	4.1086		5.6603	0.0012	269	42
341.74	400		78	400	43	0.0018	4.1102	L	5.6590	0.0018	400	42
341.89	800		85	800	43	0.0036	4.1138	L	5.6572	0.0036	800	42
342.02	700		72	700	43	0.0031	4.1168	L MINERAL POINT	5.6536	0.0032	700	42
342.15	700			700	43	0.0031	4.1199		5.6504	0.0032	700	42
342.27	600		50	600	43	0.0026	4.1226	R RC	5.6470	0.0027	600	42
342.34	400			400	43	0.0018	4.1243	TTTMPH 35	5.6446	0.0018	400	42
342.49	800		64	800	43	0.0036	4.1279	L	5.6427	0.0036	800	42
342.63	700		84	700	43	0.0031	4.1309	L	5.6391	0.0032	700	42
342.89	1400		61	1400	43	0.0062	4.1371	L	5.6360	0.0063	1400	42
342.93	200			200	43	0.0009	4.1380	REVERSE CURVE	5.6297	0.0009	200	42
343.08	800		49	800	43	0.0035	4.1415	R	5.6288	0.0036	800	42
343.14	300			191	43	0.0006	4.1421	RC	5.6252	0.0014	300	42
343.44	1600		53	509	A 43	0.0022	4.1443	L	5.6238	0.0072	1600	42
343.49	280			1516	49	0.0059	4.1502		5.6186	0.0013	288	42
343.53	215			215	49	0.0008	4.1510	[illegible]	5.6150	0.0010	215	42
343.69	800		66	707	40	0.0027	4.1537	R	5.6143	0.0036	800	42
343.69				93	D 46	0.0004	4.1541		5.6107	0.0000		
343.84	800		47	800	42	0.0036	4.1577	R	5.6107	0.0036	800	42
343.91	400			400	42	0.0018	4.1595	RC	5.6071	0.0018	400	42
344.05	700		45	700	42	0.0032	4.1627	L	5.6053	0.0032	700	42
344.18	700		42	700	42	0.0032	4.1658	L	5.6022	0.0032	700	42
344.18						0.0000	4.1658		5.5990	0.0001	25	D 44
344.20	100			100	42	0.0005	4.1663	RC	5.5989	0.0008	75	44
344.37	900		44	900	42	0.0041	4.1704	R	5.5986	0.0039	900	44
344.50	700		53	700	42	0.0032	4.1736	R	5.5947	0.0016	128	D 55
344.57	377		53	377	42	0.0017	4.1752	R	5.5942	0.0034	949	53
344.57						0.0000	4.1752	AC	5.5908	0.0002	48	D 56
344.82	1300			1300	42	0.0059	4.1811		5.5906	0.0042	1252	56
345.05	1200		56	1200	42	0.0054	4.1865	R	5.5864	0.0041	1200	58
345.05						0.0000	4.1865		5.5823	0.0002	51	D 59
345.27	1200			1200	42	0.0064	4.1919		5.5822	0.0031	1149	59
345.50	1200		56	1200	42	0.0054	4.1973	L	5.5786	0.0039	1200	58
345.69	1000		76	1000	42	0.0045	4.2018	L	5.5746	0.0032	1000	59
345.95	1400		90	1400	42	0.0063	4.2081	L	5.5714	0.0040	1400	59
345.99	200			200	42	0.0009	4.2090	RC	5.5660	0.0006	200	59
346.18	1000		58	1000	42	0.0046	4.2136	R	5.5660	0.0032	1000	59
346.18						0.0000	4.2136		5.5631	0.0006	186	D 60
346.23	246		66	246	42	0.0011	4.2147	R	5.5625	0.0002	58	69
346.34	930		100	577	42	0.0008	4.2165	R	5.5624	0.0016	600	69
346.49	800			193	A 42	0.0008	4.2193		5.5607	0.0020	800	69
346.89	2120			810	44	0.0035	4.2198	T CONEMAUGH TOWE	5.5585	0.0037	283	69
346.99	500		107	484	A 44	0.0016	4.2217	R	5.5578	0.0000		
346.99							4.2217	WEST END MINERAL	5.5578			
347.01	100			2459	50	0.0093	4.2310	RC	5.5578	0.0081	2564	A 50
347.24	1200		166	1200	50	0.0045	4.2356	L	5.5497	0.0036	1080	52
347.46	1200			1200	50	0.0045	4.2401		5.5458	0.0044	1200	52
347.55	478		131	478	50	0.0018	4.2419	L	5.5415	0.0017	476	52
347.61	300			300	50	0.0011	4.2431	RC	5.5398	0.0011	300	52
347.74	700		137	700	50	0.0027	4.2457	R	5.5387	0.0025	700	52
347.74						0.0000	4.2457	EAST END JOHNSTOWN	5.5361			
347.91	900		85	900	50	0.0034	4.2491	R	5.5361	0.0033	900	52
348.12	1100		72	868	50	0.0025	4.2516	R	5.5328	0.0040	1100	52
348.33	1000			1121	A 50	0.0039	4.2555		5.5288	0.0036	1000	52
348.50	1800		69	79	50	0.0003	4.2558	R	5.5252	0.0036	1000	52
348.70	1061		75	1607	A 59	0.0050	4.2608	R	5.5218	0.0038	1061	52
348.85	800			1456	60	0.0038	4.2646	T JOHNSTOWN	5.5177	0.0029	800	52
348.97	600		204	600	68	0.0016	4.2663	R	5.5148	0.0022	600	52
349.12	800			486	68	0.0014	4.2676	W END 3 TR	5.5126	0.0016	434	52
349.12				302	D 68	0.0009	4.2686	E END 1 TR SANG HO	5.5110			
349.16	200		52	200	52	0.0007	4.2693	R	5.5110	0.0047	1130	A 38
349.35	1000			1000	52	0.0036	4.2729		5.5063	0.0022	433	38
349.53	1000		80	1000	52	0.0036	4.2765	L	5.5041	0.0050	1000	38
349.62	438		181	438	52	0.0016	4.2780	L	5.4991	0.0022	438	38
350.00	2000			2000	52	0.0073	4.2853		5.4970	0.0100	2000	38
350.15	800		105	800	52	0.0029	4.2884	R	5.4870	0.0040	800	38
350.53	2000		86	2000	52	0.0073	4.2957	R	5.4830	0.0100	2000	38
350.82	1529		64	1529	52	0.0056	4.3012	R	5.4730	0.0076	1529	38
350.97	800			800	52	0.0029	4.3041		5.4654	0.0040	800	38
351.19	1200		57	1200	52	0.0044	4.3085	R	5.4614	0.0060	1200	38
351.35	800			230	52	0.0008	4.3094	DORNOCK	5.4555	0.0040	800	38
351.35				233	A 52	0.0008	4.3102		5.4515			
351.35				118	54	0.0004	4.3106		5.4515			
351.35				218	D 54	0.0008	4.3115		5.4515			
351.54	1000		38	1000	38	0.0050	4.3165	L	5.4515	0.0030	1000	38
351.70	889		38	889	38	0.0044	4.3209	L	5.4486	0.0044	889	38
351.70						0.0000	4.3209		5.4421	0.0002	47	D 40
351.74	200			200	38	0.0010	4.3219	RC	5.4419	0.0007	150	42

PAGE 7 COLUMN 2

READ DOWN — WESTBOUND TRAFFIC | STATION | READ UP — EASTBOUND TRAFFIC

MILES	DISTANCE ft	HWY GRADE CROSSING	MAX MPH	DISTANCE ft	SPEED mph	ET HOURS	TIME HOURS	STATION	TIME HOURS	ET HOURS	DISTANCE ft	SPEED mph
351.93	1000		42	1000	38	0.0060	4.3269	R	5.4411	0.0040	1000	42
351.93						0.0000	4.3269		5.4366	0.0043	361	D 42
352.10	900		100	900	38	0.0045	4.3314	R	5.4364	0.0010	300	46
352.29	1000		46	1000	36	0.0050	4.3364	R	5.4336	0.0020	1000	56
352.29						0.0000	4.3364		5.4309	0.0002	78	D 60
352.33	200			200	38	0.0010	4.3374	REVERSE CURVE	5.4308	0.0008	121	69
352.48	800		75	800	38	0.0040	4.3414	L	5.4300	0.0022	800	69
352.50	100			100	36	0.0005	4.3419		5.4251	0.0008	100	69
352.80	1800			1800	38	0.0080	4.3499		5.4276	0.0044	1800	69
353.11	1600		105	1600	38	0.0080	4.3578	R	5.4235	0.0044	1600	69
353.26	800			800	38	0.0040	4.3618		5.4191	0.0022	800	69
353.37	600		181	600	38	0.0030	4.3648	R	5.4169	0.0016	600	69
353.45	400		74	400	38	0.0020	4.3668	R	5.4152	0.0011	400	69
353.50	282			282	38	0.0014	4.3682	RC	5.4141	0.0008	282	69
353.63	700		89	700	38	0.0035	4.3717	L	5.4133	0.0019	700	68
353.63						0.0000	4.3717		5.4114	0.0016	750	D 38
354.39	4000			858	38	0.0041	4.3757	RC	5.4098	0.0008	441	56
354.50	586		150	205	A 38	0.0011	4.3769	L	5.4090	0.0103	4547	A 64
354.55	257		196	257	40	0.0012	4.3780	L	5.3987	0.0072	2962	A 72
355.49	5000			708	42	0.0032	4.3812		5.3915	0.0020	743	72
355.64	800			2705	A 42	0.0084	4.3896		5.3896	0.0021	800	72
356.40	4000		64	787	A 65	0.0022	4.3929	L a	5.3874	0.0105	4000	72
356.44	250			5480	69	0.0151	4.4076	RC SANG HOLLOW	5.3769	0.0005	200	72
356.48	237		86	3604	A 89	0.0096	4.4198	R b	5.3764	0.0006	237	72
356.49	20		86	478	84	0.0011	4.4179	R	5.3757	0.0001	20	72
356.49							4.4179	WEST END JOHNSTOW	5.3757			
357.25	4000			4000	84	0.0090	4.4269		5.3757	0.0106	4000	72
357.44	1000		64	725	84	0.0018	4.4288	L	5.3650	0.0026	1000	72
357.44				275	D 87	0.0007	4.4292	SEWARD	5.3625			
358.00	2966		72	2966	72	0.0079	4.4371	L	5.3625	0.0079	2966	72
358.00						0.0000	4.4371	E END 2 TR SANG HO	5.3546	0.0027	1352	D 120
358.38	2000			2000	72	0.0053	4.4424		5.3520	0.0016	848	120
358.42	214			214	72	0.0006	4.4429	L	5.3510	0.0003	214	120
359.37	5000		150	5000	72	0.0132	4.4561	R	5.3506	0.0079	5000	120
359.80	2265		150	2265	72	0.0060	4.4620	R	5.3427	0.0036	303	120
360.37	3000			1521	72	0.0040	4.4660	W END 2 TR SANG HO	5.3402	0.0000		
360.77	2139			13584	A 72	0.0276	4.4936	T NEW FLORENCE	5.3422	0.0000		
361.34	3000		120			0.0000	4.4936	R	5.3422	0.0000		
361.72	2000		120			0.0000	4.4936	R	5.3422	0.0000		
362.10	2000					0.0000	4.4936		5.3422	0.0000		
362.25	820		150			0.0000	4.4936	L	5.3422	0.0000		
362.41	824		120			0.0000	4.4936	L	5.3422	0.0000		
363.17	4000					0.0000	4.4936		5.3422	0.0219	12532	A 110
363.17				5743	A 110	0.0094	4.5033		5.3203	0.0000		
363.37	1056				120	0.0003	4.5036	T CONPIT JCT	5.3203	0.0131	4936	A 76
363.42	300	XING 296.8			120	0.0000	4.5033	PRIVATE	5.3072	0.0002	47	76
363.50	400		152	400	120	0.0006	4.5039	R REVERSE CURVE	5.3070	0.0059	3188	A 59
363.57	400		152	400	120	0.0006	4.5045	L MPH 50	5.2981	0.0005	169	59
364.09	2720			223	120	0.0004	4.5049		5.2978	0.0346	10768	59
364.09				780	D 120	0.0014	4.5063		5.2690			
364.09							4.5062	EAST END TORRANCE	5.2631			
365.01	4837		96	4065	56	0.0087	4.5150	R	5.2630	0.0022	564	A 54
365.01				477	D 96	0.0011	4.5160		5.2608			
365.57	3000		76	2936	76	0.0073	4.5233	R	5.2608	0.0065	1861	54
365.63	275				76	0.0010	4.5243	RC	5.2543	0.0010	275	54
365.63				337	D 76	0.0009	4.5252	INDIA	5.2533			
365.91	1500		99	1500	59	0.0046	4.5291	L	5.2533	0.0033	730	A 52
366.29	2000		60	2000	59	0.0064	4.5355	L LOCKPORT	5.2505	0.0019	631	52
366.61	1700			1700	59	0.0055	4.5410		5.2506	0.0062	1700	52
366.92	1619		89	1619	59	0.0052	4.5462	L	5.2444	0.0059	1619	52
367.28	1900		80	1900	59	0.0061	4.5523	L BOLIVAR	5.2385	0.0064	1900	52
367.43	805			801	59	0.0026	4.5548		5.2316	0.0029	800	52
367.60	900		78	900	59	0.0029	4.5577	R	5.2286	0.0033	900	52
367.82	1131		89	1098	59	0.0035	4.5613	R	5.2254	0.0043	1131	52
367.82				33	D 59	0.0003	4.5615		5.2211			52
368.16	1803		54	1803	54	0.0063	4.5678	R	5.2211	0.0066	1800	52
368.27	599			536	54	0.0019	4.5697	RC	5.2145	0.0021	568	52
368.27				31	D 54	0.0001	4.5698		5.2124			
368.82	2900		92	2900	52	0.0106	4.5804	L	5.2124	0.0106	2900	52
368.90	405		52	400	52	0.0015	4.5819	L	5.2019	0.0015	400	52
368.90						0.0000	4.5819		5.2004	0.0009	282	D 58
369.02	631			631	52	0.0023	4.5842		5.1995	0.0012	340	56
369.26	1300		35	1300	52	0.0047	4.5889	R	5.1985	0.0036	1300	58
369.45	1000			1000	52	0.0036	4.5926		5.1949	0.0026	1000	65
369.75	1600		134	1600	52	0.0058	4.5984	R	5.1921	0.0027	976	68
369.80	264		134	264	52	0.0010	4.5993	R	5.1894	0.0031	1051	A 60
370.34	2800			2800	52	0.0102	4.6095		5.1864	0.0081	2837	62
370.66	1700		72	1700	52	0.0062	4.6157	R	5.1783	0.0052	1700	62
370.75	500		87	500	52	0.0018	4.6175	R	5.1731	0.0016	500	62
370.90	800			800	52	0.0029	4.6204		5.1716	0.0024	800	62
371.45	2900		68	405	52	0.0015	4.6219	R	5.1691	0.0088	2900	62
371.66	1119			2360	A 52	0.0075	4.6294		5.1603	0.0034	1119	62
371.76	500		88	1771	68	0.0048	4.6342	R	5.1586	0.0015	500	62
371.81	244		86	244	68	0.0007	4.6349	R TORRANCE	5.1563	0.0007	244	62
371.92	600			486	68	0.0014	4.6362	E END 3 TR [illegible]	5.1546	0.0018	600	62
371.92				114	D 68	0.0003	4.6366	REVERSE CURVE	5.1528	0.0000		
372.30	2000		82	2000	62	0.0091	4.6407	L	5.1528	0.0061	2000	62
372.53	1200		95	1200	62	0.0037	4.6464	L	5.1467	0.0037	1200	62
372.66	700		88	700	62	0.0021	4.6485	L	5.1430	0.0021	700	62
372.75	500		72	500	62	0.0015	4.6500	L	5.1408	0.0015	500	62
372.75						0.0000	4.6500		5.1393	0.0000		
372.84	438		72	438	62	0.0013	4.6514	L	5.1393	0.0013	498	62
372.84							4.6514	WEST END TORRANCE	5.1380			
373.59	4000			3701	62	0.0113	4.6627		5.1380	0.0122	4000	62
373.73	700		82	3798	62	0.0116	4.6743	R a	5.1258	0.0021	700	62
373.84	600		82	600	62	0.0018	4.6761	R	5.1236	0.0018	600	62
373.84						0.0000	4.6781	GRAYS	5.1218	0.0006	260	D 76
373.87	177			177	62	0.0005	4.6786	RC	5.1210	0.0000		76
374.40	2800		75	2800	62	0.0066	4.6852	L b	5.1210	0.0067	1634	76
374.40		XING 301.6				0.0000	4.6852	[illegible] PA	5.1143	0.0011	477	D 96
374.71	1800			1600	62	0.0049	4.6901		5.1133	0.0022	1125	96
374.72	100	XING [illegible]	95	100	62	0.0003	4.6904	R PRIVATE c	5.1110	0.0002	100	95

READ DOWN — WESTBOUND TRAFFIC | APPENDIX B4 — STATION | READ UP — EASTBOUND TRAFFIC

PAGE 8 COLUMN 1

MILES	DISTANCE, ft	HWY GRADE CROSSING	MAX MPH	DISTANCE, ft	SPEED, mph	dT, hours	TIME, hours	STATION	TIME, hours	dT, hours	DISTANCE, ft	SPEED, mph
374.97	1269	XING 3-	95	1269	62	0.0039	4.6943	R PRIVATE	5.1106	0.0036	1269	95
375.73	4000	XING 303.6		4000	62	0.0122	4.7065	T 888 PA	5.1080	0.0080	4000	95
376.73	5278			1604	62	0.0032	4.7097		5.1009	0.0106	5278	95
377.29	3000			2743	A 62	0.0075	4.7172		5.0897	0.0060	3000	95
377.32	162		164	234	76	0.0006	4.7178	L	5.0838	0.0003	162	95
377.65	1700			5806	A 76	0.0131	4.7308		5.0834	0.0034	1700	95
377.72	400		95	0		0.0000	4.7308	R	5.0801	0.0008	400	95
377.91	1000		95			0.0000	4.7308	R MILLWOOD	5.0793	0.0020	1000	95
377.91						0.0000	4.7308		5.0773	0.0011	582	D 114
378.10	1000			0		0.0000	4.7308		5.0762	0.0006	320	114
378.23	677		134	480	95	0.0010	4.7318	L	5.0757			
378.36	700		190	700	95	0.0014	4.7332	L	5.0757			
378.72	1900			1900	95	0.0038	4.7370		5.0757			
378.81	483			483	95	0.0010	4.7379		5.0757			
378.95	700		134	700	95	0.0014	4.7393	R	5.0757			
379.72	4100			4100	95	0.0082	4.7475		5.0757	0.0161	8926	A 95
380.27	2887			1440	95	0.0029	4.7504	T DERRY	5.0596	0.0052	2813	95
380.72	2400	XING 3-	134	6789	A 95	0.0125	4.7629	R VALLEY STREET	5.0543	0.0049	2400	95
380.78	287		134			0.0000	4.7629	R	5.0496	0.0006	287	95
381.73	5000			2137	A 11	0.0036	4.7665		5.0490	0.0100	5000	95
381.86	700			326	114	0.0005	4.7670		5.0390	0.0014	700	95
381.86				567	D 11	0.0011	4.7681		5.0376			
382.30	2358		95	2359	95	0.0047	4.7728	R BRADENVILLE	5.0376	0.0047	2358	95
382.30						0.0000	4.7728		5.0329	0.0014	788	D 12
382.61	1600			1600	95	0.0032	4.7760		5.0315	0.0013	812	120
383.42	4300			4300	95	0.0086	4.7846		5.0303	0.0068	4000	120
383.61	989		134	989	95	0.0020	4.7866	L	5.0235	0.0016	989	120
384.61	5266		134	4111	95	0.0062	4.7947	L	5.0219	0.0083	5266	120
384.72	600		134	0		0.0000	4.7947	L	5.0136	0.0009	600	120
385.48	4000			8853	A 95	0.0219	4.8167	T LATROBE	5.0126	0.0063	4000	120
385.61	700		268			0.0000	4.8167	R	5.0063	0.0011	700	120
386.58	5112		268	1714	120	0.0027	4.8194	R	5.0052	0.0081	5112	120
387.00	2200		140	2200	120	0.0035	4.8228	R BEATTY	4.9972	0.0035	2200	120
387.42	2225		120	2225	120	0.0035	4.8264	R	4.9937	0.0035	2225	120
387.45	177		120	177	120	0.0000	4.8264	R	4.9902	0.0000	1886	120
388.42	5100			5100	120	0.0080	4.8347		4.9872	0.0094	6743	A 110
389.42	5278			5278	120	0.0083	4.8430		4.9778	0.0050	2926	110
389.68	1400			1400	120	0.0022	4.8452		4.9727	0.0024	1400	110
390.07	2074		190	2074	120	0.0033	4.8485	L	4.9703	0.0036	2074	110
390.41	1600			4263	120	0.0067	4.8552	DONOHOE	4.9667	0.0031	1600	110
390.94	2800			387	D 11	0.0036	4.8568		4.9636	0.0048	2800	110
391.20	1339		110	1339	110	0.0023	4.8591	L a	4.9588	0.0012	708	110
391.63	2300		110	2300	110	0.0040	4.8620	L	4.9576	0.0000		
391.76	682			682	110	0.0012	4.8632		4.9576			
392.10	1800		110	1800	110	0.0031	4.8663	R b	4.9576			
392.18	400			400	110	0.0007	4.8670		4.9576			
392.29	600			600	110	0.0010	4.8680		4.9576	0.0125	6789	A 95
392.56	1500		222	1500	110	0.0026	4.8706	L	4.9451	0.0022	1124	95
392.84	1387		328	1387	110	0.0024	4.8730	L	4.9429	0.0028	1387	95
393.18	1800			1800	110	0.0031	4.8761		4.9401	0.0036	1800	95
394.07	4700			4700	110	0.0081	4.8842	T GREENSBURG	4.9365	0.0094	4700	95
394.18	579		190	579	110	0.0010	4.8852	R	4.9271	0.0012	579	95
394.45	1400		190	949	110	0.0016	4.8869	R	4.9260	0.0036	1813	95
394.45				451	D 11	0.0008	4.8877		4.9223	0.0017	827	A 93
394.75	1584		95	1584	95	0.0032	4.8908	R	4.9207	0.0007	354	93
395.19	2300			2300	95	0.0046	4.8954		4.9200	0.0047	2300	93
395.51	1700		190	1700	95	0.0034	4.8988	L	4.9153	0.0035	1700	93
396.11	3200			3200	95	0.0064	4.9052		4.9116	0.0066	3200	93
396.18	335		268	335	95	0.0007	4.9059	L	4.9053	0.0007	335	93
396.24	311		268	311	95	0.0006	4.9065	L	4.9046	0.0006	311	93
396.77	2800			2745	95	0.0055	4.9120		4.9040	0.0057	2600	93
396.77				55	D 95	0.0001	4.9121		4.8983	0.0003	130	93
396.94	900		93	900	93	0.0018	4.9139	R	4.8980	0.0103	4847	A 78
396.96	100			100	93	0.0002	4.9141		4.8877	0.0066	2724	78
397.96	5305			5306	93	0.0108	4.9249	T RADEBAUGH	4.8811	0.0067	2538	A 66
398.28	1700			1700	93	0.0036	4.9284		4.8744	0.0112	3908	90
398.96	3686		134	3686	93	0.0073	4.9357	L JEANNETTE	4.8632	0.0008	182	A 65
399.45	2800		134	308	93	0.0006	4.9363	L	4.8627	0.0001	46	65
399.76	1562			2633	A 93	0.0056	4.9419		4.8625	0.0046	1562	65
399.76				476	100	0.0009	4.9428		4.8579	0.0000		
399.76				574	D 10	0.0012	4.9440		4.8579	0.0000		
399.96	1100		78	1100	78	0.0027	4.9467	R	4.8579	0.0032	1100	65
400.07	562		78	562	78	0.0014	4.9480	R	4.8547	0.0016	562	65
400.75	3600	XING 309.6		3347	78	0.0081	4.9562	WRIGLEYS	4.8531	0.0105	3600	65
400.75				293	D 71	0.0007	4.9568		4.8426	0.0000		
400.94	1000		86	1000	86	0.0029	4.9597	L a	4.8426	0.0029	1000	65
400.96	100			100	86	0.0003	4.9600		4.8397	0.0003	100	65
401.11	800			800	86	0.0023	4.9623		4.8394	0.0023	800	65
401.28	900		86	900	86	0.0026	4.9649	R b	4.8370	0.0026	900	65
401.53	1300			1281	86	0.0037	4.9685	T MANOR	4.8344	0.0038	1300	65
401.53				19	D 86	0.0001	4.9686		4.8306	0.0000		
401.74	1105		65	1105	65	0.0032	4.9718	L c	4.8306	0.0032	1105	65
401.74							4.9718		4.8274	0.0017	704	D 95
401.97	1200			1200	65	0.0035	4.9753		4.8257	0.0010	496	95
402.23	1400			1400	65	0.0041	4.9794	T SHAFTON	4.8247	0.0026	1400	95
402.95	3810		95	3810	65	0.0111	4.9905	R	4.8220	0.0076	3819	95
402.95							4.9905		4.8144	0.0011	816	D 115
403.85	4720		161	4690	65	0.0134	5.0038	R IRWIN	4.8132	0.0086	4104	115
403.92	378		161			0.0000	5.0038	R	4.8066	0.0006	378	115
404.77	4500			8204	A 65	0.0192	5.0230		4.8059	0.0036	2169	115
404.82	285					0.0000	5.0230		4.8023	0.0000		
405.09	1400		134			0.0000	5.0230	R LARIMER	4.8023	0.0000		
405.77	3600			9430	A 95	0.0169	5.0400		4.8023	0.0000		
405.82	285			0		0.0000	5.0400		4.8023	0.0000		
405.99	900		115			0.0000	5.0400	L ARDARA	4.8023	0.0000		
406.11	600		134		115	0.0000	5.0400	L ARA	4.8023	0.0222	11983	A 88
406.77	3500			3500	115	0.0058	5.0457		4.7801	0.0020	918	88
407.15	2000			1342	115	0.0022	5.0480		4.7781	0.0043	2000	88
407.38	1229		134				5.0480	R	4.7738			
407.38	EAST END TRAFFORD - WALL						5.0480		4.7738			
407.67	1520		134	1880	A 11	0.0031	5.0510	R	4.7738	0.0059	2746	88
407.75	433		134			0.0000	5.0510	R GP TRAFF	4.7679	0.0025	1153	88

READ DOWN — WESTBOUND TRAFFIC | APPENDIX B4 — STATION | READ UP — EASTBOUND TRAFFIC

PAGE 8 COLUMN 2

MILES	DISTANCE, ft	HWY GRADE CROSSING	MAX MPH	DISTANCE, ft	SPEED, mph	dT, hours	TIME, hours	STATION	TIME, hours	dT, hours	DISTANCE, ft	SPEED, mph
407.75	433		134			0.0000	5.0510	R GP TRAFF	4.7679	0.0025	1153	88
408.68	4900			396	118	0.0005	5.0516	TRAFFORD	4.7654	0.0164	8474	A 60
408.68				906	D 11	0.0017	5.0533		4.7490			
409.19	2700		68	2700	88	0.0058	5.0591	L	4.7490	0.0060	406	60
409.71	2743		69	2740	88	0.0059	5.0650	L	4.7430	0.0187	5915	60
409.96	1300		98	1300	88	0.0028	5.0678	L	4.7244	0.0036	1060	A 51
410.39	2300			2300	88	0.0050	5.0728	PITCAIRN YARD	4.7208	0.0060	1808	51
410.68	1500		95	1500	88	0.0032	5.0760	R WALL	4.7148	0.0025	632	A 44
410.93	1351			1351	88	0.0029	5.0789		4.7123	0.0012	300	44
411.20	1400			792	88	0.0017	5.0806		4.7110	0.0050	1400	44
411.20				808	D 86	0.0016	5.0822		4.7060	0.0006	190	A 42
411.52	1700		60	1700	60	0.0054	5.0875	R	4.7042	0.0064	1196	42
411.52	WEST END TRAFFORD WALL						5.0875		4.6968			
411.73	1093			1093	60	0.0034	5.0910	WILMERDING	4.6968	0.0049	1093	42
411.73							5.0910	EAST END WILMERDI	4.6938			
411.87	748		81	748	60	0.0024	5.0933	L PORT PERRY BR	4.6938	0.0034	748	42
411.93	300			300	60	0.0009	5.0943		4.6905	0.0014	300	42
412.17	1300			1153	60	0.0036	5.0979		4.6891	0.0069	1300	42
412.17				147	D 80	0.0005	5.0984		4.6852	0.0000		42
412.38	1080		51	1080	51	0.0040	5.1024	L	4.6832	0.0049	1080	42
412.45	400		51	400	51	0.0015	5.1039	L	4.6764	0.0018	400	42
412.87	2200			2200	51	0.0082	5.1121		4.6786	0.0099	2200	42
412.94	400		85	400	51	0.0015	5.1136	L	4.6687	0.0018	400	42
413.02	400			302	51	0.0011	5.1147		4.6649	0.0018	400	42
413.02				98	D 51	0.0004	5.1151		4.6630	0.0000		42
413.10	400		44	400	44	0.0017	5.1168	L	4.6630	0.0018	400	42
413.15	300			300	44	0.0013	5.1181	RC	4.6612	0.0014	300	42
413.29	700		44	700	44	0.0030	5.1211	R	4.6599	0.0032	700	42
413.38	480		51	480	44	0.0020	5.1231	R	4.6567	0.0022	480	42
413.38				20	D 44	0.0001	5.1232		4.6546	0.0000		42
413.49	600		42	600	42	0.0027	5.1259	R	4.6546	0.0027	600	42
413.59	500			500	42	0.0022	5.1281	RC	4.6519	0.0023	500	42
413.68	480		46	480	42	0.0021	5.1302	L	4.6496	0.0022	480	42
413.81	700			700	42	0.0031	5.1334		4.6474	0.0032	700	42
413.90	500		60	500	42	0.0022	5.1356	L	4.6443	0.0023	500	42
414.38	2500			2500	42	0.0112	5.1468	PT PERRY TUNNEL	4.6420	0.0113	2500	42
414.48	524		34	524	42	0.0023	5.1491	R PORT PERRY	4.6308	0.0024	524	42
414.48						0.0000	5.1491		4.6284	0.0003	65	D 47
414.51	200			200	42	0.0009	5.1500	REVERSE CURVE	4.6281	0.0005	135	47
414.59	400		47	400	42	0.0018	5.1518	L	4.6276	0.0016	400	47
414.59						0.0000	5.1518		4.6260	0.0018	523	D 76
414.80	1100			1100	42	0.0049	5.1567		4.6244	0.0014	577	76
414.95	800		134	800	42	0.0036	5.1603	R	4.6228	0.0020	800	76
415.08	700		81	700	42	0.0031	5.1634	R	4.6209	0.0017	700	76
415.24	858		120	858	42	0.0038	5.1672	R	4.6192	0.0021	858	76
415.45	1100			1100	42	0.0049	5.1722		4.6170	0.0027	1100	76
415.47	100		115	100	42	0.0004	5.1726	R RANKIN BRIDGE	4.6143	0.0002	100	76
415.60	700		115	700	42	0.0031	5.1757	R	4.6140	0.0010	404	76
415.76	800			800	42	0.0036	5.1793		4.6130	0.0000	0	76
415.79	180		134	180	42	0.0008	5.1801	L	4.6130	0.0050	1691	A 67
415.98	1000			1000	42	0.0045	5.1846		4.6080	0.0011	385	67
416.24	1400		134	1400	42	0.0063	5.1908	L	4.6079	0.0040	1400	67
416.47	1200			1200	42	0.0054	5.1962		4.6030	0.0034	1200	67
416.47						0.0000	5.1962		4.5996	0.0000		67
416.58	580			462	42	0.0021	5.1983		4.5996	0.0016	580	67
416.65	385		76	4983	A 42	0.0281	5.2263	L	4.5980	0.0011	385	67
416.89	1264						5.2263		4.5968	0.0036	1264	67
416.96	364		134				5.2263	R	4.5933	0.0010	364	67
417.23	1422						5.2263		4.5923	0.0040	1422	67
417.47	1264		96		76		5.2263	L	4.5883	0.0036	1254	67
417.99	2746	PRR	95	2381	76	0.0059	5.2323	L	4.5847	0.0078	2746	67
417.99		PRR/P&LE		109	D 76	0.0005	5.2328	AMITY STREET	4.5769	0.0000		
418.00	46	P&LE	67	46	67	0.0001	5.2329	L	4.5769	0.0001	46	67
418.14	758			758	67	0.0021	5.2350		4.5768	0.0021	758	67
418.17	152		96	152	67	0.0004	5.2355	L EAST, R WEST	4.5747	0.0004	152	67
418.17	19		96	19	67	0.0001	5.2355	L EAST, R WEST	4.5742	0.0001	19	67
418.20	149			149	67	0.0004	5.2359	REVERSE CURVE	4.5742	0.0004	149	67
418.23	129		96	129	67	0.0004	5.2363	R EAST, L WEST	4.5738	0.0004	129	67
418.23							5.2363	WEST END WILMERDI	4.5734			
418.36	703			703	67	0.0020	5.2383		4.5734	0.0020	703	67
418.55	1000			1994	67	0.0056	5.2439		4.5714	0.0028	1000	67
418.74	1000			1000	67	0.0028	5.2468		4.5686	0.0028	1000	67
418.86	680			680	67	0.0019	5.2486		4.5658	0.0019	680	67
418.93	340			340	67	0.0010	5.2496		4.5639	0.0010	340	67
419.12	1000			1000	67	0.0028	5.2524		4.5629	0.0028	1000	67
419.22	540			540	67	0.0015	5.2539	GLENWOOD BRIDGE	4.5601	0.0015	540	67
419.31	480		71	480	67	0.0014	5.2553	R a	4.5586	0.0013	480	67
419.42	596		71	596	67	0.0017	5.2570	R	4.5573	0.0017	596	67
419.50	415			415	67	0.0012	5.2581		4.5556	0.0012	415	67
419.62	632			632	67	0.0018	5.2599		4.5544	0.0018	632	67
419.69	368		112	368	67	0.0010	5.2610	R b	4.5526	0.0010	368	67
419.79	553		112	553	67	0.0016	5.2625	R	4.5515	0.0015	553	67
419.84	233		134	233	67	0.0007	5.2632	R	4.5500	0.0007	233	67
419.86	135		102	135	67	0.0004	5.2636	R c	4.5494	0.0004	135	67
419.88	79		102	79	67	0.0002	5.2638	R	4.5490	0.0002	79	67
419.97	499		102	499	67	0.0014	5.2652	R	4.5488	0.0014	499	67
420.03	305		90	305	67	0.0009	5.2661	R	4.5474	0.0008	305	67
420.07	196			196	67	0.0006	5.2666		4.5465	0.0008	196	67
420.12	268			268	67	0.0008	5.2674		4.5459	0.0008	268	67
420.22	539		67	539	67	0.0015	5.2689	R d	4.5452	0.0015	539	67
420.26	195			195	67	0.0006	5.2695		4.5437	0.0006	195	67
420.43	940			940	67	0.0027	5.2721		4.5431	0.0012	420	67
420.45	80		117	80	67	0.0002	5.2723	R e	4.5419			
420.64	1000		117	1000	67	0.0028	5.2751	R	4.5419			
420.79	840		117	840	67	0.0024	5.2775	R	4.5419			
420.82	160			160	67	0.0005	5.2779		4.5419	0.0078	2418	A 50
420.87	260			260	67	0.0007	5.2787		4.5341	0.0007	189	50
421.01	734		185	734	67	0.0021	5.2807	R	4.5334	0.0028	734	50
421.03	97		185	97	67	0.0003	5.2810	R	4.5307	0.0004	97	50
421.16	706			706	67	0.0020	5.2830	BECKS RUN	4.5303	0.0027	706	50
421.20	197			197	67	0.0008	5.2836		4.5276	0.0007	197	50
421.28	421			421	67	0.0012	5.2848		4.5269	0.0018	421	50

APPENDIX B4

PAGE 9 COLUMN 1

MILES	DISTANCE ft	HWY GRADE CROSSING	MAX MPH	DISTANCE ft (READ DOWN, WESTBOUND TRAFFIC)	SPEED mph	dT hours	TIME hours	STATION	TIME hours (READ UP, EASTBOUND TRAFFIC)	dT hours	DISTANCE ft	SPEED mph
421.39	579		95	579	67	0.0018	5.2884	R b	4.5253	0.0022	579	50
421.40	62		95	62	67	0.0002	5.2886	R	4.5231	0.0002	62	50
421.53	652			652	67	0.0018	5.2884		4.5228	0.0025	680	50
421.58	286		100	286	67	0.0008	5.2890	L	4.5204	0.0011	286	50
421.58						0.0000	5.2882		4.5193	0.0000		
421.63	239		100	239	67	0.0007	5.2898	L	4.5193	0.0009	239	50
421.73	550		134	550	67	0.0016	5.2915	L	4.5184	0.0021	550	50
421.77	211		95	211	67	0.0006	5.2921	L a	4.5163	0.0008	211	50
421.77						0.0000	5.2921	WILLIAMSBURG	4.5155	0.0000		
421.78	37		95	37	67	0.0003	5.2922	L	4.5155	0.0001	37	50
421.86	409			439	67	0.0012	5.2934		4.5154	0.0017	439	50
421.86						0.0000	5.2934		4.5137	0.0000		
421.96	524			524	67	0.0015	5.2946		4.5137	0.0020	524	50
422.08	634			634	67	0.0018	5.2967		4.5117	0.0024	634	50
422.13	269		75	269	67	0.0008	5.2974	L b	4.5093	0.0010	269	50
422.15	97		134	97	67	0.0003	5.2977	L	4.5083	0.0004	97	50
422.16	44		134	44	67	0.0001	5.2978	L	4.5079	0.0002	44	50
422.21	273		95	273	67	0.0008	5.2986	L c	4.5078	0.0010	273	50
422.34	583			494	67	0.0014	5.3000		4.5067	0.0019	494 A	44
422.34				1670 A	67	0.0044	5.3044		4.5048	0.0010	240	44
422.38	217						5.3044		4.5037	0.0009	217	44
422.50	619		95			0.0000	5.3044	L a	4.5028	0.0027	619	44
422.53	164					0.0000	5.3044		4.5001	0.0007	164	44
422.65	630					0.0000	5.3044		4.4994	0.0027	630	44
422.72	370		95		78	0.0000	5.3044	L b	4.4967	0.0016	370	44
422.75	175				78	0.0000	5.3044		4.4951	0.0008	175	44
422.87	632				78	0.0000	5.3044		4.4944	0.0027	632	44
422.87						0.0000	5.3044	J&L TUNNEL	4.4917	0.0000		
422.91	200					0.0000	5.3044		4.4917	0.0009	200	44
422.91				83	78	0.0002	5.3047		4.4908	0.0000		
422.91	36			498 D	78	0.0014	5.3060		4.4908	0.0002	36	44
422.97	297		50	297	50	0.0011	5.3072	L c	4.4907	0.0013	297	44
423.04	385		103	385	50	0.0015	5.3086	L	4.4894	0.0017	385	44
423.10	282			282	50	0.0011	5.3097	REVERSE CURVE	4.4877	0.0012	282	44
423.12	111			111	50	0.0004	5.3101		4.4865	0.0005	111	44
423.22	556		58	556	50	0.0021	5.3122	R d	4.4861	0.0024	556	44
423.29	338			333	50	0.0013	5.3135	REVERSE	4.4837	0.0014	333	44
423.31	111			111	50	0.0004	5.3139	CURVE	4.4822	0.0005	111	44
423.44	720		54	720	50	0.0027	5.3166	L a	4.4817	0.0031	720	44
423.48	194			194	50	0.0007	5.3174		4.4786	0.0008	194	44
423.65	896			656	50	0.0024	5.3206		4.4778	0.0039	896	44
423.67	101		186	101	50	0.0004	5.3212	L	4.4739	0.0004	101	44
423.69	103		186	103	50	0.0004	5.3216	L	4.4735	0.0004	103	44
423.79	535			535	50	0.0020	5.3236	REVERSE CURVE	4.4731	0.0025	535	44
423.84	262		186	262	50	0.0010	5.3246	R	4.4706	0.0011	262	44
423.86	100			100	50	0.0004	5.3249		4.4695	0.0004	100	44
423.87	45			45	50	0.0002	5.3251	REVERSE CURVE	4.4692	0.0002	45	44
423.92	268			268	50	0.0010	5.3261	OLIVER YARD	4.4690	0.0012	268	44
423.95	189		192	189	50	0.0007	5.3268	L	4.4679	0.0008	189	44
424.05	498			498	50	0.0018	5.3287		4.4670	0.0021	498	44
424.08	185			185	50	0.0007	5.3294		4.4649	0.0008	185	44
424.21	661		86	661	50	0.0025	5.3319	R a	4.4641	0.0028	661	44
424.24	164			164	50	0.0006	5.3325	REVERSE CURVE	4.4613	0.0007	164	44
424.26	147			147	50	0.0005	5.3331		4.4606	0.0006	147	44
424.32	285		54	505	50	0.0018	5.3350	L b	4.4600	0.0012	285	44
424.38	303			83 D	50	0.0003	5.3353	REVERSE CURVE	4.4587	0.0013	303	44
424.41	156		44	156	44	0.0007	5.3360	R a	4.4574	0.0007	156	44
424.43	109		72	109	44	0.0005	5.3365	R	4.4568	0.0011	298 D	65
424.45	122		72	122	44	0.0005	5.3370		4.4557	0.0004	122	65
424.45							5.3370	WEST END TRAFFORD	4.4553			
424.45						0.0000	5.3370	LIBERTY BRIDGE	4.4553	0.0005		
424.45	11		72	11	44	0.0000	5.3370	R d	4.4553	0.0000	11	65
424.50	270		83	270	44	0.0012	5.3382	R	4.4553	0.0006	270	83
424.62	600			600	44	0.0026	5.3408		4.4548	0.0008	98 D	98
424.68	350		134	350	44	0.0015	5.3423	R	4.4542	0.0024	854	98
424.86	914			914	44	0.0039	5.3462	SMITHFIELD STREET	4.4518	0.0025	914	98
424.87	86			86	44	0.0004	5.3466		4.4493	0.0002	86	98
424.91	220		78	220	44	0.0009	5.3475	R	4.4491	0.0005	220	66
424.93						0.0000	5.3475	MP 0 P&LE STATION	4.4484	0.0000		
425.00	450			450	44	0.0019	5.3495	REVERSE CURVE	4.4484	0.0013	450	98
425.04	200		68	200	44	0.0008	5.3503	L b	4.4472	0.0006	200	98
425.04					44	0.0000	5.3503		4.4466	0.0006	214 D	76
425.05	50			50	44	0.0002	5.3505		4.4461	0.0018	729	76
425.05						0.0000	5.3505		4.4443	0.0010	405	78
425.24	1000			1000	44	0.0043	5.3549		4.4433	0.0000	0	76
425.38	750			750	44	0.0032	5.3581		4.4433	0.0011	462 A	76
425.43	250		134	250	44	0.0011	5.3592	R FORT PITT BRIDGE	4.4422	0.0008	250	76
425.50	400		134	400	44	0.0017	5.3609	R	4.4416	0.0010	400	76
425.61	600			600	44	0.0026	5.3635		4.4406	0.0015	600	76
425.80	1000			1000	44	0.0043	5.3678		4.4391	0.0014	561	76
425.84	200			200	44	0.0009	5.3686		4.4376	0.0022	869 A	72
425.92	400		78	400	44	0.0017	5.3703	L a	4.4354	0.0004	150	72
425.97	300		78	300	44	0.0013	5.3716	L	4.4350	0.0008	350	72
425.99	100			100	44	0.0004	5.3721		4.4342	0.0003	100	72
426.12	650			650	44	0.0028	5.3749		4.4340	0.0017	850	72
426.18	350		110	350	44	0.0015	5.3764	R b	4.4323	0.0009	350	72
426.20	100		110	100	44	0.0004	5.3768	R	4.4310	0.0003	100	72
426.37	900		138	905	44	0.0039	5.3807	R WEST END BRIDGE	4.4311	0.0024	900	72
426.56	1000		136	818	44	0.0027	5.3833	R	4.4287	0.0026	1000	72
426.58	100		136				5.3833	R	4.4261	0.0003	100	72
426.75	900			2030 A	44	0.0078	5.3911		4.4264	0.0024	900	72
426.88	700			886 A	63	0.0028	5.3938		4.4239	0.0018	700	72
426.91	130		214			0.0000	5.3938	R	4.4216	0.0003	130	72
426.94	170		214	231	68	0.0006	5.3943	R	4.4213	0.0004	170	72
427.13	1000		214			0.0000	5.3943	R	4.4208	0.0026	1000	72
427.29	880		104	1882 A	68	0.0044	5.3987	R	4.4182	0.0022	880	72
427.32	150					0.0000	5.3987	REVERSE CURVE	4.4160	0.0004	150	72
427.34	100					0.0000	5.3987		4.4156	0.0002	100	72
427.42	450		76			0.0000	5.3987	L a	4.4153	0.0012	450	72
427.51	450			368	76	0.0009	5.3996		4.4141	0.0012	450	72
427.70	1000			1000	76	0.0025	5.4021		4.4129	0.0028	1000	72
427.81	600			600	76	0.0016	5.4036		4.4102	0.0016	600	72

APPENDIX B4

PAGE 9 COLUMN 2

MILES	DISTANCE ft	HWY GRADE CROSSING	MAX MPH	DISTANCE ft (READ DOWN, WESTBOUND TRAFFIC)	SPEED mph	dT hours	TIME hours	STATION	TIME hours (READ UP, EASTBOUND TRAFFIC)	dT hours	DISTANCE ft	SPEED mph
427.84	400		86	400	76	0.0010	5.4046	L b	4.4087	0.0011	400	72
427.92	190			190	76	0.0005	5.4051	REVERSE CURVE	4.4077	0.0005	190	72
427.97	280			173	76	0.0004	5.4055	CORLISS	4.4072	0.0007	280	72
427.97				87 D	76	0.0002	5.4057		4.4065	0.0000		
428.08	500		72	550	72	0.0014	5.4072	R a	4.4065	0.0014	550	72
428.18	550		72	550	72	0.0014	5.4086	R	4.4050	0.0014	550	72
428.18						0.0000	5.4086		4.4036	0.0002	87 D	76
428.27	480			480	72	0.0012	5.4098		4.4034	0.0009	383	76
428.46	1000	XING 3.4		1000	72	0.0026	5.4124	RIVER AVENUE	4.4025	0.0026	1000	76
428.64	1000			1000	72	0.0026	5.4150		4.4000	0.0025	1000	76
428.83	1000			1000	72	0.0026	5.4177		4.3975	0.0025	1000	76
428.92	480			450	72	0.0012	5.4189		4.3950	0.0011	450	76
429.02	550			550	72	0.0014	5.4203		4.3939	0.0014	550	76
429.14	600			600	72	0.0016	5.4219		4.3925	0.0015	600	76
429.21	400		107	400	72	0.0011	5.4229	R a	4.3910	0.0010	400	76
429.23	100		107	100	72	0.0003	5.4232	R	4.3900	0.0002	100	76
429.28	250			250	72	0.0007	5.4239	REVERSE CURVE	4.3898	0.0006	250	76
429.39	600		107	600	72	0.0016	5.4254	L b	4.3891	0.0015	600	76
429.40	50			50	72	0.0001	5.4256		4.3876	0.0001	50	76
429.59	1000			1000	72	0.0026	5.4282		4.3875	0.0025	1000	76
429.78	1000			1000	72	0.0026	5.4308		4.3850	0.0025	1000	76
429.82	220			220	72	0.0006	5.4314		4.3825	0.0005	220	76
429.93	550			550	72	0.0014	5.4329		4.3820	0.0014	550	76
429.97	230		78	230	72	0.0006	5.4335	L DAVIS ISLAND c	4.3806	0.0008	230	76
430.08	650		78	650	72	0.0017	5.4352	L	4.3800	0.0016	650	76
430.09						0.0000	5.4352		4.3784	0.0000		76
430.16	360			360	72	0.0008	5.4360		4.3784	0.0011	477 D	96
430.35	1000			600	72	0.0016	5.4376		4.3774	0.0017	873	96
430.54	1000			886 A	72	0.0022	5.4398	NEVILLE ISLAND	4.3756	0.0020	1000	96
430.65	600			1181	76	0.0029	5.4427		4.3736	0.0012	600	96
430.71	300		96	300	76	0.0007	5.4434	R	4.3724	0.0006	300	96
430.71						0.0000	5.4434		4.3718	0.0013	753 D	11
430.73	100			100	76	0.0002	5.4437		4.3705	0.0007	416	118
430.80	400			400	76	0.0010	5.4447		4.3698	0.0000		
430.92	600			600	76	0.0015	5.4462		4.3698	0.0000		
431.11	1000			1000	76	0.0025	5.4487		4.3698	0.0000		
431.15	240			240	76	0.0006	5.4493		4.3698	0.0000		
431.30	780		131	780	76	0.0019	5.4512	L	4.3698	0.0000		
431.31	50		131	50	76	0.0001	5.4513	L	4.3698	0.0000		
431.49	950			950	76	0.0024	5.4537		4.3698	0.0000		
431.68	1000			1000	76	0.0025	5.4562		4.3698	0.0000		
431.83	800			800	76	0.0020	5.4582		4.3698	0.0000		
431.86	200		150	200	76	0.0005	5.4587	L	4.3698	0.0000		
431.92	300		180	300	76	0.0007	5.4594	L	4.3698	0.0000		
432.04	640		152	640	76	0.0016	5.4610	L	4.3698	0.0000		
432.05	50			50	76	0.0001	5.4611	EMSWORTH DAM	4.3698	0.0000		
432.14	450			450	76	0.0011	5.4623	REVERSE CURVE	4.3698	0.0000		
432.20	300		180	200	76	0.0005	5.4628	R	4.3698	0.0000		
432.24	210					0.0000	5.4628		4.3698	0.0000		
432.24	40		182			0.0000	5.4628	L	4.3698	0.0000		
432.43	1000		182	5026 A	76	0.0131	5.4758	L	4.3698	0.0126	5026 A	96
432.57	700		152			0.0000	5.4758	L	4.3572	0.0011	600	104
432.61	250					0.0000	5.4758	REVERSE CURVE	4.3561	0.0006	350	104
432.62	50		152	8146 A	96	0.0114	5.4872	R	4.3547	0.0001	50	104
432.68	300		182			0.0000	5.4872	R	4.3546	0.0006	300	104
432.81	700					0.0000	5.4872		4.3540	0.0013	700	104
432.92	550					0.0000	5.4872		4.3528	0.0010	550	104
433.00	460					0.0000	5.4872		4.3518	0.0008	450	104
433.19	1000					0.0000	5.4872		4.3509	0.0018	1000	104
433.38	1000					0.0000	5.4872		4.3491	0.0016	1000	104
433.53	800					0.0000	5.4872		4.3473	0.0015	800	104
433.57	250		134			0.0000	5.4872	R	4.3458	0.0004	250	104
433.76	1000		134			0.0000	5.4872	R	4.3455	0.0018	1000	104
433.89	700		134			0.0000	5.4872	R	4.3437	0.0013	700	104
433.91	120	XING 8.96				0.0000	5.4872	PRIVATE	4.3424	0.0002	120	104
433.95	180	XING 9.15				0.0000	5.4872	STATION AVE	4.3422	0.0003	180	104
434.14	1000					0.0000	5.4872		4.3418	0.0018	1000	104
434.33	1000	XING 9.4				0.0000	5.4872	LEBORA AVENUE	4.3400	0.0018	1000	104
434.52	1000			332	106	0.0004	5.4876		4.3382	0.0016	1000	104
434.54	100			136 D	10	0.0003	5.4879		4.3364	0.0002	100	104
434.64	550		104	550	104	0.0010	5.4889	L a	4.3362	0.0010	550	104
434.71	350			350	104	0.0006	5.4895		4.3352	0.0001	62 D	10
434.72	90			90	104	0.0002	5.4897		4.3351	0.0006	345	106
434.89	910			910	104	0.0017	5.4913		4.3345			
435.07	950	XING 10.0		950	104	0.0017	5.4931	BROADWAY STREET	4.3345			
435.08	50		152	50	104	0.0001	5.4932	L CORAOPOLIS	4.3345			
435.12	200	XING 10.1	150	200	104	0.0004	5.4935	L MULBERRY STREET	4.3345			
435.22	500	XING 10.5		500	104	0.0009	5.4944	RC MILL STREET	4.3345			
435.27	300		150	300	104	0.0006	5.4950	R	4.3345			
435.33	300			300	104	0.0005	5.4955		4.3345			
435.38	250		152	250	104	0.0005	5.4960	R	4.3345			
435.46	450			450	104	0.0008	5.4968	REVERSE CURVE	4.3345			
435.47	50			50	104	0.0001	5.4968		4.3345			
435.53	300		160	300	104	0.0006	5.4974	L	4.3346	0.0000		
435.61	400			400	104	0.0007	5.4982	REVERSE CURVE	4.3346	0.0000		
435.62	100	XING 10.1	152	100	104	0.0002	5.4983	R MAIN STREET	4.3346	0.0000		
435.65	150			150	104	0.0003	5.4985		4.3346	0.0097	5143 A	94
435.76	550			560	104	0.0010	5.4996		4.3247	0.0007	350	94
435.80	250		180	250	104	0.0005	5.5001	R	4.3240	0.0005	250	94
435.84	200	XING 10.8		200	104	0.0004	5.5004	WATT STREET	4.3235	0.0004	200	94
435.92	400			400	104	0.0007	5.5012		4.3231	0.0008	400	94
436.03	600			600	104	0.0011	5.5023		4.3223	0.0012	600	94
436.22	1000			1000	104	0.0018	5.5041		4.3211	0.0020	1000	94
436.35	700	XING 11.4		700	104	0.0013	5.5054	LOWER THORN ST	4.3191	0.0014	700	94
436.43	300		131	300	104	0.0005	5.5059	R	4.3177	0.0006	300	94
436.60	1000		131	1000	104	0.0018	5.5077	R	4.3171	0.0020	1000	94
436.65	250	XING 11.1	131	250	104	0.0005	5.5082	R	4.3151	0.0005	250	94
436.79	750	XING 11.5		400	104	0.0007	5.5089	PRIVATE	4.3146	0.0015	750	94
436.91	650					0.0000	5.5089		4.3131	0.0013	650	94
436.91				4468 A	10	0.0078	5.5167		4.3117	0.0000		
436.98	350					0.0000	5.5167		4.3117	0.0007	350	94
437.10	650					0.0000	5.5167		4.3110	0.0013	650	94

APPENDIX B4 — PAGE 10 COLUMN 1

READ DOWN: WESTBOUND TRAFFIC — STATION — READ UP: EASTBOUND TRAFFIC

MILES	DISTANCE, ft	HWY GRADE CROSSING	MAX SPEED	DISTANCE, ft	SPEED, mph	HT, hours	TIME, hours	STATION	TIME, hours	HT, hours	DISTANCE, ft	SPEED, mph
437.17	360		164			0.0000	5.5167	R	4.3087	0.0007	360	94
437.34	900		164			0.0000	5.5167	R	4.3080	0.0018	900	94
437.36	100		121			0.0000	5.5167	R	4.3072	0.0002	100	94
437.55	1000		121			0.0000	5.5167	R	4.3070	0.0020	1000	94
437.66	600		121			0.0000	5.5167	R STOOPS FERRY	4.3050	0.0012	600	94
437.74	400			515	113	0.0000	5.5178		4.3038	0.0008	400	94
437.77	205			577 D	11	0.0011	5.5186	REVERSE CURVE	4.3030	0.0004	200	94
437.91	740		94	740	94	0.0015	5.5201	L	4.3026	0.0015	740	94
437.93	60		94	60	94	0.0001	5.5202	L a	4.3011	0.0001	60	94
438.04	600		94	600	94	0.0012	5.5214	L	4.3010	0.0012	600	94
438.11	400			400	94	0.0008	5.5222		4.2998	0.0008	415 D	108
438.29	950			950	94	0.0018	5.5241		4.2990	0.0002	98	108
438.30	50		119	50	94	0.0001	5.5242	L b	4.2988	0.0001	50	108
438.45	750		119	750	94	0.0015	5.5258	L	4.2987	0.0013	750	108
438.49	250			250	94	0.0005	5.5260		4.2974	0.0004	250	108
438.68	1000			1000	94	0.0020	5.5283		4.2970	0.0018	1000	108
438.87	1000			1000	94	0.0020	5.5303		4.2952	0.0018	1000	108
438.91	180			180	94	0.0004	5.5306		4.2936	0.0003	180	108
438.91						0.0000	5.5306		4.2931	0.0005	1406 A	105
439.06	820			820	94	0.0017	5.5320		4.2926	0.0005	250	105
439.25	1000			1000	94	0.0020	5.5340		4.2902	0.0018	1000	105
439.43	950	XING 14.5					5.5340	MCCUTCHEON WAY	4.2884	0.0017	950	105
439.44	50		106	50	94	0.0001	5.5344	R GLENWILLARD	4.2867	0.0001	50	105
439.63	1000		108	1000	94	0.0020	5.5364	R	4.2866	0.0018	1000	105
439.65	100		108	100	94	0.0002	5.5366	R	4.2848	0.0002	100	105
439.82	900			900	94	0.0018	5.5384		4.2846	0.0016	900	105
439.91	500			500	94	0.0010	5.5395		4.2830	0.0009	500	105
440.00	460			460	94	0.0009	5.5404		4.2821	0.0008	460	105
440.01	40		190	40	94	0.0001	5.5405	R	4.2813	0.0001	40	105
440.20	1000		190	1000	94	0.0020	5.5405	R	4.2812	0.0018	1000	105
440.28	450		190	450	94	0.0009	5.5434	R	4.2794	0.0008	450	105
440.39	550			100	94	0.0002	5.5436		4.2786	0.0010	550	105
440.58	1000	XING 16.6				0.0000	5.5436	PRIVATE	4.2776	0.0018	1000	105
440.77	1000			4679 A	94	0.0089	5.5525		4.2758	0.0018	1000	105
440.91	750					0.0000	5.5525	COUNTY LINE ALLEGH	4.2740	0.0014	750	105
440.96	250					0.0000	5.5525		4.2726	0.0005	250	105
441.10	750			471	105	0.0008	5.5533	SOUTH HEIGHTS	4.2721	0.0014	750	105
441.14	250		105	250	105	0.0005	5.5538	R	4.2708	0.0005	250	105
441.29	750		105	750	105	0.0014	5.5551	R	4.2703	0.0014	750	105
441.33	250			250	105	0.0005	5.5556		4.2690	0.0008	466 D	120
441.52	1000			1000	105	0.0018	5.5574		4.2682	0.0012	756	120
441.71	1000			1000	105	0.0018	5.5592		4.2670	0.0018	1000	120
441.90	1000			1000	105	0.0018	5.5610		4.2654	0.0016	1000	120
441.93	120			120	105	0.0002	5.5612		4.2638	0.0002	120	120
442.09	880			880	105	0.0016	5.5628		4.2636	0.0014	880	120
442.22	700			700	105	0.0013	5.5641		4.2622	0.0011	700	120
442.28	300		190	300	105	0.0005	5.5646	R	4.2611	0.0005	300	120
442.47	1000		190	1000	105	0.0018	5.5664	R	4.2607	0.0016	1000	120
442.65	940		190	940	105	0.0017	5.5681	R	4.2591	0.0015	940	120
442.66	60			60	105	0.0001	5.5682		4.2576	0.0001	60	120
442.85	1000			1000	105	0.0018	5.5700		4.2575	0.0016	1000	120
442.92	380			380	105	0.0007	5.5707		4.2559	0.0006	380	120
442.93	20			20	105	0.0000	5.5707		4.2553	0.0000	20	120
442.98	300		152	300	105	0.0005	5.5713	L	4.2553	0.0005	300	120
443.04	300			300	105	0.0005	5.5718	REVERSE CURVE	4.2548	0.0005	300	120
443.05	40			40	105	0.0001	5.5719	WEST ECONOMY	4.2543	0.0001	40	120
443.08	200		152	200	105	0.0004	5.5722	R	4.2543	0.0003	200	120
443.23	760			760	105	0.0014	5.5736		4.2540	0.0006	378	120
443.36	725			700	105	0.0013	5.5748		4.2534			
443.40	200		152	30	105	0.0001	5.5750	L	4.2534			
443.42	80					0.0000	5.5750	REVERSE CURVE	4.2534			
443.45	150					0.0000	5.5750		4.2534			
443.50	330		152			0.0000	5.5750	R	4.2534			
443.61	550					0.0000	5.5750		4.2534			
443.80	1000					0.0000	5.5750		4.2534			
443.91	580					0.0000	5.5750		4.2534			
443.91				8269 A	10	0.0139	5.5889		4.2534			
443.98	380					0.0000	5.5889		4.2534			
443.99	30		190			0.0000	5.5889	L	4.2534			
444.18	1000		190			0.0000	5.5889	L ALIQUIPPA	4.2534			
444.24	350		190			0.0000	5.5889	L	4.2534			
444.36	650					0.0000	5.5889		4.2534			
444.54	900					0.0000	5.5889		4.2534			
444.55	130		134			0.0000	5.5889	R	4.2534			
444.67	600		134			0.0000	5.5889	R	4.2534			
444.74	400					0.0000	5.5889		4.2534			
444.91	900					0.0000	5.5889		4.2534			
444.93	100					0.0000	5.5889		4.2534			
444.98	250			231	120	0.0004	5.5892		4.2534			
445.09	600		134	600	120	0.0008	5.5902	L	4.2534			
445.12	150			150	120	0.0002	5.5904	WEST ALIQUIPPA	4.2534			
445.31	1000			1000	120	0.0016	5.5920		4.2534	0.0216	12532 A	95
445.50	1000			1000	120	0.0016	5.5936		4.2314			
445.69	1000			1000	120	0.0016	5.5951		4.2314	0.0017	850	95
445.84	800			800	120	0.0013	5.5964		4.2297	0.0016	803	95
445.88	200		134	200	120	0.0003	5.5967	L	4.2281	0.0017	827 A	93
445.92	210			210	120	0.0003	5.5971		4.2264	0.0000	16	93
446.03	590		134	590	120	0.0009	5.5980	L	4.2264	0.0003	134	93
446.07	200		164	200	120	0.0003	5.5983	L	4.2261	0.0004	200	93
446.18	600		164	600	120	0.0009	5.5992	L	4.2257	0.0012	600	93
446.26	400			400	120	0.0006	5.5998		4.2245	0.0006	400	93
446.45	1000			1000	120	0.0016	5.6015	HOPEWELL TOWNSHI	4.2237	0.0020	1000	93
446.47	100			100	120	0.0002	5.6016		4.2218	0.0002	100	93
446.64	900		186	900	120	0.0014	5.6030	L	4.2214	0.0016	900	93
446.69	300		186	300	120	0.0005	5.6035	L	4.2198	0.0008	300	93
446.79	500			500	120	0.0008	5.6043	REVERSE CURVE	4.2190	0.0010	500	93
446.83	200		131	200	120	0.0003	5.6046	R	4.2180	0.0004	200	93
446.92	480		131	480	120	0.0008	5.6054	R	4.2176	0.0007	363	93
447.02	520		131	520	120	0.0008	5.6062	R	4.2168	0.0000		
447.03	80		131	80	120	0.0001	5.6063	R	4.2168	0.0000		
447.21	920			920	120	0.0015	5.6078		4.2168	0.0000		
447.39	1000			1000	120	0.0016	5.6093		4.2168			

APPENDIX B4 — PAGE 10 COLUMN 2

READ DOWN: WESTBOUND TRAFFIC — STATION — READ UP: EASTBOUND TRAFFIC

MILES	DISTANCE, ft	HWY GRADE CROSSING	MAX SPEED	DISTANCE, ft	SPEED, mph	HT, hours	TIME, hours	STATION	TIME, hours	HT, hours	DISTANCE, ft	SPEED, mph
447.48	400			200	120	0.0003	5.6087		4.2168			
447.58	550		134	134	120	0.0002	5.6089	R	4.2168			
447.61	150		134	28	120	0.0000	5.6099	R	4.2168			
447.61				788 D	12	0.0014	5.6113		4.2168			
447.67	300		95	300	95	0.0006	5.6119	R a	4.2168			
447.69	100			100	95	0.0002	5.6121	S COLONA	4.2168	0.0100	4647 A	75
447.73	200		95	200	95	0.0004	5.6125	R a'	4.2066			
447.77	250			250	95	0.0006	5.6130		4.2066	0.0000	0	76
447.79	100			100	95	0.0002	5.6132		4.2066	0.0000	77	78
447.81	60		133	60	95	0.0002	5.6134	R	4.2064	0.0000		
447.91	540			540	95	0.0011	5.6144		4.2064	0.0000		
447.96	280			280	95	0.0006	5.6150		4.2064	0.0000		
448.09	650			598	95	0.0012	5.6162		4.2064	0.0000		
448.09				55 D	95	0.0001	5.6163		4.2064	0.0000		
448.15	350		93	350	93	0.0007	5.6170	L b	4.2064	0.0000		
448.34	1000		93	1000	93	0.0020	5.6190	L COLONA	4.2064	0.0000		
448.51	900		93	900	93	0.0018	5.6208	L	4.2064	0.0000		
448.53	100			100	93	0.0002	5.6211		4.2064	0.0000		
448.72	1000			1000	93	0.0020	5.6231		4.2054	0.0000		
448.79	350			350	93	0.0007	5.6235		4.2064	0.0000		
448.91	650		107	650	93	0.0013	5.6251	L a	4.2064	0.0000		
448.93	60		107	60	93	0.0002	5.6253	L a'	4.2064	0.0000		
448.93						0.0000	5.6253		4.2064	0.0203	6123 A	30
448.94	70		107	70	93	0.0001	5.6255	L a'	4.1861	0.0008	1031	30
449.10	850			850	93	0.0017	5.6272	MONACA S	4.1786	0.0003	44 A	29
449.29	1000			1000	93	0.0020	5.6292		4.1783	0.0047	725	29
449.29							5.6292	EAST END MONACA-I	4.1746			
449.48	1000	XING 2	110	1000	93	0.0020	5.6313	L 4TH STREET	4.1746	0.0066	1000	29
449.50	100		110	100	93	0.0002	5.6315	L MONACA	4.1680	0.0007	100	29
449.67	900			900	93	0.0018	5.6333		4.1674	0.0059	900	29
449.78	600			600	93	0.0012	5.6345	OHIO RIVER BRANCH	4.1615	0.0039	600	29
449.86	400			24	93	0.0000	5.6346		4.1576	0.0026	400	29
449.86				376 D	62	0.0008	5.6354		4.1550			29
449.90	250		78	250	78	0.0006	5.6360	R	4.1550	0.0016	250	29
449.90							5.6360		4.1534	0.0000		29
450.05	750		78	750	78	0.0018	5.6378	R	4.1534	0.0049	750	29
450.06	60		78	60	78	0.0002	5.6380	R	4.1485	0.0006	60	29
450.12	320			320	78	0.0006	5.6386		4.1479	0.0021	320	29
450.12						0.0000	5.6386	OHIO RIVER S BANK	4.1458	0.0000		29
450.24	600			600	78	0.0015	5.6401		4.1458	0.0039	600	29
450.37	700			700	78	0.0017	5.6420		4.1419	0.0046	700	29
450.37						0.0000	5.6420	OHIO RIVER N BANK	4.1374	0.0000		29
450.43	300			300	78	0.0007	5.6427		4.1374	0.0009	300	29
450.50	400		80	400	78	0.0010	5.6437	R	4.1364	0.0026	400	29
450.61	600		80	600	78	0.0015	5.6451	R	4.1328	0.0039	600	29
450.65	200		80	200	78	0.0005	5.6456	R	4.1289	0.0013	200	29
450.73	400			400	78	0.0010	5.6466		4.1276	0.0026	400	29
450.80	400		190	400	78	0.0010	5.6475	R BEAVER P&LE STA	4.1249	0.0026	400	29
450.91	570			430	78	0.0010	5.6486		4.1223	0.0037	570	29
450.97	330					0.0000	5.6486		4.1186	0.0022	330	29
450.99	100		190			0.0000	5.6486	L	4.1165	0.0007	100	29
451.03	200		190	30	78	0.0000	5.6486	L BEAVER NEWSPAPE	4.1158	0.0013	200	29
451.03				760 D	78	0.0027	5.6513		4.1146			29
451.20	905		30	905	30	0.0057	5.6570	L	4.1145	0.0059	905	29
451.22	100			100	30	0.0006	5.6576		4.1086	0.0007	100	29
451.23	20		65	20	30	0.0001	5.6577	L	4.1079	0.0001	20	29
451.23	25			25	30	0.0001	5.6579		4.1078	0.0001	20	29
451.23	20		34	21	30	0.0001	5.6580	L	4.1077	0.0001	20	29
451.23	10			9 D	29	0.0001	5.6581	REVERSE CURVE	4.1075	0.0021	10	29
451.24	25		29	25	29	0.0002	5.6582	R	4.1075	0.0002	25	29
451.24	5		34	5	29	0.0000	5.6583	R	4.1073	0.0003	48 D	34
451.25	50		34	50	29	0.0005	5.6588	R	4.1070	0.0000	5	34
451.26	50			50	29	0.0003	5.6589	SHARON ROAD	4.1070	0.0003	50	34
451.27	80			80	29	0.0004	5.6593	REVERSE CURVE	4.1067	0.0004	65	34
451.27	14		76	14	29	0.0001	5.6594	L	4.1063	0.0001	14	34
451.28	6			6	29	0.0000	5.6595	REVERSE CURVE	4.1062	0.0000	6	34
451.28	10		88	10	29	0.0001	5.6596	R	4.1062	0.0001	10	34
451.28	5	XX		5	29	0.0000	5.6596	TRACK IN 5TH AVENUE	4.1062	0.0000	5	34
451.30	100	XX		100	29	0.0007	5.6602	SHARON ROAD	4.1062	0.0006	100	34
451.32	100	XX		100	29	0.0007	5.6609	TO BUFFALO STREET	4.1056	0.0006	100	34
451.34	100	XX		100	29	0.0007	5.6615	REVERSE CURVE	4.1050	0.0006	100	34
451.34	25	XX		25	29	0.0002	5.6617	BUFFALO STREET	4.1045	0.0001	25	34
451.35	35		34	35	29	0.0002	5.6619	L	4.1043	0.0002	35	34
451.35	40			40	29	0.0003	5.6622	REVERSE CURVE	4.1041	0.0002	40	34
451.36	15	XING		15	29	0.0001	5.6623	TUSCARAWAS ROAD	4.1039	0.0001	16	34
451.36	24		34	24	29	0.0002	5.6624	R	4.1038	0.0001	24	34
451.37	60			60	29	0.0004	5.6628		4.1037	0.0000	60	34
451.38	60			60	29	0.0004	5.6632		4.1034	0.0003	60	34
451.39	30		34	30	29	0.0002	5.6634	R	4.1030	0.0007	30	34
451.39	10	XING		10	29	0.0001	5.6635	WALNUT LANE	4.1029	0.0001	21 D	36
451.40	35			35	29	0.0002	5.6637	REVERSE CURVE	4.1028	0.0001	24	36
451.41	55		62	55	29	0.0004	5.6641	L	4.1026	0.0003	55	36
451.41	10		95	10	29	0.0001	5.6641	L	4.1023	0.0001	10	36
451.42	55		95	55	29	0.0004	5.6646	L	4.1023	0.0003	55	36
451.43	46	XING		46	29	0.0003	5.6648	SPRING LANE	4.1020	0.0002	46	36
451.45	100			100	29	0.0007	5.6654		4.1018	0.0006	100	36
451.45	25			25	29	0.0002	5.6656		4.1012	0.0001	25	36
451.46	35	XING	80	35	29	0.0002	5.6658	L DIVISION LANE	4.1011	0.0002	35	36
451.46	10		87	10	29	0.0001	5.6659	L	4.1009	0.0001	10	30
451.47	30			30	29	0.0002	5.6661		4.1008	0.0002	30	36
451.48	60			65	29	0.0004	5.6665		4.1007	0.0003	65	36
451.49	35		38	35	29	0.0002	5.6667	L	4.1004	0.0002	35	38
451.49	15	XING	38	15	29	0.0001	5.6668	L DIVISION LANE	4.1002	0.0001	15	38
451.50	30	XING		30	29	0.0002	5.6670	RD PA HWY 65	4.1001	0.0002	33 D	38
451.51	95		39	95	29	0.0004	5.6674	R	4.0999	0.0003	52	39
451.51	20		39	20	29	0.0001	5.6675	R	4.0997	0.0001	20	39
451.51	16		50	16	29	0.0001	5.6676	R	4.0996	0.0007	48 D	43
451.51	19		51	19	29	0.0001	5.6677	R	4.0994	0.0000	4	43
451.52	11		43	11	29	0.0001	5.6678	R	4.0993	0.0000		43
451.52	15			15	29	0.0001	5.6679	REVERSE CURVE	4.0993	0.0000		43
451.52	15		43	15	29	0.0001	5.6680	L	4.0993	0.0001	15	43
451.52	3	XING		3	29	0.0000	5.6680	BEAVER INDUSTRIAL	4.0993	0.0000	3	43

APPENDIX B4 — PAGE 11 — COLUMN 1

MILES	DISTANCE ft	HWY GRADE CROSSING	MAX MPH	d DISTANCE ft	SPEED mph	dT, hours	TIME, hours	STATION	TIME, hours	dT, hours	d DISTANCE ft	SPEED mph
	READ DOWN WESTBOUND TRAFFIC								READ UP EASTBOUND TRAFFIC			
451.52	10		43	10	29	0.0001	5.6881	R	4.0993	0.0000	10	43
451.52						0.0000	5.6881		4.0992	0.0036	1377	D 10(
451.52						0.0000	5.6881	VANPORT	4.0957	0.0047	2850	106
452.52	5253	17300		5253	29	0.0348	5.7024	W BEAVER	4.0910	0.0025	1380	A 105
452.65	700	XING 3.04		700	29	0.0046	5.7069	PRIVATE	4.0885	0.0010	540	103
453.09	2300		120	2300	29	0.0150	5.7220	L	4.0875	0.0042	2300	103
453.25	879			879	29	0.0087	5.7277	C	4.0833	0.0016	879	103
453.37	600		152	363	29	0.0024	5.7301	L'MPH50(TT)	4.0816	0.0011	600	103
453.52	800		152	232	A 29	0.0014	5.7315	R'RC	4.0806	0.0015	800	103
453.63	600			583	34	0.0031	5.7346		4.0791	0.0011	600	103
453.80	900		109	102	A 34	0.0006	5.7351	R	4.0780	0.0017	900	103
454.16	1900			423	36	0.0022	5.7374		4.0763	0.0003	145	103
454.33	900		106	220	A 36	0.0011	5.7385	R	4.0760	0.0000		
454.52	978			240	A 39	0.0011	5.7396		4.0760	0.0000		
454.63	600			14194	A 43	0.0347	5.7743		4.0760	0.0000		
454.77	700		103			0.0000	5.7743	R	4.0760	0.0000		
454.91	781		150			0.0000	5.7743	R	4.0760	0.0175	8476	A 79-
455.52	3200					0.0000	5.7743		4.0585	0.0011	438	79
456.13	3200	XING 6.19				0.0000	5.7743	PRIVATE	4.0575	0.0077	3200	79
456.35	1173		134			0.0000	5.7743	P/E INDUSTRY	4.0498	0.0076	3173	79
456.52	900			806	103	0.0016	5.7760		4.0422	0.0014	572	A 77-
456.92	2100	XING 7.85		1746	103	0.0032	5.7792	WABASH STREET	4.0408	0.0011	428	77
457.12	1089		124	2322	A 10	0.0042	5.7833	L W INDUSTRY	4.0398	0.0027	1089	77
457.52	2100	XING 8.08		426	108	0.0001	5.7836	PRIVATE	4.0371	0.0052	2100	77
457.52				796	D 10	0.0016	5.7851		4.0319	0.0000		
457.67	800		79	800	79	0.0019	5.7870	L	4.0319	0.0020	800	77
458.01	1800			1800	79	0.0043	5.7913		4.0300	0.0044	1800	77
458.22	1100		208	1100	79	0.0026	5.7939	L	4.0255	0.0027	1100	77
458.46	1283			1283	79	0.0031	5.7970		4.0228	0.0032	1283	77
458.52	300		94	300	79	0.0007	5.7977		4.0197	0.0007	300	77
458.69	900		94	900	79	0.0022	5.7999	R	4.0189	0.0022	900	77
458.92	1200			1164	79	0.0028	5.8027		4.0167	0.0030	1200	77
458.92				46	D 79	0.0001	5.8028		4.0138			
459.13	1100		77	1100	77	0.0027	5.8055	R	4.0138	0.0051	2078	77
459.13						0.0000	5.8055		4.0087	0.0017	665	A 74-
459.30	900			900	77	0.0022	5.8077		4.0070	0.0022	869	74
459.52	1175		119	1175	77	0.0029	5.8106	R	4.0048	0.0030	1175	74
460.52	5278		100	5278	77	0.0130	5.8236	R'MIDLAND	4.0018	0.0136	5278	74
460.57	290		153	290	77	0.0007	5.8243	R	3.9882	0.0022	847	A 70-
461.31	3900			3357	77	0.0083	5.8325		3.9860	0.0021	791	70
461.31				1401	A 77	0.0033	5.8359		3.9839			
461.52	1100		99	59	82	0.0001	5.8360	L	3.9839	0.0058	1987	A 58-
461.52				183	D 82	4E-04	5.8365		3.9780			
461.69	900		74	900	74	0.0023	5.8388	L'MPH45	3.9780	0.0031	993	58
461.90	1100			1100	74	0.0028	5.8416		3.9749	0.0036	1100	58
462.01	600		134	600	74	0.0015	5.8431	L	3.9713	0.0020	600	58
462.16	800			800	74	0.0020	5.8452	MIDL YG LEAD	3.9694	0.0026	800	58
462.28	600		101	600	74	0.0015	5.8467	L	3.9668	0.0020	600	58
462.39	569			569	74	0.0015	5.8481		3.9648	0.0018	569	58
462.52	700		134	700	74	0.0018	5.8499	L	3.9630	0.0023	700	58
463.07	3186			3084	74	0.0079	5.8578		3.9607	0.0095	2900	58
463.07				84	D 74	0.0002	5.8581	WEST END MONACA-I	3.9512	0.0000		
463.17	633		70	633	70	0.0017	5.8598	L EAST END SMITHS F	3.9512	0.0017	531	58
463.22	0			0	70	0.0000	5.8598		3.9496	0.0009	269	58
463.39	844	XING 13.96		844	70	0.0023	5.8620	CHURCH STREET	3.9486	0.0029	900	58
463.52	633		107	633	70	0.0017	5.8636	R'MPH45	3.9456	0.0022	678	58
463.56	200		107		70	0.0000	5.8638	R	3.9434	0.0007	200	58
463.59	200			175	70	0.0005	5.8642	MPH 45	3.9428	0.0007	200	58
463.59				225	D 70	0.0007	5.8649	REVERSE CURVE	3.9421	0.0000		
463.71	631		58	631	58	0.0021	5.8670	L'RC	3.9421	0.0021	631	58
463.71						0.0000	5.8670	LITTLE BEAVER	3.9401	0.0019	748	D 92-
463.77	300			300	58	0.0010	5.8679		3.9382	0.0007	358	92
463.88	600	XING 1	152	600	58	0.0020	5.8699	R PRIVATE	3.9374	0.0000		
464.00	600		152	600	58	0.0020	5.8719	L REVERSE CURVE	3.9374	0.0000		
464.51	2700			2700	58	0.0088	5.8807	OHIO/PA LINE	3.9374	0.0197	8300	A 56-
465.52	5332	XING 15.10		5332	56	0.0174	5.8981	PUTNAM SATREET	3.9177	0.0004	124	56
465.59	400	XING 15.29		400	58	0.0013	5.8994	MYRTLE STREET	3.9173	0.0014	400	56
465.64	256	XING 1	152	256	56	0.0008	5.9002	L WALTERS STREET	3.9159	0.0008	256	56
465.74	500		152	500	58	0.0016	5.9018	R REVERSE CURVE	3.9151	0.0017	500	56
465.95	1100	XING 16.5		312	58	0.0010	5.9029	MPH50(TT) VIRGINIA S	3.9134	0.0037	1100	56
466.14	1000		129			0.0000	5.9029	L	3.9097	0.0034	1000	56
466.34	1100		142			0.0000	5.9029	L	3.9063	0.0037	1100	56
466.51	900			2616	A 56	0.0117	5.9145		3.9026	0.0030	900	56
467.01	2600	OHIO		4682	A 73	0.0108	5.9254	US 30	3.8995	0.0088	2600	56
467.20	1000		115			0.0000	5.9254	L	3.8907	0.0034	1000	56
467.27	400			82	90	0.0002	5.9256		3.8873	0.0014	400	56
467.27				406	D 90	0.0009	5.9265	REVERSE CURVE	3.8860	0.0000		
467.37	500		73	500	73	0.0013	5.9278	R	3.8860	0.0017	500	56
467.46	500		112	500	73	0.0013	5.9291	R'MPH50(TT)	3.8843	0.0017	500	56
467.52	320				73	0.0000	5.9291		3.8826	0.0011	320	56
467.58	300			298	73	0.0008	5.9299		3.8815	0.0010	300	56
467.58				322	D 73	0.0009	5.9308		3.8805	0.0000		
467.62	200	XING 1	56	200	56	0.0007	5.9315	R PRIVATE	3.8805	0.0007	200	56
467.62						0.0000	5.9315		3.8798	0.0002	68	D 60-
467.75	700			700	56	0.0024	5.9339		3.8796	0.0020	632	60
467.80	252	XING 1	60	252	56	0.0009	5.9347	R BROADWAY	3.8776	0.0008	252	60
467.80						0.0000	5.9347		3.8768	0.0002	54	D 60-
467.87	400	XING 16.40		400	56	0.0014	5.9361	TT MPH 35 WASHINGT	3.8767	0.0000		63
467.99	600	XING 1	63	600	56	0.0020	5.9381	R MARKET STREET	3.8767	0.0026	949	63
467.99		XING 16.60				0.0000	5.9381	JACKSON STREET	3.8738	0.0006	178	D 72-
468.10	600			600	56	0.0020	5.9401	E LIVERPOOL	3.8733	0.0011	422	72
468.21	600		92	600	56	0.0020	5.9421	R	3.8722	0.0016	600	72
468.33	600			600	56	0.0020	5.9442	MPH35	3.8706	0.0016	600	72
468.44	600		95	600	56	0.0020	5.9462	R	3.8691	0.0016	600	72
468.52	400			400	56	0.0014	5.9476		3.8675	0.0011	400	72
468.57	300			300	56	0.0010	5.9486		3.8664	0.0008	300	72
468.67	500		162	500	56	0.0017	5.9503	L	3.8656	0.0013	500	72
468.84	900			900	56	0.0030	5.9533		3.8643	0.0024	900	72
468.93	500		72	500	56	0.0017	5.9550	L	3.8620	0.0013	500	72
468.93						0.0000	5.9550		3.8606	0.0010	408	D 90-
469.05	600		96	600	56	0.0020	5.9570	L	3.8596	0.0004	172	90
469.26	1100			1100	56	0.0037	5.9607		3.8593	0.0025	1100	90

APPENDIX B4 — PAGE 11 — COLUMN 2

MILES	DISTANCE ft	HWY GRADE CROSSING	MAX MPH	d DISTANCE ft	SPEED mph	dT, hours	TIME, hours	STATION	TIME, hours	dT, hours	d DISTANCE ft	SPEED mph
	READ DOWN WESTBOUND TRAFFIC								READ UP EASTBOUND TRAFFIC			
469.43	900		94	900	58	0.0030	5.9638	L	3.8570	0.0019	900	90
469.47	215			215	56	0.0007	5.9645		3.8561	0.0013	607	90
470.03	3000			1233	56	0.0042	5.9687		3.8638	0.0067	2996	A 80
470.26	1200		95	509	A 56	0.0017	5.9704	R	3.8471	0.0006	257	80
470.51	1337			443	60	0.0014	5.9718	MPH 50(TT)	3.8465	0.0032	1337	80
470.78	1400			541	A 60	0.0017	5.9734	BELT LINE	3.8433	0.0075	2921	A 67
471.12	1800		95	459	63	0.0014	5.9748	L	3.8358	0.0024	834	67
471.29	900		90	1691	A 63	0.0047	5.9795	L	3.8335	0.0025	900	67
471.53	1131		101	3309	72	0.0087	5.9887	L'OKWB	3.8309	0.0032	1131	67
471.66	500			4896	A 72	0.0114	5.9995		3.8277	0.0023	800	67
471.83	900		252			0.0000	5.9996	R	3.8255	0.0025	900	67
471.96	700			88	90	0.0002	5.9998		3.8229	0.0020	700	67
472.07	600		110	800	90	0.0013	6.0011	L	3.8210	0.0017	600	67
472.23	800			551	90	0.0012	6.0022	MPH 50(TT)	3.8193	0.0018	620	67
472.23				246	D 90	0.0006	6.0028		3.8175			
472.46	1255		80	1256	80	0.0030	6.0056	R	3.8175	0.0000		
472.52	300			300	80	0.0007	6.0065		3.8175	0.0000		
472.61	500			500	80	0.0012	6.0076		3.8175	0.0094	2837	A 45
472.78	900		96	900	80	0.0021	6.0098	R	3.8081	0.0013	298	45
473.05	1380	XING 23.47		1100	80	0.0026	6.0124	18TH STREET	3.8068	0.0058	1380	45
473.05				280	D 80	0.0007	6.0131	WELLSVILLE	3.8010			
473.22	900		67	900	67	0.0025	6.0156	L	3.8010	0.0038	900	45
473.35	700			700	67	0.0020	6.0176		3.7972	0.0029	700	45
473.48	700		120	700	67	0.0020	6.0196	L	3.7943	0.0029	700	45
473.52	200			200	67	0.0006	6.0202		3.7913	0.0008	200	45
473.67	800			800	67	0.0023	6.0224		3.7905	0.0034	800	45
473.84	900		81	900	67	0.0025	6.0250	L	3.7871	0.0036	900	45
474.11	1400			1400	67	0.0040	6.0289		3.7833	0.0059	1400	45
474.30	1000		67	1000	67	0.0028	6.0318	L	3.7774	0.0042	1000	40
474.53	1222			1222	67	0.0036	6.0352		3.7732	0.0051	1222	45
474.62	500			500	67	0.0014	6.0366		3.7681	0.0021	500	45
474.74	600		107	600	67	0.0017	6.0383	L	3.7660	0.0025	600	45
474.81	400			39	67	0.0001	6.0384	RC	3.7636	0.0017	400	45
474.81				361	D 67	0.0012	6.0397		3.7618			
474.94	700		45	700	45	0.0029	6.0426	R	3.7618	0.0029	700	40
474.94						0.0000	6.0426		3.7588	0.0003	70	D 50
475.08	700	XING 25.73		700	45	0.0029	6.0455	TT MPH 35 PRIVATE	3.7586	0.0024	630	50
475.23	800		50	800	45	0.0034	6.0489	L YELLOW CREEK	3.7562	0.0030	800	50
475.34	600			600	45	0.0025	6.0514	MPH 35	3.7531	0.0002	61	D 54
475.52	952		54	952	45	0.0040	6.0554	R RC	3.7529	0.0052	1491	54
475.79	1400		54	1400	45	0.0059	6.0613	R	3.7477	0.0049	1400	54
476.00	1100			1100	45	0.0046	6.0660		3.7428	0.0003	100	D 60
476.09	496		86	496	45	0.0021	6.0680	L	3.7424	0.0076	2410	60
476.35	1400		95	1400	45	0.0059	6.0739	L'DIV POST	3.7348	0.0078	2060	A 39
476.49	700			700	45	0.0029	6.0769		3.7271	0.0006	130	39
476.68	1000			1000	45	0.0042	6.0811		3.7264	0.0048	1000	39
476.90	1200		78	1200	45	0.0051	6.0861	L	3.7216	0.0058	1200	39
477.13	1200			850	45	0.0027	6.0889		3.7157	0.0068	1200	39
477.13		XING 27.88		419	A 45	0.0017	6.0906	DANDOS CROSSING	3.7089	0.0000		
477.55	2195		72	1081	50	0.0041	6.0947	R	3.7089	0.0107	2195	39
477.70	800		92	467	A 50	0.0017	6.0963	R	3.6983	0.0039	800	39
477.85	800		63	2346	54	0.0084	6.1047	R	3.6954	0.0039	800	39
478.00	800		60	110	54	0.0004	6.1051	R	3.6915	0.0039	800	39
478.30	1800	XING 28.05		1292	A 54	0.0042	6.1092	HAMMONDSVILLE WA	3.6876	0.0078	1600	39
478.44	700		105	700	63	0.0021	6.1113	L	3.6798	0.0034	700	39
478.55	575		63	309	D 62	0.0013	6.1127	L	3.6764	0.0028	575	39
478.72	900		39	900	39	0.0044	6.1170	L	3.6736	0.0044	900	39
478.76	225			225	39	0.0011	6.1181	REVC	3.6693	0.0001	23	D 41
478.97	1100		41	1100	39	0.0053	6.1235	R	3.6692	0.0050	1077	41
479.14	900			900	39	0.0044	6.1279		3.6642	0.0005	120	D 50
479.40	1400		57	1400	39	0.0068	6.1347	L	3.6637	0.0053	1400	50
479.55	800	XING 30.04		800	39	0.0039	6.1385	EAST AVENUE	3.6584	0.0030	800	50
479.74	1000		82	1000	39	0.0049	6.1434	R	3.6553	0.0038	1000	50
479.95	1100			1100	39	0.0053	6.1487	IRONDALE	3.6516	0.0042	1100	50
480.20	1300		84	1300	39	0.0063	6.1550	L	3.6474	0.0049	1300	50
480.44	1300		57	1300	39	0.0063	6.1614	L	3.6425	0.0049	1300	50
480.56	590			590	39	0.0029	6.1642		3.6375	0.0104	2746	50
480.61	300			300	39	0.0015	6.1657		3.6271	0.0005	144	A 48
481.01	2100		50	985	39	0.0048	6.1705	R	3.6266	0.0004	100	48
481.04	145			117	A 39	0.0006	6.1710	REVC	3.6262	0.0006	140	48
481.34	1800		50	1296	41	0.0056	6.1766	L	3.6256	0.0063	1600	48
481.57	1200		62	735	A 41	0.0031	6.1797	L	3.6193	0.0047	1200	48
481.87	1600			4814	50	0.0182	6.1979	NEW SALISBURY	3.6146	0.0063	1600	48
482.04	900		91	900	50	0.0034	6.2013	R	3.6083	0.0036	900	48
482.17	700	XING 32.55		700	50	0.0027	6.2040	PRIVATE	3.6047	0.0028	700	48
482.40	1200		74	1200	50	0.0045	6.2085	L	3.6019	0.0047	1200	48
482.56	856			855	50	0.0032	6.2117		3.5972	0.0034	855	48
482.73	900			900	50	0.0034	6.2151		3.5938	0.0036	900	48
482.91	900		85	900	50	0.0034	6.2186	L	3.5903	0.0036	900	48
483.08	900			871	50	0.0033	6.2219		3.5867	0.0035	900	48
483.08				29	D 50	0.0001	6.2220		3.5832			
483.32	1300		48	1300	48	0.0051	6.2271	R	3.5832	0.0051	1300	48
483.32						0.0000	6.2271		3.5781	0.0012	380	D 69
483.32						0.0000	6.2271		3.5769	0.0001	49	69
483.36	190			190	48	0.0007	6.2278	REVC	3.5768	0.0086	2119	A 56
483.49	700		66	700	48	0.0028	6.2306	L	3.5681	0.0024	701	56
483.55	300			300	48	0.0012	6.2318		3.5658	0.0025	899	A 50
483.55							6.2318		3.5633			
484.08	2800			2800	48	0.0110	6.2428		3.5633	0.0070	1656	50
484.29	1100		97	1100	48	0.0049	6.2477	R	3.5562	0.0036	144	A 48
484.56	1471			1471	48	0.0058	6.2530		3.5557	0.0025	633	45
484.66	500			500	48	0.0020	6.2550		3.5532	0.0020	500	48
484.83	900	XING 3:	73	900	48	0.0036	6.2585	L McGARY	3.5512	0.0036	900	46
485.30	2500	XING 36.71		2500	48	0.0099	6.2684	NOLAN	3.5477	0.0099	2500	48
485.54	1267		153	539	48	0.0021	6.2705	R	3.5378	0.0050	1267	46
485.56	100			843	A 48	0.0031	6.2736		3.5328	0.0004	100	48
485.88	1700	XING 36.11		1685	58	0.0067	6.2793	PRIVATE	3.5324	0.0067	1700	46
486.09	1100		56	1100	56	0.0037	6.2830	R	3.5257	0.0043	1100	46
486.15	300	XING 36.50		207	56	0.0007	6.2837	REV C HAYTI	3.5214	0.0012	300	46
486.15				93	D 56	0.0003	6.2840		3.5202			
486.51	1900		50	1900	50	0.0072	6.2912	L'ROGERS	3.5202	0.0075	1900	46
486.33	100			71	50	0.0000	6.2919	RC	3.5127	0.0004	100	46

READ DOWN — WESTBOUND TRAFFIC | APPENDIX B4 — STATION | READ UP — EASTBOUND TRAFFIC

PAGE 12 COLUMN 1

MILES	DISTANCE, ft	HWY GRADE CROSSING	MAX SPEED	d DISTANCE, feet	SPEED, mph	dT, hours	TIME, hours	STATION	TIME, hours	dT, hours	d DISTANCE, feet	SPEED, mph
486.53				20	D 50	0.0001	6.2918		3.5123			
486.55	100		48	100	48	0.0004	6.2920	R	3.5123	0.0004	100	48
486.82	1431		48	1431	48	0.0056	6.2976	R	3.5119	0.0056	1431	48
486.82		XING 37.39				0.0000	6.2976	MAIN STREET	3.5062	0.0003	90	D 54
486.93	500			500	48	0.0020	6.2996	RC SALINEVILLE	3.5059	0.0014	390	54
487.06	800		55	800	48	0.0032	6.3028	L	3.5045	0.0028	800	54
487.23	900			900	48	0.0036	6.3063		3.5017	0.0032	900	54
487.40	900		95	900	48	0.0036	6.3099	L	3.4986	0.0032	900	54
487.56	800			800	48	0.0032	6.3130		3.4954	0.0028	800	54
487.84	1500			1500	48	0.0059	6.3189		3.4925	0.0053	1500	54
487.97	700	XING 3	69	700	48	0.0028	6.3217	R FOUNDRY HILL ROA	3.4874	0.0025	700	54
488.03	300			300	48	0.0012	6.3229	REVC	3.4849	0.0011	300	54
488.14	600		76	600	48	0.0024	6.3252	L	3.4838	0.0021	600	54
488.22	400			400	48	0.0016	6.3268	REVC	3.4817	0.0014	400	54
488.41	1000		54	1000	48	0.0039	6.3308	R	3.4803	0.0035	1000	54
488.43	116			116	48	0.0005	6.3312	REV C	3.4768	0.0004	136	D 62
488.54	600		62	600	48	0.0024	6.3336	L	3.4764	0.0018	600	62
488.75	1100		62	1100	48	0.0043	6.3379	L MPH 35	3.4746	0.0034	1100	62
488.75		XING 39.33				0.0000	6.3379	DOBSON	3.4712	0.0006	197	D 72
488.79	208			208	48	0.0006	6.3388	REV C	3.4706	0.0000	11	72
488.92	700		62	575	48	0.0023	6.3410	R	3.4706	0.0018	700	72
489.08	800		86			0.0000	6.3410	R	3.4688	0.0036	1341	72
489.25	900			781	A 48	0.0022	6.3433		3.4652	0.0011	418	A 70-
489.55	1600		152	2663	54	0.0093	6.3526	L	3.4641	0.0000		
489.64	500		152	500	54	0.0016	6.3543	L	3.4641	0.0000		
489.78	700			700	54	0.0025	6.3568		3.4641	0.0094	3022	A 50-
489.93	800	XING 4	134	800	54	0.0028	6.3596	R HULL ROAD	3.4547	0.0020	919	50
490.04	600			600	54	0.0021	6.3617		3.4527	0.0023	600	50
490.54	2613		95	2376	54	0.0083	6.3700	L MPH 35(TT)	3.4504	0.0099	2613	50
490.68	788		95	1109	A 54	0.0036	6.3737	L	3.4405	0.0029	788	50
490.83	800			797	62	0.0022	6.3759		3.4378	0.0030	800	50
490.83				1874	A 62	0.0053	6.3811		3.4046			
491.16	1700		72			0.0000	6.3811	R	3.4346	0.0064	1700	50
491.27	600			415	72	0.0011	6.3822		3.4282	0.0023	600	50
491.38	600		108	600	72	0.0016	6.3838	R	3.4259	0.0023	600	50
491.53	800			800	72	0.0021	6.3859		3.4236	0.0030	800	50
491.59	300			258	72	0.0000	6.3859		3.4206	0.0011	300	50
491.59				42	D 72	0.0001	6.3860		3.4195			
491.84	1300		70	1300	70	0.0035	6.3895	L	3.4195	0.0049	1300	50
491.91	400			48	70	0.0001	6.3896	MPH 35	3.4146	0.0015	400	50
491.91				352	D 70	0.0011	6.3907	REVERSE CURVE	3.4130			
492.06	800		50	800	50	0.0030	6.3936	FOK	3.4130	0.0030	800	50
492.06						0.0000	6.3936	REVERSE CURVE	3.4100	0.0002	91	D 54
492.20	700		54	700	50	0.0027	6.3964	L	3.4098	0.0022	639	54
492.20						0.0000	6.3964		3.4075	0.0023	895	D 95
492.31	800			800	50	0.0023	6.3987	SUMMITVILLE	3.4052	0.0000	14	95
492.37	310		95	310	50	0.0012	6.3999	L W END 2 TRACKS	3.4052	0.0000	0	95
492.52	800			800	50	0.0030	6.4029	MPH 40(TT)	3.4052	0.0018	800	95
492.53	32			32	50	0.0001	6.4030		3.4036	0.0001	32	95
492.62	500	XING 4	95	500	50	0.0018	6.4049	R PRIVATE	3.4036	0.0103	5146	95
493.08	2400			2400	50	0.0091	6.4140		3.3933			
493.53	2400			2400	50	0.0091	6.4231		3.3933	0.0096	4800	95
493.53						0.0000	6.4231		3.3837	0.0017	827	A 93-
494.53	5280			3298	50	0.0123	6.4354		3.3821	0.0050		
495.25	3800	XING 46 30		487	A 50	0.0017	6.4371		3.3821	0.0103	4647	A 78-
495.46	1100		95	248	54	0.0009	6.4380	L	3.3718	0.0001	50	78
495.55	460			9778	A 54	0.0242	6.4621		3.3716	0.0011	460	78
496.53	5200			3049	95	0.0061	6.4682		3.3705	0.0126	5200	78
496.72	1000			55	D 95	0.0001	6.4683		3.3579	0.0024	1000	78
496.87	800		93	800	93	0.0000	6.4683	L	3.3556	0.0019	800	78
497.53	3480			3104	93	0.0065	6.4748		3.3535	0.0084	3480	78
497.53				376	D 93	0.0008	6.4755		3.3451	0.0000		
497.76	1200		78	1200	78	0.0029	6.4784	L KENSINGTON	3.3451	0.0029	1200	78
497.76						0.0000	6.4784		3.3422	0.0009	491	D 95
497.95	1000			1000	78	0.0024	6.4808	MPH 40(TT)	3.3412	0.0011	569	95
498.12	900		95	900	78	0.0022	6.4830	R	3.3401	0.0018	900	95
498.42	1600			1600	78	0.0039	6.4869	WEST KENSINGTON	3.3383	0.0032	1600	95
498.52	524		95	524	78	0.0013	6.4881	R	3.3351	0.0010	524	95
498.56	185		95	185	78	0.0004	6.4886	R	3.3341	0.0004	185	95
499.13	3039			3039	78	0.0074	6.4960		3.3337	0.0014	786	D 120
499.13							6.4960		3.3323	0.0014	264	120
499.52	2061			2061	78	0.0050	6.5010		3.3309	0.0033	2061	120
500.37	4500	XING 51.18		1681	78	0.0041	6.5051	BROGAN ROAD	3.3277	0.0071	4500	120
500.43	300	XING 5	164	5474	A 78	0.0119	6.5170	L LIPPINCOT ROAD	3.3206	0.0005	300	120
500.52	476	XING 51.85		12532	A 95	0.0219	6.5390	PRIVATE	3.3201	0.0008	476	120
501.53	5342	XING 52.28				0.0000	6.5390	PRIVATE	3.3193	0.0084	5342	120
501.68	800					0.0000	6.5390		3.3109	0.0013	800	120
501.78	500		268			0.0000	6.5390	R	3.3096	0.0008	500	120
502.49	3752					0.0000	6.5390		3.3058	0.0056	3752	120
502.66	900					0.0000	6.5390		3.3039	0.0014	900	120
502.96	1600	XING 5	134			0.0006	6.5390	R US HWY 30	3.3015	0.0026	1600	120
503.11	800					0.0000	6.5390	E ROCHESTER	3.2990	0.0013	800	120
503.51	2073		134	1346	120	0.0021	6.5411	L	3.2977	0.0033	2073	120
503.56	300		134	300	120	0.0005	6.5416	L	3.2944	0.0036	300	120
503.70	714						6.5416		3.2940	0.0011	714	120
503.70		JCT 54.6			120	0.0000	6.5416	JCT MAHONING SECC	3.2928	0.0000		120
503.74	213			213	120	0.0003	6.5418	BAYARD	3.2928	0.0003	213	120
503.82	400	XING 5	233	400	120	0.0006	6.5425	R BAYARD STREET	3.2925	0.0006	400	120
503.89	416			416	120	0.0007	6.5432		3.2919	0.0007	416	120
503.96	370			370	120	0.0006	6.5438		3.2912	0.0006	370	120
504.89	4900	XING 56.88		4900	120	0.0077	6.5515	PRIVATE	3.2906	0.0077	4900	120
505.89	5276	XING 57.57		5276	120	0.0083	6.5598	PRIVATE	3.2829	0.0083	5276	120
506.00	600			600	120	0.0009	6.5608		3.2746	0.0009	600	120
506.28	1472		95	1472	120	0.0023	6.5631	R	3.2738	0.0023	1472	120
506.89	3200			3200	120	0.0051	6.5682		3.2713	0.0051	3200	120
507.89	5280			5280	120	0.0081	6.5765		3.2583	0.0083	5280	120
508.06	900			900	120	0.0014	6.5779	HUGLE RUN	3.2579	0.0014	900	120
508.26	1074	XING 5	134	1074	120	0.0017	6.5796	L PRIVATE	3.2565	0.0017	1074	120
508.89	3300			3300	120	0.0052	6.5848	MP	3.2548	0.0052	3300	120
508.95	300			300	120	0.0005	6.5853		3.2496	0.0005	300	120
509.08	700		190	700	120	0.0011	6.5864	R/REVC	3.2491	0.0011	700	120
509.21	700		190	700	120	0.0011	6.5875	L/REVC	3.2480	0.0011	700	120

READ DOWN — WESTBOUND TRAFFIC | APPENDIX B4 — STATION | READ UP — EASTBOUND TRAFFIC

PAGE 12 COLUMN 2

MILES	DISTANCE, ft	HWY GRADE CROSSING	MAX SPEED	d DISTANCE, feet	SPEED, mph	dT, hours	TIME, hours	STATION	TIME, hours	dT, hours	d DISTANCE, feet	SPEED, mph
509.38	893			893	120	0.0014	6.5889		3.2469	0.0014	893	120
509.72	1800		134	1800	120	0.0026	6.5917	L	3.2455	0.0026	1800	120
509.89	900			900	120	0.0014	6.5932		3.2427	0.0014	900	120
510.88	5230			5230	120	0.0083	6.6014	PARISONOARD	3.2413	0.0083	5230	120
511.22	1800			1800	120	0.0026	6.6043		3.2330	0.0026	1800	120
511.39	900		134	900	120	0.0014	6.6057	R PARKS RD	3.2302	0.0014	900	120
511.90	2660			2660	120	0.0042	6.6099		3.2287	0.0042	2660	120
512.05	800			800	120	0.0013	6.6111		3.2245	0.0015	800	120
512.66	3229		190	3229	120	0.0051	6.6162	R	3.2233	0.0051	3229	120
512.91	1300	XING 64.88		1300	120	0.0021	6.6183	PRIVATE	3.2182	0.0021	1300	120
513.87	5086	XING 65.08		5086	120	0.0080	6.6263	PRIVATE	3.2161	0.0080	5086	120
514.89	5383	XING 65.71		5383	120	0.0085	6.6348	PRIVATE	3.2081	0.0085	5383	120
515.06	900			900	120	0.0014	6.6362	NKP RD	3.1996	0.0014	900	120
515.34	1473		134	1473	120	0.0023	6.6386	R	3.1982	0.0023	1473	120
515.89	2900			2900	120	0.0046	6.6431		3.1959	0.0046	2900	120
516.15	1364			1364	120	0.0022	6.6453	LOUISVILLE	3.1913	0.0022	1364	120
516.52	2000		129	2000	120	0.0032	6.6484	L	3.1891	0.0032	2000	120
516.94	2200			2200	120	0.0035	6.6519	FAIRHOPE	3.1860	0.0035	2200	120
517.30	1900			1900	120	0.0030	6.6549		3.1825	0.0030	1900	120
517.51	1124		134	1124	120	0.0018	6.6567	R	3.1795	0.0018	1124	120
517.87	1900			1900	120	0.0030	6.6597		3.1777	0.0030	1900	120
517.87		CPFAIRHOPE				0.0000	6.6597	JCT BAYARD BR	3.1747	0.0000		120
517.97	500	XING 96.56		500	120	0.0008	6.6605	BEK ROAD	3.1747	0.0008	500	120
518.08	600	XING 97.13		600	120	0.0009	6.6614	BROADWAY	3.1739	0.0009	600	120
518.14	300		134	300	120	0.0005	6.6619	L	3.1730	0.0005	300	120
518.14						0.0000	6.6619		3.1725	0.0000		120
518.20	300		134	300	120	0.0005	6.6624	L	3.1725	0.0005	300	120
518.53	1753			1753	120	0.0028	6.6651		3.1720	0.0028	1753	120
518.74	1100		129	1100	120	0.0017	6.6669	L	3.1693	0.0017	1100	120
518.98	1300			1300	120	0.0021	6.6689		3.1675	0.0021	1300	120
519.98	5285			5288	120	0.0083	6.6773		3.1655	0.0083	5288	120
520.98	5280			5280	120	0.0083	6.6856		3.1571	0.0083	5280	120
521.97	5236			5236	120	0.0083	6.6939		3.1488	0.0083	5236	120
522.28	1600			1600	120	0.0025	6.6964		3.1405	0.0025	1600	120
522.62	1800			1800	120	0.0028	6.6992	CANTON	3.1380	0.0028	1800	120
522.88	1400		185	1400	120	0.0022	6.7014	R	3.1352	0.0022	1400	120
523.00	610			610	120	0.0010	6.7024		3.1330	0.0010	610	120
523.12	634			634	120	0.0010	6.7034		3.1320	0.0000		
523.12						0.0000	6.7034		3.1320	0.0058	3605	A 11-
523.99	4600			4600	120	0.0073	6.7107		3.1262	0.0000		
525.00	5306			5306	120	0.0084	6.7190		3.1262	0.0192	10428	A 91
525.00		XING 103.57				0.0000	6.7190	MARYLAND AVENUE	3.1070	0.0016	756	91
525.28	1500	XING 1[illegible]107		1500	120	0.0024	6.7214	R RAFF RD	3.1054	0.0061	2798	A 82
526.00	3780			3780	120	0.0060	6.7274	WEST CANTON	3.0993	0.0005	220	82
526.56	2966	XING 105.56		2966	120	0.0073	6.7346	PERRY DRIVE	3.0988	0.0083	3600	82
526.56							6.7346	EAST END MASSILLO	3.0905			
526.68	634			634	120	0.0073	6.7419		3.0905			
526.83	800		190	800	120	0.0073	6.7492	L	3.0905	0.0018	800	82
527.00	897					0.0000	6.7492		3.0887	0.0021	897	82
527.15	800			1514	120	0.0024	6.7516		3.0866	0.0026	1101	82
527.15				897	D 12	0.0016	6.7532		3.0840	0.0019	809	A 79
527.43	1500	XING 1[illegible]	91	1500	91	0.0031	6.7563	L GENOA STREET	3.0821	0.0094	3540	A 62
528.00	2991			2995	91	0.0061	6.7624		3.0727	0.0046	1516	62
528.04	192				91	0.0000	6.7624		3.0680	0.0006	192	62
528.04				228	D 91	0.0005	6.7629	PERRY HEIGHTS	3.0674			
528.30	1400		82	1400	82	0.0032	6.7662	R	3.0674	0.0022	701	A 57
528.62	1700			1700	82	0.0039	6.7701		3.0652	0.0023	704	57
528.70	400		134	400	82	0.0009	6.7710	L	3.0629	0.0013	400	57
529.00	1600			1600	82	0.0037	6.7747		3.0615	0.0053	1600	57
529.23	1187			1187	82	0.0027	6.7774		3.0562	0.0039	1187	57
529.32	500		134	500	82	0.0012	6.7786	L	3.0523	0.0017	500	57
529.46	700			629	82	0.0015	6.7801		3.0506	0.0023	700	57
529.46				71	D 82	0.0002	6.7802		3.0483			
529.76	1600	XING 1[illegible]	79	1600	79	0.0038	6.7841	R TTMPH 50 HIGH STF	3.0483	0.0053	1600	57
530.00	1300			1300	79	0.0031	6.7872	REED	3.0430	0.0043	1300	57
530.12	600			248	79	0.0006	6.7878		3.0387	0.0020	600	57
530.12				352	D 79	0.0009	6.7887		3.0367			
530.48	1909		62	1909	62	0.0058	6.7945	L	3.0367	0.0053	1909	57
530.57	500			413	62	0.0013	6.7958	REVERSE CURVE	3.0303	0.0017	500	57
530.57				87	D 62	0.0003	6.7961		3.0287			
530.99	2200		57	2200	57	0.0073	6.8034	R	3.0287	0.0073	2200	57
530.99					57	0.0000	6.8034		3.0214	0.0001	17	D 58
531.07	400		58	400	57	0.0013	6.8047	R	3.0213	0.0013	383	58
531.07						0.0000	6.8047		3.0200	0.0017	642	D 88
531.28	1100			1100	57	0.0037	6.8084	MASSILLON	3.0184	0.0009	438	88
531.41	700		234	700	57	0.0023	6.8107	R	3.0174	0.0015	700	88
531.75	1800			1800	57	0.0060	6.8167		3.0159	0.0039	1800	88
531.95	1081		92	1081	57	0.0035	6.8202	L	3.0121	0.0020	1081	88
531.95				B&C		0.0000	6.8202	MACE TOWER	3.0087			
532.03	400			400	57	0.0013	6.8215		3.0087	0.0009	400	88
532.44	2200	XING 111.36		2200	57	0.0073	6.8289	EARL ROAD	3.0089	0.0047	2200	88
532.76	1700		88	1700	57	0.0056	6.8345	L TTMPH(70/60/50)	3.0041	0.0037	1700	88
532.76						0.0000	6.8345		3.0005	0.0007	380	D 10
532.88	594	XING 111.92		594	57	0.0020	6.8365	RC BOYDS CHN	2.9998	0.0004	234	101
533.01	700		101	700	57	0.0023	6.8388	R	2.9993	0.0013	700	101
533.18	900		101	345	57	0.0011	6.8399	R	2.9980	0.0017	900	101
533.37	1009			175	A 57	0.0006	6.8405		2.9963	0.0003	183	D10
533.37		XING 112.36					6.8405	BEAUMONT AVENUE	2.9960	0.0015	826	107
533.39	94			230	58	0.0008	6.8413		2.9945	0.0002	94	107
533.82	2300		107			0.0000	6.8413	L	2.9944	0.0041	2300	107
534.01	1000			8647	A 58	0.0169	6.8582		2.9903	0.0016	1000	107
534.05	187					0.0000	6.8582		2.9886	0.0003	187	107
534.31	1400		107			0.0000	6.8582	R	2.9882	0.0025	1400	107
534.31							6.8582	WEST END MASSILLO	2.9857			
535.01	3700					0.0000	6.8582		2.9857	0.0007	430	D 12
535.79	4100	XING 114.05		5001	A 88	0.0109	6.8682	KENYON AVENUE	2.9850	0.0118	7987	120
535.98	1000		129			0.0000	6.8682	L	2.9734	0.0016	1000	120
536.01	135			3190	A 10	0.0050	6.8732		2.9718	0.0002	135	120
537.01	5282	XING 115.19				0.0000	6.8732	SR 93	2.9716	0.0083	5282	120
537.76	4000	XING 116.15		7336	A 10	0.0122	6.8854	CLERMONT	2.9632	0.0063	4000	120
537.76		XING 116.73		2168	120	0.0036	6.8889	ALABAMA AVE	2.9569			
537.92	800		134	800	120	0.0013	6.8901	L	2.9569	0.0013	800	120

APPENDIX B4

PAGE 13 COLUMN 1

MILES	DISTANCE FT	HWY GRADE CROSSING	MAX MPH	DISTANCE FT	SPEED MPH	ET HOURS	TIME HOURS	STATION	TIME HOURS	ET HOURS	DISTANCE FT	SPEED MPH
		READ DOWN WESTBOUND TRAFFIC						STATION	READ UP EASTBOUND TRAFFIC			
538.01	511	XING 117.0x		511	120	0.0008	6.8909	RAILROAD STREET N	2.9657	0.0008	511	120
539.02	5298			5298	120	0.0084	6.8993	NORTH LAWRENCE	2.9646	0.0084	5298	120
540.03	5377	XING 119.2x		5377	120	0.0085	6.9078	ANDERSON ROAD	2.9465	0.0085	5377	120
540.75	3800	XING 118.8x		3800	120	0.0060	6.9138	PRIVATE	2.9380	0.0060	3800	120
541.02	1422	XING 1	126	1422	120	0.0022	6.9160	L PRIVATE	2.9320	0.0022	1422	120
541.04	98	XING 1	126	98	120	0.0002	6.9162	L PRIVATE	2.9298	0.0002	98	120
542.03	5200	XING 126.9x		5200	120	0.0082	6.9244	SR 94	2.9298	0.0069	4363	120
543.03	5297	XING 121.8x		5297	120	0.0084	6.9328	TANNERVILLE	2.9227	0.0000		
543.75	3800			3800	120	0.0060	6.9388		2.9227			
543.75						0.0000	6.9388		2.9227	0.0219	12533	A 95
543.97	900		130	900	120	0.0014	6.9402	R	2.9006			
544.03	598	XING 123.6x		598	120	0.0009	6.9411	WALNUT	2.9006	0.0000		
544.32	1500	XING 125.8x		1500	120	0.0024	6.9435	MARKET STREET	2.9008	0.0008	411	96
545.04	3789		128	3789	120	0.0060	6.9495	L ORRVILLE	2.9000	0.0076	3789	96
545.53	2600		134	2600	120	0.0041	6.9536	L	2.8924	0.0052	2600	96
546.04	2700			2700	120	0.0043	6.9579	(WLE OHIO)	2.8872	0.0054	2700	96
546.32	1500	XING 125.1x		712	120	0.0011	6.9590	CROWN HILL	2.8816	0.0030	1500	96
546.32		XING 125.2x		788	D 12	0.0014	6.9603	SCHOOLHOUSE ROAD	2.8786			
546.59	1400		95	1400	96	0.0028	6.9631	R WEST ORRVILLE	2.8738	0.0028	1400	96
546.85	1400		146	1400	96	0.0028	6.9659	R	2.8761	0.0002	114	D 99
547.05	1029		130	1029	96	0.0021	6.9680	R	2.8756	0.0006	328	99
547.28	1300	XING 1	132	1300	96	0.0026	6.9706	R MUSKRAT ROAD	2.8782	0.0008		
547.55	1368	XING 1	132	1368	96	0.0027	6.9733	R BACK ORRVILLE	2.8752	0.0000		
548.05	2600	XING 127.9x		2600	96	0.0052	6.9785	ROHRER ROAD	2.8752	0.0283	11664	A 51
549.05	5302	XING 128.0x		3306	96	0.0066	6.9851	PRIVATE	2.8469	0.0030	900	51
549.09	200	XING 128.7x		7786	A 96	0.0142	6.9992	APPLEGATE	2.8439	0.0007	200	51
550.05	5100		124			0.0000	6.9992	L	2.8426	0.0189	5100	51
550.36	1600					0.0000	6.9992		2.8239	0.0060	1600	51
550.53	900		126			0.0000	6.9992	L MPH 56	2.8179	0.0030	906	51
550.66	700	XING 129.6x				0.0000	6.9992	CO RD 75	2.8146	0.0026	700	51
550.66							6.9992	EAST END SMITHVILLE	2.8120			
550.85	1000		112	2630	112	0.0048	7.0040	R a	2.8120	0.0037	1006	51
550.96	600			42	112	0.0001	7.0041	REVERSE CURVE	2.8083	0.0022	600	51
550.96				1468	D 11	0.0034	7.0075		2.8060			
551.06	518		51	518	51	0.0019	7.0094	L SMITHVILLE b	2.8060	0.0019	518	51
551.27	1120		51	1120	51	0.0040	7.0136	L	2.8041	0.0042	1120	51
552.03	4000			4000	51	0.0149	7.0284		2.7999	0.0002	60	D 56
552.07	187			187	51	0.0007	7.0291	L MPH 20	2.7997	0.0142	4126	55
552.37	1600		67	1600	51	0.0059	7.0351	R c	2.7855	0.0056	1600	55
552.62	1300			1300	51	0.0048	7.0399		2.7800	0.0045	1300	56
553.03	2200		80	2200	51	0.0082	7.0481	L	2.7755	0.0076	2200	56
553.05	98		60	98	51	0.0004	7.0484	L	2.7679	0.0003	98	55
553.20	800			800	51	0.0030	7.0514		2.7676	0.0028	800	55
553.51	1600		85	815	51	0.0030	7.0544	R c	2.7649	0.0065	1600	55
553.51							7.0544	WEST END SMITHVILLE	2.7583			
553.62	600			1207	A 51	0.0042	7.0586		2.7583	0.0021	600	55
553.62				45	51	0.0001	7.0587		2.7573			
553.62				100	D 51	0.0003	7.0591		2.7573			
553.62							7.0591	EAST END WOOSTER	2.7573			
554.02	2100		50	2100	55	0.0072	7.0663	R a	2.7573	0.0072	2100	55
554.15	700		55	700	56	0.0024	7.0687	R	2.7500	0.0024	706	55
554.15						0.0000	7.0687		2.7476	0.0002	67	D 56
554.19	236			236	56	0.0006	7.0696	REVERSE CURVE	2.7474	0.0006	169	59
554.69	2000	XING 1	59	2600	56	0.0090	7.0785	L a BEGIN 2 TR PRIVA	2.7469	0.0083	2606	59
554.91	1200			1200	56	0.0041	7.0826		2.7385	0.0039	1200	59
555.05	700		65	700	56	0.0024	7.0850	L a	2.7347	0.0022	700	59
555.18	700		65	700	56	0.0024	7.0874	L	2.7324	0.0022	706	59
555.22	218			218	56	0.0008	7.0882	REVERSE CURVE	2.7302	0.0007	218	59
555.31	500		59	500	56	0.0017	7.0898	R 4	2.7295	0.0016	500	59
555.31						0.0000	7.0898		2.7279	0.0006	314	D 75
555.75	2300			2300	56	0.0079	7.0978		2.7273	0.0050	1986	75
555.90	800		76	800	56	0.0026	7.1006	L a	2.7220	0.0020	800	75
556.02	600			600	56	0.0021	7.1026		2.7200	0.0015	600	75
556.11	500			500	56	0.0017	7.1044		2.7184	0.0013	500	75
556.34	1207	XING 1	75	646	56	0.0022	7.1066	L / WOOSTER LIBERT	2.7172	0.0030	1207	75
556.34		XING 135.24				0.0000	7.1066	SOUTH STREET	2.7141	0.0016	781	D 104
556.55	1100			504	A 56	0.0017	7.1082		2.7125	0.0006	339	104
556.62	400		104	400	59	0.0015	7.1096	R g	2.7119	0.0007	400	104
556.93	1600			1600	59	0.0051	7.1147		2.7112	0.0003	188	D 110
557.04	600	XING 1	110	600	59	0.0019	7.1166	L a PRIVATE	2.7108	0.0068	3661	110
557.05	98	XING 1	110	98	59	0.0002	7.1168	L PRAIRIE ROAD	2.7040	0.0036	2062	A 106
557.05							7.1168	WEST END WOOSTER	2.7064			
557.94	4700			4700	59	0.0151	7.1319		2.7064	0.0014	788	106
558.06	700			700	A 59	0.0083	7.1402		2.6991	0.0013	700	106
558.19	800		141	800	75	0.0015	7.1417	R	2.6978	0.0011	600	106
558.20	36			36	75	0.0001	7.1418	REVERSE CURVE	2.6967	0.0001	36	106
558.31	600		141	600	75	0.0015	7.1433	L OK	2.6967	0.0011	600	106
559.03	3800			1185	75	0.0030	7.1464		2.6956	0.0068	3800	106
559.88	4500			9850	A 75	0.0208	7.1672		2.6888	0.0080	4500	106
560.02	714		106			0.0000	7.1672	KILLBUCK	2.6808	0.0013	714	106
560.77	4000			924	A 10	0.0017	7.1689		2.6795	0.0008	464	D 120
560.98	1100		134	3202	106	0.0057	7.1746	R FISH	2.6787	0.0056	3557	120
561.01	134	XING 140.1x		134	106	0.0002	7.1748	MESSNER ROAD	2.6731	0.0000		
561.01						0.0000	7.1748		2.6731	0.0158	12546	A 94
562.03	5405	XING 141.0x		5405	106	0.0097	7.1845	WILLOW ROAD	2.6573	0.0000		
563.04	5350	XING 142.9x		361	106	0.0006	7.1851	FORCE ROAD	2.6573	0.0078	3642	A 63
564.04	5232			7805	A 10	0.0131	7.1982		2.6495	0.0044	1669	A 76
564.41	2000	XING 143.3		2000	120	0.0032	7.2013	CEMETERY ROAD	2.6450	0.0038	1481	A 89
564.64	1216		181	1216	120	0.0019	7.2033	L	2.6412	0.0015	469	89
565.02	2000			2000	120	0.0032	7.2064		2.6389	0.0056	2000	89
565.25	1200		180	1200	120	0.0019	7.2083	L	2.6344	0.0033	1200	69
565.38	700	XING 144.3x		2302	120	0.0036	7.2119	WOOD	2.6311	0.0019	700	69
565.38		XING 144.4x		816	D 12	0.0014	7.2134	ROBINSON	2.6292			
565.71	1700	XING 1	94	1700	94	0.0034	7.2166	R a MARKET STREET	2.6292	0.0047	1700	69
565.89	1000	XING 144.6x		1000	94	0.0020	7.2186	SHREVE MAIN STREE	2.6240	0.0027	1000	69
566.04	750		104	467	94	0.0008	7.2196	R	2.6218	0.0021	751	66
566.04				286	D 94	0.0008	7.2204		2.6197			
566.28	1300		83	1300	83	0.0030	7.2233	R b	2.6197	0.0036	1300	69
566.38	500			387	83	0.0008	7.2241	REVERSE CURVE	2.6161	0.0014	505	69
566.38				163	D 83	0.0004	7.2246		2.6147			
566.64	1378		74	1229	76	0.0031	7.2276	L c	2.6147	0.0008	1376	69
566.64				149	D 76	0.0004	7.2280		2.6110			

APPENDIX B4

PAGE 13 COLUMN 2

MILES	DISTANCE FT	HWY GRADE CROSSING	MAX MPH	DISTANCE FT	SPEED MPH	ET HOURS	TIME HOURS	STATION	TIME HOURS	ET HOURS	DISTANCE FT	SPEED MPH
		READ DOWN WESTBOUND TRAFFIC						STATION	READ UP EASTBOUND TRAFFIC			
566.90	1400		76	1400	69	0.0038	7.2318	L	2.6110	0.0038	1400	69
566.90						0.0000	7.2318		2.6071	0.0004	149	D 76
567.00	500			500	69	0.0014	7.2332	RC	2.6067	0.0008	351	76
567.04	200		76	200	69	0.0006	7.2337	R J	2.6059	0.0005	200	76
567.34	1601	XING 1	76	1601	69	0.0044	7.2381	R BROWNS	2.6054	0.0040	1601	76
567.34						0.0000	7.2381		2.6014	0.0011	477	D 96
567.34						0.0000	7.2381		2.6003	0.0006	321	96
568.04	3700	XING 147.1x		3700	69	0.0102	7.2483	PRIVATE	2.5997	0.0088	4356	A 83
568.87	4392	XING 146.5x		4392	69	0.0121	7.2603	SCRUBBYS	2.5911	0.0074	3236	83
569.21	1800	XING 148.5x		607	69	0.0017	7.2620	BIG PRAIRIE RTE 226	2.5837	0.0041	1800	83
569.38	887		134	3359	A 69	0.0063	7.2683	L	2.5796	0.0020	887	83
570.04	3600			3201	A 83	0.0069	7.2773		2.5775	0.0080	3600	83
570.10	377			305	93	0.0006	7.2779		2.5696	0.0006	277	83
570.34	1300		134	258	D 93	0.0006	7.2785	L	2.5689	0.0030	1300	83
570.80	2400		83	2400	83	0.0055	7.2839	L a	2.5660	0.0055	2400	83
570.80						0.0000	7.2839		2.5605	0.0004	204	D 91
570.85	300			300	85	0.0007	7.2846	REVERSE CURVE	2.5600	0.0002	96	91
571.04	1000		98	1000	85	0.0023	7.2869	R b	2.5598	0.0021	1000	91
571.25	1100		110	1100	85	0.0025	7.2894	R	2.5576	0.0015	734	91
571.33	400			400	85	0.0009	7.2903		2.5562	0.0000		
571.57	1300	XING 1	80	1300	85	0.0030	7.2933	R a LAKEVILLE CR LS	2.5562	0.0111	4884	A 74
572.05	2500	XING 150.0x		2500	85	0.0057	7.2990	STATE ROUTE 179	2.5451	0.0006	246	74
572.29	1300			1300	85	0.0030	7.3020		2.5445	0.0033	1300	74
572.52	1217		91	1217	85	0.0028	7.3047	L d	2.5412	0.0044	1546	A 66
573.05	2800			1863	85	0.0043	7.3090		2.5367	0.0053	1883	66
573.05				2556	A 63	0.0056	7.3146		2.5315	0.0000		
574.05	5281			3935	91	0.0082	7.3228		2.5315	0.0152	5281	66
574.19	700			411	D 91	0.0009	7.3237		2.5163	0.0020	700	66
574.74	2900	XING 1	74	2900	74	0.0074	7.3311	R TR 385 a	2.5143	0.0083	2900	66
574.79	288			122	74	0.0003	7.3314	REVERSE CURVE	2.5060	0.0008	288	66
574.79				164	D 74	0.0004	7.3319		2.5052			
575.06	1400		56	1400	56	0.0040	7.3359	L b	2.5052	0.0040	1400	56
575.19	700			700	56	0.0020	7.3379		2.5012	0.0009	347	D 80
575.19						0.0000	7.3379		2.5003	0.0008	353	80
575.36	900		82	900	56	0.0026	7.3405	L c	2.4994	0.0021	900	80
575.36						0.0000	7.3405		2.4974	0.0002	96	D 86
575.51	804		101	804	56	0.0023	7.3428	L	2.4971	0.0016	706	86
576.06	2900	XING 155.2x		2900	56	0.0083	7.3511	TR 5	2.4956	0.0064	2900	86
576.95	4700			4700	86	0.0135	7.3646		2.4892	0.0104	4700	86
577.06	564	XING 1	88	564	56	0.0016	7.3662	R a LOUDONVILLE JCT	2.4789	0.0012	564	88
577.60	2848	XING 1	86	442	56	0.0013	7.3675	R SPRING STREET	2.4776	0.0063	2548	86
577.60				3712	A 56	0.0084	7.3759		2.4714	0.0010	502	D 104
578.07	2500			1110	A 82	0.0025	7.3784	OP MOHICAN	2.4704	0.0036	1990	104
578.90	4400			4484	96	0.0096	7.3883		2.4667	0.0080	4400	104
579.07	893	XING 1	104	893	96	0.0020	7.3903	L PRIVATE	2.4587	0.0018	893	104
579.19	610		104	610	96	0.0013	7.3916	L	2.4571	0.0011	610	104
580.08	4700			2587	96	0.0057	7.3983		2.4560	0.0003	166	D 110
580.32	1303	XING 1	110	6963	A 96	0.0139	7.4122	LE PERRY PRIVATE	2.4556	0.0160	3815	110
581.06	4000			413	104	0.0008	7.4130		2.4496	0.0004	233	D 117
581.08						0.0000	7.4130		2.4452	0.0004	273	117
581.27	1000	XING 160.0x		967	104	0.0018	7.4148	BRIDGE STREET	2.4448	0.0064	3631	A 11x
582.08	4264	XING 1	140	2986	A 10	0.0053	7.4201	R PRIVATE	2.4384	0.0111	5064	A 87
582.67	3100		128	3047	110	0.0062	7.4262	R PERRYSVILLE	2.4273	0.0001	15	87
583.09	2260	XING 162.5x		5740	A 11	0.0084	7.4342	PRIVATE	2.4273	0.0044	2260	87
583.87	4100	XING 162.8x		1881	120	0.0031	7.4378	TOWNSHIP ROAD 10x	2.4229	0.0108	5002	A 82
584.10	1197		128	1197	120	0.0019	7.4397	L	2.4123	0.0022	958	82
584.19	500		128	500	120	0.0008	7.4405	L	2.4101	0.0010	500	82
584.38	1000			268	120	0.0004	7.4408		2.4090	0.0023	1000	82
584.38				732	D 12	0.0013	7.4420		2.4067	0.0050	2086	A 74
584.68	1600		97	1600	97	0.0031	7.4453	L PERRYSV	2.4017	0.0021	806	74
584.95	1390	XING 163.3x		1390	97	0.0027	7.4480	COUNTY ROAD 1075	2.3996	0.0036	1390	74
585.10	800	XING 1	129	800	97	0.0016	7.4496	L PRIVATE	2.3960	0.0020	800	74
585.25	800		129	800	97	0.0016	7.4512	L	2.3940	0.0020	800	74
585.76	2700			2306	97	0.0045	7.4557		2.3919	0.0069	2700	74
585.76				394	D 97	0.0006	7.4565		2.3850			
585.97	1104		82	1104	82	0.0026	7.4590	R a	2.3850	0.0028	1104	74
586.10	700			700	82	0.0016	7.4607		2.3822	0.0018	700	74
586.61	2700			2517	82	0.0058	7.4665		2.3804	0.0069	2700	74
586.61				183	D 82	0.0004	7.4669		2.3735			
586.90	1500		74	1500	74	0.0030	7.4706	L b	2.3735	0.0038	1500	74
586.90						0.0000	7.4706		2.3697	0.0006	259	D 86
587.11	1102			1102	74	0.0028	7.4736		2.3690	0.0016	743	86
587.45	1800			1800	74	0.0046	7.4782		2.3672	0.0038	1800	86
587.60	800		93	800	74	0.0020	7.4802	L c	2.3634	0.0017	800	86
587.80	1067	XING 168.6x		1067	74	0.0027	7.4830	PRIVATE	2.3617	0.0023	1067	86
588.03	1200		90	1200	74	0.0031	7.4860	R d	2.3594	0.0026	1200	86
588.10	400			400	74	0.0010	7.4871		2.3569	0.0009	400	86
588.26	800			800	74	0.0020	7.4891	COULTER	2.3560	0.0017	800	86
588.39	700		97	700	74	0.0018	7.4909	R e	2.3543	0.0015	700	86
588.94	2900	XING 167.8x		2900	74	0.0074	7.4983	PRIVATE	2.3528	0.0060	2900	86
589.01	400	XING 1	89	231	74	0.0006	7.4989	R LUCAS 1	2.3467	0.0009	400	86
589.10	467			4205	A 74	0.0097	7.5086		2.3458	0.0010	467	86
589.16	295						7.5086		2.3448	0.0006	295	86
589.33	900		89				7.5086	L g	2.3442	0.0019	900	86
589.33		XING 1168.5					7.5086	UNION STREET	2.3423	0.0003	162	D 95
589.33		XING 166.46					7.5086	DEKROM	2.3420	0.0017	831	95
589.72	2100	XING 166.68				0.0000	7.5086	LUCAS HANLEY BRET	2.3403	0.0022	1107	95
590.10	2000		107	1718	89	0.0037	7.5122	R h	2.3381	0.0040	2000	95
590.52	2200		134	2200	89	0.0047	7.5170	R	2.3341	0.0044	2200	95
590.80	1500			1500	89	0.0032	7.5202	OP LUCAS	2.3297	0.0030	1500	95
590.94	693		110	690	86	0.0015	7.5216	R i	2.3267	0.0014	693	95
591.07	700		98	700	89	0.0015	7.5231	R	2.3254	0.0014	700	95
591.11	200			200	89	0.0004	7.5236		2.3240	0.0004	200	95
591.60	2600			1607	89	0.0034	7.5270		2.3236	0.0052	2600	95
592.03	2300		95	2183	A 89	0.0044	7.5314	L i	2.3184	0.0046	2300	95
592.03						0.0000	7.5314		2.3138	0.0012	650	D 116
592.10	377	XING 171.5x		1507	95	0.0030	7.5344	PRIVATE	2.3126	0.0027	1673	116
593.11	5281	XING 171.9x		5281	95	0.0105	7.5450	SITTS ROAD	2.3099	0.0000		
594.05	5000			5000	95	0.0100	7.5549		2.3099	0.0064	3318	A 11x
594.10	279		152	279	95	0.0006	7.5555	L	2.3043	0.0184	8989	A 9x
594.16	300	XING	152	63	95	0.0001	7.5556	URC ILLINOIS AVENUE	2.2860	0.0000		
594.25	461		152			0.0000	7.5556	R 1 MPH	2.2860	0.0007		

APPENDIX B4

READ DOWN WESTBOUND TRAFFIC — STATION — READ UP EASTBOUND TRAFFIC

PAGE 14 COLUMN 1

MILES	DISTANCE ft	HWY GRADE CROSSING	MAX MPH	d DISTANCE ft	SPEED mph	dT hours	TIME hours	STATION	TIME hours	dT hours	d DISTANCE ft	SPEED mph
595.11	4600			8261	A 95	0.0150	7.5706		2.2860	0.0040	1888	90
595.48	2000					0.0000	7.5706		2.2840	0.0042	2000	90
595.60	600		116	50	113	0.0001	7.5707	R a	2.2796	0.0013	600	90
595.71	600	XING 1	134	99	D 11	0.0002	7.5709	R 4TH STREET	2.2785	0.0013	600	90
595.82	600	XING 1	110	600	110	0.0010	7.5719	R 5TH STREET b	2.2773	0.0013	600	90
596.11	1512	XING 175.0C		1512	110	0.0026	7.5745	MANSFIELD ORANGE	2.2760	0.0032	1512	90
596.83	3800	XING 175.3C		3213	110	0.0059	7.5800	SURREY ROAD	2.2728	0.0080	3800	90
596.83				587	D 11	0.0011	7.5811		2.2646	0.0000		
596.83			90		90	0.0000	7.5811	B&O RR XING c	2.2646	0.0000		90
596.83		XING 175.57				0.0000	7.5811	MAIN STREET	2.2646	0.0000		
596.89	300	XING 1	190	300	90	0.0006	7.5818	LW MANS LONGVIEW	2.2646	0.0010	455	90
597.13	1274			1274	90	0.0027	7.5845		2.2639	0.0000		
597.30	900			900	90	0.0019	7.5864		2.2639	0.0081	3574	A 78-
597.38	400		190	400	90	0.0006	7.5872	L	2.2556	0.0000		
597.62	1268			1268	90	0.0027	7.5899		2.2558	0.0003	113	78
597.96	1800	XING 1	124	1800	90	0.0038	7.5937	L EMP STL MILL	2.2555	0.0044	1800	78
598.13	900	XING 177.2		900	90	0.0019	7.5956	PRIVATE	2.2512	0.0022	900	78
598.75	3300			1500	90	0.0032	7.5987		2.2490	0.0080	3300	78
599.13	1987		134	0	90	0.0000	7.5987	L	2.2410	0.0046	1987	78
599.68	2900			5825	A 90	0.0126	7.6115		2.2361	0.0070	2900	78
599.68				108	108	0.0002	7.6117		2.2291	0.0000		
599.68				758	D 10	0.0016	7.6132		2.2291	0.0000		
599.89	1100		78	1100	78	0.0027	7.6159	R d	2.2291	0.0027	1100	78
599.89						0.0000	7.6159		2.2264	0.0014	594	D 104
600.00	593			593	78	0.0014	7.6173	REVERSE CURVE	2.2250	0.0006	277	104
600.13	700		107	700	78	0.0017	7.6190	L e	2.2245	0.0000		
600.32	997		107	997	78	0.0024	7.6216	L	2.2245	0.0000		
600.47	800			800	76	0.0019	7.6234		2.2245	0.0000		
600.58	600		178	600	78	0.0015	7.6249	L	2.2245	0.0000		
600.72	700	XING 179.5		700	78	0.0017	7.6266	LEPPO ROAD	2.2245	0.0000		
600.87	800		135	800	78	0.0019	7.6285	R	2.2245	0.0088	4630	A 93-
601.11	1300	XING 180.5		1300	78	0.0032	7.6317	SPRING MILL	2.2156	0.0018	889	93
601.40	1510		134	1510	78	0.0037	7.6353	L	2.2138	0.0031	1510	93
602.14	3900			3000	78	0.0073	7.6426		2.2107	0.0079	3900	93
602.65	2700			4847	A 78	0.0103	7.6529		2.2028	0.0055	2700	93
602.78	701		167	0		0.0000	7.6529	L f	2.1973	0.0014	701	93
603.03	1300	XING 181.9		954	93	0.0019	7.6548	PRIVATE	2.1958	0.0026	1300	93
603.14	600		93	606	93	0.0012	7.6560	L g	2.1932	0.0012	600	93
603.32	927	XING 1	93	927	93	0.0019	7.6579	L ROCK ROAD	2.1920	0.0019	927	93
603.32							7.6579		2.1901	0.0015	643	D 12
603.32						0.0000	7.6579	TOLEDO JCT	2.1886	0.0001	57	120
604.15	4400	XING 183.8		4400	93	0.0090	7.6669	PRIVATE	2.1885	0.0065	3600	120
605.08	4912	XING 184.2		4912	93	0.0100	7.6769	SHELBY ROAD	2.1830	0.0078	4912	120
606.16	5710			1668	60	0.0034	7.6803		2.1752	0.0072	4553	120
607.17	5295	XING 186.4		13359	A 93	0.0236	7.7039	HORNING ROAD	2.1680	0.0000	0	
608.17	5308			200	120	0.0003	7.7043		2.1680	0.0000	0	69
609.18	5302			5302	120	0.0084	7.7126		2.1680	0.0000	0	69
609.52	1800			2866	120	0.0046	7.7172		2.1680	0.0361	18898	A 74
609.62	554		163	554	120	0.0009	7.7180	R	2.1319	0.0013	516	74
609.79	900			900	120	0.0014	7.7195	CRESTLINE	2.1306	0.0023	900	74
609.83	200	OH		200	120	0.0003	7.7198	SR 61	2.1283	0.0005	200	74
609.94	600		163	600	120	0.0009	7.7207	R c	2.1278	0.0015	600	74
610.10	800			800	120	0.0013	7.7220		2.1262	0.0020	800	74
610.17	400		163	400	120	0.0006	7.7226	R c	2.1242	0.0010	400	74
610.30	700		163	700	120	0.0011	7.7237	R	2.1232	0.0018	700	74
610.34	204			204	120	0.0003	7.7240		2.1214	0.0005	204	74
610.42	400			400	120	0.0006	7.7247		2.1208	0.0010	400	74
611.18	4000			4335	120	0.0068	7.7315		2.1198	0.0102	4000	74
611.48	1600			1285	D 15	0.0024	7.7340		2.1096	0.0041	1600	74
611.59	600		76	634	76	0.0016	7.7356	L CREST WEST	2.1055	0.0015	600	74
611.61	78			44	D 76	0.0001	7.7356	REV CUR TT40MPH	2.1040	0.0002	78	74
611.76	800		74	800	74	0.0020	7.7377	R CREST WEST WEST	2.1038	0.0020	800	74
611.76						0.0000	7.7377		2.1017	0.0026	1309	D 120
612.44	3600			3318	74	0.0085	7.7462		2.0991	0.0036	2291	120
612.57	700	XING 1	176	1051	74	0.0026	7.7488	R SR 598	2.0955	0.0011	700	120
612.74	900	XING 1	176	831	74	0.0021	7.7511	R CRESTLINE ROAD	2.0944	0.0014	900	120
613.57	4380	XING 192.6		4380	74	0.0112	7.7623	BIDDLE ROAD	2.0930	0.0069	4380	120
614.56	5192	XING 193.8		1426	74	0.0036	7.7659	BECK ROAD	2.0861	0.0082	5192	120
615.56	5288	XING 194.62				0.0000	7.7659	MAIN STREET	2.0779	0.0083	5288	120
616.56	5284	XING 194.7		18898	A 74	0.0361	7.8020	WESTERN AVENUE	2.0696	0.0083	5284	120
617.56	5269	XING 195.		735	120	0.0012	7.8032	OLENTANGY ROAD	2.0612	0.0083	5269	120
618.53	5092	XING 195.8		5092	120	0.0080	7.8112	PRIVATE	2.0529	0.0080	5092	120
619.56	5470	XING 196.2		5470	120	0.0086	7.8199	PRIVATE	2.0449	0.0086	5470	120
620.56	5275	XING 197.0		5275	120	0.0083	7.8282	LWR LEESVILLE	2.0362	0.0083	5275	120
621.56	5254	XING 197.6		5254	120	0.0083	7.8365	PRIVATE	2.0279	0.0083	5254	120
621.76	1100	XING 197.7		1100	120	0.0017	7.8382	PRIVATE	2.0196	0.0017	1100	120
621.95	985	XING 198.2		985	120	0.0015	7.8398	PRIVATE	2.0178	0.0015	985	120
622.14	1000	XING 198.7		1000	120	0.0016	7.8413	ALBRIGHT	2.0163	0.0016	1000	120
622.34	1100			1100	120	0.0017	7.8431		2.0147	0.0017	1100	120
622.44	500			500	120	0.0008	7.8439	BUCYRUS	2.0130	0.0008	500	120
622.46	100			100	120	0.0002	7.8440		2.0122	0.0002	100	120
622.55	500	UG 201.20		500	120	0.0008	7.8448	SANDUSKY RIVER	2.0121	0.0008	500	120
622.76	1100		163	1100	120	0.0017	7.8465	R	2.0113	0.0017	1100	120
623.55	4186			4186	120	0.0066	7.8531		2.0095	0.0066	4186	120
624.55	5286	XING 200.1		5286	120	0.0083	7.8615	McCRACKEN	2.0029	0.0083	5286	120
625.55	5280	XING 203.8		5280	120	0.0083	7.8698	SIMMS CR	1.9946	0.0083	5280	120
626.58	5390	XING 204.7		5390	120	0.0085	7.8783	PRIVATE	1.9863	0.0085	5390	120
627.57	5272	XING 204.9		5272	120	0.0083	7.8866	GLENVILLE CR	1.9777	0.0083	5272	120
628.58	5298	XING 205.6		5298	120	0.0084	7.8950	PRIVATE	1.9694	0.0084	5298	120
629.45	4593	XING 206.1		4593	120	0.0072	7.9023	STRIEB CR KN	1.9611	0.0072	4593	120
630.57	5928	XING 207.1		5928	120	0.0094	7.9116	MARION-MELMO	1.9538	0.0094	5928	120
630.57		XING 207.70				0.0000	7.9116	PRIVATE	1.9445	0.0000		120
630.57		XING 208.13				0.0000	7.9116	COUNTY LINE	1.9445	0.0000		120
630.57		XING 208.60				0.0000	7.9116	PRIVATE	1.9445	0.0000		120
631.60	5440	XING 208.9		5440	120	0.0086	7.9202	NEVADA AYERS STRE	1.9445	0.0086	5440	120
631.60		XING 209.10				0.0000	7.9202	MORRISON STREET	1.9359	0.0000		120
631.60		XING 209.27				0.0000	7.9202	GOODBREAD STREET	1.9359	0.0000		120
632.60	5270	XING 210.61		5270	120	0.0083	7.9285	EDENVILLE ROAD	1.9359	0.0083	5270	120
633.60	5290	XING 211.6		5290	120	0.0083	7.9369	PRIVATE	1.9276	0.0083	5290	120
634.60	5289	XING 212.1		5289	120	0.0083	7.9452	DOUGLAS RD	1.9192	0.0083	5289	120
635.60	5269	XING 213.2		5269	120	0.0083	7.9535	PRIVATE	1.9109	0.0083	5269	120
636.60	5280	XING213.80		5280	120	0.0083	7.9619	ROCK RUN ROAD	1.9025	0.0083	5280	120

APPENDIX B4

READ DOWN WESTBOUND TRAFFIC — STATION — READ UP EASTBOUND TRAFFIC

PAGE 14 COLUMN 2

MILES	DISTANCE ft	HWY GRADE CROSSING	MAX MPH	d DISTANCE ft	SPEED mph	dT hours	TIME hours	STATION	TIME hours	dT hours	d DISTANCE ft	SPEED mph
637.60	5284	XING 214.3		5284	120	0.0083	7.9702	PRIVATE	1.8942	0.0083	5284	120
637.75	787	XING 2	190	787	120	0.0012	7.9714	L WILLIAMS CR	1.8859	0.0012	787	120
637.75		XING 215.50			120	0.0000	7.9714	PRIVATE	1.8846	0.0000		120
637.75		XING 215.70			120	0.0000	7.9714	PRIVATE	1.8846	0.0000		120
638.60	4500	XING 216.1		4500	120	0.0071	7.9785	PRIVATE	1.8846	0.0071	4500	120
639.09	2597	XING 217.0		2597	120	0.0041	7.9826	UPPER SANDUSKY	1.8775	0.0041	2597	120
639.09	6			6	120	0.0000	7.9827		1.8734	0.0000	6	120
639.19	476		163	476	120	0.0008	7.9834	L	1.8734	0.0008	476	120
639.60	2200			2200	120	0.0035	7.9869		1.8727	0.0035	2200	120
640.60	5061	XING 219.1		5281	120	0.0083	7.9952	WHITE CR	1.8692	0.0083	5281	120
641.58	5185	XING 218.9		5185	120	0.0082	8.0034	PRIVATE	1.8609	0.0082	5185	120
641.58		XING 220.20			120	0.0000	8.0034	MP 220 PRIVATE	1.8527	0.0000		120
642.58	5281	XING 220.7		5281	120	0.0083	8.0117	WILL CR	1.8527	0.0083	5281	120
643.58	5281	XING 220.9		5261	120	0.0083	8.0201	MIGRET CR	1.8443	0.0083	5281	120
644.58	5283	XING 221.3		5283	120	0.0083	8.0284	PRIVATE	1.8360	0.0083	5283	120
645.58	5280	XING 222.2		5280	120	0.0083	8.0367	GAMBER CR	1.8277	0.0083	5280	120
646.58	5279	XING 220.5		5279	120	0.0083	8.0451	PRIVATE	1.8193	0.0083	5279	120
647.59	5282	XING 223.2		5282	120	0.0083	8.0534	KRAUS CR	1.8110	0.0083	5282	120
648.59	5284	XING 223.6		5264	120	0.0083	8.0617	PRIVATE	1.8027	0.0083	5264	120
649.58	5275	XING 223.9		5275	120	0.0083	8.0701	PRIVATE	1.7943	0.0083	5275	120
650.58	5279	XING 224.2		5279	120	0.0083	8.0784	MAIN STREET SR	1.7860	0.0083	5279	120
650.58		XING 224.30				0.0000	8.0784	ALLEY STREET	1.7777			
650.58		XING 225.80				0.0000	8.0784	LILES CR	1.7777			
650.58		XING 226.84				0.0000	8.0784	HELLER CR	1.7777			
650.58		XING 227.20				0.0000	8.0784	PRIVATE	1.7777			
650.58		XING 227.50				0.0000	8.0784	PRIVATE	1.7777			
650.58		XING 227.80				0.0000	8.0784	PRIVATE	1.7777			
650.58		XING 227.94				0.0000	8.0784	ANGLING ROAD	1.7777			
650.58		XING 228.38				0.0000	8.0784	COUNTY LINE	1.7777			
650.58		XING 228.90				0.0000	8.0784	LOUISA STREET	1.7777			
651.59	5296	XING 228.9		5296	120	0.0084	8.0868	FOREST MARY STREE	1.7777	0.0084	5296	120
652.59	5275	XING 229.2		5275	120	0.0083	8.0951	GORMLY STREET	1.7693	0.0083	5275	120
653.59	5273	XING 229.3		5273	120	0.0083	8.1034	MARTIN STREET SR	1.7610	0.0083	5273	120
654.59	5278	XING 229.7		5278	120	0.0083	8.1117	CAMPBELL STREET	1.7527	0.0083	5278	120
655.59	5289	XING 230.5		5289	120	0.0083	8.1201	BERLIN CR	1.7443	0.0083	5289	120
656.58	5289	XING 231.0		5289	120	0.0083	8.1284	PRIVATE	1.7360	0.0083	5289	120
656.58		XING 231.70			120	0.0000	8.1284	PRIVATE	1.7277	0.0000		120
657.58	5278	XING 234.1		5278	120	0.0083	8.1367	GROAT ROAD	1.7277	0.0083	5278	120
657.87	1500	XING 234.5		1500	120	0.0024	8.1391	WYKES CR	1.7193	0.0024	1500	120
657.87					120	0.0000	8.1391	DUNKIRK	1.7170	0.0000		120
658.58	3771			3771	120	0.0060	8.1451		1.7170	0.0060	3771	120
659.58	5278			5278	120	0.0083	8.1534		1.7110	0.0083	5278	120
660.58	5280			5280	120	0.0083	8.1617		1.7027	0.0083	5280	120
661.58	5278			5278	120	0.0083	8.1700		1.6944	0.0083	5278	120
662.58	5280	XING 241.6		5280	120	0.0083	8.1784	TRAVERSE PK	1.6860	0.0083	5280	120
663.58	5277	XING 241.8		5277	120	0.0083	8.1867	PRIVATE	1.6777	0.0083	5277	120
664.58	5278	XING 242.8		5278	120	0.0083	8.1950	PETERSON CR	1.6694	0.0083	5278	120
665.58	5273	XING 243.6		5273	120	0.0083	8.2034	SCOTT XING	1.6610	0.0083	5273	120
666.58	5285	XING 244.6		5285	120	0.0083	8.2117	1 MILE XING	1.6527	0.0083	5285	120
667.60	5394	XING 245.5		5394	120	0.0085	8.2202	ADA JOHNSON STREE	1.6444	0.0085	5394	120
668.60	5261	XING 245.0		5281	120	0.0083	8.2285	MAIN STREET	1.6359	0.0083	5281	120
669.60	5278	XING 245.7		5278	120	0.0083	8.2368	GILBERT STREET	1.6276	0.0083	5278	120
670.62	5377	XING 246.6		5377	120	0.0085	8.2453	KLINGLER RD	1.6192	0.0085	5377	120
671.62	5303	XING 247.6		5303	120	0.0084	8.2537	ST PAUL RD	1.6107	0.0084	5303	120
671.62		XING 248.10				0.0000	8.2537	PRIVATE	1.6024			
671.62		XING 248.85				0.0000	8.2537	COUNTY LINE	1.6024			
671.62		XING 249.56				0.0000	8.2537	BENTLEY ROAD	1.6024			
671.62		XING 250.13				0.0000	8.2537	PEVEE ROAD	1.6024			
672.62	5289	XING 250.4		5289	120	0.0083	8.2621	PRIVATE	1.6024	0.0083	5289	120
673.62	5286	XING 250.7		5286	120	0.0083	8.2704	LAFAYETTE RD	1.5940	0.0083	5286	120
674.63	5297	XING 250.9		5297	120	0.0084	8.2788	SHRIDER ROAD	1.5857	0.0084	5297	120
675.63	5304	XING 251.3		5304	120	0.0084	8.2871	VINT RD	1.5773	0.0084	5304	120
676.65	5371	XING 251.6		5371	120	0.0085	8.2956	PHILLIPS RD	1.5689	0.0085	5371	120
677.63	5202	XING 252.2		5202	120	0.0082	8.3038	S HIGH STREET	1.5605	0.0082	5202	120
678.64	5298	XING 252.3		5298	120	0.0084	8.3122	CHURCH ST	1.5523	0.0084	5298	120
679.09	2400	XING 252.4		2400	120	0.0038	8.3160	WASHINGTON	1.5439	0.0038	2400	120
679.15	293	XING 2	149	293	120	0.0005	8.3164	R RUMBAUGH RD	1.5401	0.0005	293	120
679.64	2600	XING 253.6		2600	120	0.0041	8.3205	FISHER ROAD	1.5396	0.0041	2600	120
679.64		XING 254.17			120	0.0000	8.3205	MCCLURE XING	1.5355	0.0000		120
679.64		XING 254.66			120	0.0000	8.3205	COOL ROAD	1.5355	0.0000		120
680.65	5336	XING 255.6		5336	120	0.0084	8.3289	THAYER ROAD	1.5355	0.0084	5336	120
681.22	3000	XING 256.9		3000	120	0.0047	8.3337	FETTER ROAD	1.5271	0.0047	3000	120
681.67	2373	XING 257.6		2373	120	0.0037	8.3374	METZGER ROAD	1.5224	0.0037	2373	120
681.88	1100	XING 258.4		1100	120	0.0017	8.3392	ROUSH ROAD	1.5188	0.0017	1100	120
681.97	500			500	120	0.0008	8.3400	LIMA	1.5169	0.0008	500	120
682.06	485		163	485	120	0.0008	8.3407	R	1.5161	0.0008	485	120
682.06					120	0.0000	8.3407		1.5154	0.0000		120
682.21	800			800	120	0.0013	8.3420		1.5154	0.0013	800	120
682.65	2300		134	2300	120	0.0036	8.3456	R	1.5141	0.0036	2300	120
682.76	576		134	576	120	0.0009	8.3465	R	1.5105	0.0009	576	120
683.65	4700	XING 262.9		4700	120	0.0074	8.3539	CABLE ROAD	1.5095	0.0074	4700	120
684.65	5296	XING 263.5		5296	120	0.0084	8.3623	HARTZLER RD	1.5021	0.0084	5296	120
685.65	5285	XING 264.0		5288	120	0.0083	8.3706	EASTTOWN ROAD	1.4938	0.0083	5288	120
686.67	5393	XING 264.8		5393	120	0.0085	8.3792	EAST ROAD	1.4854	0.0085	5393	120
687.67	5280	XING 265.2		5280	120	0.0083	8.3875	BATY ROAD	1.4769	0.0083	5280	120
688.68	5286	XING 265.8		5286	120	0.0083	8.3958	PRIVATE	1.4686	0.0083	5286	120
689.68	5285	XING 266.4		5285	120	0.0083	8.4042	GREENLAWN AVE	1.4602	0.0083	5285	120
690.68	5282	XING 266.8		5282	120	0.0083	8.4125	MAIN ROAD	1.4519	0.0083	5282	120
691.68	5281	XING 267.2		5281	120	0.0083	8.4208	WAPAK ROAD	1.4436	0.0083	5281	120
691.68		XING 267.40			120	0.0000	8.4208	PRIVATE	1.4352	0.0000		
691.68		XING 267.65				0.0000	8.4208	KEMP ROAD	1.4352	0.0000		
691.68		XING 267.90				0.0000	8.4208	PRIVATE	1.4352	0.0000		
691.68		XING 268.10				0.0000	8.4208	PRIVATE	1.4352	0.0000		
691.68		XING 268.83				0.0000	8.4208	GRUBB ROAD	1.4352	0.0000		
691.68		XING 269.20				0.0000	8.4208	PRIVATE	1.4352	0.0000		
692.68	5292	XING 270.4		5292	120	0.0084	8.4292	STATE RD	1.4352	0.0084	5292	120
693.68	5283	XING 270.6		5283	120	0.0083	8.4375	OLD DELPHOS	1.4268	0.0083	5283	120
694.68	5290	XING 271.2		5290	120	0.0083	8.4459	DEFIANCE TRA	1.4185	0.0083	5290	120
695.27	3100	XING 271.7		3100	120	0.0049	8.4508	BAUGH RD	1.4102	0.0049	3100	120
695.68	2172	XING 2	134	2172	120	0.0034	8.4542	L PELTIER RD	1.4053	0.0034	2172	120
695.87	1000	XING 2	134	1000	120	0.0016	8.4558	L LEHMAN RD	1.4019	0.0016	1000	120
695.87		XING 273.69				0.0000	8.4556	PRIVATE	1.4003			

APPENDIX B4

READ DOWN — WESTBOUND TRAFFIC | STATION | READ UP — EASTBOUND TRAFFIC

PAGE 15 COLUMN 1

MILES	DISTANCE ft	HWY GRADE CROSSING	MAX MPH	DISTANCE ft	SPEED mph	ST. HOURS	TIME HOURS	STATION	TIME HOURS	ST. HOURS	DISTANCE ft	SPEED mph
696.08	1100	XING	274.37	1100	120	0.0017	8.4575	DELPHOS PIERCE ST	1.4003	0.0017	1100	120
696.44	1900	XING	274.38	1900	120	0.0030	8.4605	S MAIN STREET	1.3868	0.0030	1900	120
696.63	1000	XING	274.34	1000	120	0.0016	8.4621	R CANAL STREET	1.3866	0.0016	1000	120
696.68	284	XING	274.71	284	120	0.0004	8.4625	JEFFERSON STREET	1.3840	0.0004	284	120
697.68	5277	XING	274.76	5277	120	0.0083	8.4708	S CLAY STREET	1.3836	0.0083	5277	120
698.68	5288	XING	274.66	5288	120	0.0083	8.4792	S BREDICK STREET	1.3802	0.0083	5280	120
699.68	5288	XING	275.06	5288	120	0.0083	8.4875	STATE STREET	1.3769	0.0083	5290	120
700.68	5289	XING	276.65	5289	120	0.0083	8.4959	BRICKNER RD	1.3686	0.0083	5289	120
701.68	5284	XING	277.28	5284	120	0.0083	8.5042	PRIVATE	1.3602	0.0083	5284	120
701.68		XING	277.98			0.0000	8.5042	BOOKEY RD	1.3519	0.0000		
701.68		XING	278.88			0.0000	8.5042	CONVERSE RD	1.3519	0.0000		
701.68		XING	279.71			0.0000	8.5042	MIDDLEPOINT	1.3519	0.0000		
702.68	5281	XING	280.05	5281	120	0.0083	8.5126	MAIN STREET	1.3519	0.0083	5281	120
703.68	5281	XING	280.22	5281	120	0.0083	8.5209	ADAMS STREET	1.3436	0.0083	5281	120
704.68	5285	XING	280.31	5285	120	0.0083	8.5292	MASON STREET	1.3353	0.0030	1886	120
705.68	5282	XING	280.76	5282	120	0.0083	8.5376	DOG CREEK RD	1.3322	0.0000		
706.68	5281	XING	281.06	5281	120	0.0083	8.5459	RINGWALD RD	1.3322	0.0300	18846	A 84
707.69	5297	XING	282.30	5297	120	0.0084	8.5543	CHENOWETH RD	1.3022	0.0065	2408	84
708.60	4805	XING	283.54	4805	120	0.0076	8.5618	GAMBLE RD	1.2967	0.0119	5286	84
708.69	486	XING	2 131	486	120	0.0008	8.5626	L HOAGLIN CTR	1.2848	0.0011	486	84
708.93	1300	XING	2 131	1300	120	0.0021	8.5647	L GILLAND RD	1.2837	0.0026	1300	84
708.93		XING	285.68			0.0000	8.5647	MENDON RD	1.2808	0.0000		
709.22	1600	XING	286.75	403	120	0.0007	8.5650	VAN WERT WAYNE ST	1.2806	0.0034	1000	84
709.22		XING	288.86	1077	D 10	0.0020	8.5673	VINE STREET	1.2774	0.0000		
709.39	900	XING	2 84	900	84	0.0020	8.5693	R FRANKLIN STREET	1.2774	0.0029	900	84
709.56	900	XING	2 85	900	84	0.0020	8.5714	R CHESTNUT STREET	1.2754	0.0003	153	D 90
709.58	100	XING	287.04	100	84	0.0002	8.5716	RACE STREET	1.2750	0.0018	847	90
709.58		XING	287.15			0.0000	8.5716	N TYLER STREET	1.2733	0.0000		
709.58		XING	287.17			0.0000	8.5716	HARRISON STREET	1.2730	0.0000		
709.58		XING	287.24			0.0000	8.5716	CHERRY STREET	1.2733	0.0000		
709.58		XING	287.30			0.0000	8.5716	WALNUT STREET	1.2733	0.0000		
709.58		XING	287.37			0.0000	8.5716	MARKET STREET	1.2730	0.0000		
709.58		XING	287.48			0.0000	8.5716	WASHINGTON STREE	1.2730	0.0000		
709.58		XING	287.55			0.0000	8.5716	JEFFERSON STREET	1.2733	0.0000		
709.58		XING	287.67			0.0000	8.5716	N SHANNON	1.2733	0.0000		
709.58		XING	287.74			0.0000	8.5716	WALL STREET	1.2730			
709.58			90				8.5716	CR VAN WERT	1.2733			
709.58		XING	288.07			0.0000	8.5716	BURT STREET	1.2733	0.0017	924	D 120
709.70	642	XING	288.37	642	84	0.0014	8.5731	FISHER AVENUE	1.2716	0.0000	0	120
710.69	5241	XING	288.06	5241	84	0.0118	8.5849	JOHN BROWN RD	1.2716	0.0078	4800	120
711.69	5264	XING	289.56	4117	84	0.0093	8.5942	PRIVATE	1.2638	0.0080	5084	120
712.70	5288	XING	290.11	1913	A 84	0.0042	8.5983	LIBERTY UNION	1.2564	0.0083	5288	120
713.70	5287	XING	291.27	4402	A 90	0.0058	8.6042	RICHEY CR	1.2471	0.0083	5287	120
714.70	5287	XING	292.38	685	120	0.0011	8.6052	MACE XING	1.2387	0.0080	5287	120
715.70	5284	XING	293.55	5284	120	0.0083	8.6336	ALT RT US 30	1.2304	0.0080	5284	120
715.70		XING	293.59			0.0000	8.6336	BERFSNER RD	1.2221	0.0000		
715.70		XING	294.71			0.0000	8.6336	MAIN ST CONVOY	1.2221	0.0000		
715.70		XING	294.86			0.0000	8.6336	TULLY ST	1.2221	0.0000		
716.70	5287	XING	295.46	5287	120	0.0083	8.6419	PRIVATE	1.2221	0.0083	5287	120
717.70	5288	XING	295.65	5288	120	0.0080	8.6505	PAYNE CR	1.2137	0.0080	5288	120
718.70	5286	XING	296.38	5286	120	0.0083	8.6586	SHANER CR	1.2054	0.0080	5286	120
719.70	5280	XING	296.94	5280	120	0.0083	8.6669	LARE CR	1.1970	0.0083	5280	120
720.70	5292	XING	297.16	5292	120	0.0084	8.6753	MENTZER	1.1887	0.0084	5290	120
721.71	5288	XING	297.66	5288	120	0.0083	8.6836	PRIVATE	1.1803	0.0080	5288	120
721.71		XING	298.12			0.0000	8.6836	MENTZER CURC	1.1720			
721.71		XING	298.83			0.0000	8.6836	CLEM XING	1.1720			
721.71		XING	299.46			0.0000	8.6836	SCHOOLHOUSE	1.1720			
722.71	5282	XING	300.38	5282	120	0.0083	8.6920	OH IN ST LINE RD	1.1720	0.0083	5282	120
723.71	5281	XING	301.5K	5281	120	0.0083	8.7003	MORGAN RD	1.1637	0.0083	5281	120
724.71	5289	XING	301.8K	5289	120	0.0083	8.7087	PRIVATE	1.1553	0.0083	5089	120
725.71	5285	XING	302.64	5285	120	0.0083	8.7170	LORTIE RD	1.1470	0.0083	5285	120
726.71	5294	XING	303.9K	5294	120	0.0084	8.7254	OHIO STREET	1.1386	0.0084	5294	120
727.71	5289	XING	304.05	5289	120	0.0083	8.7337	MAIN STREET	1.1303	0.0083	5089	120
728.72	5287	XING	304.3	5287	120	0.0084	8.7421	WASHINGTON	1.1219	0.0084	5287	120
729.72	5274	XING	306.5K	5274	120	0.0083	8.7504	SNYDER RD	1.1136	0.0083	5274	120
730.72	5292	XING	306.86	5292	120	0.0084	8.7587	HOFFMAN RD	1.1052	0.0084	5092	120
731.72	5281	XING	307.11	5281	120	0.0083	8.7671	GROTRAIN RD	1.0969	0.0083	5081	120
731.72		XING	307.71			0.0000	8.7671	WILSON RD	1.0886			
731.72		XING	308.30			0.0000	8.7671	FACKLER RD	1.0886			
731.72		XING	308.85			0.0000	8.7671	GARADOT RD	1.0886			
731.72		XING	308.97			0.0000	8.7671	HOUK RD	1.0886			
731.72		XING	308.71			0.0000	8.7671	WASHINGTON	1.0886			
731.72		XING	309.94			0.0000	8.7671	FRANKE RD	1.0886			
732.72	5281	XING	310.66	5281	120	0.0083	8.7754	PRIVATE	1.0886	0.0083	5281	120
733.72	5281	XING	310.96	5281	120	0.0080	8.7837	PRIVATE	1.0802	0.0083	5281	120
734.73	5287	XING	311.07	5287	120	0.0083	8.7921	TILLMAN RD	1.0718	0.0083	5287	120
734.72		XING	311.75		120	0.0000	8.7921	MINNICH ROAD	1.0636	0.0000		120
734.72		XING	312.50		120	0.0000	8.7921	PRIVATE	1.0636	0.0000		120
734.72		XING	312.80		120	0.0000	8.7921	PRIVATE	1.0636	0.0000		120
734.72		XING	313.09		120	0.0000	8.7921	PRIVATE	1.0636	0.0000		120
734.72		XING	313.26		120	0.0000	8.7921	PAULDING ROAD	1.0636	0.0000		120
735.72	5293	XING	313.47	5293	120	0.0084	8.8004	HARTZELL ROAD	1.0636	0.0084	5293	120
736.72	5283	XING	315.13	5283	120	0.0083	8.8087	ADAMS CENTER	1.0552	0.0083	5283	120
737.72	5288	XING	316.26	5288	120	0.0083	8.8171	MEYER ROAD	1.0469	0.0080	5286	120
738.73	5329			5329	120	0.0084	8.8255		1.0385	0.0084	5329	120
739.71	5192			5192	120	0.0082	8.8337		1.0301	0.0082	5192	120
740.58	4600			4600	120	0.0073	8.8410		1.0219	0.0073	4600	120
740.73	740		106	748	120	0.0012	8.8421	L	1.0147	0.0012	749	120
740.88	800			800	120	0.0013	8.8434		1.0135	0.0013	800	120
741.14	1400			1400	120	0.0022	8.8456	FORT WAYNE	1.0122	0.0022	1400	120
741.37	1200			1200	120	0.0019	8.8475		1.0100	0.0019	1200	120
741.65	1500			1500	120	0.0024	8.8499		1.0081	0.0024	1500	120
741.75	481			481	120	0.0008	8.8506		1.0058	0.0008	481	120
741.84	489			489	120	0.0008	8.8514		1.0050	0.0008	489	120
742.73	4700			4700	120	0.0074	8.8588		1.0042	0.0074	4700	120
742.75	100		108	100	120	0.0002	8.8590	R	0.9968	0.0002	100	120
742.82	400		108	400	120	0.0006	8.8596	R	0.9966	0.0006	400	120
742.91	462		108	462	120	0.0007	8.8603	R	0.9960	0.0007	462	120
743.76	4500	XING	323.64	4500	120	0.0071	8.8674	THOMAS ROAD	0.9953	0.0071	4500	120
744.77	5348	XING	324.86	5348	120	0.0084	8.8759	HADLEY ROAD	0.9882	0.0084	5348	120
745.78	5322	XING	325.04	5322	120	0.0084	8.8843	BASS ROAD	0.9798	0.0084	5322	120
746.77	5180	XING	326.46	5180	120	0.0082	8.8925	PRIVATE	0.9714	0.0080	5180	120

APPENDIX B4

READ DOWN — WESTBOUND TRAFFIC | STATION | READ UP — EASTBOUND TRAFFIC

PAGE 15 COLUMN 2

MILES	DISTANCE ft	HWY GRADE CROSSING	MAX MPH	DISTANCE ft	SPEED mph	ST. HOURS	TIME HOURS	STATION	TIME HOURS	ST. HOURS	DISTANCE ft	SPEED mph
747.76	5277	XING	326.14	5277	120	0.0083	8.9008	MAIN STREET	0.9632	0.0083	5277	120
748.76	5278	XING	328.36	5278	120	0.0083	8.9091	ARCOLA COLUMBIA S	0.9548	0.0083	5278	120
749.77	5290	XING	328.6K	5290	120	0.0083	8.9175	PRIVATE	0.9465	0.0083	5290	120
750.78	5363	XING	329.06	5363	120	0.0085	8.9260	PRIVATE	0.9382	0.0085	5363	120
751.77	5224	XING	329.36	5224	120	0.0082	8.9342	PRIVATE	0.9297	0.0082	5024	120
752.77	5298	XING	329.6K	5298	120	0.0084	8.9426	PRIVATE	0.9214	0.0084	5296	120
753.78	5297	XING	330.61	5297	120	0.0084	8.9509	COUNTY LINE	0.9131	0.0084	5297	120
754.78	5294	XING	331.56	5294	120	0.0084	8.9593	BEECHING ROAD	0.9047	0.0084	5294	120
755.78	5293	XING	332.87	5293	120	0.0084	8.9676	800E	0.8964	0.0084	5293	120
756.79	5293	XING	333.85	5293	120	0.0084	8.9760	MILL STREET	0.8880	0.0084	5293	120
757.79	5292	XING	334.77	5292	120	0.0084	8.9843	MOWERY ROAD	0.8797	0.0084	5292	120
758.79	5269	XING	336.5K	5269	120	0.0083	8.9926	PAGES CR	0.8713	0.0083	5269	120
759.79	5290	XING	338.45	5290	120	0.0083	9.0010	SR 9	0.8630	0.0083	5290	120
760.82	5454	XING	338.91	5454	120	0.0086	9.0096	LINE STREET SR	0.8546	0.0086	5454	120
761.80	5185	XING	339.8K	5185	120	0.0082	9.0177	POOR FARM CR	0.8461	0.0082	5185	120
762.80	5283	XING	341.3K	5283	120	0.0083	9.0260	WOLF ROAD	0.8379	0.0074	4716	120
763.80	5298	XING	342.08	5298	120	0.0084	9.0344	PEARL CR 300E	0.8304	0.0122	7336	A 107
764.10	1600	XING	343.06	1600	120	0.0025	9.0369	BUTLER CR 40	0.8182	0.0020	107	107
764.39	1503	XING	3 147	1503	120	0.0024	9.0393	L PRIVATE	0.8180	0.0027	1503	107
764.80	2200			2200	120	0.0035	9.0428		0.8154	0.0039	2200	107
765.80	5290			5290	120	0.0083	9.0511		0.8115	0.0094	5290	107
766.16	1900			483	D 120	0.0007	9.0518		0.8021	0.0034	1900	107
766.45	1491		107	1491	107	0.0008	9.0526	R	0.7987	0.0026	1491	107
766.81	1900			1900	107	0.0034	9.0578		0.7961	0.0007	403	D 120
767.72	4800			4800	107	0.0085	9.0663		0.7954	0.0089	6267	120
767.81	486	XING	3 164	486	107	0.0008	9.0672	L BRD 5 CENTER ST	0.7865	0.0008	486	120
767.84	164	XING	3 164	3812	107	0.0067	9.0740	L LARWILL DEPOT ST	0.7847	0.0003	164	120
768.81	5100	XING	347.3K	7336	A 10	0.0122	9.0862	BRINKLEY ROAD	0.7844	0.0090	5100	120
769.81	5294	XING	348.61	897	120	0.0014	9.0876	BLACK CR	0.7754	0.0084	5294	120
770.81	5286	XING	349.3K	5080	105	0.0083	9.0959	PRIVATE	0.7680	0.0083	5286	120
771.81	5287	XING	348.8K	5287	120	0.0083	9.1043	PRIVATE	0.7597	0.0083	5287	120
772.81	5282	XING	350.38	5282	120	0.0083	9.1126	MARKET STREET	0.7513	0.0083	5282	120
773.82	5283	XING	350.61	5283	120	0.0083	9.1210	PIERCETON FIRST ST	0.7430	0.0083	5283	120
774.81	5279	XING	352.5K	5279	120	0.0083	9.1293	VAN NESS ROAD	0.7347	0.0083	5279	120
775.81	5278	XING	353.4K	5278	120	0.0083	9.1376	WOOSTER ROAD	0.7263	0.0083	5278	120
776.82	5293	XING	354.78	5293	120	0.0084	9.1460	SCHADE 250E	0.7180	0.0084	5293	120
777.83	5323			5323	120	0.0084	9.1544		0.7097	0.0084	5323	120
778.82	5233			5233	120	0.0083	9.1626		0.7013	0.0083	5233	120
779.54	3800			3800	120	0.0060	9.1686		0.6930	0.0060	3800	120
779.82	1500	XING	3 164	1500	120	0.0024	9.1710	L	0.6870	0.0024	1500	120
780.01	996	XING	3 164	996	120	0.0018	9.1728	L	0.6846	0.0016	996	120
780.52	2700	XING	358.24	2700	120	0.0043	9.1758	WARSAW	0.6831	0.0043	2700	120
780.82	1600			1600	120	0.0025	9.1784		0.6788	0.0025	1600	120
781.82	5282			5282	120	0.0083	9.1867		0.6763	0.0083	5282	120
781.86	178	XING	360.0K	178	120	0.0003	9.1870	JACKSON CR	0.6679	0.0003	178	120
782.22	1900		134	1900	120	0.0030	9.1910	R BURKET	0.6677	0.0030	1900	120
782.82	3200	XING	361.75	3200	120	0.0051	9.1960	PRIVATE	0.6647	0.0051	3200	120
783.83	5318	XING	362.0K	5318	120	0.0084	9.2044	PRIVATE	0.6596	0.0084	5318	120
784.83	5246	XING	362.26	5246	120	0.0083	9.2127	THOMAS CR 19	0.6512	0.0083	5246	120
785.83	5291	XING	363.28	5291	120	0.0084	9.2210	PRIVATE	0.6429	0.0084	5291	120
785.83		XING	363.72		120	0.0000	9.2210	SELBY CR 525	0.6346			
786.84	5362	XING	364.5K	5350	120	0.0084	9.2295	PRIVATE	0.6346	0.0084	5362	120
787.83	5220	XING	366.36	5220	120	0.0082	9.2377	PRAIRIE STREET	0.6261	0.0082	5220	120
788.83	5291	XING	366.48	5291	120	0.0084	9.2461	ATWOOD HARRISON	0.6179	0.0084	5291	120
789.83	5279	XING	366.48	5279	120	0.0083	9.2544	ROBINSON ROAD	0.6096	0.0083	5279	120
790.82	5227	XING	367.46	5227	120	0.0082	9.2627	SNYDER CR 87	0.6012	0.0082	5227	120
791.83	5307	XING	368.15	5307	120	0.0084	9.2711	PRIVATE	0.5930	0.0084	5307	120
791.83		XING	368.77			0.0000	9.2711	COOK CR	0.5846			
791.83		XING	369.29			0.0000	9.2711	WALNUT STREET CR	0.5846			
791.83		XING	369.36			0.0000	9.2711	ETNA GREEN	0.5846			
792.85	5379	XING	370.36	5378	120	0.0085	9.2796	S BEECH ROAD	0.5846	0.0085	5379	120
793.83	5183	XING	372.96	5180	120	0.0082	9.2878	HARRIS STREET	0.5761	0.0082	5183	120
794.86	5439	XING	373.06	5439	120	0.0086	9.2964	BOURBON MAIN STRE	0.5679	0.0086	5439	120
794.86		XING	373.17			0.0000	9.2964	BOURBON STREET	0.5593			
794.86		XING	373.26			0.0000	9.2964	CENTER STREET	0.5593			
794.86		XING	374.13			0.0000	9.2964	ISCHR ST SE	0.5593			
795.84	5182	XING	375.3K	5182	120	0.0081	9.3045	STOCKMAN ROAD	0.5593	0.0081	5182	120
796.84	5262	XING	376.24	5262	120	0.0083	9.3128	WYMER CR	0.5511	0.0083	5262	120
797.83	5270	XING	377.04	5270	120	0.0083	9.3211	COMPTON CR	0.5428	0.0083	5270	120
798.83	5279	XING	377.36	5278	120	0.0083	9.3294	HAWTHORNE ROAD	0.5345	0.0083	5278	120
799.83	5279	XING	378.66	5279	120	0.0083	9.3378	PRIVATE	0.5262	0.0083	5279	120
800.83	5283	XING	379.15	5283	120	0.0083	9.3461	S IRIS ROAD	0.5179	0.0083	5283	120
801.83	5277	XING	379.9K	5277	120	0.0083	9.3544	PRIVATE	0.5096	0.0083	5277	120
802.84	5315	XING	380.2K	5315	120	0.0084	9.3628	PRIVATE	0.5012	0.0084	5315	120
803.82	5158	XING	380.5K	5158	120	0.0081	9.3710	PRIVATE	0.4928	0.0081	5158	120
804.83	5367	XING	381.05	5367	120	0.0085	9.3794	KING ROAD	0.4847	0.0085	5367	120
805.67	4400	XING	383.26	4400	120	0.0069	9.3864	LIBERTY STREET	0.4762	0.0069	4400	120
805.84	944	XING	384.06	944	120	0.0015	9.3879	PLYMOUTH FIFTH ST	0.4692	0.0015	944	120
806.83	5216	XING	384.61	5216	120	0.0082	9.3961	OAK ROAD	0.4678	0.0082	5216	120
807.83	5280	XING	385.71	5280	120	0.0083	9.4044	PRIVATE	0.4595	0.0083	5280	120
808.83	5267	XING	385.9K	5267	120	0.0083	9.4128	PRIVATE	0.4512	0.0083	5087	120
809.84	5293	XING	386.36	5293	120	0.0084	9.4211	ANGLING ROAD	0.4428	0.0084	5293	120
810.83	5289	XING	386.81	5289	120	0.0083	9.4294	QUINCE ROAD	0.4345	0.0083	5289	120
811.83	5281	XING	387.77	5281	120	0.0083	9.4378	REDWOOD ROAD	0.4262	0.0083	5281	120
811.83		XING	388.14		120	0.0000	9.4378	PRIVATE	0.4178			
812.83	5268	XING	390.81	5268	120	0.0083	9.4461	UNION ROAD	0.4178	0.0083	5268	120
813.84	5298			5298	120	0.0084	9.4545	DONALDSON	0.4095	0.0084	5298	120
814.84	5298	XING	393.5K	5298	120	0.0084	9.4628	PRIVATE	0.4012	0.0084	5298	120
815.86	5411	XING	394.96	5411	120	0.0085	9.4714	SR 30	0.3928	0.0085	5411	120
816.84	5164	XING	395.54	5164	120	0.0082	9.4796	WALSH CR	0.3843	0.0082	5164	120
817.84	5285	XING	396.54	5285	120	0.0083	9.4879	786E	0.3761	0.0083	5285	120
818.84	5288			5288	120	0.0083	9.4962		0.3678	0.0083	5288	120
819.85	5295	XING	398.06	5295	120	0.0083	9.5046	STARKE STREET	0.3594	0.0083	5295	120
820.85	5279	XING	398.56	5279	120	0.0083	9.5129	HAMLET JEFFERSON	0.3511	0.0083	5279	120
821.65	5301	XING	399.11	5301	120	0.0084	9.5212	500E	0.3428	0.0084	5301	120
822.85	5282	XING	401.17	5282	120	0.0083	9.5296	300E	0.3344	0.0083	5282	120
823.85	5282	XING	401.66	5282	120	0.0083	9.5379	PRIVATE	0.3261	0.0083	5282	120
824.85	5270	XING	402.3K	5270	120	0.0083	9.5462	PRIVATE	0.3177	0.0083	5270	120
825.85	5287	XING	403.36	5287	120	0.0083	9.5546	50E	0.3094	0.0083	5287	120
826.85	5292	XING	405.06	5292	120	0.0084	9.5629	100W	0.3011	0.0084	5292	120
827.85	5280	XING	405.76	5280	120	0.0083	9.5713	140W	0.2927	0.0083	5280	120
828.88	5422	XING	406.36	5422	120	0.0086	9.5798	SR 39	0.2844	0.0086	5422	120

APPENDIX B4 — PAGE 16 COLUMN 1

	READ DOWN WESTBOUND TRAFFIC							STATION	READ UP EASTBOUND TRAFFIC			
MILES	DISTANCE, ft	HWY GRADE CROSSING	MAX MPH	DISTANCE, ft	SPEED mph	dT, hours	TIME, hours	PAGE 16 COLUMN 1	TIME, hours	dT, hours	DISTANCE, ft	SPEED mph
829.83	5046	XING	407.26	5046	120	0.0083	9.5878	LONG LANE	0.2785	0.0080	5046	120
829.83		XING	407.70		120	0.0000	9.5878	PRIVATE	0.2678	0.0000		120
830.67	4400	XING	408.61	4400	120	0.0069	9.5947	THOMPSON STREET	0.2678	0.0069	4400	120
830.67		XING	408.74		120	0.0000	9.5947	OHIO STREET	0.2608	0.0000		120
830.67		XING	408.83		120	0.0000	9.5947	HANNA 450W	0.2600	0.0000		120
830.85	986	XING	409.28	986	120	0.0015	9.5962	PRIVATE	0.2608	0.0015	986	120
831.85	5281	XING	410.38	5281	120	0.0083	9.6045	SCHULL 600W	0.2594	0.0083	5281	120
832.85	5285	XING	411.26	5285	120	0.0083	9.6128	FARRESTER ROAD	0.2510	0.0083	5285	120
833.85	5273	XING	411.65	5273	120	0.0083	9.6212	TILDON 750W	0.2427	0.0083	5273	120
835.02	6182	XING	412.86	6182	120	0.0098	9.6310	PRIVATE	0.2344	0.0098	6182	120
835.85	4367	XING	413.40	4367	120	0.0069	9.6379	KLEINS 900W	0.2246	0.0069	4367	120
836.85	5280	XING	414.25	5280	120	0.0083	9.6462	N MAIN STREET	0.2177	0.0083	5280	120
837.85	5272	XING	414.81	5272	120	0.0083	9.6545	WINNATAH S ILLINOIS	0.2094	0.0083	5272	120
838.84	5251	XING	415.16	5251	120	0.0083	9.6626	S WASHINGTON	0.2011	0.0083	5251	120
839.84	5293	XING	415.15	5293	120	0.0084	9.6712	LINCOLN STREET	0.1928	0.0084	5293	120
840.84	5270	XING	415.48	5270	120	0.0083	9.6795	1100W	0.1844	0.0083	5270	120
841.45	3200	XING	416.50	3200	120	0.0051	9.6846	OSBRN GR CO	0.1761	0.0083	5270	120
841.56	570	XING	4 176	570	120	0.0009	9.6855	R FISHER 575E	0.1678			
841.84	1500	XING	418.01	1500	120	0.0024	9.6878	DOLSON 450E	0.1678			
841.84		XING	418.79			0.0000	9.6878	BICKEL 400E	0.1678			
841.84		XING	420.00			0.0000	9.6878	MP 420 PRIVATE	0.1678			
842.84	5280	XING	420.23	5280	120	0.0083	9.6962	MONTDALE ROAD	0.1678	0.0083	5280	120
843.84	5283	XING	421.05	5283	120	0.0083	9.7045	PRIVATE	0.1595	0.0083	5283	120
843.97	700	XING	422.29	700	120	0.0011	9.7056	CEMETERY ROAD	0.1512	0.0011	700	120
844.76	4200	XING	4 134	4200	120	0.0066	9.7122	R GREENWICH STREE	0.1501	0.0066	4200	120
844.84	394	XING	423.42	394	120	0.0006	9.7128	AXE STREET	0.1434	0.0006	394	120
845.37	2800	XING	423.86	2800	120	0.0044	9.7172	FRANKLIN STREET	0.1428	0.0044	2800	120
845.46	486	XING	4 134	486	120	0.0008	9.7180	L VALPARAISO WASH	0.1384	0.0008	486	120
845.84	2000	XING	423.96	2000	120	0.0032	9.7212	LAFAYETTE STREET	0.1376	0.0032	2000	120
846.84	5279	XING	423.81	5279	120	0.0083	9.7295	NAPOLEON STREET	0.1345	0.0083	5279	120
847.84	5274	XING	424.81	5274	120	0.0083	9.7378	STEVENS XING	0.1261	0.0083	5274	120
848.67	4400	XING	425.78	4400	120	0.0069	9.7448	150 W	0.1178	0.0069	4400	120
848.84	892			892	120	0.0014	9.7462		0.1108	0.0014	892	120
849.84	5299			5299	120	0.0084	9.7546		0.1095	0.0084	5299	120
850.86	5340			5340	120	0.0084	9.7630		0.1011	0.0084	5340	120
851.84	5206	XING	430.21	5206	120	0.0082	9.7712	PRIVATE	0.0927	0.0082	5206	120
852.84	5291	XING	430.81	5291	120	0.0084	9.7796	WHEELER PARK AVE	0.0844	0.0084	5291	120
853.85	5289	XING	431.02	5289	120	0.0083	9.7879	GOTT RD 7TH	0.0761	0.0083	5289	120
854.83	5212	XING	432.36	5212	120	0.0082	9.7961	TREAGER RD	0.0677	0.0041	2621	120
855.85	5354	XING	433.05	5354	120	0.0085	9.8046	COUNTY LINE	0.0636	0.0219	12632	A 95
856.85	5316	XING	434.57	5316	120	0.0084	9.8129	HOBART ILLINOIS STF	0.0417	0.0000		95
856.99	700	XING	434.67	700	120	0.0011	9.8140	LINDA STREET	0.0417	0.0000		
857.15	844	XING	4 95	844	120	0.0013	9.8154	R CLEVELAND STREE	0.0417	0.0000		95
857.85	3700	XING	435.22	3700	120	0.0058	9.8212	LAKE PARK AVE	0.0417	0.0000		
858.85	5326	XING	435.94	5326	120	0.0084	9.8296	WISCONSIN STREET	0.0417	0.0000		
859.06	1080			1080	120	0.0017	9.8313		0.0417	0.0417	12353	A0-9
859.06		XING	437.16		120	0.0000	9.8313	LIVERPOOL RD E END LIV TE EAST END LIVERPO				
859.85	4200	XING	436.96	4200	120	0.0066	9.8379	INDIANA AVENUE				
859.96	600	XING	4 5	600	120	0.0009	9.8389	VIRGINIA STREET				
859.96					120	0.0000	9.8389	I-65				
860.18	1122				120	0.0000	9.8389					
860.18					120	0.0000	9.8389	I-90				
860.86	3800			4507	120	0.0071	9.8460					
861.22	1697			2112	[illegible]	0.0067	9.8526					
861.22					0		9.8526	WEST END LIVERPOOL TERMINAL				
861.67	2363											
861.67		XING	440.28					21ST STREET				
862.66	5263	XING	440.36					BROADWAY				
862.66		XING	440.47					WASHINGTON ST				
862.66		XING	440.60					19TH STREET				
862.66		XING	440.70					MADISON STREET				
862.66		XING	440.84					JACKSON AVE				
862.66		XING	441.00					HARRISON STREET				
862.66		XING	441.14					15TH AVENUE				
862.66		XING	441.23					13TH AVENUE				
862.66		XING	441.65					GRANT STREET				
862.66		XING	441.74					10TH AVENUE				
862.66			90					TOLLESTON CSX CROSSING PRR CUT				
862.96		XING	441.90					9TH AVENUE				
863.67	5298	OHD	442.02					IHB RR				
864.67	5293	XING	442.25					TAFT STREET				
864.67		XING	442.76					5TH AVENUE				
864.87	1050											
865.18	1500	OH	443.164					R CSSB OVERHEAD				
865.67	2700	XING	444.10					CLARKE ROAD				
866.67	5325											
867.08	2115											
867.06		OHD	445.40					EJ&E OVERHEAD				
867.28	793											
867.23		XING	445.55					PRIVATE				
867.35	651											
867.35			90					CSX CLARKE JCT				
867.85	1718											
868.86	5304							VIA PRR				
869.68	5296											
870.66	5154							PRR ROW TO LANE ST				
871.42	4030							CLARKE JCT-HAMMOND				
871.56	740		190					L				
871.71	800											
872.43	3630		70					INDIANA HARBOR CANAL				
872.50	920		152					R				
872.51	60							REVERSE CURVE				
872.60	480		150					L				
872.59	495							CP 506				
873.66	8227							PRR "IN PLACE" TO WESTERN AVENUE				
874.02	1900											
874.25	1200		76					R REVERSE CURVE				
874.48	1200		76					L				
874.57	500											
874.72	780		94					L				
874.73	70		94					L				
874.77	200											

APPENDIX B4 — PAGE 16 COLUMN 2

	READ DOWN WESTBOUND TRAFFIC							STATION	READ UP EASTBOUND TRAFFIC			
MILES	DISTANCE, ft	HWY GRADE CROSSING	MAX MPH	DISTANCE, ft	SPEED mph	dT, hours	TIME, hours	PAGE 16 COLUMN 2	TIME, hours	dT, hours	DISTANCE, ft	SPEED mph
874.88	1100		50					R				
875.57	3100											
875.70	711		110					L				
876.05	1860		107					L				
876.09	210							REVERSE CURVE				
876.12	160		107					R				
876.32	1065											
876.40	400	UG 454	50					CALUMET RIVER				
876.70	1600											
876.86	860											
877.02	1200		107					L REVERSE CURVE				
877.19	500		107					R				
877.32	700											
877.46	734		120					R				
877.68	1200											
878.71	5421											
879.71	5050											
880.66	5184											
881.58	5275											
881.90	1800											
882.05	303		35					L ENGLEWOOD				
882.69	3400							GRIP (METRA) OHC				
882.72	164											
882.95	1200		58					R 59TH STREET				
883.66	3770											
883.67	30		68					R 54TH STREET				
883.73	310		86									
884.02	1520											
884.10	400		67					L 51ST STREET				
884.30	1080											
884.40	516		64					L				
884.62	1164		81									
884.62												
884.67	291		89					R 47TH STREET				
884.70	120		89					R				
885.73	5456											
886.77	5806											
887.38	3300											
887.45	380		86					L REVERSE CURVE				
887.51	340		88					R 23RD STREET				
887.60	700											
887.67	146		68					L				
887.72	230		68					L				
887.78	323		85					R REVERSE CURVE				
887.75			30					IC (CN) RR CROSSING				
887.84	310		70					S BRANCH CHICAGO RIVER				
887.89	300		107					L				
888.07	950							LUMBER STREET				
888.13	320		107					R				
888.23	520							REVERSE CURVE				
888.31	420		107					L				
888.44	677	OH 468.57						ST CHARLES AIR LINE				
888.56	600		134					L				
888.60	240											
888.84	1250											
888.94	555		55					R				
889.12	950											
889.20	560		30					L				
889.31	423		15					CHICAGO UNION STATION				
889.42	550		190					R				
889.50	750		15									
889.56	120											
889.60	200		41					R REVERSE CURVE				
889.65	135		44					L				
889.72	350											
889.74	120		62					L				
889.77	150							REVERSE CURVE				
889.80	150		30					R				
889.83	200											
889.86	250		41					R				
890.11	1200											
890.11								LAKE STREET				
890.15	240		37					L				
890.20	240							CANAL STREET AT GRADE				
890.22	100		169					L				
890.22								EAST END DOWNTOWN CHICAGO TERMINAL				
890.46	1300											
890.56	500		72					R				
890.66	580											
890.77	450		78					L				
891.00	1200											
891.20	1100		35					R				
891.22	81		76					R				
891.26	250							REVERSE CURVE				
891.34	400		62					L				
891.53	1000											
891.81	450		76					R REVERSE CURVE				
891.71	300		76					L PAULINA STREET				
892.21	2650											
892.21								WESTERN AVE TOWER A2				
892.24	160											
892.62	2002	NOTE: USE PRR (PCC&STL) ROW, NOT RTA (MILW)										
892.62								WEST END DOWNTOWN CHICAGO TERMINAL				
892.74	850		75					L				
892.74								WESTERN AVENUE				
892.78	200											
892.88	550		48					L				
892.99	580		46					L				
893.22	1210							FULTON ST				
893.22								MILEPOST 311				
893.30	400											
893.30			60					R MADISON ST				

Appendix B5 - Improved Right of Way on Partially Rebuilt Alignment per Appendix B4. Streamlined high-speed passenger train operating at 200 mph maximum speed.

Maximum allowable speed on curves as shown for high speed equipment only.

FRA Class 9 track where speeds over 180 mph are possible.

Other tracks classified per maximum possible speed.

Track re-built for maximum speed up to 200 mph. E_a = 8.635 inches, E_u = 4 inches

Note: for HS Freight (TOFC) E_a = 5.84 inches; for HS passenger if E_a = 5.84 inches E_u = 6.795 inches.

Wheel load 0.6 minimum per FRA 213.307d1

Assumed c.g. of moving equipment 48 inches above top of rail

Train length 1,360 feet: 10 non-articulated streamlined passenger cars, 8 locomotive units @ 10,000 hp each. Maximum train braking rate 5 mph/second.

20% speed reduction for reverse curves with 600 feet or less between curves.

Gradients shown in Appendix B2.

Highway grade crossing locations shown are approximate. See Appendix B2 and Appendix B3 for more detailed right of way information.

Distance New York Penn Station to Chicago Union Station 892.3 miles

Westbound trip	6.737 hours
Eastbound trip	6.795 hours

APPENDIX B5

READ DOWN — WESTBOUND TRAFFIC | STATION | READ UP — EASTBOUND TRAFFIC

PAGE 1 COLUMN 1

MILES	DISTANCE	HWY GRADE CROSSING	MAX MPH	DISTANCE	SPEED	ET, HOURS	TIME, HOURS	STATION	TIME, HOURS	ET, HOURS	DISTANCE	SPEED
0								PENN STATION NEW YORK				
-0.258	1360							EAST END PLATFORM	6.7822	0.0033	528	D 6C
0.00	4510		60	1086	A 04	0.0069	0.0069	WEST END PLATFORM	6.7756	0.0162	5773	60
1.00	778		226	4202	60	0.0133	0.0201	R	6.7613	0.0025	778	60
2.00	5288		60	5288	60	0.0167	0.0368		6.7589	0.0167	5288	60
2.92	4860		60	4860	60	0.0153	0.0522		6.7422	0.0153	4860	60
2.92							0.0522		6.7266	0.0032	1480	D 15
3.00	427		117	427	60	0.0013	0.0535	L RTE 1&9	6.7237	0.0000	22	117
3.38	2006		117	930	60	0.0029	0.0565	L	6.7237	0.0006	500	117
3.38					60	0.0000	0.0565	E MEADOWS	6.7228	0.0023	1700	D 15
4.00	3274			4192	A 60	0.0087	0.0651		6.7205	0.0000	12	158
5.00	5281			7570	A 11	0.0106	0.0756		6.7206	0.0000		
5.45	2377						0.0756		6.7206	0.0000		
5.73	1479		226				0.0756	R	6.7205	0.0000		
5.85	589						0.0756		6.7205	0.0188	11593	A 60
6.01	838						0.0756	A	6.7017	0.0026	834	60
6.01				101	161	0.0001	0.0759	W MEADOWS	6.6991	0.0000		60
6.10	526			3274	D 16	0.0056	0.0815		6.6991	0.0017	526	60
6.17	330		60	330	60	0.0010	0.0825	HACKENSACK R	6.6975	0.0010	330	60
6.17							0.0825		6.6964	0.0054	3133	D 15
6.17				1360	60	0.0043	0.0868		6.6910	0.0001	55	158
7.00	4409			11360	A 60	0.0185	0.1053		6.6909	0.0000		
7.39	2022					0.0000	0.1053	HUDSON	6.6908	0.0000		
8.04	3443					0.0000	0.1053	R	6.6909	0.0000		
8.28	1273		233			0.0000	0.1053	R	6.6909	0.0000		
8.37	487					0.0000	0.1053		6.6909	0.0169	10824	A 73
9.13	4035					0.0000	0.1053		6.6740	0.0035	1360	73
9.13				59	158	0.0001	0.1054		6.6704	0.0000		
9.13				2880	D 15	0.0047	0.1101		6.6704			
9.36	1205		73	1205	73	0.0031	0.1132	L HARRISON	6.6704	0.0031	1205	73
9.36							0.1132		6.6673	0.0019	929	D 10
9.43	371			371	73	0.0010	0.1142	REVERSE CURVE	6.6654	0.0001	47	108
9.61	955		108	955	73	0.0025	0.1167	R	6.6653	0.0006	350	108
9.66	265			34	73	0.0001	0.1168	REVERSE CURVE	6.6647	0.0004	239	108
9.76	531		162	2198	A 73	0.0047	0.1214	L	6.6643	0.0003		
9.85	424					0.0000	0.1214		6.6640	0.0018	1054	A 97
10.07	1168		233	217	104	0.0004	0.1218	L NEWARK	6.6623	0.0021	1095	97
10.12	265			206	D 10	0.0004	0.1222		6.6602	0.0005	285	97
10.24	637		97	637	97	0.0012	0.1234	L	6.6597	0.0012	637	97
10.24							0.1234		6.6584	0.0021	1290	D 13
10.35	584			584	97	0.0011	0.1246		6.6563	0.0001	58	135
10.48	686			686	97	0.0013	0.1259		6.6562	0.0000	0	135
10.59	581		136	90	97	0.0002	0.1261	R HUNTER	6.6562	0.0007	500	136
11.35	4013			4080	A 97	0.0076	0.1337		6.6556	0.0017	1320	D 16
11.55	1054			5042	A 13	0.0079	0.1416		6.6538	0.0043	3747	165
11.89	1832		165			0.0000	0.1416	L	6.6495	0.0021	1830	165
12.38	2553					0.0000	0.1416		6.6474	0.0010	919	D 20
13.37	5262			11149	A 16	0.0115	0.1531		6.6464	0.0001	97	183
13.56	1002					0.0000	0.1531		6.6463	0.0000		
13.94	2004		272			0.0000	0.1531	R	6.6463	0.0000		
14.37	2267						0.1531		6.6463	0.0000		
14.39	106					0.0000	0.1531		6.6463	0.0216	16393	A 94
14.53	742		272			0.0000	0.1531	L	6.6247	0.0009	436	94
15.37	4399					0.0000	0.1531		6.6239	0.0013	594	A 86
15.45	434			717	200	0.0007	0.1538		6.6226	0.0007	330	86
15.45				4571	D 20	0.0059	0.1597		6.6219			
15.61	867		94	854	94	0.0017	0.1614	L ELIZABETH	6.6219	0.0019	867	86
15.64	163				94	0.0000	0.1614	REVERSE CURVE	6.6200	0.0004	163	86
15.64				211	D 84	0.0004	0.1619	TT 55 MPH	6.6196			
16.09	2330		86	2330	86	0.0051	0.1670	R	6.6196	0.0051	2330	86
16.09							0.1670	ELMORA	6.6145	0.0063	4762	D 20
16.37	1517			1360	86	0.0030	0.1700	E END 6 TRACKS	6.6082	0.0064	5670	200
17.37	5283				86	0.0000	0.1700		6.6025	0.0000		
18.38	5289			10319	A 86	0.0228	0.1928		6.6025	0.0063	6309	A 15
19.37	5258					0.0000	0.1928		6.5965	0.0006	586	182
19.52	774			406	182	0.0004	0.1932		6.5959	0.0008	774	182
19.74	1187		182	1187	182	0.0012	0.1945	R	6.5951	0.0012	1187	182
20.12	2014			1360	182	0.0014	0.1959		6.5939	0.0010	1008	D 20
20.38	1342		300	6309	A 18	0.0063	0.2021	L	6.5929	0.0024	2466	200
20.51	695					0.0000	0.2021		6.5905	0.0000		
20.68	909		222			0.0000	0.2021	R	6.5905	0.0000		
20.87	1016					0.0000	0.2021	RAHWAY	6.5905	0.0000		
20.90	160					0.0000	0.2021		6.5905	0.0000		
20.95	214					0.0000	0.2021		6.5905	0.0000		
21.14	1016		300			0.0000	0.2021	L	6.5905	0.0049	5027	A 18
21.37	1230			134	200	0.0001	0.2023		6.5856	0.0001	65	186
21.62	1295			793	D 20	0.0008	0.2031		6.5855	0.0010	1018	186
21.95	1781		186	1781	186	0.0018	0.2049	R	6.5845	0.0000		
22.02	378			378	186	0.0004	0.2053	REVERSE CURVE	6.5845	0.0013	1282	A 18
22.13	540		218	540	186	0.0006	0.2058	L	6.5832	0.0000		
22.37	1295			442	186	0.0006	0.2063		6.5832	0.0000		
22.96	3108			4785	A 18	0.0048	0.2110	COLONIA	6.5832	0.0000		
22.96				131	196	0.0001	0.2112		6.5832	0.0000		
22.96				340	D 16	0.0003	0.2115		6.5832	0.0000		
23.08	621		190	541	190	0.0005	0.2120	R	6.5832	0.0000		
23.08				265	190	0.0003	0.2123		6.5832	0.0000		
23.20	621			436	D 16	0.0004	0.2128		6.5832	0.0000		
23.35	828		182	828	182	0.0009	0.2136	L	6.5832	0.0000		
23.38	158		182	158	182	0.0002	0.2138	L	6.5832	0.0105	9123	A 14
23.79	2158			425	182	0.0004	0.2142		6.5727	0.0018	1360	146
23.79				1732	D 17	0.0020	0.2162		6.5709	0.0000		
24.06	1421		146	1421	146	0.0018	0.2181	R	6.5709	0.0018	1421	146
24.15	474			474	146	0.0006	0.2187	REVERSE CURVE	6.5691	0.0006	474	146
24.36	1105		146	1105	146	0.0014	0.2201	ISELIN	6.5685	0.0009	719	146
24.66	1602		146	1602	146	0.0021	0.2222	L	6.5675	0.0000		
24.80	701		146	248	146	0.0003	0.2225	L	6.5675	0.0027	2010	A 13
24.92	681			453	D 14	0.0006	0.2231	METROPARK	6.5648	0.0031	2176	135
25.18	1361		136	1301	136	0.0019	0.2250	R	6.5618	0.0011	778	A 13
25.18				1360	136	0.0019	0.2269		6.5607			
25.18				500	A 13	0.0007	0.2276		6.5607			
25.37	995			100	136	0.0001	0.2278	MENLO PARK	6.5607	0.0011	782	130
25.48	598			314	D 13	0.0004	0.2282		6.5596	0.0009	598	130
25.92	2284		130	2284	130	0.0033	0.2315	L	6.5587	0.0033	2284	130
26.10	979			979	130	0.0014	0.2330		6.5553	0.0009	891	D 14
26.39	1522		147	381	130	0.0006	0.2335	R	6.5544	0.0005	404	147
26.84	2398		147	3003	A 13	0.0042	0.2377	R	6.5539	0.0075	5199	A 11
27.37	2755			1896	147	0.0024	0.2401	METUCHEN	6.5464	0.0023	1360	114
27.37				509	A 14	0.0008	0.2409		6.5441			
27.53	860			282	150	0.0003	0.2412		6.5441	0.0014	860	114
27.53				1384	D 14	0.0020	0.2432		6.5427			
27.79	1397		114	1397	114	0.0023	0.2455	R SOUTH METUCHEN	6.5427	0.0023	1397	114
27.79				1360	114	0.0023	0.2478		6.5404	0.0032	2433	D 17
28.19	2095			11121	A 11	0.0145	0.2623		6.5372	0.0001	75	172
28.36	913		233			0.0000	0.2623	L	6.5371	0.0006	500	172
28.47	573		233			0.0000	0.2623	L	6.5365	0.0006	573	172
28.65	937					0.0000	0.2623		6.5359	0.0010	937	172
29.00	1874		172			0.0000	0.2623	L	6.5349	0.0021	1874	172
29.36	1874					0.0000	0.2623		6.5328	0.0016	1528	D 20
30.14	4135					0.0000	0.2623		6.5312	0.0042	4481	200
30.35	1074		230	9295	A 17	0.0094	0.2718	R	6.5270	0.0010	1074	200
30.37	107					0.0000	0.2718		6.5260	0.0001	107	200
31.36	5273					0.0000	0.2718		6.5259	0.0050	5273	200

APPENDIX B5

READ DOWN — WESTBOUND TRAFFIC | STATION | READ UP — EASTBOUND TRAFFIC

PAGE 1 COLUMN 2

MILES	DISTANCE	HWY GRADE CROSSING	MAX MPH	DISTANCE	SPEED	ET, HOURS	TIME, HOURS	STATION	TIME, HOURS	ET, HOURS	DISTANCE	SPEED
31.53	875					0.0000	0.2718		6.5209	0.0008	875	200
32.04	2676		229	630	200	0.0006	0.2724	L	6.5201	0.0025	2676	200
32.36	1698			1698	200	0.0016	0.2740		6.5175	0.0016	1698	200
32.44	420			420	200	0.0004	0.2744		6.5159	0.0004	420	200
32.65	1103		226	1103	200	0.0010	0.2754	L	6.5155	0.0010	1103	200
33.35	3727			3727	200	0.0035	0.2789	NEW BRUNSWICK	6.5145	0.0035	3727	200
34.36	5290			5290	200	0.0050	0.2840		6.5109	0.0050	5290	200
34.93	3029			3029	200	0.0029	0.2868		6.5059	0.0029	3029	200
35.35	2208		204	2208	200	0.0021	0.2889	L	6.5031	0.0021	2208	200
36.37	5404			5404	200	0.0051	0.2940		6.5010	0.0051	5404	200
37.36	5202			5202	200	0.0049	0.2990		6.4959	0.0049	5202	200
38.36	5303			5303	200	0.0050	0.3040		6.4909	0.0050	5303	200
39.36	5259			5259	200	0.0050	0.3090		6.4859	0.0049	5133	200
40.36	5285			5285	200	0.0050	0.3140		6.4810	0.0000		
40.44	430			430	200	0.0004	0.3144	DEANS	6.4810	0.0063	6309	A 18
40.66	1183		272	175	200	0.0002	0.3145	L	6.4748	0.0007	715	182
40.78	645			1008	D 20	0.0010	0.3155		6.4741	0.0007	645	182
41.36	3013		182	3013	182	0.0031	0.3187	R	6.4734	0.0031	3013	182
42.35	5253			1360	182	0.0014	0.3201	MONMOUTH JCT	6.4702	0.0010	1308	D 20
43.36	5314			6309	A 18	0.0087	0.3287		6.4692	0.0091	9559	200
44.36	5282			8825	200	0.0084	0.3371		6.4602	0.0050	5282	200
45.36	5286			5286	200	0.0050	0.3421		6.4552	0.0050	5286	200
46.36	5284			5284	200	0.0050	0.3471		6.4502	0.0050	5284	200
47.36	5276			5276	200	0.0050	0.3521		6.4452	0.0050	5276	200
48.36	5274			5274	200	0.0050	0.3571		6.4402	0.0050	5274	200
48.42	317			317	200	0.0003	0.3574	PRINCETON JUNCTION	6.4352	0.0003	317	200
49.36	4968			4968	200	0.0047	0.3621		6.4349	0.0047	4968	200
50.36	5268			5268	200	0.0050	0.3671		6.4302	0.0050	5268	200
51.36	5295			5295	200	0.0050	0.3721		6.4252	0.0050	5295	200
51.68	1671			1671	200	0.0016	0.3737		6.4202	0.0016	1671	200
51.89	1148		261	1148	200	0.0011	0.3748	R LAWRENCE STATION	6.4186	0.0011	1148	200
52.36	2453			2453	200	0.0023	0.3771		6.4175	0.0023	2453	200
53.36	5291			5291	200	0.0050	0.3821		6.4152	0.0050	5291	200
54.36	5277			5277	200	0.0050	0.3871		6.4102	0.0073	7744	200
55.36	5280			5280	200	0.0050	0.3921		6.4028	0.0063	6309	A 18
56.36	5276			5276	200	0.0050	0.3971		6.3966	0.0016	1514	A 17
57.17	4273			4273	200	0.0040	0.4011		6.3950	0.0000		
57.35	990		286	3652	200	0.0035	0.4046	L	6.3950	0.0000		
57.46	532		286		200	0.0000	0.4046	L	6.3950	0.0000		
58.10	3402			1272	D 20	0.0013	0.4059	TRENTON	6.3950	0.0000		
58.23	691		182	691	177	0.0007	0.4066	R	6.3950	0.0000		
58.35	638			638	177	0.0007	0.4073	A	6.3950	0.0000		
58.96	3236		177	2600	177	0.0028	0.4101	L	6.3950	0.0000		
59.27	1600			0	177	0.0000	0.4101	MORRISVILLE PA	6.3950	0.0226	15868	A 73
59.36	500		233		177	0.0000	0.4101	R	6.3724	0.0007	260	73
59.57	1100			23	177	0.0000	0.4101		6.3717	0.0029	1100	73
59.57				3813	D 17	0.0058	0.4159		6.3688			
59.72	800		73	800	73	0.0021	0.4180	R STA	6.3686	0.0021	800	73
59.78	300			300	73	0.0008	0.4187	REVERSE CURVE	6.3668	0.0002	68	D 76
59.96	980		76	980	73	0.0025	0.4213	L	6.3666	0.0015	613	76
60.02	300		76	80	73	0.0002	0.4215	L	6.3651	0.0007	300	76
60.04	100			164	A 73	0.0004	0.4219	REVERSE CURVE	6.3643	0.0020	721	A 59
60.19	800		76	958	76	0.0024	0.4243	R	6.3623	0.0025	780	59
60.30	500			243	76	0.0006	0.4249		6.3598	0.0019	560	59
60.30				397	D 76	0.0009	0.4258		6.3579			
60.43	700		59	700	59	0.0022	0.4281	R	6.3579	0.0022	700	59
60.43							0.4281		6.3557	0.0013	500	D 83
60.53	500			500	59	0.0016	0.4297	REVERSE CURVE	6.3544	0.0011	500	83
60.72	1000		83	880	59	0.0028	0.4325	L	6.3532	0.0011	500	83
60.78	300			1101	A 59	0.0029	0.4354	REVERSE CURVE	6.3521	0.0009	427	D 15
60.96	1000		186	339	83	0.0008	0.4361	R	6.3512	0.0000	26	99
61.17	1100			60	83	0.0001	0.4363		6.3511	0.0008	405	A 94
61.17				780	A 83	0.0017	0.4379		6.3503			
61.17				300	94	0.0005	0.4385		6.3503	0.0038	1529	A 96
61.40	1200		94	543	94	0.0011	0.4395	R	6.3465	0.0030	1060	66
61.46	300			657	D 90	0.0015	0.4411	REVERSE CURVE	6.3434	0.0009	300	66
61.61	900		66	800	66	0.0023	0.4434	L	6.3426	0.0023	800	66
61.61				1360	66	0.0039	0.4473		6.3403	0.0028	1369	D 11
61.96	1600			4309	A 90	0.0085	0.4559		6.3374	0.0001	52	117
							0.4559		6.3374	0.0047	2570	A 91
62.79	4380			103	120	0.0002	0.4560		6.3327	0.0028	1360	91
62.79				786	D 12	0.0014	0.4574		6.3299			
62.95	800		95	800	95	0.0016	0.4590	R	6.3299	0.0017	800	91
62.96	100			91	95	0.0002	0.4592	W YARD	6.3282	0.0021	1009	91
62.96	100			106	D 95	0.0002	0.4594		6.3261	0.0002	100	91
63.08	500		91	500	91	0.0010	0.4605	R	6.3259	0.0010	500	91
63.08							0.4605		6.3249	0.0034	2219	D 15
63.08	0			1360	91	0.0028	0.4633		6.3214	0.0002	177	183
63.08	0			4674	A 91	0.0074	0.4706		6.3212	0.0000		
63.96	4680					0.0000	0.4706		6.3212	0.0032	3924	A 13
64.53	3000					0.0000	0.4706		6.3180	0.0015	1024	132
64.96	2280		132	5846	132	0.0084	0.4790	R	6.3145	0.0000		
65.09	680		132		132	0.0000	0.4790	R	6.3145	0.0074	4436	A 94
65.57	2500			1260	D 13	0.0021	0.4811	REVERSE CURVE	6.3072	0.0027	1360	94
65.96	2100		94	2100	94	0.0042	0.4854	L WOODBOURNE	6.3044	0.0042	2100	94
65.96				1360	94	0.0027	0.4881		6.3002	0.0058	3643	D 18
66.96	5280			12508	A 94	0.0174	0.5055		6.2949	0.0004	371	188
67.38	2200					0.0000	0.5055		6.2945	0.0000		
67.62	1280		233			0.0000	0.5055	R	6.2945	0.0000		
67.72	500					0.0000	0.5055		6.2945	0.0000		
67.96	1300		286			0.0000	0.5055	R	6.2945	0.0000		
68.11	780		286	526	170	0.0006	0.5061	R	6.2945	0.0076	7153	A 16
68.74	3300			248	D 17	0.0003	0.5064		6.2869	0.0000		
68.96	1200		165	1200	165	0.0014	0.5078	R	6.2869	0.0000		
69.09	680		165	680	165	0.0008	0.5086	R	6.2869	0.0079	6293	A 13
69.59	2900			1280	165	0.0015	0.5100		6.2790	0.0019	1360	136
69.59				1320	D 16	0.0017	0.5117		6.2771			
69.83	1300		135	1300	135	0.0018	0.5135	L	6.2771	0.0018	1300	135
69.83				1360	135	0.0019	0.5154	HOLLAND	6.2753	0.0000		
69.96	700			1427	A 13	0.0019	0.5174		6.2753	0.0010	700	135
70.06	500					0.0000	0.5174		6.2743	0.0007	500	135
70.19	700		165	367	143	0.0005	0.5179	R	6.2736	0.0010	700	135
70.49	1580			326	D 14	0.0004	0.5183		6.2726	0.0022	1580	135
70.70	1100		135	1100	135	0.0015	0.5199	L	6.2704	0.0015	1100	135
70.96	1400			1360	135	0.0019	0.5218		6.2689	0.0016	1272	D 16
70.98	80			6052	A 13	0.0076	0.5294		6.2672	0.0001	76	164
71.38	2100		190		0	0.0000	0.5294	R	6.2671	0.0000		
71.70	1700				0	0.0000	0.5294		6.2671	0.0000		
71.87	900	XING 12.91			0	0.0000	0.5294	L BUSTLETON	6.2671	0.0000		
71.96	500			76	164	0.0001	0.5295		6.2671	0.0076	6052	A 13
72.36	2080			1272	D 16	0.0016	0.5311		6.2595	0.0019	1360	135
72.89	2800		135	2800	135	0.0039	0.5350	R STREET RD	6.2576	0.0039	2800	135
72.96	400			400	135	0.0006	0.5356		6.2537	0.0036	3194	D 20
73.66	3680			900	135	0.0013	0.5369		6.2500	0.0033	3537	200
73.91	1300		202	17442	A 13	0.0194	0.5564	L	6.2467	0.0012	1300	200
73.96	300	XING 15.95				0.0000	0.5564	COUNTY LINE	6.2455	0.0003	300	200
74.96	5280					0.0000	0.5564		6.2452	0.0050	5280	200
75.96	5280					0.0000	0.5564		6.2402	0.0000		
76.96	5280			5444	200	0.0052	0.5615		6.2402	0.0115	11149	A 165 200
77.84	4800			1874	D 20	0.0019	0.5635		6.2287	0.0016	1360	165
77.96	580		165	580	165	0.0008	0.5643	L	6.2271	0.0008	580	165
78.24	1480		165	1480	165	0.0017	0.5660	L	6.2263	0.0017	1480	165

APPENDIX B5

PAGE 2 COLUMN 1

READ DOWN WESTBOUND TRAFFIC | STATION | READ UP EASTBOUND TRAFFIC

MILES	DISTANCE	XING	MAX SPD	DISTANCE	SPEED	ST. HOURS	TIME HOURS	STATION	TIME HOURS	ST. HOURS	DISTANCE	SPEED	MILES
78.95	3700			1380	185	0.0018	0.5875	YORK RD	6.2248	0.0019	1874	D 20	115.30
78.96	100		233	11149	A 16	0.0130	0.5805	L	6.2227	0.0006	587	200	115.52
78.64	2580		233			0.0000	0.5806	L	6.2221	0.0034	3580	200	115.56
78.96	1700					0.0000	0.5805		6.2187	0.0000			115.71
80.15	880			0	200	0.0000	0.5805		6.2187	0.0000			116.39
80.76	3200		233	751	200	0.0007	0.5812	R	6.2187	0.0000			116.39
80.96	1100			3428	200	0.0114	0.5926		6.2187	0.0115	11149	A 16	116.39
81.76	4200			1874	D 20	0.0018	0.5946		6.2072	0.0016	1960	165	116.39
81.96	1080		165	1080	165	0.0012	0.5958	R	6.2056	0.0012	1080	165	116.52
82.12	800		165	600	165	0.0008	0.5967	R CAMP HILL	6.2044	0.0008	800	165	116.98
82.65	2800			1360	165	0.0016	0.5983		6.2036	0.0013	1191	D 18	117.10
82.96	1680		190	7801	A 18	0.0080	0.6063	L	6.2002	0.0001	68	188	117.46
83.29	1700		190	3648	A 19	0.0035	0.6098	L	6.2021	0.0000			117.52
83.96	3520					0.0000	0.6098		6.2021	0.0162	10191	A 13	117.52
84.96	5280			17	200	0.0000	0.6098		6.1869	0.0010	980	135	117.94
85.09	680			3194	D 35	0.0036	0.6134		6.1860	0.0010	880	135	118.52
85.49	2100		135	2100	135	0.0029	0.6164	R STENTON	6.1850	0.0029	2100	135	118.60
85.96	2500			1380	135	0.0019	0.6180		6.1821	0.0036	3194	D 35	118.87
86.96	5280			17440	A 13	0.0194	0.6377		6.1784	0.0052	5304	200	119.34
87.96	5280	XING 28.18				0.0000	0.6377	CONSHOHOCKEN	6.1732	0.0000			119.34
88.96	5280	XING 29.52				0.0000	0.6377	GRAVERS	6.1732	0.0088	5642	A 18	119.34
88.96	0			638	200	0.0006	0.6383		6.1696	0.0005	0	0	119.52
89.38	2180			1062	D 35	0.0011	0.6394		6.1868	0.0000			120.26
89.85	2500		181	2500	181	0.0026	0.6420	L	6.1686	0.0000			120.52
89.96	600			300	181	0.0006	0.6426	EARNEST	6.1686	0.0128	10850	A 13	120.61
90.70	3880			1748	181	0.0018	0.6445		6.1539	0.0059	1380	135	120.88
90.70				2130	D 18	0.0026	0.6470		6.1519	0.0000			121.27
90.96	1400		135	1400	135	0.0020	0.6490	L YRAMBO	6.1519	0.0020	1402	135	121.52
91.02	250		135	280	135	0.0004	0.6494	L	6.1499	0.0004	280	135	122.52
91.28	1400			1380	135	0.0019	0.6513		6.1495	0.0045	760	D 15	123.00
91.96	3600		153	3435	A 13	0.0045	0.6558	R HENDERSON	6.1488	0.0052	4240	153	123.19
92.07	580		153	786	153	0.0010	0.6568	R	6.1433	0.0007	980	153	123.52
92.96	4700			1380	153	0.0017	0.6585		6.1426	0.0007	980	D 16	124.52
93.13	880			2858	A 15	0.0034	0.6619		6.1419	0.0058	5020	165	124.61
93.47	1800		165	2160	165	0.0036	0.6655	L	6.1361	0.0021	1800	165	124.85
93.96	2600			1360	165	0.0016	0.6671	KING of PRUSSIA	6.1341	0.0019	1874	D 20	125.52
94.96	5250			11149	A 16	0.0115	0.6786		6.1321	0.0067	4006	200	126.28
95.96	5280					0.0000	0.6786		6.1264	0.0050	5280	200	126.43
96.07	580					0.0000	0.6786		6.1214	0.0005	580	200	126.52
96.38	1600		200	2831	200	0.0027	0.6813	R	6.1209	0.0015	1600	200	126.83
96.96	3100			3100	200	0.0029	0.6842		6.1194	0.0045	4793	200	126.93
97.96	5280			7387	200	0.0069	0.6911	HOWELLVILLE	6.1149	0.0048	5007	A 16	127.52
98.49	2800			793	D 28	0.0006	0.6916		6.1098	0.0014	1360	188	127.52
98.96	2480		186	2480	186	0.0025	0.6944	R	6.1085	0.0025	2430	186	127.68
98.96	317		186	317	186	0.0003	0.6947		6.1060	0.0003	317	186	127.72
99.11	483			483	186	0.0006	0.6952	CR ERROR	6.1057	0.0006	780	188	128.18
99.60	2606		186	2800	186	0.0026	0.6978	L	6.1049	0.0029	2800	186	128.26
99.96	1600			1380	188	0.0014	0.6992		6.1022	0.0008	793	D 28	128.52
100.85	4700			5027	A 19	0.0049	0.7042		6.1014	0.0088	5807	200	128.72
100.96	580		404	780	200	0.0006	0.7048	R	6.0959	0.0005	580	200	128.72
101.15	980		404	980	200	0.0009	0.7058	R	6.0954	0.0009	980	200	129.33
101.96	4300			4300	200	0.0041	0.7099		6.0945	0.0041	4300	200	129.52
102.17	1100			1100	200	0.0010	0.7110		6.0904	0.0010	1100	200	129.52
102.28	580		200	580	200	0.0006	0.7116	L	6.0894	0.0005	580	200	129.62
102.75	2442	UG 2.07		2442	200	0.0023	0.7138	STREAM	6.0888	0.0023	2442	200	129.73
102.82	370			370	200	0.0004	0.7142		6.0865	0.0004	370	200	129.73
102.96	788			788	200	0.0007	0.7149		6.0861	0.0007	788	200	130.12
103.22	1373		233	1373	200	0.0013	0.7162		6.0854	0.0013	1373	200	130.28
103.49	1425	OHD 1.46		1425	200	0.0013	0.7176	PARTE 352	6.0841	0.0013	1425	200	130.28
103.63	740			740	200	0.0007	0.7183		6.0827	0.0007	740	200	130.52
103.81	954		200	954	200	0.0009	0.7192	L	6.0820	0.0009	954	200	130.52
103.96	788		200	788	200	0.0007	0.7199	L	6.0811	0.0007	788	200	130.83
104.11	802	XING 45.18		802	200	0.0008	0.7207	W HENDERSON	6.0804	0.0008	802	200	131.19
104.16	245			240	200	0.0002	0.7209		6.0796	0.0000	245	200	131.19
104.41	1313		200	1313	200	0.0012	0.7221	R	6.0784	0.0012	1313	200	131.52
104.46	260	UG 0.35		260	200	0.0002	0.7224	RAVINE ROAD	6.0782	0.0002	260	200	131.58
104.60	750			750	200	0.0007	0.7231		6.0778	0.0007	750	200	131.58
104.69	453		200	453	200	0.0004	0.7235	R	6.0772	0.0004	453	200	131.95
104.71	101			101	200	0.0001	0.7236		6.0768	0.0001	101	200	131.98
104.76	274			274	200	0.0003	0.7239		6.0767	0.0003	274	200	132.27
104.81	257			257	200	0.0002	0.7241	GLEN LOCH	6.0764	0.0002	257	200	132.52
104.99	980			980	200	0.0009	0.7251	T	6.0762	0.0009	980	200	132.72
105.28	1524			1524	200	0.0014	0.7265		6.0752	0.0014	1524	200	132.72
105.52	1252			1252	200	0.0012	0.7277	T	6.0738	0.0012	1252	200	133.20
105.65	673			673	200	0.0006	0.7283	T	6.0726	0.0006	673	200	133.52
105.87	1180		233	1180	200	0.0011	0.7295	L	6.0720	0.0011	1180	200	134.00
106.10	1191		201	1191	200	0.0011	0.7306	R REVERSE CURVE	6.0709	0.0011	1191	200	134.18
106.52	2226			2226	200	0.0021	0.7327	SHIP ROAD	6.0697	0.0021	2226	200	134.52
107.52	5280			5280	200	0.0050	0.7377		6.0678	0.0034	456	200	135.32
107.75	1250			1250	200	0.0012	0.7389		6.0658				135.52
108.52	4050		426	4050	200	0.0038	0.7427	L	6.0658				136.13
108.68	845		426	845	200	0.0008	0.7435	L	6.0658				136.52
109.28	1715			1715	200	0.0016	0.7451	T	6.0658				136.89
109.52	1214			1214	200	0.0011	0.7463	R ACKWORTH	6.0658				137.16
109.70	970		200	970	200	0.0009	0.7472	R	6.0658				137.52
109.96	1024			1024	200	0.0010	0.7481		6.0658				137.96
110.26	1540		200	880	200	0.0008	0.7490	L	6.0658				138.44
110.26					200	0.0000	0.7490		6.0658	0.0217	18966	A 12	138.52
110.52	1847			4	200	0.0000	0.7490	T	6.0441	0.0009	167	125	138.52
110.74	1173			3625	D 35	0.0042	0.7532	T	6.0436	0.0018	1174	125	139.02
111.14	2080		125	2080	125	0.0032	0.7563	R	6.0420	0.0030	2080	125	139.22
111.14				1380	125	0.0021	0.7584		6.0389	0.0022	1701	D 16	139.47
111.52	2027			7820	A 12	0.0101	0.7685	T	6.0367	0.0004	428	185	139.47
111.75	1207					0.0000	0.7685	DOWNINGTOWN	6.0363	0.0014	1202	185	139.52
112.08	1777		165			0.0000	0.7685	L	6.0349	0.0020	1777	185	140.12
112.28	1046					0.0000	0.7685	T	6.0329	0.0007	602	D 17	140.12
112.52	1268		177			0.0000	0.7685	R	6.0322	0.0018	1699	177	140.48
112.73	1109		177	3326	A 16	0.0037	0.7722	R	6.0304	0.0019	1788	177	140.52
113.18	2379					0.0000	0.7722	T	6.0285	0.0037	3326	A 16	140.97
113.52	1796		177	0		0.0000	0.7722	L	6.0248	0.0002	169	185	140.97
113.61	466		177	870	177	0.0007	0.7729	L	6.0246	0.0006	466	185	141.18
113.75	725			602	D 17	0.0007	0.7736	T	6.0241	0.0008	725	185	141.52
113.95	1087		185	1087	185	0.0012	0.7748	R	6.0232	0.0012	1087	185	141.63
114.16	1087		300	1087	185	0.0012	0.7761	R	6.0220	0.0014	1300	D 16	142.44
114.35	1035			273	165	0.0003	0.7764	T	6.0206	0.0008	820	190	142.44
114.52	880		233	6522	A 18	0.0066	0.7830	L	6.0196	0.0009	880	190	142.52
114.56	216					0.0000	0.7830	T	6.0189	0.0002	216	190	142.72
114.70	754		190		186	0.0000	0.7830	L	6.0187	0.0008	903	190	143.02

PAGE 2 COLUMN 2

READ DOWN WESTBOUND TRAFFIC | STATION | READ UP EASTBOUND TRAFFIC

DISTANCE	XING	MAX SPD	DISTANCE	SPEED	ST. HOURS	TIME HOURS	STATION	TIME HOURS	ST. HOURS	DISTANCE	SPEED
3610			100	186	0.0001	0.7831	T	6.0179	0.0014	1379	A 186 190
700		186	700	186	0.0007	0.7838	L	6.0165	0.0000		
209		186	209	186	0.0002	0.7840	L	6.0165	0.0000		
764		186	764	186	0.0008	0.7848	R REVERSE CURVE	6.0165	0.0066	6122	A 16
3697			1360	186	0.0014	0.7862	T	6.0099	0.0016	1360	165
			695	A 18	0.0007	0.7869		6.0083			
			370	188	0.0004	0.7873		6.0083			
			1191	D 18	0.0013	0.7885		6.0083			
680		165	680	165	0.0008	0.7893	R	6.0083	0.0005	680	165
2485		165	2463	165	0.0028	0.7921	R	6.0078	0.0028	2463	165
621			621	165	0.0007	0.7928	T	6.0048	0.0007	621	165
1915		165	1915	165	0.0022	0.7950	L	6.0043	0.0022	1915	165
301			311	165	0.0004	0.7954	T	6.0019	0.0019	1874	D 19
			1049	159	0.0012	0.7966	COATESVILLE	5.9986	0.0005	573	200
2790			7501	A 16	0.0074	0.8040	T	5.9964	0.0000		
3040		233			0.0000	0.8040	L	5.9894	0.0000		
414		233			0.0000	0.8040	L	5.9894	0.0035	5648	A 19
1449					0.0000	0.8040	T	5.9858	0.0014	1360	190
2485		190	1078	190	0.0011	0.8050	R	5.9845	0.0025	2485	190
			1360	190	0.0014	0.8064		5.9820	0.0003	340	D 19
			3211	A 19	0.0028	0.8091		5.9817	0.0005	549	190
952			116	235	0.0001	0.8093	T	5.9813	0.0007	2807	A 18
3984			279	D 19	0.0003	0.8095	T	5.9806	0.0014	1360	188
1346		188	1346	188	0.0014	0.8108	L	5.9893	0.0014	1346	188
485		188	485	188	0.0005	0.8114	L	5.9879	0.0005	485	188
1455			1455	188	0.0015	0.8128	T	5.9874	0.0015	1455	188
2047		188	2047	188	0.0021	0.8149	R	5.9859	0.0021	2047	188
1295			1360	188	0.0014	0.8163	T	5.9838	0.0007	593	D 20
5280			4341	A 18	0.0042	0.8205		5.9832	0.0001	80	200
2506			2408	200	0.0032	0.8237	T	5.9831	0.0024	2506	200
984		200	984	200	0.0009	0.8247	L	5.9807	0.0009	984	200
1780			1760	200	0.0017	0.8263	T	5.9798	0.0017	1780	200
5280			5280	200	0.0050	0.8313	PARKESBURG	5.9781	0.0021	2200	200
495			495	200	0.0005	0.8318	T	5.9781	0.0000		
1265		233	1265	200	0.0012	0.8330	R	5.9761	0.0002	200	200
3520			3520	200	0.0033	0.8363	T	5.9759	0.0000		
4015			3964	200	0.0034	0.8397	T	5.9759	0.0000		
770		233	6	200	0.0000	0.8397	R	5.9759	0.0233	20035	A 11
495			0	200	0.0000	0.8397	T	5.9526	0.0013	903	117
2145			0	200	0.0000	0.8397	T	5.9512	0.0042	2332	94
			3859	D 20	0.0046	0.8443		5.9471	0.0000		
3135		117	2425	117	0.0039	0.8483	R	5.9471	0.0025	1300	94
			712	D 11	0.0012	0.8495	ATGLEN	5.9448	0.0000		
834		94	834	94	0.0017	0.8512	R	5.9446	0.0017	834	94
500			500	94	0.0010	0.8522	REVERSE CURVE	5.9429	0.0021	1290	D 13
2158		132	820	94	0.0017	0.8539	L	5.9408	0.0000	1406	132
389			3436	A 94	0.0071	0.8610	REVERSE CURVE	5.9408	0.0006	389	132
1989		132			0.0000	0.8610	R	5.9402	0.0020	1386	132
1056		132		132	0.0000	0.8610	R CHRISTIANA	5.9382	0.0004	290	132
			1066	132	0.0015	0.8626		5.9378	0.0000		
3124			1916	A 13	0.0026	0.8652	T	5.9378	0.0080	4740	A 90
1000		105	257	145	0.0003	0.8655	L	5.9298	0.0005	300	90
			444	D 14	0.0006	0.8661		5.9293	0.0000		
550		132	0	132	0.0000	0.8661	L	5.9293	0.0045	2119	90
550			0	132	0.0000	0.8661	REVERSE CURVE	5.9248	0.0012	594	A 82
			1358	D 13	0.0023	0.8685		5.9236	0.0000		
2090		90	2090	90	0.0044	0.8729	R	5.9236	0.0012	535	82
625			625	90	0.0013	0.8742	T	5.9224	0.0019	835	82
			202	D 90	0.0004	0.8746		5.9205	0.0000		
1265		82	369	82	0.0009	0.8755	L	5.9205	0.0029	1265	82
			900	82	0.0021	0.8776	GAP	5.9176	0.0000		
1650		82	1650	82	0.0038	0.8814	L	5.9176	0.0038	1650	82
1670		206	1360	82	0.0031	0.8845	L	5.9138	0.0043	1870	82
			1904	A 82	0.0003	0.8848		5.9096	0.0000		
1760			34	106	0.0001	0.8849	T	5.9095	0.0041	1760	82
580				106	0.0000	0.8849		5.9054	0.0006	300	82
			982	D 10	0.0013	0.8862		5.9048	0.0000		
2090		82	2090	82	0.0048	0.8910	R	5.9046	0.0048	2090	82
			1360	83	0.0021	0.8942		5.8998	0.0017	807	D 11
1540			2866	A 83	0.0021	0.8963	T	5.8981	0.0011	583	113
1920		113	0		0.0000	0.8963	L	5.8970	0.0022	1526	113
1045		113		113	0.0000	0.8963	L	5.8948	0.0018	1046	113
			1289	113	0.0021	0.9014		5.8930	0.0044	3580	D 19
2580			9397	A 113	0.0126	0.9140	T	5.8886	0.0003	245	103
1700		233	0	165	0.0000	0.9140	L	5.8883	0.0000		
2550			0	165	0.0000	0.9140	T	5.8883	0.0000		
395		233	1435	A 162	0.0027	0.9166	L	5.8883	0.0000		
1835			126	174	0.0001	0.9169	T	5.8883	0.0080	8605	A 16
4235			447	D 17	0.0005	0.9174	T	5.8760	0.0015	1360	165
1046		165	1046	165	0.0012	0.9186	R	5.8777	0.0012	1046	165
3245		165	3245	165	0.0037	0.9223	R	5.8765	0.0037	3245	165
2085			1360	165	0.0016	0.9239	T	5.8728	0.0003	296	D 17
1080			5782	A 16	0.0063	0.9301	T	5.8725	0.0001	98	171
1468		202			0.0000	0.9301	L	5.8724			
1815			160	165	0.0002	0.9303	T	5.8724	0.0097	7806	A 13
2334			2347	D 16	0.0028	0.9331	T	5.8627	0.0019	1360	135
2501		135	2501	135	0.0035	0.9366	L	5.8608	0.0035	2501	135
445			445	135	0.0006	0.9372	T	5.8572	0.0014	991	135
			915	135	0.0013	0.9385		5.8559	0.0000		
2640	XING 59.18		2010	A 13	0.0027	0.9412	IRISHTOWN ROAD	5.8559	0.0047	3089	A 11
1648		146	820	146	0.0011	0.9423	R	5.8511	0.0001	40	113
1220			6	146	0.0000	0.9423	T	5.8511	0.0022	1920	113
			1254	D 14	0.0018	0.9441		5.8488			
275		113	275	113	0.0005	0.9446	R	5.8488	0.0005	275	113
3190		113	3190	113	0.0063	0.9469	R	5.8484	0.0057	3190	113
			1360	113	0.0023	0.9522		5.8430			113
1870			506	A 113	0.0008	0.9530	T	5.8430	0.0002	130	D 11
720		117	224	117	0.0004	0.9534	L BIRD IN HAND	5.8428	0.0032	1959	117
2590		117	2390	117	0.0039	0.9572	L	5.8397	0.0039	2390	117
			1360	117	0.0022	0.9594		5.8358			
1112			8866	A 11	0.0118	0.9713	T	5.8358	0.0027	1685	D 18
1778		165		165	0.0000	0.9713	L	5.8331	0.0010	900	165
605		165		165	0.0000	0.9713	L	5.8321	0.0007	605	165
4235				165	0.0000	0.9713	T	5.8314	0.0009	985	D 18
					0.0000	0.9713		5.8304	0.0003	250	182
440		233			0.0000	0.9713	L	5.8302	0.0000		
1045		233			0.0000	0.9713	L	5.8302	0.0052	4640	A 16
1595			505	185	0.0006	0.9719	T	5.8249	0.0016	1360	185

APPENDIX B5

PAGE 3 COLUMN 1

MILES	DISTANCE, ft	HWY GRADE CROSSING	MAX MPH	READ DOWN WESTBOUND TRAFFIC DISTANCE, ft	SPEED, mph	dT, hours	TIME, hours	STATION	READ UP EASTBOUND TRAFFIC TIME, hours	dT, hours	DISTANCE, ft	SPEED, mph
143.32	1595		165	1595	165	0.0018	0.9737 R		5.8233	0.0018	1595	165
143.32							0.9737		5.8215	0.0019	1815	D 19
143.52	1045			1045	165	0.0012	0.9749 T		5.8196	0.0001	122	198
143.52				315	165	0.0004	0.9753		5.8196	0.0000		
144.32	4224					0.0000	0.9753 T		5.8195	0.0000		
144.52	1056		286	6122	A 16	0.0105	0.9858 R		5.8190	0.0000		
144.52						0.0000	0.9858		5.8195	0.0048	4661	A 18
145.02	2640		280			0.0000	0.9858 R		5.8149	0.0000		
145.52	2640			3003	A 18	0.0035	0.9893 T		5.8149	0.0000		
145.52				305	186	0.0003	0.9896		5.8148	0.0000		
145.57	275			580	D 15	0.0006	0.9901 T		5.8149	0.0000		
145.67	495		186	495	186	0.0005	0.9906 R		5.8149	0.0086	6127	A 16
145.75	440			634	186	0.0008	0.9913	REVERSE CURVE	5.8063	0.0001	95	195
145.94	990		186	0	186	0.0000	0.9913 L		5.8062	0.0011	990	185
145.99	275			1081	D 18	0.0012	0.9924 T		5.8071	0.0003	275	165
146.18	990		165	990	165	0.0011	0.9936 L		5.8098	0.0011	990	165
146.52	1815			1360	165	0.0016	0.9951 T		5.8056	0.0010	919	D 18
146.52							0.9951		5.8046	0.0003	277	183
146.90	2006			6464	A 16	0.0069	1.0021	LANCASTER STATION	5.8040	0.0000		
147.52	2274						1.0021		5.8043	0.0079	7124	A 15
148.01	2585			342	187	0.0003	1.0024 T		5.7964	0.0018	1360	157
148.01				1514	D 16	0.0017	1.0041	DILLERVILLE - COL BR	5.7948	0.0000		
148.52	2695		157	2695	157	0.0033	1.0073 R		5.7948	0.0033	2695	157
148.52					157	0.0000	1.0073		5.7915	0.0000		
148.72	1045		157	1045	157	0.0013	1.0086 R		5.7915	0.0013	1045	157
							1.0086		5.7903	0.0014	1243	D 18
149.52	4225			1360	157	0.0016	1.0102 T		5.7889	0.0003	257	182
149.52				5535	A 15	0.0083	1.0185		5.7886	0.0063	4840	A 16
150.18	3465			151	178	0.0002	1.0187 T		5.7830	0.0016	1360	165
150.18				654	D 17	0.0007	1.0194		5.7818			
150.52	1815		165	1455	165	0.0021	1.0196 L		5.7618	0.0021	1815	165
150.56	220		165	220	165	0.0003	1.0197 L		5.7797	0.0003	220	165
150.56				1360	165	0.0016	1.0213		5.7794	0.0003	296	D 18
150.56				1602	A 16	0.0018	1.0231		5.7791	0.0002	152	171
151.21	3410	XING 71.06		152	171	0.0002	1.0232	E PETER ROAD	5.7789	0.0015	1602	A 16
151.21				296	D 17	0.0003	1.0236		5.7775	0.0016	1360	165
151.52	1650		165	1650	165	0.0019	1.0255 L		5.7758	0.0019	1650	165
151.64	660		165	660	165	0.0008	1.0262 L		5.7740	0.0008	660	165
151.93	1485			1485	165	0.0017	1.0279 T		5.7730	0.0017	1485	165
152.31	2035		165	2035	165	0.0023	1.0303 R		5.7716	0.0065	5705	165
152.52	1100			1100	165	0.0013	1.0315 T		5.7690	0.0028	2406	A 15
153.52	5280			260	165	0.0003	1.0318		5.7622	0.0004	304	156
153.52				4507	A 16	0.0049	1.0368		5.7618	0.0003		
153.72	1056			288	181	0.0003	1.0371 T		5.7618	0.0013	1056	156
153.72				1281	D 18	0.0014	1.0386		5.7605			
154.23	2668		155	2668	156	0.0033	1.0418 L		5.7605	0.0033	2668	156
154.37	778			778	155	0.0010	1.0427 T		5.7572	0.0006	480	D 18
154.37				582	155	0.0007	1.0434		5.7567	0.0004	308	165
154.37				2406	A 15	0.0028	1.0463		5.7563			
154.52	778		165			0.0000	1.0463 R		5.7563	0.0009	776	165
154.72	1056		165	306	165	0.0002	1.0465 R		5.7554	0.0012	1084	165
155.43	3724	XING 75.82		6808	A 16	0.0073	1.0538	PRIVATE	5.7542	0.0012	1081	D 18
155.52	500		369			0.0000	1.0538 L		5.7551	0.0001	114	188
156.03	2668		369			0.0000	1.0538 L		5.7529	0.0000		
156.52	2612					0.0000	1.0538 T		5.7529	0.0008	6122	A 16
156.52				585	188	0.0006	1.0544		5.7463			
156.60	440			1191	D 16	0.0013	1.0557 T		5.7463	0.0000		
156.83	1250		165	1250	165	0.0014	1.0571 L		5.7463	0.0049	4066	A 14
157.13	1595			771	165	0.0008	1.0579 T		5.7414	0.0018	1360	147
157.13				824	D 16	0.0010	1.0589		5.7397	0.0000		
157.33	1045		147	1045	147	0.0013	1.0603 L		5.7397	0.0013	1045	147
							1.0603		5.7383	0.0010	824	D 18
157.52	990			990	147	0.0013	1.0616 T		5.7373	0.0000	18	165
							1.0616		5.7373	0.0049	4066	A 14
158.52	5280			370	147	0.0005	1.0620		5.7324	0.0018	1360	147
158.52				4626	A 14	0.0047	1.0667		5.7306			
158.67	778	XING 79.13		288	164	0.0003	1.0670	MOORES MILL ROAD	5.7306	0.0010	778	147
158.67				775	D 16	0.0009	1.0680		5.7296			
158.87	1056		147	1056	147	0.0014	1.0693 R		5.7296	0.0014	1056	147
159.28	2168			1360	147	0.0018	1.0711 T		5.7283	0.0024	2181	D 19
159.28				8743	A 14	0.0106	1.0816		5.7258	0.0003	281	191
159.52	1276		202	0		0.0000	1.0816 R		5.7265	0.0000		
159.52						0.0000	1.0816		5.7255			
159.61	500		202			0.0000	1.0816 R		5.7255	0.0000		
160.38	4057					0.0000	1.0816 T		5.7255	0.0000		
160.52	723		165			0.0000	1.0816 R		5.7255			
160.68	834	XING 8	165	150	188	0.0002	1.0818 R	MARKET ST MT JOY	5.7255	0.0083	7850	A 18
161.08	2112			919	D 18	0.0010	1.0828 T		5.7172	0.0016	1360	186
161.52	2334		165	2334	165	0.0027	1.0855 R		5.7156	0.0027	2334	165
161.55	165		165	165	165	0.0002	1.0857 R		5.7130	0.0002	165	165
162.30	3960			1360	165	0.0016	1.0872 T		5.7128	0.0009	812	D 18
162.52	1135		233	7579	A 16	0.0067	1.0940 L		5.7119	0.0001	136	181
162.75	1210		233	0		0.0000	1.0940 L		5.7117	0.0000		
163.23	2530			0		0.0000	1.0940 T		5.7117	0.0000		
163.52	1540		186	1393	186	0.0016	1.0956 L		5.7117	0.0000		
163.52				100	D 18	0.0001	1.0957		5.7117			
163.76	1278		184	1278	184	0.0013	1.0970 L		5.7117	0.0000		
164.27	2668			1360	184	0.0014	1.0984 T		5.7117	0.0167	13893	A 11
164.27				676	A 18	0.0007	1.0991		5.6950			
164.27				325	186	0.0003	1.0994		5.6950			
164.27				305	D 18	0.0005	1.0998		5.6950			
164.52	1334		117	1334	117	0.0022	1.1021 R	ELIZABETHTOWN	5.6950	0.0022	1334	117
164.86	1779		117	1779	117	0.0029	1.1049 R		5.6928	0.0029	1779	117
164.99	723			723	117	0.0012	1.1061 T		5.6900	0.0012	723	117
164.99				637	117	0.0010	1.1071		5.6888			
165.29	1596		165	1066	A 11	0.0017	1.1088 L		5.6888	0.0025	1596	117
165.40	611			180	125	0.0003	1.1091 T		5.6863	0.0010	611	117
165.40				264	D 15	0.0004	1.1095		5.6853			
165.52	611		117	611	117	0.0010	1.1105 L		5.6853	0.0010	611	117
165.71	1000		117	1000	117	0.0016	1.1121 L		5.6843	0.0016	1000	117
165.71				1360	117	0.0022	1.1143		5.6827	0.0046	3856	D 20
166.52	4280					0.0000	1.1143 T		5.6781	0.0004	421	200
167.24	3779			20035	A 11	0.0233	1.1377 T		5.6777	0.0036	3779	200
167.49	1334		233			0.0000	1.1377		5.6741	0.0013	1334	200
167.52	167					0.0000	1.1377 T		5.6726	0.0002	167	200
168.52	5280					0.0000	1.1377		5.6727	0.0011	1117	200
169.52	5280					0.0000	1.1377		5.6716	0.0127	12165	A 16
169.52				742	200	0.0007	1.1384		5.6589			

PAGE 3 COLUMN 2

MILES	DISTANCE, ft	HWY GRADE CROSSING	MAX MPH	READ DOWN WESTBOUND TRAFFIC DISTANCE, ft	SPEED, mph	dT, hours	TIME, hours	STATION	READ UP EASTBOUND TRAFFIC TIME, hours	dT, hours	DISTANCE, ft	SPEED, mph
170.29	4087			2065	D 20	0.0022	1.1405 T		5.6589	0.0016	1380	161
170.52	1198		161	1198	161	0.0014	1.1419 L		5.6573	0.0002	195	161
170.68	825		161	825	161	0.0013	1.1429 L		5.6571	0.0000		
171.17	2585			1360	161	0.0016	1.1445 T		5.6571	0.0000		
171.52	1870		165	1016	A 161	0.0012	1.1457 R		5.6571	0.0106	7870	A 11
171.52				2079	165	0.0024	1.1481		5.6464			
171.81	1556		165	1556	165	0.0015	1.1499 R		5.6464	0.0042	2312	A 94
172.28	2445			490	165	0.0005	1.1504 T		5.6423	0.0002	82	94
172.28				1955	D 18	0.0027	1.1531		5.6421	0.0000		
172.52	1276		117	584	117	0.0009	1.1540 R		5.6421	0.0026	1278	94
172.52				712	D 11	0.0013	1.1553		5.6395	0.0000		
172.87	1870		94	1870	94	0.0038	1.1590 R		5.6395	0.0038	1870	94
172.87				1360	94	0.0027	1.1618	REVERSE CURVE	5.6358	0.0000		
172.87				1289	A 94	0.0024	1.1642		5.6356	0.0000	1711	D 10
173.46	3080		108	421	108	0.0007	1.1649 L		5.6258	0.0024	1369	108
173.52	330			330	108	0.0008	1.1655 T		5.6274	0.0027	1659	D 15
173.52				1080	108	0.0018	1.1673	ROYALTON BRANCH	5.6247	0.0003	281	156
174.12	3168			5636	A 108	0.0083	1.1757 T		5.6244	0.0026	2105	A 14
174.26	723		281			0.0000	1.1757 L		5.6218	0.0000	26	146
174.33	389					0.0000	1.1757 T		5.6217	0.0005	389	146
174.49	834		571				1.1757 L		5.6212	0.0011	834	146
174.51	111					0.0000	1.1757 T		5.6202	0.0001	111	146
174.52	56		146			0.0000	1.1757 L	MIDDLETOWN	5.6200	0.0001	56	146
174.81	1556		146	171	146	0.0002	1.1759 L		5.6199	0.0000	1556	146
174.89	389			389	146	0.0005	1.1764	REVERSE CURVE	5.6179	0.0018	1521	D 17
175.09	1056		178	971	146	0.0013	1.1776 R		5.6161	0.0000	5	178
175.29	1056			6453	A 14	0.0076	1.1853 T		5.6161	0.0008	780	178
175.49	1056		190			0.0000	1.1853 L		5.6153	0.0000		
175.52	167					0.0000	1.1853 T		5.6150	0.0000		
175.81	1556					0.0000	1.1853 T		5.6153	0.0000		
175.96	778		571	55	173	0.0001	1.1853 L		5.6153	0.0054	4807	A 16
176.43	2445			635	D 17	0.0007	1.1861 T	FLYOVER	5.6099	0.0016	1360	160
176.52	500		160	500	160	0.0006	1.1867 R	HIGHSPIRE	5.6083	0.0006	500	160
176.81	1556		160	1556	160	0.0018	1.1885 R		5.6077	0.0018	1556	160
177.41	3168			1360	160	0.0016	1.1901 T		5.6059	0.0002	2112	D 20
177.52	556		369			0.0000	1.1901 L		5.6036	0.0015	1612	200
177.72	1045		369			0.0000	1.1901 L		5.6021	0.0010	1045	200
178.46	3906			12400	A 18	0.0130	1.2031 T		5.6011	0.0087	3906	200
178.52	330		219			0.0000	1.2031 R		5.5974	0.0003	330	200
178.97	2385		219			0.0000	1.2031 R		5.5971	0.0023	2385	200
179.52	2885			524	200	0.0005	1.2036 T		5.5948	0.0027	2885	200
180.52	5280			5280	200	0.0056	1.2088	STEELTON	5.5921	0.0004	385	200
181.14	3300			3300	200	0.0031	1.2117 T		5.5917	0.0115	11140	A 18
181.49	1815		272	1815	200	0.0017	1.2134 R		5.5802	0.0000		
181.52	165			165	200	0.0002	1.2136 T		5.5802	0.0000		
181.52				460	200	0.0004	1.2140		5.5802	0.0000		
181.96	2334			1874	D 20	0.0019	1.2160 T		5.5802	0.0016	1360	165
182.20	1278		165	1278	165	0.0015	1.2174 R		5.5787	0.0015	1278	165
182.42	1167			1167	165	0.0013	1.2188 T		5.5772	0.0019	1874	D 20
182.50	389		202	193	165	0.0002	1.2190 L		5.5752	0.0001	82	200
182.52	111			6122	A 18	0.0066	1.2256 T		5.5752	0.0132	111	200
182.52				5027	A 18	0.0049	1.2305		5.5620	0.0000		
182.85	1760					0.0000	1.2305 T		5.5620	0.0013	1360	200
182.98	660		200			0.0000	1.2305 R		5.5607	0.0006	660	200
183.52	2860			0	200	0.0000	1.2305 T		5.5601	0.0027	2860	200
183.86	1778				200	0.0000	1.2305 T		5.5573	0.0016	1685	200
183.91	269			0	200	0.0000	1.2305		5.5557			
183.91		STATE TOWER 104.18			200	0.0000	1.2305	HARRISBURG	5.5557			
183.92	54			0	200	0.0000	1.2305		5.5557			
183.96	215			0	200	0.0000	1.2305 T		5.5557			
184.08	377			0	200	0.0000	1.2305		5.5557			
184.14	593			0	200	0.0000	1.2305		5.5557			
184.27	700		200		200	0.0000	1.2305 L		5.5557			
184.42	754			0	200	0.0000	1.2305 T		5.5557			
184.52	539			0	200	0.0000	1.2305 T		5.5557			
184.56	218			0	200	0.0000	1.2305 T		5.5557			
184.59	163		200	98	200	0.0001	1.2306 R		5.5557			
184.73	707			707	200	0.0007	1.2313 T		5.5557			
184.73	0	HERR STR6		0	200	0.0000	1.2313	WEST END AMTRAK - EA	5.5557			
184.98	1361			1361	200	0.0013	1.2326 T		5.5557			
185.07	435			435	200	0.0004	1.2330		5.5557			
185.15	436			435	200	0.0004	1.2334		5.5557			
185.27	653			653	200	0.0006	1.2340 T		5.5557			
185.46	980		200	980	200	0.0009	1.2349 R		5.5557			
185.52	327			327	200	0.0003	1.2353 T		5.5557			
185.65	705			701	200	0.0007	1.2359 T		5.5557			
155.68	162		200	162	200	0.0002	1.2361 L		5.5557			
186.52	4419			4419	200	0.0042	1.2403 T		5.5557	0.0130	12400	A 18
187.52	5281			5281	200	0.0060	1.2453 T		5.5428	0.0000		
187.52				5	200	0.0000	1.2453		5.5428	0.0154	10400	A 88
188.42	4736			4731	D 20	0.0062	1.2515 T		5.5273	0.0029	1360	88
188.45	163		88	163	88	0.0004	1.2516 R		5.5244	0.0004	163	88
188.48	163			163	88	0.0004	1.2522	REVERSE CURVE	5.5241	0.0004	163	88
188.52	218		88	218	88	0.0006	1.2527 L		5.5237	0.0005	218	88
188.52						0.0000	1.2527	LINGLESTOWN	5.5232	0.0003	160	D 94
188.60	437			437	88	0.0009	1.2536 T		5.5229	0.0037	1514	94
188.84	1257		94	1360	88	0.0028	1.2565 L		5.5193	0.0026	1257	94
188.84				1734	A 88	0.0033	1.2599		5.5167	0.0000		
189.39	2897			8	108	0.0000	1.2599 T		5.5167	0.0038	1360	67
189.39				1052	D 10	0.0023	1.2621	ROCKVILLE	5.5129	0.0000		
189.43	219		67	219	67	0.0006	1.2628 L		5.5129	0.0006	219	67
189.53	547		67	547	67	0.0015	1.2643 L		5.5123	0.0015	547	67
189.58	218		67	218	67	0.0006	1.2649 L		5.5107	0.0006	218	67
189.58				1360	67	0.0038	1.2688		5.5101	0.0038	2015	D 13
190.29	3796					0.0000	1.2688 T		5.5063	0.0024	1741	135
190.29				6505	A 67	0.0117	1.2805	SUSQUEHANNA RIVER	5.5039	0.0000		
190.46	925		135			0.0000	1.2805 R		5.5039	0.0013	925	135
190.53	381					0.0000	1.2805	MARYSVILLE	5.5026	0.0006	381	135
190.71	916					0.0000	1.2805 T		5.5020	0.0013	916	135
191.26	2905		135	1022	135	0.0014	1.2819 R		5.5008	0.0023	1672	135
191.53	1455			1360	135	0.0019	1.2839 T		5.4984	0.0056	4129	A 10
191.53				1427	A 13	0.0019	1.2858		5.4919			
191.53				32	143	0.0000	1.2858		5.4919	0.0066	3139	103
192.06	2797			1443	D 14	0.0022	1.2880 T		5.4861	0.0014	737	A 95
192.55	2581		103	2581	103	0.0047	1.2928 L		5.4847	0.0001	75	95
192.62	335		103	1063	103	0.0019	1.2947 L		5.4846	0.0007	335	95
192.80	950			232	D 10	0.0004	1.2952 T		5.4839	0.0028	1363	95
192.87	391		95	391	95	0.0008	1.2960 R		5.4810	0.0001	93	A 94-06

APPENDIX B5

PAGE 4 COLUMN 1

READ DOWN — WESTBOUND TRAFFIC

MILES	DISTANCE	XING / BRIDGE CROSSING	MAX MPH	DISTANCE	SPEED	ACC.	TIME	STATION
193.00	671			671	95	0.0013	1.2973	
193.00		BANKS			CW	0.0000	1.2973	BANKS
193.13	700			672	95	0.0013	1.2986	
193.13		XING 114.2X		29	D 95	0.0001	1.2987	PRIVATE
193.35	1160		94	1163	94	0.0023	1.3010	L
193.35							1.3010	
193.60	1300			1300	94	0.0026	1.3036	
194.15	2900			98	94	0.0001	1.3038	
194.37	1176		179	7748	A 94	0.0118	1.3156	R
194.60	1200	XING 115.3X		76	150	0.0001	1.3157	PRIVATE
195.45	4500	XING 115.8X		1891	D 15	0.0029	1.3186	PRIVATE
195.60	791		98	791	98	0.0036	1.3222	R
195.80	1084		96	1584	96	0.0021	1.3243	R
195.80							1.3243	
196.37	3000			1360	98	0.0026	1.3269	
196.60	1200		147	5311	A 98	0.0064	1.3333	R
197.17	3000		147	39	139	0.0001	1.3353	R
197.28	885			1156	D 10	0.0018	1.3371	
197.44	500		107	890	107	0.0014	1.3385	R
197.60	800		136	800	107	0.0014	1.3400	R
197.80	1100		110	587	107	0.0010	1.3410	R
197.86	300			0	107	0.0000	1.3410	
197.96				1313	D 10	0.0032	1.3442	
198.16	1100		90	1160	80	0.0042	1.3483	L
198.16							1.3483	
198.21	270			270	50	0.0010	1.3494	TRC TT MPH35
198.37	800		54	800	50	0.0030	1.3524	R
198.37				290	50	0.0011	1.3535	
198.57	1100			114	A 50	0.0004	1.3539	
198.59	100			758	54	0.0028	1.3567	
199.37	4098		125	160	54	0.0006	1.3573	R DUNCANNON
199.37						0.0000	1.3573	
199.46	500			5468	A 54	0.0110	1.3683	
199.82	2400			2500	A 12	0.0037	1.3720	
200.49	3000		141	1264	141	0.0024	1.3744	L
200.73	1300		141	1306	141	0.0017	1.3761	L
201.09	1880			1300	141	0.0018	1.3779	
201.34	1300		160	0		0.0000	1.3779	R
201.48	800			5293	A 14	0.0065	1.3844	
201.64	800			1402	A 16	0.0018	1.3862	
202.00	1898	XING 1 171				0.0000	1.3862	L AQUEDUCT
202.00							1.3862	
202.49	2600	XING 123.3X		0	171	0.0000	1.3862	PRIVATE
203.49	5265	XING 125.8X		8023	A 17	0.0083	1.3945	PRIVATE
204.49	5279					0.0000	1.3945	
204.72	1200			1164	190	0.0011	1.3956	
204.72				3981	D 15	0.0044	1.4000	
205.11	2083	XING 1 116		2083	116	0.0034	1.4034	L PRIVATE
205.45	1800			1805	116	0.0029	1.4064	
205.49	182			182	116	0.0003	1.4067	
206.06	3000		116	3000	116	0.0049	1.4116	L
206.28	1200			1200	116	0.0020	1.4135	
206.43	900		127	1561	116	0.0025	1.4161	R
206.60	752		189	101	D 11	0.0002	1.4162	R
206.81	1100		113	1100	113	0.0019	1.4181	R
206.81		XING 127.30					1.4181	PRIVATE
207.03	1200		155	1360	113	0.0023	1.4204	R
207.26	1200	XING 1 120		2610	A 11	0.0040	1.4244	R PRIVATE
207.49	1200			81	132	0.0001	1.4245	
207.70	1100			648	D 12	0.0010	1.4256	
208.00	1600	XING 1 114		1600	114	0.0027	1.4282	L PRIVATE
208.24	1267			877	114	0.0015	1.4296	
208.24				410	D 11	0.0007	1.4303	
208.49	1300		101	1303	101	0.0024	1.4328	L
208.66	930		101	930	101	0.0017	1.4345	L
208.66				1380	101	0.0026	1.4371	
208.66				3683	A 10	0.0060	1.4430	
209.50	4400		180			0.0000	1.4430	L
209.50				181	131	0.0002	1.4433	MP 130
209.75	1313			509	D 13	0.0008	1.4441	T 201.5
210.47	3800		117	3440	117	0.0056	1.4496	R
210.47				380	D 11	0.0006	1.4502	
211.07	3200		106	3230	106	0.0057	1.4560	R
211.07						0.0000	1.4560	
211.27	1024			1024	106	0.0018	1.4578	
211.44	900		160	336	106	0.0006	1.4584	L
211.53	500		160			0.0000	1.4584	L
211.80	1400			8908	A 10	0.0125	1.4709	T NEWPORT
212.14	1800	XING 1 160				0.0000	1.4709	R PRIVATE
212.33	1000					0.0000	1.4709	
212.40	373	XING 1 160				0.0000	1.4709	R PRIVATE
212.87	2470		160	572	160	0.0007	1.4716	R
213.40	2800					0.0000	1.4716	
213.55	825			3860	A 16	0.0045	1.4761	
214.39	4400		175	2869	175	0.0029	1.4790	L
214.99	3176			1360	175	0.0015	1.4804	
214.99				1431	A 17	0.0015	1.4820	
215.39	2100		164	136	160	0.0001	1.4821	L
215.39				2349	D 16	0.0029	1.4850	
215.73	1800		126	1500	126	0.0027	1.4877	L
215.80	400			400	126	0.0006	1.4882	TRC MPH70(TT)
215.94	716		126	719	126	0.0011	1.4893	R
216.37	2300			1380	126	0.0020	1.4913	
216.90	2800			4666	A 12	0.0067	1.4980	
217.04	713		155			0.0000	1.4980	R
217.38	1800			1257	155	0.0015	1.4995	T MILLERSTOWN
217.84	2414		283	242	155	0.0003	1.4998	L
217.84				2172	D 15	0.0033	1.5031	
218.37	2800		96	2800	96	0.0055	1.5086	L
218.73	1900			1360	96	0.0027	1.5113	
218.90	902	XING 1 156		8388	A 96	0.0098	1.5212	R PRIVATE
219.15	1300					0.0000	1.5212	
219.37	1200		157			0.0000	1.5212	R
219.39	100		158	19	144	0.0000	1.5213	R 211.5
220.04	3400			1084	D 14	0.0015	1.5228	
220.34	1620		117	1620	117	0.0028	1.5254	L OURWARD CURVE
220.39	249		117	249	117	0.0004	1.5258	L
220.98	3000			1380	117	0.0022	1.5280	
221.08	700		165	1625	A 11	0.0025	1.5305	R
221.09				70	129	0.0001	1.5306	
221.18	500			1146	D 12	0.0019	1.5325	
221.34	800		94	419	94	0.0008	1.5334	R
221.34				381	D 94	0.0008	1.5342	REVERSE CURVE
221.54	1100		78	1190	79	0.0026	1.5368	L
221.54							1.5369	
221.66	600			600	78	0.0014	1.5383	REVERSE CURVE
221.83	1000		101	760	78	0.0018	1.5401	R THOMPSONTOWN
221.90	282			177	A 79	0.0004	1.5405	TRC MPH95
222.11	1100		82	1445	82	0.0033	1.5439	L
222.11							1.5439	
222.34	1200			1200	82	0.0028	1.5466	
222.37	164			160	82	0.0004	1.5470	
222.58	1100		180	8610	A 82	0.0139	1.5609	R
223.34	4000			167	151	0.0002	1.5611	

READ UP — EASTBOUND TRAFFIC

TIME	ACC.	DISTANCE	SPEED
5.4612	0.0013	660	94
5.4790		BANKS	TOW
5.4798	0.0014	700	94
5.4785			
5.4780	0.0023	1163	94
5.4761	0.0032	2048	D 15
5.4730	0.0000	15	131
5.4729	0.0000		
5.4729	0.0000		
5.4729	0.0117	7083	A 98
5.4612	0.0021	1363	98
5.4588	0.0015	791	98
5.4570	0.0021	1084	98
5.4546	0.0024	1507	D 14
5.4526	0.0005	191	141
5.4523	0.0073	4607	A 12
5.4450	0.0002	137	107
5.4448	0.0000		
5.4448	0.0000		
5.4448	0.0078	3363	A 50
5.4368	0.0021	500	50
5.4348	0.0030	600	50
5.4318	0.0000		
5.4318	0.0042	1100	50
5.4276	0.0002	81	D 54
5.4274	0.0007	208	54
5.4267	0.0028	800	54
5.4238	0.0029	1864	D 12
5.4198	0.0002	134	125
5.4197	0.0000	0	125
5.4192	0.0050	3300	125
5.4147	0.0006	824	D 14
5.4138	0.0012	918	141
5.4126	0.0016	1380	141
5.4107	0.0040	3000	141
5.4067	0.0017	1300	141
5.4050	0.0043	1077	D 16
5.4036	0.0024	2103	188
5.4012	0.0000	298	D 17
5.4009	0.0014	1304	171
5.3994	0.0021	1896	171
5.3970	0.0005	463	D 18
5.3968	0.0002	173	180
5.3967	0.0168	13190	A 11
5.3801	0.0023	150	118
5.3798	0.0020	1200	116
5.3778	0.0000		
5.3778	0.0034	2083	116
5.3744	0.0029	1800	116
5.3715	0.0003	182	116
5.3712	0.0038	3000	116
5.3675	0.0006	378	A 11
5.3670	0.0015	900	113
5.3655	0.0013	752	113
5.3542	0.0018	1100	113
5.3524	0.0011	660	D 10
5.3510	0.0000	172	132
5.3511	0.0036	2485	A 11
5.3472	0.0026	1544	114
5.3447	0.0024	1343	A 10
5.3423	0.0001	73	101
5.3422	0.0024	1287	101
5.3398	0.0000		
5.3398	0.0024	1300	101
5.3373	0.0017	930	101
5.3356	0.0020	1287	D 13
5.3336	0.0002	184	187
5.3334	0.0044	2942	A 11
5.3290	0.0041	2032	117
5.3249	0.0022	1266	A 19
5.3227	0.0024	1360	106
5.3203	0.0000		
5.3203	0.0057	3000	106
5.3148	0.0030	2107	D 16
5.3118	0.0000	17	160
5.3116	0.0000	0	180
5.3116	0.0004	300	180
5.3112	0.0017	1400	180
5.3090	0.0021	1800	160
5.3074	0.0012	1000	160
5.3062	0.0004	373	180
5.3058	0.0029	3475	180
5.3029	0.0008	737	D 15
5.3020	0.0011	1082	175
5.3009	0.0000		
5.3009	0.0000		
5.3009	0.0126	10147	A 12
5.2983	0.0026	1401	126
5.2865	0.0004	775	A 12
5.2861	0.0014	960	126
5.2806	0.0006	400	126
5.2830	0.0016	1037	126
5.2815	0.0000		91
5.2815	0.0000		91
5.2815	0.0006		91
5.2815	0.0057	3383	A 96
5.2758	0.0027	1380	96
5.2731			
5.2731	0.0134	7786	96
5.2577	0.0030	1348	D 15
5.2547	0.0001	77	150
5.2546	0.0000		
5.2546	0.0000		
5.2546	0.0076	5417	A 11
5.2470	0.0022	1380	117
5.2446	0.0035	2160	117
5.2410	0.0040	2082	A 94
5.2371	0.0022	1017	A 79
5.2349	0.0001	80	79
5.2348			
5.2348	0.0012	500	79
5.2336	0.0018	802	79
5.2318			
5.2318	0.0026	1100	79
5.2290	0.0002	71	D 82
5.2288	0.0012	920	82
5.2276	0.0023	1000	82
5.2253	0.0007	282	82
5.2247	0.0025	1100	82
5.2221	0.0046	2959	D 16
5.2176	0.0002	180	164
5.2174	0.0000		
5.2174	0.0000		
5.2174	0.0066	9080	A 13

APPENDIX B5

PAGE 4 COLUMN 2

READ DOWN — WESTBOUND TRAFFIC

MILES	DISTANCE	XING / BRIDGE CROSSING	MAX MPH	DISTANCE	SPEED	ACC.	TIME	STATION
224.23	4700			827	D 15	0.0011	1.5822	
224.34	596	XING 1 131		596	131	0.0009	1.5831	R PRIVATE
224.43	496		131	495	131	0.0007	1.5838	R
225.34	4800			1360	131	0.0153	1.5791	
225.98	3340			8028	A 13	0.0106	1.5897	
226.35	2000		100		0	0.0000	1.5897	L
226.37	100		199		0	0.0000	1.5897	L
226.83	2500			66	171	0.0001	1.5898	
226.85				2783	D 17	0.0036	1.5937	
227.32	2500		102	2500	102	0.0046	1.5983	R TUSCARORA
227.41	480		102	485	102	0.0009	1.5992	R
227.41		XING 148.5					1.5992	PRIVATE
227.41				1360	102	0.0025	1.6017	
227.90	2090			1608	A 10	0.0030	1.6047	
228.32	2200		119	1591	119	0.0024	1.6074	L
228.72	2100		119	2100	119	0.0033	1.6107	L
228.97	1350			1350	119	0.0021	1.6128	
229.24	1400		132	16	119	0.0000	1.6128	R
229.52	1500			3396	A 11	0.0048	1.6176	
229.52	0		132	157	141	0.0002	1.6180	R PORT ROYAL
229.86	1000			1237	D 14	0.0019	1.6199	
230.21	1787		107	1762	107	0.0031	1.6231	R
230.27	270		107	270	107	0.0005	1.6235	R
231.21	5000			1360	107	0.0024	1.6258	
231.21				4862	A 10	0.0073	1.6332	
231.74	2800			13	141	0.0000	1.6332	
231.74		XING 152.0		1820	D 14	0.0025	1.6358	PRIVATE
231.89	800		94	288	94	0.0006	1.6364	L TRC
231.89				514	D 94	0.0012	1.6376	
232.22	1751		73	1751	73	0.0045	1.6421	R TT MPH 50
232.30	400			400	73	0.0010	1.6431	
232.56	1400		73	1124	73	0.0029	1.6461	R TRC
232.56				271	D 73	0.0008	1.6468	
232.68	600		59	600	59	0.0019	1.6488	L MPH40
232.68		XING 153.9X		1380	59	0.0044	1.6531	JUNIATA STREET
233.10	2210			961	A 59	0.0026	1.6558	T MIFFLIN
233.21	600		91	469	81	0.0011	1.6569	L
233.21				1380	81	0.0032	1.6600	
233.46	1300		125	396	A 81	0.0013	1.6614	L MPH 45
233.55	500					0.0000	1.6614	TRC
233.67	1980		90	1524	90	0.0032	1.6646	R
233.67							1.6646	
234.00	700			700	90	0.0015	1.6660	
234.21	1100		117	862	90	0.0014	1.6674	L
234.83	3295		117	2636	A 90	0.0048	1.6722	L
235.06	1200			2298	117	0.0037	1.6759	
235.21	800		117	800	117	0.0013	1.6772	R
235.36	600		117	800	117	0.0013	1.6785	R
235.57	1100			1100	117	0.0018	1.6803	
235.70	700		106	700	117	0.0011	1.6814	R
236.22	2885			2885	117	0.0044	1.6858	
236.58	1900			1002	117	0.0016	1.6874	T DENHOLM
236.58				666	D 11	0.0017	1.6891	
236.88	1600		87	1600	87	0.0041	1.6932	L
236.93							1.6932	
237.07	700			700	87	0.0015	1.6948	
237.22	781		156	680	87	0.0014	1.6962	L
237.33	581		156	9671	A 87	0.0145	1.7107	L
238.08	4000					0.0000	1.7107	
238.22	700		108			0.0000	1.7107	R
238.26	235		156			0.0000	1.7107	R
238.62	1900					0.0000	1.7107	
238.85	1200		150			0.0000	1.7107	L
239.21	1900			1066	156	0.0013	1.7120	OP HAWSTONE
239.21						0.0000	1.7120	
239.36	800			800	156	0.0010	1.7130	
239.53	1014		199	1014	156	0.0012	1.7142	L
240.21	3600			3600	156	0.0042	1.7185	
240.36	600			606	156	0.0040	1.7194	
240.64	1471		190	1471	156	0.0016	1.7212	L
240.97	1700			1700	156	0.0021	1.7233	
241.23	1400		156		156	0.0000	1.7233	R
241.42	1000		156		156	0.0000	1.7233	R
241.57	800			819	156	0.0010	1.7243	
241.57				2381	D 15	0.0037	1.7279	
241.87	1586		90	1586	90	0.0033	1.7313	L
241.99	600			327	90	0.0007	1.7320	REVERSE CURVE
241.99				273	D 90	0.0006	1.7326	
242.18	1000		79	1000	79	0.0024	1.7350	R
242.33	800		79	800	79	0.0019	1.7369	R
242.73	2114			1380	79	0.0033	1.7402	SHAWNEE
243.16	2300		98	1442	A 79	0.0031	1.7432	R
243.30	700		98	2312	98	0.0044	1.7476	R
243.30				1431	98	0.0027	1.7504	
243.60	1600			166	D 93	0.0003	1.7507	
243.84	1291		93	1291	93	0.0026	1.7533	L
244.01	900			900	93	0.0018	1.7550	
244.16	800		96	493	93	0.0009	1.7561	R
244.32	800		98	987	A 93	0.0005	1.7566	R
244.49	900		102	985	102	0.0007	1.7572	R
244.56	400				102	0.0014	1.7586	TRC MPH35
244.56				1088	D 10	0.0027	1.7613	
244.77	1100		54	1100	54	0.0038	1.7652	L
244.77							1.7652	
244.97	1064			1064	54	0.0037	1.7688	
245.16	1000		79	308	54	0.0011	1.7699	L
245.44	1500		79	1051	A 54	0.0027	1.7726	L LEWISTOWN
245.44				1140	79	0.0027	1.7754	
246.05	3200			1380	79	0.0033	1.7786	
246.16	567		144	7560	A 79	0.0125	1.7911	R
247.08	4607		144			0.0000	1.7911	R
247.18	500		148	229	144	0.0003	1.7914	R
247.62	2300			1380	144	0.0016	1.7932	
247.78	872		230	4414	A 14	0.0054	1.7986	R
248.08	1600			108	164	0.0001	1.7988	
248.33	1300	XING 169.4X		190	D 16	0.0002	1.7990	GRANVILLE
248.64	2880		160	2880	160	0.0030	1.8022	L
249.08	1300			1300	180	0.0131	1.8152	
249.08				83	180	0.0001	1.8153	
249.08				1817	A 18	0.0003	1.8156	
249.94	4500	XING 170.8X		236	181	0.0003	1.8158	PRIVATE
249.94		XING 170.84		2587	D 18	0.0039	1.8197	ANDERSON
250.08	771		91	771	91	0.0016	1.8213	L
250.44	1900		91	1900	91	0.0040	1.8253	L
250.44				660	91	0.0014	1.8267	
250.57	700			700	91	0.0015	1.8281	
250.82	1300		104	1110	A 88	0.0022	1.8303	R
250.82				690	104	0.0016	1.8319	
251.05	1000		127	2804	A 10	0.0046	1.8365	R
251.58	178	XING 172.20				0.0000	1.8365	PRIVATE
251.61	2774	XING 173.9X		1998	A 12	0.0023	1.8388	PRIVATE
252.08	2900		158	148	137	0.0002	1.8390	R
252.08				746	D 13	0.0011	1.8401	
252.30	1174		117	1174	117	0.0018	1.8420	R

READ UP — EASTBOUND TRAFFIC

	TIME	ACC.	DISTANCE	SPEED
	5.2048	0.0020	1360	131
	5.2068	0.0009	596	131
	5.2080	0.0007	496	131
	5.2053	0.0017	1332	D 18
	5.2035	0.0003	292	160
	5.2032	0.0000		
	5.2032	0.0138	8708	A 10
	5.1885	0.0025	1360	102
	5.1868			
	5.1868	0.0048	2500	102
	5.1823	0.0009	485	102
	5.1814	0.0008	851	D 11
	5.1805	0.0001	40	119
	5.1804	0.0032	2000	119
	5.1772	0.0036	2200	118
	5.1737	0.0033	2100	119
	5.1704	0.0007	479	D 76
	5.1697	0.0001	56	132
150	5.1666	0.0014	1000	132
	5.1651	0.0051	3255	A 10
	5.1630	0.0024	1380	107
	5.1606	0.0118	4600	107
	5.1488	0.0002	1181	A 94
	5.1486	0.0030	1324	A 73
	5.1436			
	5.1436	0.0015	500	73
	5.1421			
	5.1421	0.0021	900	73
	5.1401			
	5.1401	0.0045	1731	73
	5.1356	0.0016	587	A 59
	5.1340	0.0044	1380	59
	5.1296			
	5.1296	0.0019	800	59
	5.1277	0.0012	452	D 78
	5.1265	0.0041	1758	81
	5.1224	0.0014	600	81
	5.1210	0.0005	226	D 90
	5.1205	0.0023	1074	90
	5.1182	0.0011	500	90
	5.1171	0.0035	1660	90
	5.1136	0.0015	820	D 11
	5.1121	0.0000	20	117
	5.1121	0.0064	450	117
	5.1057	0.0053	3295	117
	5.1004	0.0019	1200	117
	5.0984	0.0022	1080	117
	5.0962	0.0013	800	117
	5.0949	0.0018	1100	117
	5.0931	0.0017	1072	117
	5.0914	0.0083	2863	A 87
	5.0881	0.0030	1380	87
	5.0852			
	5.0852	0.0041	1900	87
	5.0796	0.0036	2459	D 15
	5.0752	0.0027	2243	156
	5.0725	0.0000		
	5.0725	0.0000		
	5.0725	0.0017	1360	156
	5.0708	0.0008	700	156
	5.0700	0.0003	236	156
	5.0697	0.0002	185	D 15
	5.0695	0.0035	2816	160
	5.0660	0.0006	404	D 16
	5.0650	0.0002	28	169
	5.0653	0.0028	2282	A 15
	5.0626	0.0011	827	A 15
	5.0617	0.0000		75
	5.0617	0.0000		75
	5.0617	0.0000		75
	5.0617	0.0000		75
	5.0617	0.0140	9344	A 90
	5.0477	0.0031	1473	90
	5.0446	0.0016	713	A 79
	5.0430			
	5.0430	0.0018	780	79
	5.0412	0.0014	600	79
	5.0397			
	5.0397	0.0024	1000	79
	5.0373	0.0019	800	79
	5.0354	0.0011	622	D 98
	5.0343	0.0083	4338	98
	5.0260	0.0010	434	A 93
	5.0250			
	5.0250	0.0028	1380	93
	5.0225	0.0036	1722	93
	5.0187	0.0000		
	5.0187	0.0051	1898	A 54
	5.0137	0.0002	50	54
	5.0135	0.0032	600	54
	5.0103	0.0014	400	54
	5.0089			
	5.0086	0.0036	1100	54
	5.0051	0.0014	468	D 75
	5.0037	0.0014	588	79
	5.0023	0.0024	1000	79
	4.9999	0.0036	1500	79
	4.9963	0.0036	2126	D 14
	4.9927	0.0014	1074	144
	4.9913	0.0007	902	144
	4.9905	0.0065	4607	144
	4.9841	0.0007	500	148
	4.9834	0.0013	1049	D 15
	4.9822	0.0225	200	167
	4.9597	0.0000		
	4.9597	0.0000		
	4.9597	0.0020	1736	A 10
	4.9576	0.0150	10205	A 91
	4.9426			
	4.9426			
	4.9426	0.0028	1380	91
	4.9398			
	4.9398	0.0016	771	91
	4.9382	0.0040	1900	91
	4.9343	0.0007	372	D 10
	4.9335	0.0006	328	104
	4.9329	0.0024	1300	104
	4.9306	0.0013	778	D 12
	4.9293	0.0009	421	127
	4.9287	0.0002	113	D 13
	4.9285	0.0032	2185	130
	4.9253	0.0028	1815	A 11
	4.9225	0.0051	3141	117
	4.9175	0.0014	832	A 110-117

APPENDIX B5

PAGE 5 COLUMN 1

MILES	DISTANCE FT	XING GRADE CROSSING	MAX MPH	DISTANCE FT (READ DOWN WESTBOUND TRAFFIC)	SPEED	ET HOURS	TIME HOURS	STATION	TIME HOURS (READ UP EASTBOUND TRAFFIC)	ET HOURS	DISTANCE FT	SPEED
252.30				1380	117	0.0022	1.8442		4.9161			
252.30				929	A 11	0.0015	1.8456		4.9161			
252.83	2800			31	124	0.0000	1.8457		4.9161	0.0025	1380	110
252.83				480	D 12	0.0008	1.8465		4.9137			
253.08	1300		110	1300	110	0.0022	1.8487	R	4.9137	0.0022	1300	110
253.08		XING 174.6		1360	110	0.0023	1.8510	HORNINGFORD	4.9115	0.0019	1287	D 14
253.44	2186		286	5064	A 11	0.0075	1.8585	R	4.9096	0.0004	283	144
254.08	3100	XING 175.3		55	144	0.0001	1.8586	PRIVATE	4.9092	0.0066	4737	A 11
254.53	2379			1166	D 14	0.0017	1.8603		4.9025	0.0023	1360	113
255.06	2800	XING 1	113	2800	113	0.0049	1.8652	L WINGARTS	4.9000	0.0049	2800	113
256.08	5286	XING 172.1		1360	113	0.0023	1.8675	PRIVATE	4.8952	0.0019	1297	D 14
256.12	191	XING 177.8		5486	A 11	0.0079	1.8754	HWY 103 McVEYTOWN	4.8933	0.0003	205	147
256.63	2700		165	0	148	0.0000	1.8754	L	4.8930	0.0077	5324	A 11
256.63				1340	D 14	0.0019	1.8773		4.8853	0.0023	1360	113
257.06	2300		113	2300	113	0.0039	1.8812	L MATAWANNA	4.8830	0.0063	3735	113
257.08	100			106	113	0.0002	1.8814		4.8765	0.0003	208	A 11
257.08				1280	113	0.0021	1.8836		4.8764			
257.08				1120	A 11	0.0018	1.8853		4.8764			
257.83	2800			144	122	0.0002	1.8855		4.8764	0.0023	1380	111
257.63				378	D 12	0.0006	1.8861		4.8741			
257.95	1700		111	1700	111	0.0029	1.8890	R	4.8741	0.0029	1700	111
258.08	680			680	111	0.0017	1.8902		4.8712	0.0019	1319	D 14
258.95	6100			980	111	0.0012	1.8913		4.8692	0.0000	17	146
259.05				3854	A 11	0.0056	1.8971		4.8692			
259.05				78	138	0.0001	1.8972		4.8692			
259.05				536	D 12	0.0008	1.8980		4.8692			
259.08	148	XING 1	124	148	124	0.0002	1.8982	R PRIVATE	4.8692	0.0076	5232	A 11
259.23	800		124	800	124	0.0012	1.8994	R RYDE	4.8616	0.0000	188	112
259.30	400			400	124	0.0006	1.9000		4.8613	0.0007	400	112
259.46	800		150	385	124	0.0006	1.9006	R	4.8606	0.0014	800	112
259.46				415	D 12	0.0007	1.9013		4.8593			
259.58	711		112	711	112	0.0012	1.9025	R	4.8593	0.0012	711	112
							1.9025		4.8581	0.0017	1159	D 14
260.08	2800			1360	112	0.0023	1.9048		4.8564	0.0003	227	143
261.07	5222			6434	A 11	0.0092	1.9140	T VINEYARD	4.8561	0.0078	8276	A 58
261.68	3200			333	152	0.0004	1.9144	T Y'S & CURVE	4.8482	0.0044	1360	56
261.68				2895	D 14	0.0032	1.9190		4.8436			
261.89	1137		58	1137	58	0.0037	1.9233	PRC	4.8436	0.0037	1137	58
261.89				1360	58	0.0044	1.9278		4.8401			58
261.89				43	A 58	0.0001	1.9279		4.8401	0.0001	17	D 58
262.06	900		59	0	59	0.0000	1.9279	L	4.8400	0.0008	880	59
262.24	917		59	414	59	0.0013	1.9292	L	4.8372	0.0029	917	59
262.24						0.0005	1.9292		4.8342	0.0032	1487	D 11
262.60	1900		185	1360	58	0.0044	1.9336	L	4.8310	0.0000	3	117
262.78	1030			4213	A 58	0.0088	1.9423		4.8310	0.0023	1400	117
262.92	700		117			0.0000	1.9423	R	4.8287	0.0011	700	117
262.92							1.9423		4.8276	0.0010	985	D 12
263.07	800		135	2583	A 11	0.0038	1.9462	R	4.8266	0.0000	135	135
263.16	500		185	1380	135	0.0019	1.9481	R	4.8264	0.0031	2678	D 13
263.94	4085			10800	A 13	0.0128	1.9610		4.8253	0.0002	158	191
264.07	700		288			0.0000	1.9610	R	4.8231	0.0000		
264.62	2874		288			0.0000	1.9610	R	4.8231	0.0000		
265.07	2450		385			0.0000	1.9610	R	4.8231	0.0000		
265.50	2280		385			0.0000	1.9610	R	4.8231	0.0000		
266.07	3000			81	181	0.0001	1.9611		4.8231	0.0183	14143	A 13
266.31	2300			2132	D 16	0.0026	1.9636		4.8069	0.0019	1380	135
266.85	1834		135	1834	135	0.0026	1.9662	L NEWTON HAMILTON	4.8050	0.0023	1834	135
267.04	1000			1000	135	0.0014	1.9676		4.8027	0.0007	486	D 14
267.12	402			380	135	0.0005	1.9681		4.8020	0.0012	906	147
267.97	4500		147	2225	A 13	0.0030	1.9711	R	4.8009	0.0058	4500	147
268.48	2700		147	5017	147	0.0065	1.9776	R	4.7951	0.0018	1377	147
268.97	2591			1380	147	0.0018	1.9790		4.7933	0.0030	2225	A 13
268.97				3080	A 14	0.0040	1.9830		4.7903			
268.97				53	162	0.0001	1.9834		4.7903	0.0056	3636	A 12
269.60	3292			1178	D 12	0.0015	1.9848		4.7847	0.0024	1360	108
269.98	2000		135	1036	135	0.0015	1.9864	L	4.7823	0.0035	2000	108
269.98		XING 181.6		982	D 12	0.0015	1.9878	PRIVATE	4.7788			
270.18	1100		108	1100	108	0.0019	1.9898	L MT UNION	4.7788	0.0019	1100	108
270.27	477			477	108	0.0005	1.9906	REVERSE CURVE	4.7769	0.0008	472	D 12
270.27				883	108	0.0015	1.9922		4.7761	0.0000	5	122
270.77	2800		122	1947	A 10	0.0027	1.9949	R	4.7761	0.0040	2800	122
271.07	1600		122	1670	122	0.0026	1.9975	R	4.7720	0.0009	584	122
271.33	1382			1380	122	0.0021	1.9996		4.7711	0.0069	3788	A 62
271.39				870	A 12	0.0013	2.0009		4.7642			
271.77	2000		135	185	128	0.0003	2.0012	L	4.7642	0.0021	912	62
271.77				1257	D 10	0.0022	2.0034		4.7621			
271.82	300		88	298	88	0.0006	2.0041	L	4.7621	0.0007	300	82
271.85	148				88	0.0000	2.0041	TRC	4.7614	0.0003	148	82
271.85				150	D 88	0.0003	2.0044	MAPLETON	4.7610			
272.76	4800		82	4800	82	0.0111	2.0155	R MP 160	4.7610	0.0111	4800	82
272.99	1200		82	1200	82	0.0028	2.0182	R	4.7500	0.0028	1200	82
272.99							2.0182		4.7472	0.0022	1197	D 12
272.99							2.0182		4.7450	0.0002	100	122
273.42	2300			1360	82	0.0031	2.0214		4.7448	0.0018	1120	A 11
273.70	1477		113	2686	A 82	0.0051	2.0265	R	4.7430	0.0023	1360	113
273.76	300			51	113	0.0001	2.0266		4.7407	0.0031	2387	D 16
274.14	2000			1060	113	0.0016	2.0284		4.7377	0.0001	117	188
274.75	3248		185	6011	A 11	0.0090	2.0373	L	4.7375	0.0000		
274.96	1085		165	171	152	0.0002	2.0376	L	4.7375	0.0089	5660	A 14
275.26	1600			369	D 15	0.0006	2.0381		4.7307	0.0019	1427	A 13
275.53	1400		143	1074	140	0.0014	2.0386	L	4.7287	0.0002	160	130
275.76	1200		147	328	D 14	0.0004	2.0399	L MILL CREEK	4.7285	0.0017	1200	135
276.28	2700	XING 1	135	2700	135	0.0039	2.0438	L GAISVILLE	4.7268	0.0039	2790	135
276.28							2.0438		4.7229	0.0016	1224	D 14
276.28							2.0438		4.7213	0.0000	16	183
276.76	2500			1360	135	0.0019	2.0457		4.7213	0.0074	5812	A 13
276.79	178			7583	A 13	0.0094	2.0551		4.7140	0.0189	1527	A 12
277.43	3800		233			0.0000	2.0551	R	4.6951	0.0000		86
277.76	1800			271	170	0.0003	2.0554		4.6951	0.0000		98
278.27	2700			2464	D 17	0.0030	2.0586		4.6951	0.0031	1600	A 11
278.49	1179	XING 1	110	1179	110	0.0020	2.0606	R JACOBS	4.6920	0.0081	4713	110
278.76	1400			1360	110	0.0023	2.0631		4.6836	0.0034	1803	A 90
278.76				3075	A 11	0.0055	2.0686		4.6805			
278.76	5297			237	108	0.0000	2.0689		4.6805	0.0029	1380	90
279.76				1525	D 12	0.0026	2.0715		4.6776			
280.01	1300		90	1300	90	0.0027	2.0742	L	4.6776	0.0042	1983	90
280.05	240			240	90	0.0005	2.0747	REVERSE CURVE	4.6734	0.0006	290	A 80
280.45	2100		100	1987	90	0.0042	2.0789	R	4.6728	0.0030	1395	86
280.45				103	D 90	0.0002	2.0792	REVERSE CURVE	4.6698			
280.73	1500		86	1500	86	0.0033	2.0825	L	4.6698	0.0033	1500	98
280.92	1000	XING 2	98	1000	86	0.0022	2.0847	L MPH 55 4TH STREET	4.6665	0.0022	1000	98
281.07	800			800	86	0.0018	2.0864	T HUNTINGDON	4.6643	0.0013	690	D 11
281.26	1000		199	366	86	0.0012	2.0877	R	4.6630	0.0019	1110	110
281.40	700			2593	A 86	0.0040	2.0917		4.6611	0.0012	700	110
281.68	1500		110	547	155	0.0006	2.0923	R	4.6586	0.0060	2800	110
281.73	236			236	159	0.0004	2.0931		4.6548	0.0040	2060	A 66
282.23	3197			1171	110	0.0019	2.0950		4.6508	0.0001	60	86
282.33				2538	A 11	0.0033	2.0982		4.6507			
282.58	1300		177	94	126	0.0001	2.0984	R	4.6507	0.0029	1300	86
282.58				1244	D 15	0.0022	2.1006		4.6478			
282.73	800		86	800	86	0.0018	2.1024	R	4.6478	0.0016	800	86
282.88	800		98	800	86	0.0018	2.1041	R	4.6460	0.0014	653	86
282.94	300			300	98	0.0007	2.1048	TYRONE	4.6446	0.0030	1790	D 14

PAGE 5 COLUMN 2

MILES	DISTANCE FT	HWY GRADE CROSSING	MAX MPH	DISTANCE FT (READ DOWN WESTBOUND TRAFFIC)	SPEED	ET HOURS	TIME HOURS	STATION	TIME HOURS (READ UP EASTBOUND TRAFFIC)	ET HOURS	DISTANCE FT	SPEED
283.13	1000		218	1000	88	0.0022	2.1070	L RC MPH 55	4.6416	0.0002	156	140
283.30	900		177	60	86	0.0001	2.1071	L	4.6414	0.0000		
283.60	1600			8846	A 86	0.0113	2.1184		4.6414	0.0000		
283.78	970		151			0.0000	2.1184	R	4.6414	0.0000		
283.78	277		151			0.0000	2.1184	R	4.6414	0.0108	5436	A 85
284.73	5000					0.0000	2.1184		4.6306	0.0030	1360	85
284.84	600			101	143	0.0001	2.1185		4.6277	0.0013	600	86
284.84				1940	D 14	0.0032	2.1217		4.6264	0.0000		
285.05	1100		65	1100	86	0.0025	2.1242	L	4.6264	0.0039	1787	65
285.23	977		113	977	86	0.0022	2.1264	L	4.6225	0.0008	300	A 70
285.50	1400	XING 206.6		1256	85	0.0028	2.1292	T WARRIOR RIDGE	4.6218	0.0033	1380	79
285.58				144	D 85	0.0003	2.1295		4.6184	0.0000		
285.73	1200		79	1200	79	0.0029	2.1324	R	4.6184	0.0029	1200	79
285.97	1300		79	1300	79	0.0031	2.1355	R	4.6155	0.0031	1300	79
285.97							2.1355		4.6124	0.0005	320	D 83
285.98	87			87	79	0.0002	2.1357	REVERSE CURVE	4.6119	0.0000	0	98
286.22	1200		98	1200	79	0.0029	2.1386	L	4.6119	0.0023	1067	98
286.22				73	79	0.0002	2.1387		4.6096	0.0006	302	D 94
286.44	1200		96	582	A 76	0.0013	2.1401	L	4.6090	0.0017	898	96
286.44				546	88	0.0012	2.1412		4.6073	0.0014	818	D 96
286.73	1500			350	88	0.0000	2.1416		4.6059	0.0002	120	124
286.95	1190			950	A 88	0.0017	2.1433		4.6057	0.0040	2380	A 10
286.95				340	98	0.0007	2.1440		4.6017	0.0003		
287.33	2000		153	3866	A 99	0.0045	2.1485	L PETERSBURG	4.6017	0.0026	1380	104
287.33				40	123	0.0001	2.1486		4.5993	0.0000		
287.33				580	D 12	0.0011	2.1496		4.5993	0.0000		
287.73	2100		104	2100	104	0.0038	2.1534	L	4.5993	0.0039	2100	104
287.73							2.1534		4.5954	0.0022	1450	D 14
287.90	900			900	104	0.0016	2.1551		4.5932	0.0000	30	144
288.26	1985		144	480	104	0.0008	2.1559	R	4.5932	0.0017	1300	144
288.73	2500			5991	A 10	0.0086	2.1645		4.5915	0.0019	1808	D 17
288.73		XING 210.35					2.1645	PRIVATE	4.5896	0.0002	173	176
289.73	5758	XING 210.55		6862	A 14	0.0100	2.1745	PRIVATE	4.5894	0.0068	6862	A 15
290.64	4600					0.0000	2.1748		4.5826	0.0017	1360	155
290.70	322		155			0.0000	2.1748	R	4.5809	0.0004	322	155
291.04	1800	XING 2	155	3094	155	0.0039	2.1787	R BARREE	4.5806	0.0100	7313	A 11
291.63	3100			1749	D 15	0.0025	2.1812		4.5702	0.0016	1045	110
291.69	315		110				2.1812	L	4.5684	0.0005	315	110
291.84	805		110	1300	110	0.0022	2.1836	L	4.5678	0.0014	805	110
291.96	800			1360	110	0.0023	2.1858		4.5664	0.0010	600	110
292.37	2130			1181	A 11	0.0019	2.1877	T MPH 50WB	4.5654	0.0030	1745	110
292.52	810			542	120	0.0008	2.1885	T TUNNEL	4.5624	0.0024	1320	A 96
292.52				337	D 12	0.0006	2.1891		4.5600	0.0000		
292.64	630		110	630	110	0.0011	2.1902	R	4.5600	0.0011	590	96
292.64				220	110	0.0004	2.1905		4.5529	0.0000		
292.80	900			157	110	0.0003	2.1908	T SPRUCE CREEK	4.5589	0.0004	327	96
292.80				423	D 11	0.0008	2.1917		4.5584	0.0000		
292.98	1000		96	1000	96	0.0020	2.1936	R	4.5584	0.0042	1803	A 66
293.17	1000		103	800	96	0.0016	2.1952	R	4.5543	0.0022	720	66
293.29	580			77	96	0.0002	2.1954	REVERSE CURVE	4.5521	0.0047	2081	66
293.29				733	D 96	0.0017	2.1970		4.5473			
293.63	1600		96	1800	66	0.0052	2.2022	L	4.5473	0.0052	1800	96
293.63							2.2022		4.5422	0.0006	208	D 76
293.66	200		175	200	66	0.0006	2.2028	L	4.5416	0.0003	732	76
294.06	2100			1160	66	0.0033	2.2061		4.5413	0.0067	2706	76
294.31	1300		76	486	A 66	0.0013	2.2073	L	4.5346	0.0010	371	A 68
294.51	1077			3379	78	0.0084	2.2158		4.5336	0.0018	660	68
294.64	700		128	369	D 76	0.0004	2.2162	R	4.5318	0.0019	700	66
294.85	1100		66	1100	68	0.0031	2.2193	R	4.5298	0.0031	1100	66
295.02	900		104	900	68	0.0025	2.2218	R	4.5268	0.0019	978	D 10
295.22	1062	XING 218.6		460	58	0.0013	2.2231	PEMBERTON	4.5248	0.0000	104	100
295.43	1100		185	2403	A 66	0.0052	2.2283	L	4.5246	0.0063	2720	A 58
295.62	1000			0		0.0000	2.2283		4.5184	0.0052	1800	58
295.81	1000					0.0000	2.2283		4.5132	0.0006	163	A 53
295.81				208	104	0.0004	2.2286		4.5126			
295.81				1093	D 10	0.0028	2.2317	UNION FURNACE	4.5126			
296.06	1300		96	1300	58	0.0042	2.2364	L	4.5126	0.0032	807	53
296.14	453			372	58	0.0012	2.2367	TRC MPH 35	4.5094	0.0016	463	53
296.14				81	D 58	0.0009	2.2368		4.5078			
296.37	1200		53	1200	53	0.0043	2.2412	R	4.5078	0.0156	4345	53
296.64	1400			1380	53	0.0049	2.2461		4.4922	0.0003	95	A 50
297.07	2300			2158	A 53	0.0054	2.2515		4.4919	0.0017	460	50
297.13	300		219			0.0000	2.2515	R	4.4902	0.0011	300	50
297.24	600			127	96	0.0003	2.2518		4.4890	0.0023	600	50
297.24				987	D 96	0.0025	2.2543		4.4867			
297.28	200		50	200	50	0.0008	2.2560	L	4.4867	0.0008	200	50
297.28							2.2560		4.4860	0.0042	1962	D 12
297.58	1600		54	1360	50	0.0052	2.2603	R RC	4.4818	0.0001	87	126
297.63	247			114	A 50	0.0004	2.2606		4.4816	0.0000		
297.71	400			770	54	0.0027	2.2633		4.4816	0.0000		
297.90	1000		110	713	54	0.0025	2.2658	L	4.4816	0.0000		
298.16	1400			4145	A 54	0.0090	2.2748	T BIRMINGHAM	4.4816	0.0113	5800	A 54
298.63	2487		155			0.0000	2.2748	L	4.4704	0.0041	1181	54
298.80	900		155	367	115	0.0007	2.2754	L	4.4662	0.0001	19	A 53
298.99	1000			1512	D 11	0.0034	2.2788		4.4662	0.0006	310	53
299.16	900		54	1184	64	0.0042	2.2828	R	4.4656	0.0032	900	53
299.21	300			16	D 54	0.0001	2.2830	REVERSE CURVE	4.4624	0.0011	300	53
299.42	1085		53	1085	53	0.0036	2.2866	L	4.4613	0.0039	1085	53
299.42							2.2866		4.4574	0.0010	307	D 71
299.63	1100			1100	53	0.0038	2.2906		4.4564	0.0021	773	71
299.78	800			260	53	0.0009	2.2917		4.4544	0.0021	800	71
299.97	1000		71	684	A 53	0.0021	2.2938	L	4.4522	0.0027	1000	71
300.03	300			1156	71	0.0031	2.2969	REVERSE CURVE	4.4496	0.0040	1518	71
300.23	1100	XING	72	1080	71	0.0028	2.2997	R PRIVATE	4.4455	0.0017	594	A 56
300.39	800			76	A 71	0.0002	2.2999		4.4438	0.0003	96	56
300.63	1262		90	1724	72	0.0045	2.3045	L T MPH 35	4.4435	0.0043	1262	56
300.63				300	D 72	0.0008	2.3054		4.4392	0.0000		
300.79	860		56	860	56	0.0029	2.3083	L TYRONE	4.4380	0.0029	860	56
300.79		XING 222.18					2.3083	WILSON ROAD	4.4363	0.0014	560	D 81
300.88	500			500	56	0.0017	2.3100	RC	4.4349	0.0001	48	81
301.11	1200		96	860	56	0.0029	2.3129	R	4.4348	0.0027	1150	81
301.32	1100		81	1087	A 56	0.0030	2.3159	R	4.4321	0.0026	1100	81
301.32				1713	81	0.0040	2.3199		4.4295	0.0001	24	D 80
301.43	400			80	A 81	0.0001	2.3200	RC	4.4295	0.0013	678	82
301.62	1000		82	180	82	0.0003	2.3204	L	4.4281	0.0025	1000	82
301.62							2.3204		4.4256	0.0012	970	D 10
301.67	273		103	273	82	0.0006	2.3210	L	4.4247	0.0000	0	103
301.84	900			900	82	0.0021	2.3231		4.4247	0.0014	866	D 12
302.17	1700		100	387	82	0.0004	2.3235	R	4.4232	0.0003	188	128
302.62	2400			4427	A 82	0.0098	2.3334		4.4229	0.0073	4261	A 90
303.00	2000			1	135	0.0000	2.3334		4.4156	0.0028	1360	90
303.00				1485	D 13	0.0025	2.3359	GRV E END 3 TR	4.4128	0.0000		
303.36	1900		90	1900	90	0.0040	2.3399	L	4.4128	0.0040	1900	90
303.36		XING 224.80					2.3399	RC PRIVATE	4.4088	0.0000		
303.47	600	XING 225.06		600	90	0.0013	2.3411	T T MPH 50 PRIVATE	4.4088	0.0013	600	90
303.62	782	XING 2	90	782	90	0.0016	2.3426	R TIPTON ROAD	4.4075	0.0016	782	90
303.90	1473		90	1473	90	0.0033	2.3431	R	4.4058	0.0031	1473	90
304.62	3800			1360	90	0.0056	2.3486	TIPTON	4.4028	0.0027	1648	D 13
304.62				5932	A 90	0.0087	2.3586		4.4000	0.0002	140	130
304.62				140	130	0.0002	2.3588		4.3998	0.0087	5932	A 90
305.62	5278			1646	D 13	0.0027	2.3615		4.3912	0.0029	1360	90
306.43	4300		90	4300	90	0.0090	2.3706	L	4.3873	0.0090	4300	90
306.43		XING 227.90					2.3706	PA RTE 500	4.3783	0.0003	136	D 95.90

APPENDIX B5

PAGE 6

COLUMN 1

MILES	DISTANCE	HWY GRADE CROSSING	MAX MPH	DISTANCE (READ DOWN WESTBOUND TRAFFIC)	SPEED	TIME	CUM. TIME	STATION	TIME (READ UP EASTBOUND TRAFFIC)	CUM. TIME	DISTANCE	SPEED
306.49	300			300	90	0.0033	2.3739	TFOSTORIA RC	4.3780	0.0003	154	95
306.56	358		96	358	90	0.0008	2.3746	R'TTMPH 60	4.3777	0.0007	358	95
306.56				702	90	0.0015	2.3761		4.3770	0.0000		
307.01	2409		95	373	A 90	0.0008	2.3769	R	4.3770	0.0046	2409	95
307.01				1334	95	0.0027	2.3796		4.3722	0.0046	3271	D 17
307.54	2800			1369	95	0.0027	2.3823		4.3676	0.0005	245	177
308.55	5289	XING 230.7		6898	A 95	0.0144	2.3966	PRIVATE	4.3673	0.0048	4577	A 16
309.08	2800					0.0000	2.3966		4.3626	0.0033	2800	180
309.32	1279		100	913	180	0.0011	2.3977	L	4.3592	0.0015	1276	180
309.55	1200			1200	180	0.0014	2.3991		4.3577	0.0018	1093	180
310.54	5258			160	180	0.0002	2.3993		4.3561	0.0101	7505	A 11
310.54				5758	A 16	0.0068	2.4061	E END 4 TR	4.3459			
311.26	3800			377	181	0.0147	2.4209	T'ANTIS	4.3459	0.0022	1360	118
311.26				2763	D 18	0.0035	2.4244		4.3438			
311.51	1300		118	1300	118	0.0021	2.4264	R	4.3438	0.0021	1300	118
311.55	202			202	115	0.0003	2.4258	T'MPH90(TT)	4.3417	0.0027	1909	D 15
311.57	108			108	118	0.0002	2.4269	RC	4.3390	0.0002	154	166
312.06	2600		166	1050	118	0.0017	2.4286	L	4.3388	0.0014	1200	166
312.53	2500			8069	A 11	0.0119	2.4405		4.3374	0.0000		
312.57	200					0.0000	2.4405		4.3374	0.0000		
313.04	2500		202			0.0000	2.4405	R	4.3374	0.0000		
313.53	2549					0.0000	2.4405		4.3374	0.0000		
314.13	3200			2182	A 16	0.0024	2.4429		4.3374	0.0154	11036	A 98
314.27	700		288			0.0000	2.4429	L	4.3220	0.0003	162	99
314.49	1200			215	174	0.0002	2.4432	T'TTMPH30	4.3217	0.0023	1200	99
314.49				3003	D 17	0.0040	2.4473	ALTOONA	4.3194			
314.52	142		99	142	99	0.0003	2.4476	R	4.3194	0.0003	142	99
314.62	520		99	520	99	0.0010	2.4486	R	4.3192	0.0010	520	99
315.53	4800			1360	99	0.0026	2.4512	T'ALTO	4.3182	0.0030	1998	D 15
316.24	3787			8725	A 99	0.0102	2.4614	T'SLOPE	4.3152	0.0002	191	153
316.24				6	147	0.0000	2.4614	E END 3 TR	4.3149			
316.24				498	D 14	0.0007	2.4621		4.3148			
316.51	1400		135	1400	135	0.0020	2.4640	L BRICKYARD	4.3149	0.0045	3435	A 13
316.77	1400	XING 2	135	100	135	0.0001	2.4647	L COBURN ROAD	4.3104	0.0117	6805	A 67
317.17	2100			9	135	0.0000	2.4642	T'TTMPH 35	4.2987	0.0036	1360	67
317.17				1386	135	0.0019	2.4661		4.2948			
317.17				2015	D 13	0.0038	2.4699		4.2948			
317.46	1500		67	1500	67	0.0042	2.4741	R	4.2948	0.0042	1500	67
317.53	372			372	67	0.0011	2.4752		4.2906	0.0004	145	D 74
317.83	1574			988	67	0.0028	2.4780		4.2902	0.0046	1601	74
318.01	1000		99	2005	A 67	0.0045	2.4825	R	4.2856	0.0017	1000	74
318.22	1100			47	99	0.0001	2.4826	T'TTMPH35	4.2839	0.0034	1347	74
318.22				634	D 96	0.0014	2.4840		4.2805			
318.39	900		74	900	74	0.0023	2.4863	L	4.2805	0.0021	796	A 54
318.53	700			700	74	0.0018	2.4881		4.2784	0.0056	1962	54
318.68	800			425	74	0.0011	2.4892		4.2728	0.0004	114	A 50
318.68				375	D 74	0.0011	2.4903		4.2727			
318.96	1500		54	1500	54	0.0053	2.4955	R	4.2722	0.0041	1084	50
319.01	276			215	54	0.0008	2.4963	T'TTMPH 35 RC	4.2681	0.0010	276	50
319.01		UG 240.69		81	D 54	0.0002	2.4965	SCOTTS RUN	4.2670			
319.24	1200		50	1200	50	0.0046	2.5011	L	4.2670	0.0045	1205	50
319.53	1500			1060	50	0.0032	2.5082		4.2629	0.0027	1071	D 59
319.68	800			1644	A 50	0.0051	2.5113		4.2597	0.0002	97	59
							2.5113		4.2596	0.0042	1080	A 70
319.83	800		105			0.0000	2.5113	L	4.2554	0.0007	884	A 44
320.00	900			81	81	0.0002	2.5115		4.2547	0.0003	73	44
320.11	600		99	1215	D 96	0.0051	2.5166	R	4.2544	0.0026	600	44
320.23	600		70			0.0000	2.5166	R	4.2518	0.0026	600	44
320.24	87			252	70	0.0007	2.5173	RC	4.2492	0.0004	87	44
320.24				435	D 70	0.0014	2.5187		4.2489			
320.41	900		44	900	44	0.0039	2.5226	L	4.2489	0.0039	900	44
320.53	600		54	600	44	0.0026	2.5252	L'HORSESHOE	4.2450	0.0026	600	44
320.70	900		44	900	44	0.0039	2.5290	L	4.2424	0.0036	900	44
320.70							2.5290		4.2388	0.0012	330	D 65
320.73	182			182	44	0.0008	2.5298	RC	4.2374	0.0001	46	65
320.81	400		64	400	44	0.0017	2.5315	R	4.2372	0.0010	300	65
320.90	500			500	44	0.0022	2.5337		4.2362	0.0039	500	65
321.05	800		74	278	44	0.0012	2.5349	R	4.2323	0.0039	800	65
321.15	500		105	702	A 44	0.0024	2.5373	R	4.2284	0.0015	500	65
321.26	500		99	920	65	0.0027	2.5400	R	4.2268	0.0017	600	65
321.41	800		70	800	65	0.0023	2.5423	R	4.2252	0.0023	800	65
321.53	600			580	65	0.0016	2.5440		4.2229	0.0027	1286	D 11
321.72	1000			195	A 65	0.0006	2.5445		4.2201	0.0003	158	114
321.89	900		117	602	70	0.0016	2.5461	L'MG	4.2198	0.0000		
322.20	1670			3406	A 70	0.0069	2.5530		4.2198	0.0058	3026	A 78
322.53	1700		117	90	114	0.0001	2.5532	R	4.2139	0.0058	2373	78
322.53				1014	D 11	0.0020	2.5552		4.2081			
322.73	1100		78	1000	78	0.0024	2.5576	R	4.2081	0.0026	912	A 56
322.85	600			1388	78	0.0033	2.5609		4.2068	0.0015	400	A 43
323.06	1100		233	432	D 76	0.0012	2.5621	R	4.2041	0.0021	475	43
323.21	800		56	811	56	0.0021	2.5642	R	4.2020	0.0036	800	43
323.22	66				56	0.0000	2.5642	REVERSE CURVE	4.1984	0.0004	86	43
323.41	1000		43	189	D 54	0.0007	2.5649	L	4.1981	0.0044	1000	43
323.41				1000	43	0.0044	2.5693		4.1937	0.0002	30	D 46
323.45	200			200	43	0.0009	2.5702	RC	4.1935	0.0001	13	46
323.54	500		70	500	43	0.0022	2.5724	R	4.1934	0.0012	300	46
323.64	500		70	500	43	0.0022	2.5746	R	4.1922	0.0021	500	46
323.67	149			149	43	0.0007	2.5753	TRC	4.1901	0.0006	149	46
323.86	1000		46	11	43	0.0000	2.5753	L'MPH 30(TT)	4.1895	0.0041	1000	46
323.97	600			81	A 43	0.0003	2.5757		4.1854	0.0012	348	D 67
324.10	700		94	2283	46	0.0084	2.5851	R	4.1842	0.0046	1625	67
324.25	800		67	60	46	0.0002	2.5854	R	4.1797	0.0024	726	A 46
324.41	800			725	A 46	0.0028	2.5881		4.1772	0.0023	548	46
324.48	400		104	1214	67	0.0034	2.5916	R	4.1750	0.0016	400	46
324.52	200		68	60	67	0.0002	2.5917	R	4.1733	0.0008	200	46
324.56	211			3	67	0.0000	2.5917	RC	4.1725	0.0009	211	46
324.56				349	D 60	0.0012	2.5929		4.1718			
324.81	1300		46	1300	46	0.0054	2.5983	L	4.1716	0.0064	1300	46
324.81							2.5983		4.1653	0.0043	1945	D 12
324.90	500			500	46	0.0021	2.6003	T'BENNY	4.1619	0.0003	173	124
325.00	500		186	500	46	0.0021	2.6024	L'EB -2.27%	4.1617	0.0000		
325.11	600			360	46	0.0015	2.6039	T(tr 1) RC	4.1617	0.0015	929	A 11
325.53	2200		132	4638	A 46	0.0097	2.6135	R	4.1602	0.0012	753	117
325.64	607		165			0.0000	2.6135	R	4.1590	0.0010	607	117
325.75	600		117			0.0000	2.6135	R	4.1580	0.0006	361	117
325.93	900			1815	A 11	0.0040	2.6175		4.1574	0.0083	4079	A 63
326.53	3200					0.0000	2.6175	T'TUNNELS	4.1491	0.0008	280	63
326.61	400			497	130	0.0007	2.6183	T'TUNNELS	4.1483	0.0012	400	63
326.74	700			1897	D 13	0.0037	2.6220	GALLITZEN	4.1471	0.0021	700	63
326.97	1200		63	1200	63	0.0036	2.6256	L@UN E END 4 TR	4.1450	0.0036	1200	63
326.97							2.6256		4.1414	0.0023	1096	D 10
327.20	1246			1246	63	0.0037	2.6293	EAST END 5 TR	4.1391	0.0004	194	105
327.41	1100		117	114	63	0.0003	2.6297	R	4.1387	0.0047	2257	A 74
327.75	1800		117	1576	A 63	0.0038	2.6335	R	4.1340	0.0017	680	74
327.88	700			1472	92	0.0030	2.6365		4.1323	0.0018	700	74
327.88				438	D 10	0.0010	2.6375		4.1305	0.0000		
328.15	1400		74	1400	74	0.0036	2.6411	L	4.1305	0.0036	1400	74
328.15							2.6411		4.1269	0.0059	3848	D 15
328.40	1338			1338	74	0.0034	2.6445		4.1210	0.0002	218	180
328.40				22	74	0.0001	2.6446		4.1208	0.0000		
329.41	5304			3802	A 74	0.0070	2.6516	WEST END 5 TRACKS	4.1208	0.0119	3382	A 11
329.86	2400			11056	A 11	0.0142	2.6658	T'CRESSON	4.1089	0.0000		
330.15	1536		230			0.0000	2.6658	R	4.1089	0.0000		
330.38	1200					0.0000	2.6658		4.1089	0.0000		

APPENDIX B5

PAGE 6

COLUMN 2

MILES	DISTANCE	HWY GRADE CROSSING	MAX MPH	DISTANCE (READ DOWN WESTBOUND TRAFFIC)	SPEED	TIME	CUM. TIME	STATION	TIME (READ UP EASTBOUND TRAFFIC)	CUM. TIME	DISTANCE	SPEED
331.39	5311	XING 253.3		189	173	0.0002	2.6661	RISBINS ROAD	4.1089	0.0127	9680	A 11
331.67	1500			2382	D 17	0.0031	2.6692		4.0962	0.0022	1360	117
332.08	2162		117	2162	117	0.0035	2.6727	L	4.0940	0.0035	2162	117
332.36	1500			1500	117	0.0024	2.6751		4.0905	0.0024	1500	117
332.55	982			982	117	0.0016	2.6767		4.0880	0.0016	982	117
332.72	900		117	900	117	0.0015	2.6781	R'LILLY	4.0864	0.0015	900	117
332.72						0.0000	2.6781		4.0850	0.0029	2181	D 18
333.25	2800			1360	117	0.0022	2.6803		4.0821	0.0000	12	169
333.59	1794			6896	A 11	0.0118	2.6921		4.0821	0.0012	1041	A 16
333.97	2000		165			0.0000	2.6921	R	4.0809	0.0035	3004	165
334.12	800					0.0000	2.6921	CASSANDRA	4.0774	0.0000	0	
334.98	4519			4176	A 16	0.0046	2.6967		4.0774	0.0116	9886	A 11
335.85	4631					0.0000	2.6967		4.0658	0.0022	1360	117
336.01	600			178	180	0.0002	2.6969		4.0634	0.0013	800	117
336.01				2744	D 18	0.0035	2.7004		4.0621			
336.10	500		117	500	117	0.0008	2.7012	R	4.0621	0.0008	500	117
336.19	500			500	117	0.0008	2.7020	T'PORTAGE	4.0613	0.0009	587	D 17
336.46	1400		135	880	117	0.0014	2.7034	R	4.0604	0.0001	88	133
336.46				2445	A 11	0.0037	2.7071		4.0603			
336.71	1346		147	114	134	0.0002	2.7073	R	4.0603	0.0040	2830	A 11
336.98	1400			727	D 13	0.0011	2.7084		4.0563	0.0023	1390	114
337.24	1400		114	1400	114	0.0023	2.7107	R	4.0540	0.0023	1400	114
337.51	1400			1390	114	0.0023	2.7130		4.0517	0.0002	102	D 11
337.59	426		117	381	A 11	0.0006	2.7136	L	4.0515	0.0028	1724	117
338.01	2200		117	1380	117	0.0022	2.7158	L	4.0487	0.0036	2200	117
338.50	2626					0.0000	2.7158	WILMORE	4.0451	0.0034	2636	D 17
339.08	3100			3656	A 11	0.0053	2.7211	WEST END POP RUN	4.0418	0.0004	376	175
339.43	1774		233	10674	A 14	0.0124	2.7335	R	4.0414	0.0000		
340.06	3361		233			0.0000	2.7335	R	4.0414	0.0000		
340.35	1500					0.0000	2.7335		4.0414	0.0104	6816	A 14
340.47	665			2406	A 15	0.0024	2.7359		4.0309	0.0000		
340.97	2600		255			0.0000	2.7359	L	4.0309	0.0000		
341.27	1600			347	191	0.0003	2.7363	SUMMERHILL	4.0308	0.0137	7735	A 63
341.76	2800			2438	D 19	0.0028	2.7391		4.0173	0.0006	260	63
341.97	1100		141	0	141	0.0000	2.7391	R	4.0165	0.0033	1100	63
341.97				2334	D 14	0.0043	2.7434	E END MIN POINT	4.0132			
342.16	1000		63	1000	63	0.0030	2.7464	R	4.0132	0.0030	1000	63
342.22	300			300	63	0.0009	2.7473	REVERSE CURVE	4.0102	0.0002	76	D 67
342.26	248		67	248	63	0.0007	2.7480	L	4.0099	0.0013	472	67
342.36	600		67	600	63	0.0018	2.7498	L	4.0086	0.0017	600	67
342.53	800		99	212	63	0.0006	2.7505	L	4.0069	0.0015	658	D 94
342.91	2000			187	A 63	0.0005	2.7510	T'TTMPH 40	4.0054	0.0002	83	94
342.91				961	67	0.0028	2.7537		4.0052	0.0015	719	A 84
342.91				861	A 67	0.0022	2.7559		4.0037			
342.91				578	84	0.0013	2.7572		4.0037	0.0031	1390	84
343.26	1866		84	1866	84	0.0042	2.7614	R'SOUTH FORK	4.0007	0.0080	3535	84
343.62	1871			1360	84	0.0031	2.7645		3.9927	0.0010	441	A 76
343.77	800		112	1036	A 84	0.0022	2.7668	L	3.9916	0.0024	945	76
343.92	600		98	514	98	0.0010	2.7678	L	3.9893	0.0018	655	A 81
343.92				561	D 98	0.0012	2.7688		3.9875			
344.07	800		76	800	76	0.0020	2.7708	L'MPH 35	3.9875	0.0017	960	81
344.22	800			489	76	0.0012	2.7721		3.9857	0.0039	1268	61
344.22				301	D 74	0.0008	2.7729		3.9818			
344.26	200		61	200	61	0.0006	2.7735	R	3.9818	0.0009	272	A 53
344.41	800		61	800	61	0.0025	2.7760	R	3.9809	0.0009	260	53
344.62	1100		77	986	61	0.0030	2.7790	R REV CURVE	3.9800	0.0039	1100	53
344.62				134	D 61	0.0004	2.7795		3.9760			
344.79	900		53	900	53	0.0032	2.7827	L'TTMPH 30	3.9760	0.0032	900	53
344.79							2.7827		3.9728	0.0013	458	D 57
344.96	900		77	900	53	0.0032	2.7859	L	3.9715	0.0011	442	77
345.00	198			198	53	0.0007	2.7866	RC	3.9704	0.0007	196	57
345.11	600		66	261	53	0.0009	2.7875	R	3.9697	0.0017	600	66
345.26	800		96	498	A 53	0.0015	2.7890	R	3.9680	0.0012	497	D 53
345.43	870			1241	66	0.0036	2.7926		3.9668	0.0025	1173	88
345.50	400		96	1381	A 66	0.0033	2.7959	L	3.9643	0.0009	400	86
345.65	900		104	40	91	0.0001	2.7960	L	3.9634	0.0026	1200	86
345.65				79	D 91	0.0002	2.7961		3.9608			
345.79	700		98	700	86	0.0015	2.7975	L'MINERAL POINT	3.9608	0.0015	840	A 78
345.92	700			128	86	0.0003	2.7979		3.9594	0.0009	300	78
345.92				572	D 58	0.0014	2.7994		3.9585			
346.03	600		62	600	62	0.0018	2.8012	R'RC	3.9585	0.0015	600	78
346.11	400			400	62	0.0012	2.8024	T'TTMPH 35	3.9570	0.0010	400	78
346.26	800		78	800	62	0.0024	2.8049	L	3.9561	0.0023	956	78
346.39	700		78	160	62	0.0005	2.8053	L	3.9538	0.0022	765	A 60
346.39				141	A 62	0.0004	2.8056		3.9518			
346.39				342	98	0.0010	2.8067		3.9510			
346.39				57	D 98	0.0002	2.8069		3.9516			
346.66	1400		63	1546	63	0.0048	2.8116	L	3.9516	0.0037	1160	80
346.70	200			54	D 63	0.0002	2.8117	REVERSE CURVE	3.9479	0.0006	200	80
346.85	800		80	800	80	0.0025	2.8143	R	3.9473	0.0025	800	80
346.90	300			300	80	0.0008	2.8152	RC	3.9448	0.0049	1537	60
347.21	1600		65	1066	80	0.0030	2.8185	L	3.9399	0.0010	304	A 51
347.26	286			207	A 80	0.0006	2.8192		3.9389	0.0013	345	51
347.30	215			834	85	0.0024	2.8218	RC	3.9376	0.0008	215	51
347.45	800		85	562	85	0.0016	2.8232	R	3.9368	0.0030	800	51
347.45				298	D 85	0.0008	2.8240		3.9338			
347.61	800		51	800	51	0.0030	2.8270	R	3.9336	0.0030	800	51
347.61							2.8270		3.9309	0.0001	31	D 53
347.68	400			400	51	0.0015	2.8285	RC	3.9308	0.0013	389	53
347.81	700		55	700	51	0.0026	2.8311	L	3.9294	0.0026	700	53
347.95	700		53	260	51	0.0010	2.8320	L	3.9269	0.0025	700	53
347.95				36	A 51	0.0003	2.8323		3.9244	0.0001	16	D 54
347.97	100			401	53	0.0014	2.8337	RC	3.9244	0.0003	64	54
348.14	900		54	900	53	0.0032	2.8370	R	3.9241	0.0032	900	54
348.27	700		65	480	53	0.0016	2.8386	R	3.9209	0.0006	192	D 65
348.34	377		65	19	A 53	0.0001	2.8387	R	3.9203	0.0026	885	65
348.34							2.8387	AC	3.9177	0.0002	79	D 69
348.59	1300			981	54	0.0034	2.8421		3.9175	0.0034	1271	69
348.81	1200		69	415	A 54	0.0013	2.8434	R	3.9142	0.0033	1200	69
348.81				802	65	0.0019	2.8454		3.9109	0.0002	62	D 72
							2.8454		3.9107	0.0041	1540	72
349.04	1200			275	A 65	0.0008	2.8461		3.9067	0.0013	438	A 61
349.27	1200		72	2224	72	0.0068	2.8530	L	3.9054	0.0011	380	61
349.27				1308	A 72	0.0030	2.8550		3.9043			
349.46	1000		95	9	93	0.0000	2.8550	L	3.9043	0.0031	1000	61
349.46				723	D 93	0.0018	2.8568		3.9012			
349.72	1400		81	1400	81	0.0043	2.8611	L	3.9012	0.0043	1400	61
349.76	200			330	81	0.0006	2.8617	RC	3.8968	0.0006	215	D 72
349.95	1000		72	1000	81	0.0031	2.8648	R	3.8962	0.0026	985	72
349.95				180	81	0.0005	2.8653		3.8936	0.0007	299	D 85
350.00	248		85	438	A 81	0.0013	2.8666	R	3.8929	0.0000	0	85
350.11	900		237	246	72	0.0007	2.8672	R	3.8929	0.0026	1496	D 13
350.26	900			514	72	0.0014	2.8686		3.8903	0.0009	611	132
350.66	2120			720	A 72	0.0017	2.8703	CONEMAUGH TOWER	3.8894	0.0007	505	132
350.76	500		132	5089	A 85	0.0088	2.8791	R	3.8887	0.0000		
350.78	100					0.0000	2.8791	RC	3.8887	0.0000		
351.00	1200		204			0.0000	2.8791	L	3.8887	0.0000		
351.23	1200					0.0000	2.8791		3.8887	0.0000		
351.32	478		162			0.0000	2.8791	L	3.8887	0.0057	3564	A 10
351.38	300					0.0000	2.8791	RC	3.8830	0.0026	1307	A 88
351.51	700		168	100	132	0.0001	2.8792	R	3.8804	0.0039	1816	88
351.51				969	D 13	0.0016	2.8808		3.8766			
351.68	900		104	449	104	0.0008	2.8816	R	3.8766	0.0006	284	A 84
351.88				451	D 10	0.0009	2.8825	E END 4 TR	3.8759			

APPENDIX B5 — PAGE 7 — COLUMN 1 / COLUMN 2

READ DOWN WESTBOUND TRAFFIC — STATION — READ UP EASTBOUND TRAFFIC

MILES	DISTANCE, ft	HWY GRADE CROSSING	MAX MPH	W DISTANCE	SPEED	ET, hours	TIME, hours	STATION (COLUMN 1)	TIME, hours	ET, hours	E DISTANCE	SPEED	MILES	DISTANCE, ft	HWY GRADE CROSSING	MAX MPH	W DISTANCE	SPEED	ET, hours	TIME, hours	STATION (COLUMN 2)	TIME, hours	ET, hours	E DISTANCE	SPEED
351.89	1100		88	1369	88	0.0043	2.8868	R	3.8759	0.0008	360	84	376.60				1360	88	0.0029	3.1607		3.6131			
352.08	1000			101 D	86	0.0002	2.8870	W END 4 TR	3.8751	0.0023	1000	84	376.60				1615 A	89	0.0033	3.1641		3.6131			
352.27	1000		84	1000	84	0.0023	2.8893	R	3.8728	0.0023	1000	84	377.36	4000			7	107	0.0000	3.1641	GRAYS	3.6131	0.0034	1360	76
352.27							2.8893		3.8706	0.0034	207 D	92	377.36				832 d	107	0.0017	3.1658		3.6097			
352.47	1061		92	1061	84	0.0024	2.8917	R	3.8701	0.0003	164	92	377.49	700		76	700	76	0.0017	3.1675	R	3.6097	0.0017	700	76
352.62	800			299	84	0.0007	2.8923	T JOHNSTOWN	3.8698	0.0037	1530 A	84	377.61	600		76	600	75	0.0015	3.1690	R	3.6080	0.0015	600	76
352.73	600		251	1036 A	64	0.0022	2.8945	R	3.8681	0.0017	560	84	377.61							3.1690		3.6065	0.0010	448 D	94
352.89	600			57	88	0.0001	2.8946	W END 3 TR	3.8645	0.0024	600	84	377.64	177			177	78	0.0004	3.1695	RC	3.6055	0.0001	25	94
352.89				808 D	96	0.0019	2.8965	E END 1 TR SANG HOL	3.8621				378.17	2800		94	1160	78	0.0020	3.1724	L	3.6054	0.0050	2500	94
352.92	200		64	200	64	0.0006	2.8971	R	3.8621	0.0006	200	64	378.17		XING 301.8					3.1724	T 589 PA	3.8004	0.0013	712 D	11
353.11	1000			1000	64	0.0030	2.9000		3.8615	0.0019	837 D	96	378.47	1600			3492 A	75	0.0087	3.1792		3.5991	0.0014	888	117
353.30	1000		96	360	64	0.0011	2.9011	L	3.8586	0.0022	1183	96	378.49	100	XING 3	117			0.0000	3.1792	R PRIVATE	3.5977	0.0002	100	117
353.39	436		222	2126 A	64	0.0049	2.9060	L	3.8573	0.0010	570 D	8X	378.74	1289	XING 3	117	1114	117	0.0018	3.1810	R PRIVATE	3.5975	0.0021	1289	117
353.76	2000					0.0000	2.9060		3.8563	0.0001	40	117	379.49	4000	XING 303.6		1300	117	0.0022	3.1832	T 589 PA	3.5954	0.0030	2261 D	17
353.92	800		190	630 A	86	0.0001	2.9061	R	3.8563	0.0022	1266 A	10	380.49	5276			10439 A	11	0.0130	3.1962		3.5924	0.0000	11	171
354.29	2000		106	3113	106	0.0056	2.9116	R	3.8541	0.0010	560	106	381.06	3000			166	169	0.0002	3.1963		3.5924	0.0000		
354.29				1360	106	0.0024	2.9141		3.8531	0.0033	1865	106	381.09	162		202		169	0.0000	3.1963	L	3.5924	0.0136	10466 A	11
354.29				117 A	100	0.0002	2.9143		3.8498	0.0036	1904 A	82	381.42	1700			2181 D	169	0.0029	3.1992		3.5786	0.0022	1360	117
354.38	1520		115	159	107	0.0003	2.9146	R	3.8460	0.0012	534	82	381.49	400		117	400	117	0.0006	3.1999	R	3.5766	0.0006	400	117
354.74	800			683 D	10	0.0014	2.9160		3.8447	0.0043	1466 A	46	381.66	1000		117	1000	117	0.0016	3.2015	R MILLWOOD	3.5760	0.0016	1000	117
354.96	1200		82	1200	82	0.0026	2.9187	R	3.8404	0.0023	560	46	381.68							3.2015		3.5744	0.0027	1985 D	16
355.11	800			124	82	0.0003	2.9190	DORNOCK	3.8381	0.0033	600	46	381.87	1000			1000	117	0.0016	3.2031		3.5717	0.0000	7	165
355.11				676 D	82	0.0020	2.9210		3.8348				382.00	677		165	360	117	0.0006	3.2037	L	3.5717	0.0004	986	165
355.30	1000		46	1000	46	0.0041	2.9251	L	3.8348	0.0041	1000	46	382.13	700		233	0		0.0000	3.2037	L	3.5712			
355.47	889		46	889	46	0.0037	2.9286	L	3.8307	0.0037	889	48	382.49	1900			0		0.0000	3.2037		3.5712	0.0022	1900	165
355.47							2.9286		3.8275	0.0002	56 D	50	382.58	483			0		0.0000	3.2037		3.5691	0.0006	483	165
355.51	200			200	46	0.0006	2.9296	RC	3.8066	0.0005	144	50	382.71	700		165	0	0	0.0000	3.2037	R	3.5685	0.0008	700	165
355.70	1000		52	1000	46	0.0041	2.9337	R	3.8267	0.0036	1000	52	383.49	4100			8886 A	11	0.0118	3.2156	2 TR (W OF SIDING)	3.5677	0.0008	2241	165
355.70							2.9337		3.8228	0.0018	542 D	8X	384.04	2687					0.0000	3.2156	T DERRY	3.5651	0.0033	2687	165
355.87	900		233	160	46	0.0007	2.9344	R	3.8211	0.0006	359	80	384.49	2400	XING 3	165			0.0000	3.2156	R VALLEY STREET	3.5618	0.0000		
356.06	1000		80	180	46	0.0007	2.9351	R	3.8202	0.0024	1000	68	384.55	287		165	4686	165	0.0056	3.2211	R	3.5616	0.0118	8866 A	11
356.06				1020	52	0.0037	2.9388		3.8179	0.0023	1206 D	12	385.49	5000			1360	165	0.0016	3.2227		3.5500	0.0011	660	117
356.10	200			1167 A	52	0.0033	2.9421	REVERSE CURVE	3.8150	0.0000	15	121	385.49				1854 A	16	0.0021	3.2248		3.5489			
356.25	800		93	373	80	0.0009	2.9430	L	3.8156	0.0000			385.63	700			155	172	0.0002	3.2249		3.5489	0.0011	700	117
356.27	100			100	80	0.0002	2.9433		3.8156	0.0056	3116 A	90	385.63				2331 D	17	0.0031	3.2280		3.5478			
356.57	1600			260	80	0.0006	2.9439		3.8100	0.0007	329	90	386.07	2359		117	2359	117	0.0038	3.2318	R BRADENVILLE	3.5478	0.0038	2359	117
356.87	1600		126	890 A	80	0.0019	2.9458	R	3.8093	0.0008	353 A	65	386.07							3.2318		3.5440	0.0041	3287 D	12
356.87				110	93	0.0002	2.9460		3.8085				386.38	1600			1360	117	0.0022	3.2340		3.5399	0.0000	7	190
356.87				2401 A	93	0.0043	2.9503		3.8085				387.19	4300			8886 A	11	0.0115	3.2458		3.5399	0.0080	7504 A	16
357.03	800			119	117	0.0002	2.9505		3.8085	0.0002	78	85	387.38	989		165			0.0000	3.2458	L	3.5319	0.0000		
357.14	600		222	820 D	11	0.0015	2.9520	R	3.8084	0.0013	600	85	388.38	5266		165	1909	165	0.0022	3.2480	L	3.5319	0.0016	1360	165
357.21	400		90	400	90	0.0009	2.9529	R	3.8070	0.0009	400	85	388.49	600		165	600	165	0.0007	3.2487	L	3.5304	0.0007	600	165
357.27	282			154	90	0.0003	2.9532	RC	3.8061	0.0006	282	85	389.25	4000			1360	165	0.0016	3.2503	T LATROBE	3.5297	0.0014	1302 D	19
357.27				126 D	90	0.0003	2.9535		3.8055				389.38	700		330	7153 A	18	0.0076	3.2579	R	3.5283	0.0000	8	190
357.40	700		85	700	85	0.0018	2.9550	L	3.8065	0.0016	700	85	390.35	5112		330	339	189	0.0004	3.2583	R	3.5283	0.0059	5647 A	17
357.40				1360	85	0.0030	2.9581	W END JOHNSTOWN	3.8040	0.0036	2240 D	15	390.35				900 D	18	0.0009	3.2592		3.5224			
358.16	4000					0.0000	2.9581	SO	3.8003	0.0001	70	150	390.76	2200		172	1030	172	0.0023	3.2616	R BEATTY	3.5224	0.0070	5922 A	14
358.27	586		233	7592 A	85	0.0122	2.9702	L	3.8003	0.0000			390.76				1170 D	17	0.0014	3.2630		3.5154			
358.31	257		233			0.0000	2.9702	L	3.8003	0.0103	6963 A	10	391.19	2225		147	2225	147	0.0029	3.2658	R	3.5154	0.0018	1360	147
359.26	5000			101	146	0.0001	2.9704		3.7900	0.0010	560	103	391.22	177		147	177	147	0.0002	3.2661	R	3.5136	0.0002	177	147
359.41	800			1570 D	14	0.0024	2.9727		3.7890	0.0015	600	103	392.18	5100			1360	147	0.0018	3.2678		3.5134	0.0025	2237 D	18
360.17	4000		103	4000	103	0.0074	2.9801	L	3.7875	0.0074	4000	103	393.18	5276			14053 A	14	0.0153	3.2832		3.5109	0.0004	360	192
							2.9801		3.7801	0.0002	92 D	10	393.45	1400					0.0000	3.2832		3.5105	0.0136	12279 A	14
360.21	200			100	103	0.0144	2.9945	RC SANG HOLLOW	3.7800	0.0045	2506	106	393.84	2074		344			0.0000	3.2832	L	3.4968	0.0000		
360.25	237		106	237	103	0.0004	2.9950	R	3.7756	0.0025	298 A	10	394.18	1800			20	167	0.0000	3.2832	DONOHOE	3.4969	0.0030	2225 A	13
360.26	20		106	20	103	0.0000	2.9950	R	3.7750	0.0024	1231 A	88	394.71	2800			3019 D	16	0.0034	3.2866		3.4939	0.0019	1360	135
361.01	4000			1003	103	0.0016	2.9969		3.7726	0.0008	360	86	394.97	1339		135	1339	135	0.0019	3.2885	L	3.4920	0.0018	1339	135
361.01				2284 A	10	0.0038	3.0007		3.7719				395.40	2300		135	2300	135	0.0032	3.2917	L	3.4902	0.0032	2300	135
361.01				150	123	0.0002	3.0009		3.7718				395.53	682			682	135	0.0010	3.2927		3.4669	0.0010	682	135
361.01				683 D	12	0.0011	3.0020		3.7718				395.87	1800		135	1800	135	0.0025	3.2952	R	3.4880	0.0025	1800	135
361.20	1000		103	504	103	0.0009	3.0029	L	3.7718	0.0022	1000	88	395.95	400			400	135	0.0006	3.2958		3.4854	0.0018	1417 D	16
361.20				496 D	10	0.0010	3.0039		3.7690				395.95						0.0000	3.2958		3.4817	0.0003	226	167
361.77	2996		88	2996	85	0.0067	3.0106	L SEWARD	3.7696	0.0064	2996	88	396.06	600			800	135	0.0008	3.2966		3.4814			167
361.77							3.0106	E END 2 TR SANG HOL	3.7632	0.0053	3775 D	18	396.35	1500		272	360	135	0.0005	3.2971	L	3.4814	0.0000		
362.15	2000			1360	88	0.0020	3.0135		3.7579	0.0002	236	183	396.61	1307		404			0.0000	3.2971	L	3.4614	0.0085	6786 A	13
362.19	214			12140 A	85	0.0174	3.0310	T	3.7577	0.0000			396.95	1800			9463 A	13	0.0103	3.3075		3.4728	0.0000		
363.14	5000		233			0.0000	3.0310	R	3.7577	0.0000			397.84	4700					0.0000	3.3075	T GREENSBURG	3.4729	0.0000		
363.57	2285		233			0.0000	3.0310	R	3.7577	0.0000			397.95	579		233			0.0000	3.3075	R	3.4726	0.0039	2593 A	11
364.13	3000			107	187	0.0002	3.0312	W END 2 TR SANG HOL	3.7577	0.0106	9243 A	14	398.22	1400		233	171	173	0.0002	3.3076	R	3.4691	0.0022	1360	117
364.34	2139			921 D	18	0.0011	3.0323	T NEW FLORENCE	3.7471	0.0016	1360	147	398.22				2382 D	17	0.0031	3.3106		3.4669			
365.11	3000		147	3000	147	0.0039	3.0362	R	3.7454	0.0039	3000	147	398.52	1594		117	1594	117	0.0026	3.3133	R	3.4669	0.0026	1594	117
365.49	2000		147	2000	147	0.0026	3.0388	R	3.7415	0.0026	2000	147	398.95	2300			1360	117	0.0022	3.3155		3.4643	0.0023	1654 D	15
365.87	2000			2000	147	0.0026	3.0413		3.7389	0.0026	2000	147	399.28	1700		233	7401 A	11	0.0101	3.3256	L	3.4620	0.0001	63	158
366.02	820		233	820	147	0.0011	3.0424	L	3.7363	0.0011	820	147	399.88	3200					0.0000	3.3256		3.4619	0.0098	7188 A	11
366.18	824		147	824	147	0.0011	3.0435	L	3.7353	0.0011	824	147	399.95	335		330			0.0000	3.3256	L	3.4521	0.0000		
366.93	4000			1360	147	0.0016	3.0452		3.7342	0.0006	448 D	15	400.00	311		330			0.0000	3.3256	L	3.4521	0.0006	361 A	11
366.93						0.0000	3.0452		3.7337	0.0002	133	157	400.53	2800			63	158	0.0001	3.3257		3.4515	0.0023	1360	114
367.13	1056					0.0000	3.0452	T CONNPIT JCT 3 TR	3.7335				400.53				1802 D	15	0.0026	3.3282		3.4492			
367.13				5109 A	14	0.0061	3.0513		3.7335				400.71	900		114	900	114	0.0015	3.3297	R	3.4402	0.0015	900	114
367.19	300	XING 290.8				0.0000	3.0513	PRIVATE	3.7335	0.0095	6932 A	11	400.72	100			100	114	0.0002	3.3299		3.4477	0.0028	2067 D	16
367.27	400		186				3.0513	R REVERSE CURVE	3.7240	0.0041	2505	117	401.73	5305			1250	114	0.0021	3.3320	T RADEBAUGH	3.4449	0.0003	287	165
367.34	400		186			0.0000	3.0513	L MPH 60	3.7199	0.0042	2332 A	94	402.05	1700			10057 A	11	0.0139	3.3458		3.4446	0.0000		
367.86	2720			236	169	0.0003	3.0516		3.7158	0.0039	1951	94	402.73	3586		165			0.0000	3.3458	L JEANNETTE	3.4446			
367.86				2181 D	16	0.0029	3.0545		3.7118				403.22	2600		165			0.0000	3.3458	L	3.4446	0.0340	11149 A	96
368.77	4837		117	4125	117	0.0067	3.0611	R	3.7118	0.0030	1324 A	73	403.52	1592					0.0000	3.3458		3.4105	0.0027	1360	96
368.77				712 D	11	0.0013	3.0624		3.7088				403.52				51	170	0.0001	3.3459		3.4078			
369.34	3000		94	2761	94	0.0056	3.0680	R	3.7088	0.0028	1085	73	403.52				2919 D	17	0.0042	3.3501		3.4078			
369.39	275				94	0.0000	3.0680	RC	3.7060	0.0007	275	73	403.73	1100		95	1100	95	0.0022	3.3523	R	3.4078	0.0060	2893	95
369.39				514 D	94	0.0012	3.0691	INDIA	3.7053				403.84	562		95	562	95	0.0011	3.3534	R	3.4019	0.0019	909 A	82
369.68	1500		73	1500	73	0.0039	3.0730	L	3.7053	0.0036	1500	73	404.52	3600	XING 328.6		1360	95	0.0027	3.3561	WRIGLEYS	3.3986	0.0042	1640	82
370.06	2000		73	2000	73	0.0052	3.0782	L LOCKPORT	3.7014	0.0052	2000	73	404.52				1310 A	95	0.0024	3.3585		3.3957			
370.38	1700			1360	73	0.0035	3.0817		3.6962	0.0028	1474 D	11	404.52				174	109	0.0003	3.3588		3.3957			
370.69	1619		117	3656 A	73	0.0072	3.0889	L	3.6934	0.0001	47	124	404.52				756 D	10	0.0015	3.3603		3.3957	0.0001	60 A	81
371.05	1900		117	319	117	0.0005	3.0894	L BOLIVAR	3.6933	0.0066	4880 A	73	404.71	1000		82	1000	82	0.0023	3.3626	L	3.3955	0.0011	460	81
371.20	800			664 D	11	0.0012	3.0906		3.6847	0.0006	329 A	66	404.73	100			100	82	0.0002	3.3628		3.3944	0.0036	1881	81
371.37	900		66	900	95	0.0018	3.0924	R	3.6839	0.0066	3297	66	404.88	800			24 D	82	0.0001	3.3629		3.3906	0.0001	59 A	80
371.59	1181		117	496	95	0.0010	3.0934	R	3.6773	0.0007	72 A	64	405.05	900		81	900	81	0.0021	3.3650	R	3.3904	0.0001	80	80
371.59				685 D	95	0.0016	3.0950		3.6771				405.30	1300			1276	81	0.0030	3.3680	T MANOR	3.3902	0.0031	1300	86
371.93	1800		86	1800	66	0.0052	3.1002	R	3.6771	0.0025	791	64	405.30				24 D	81	0.0001	3.3681		3.3872			
372.04	569			531	86	0.0015	3.1017	RC	3.6747	0.0017	569	64	405.51	1105		80	1105	80	0.0026	3.3707	L	3.3872	0.0028	1105	80
372.04				38 D	86	0.0001	3.1018		3.6731				405.51							3.3707		3.3845	0.0021	1069 D	11
372.59	2900		64	2900	64	0.0086	3.1104	L	3.6731	0.0086	2900	64	405.73	1200			1200	80	0.0028	3.3735		3.3825	0.0009	531	117
372.67	400		64	400	64	0.0012	3.1116	L	3.6645	0.0012	400	64	405.73				160	80	0.0004	3.3739		3.3818			
372.67							3.1116		3.6633	0.0029	1407 D	11	406.00	1400			3291 A	80	0.0063	3.3802	T SHAFTON	3.3816	0.0016	1000	117
372.78	631			631	64	0.0019	3.1135		3.6604	0.0000	24	117	406.72	3810		117	2535	117	0.0041	3.3843	R	3.3800	0.0062	3810	117
373.03	1300		117	726	64	0.0022	3.1156	R	3.6603	0.0008	500	117	406.72				1360	117	0.0022	3.3865		3.3738	0.0037	2958 D	18
373.22	1000			4033 A	84	0.0062	3.1236		3.6595	0.0026	1602	117	407.61	4720		222	8686 A	11	0.0118	3.3983	R ERWIN	3.3701	0.0002	174	184
373.52	1600		165	929 A	11	0.0015	3.1253	R	3.6569	0.0000			407.69	376		222			0.0000	3.3983	R	3.3699	0.0000		
373.57	264		190	180	124	0.0003	3.1256	R	3.6569	0.0048	2702 A	89	408.54	4600					0.0000	3.3983		3.3699	0.0059	5444 A	16
374.10	2800			1003 D	12	0.0019	3.1275		3.6520	0.0056	2651	89	408.59	285					0.0000	3.3983		3.3640	0.0000		
374.43	1700		89	1700	89	0.0036	3.1311	R	3.6463	0.0008	349 A	84	408.86	1400		165	1035	165	0.0012	3.3995	R LARIMER	3.3640	0.0065	5230 A	14
374.52	500		101	373	89	0.0006	3.1319	R	3.6456	0.0013	560	84	408.86				1360	165	0.0015	3.4010		3.3576			
374.67	800			127 D	85	0.0003	3.1322		3.6443	0.0066	4362	84	408.86				1041 A	16	0.0017	3.4022		3.3576			
375.22	2900		84	2900	84	0.0065	3.1387	R	3.6346	0.0019	441 A	78	409.54	3600			211	169	0.0002	3.4024		3.3576	0.0014	1075	141
375.43	1119			1119	84	0.0025	3.1412		3.6334	0.0000	16	76	409.59	285			1273 D	16	0.0016	3.4040		3.3561	0.0004	255	141
375.53	500		108	241	84	0.0005	3.1418	R	3.6334	0.0012	500	76	409.76	900		141	900	141	0.0012	3.4052	L ARDARA	3.3557	0.0012	900	141
375.57	244		108	1217 A	84	0.0025	3.1443	R TORRANCE	3.6322	0.0006	244	76	409.88	600		165	600	141	0.0008	3.4060	L	3.3546	0.0018	1323 D	17
375.69	600			66	100	0.0001	3.1444	E END 3 TR (SIDING)	3.6315	0.0015	600	76	410.54	3500			780	141	0.0010	3.4070		3.3529	0.0003	290	170
375.69				820 D	10	0.0013	3.1458	REVERSE CURVE	3.6301				410.92	2000			5230 A	14	0.0065	3.4135		3.3526	0.0000		
376.07	2000		76	2000	76	0.0050	3.1507	L	3.6301	0.0050	2000	76	411.44	2749		165	2259	165	0.0026	3.4161	R	3.3526	0.0000		
376.29	1200		117	1360	76	0.0034	3.1541	L	3.6251	0.0007	289 D	17	411.52	433		165	1793	165	0.0021	3.4181	R DP TRAFF	3.3526	0.0150	11209 A	108-170
376.43	700		110	725 A	76	0.0017	3.1558	L	3.6244	0.0035	1611	88	411.52				1041 A	16	0.0012	3.4193	TRAFFORD	3.3376			
376.52	500		86	500	88	0.0011	3.1569	L	3.6209	0.0061	2851	88	412.45	4600			21	169	0.0000	3.4193		3.3376	0.0024	1360	108
376.60	436		88	436	88	0.0009	3.1578	L	3.6148	0.0017	725 A	76	412.45				2478 D	16	0.0034	3.4227		3.3352			

APPENDIX B5

PAGE 8 COLUMN 1

READ DOWN WESTBOUND TRAFFIC — STATION — READ UP EASTBOUND TRAFFIC

MILES	DISTANCE	MAX SPD	DISTANCE / SPEED	HOURS	TIME	STATION	TIME	HOURS	DISTANCE / SPEED	MILES	DISTANCE	MAX SPD
412.86	2700	106	2700 106	0.0047	3.4275 L		3.3352	0.0047	2700 106	424.79	44	186
413.48	2740	102	1980 108	0.0024	3.4298 L		3.3306	0.0008	472 D 12	424.85	273	117
413.73	1300	122	1647 A 10	0.0027	3.4326 L		3.3297	0.0046	2867 122	424.98	680	
413.73			1300 122	0.0021	3.4347		3.3251	0.0010	614 A 11	425.02	217	
413.73			1340 A 12	0.0020	3.4367		3.3241			425.13	618	117
414.16	2300		122 131	0.0002	3.4369		3.3241	0.0025	1584 117	425.17	164	
414.16			909 D 13	0.0008	3.4378	PITCAIRN YARD	3.3216			425.28	630	
414.45	1505	117	1500 117	0.0024	3.4401	SWALL	3.3216	0.0000	117	425.35	370	117
414.70	1351		1351 117	0.0022	3.4422		3.3216	0.0070	9902 A 74	425.35		
414.97	1400		185 117	0.0003	3.4426		3.3146	0.0035	1380 74	425.49	175	
414.97			1205 D 15	0.0024	3.4450		3.3111			425.51	602	
415.29	1700	74	1700 74	0.0044	3.4493 R		3.3111	0.0044	1700 74	425.51		
415.50	1093		1093 74	0.0026	3.4521	WILMERDING	3.3067	0.0009	411 D 91	425.54	203	
415.50				0.0000	3.4521	CP WING	3.3058	0.0008	138 91	425.54		
415.64	748	58	287 74	0.0007	3.4528 L		3.3050	0.0036	1502 A 82	425.55	36	
415.69	300		1578 A 74	0.0028	3.4536		3.3017	0.0002	90 82	425.61	297	82
415.94	1300		79 94	0.0002	3.4558		3.3016	0.0040	1300 82	425.68	385	131
415.94			752 D 94	0.0018	3.4578		3.2976	0.0006		425.73	282	
416.15	1080	62	1086 82	0.0033	3.4608 L		3.2976	0.0066	2684 82	425.73		
416.22	400	62	400 62	0.0012	3.4621 L		3.2905	0.0012	400 82	425.75	111	
416.64	2200		1300 62	0.0040	3.4660		3.2876	0.0010	296 A 54	425.86	606	70
416.71	400	104	982 A 62	0.0026	3.4686 L		3.2868	0.0034	900 54	425.92	333	
416.79	400		62 82	0.0001	3.4690		3.2832	0.0014	400 54	425.92		
416.79			580 D 82	0.0016	3.4705		3.2818			425.94	111	
416.86	400	54	400 54	0.0014	3.4720 L		3.2818	0.0002	58 A 52	426.08	720	60
416.92	300		300 54	0.0011	3.4730 RC		3.2816	0.0007	160 52	426.12	194	
417.05	700	54	700 54	0.0025	3.4755 R		3.2810	0.0025	700 52	426.29	808	
417.15	480	62	480 54	0.0017	3.4772 R		3.2784	0.0017	480 52	426.31	101	229
417.15			31 D 54	0.0001	3.4773		3.2767			426.33	103	229
417.26	600	52	600 52	0.0022	3.4796 R		3.2767	0.0022	600 52	426.43	530	
417.35	500		500 52	0.0016	3.4813 RC		3.2745	0.0004	114 D 56	426.48	262	229
417.44	480	59	480 52	0.0017	3.4830 L		3.2741	0.0028	808 56	426.50	100	
417.58	700		380 50	0.0014	3.4844		3.2713	0.0058	1800 56	426.50	45	
417.67	500	74	245 A 52	0.0006	3.4852 L		3.2656	0.0020	537 A 42	426.56	286	
418.15	2500		735 56	0.0024	3.4875		3.2635	0.0061	1980 42	426.59	189	180
418.15			511 A 56	0.0017	3.4890		3.2574			426.69	498	
418.15			1153 74	0.0030	3.4922	PT PERRY TUNNEL	3.2574			426.72	195	
418.15			944 D 74	0.0018	3.4940	JCT PORT PERRY BR	3.2574	0.0000		426.84	651	106
418.24	524	42	524 42	0.0024	3.4964 R		3.2574	0.0024	524 42	426.87	164	
418.24				0.0000	3.4964		3.2550	0.0002	52 D 46	426.90	147	
418.26	200		200 42	0.0009	3.4973	REVERSE CURVE	3.2548	0.0006	148 46	426.96	285	86
418.36	400	46	400 42	0.0018	3.4991 L		3.2542	0.0016	400 46	427.03	303	
418.36					3.4991		3.2525	0.0029	1107 D 96	427.04	156	54
418.57	1100		760 42	0.0034	3.5025		3.2496	0.0000	22 96	427.06	105	86
418.72	800	185	112 A 42	0.0005	3.5030 R		3.2496	0.0014	750 96	427.06		
418.85	700	56	488 46	0.0020	3.5050 R		3.2481	0.0013	700 99	427.09	122	68
419.01	858	147	2731 A 46	0.0069	3.5119 R		3.2468	0.0019	1157 D 12	427.09	11	86
419.22	1100		4849 A 96	0.0077	3.5197		3.2449	0.0003	208 131	427.14	270	78
419.24	100	141		0.0000	3.5197 R		3.2446	0.0000	59	427.25	600	
419.37	700	141		0.0000	3.5197 R		3.2446	0.0000	88	427.32	300	180
419.52	800			0.0000	3.5197		3.2446	0.0000	86	427.49	914	
419.56	186	166		0.0000	3.5197 L		3.2446	0.0000	89	427.51	86	
419.75	1000			0.0000	3.5197		3.2446	0.0000	89	427.55	220	94
420.01	1400	165		0.0000	3.5197 L		3.2446	0.0097	5540 A 89	427.55		
420.24	1200			0.0000	3.5197		3.2348	0.0010	433 89	427.64	450	
420.29	245	295	111 137	0.0002	3.5198 L		3.2338	0.0005	245 89	427.67	300	80
420.42	682		1814 D 13	0.0032	3.5230		3.2333	0.0016	682 86	427.67		
420.43	73	80	73 80	0.0002	3.5231 L		3.2317	0.0002	73 58	427.68	30	
420.47	233	80	233 80	0.0006	3.5237 L		3.2315	0.0006	233 80	427.87	1000	
420.48	31		31 80	0.0001	3.5238		3.2310	0.0001	48 D 82	428.02	750	
420.48				0.0000	3.5238		3.2308	0.0014	903 82	428.06	250	165
420.60	620		620 80	0.0015	3.5253		3.2295	0.0000		428.14	400	165
420.62	116	82	116 80	0.0003	3.5256 L		3.2295	0.0003	116 82	428.25	600	
420.63	44	87	44 80	0.0001	3.5256 L		3.2292	0.0001	44 82	428.44	1000	
420.63				0.0000	3.5256 L	AMITY STREET	3.2291	0.0000		428.48	200	
420.64	46	82	46 80	0.0001	3.5257 L		3.2291	0.0001	46 82	428.56	400	95
420.76	758		500 80	0.0017	3.5266		3.2290	0.0029	1086 D 11	428.61	300	95
420.81	152	119	119 A 80	0.0008	3.5272 L	EAST R WEST	3.2270	0.0000	22 116	428.63	100	
420.81	18	116	18 87	0.0000	3.5272 L	EAST R WEST	3.2270	0.0000		428.75	650	
420.84	149				3.5272	REVERSE CURVE	3.2270	0.0000		428.82	360	135
420.86	129	118	96 82	0.0002	3.5275 R	EAST L WEST	3.2270	0.0000	129 116	428.64	100	136
421.00	705		820 82	0.0014	3.5289		3.2268	0.0011	703 116	429.01	900	169
421.00					3.5289		3.2256	0.0000		429.20	1000	169
421.19	1000		3302 A 82	0.0040	3.5329		3.2256	0.0003	187 116	429.22	100	169
421.38	1000			0.0000	3.5329		3.2253	0.0000		429.39	900	
421.50	880			0.0000	3.5329		3.2253	0.0000		429.52	700	
421.57	340			0.0000	3.5329		3.2253	0.0055	2993 A 87	429.55	130	283
421.76	1000		616 A 11	0.0010	3.5339		3.2198	0.0018	829 87	429.58	170	283
421.76			81 123	0.0001	3.5340		3.2181	0.0000		429.77	1000	283
421.86	540		1109 D 12	0.0000	3.5380		3.2181	0.0012	540 87	429.83	930	152
421.94	460	67	460 87	0.0010	3.5370 R		3.2169	0.0010	460 87	429.96	150	
422.06	586	87	586 67	0.0013	3.5383 R		3.2159	0.0013	586 87	429.96	100	
422.14	415		415 87	0.0009	3.5392		3.2146	0.0009	476 D 10	430.06	400	94
422.14					3.5392		3.2136	0.0002	133 104	430.13	450	
422.26	632		632 87	0.0014	3.5406		3.2134	0.0000		430.34	1000	
422.33	368	130	313 87	0.0007	3.5413 R		3.2134	0.0000		430.45	900	
422.45	585	138	1595 A 87	0.0059	3.5472 R		3.2134	0.0000		430.52	400	106
422.47	233	165		0.0000	3.5472 R		3.2134	0.0036	1717 A 82	430.56	190	
422.50	135	120		0.0000	3.5472 R		3.2099	0.0000	15 82	430.61	200	
422.51	79	125		0.0000	3.5472 R		3.2098	0.0002	79 82	430.61		
422.61	400	128		0.0000	3.5472 R		3.2097	0.0012	460 82	430.71	550	96
422.67	305	110		0.0000	3.5472 R		3.2086	0.0007	306 82	430.82	500	89
422.70	156			0.0000	3.5472		3.2079	0.0005	196 82	430.82		
422.70			94 108	0.0001	3.5473		3.2074	0.0000		430.90	450	
422.75	296		982 D 10	0.0013	3.5486		3.2074	0.0006	286 82	431.09	1000 XING 3.4	
422.86	539	82	539 82	0.0012	3.5498 R		3.2068	0.0012	539 82	431.28	1000	
422.89	106		185 82	0.0005	3.5503		3.2050	0.0026	1847 D 11	431.47	1000	
422.89				0.0000	3.5503		3.2026	0.0002	156 104	431.56	450	
423.07	940		940 82	0.0022	3.5525		3.2024	0.0000		431.66	550	
423.08	80	144	60 82	0.0001	3.5526 R		3.2024	0.0000		431.77	400	
423.27	1000	144	185 82	0.0004	3.5530 R		3.2024	0.0000		431.85	400	132
423.43	840	144	3920 A 82	0.0072	3.5601 R		3.2024	0.0000		431.87	100	132
423.46	180			0.0000	3.5601		3.2024	0.0000		431.92	250	
423.51	260			0.0000	3.5601		3.2024	0.0000		432.03	800	132
423.65	734	227		0.0000	3.5601 R		3.2024	0.0037	2445 A 11	432.04	50	
423.67	97	227		0.0000	3.5601 R		3.1987	0.0001	36 117	432.23	1000	
423.80	706			0.0000	3.5601	BECKS RUN	3.1987	0.0011	706 117	432.42	1000	
423.84	197			0.0000	3.5601		3.1975	0.0003	197 117	432.46	220	
423.84			119 123	0.0002	3.5603		3.1972	0.0000		432.56	550	
423.92	421		211 D 15	0.0003	3.5607		3.1972	0.0007	421 117	432.56		
424.03	579	117	579 117	0.0009	3.5616 R		3.1965	0.0008	579 117	432.61	730	94
424.04	62	117	62 117	0.0001	3.5617 R		3.1956	0.0000	11 117	432.73	650	94
424.16	652		652 117	0.0011	3.5628		3.1956	0.0000		432.73		
424.22	286	147	286 117	0.0005	3.5633 L		3.1956	0.0000		432.80	350	
424.26	239	147	209 117	0.0004	3.5636 L		3.1956	0.0000		432.99	1000	
424.37	580	165	580 117	0.0009	3.5645 L		3.1956	0.0000		433.18	1000	
424.41	211	117	211 117	0.0003	3.5648 L		3.1956	0.0000		433.29	600	
424.41	37	117	37 117	0.0001	3.5649 L		3.1956	0.0040	2202 A 95	433.33	300	117
424.50	439		439 117	0.0007	3.5656		3.1916	0.0004	202 95	433.35		
424.60	524		524 117	0.0008	3.5665		3.1912	0.0010	524 95	433.37	100	
424.72	634		367 117	0.0006	3.5671		3.1901	0.0013	684 95	433.44	400	
424.77	369	86	348 A 11	0.0005	3.5676 L		3.1889	0.0005	269 95	433.56	600	
424.78	97	165		0.0000	3.5676 L		3.1883	0.0008	451 D 11	433.74	1000	
424.79			185 120	0.0003	3.5680		3.1875	0.0003	160 159	433.79	240	

APPENDIX B5

PAGE 8 COLUMN 2

READ DOWN WESTBOUND TRAFFIC — STATION — READ UP EASTBOUND TRAFFIC

DISTANCE / SPEED	HOURS	TIME	STATION	TIME	HOURS	DISTANCE / SPEED
104 D 12	0.0002	3.5681 L		3.1872	0.0000	
272 117	0.0004	3.5686 L		3.1872	0.0000	
650 117	0.0011	3.5696		3.1872	0.0000	
		3.5696		3.1872	0.0000	
	0.0000	3.5696 L		3.1872	0.0000	
	0.0000	3.5696		3.1872	0.0000	
690 117	0.0010	3.5704		3.1872	0.0060	2330 A 82
370 117	0.0008	3.5713 L		3.1812	0.0003	144 82
	0.0000	3.5713		3.1809	0.0002	72 82
170 117	0.0003	3.5715		3.1807	0.0015	815 A 70
622 117	0.0010	3.5725		3.1792	0.0008	250 70
	0.0000	3.5725	J&L TUNNEL	3.1784	0.0000	
203 117	0.0003	3.5729		3.1784	0.0006	303 70
34 117	0.0001	3.5729		3.1778	0.0000	
1022 D 11	0.0018	3.5749		3.1778	0.0001	36 70
297 82	0.0007	3.5756 L		3.1778	0.0008	297 70
385 82	0.0009	3.5765 L		3.1770	0.0005	174 A 66
125 82	0.0003	3.5767	REVERSE CURVE	3.1765	0.0007	249 66
82	0.0000	3.5767		3.1758	0.0000	
288 D 11	0.0007	3.5774		3.1758	0.0003	111 66
568 70	0.0015	3.5789 R		3.1754	0.0016	558 66
333 70	0.0009	3.5798	REVERSE	3.1738	0.0010	303 66
31 70	0.0001	3.5799		3.1729	0.0000	
60 D 70	0.0002	3.5801	CURVE	3.1729	0.0003	111 66
720 66	0.0021	3.5822 L		3.1726	0.0021	720 66
194 66	0.0006	3.5827		3.1705	0.0006	194 66
880 66	0.0026	3.5853		3.1700	0.0026	880 66
101 66	0.0003	3.5856 L		3.1674	0.0003	101 66
103 66	0.0003	3.5859 L		3.1671	0.0003	103 66
93 66	0.0002	3.5861	REVERSE CURVE	3.1668	0.0015	530 66
2808 A 66	0.0058	3.5919 R		3.1653	0.0008	262 66
	0.0000	3.5919		3.1645	0.0003	100 66
	0.0000	3.5919	REVERSE CURVE	3.1642	0.0001	45 66
	0.0000	3.5919		3.1641	0.0008	268 66
	0.0000	3.5919 L		3.1633	0.0006	188 66
	0.0000	3.5919		3.1626	0.0014	498 66
	0.0000	3.5919		3.1613	0.0014	435 A 54
	0.0000	3.5919 R		3.1600	0.0016	461 54
164 106	0.0003	3.5922	REVERSE CURVE	3.1583	0.0006	164 54
147 106	0.0003	3.5925		3.1578	0.0005	147 54
364 96	0.0010	3.5935 L		3.1572	0.0010	285 54
211 D 96	0.0007	3.5942	REVERSE CURVE	3.1562	0.0011	303 54
158 54	0.0005	3.5947 R		3.1553	0.0005	158 54
106 54	0.0004	3.5951 R		3.1546	0.0013	465 D 78
	0.0000	3.5951		3.1533	0.0001	47 78
122 54	0.0004	3.5955	LIBERTY BRIDGE	3.1532		
11 54	0.0000	3.5956		3.1532		
270 54	0.0009	3.5966 R		3.1532		
610 54	0.0021	3.5986		3.1532	0.0046	1010 D 83
240 54	0.0009	3.5996 R		3.1486	0.0016	304 83
993 A 54	0.0028	3.6023	SMITHFIELD STREET	3.1486		
840 A 78	0.0015	3.6038		3.1468	0.0002	85 83
	0.0000	3.6038 R		3.1464	0.0005	220 83
14 88	0.0000	3.6038	MP 0 P&LE STATION	3.1459	0.0000	
125 D 88	0.0003	3.6041	REVERSE CURVE	3.1459	0.0010	450 83
300 83	0.0005	3.6046 L		3.1449	0.0005	200 83
	0.0001	3.6046		3.1445	0.0018	983 D 11
50 83	0.0001	3.6047		3.1428	0.0001	42 116
1000 83	0.0023	3.6069		3.1425	0.0000	
310 83	0.0007	3.6076		3.1425	0.0000	
2279 A 83	0.0044	3.6121 R	FORT PITT BRIDGE	3.1425	0.0000	
	0.0000	3.6121 R		3.1425	0.0038	2135 A 95
	0.0000	3.6121		3.1387	0.0002	100 95
160 110	0.0003	3.6124		3.1384	0.0020	1000 95
451 D 11	0.0008	3.6132		3.1364	0.0004	200 95
400 95	0.0008	3.6140 L		3.1360	0.0003	100 95
300 95	0.0006	3.6146 L		3.1357	0.0006	300 95
100 95	0.0002	3.6148		3.1351	0.0021	1232 D 132-95
650 95	0.0013	3.6161		3.1331	0.0002	172 132
350 95	0.0007	3.6168 R		3.1328	0.0000	
100 95	0.0002	3.6170 R		3.1328	0.0000	
160 95	0.0003	3.6173 R	WEST END BRIDGE	3.1328	0.0000	
	0.0000	3.6173 R		3.1328	0.0000	
		3.6173 R		3.1328	0.0000	
4812 A 95	0.0074	3.6247		3.1328	0.0000	
		3.6247		3.1328	0.0000	
		3.6247 R		3.1328	0.0000	
		3.6247 R		3.1328	0.0081	4436 A 94-132
		3.6247 R		3.1248	0.0005	200 94
		3.6247 R		3.1242	0.0017	850 94
30 133	0.0000	3.6248	REVERSE	3.1225	0.0003	160 94
1098 D 13	0.0022	3.6269	CURVE	3.1222	0.0002	100 94
450 94	0.0009	3.6279 L		3.1220	0.0008	450 94
400 94	0.0009	3.6288		3.1211	0.0024	1170 94
910 94	0.0018	3.6306		3.1188	0.0008	370 A 89-94
987 A 94	0.0019	3.6325		3.1180	0.0011	510 89
	0.0000	3.6325 L		3.1169	0.0008	400 89
	0.0000	3.6325	REVERSE CURVE	3.1161	0.0004	190 89
1458 105	0.0026	3.6351	CORLISS	3.1157	0.0006	280 89
456 D 10	0.0009	3.6360		3.1151		
550 89	0.0012	3.6372 R		3.1151	0.0012	550 89
500 89	0.0012	3.6383 R		3.1138	0.0012	550 89
		3.6383		3.1126	0.0024	1394 D 132-89
450 89	0.0010	3.6393		3.1104	0.0001	56 132
910 89	0.0019	3.6412	RIVER AVENUE	3.1103	0.0000	
4586 A 89	0.0072	3.6484		3.1103	0.0014	1800 132
	0.0000	3.6484		3.1089	0.0014	104 132
	0.0000	3.6484		3.1075	0.0000	
	0.0000	3.6484		3.1075	0.0000	
	0.0000	3.6484		3.1075	0.0000	
	0.0000	3.6484 R		3.1075	0.0000	
	0.0000	3.6484 R		3.1075	0.0000	
	0.0000	3.6484	REVERSE CURVE	3.1075	0.0000	
	0.0000	3.6484 L		3.1075	0.0000	
391 A 12	0.0014	3.6498		3.1075	0.0000	
	0.0000	3.6498		3.1075	0.0074	4436 A 94-132
	0.0000	3.6498		3.1001	0.0012	500 94
	0.0000	3.6498		3.0989	0.0004	200 94
15 134	0.0000	3.6498		3.0985	0.0011	550 94
1338 D 14	0.0022	3.6521		3.0974	0.0000	
230 94	0.0005	3.6526 L	DAVIS SCHOOL	3.0974	0.0006	230 94
650 94	0.0013	3.6539 L		3.0969	0.0013	650 94
		3.6539		3.0956	0.0013	712 D 117-94
350 94	0.0007	3.6546		3.0943	0.0001	38 117
1000 94	0.0020	3.6566		3.0942	0.0016	600 117
10 94	0.0000	3.6566	NEVILLE ISLAND	3.0938	0.0016	1000 117
2232 A 94	0.0042	3.6608		3.0917	0.0010	500 117
	0.0000	3.6608 R		3.0907	0.0005	300 117
		3.6608		3.0902	0.0024	1794 D 161-117
	0.0000	3.6608		3.0878	0.0000	8 161
	0.0000	3.6608		3.0877	0.0000	161
656 117	0.0011	3.6619		3.0877	0.0000	161
280 117	0.0004	3.6622		3.0877	0.0004	300 161
7870 A 11	0.0108	3.6729		3.0874	0.0000	240 161

APPENDIX 85 — PAGE 9 — COLUMN 1

READ DOWN — WESTBOUND TRAFFIC | STATION | READ UP — EASTBOUND TRAFFIC

MILES	DISTANCE FT	HWY GRADE CROSSING	MAX MPH	DISTANCE FT	SPEED MPH	ET HOURS	TIME HOURS	STATION	TIME HOURS	ET HOURS	DISTANCE FT	SPEED MPH
433.93	760		161			0.0000	3.6729	L	3.0871	0.0009	760	161
433.94	50		161			0.0000	3.6729	L	3.0862	0.0001	50	161
434.12	950					0.0000	3.6729		3.0862	0.0014	1272	D 18
434.31	1000					0.0000	3.6729		3.0848	0.0007	678	186
434.46	800					0.0000	3.6729		3.0841	0.0002	216	186
434.50	200		186			0.0000	3.6729	L	3.0839	0.0000		
434.58	300		186			0.0000	3.6729	L	3.0839	0.0000		
434.66	640		186			0.0000	3.6729	L	3.0839	0.0000		
434.69	60					0.0000	3.6729	EMSWORTH DAM	3.0839	0.0000		
434.78	450					0.0000	3.6729	REVERSE CURVE	3.0839	0.0000		
434.83	300		186			0.0000	3.6729	R	3.0839	0.0000		
434.87	210					0.0000	3.6729	REVERSE CURVE	3.0839	0.0000		
434.88	40		186			0.0000	3.6729	L	3.0839	0.0000		
435.07	1000		186			0.0000	3.6729	L	3.0839	0.0000		
435.20	700		186			0.0000	3.6729	L	3.0839	0.0000		
435.25	250			4929	A 16	0.0055	3.6784	REVERSE CURVE	3.0839	0.0000		
435.26	50		186			0.0000	3.6784	R	3.0839	0.0000		
435.32	300		186			0.0000	3.6784	R	3.0839	0.0000		
435.43	700					0.0000	3.6784		3.0839	0.0000		
435.55	550					0.0000	3.6784		3.0839	0.0066	6122	A 16
435.64	450					0.0000	3.6784		3.0773	0.0000		
435.83	1000					0.0000	3.6784		3.0773	0.0000		
436.02	1000			35	179	0.0000	3.6784		3.0773	0.0000		
436.17	800			706	D 17	0.0008	3.6792		3.0773	0.0000		
436.21	200		165	200	165	0.0002	3.6794	R	3.0773	0.0000		
436.40	1000		165	1000	165	0.0011	3.6805	R	3.0773	0.0000		
436.53	700		165	700	165	0.0008	3.6814	R	3.0773	0.0000		
436.55	120	XING 8.96		1360	165	0.0016	3.6829	PRIVATE	3.0773	0.0000		
436.55				243	A 16	0.0003	3.6832		3.0773	0.0000		
436.59	180					0.0000	3.6832		3.0773	0.0000		
436.77	1000					0.0000	3.6832		3.0773	0.0096	7402	A 12
436.96	1000	XING 9.4				0.0000	3.6832	LENORA AVENUE	3.0677	0.0004	260	128
437.15	1000			158	186	0.0002	3.6834		3.0674	0.0015	1000	128
437.17	100			1639	D 18	0.0021	3.6855		3.0659	0.0001	100	128
437.28	550		128	550	128	0.0008	3.6863	L	3.0657	0.0008	550	128
437.34	350			350	128	0.0005	3.6868		3.0649	0.0028	2298	D 17
437.36	90			90	128	0.0001	3.6869		3.0621	0.0001	95	179
437.53	910			910	128	0.0013	3.6883		3.0620	0.0000		
437.71	950	XING 10.3		10	128	0.0000	3.6883	BROADWAY STREET	3.0620	0.0000		
437.72	50		186				3.6883	L CORAOPOLIS	3.0620	0.0000		
437.76	200	XING 1	186				3.6883	L MULBERRY STREET	3.0620	0.0000		
437.85	500	XING 10.5					3.6883	MILL STREET REV CUR	3.0620	0.0000		
437.91	300		186				3.6883	R	3.0620	0.0000		
437.97	300			6388	126	0.0003	3.6886		3.0620	0.0000		
438.02	350		186				3.6886	R	3.0620	0.0000		
438.10	450						3.6886	REVCURVE	3.0620	0.0000		
438.11	50						3.6886		3.0620	0.0000		
438.17	300		186				3.6886	L	3.0620	0.0000		
438.24	400						3.6886	REVERSE CURVE	3.0620	0.0000		
438.26	100	XING 1	186				3.6886	R MAIN STREET	3.0620	0.0000		
438.29	150						3.6886		3.0620	0.0000		
438.39	550						3.6886		3.0620	0.0000		
438.44	250		233				3.6886	R	3.0620	0.0000		
438.48	200	XING 10.9					3.6886	WATT STREET	3.0620	0.0000		
438.56	400						3.6886		3.0620	0.0000		
438.67	600						3.6886		3.0620	0.0055	4929	A 16
438.86	1000			0	161	0.0000	3.6886		3.0565	0.0000		
438.99	700	XING 11.4		0	0	0.0000	3.6886	LOWER THORN ST	3.0565	0.0000		
439.05	300		161	1804	161	0.0019	3.6905	R	3.0565	0.0000		
439.24	1000		161	1000	161	0.0012	3.6987	R	3.0565	0.0035	2890	A 14
439.28	250	XING 1	161	250	161	0.0003	3.7000	R	3.0530	0.0000		
439.43	750	XING 11.5		750	161	0.0009	3.7009	PRIVATE	3.0530	0.0000		
439.55	650			610	161	0.0007	3.7016		3.0530	0.0000		
439.62	350			350	161	0.0004	3.7020		3.0530	0.0000		
439.74	650			650	161	0.0008	3.7028		3.0530	0.0000		
439.81	350		200	701	161	0.0008	3.7036	R	3.0530	0.0000		
439.88	400		200	588	D 18	0.0007	3.7043	R	3.0530	0.0000		
439.99	100		148	50	148	0.0001	3.7044	R	3.0530	0.0071	4980	A 11
440.18	1000		148			0.0000	3.7044	R	3.0459	0.0000	160	117
440.30	600		148			0.0000	3.7044	R STOOPS FERRY	3.0459	0.0010	600	117
440.37	400			19	170	0.0000	3.7044		3.0446	0.0006	400	117
440.41	200			2231	D 17	0.0029	3.7074	REVERSE CURVE	3.0440	0.0003	200	117
440.55	740		117	740	117	0.0012	3.7086	L	3.0437	0.0012	740	117
440.56	60		117	60	117	0.0001	3.7086	L	3.0425	0.0001	60	117
440.68	600		117	600	117	0.0010	3.7096	L	3.0424	0.0010	600	117
440.75	400			400	117	0.0006	3.7103		3.0414	0.0016	1119	D 14
440.93	950			950	117	0.0015	3.7118		3.0398	0.0003	231	146
440.94	50		146	10	117	0.0000	3.7118	L	3.0395	0.0001	50	146
441.08	750		146	4803	A 11	0.0066	3.7184	L	3.0394	0.0001	102	146
441.13	250					0.0000	3.7184		3.0393	0.0000		
441.32	1000					0.0000	3.7184		3.0393	0.0000		
441.51	1000					0.0000	3.7184		3.0393	0.0000		
441.54	180					0.0000	3.7184		3.0393	0.0000		
441.70	800					0.0000	3.7184		3.0393	0.0012	2384	A 13
441.89	1000			855	146	0.0011	3.7195		3.0381	0.0005	410	133
442.07	950	XING 14.5		532	D 14	0.0007	3.7202	PRIVATE?	3.0376	0.0014	950	133
442.08	50		133	50	133	0.0001	3.7203	R GLENWILLARD	3.0342	0.0001	50	133
442.27	1000		133	1000	133	0.0014	3.7217	R	3.0341	0.0014	1000	133
442.29	100		133	100	133	0.0001	3.7219	R	3.0327	0.0001	100	133
442.46	900			900	133	0.0013	3.7232		3.0325	0.0017	1350	D 16
442.55	500			490	133	0.0007	3.7238		3.0308	0.0001	54	164
442.64	480					0.0000	3.7238		3.0307	0.0000		
442.65	40		233			0.0000	3.7238	L	3.0307	0.0000		
442.84	1000		233			0.0000	3.7238	L	3.0307	0.0000		
442.92	450		233	4919	A 13	0.0064	3.7302	L	3.0307	0.0000		
443.02	550					0.0000	3.7302		3.0307	0.0000		
443.21	1000	XING 15.6				0.0000	3.7302	PRIVATE	3.0307	0.0000		
443.40	1000					0.0000	3.7302		3.0307	0.0108	7020	A 12
443.55	750					0.0000	3.7302	COUNTY LINE ALLE-BE.	3.0200	0.0005	380	129
443.59	250			100	158	0.0002	3.7304		3.0195	0.0004	250	129
443.74	750			1221	D 15	0.0016	3.7320	SOUTH HEIGHTS	3.0191	0.0011	750	129
443.78	250		129	250	129	0.0004	3.7324	R	3.0180	0.0004	250	129
443.92	750		129	750	129	0.0011	3.7335	R	3.0176	0.0011	750	129
443.97	250			250	129	0.0004	3.7338		3.0165	0.0038	3251	D 16
444.16	1000			1000	129	0.0015	3.7353		3.0127	0.0002	178	197
444.35	1000			110	129	0.0002	3.7355		3.0126	0.0000		
444.54	1000			13383	A 12	0.0151	3.7505		3.0126	0.0000		
444.56	120					0.0000	3.7505		3.0126	0.0000		
444.73	880					0.0000	3.7505		3.0126	0.0000		
444.86	700					0.0000	3.7505		3.0126	0.0000		
444.92	300		233			0.0000	3.7505	R	3.0126	0.0038	2863	A 16
445.11	1000		233			0.0000	3.7505	R	3.0088	0.0000		
445.29	940		233			0.0000	3.7505	R	3.0088	0.0000		
445.30	50					0.0000	3.7505		3.0088	0.0000		
445.49	1000					0.0000	3.7505		3.0088	0.0010	960	186
445.56	380					0.0000	3.7505		3.0078	0.0004	380	186
445.58	20					0.0000	3.7505		3.0074	0.0000	20	186
445.62	300		186			0.0000	3.7505	L	3.0074	0.0003	300	186
445.68	300					0.0000	3.7505	REVERSE CURVE	3.0071	0.0003	300	186
445.68	40					0.0000	3.7505	WEST ECONOMY	3.0068	0.0000	40	186
445.72	200		186			0.0000	3.7505	R	3.0067	0.0008	818	186
445.87	760					0.0000	3.7505		3.0059			
446.00	720					0.0000	3.7505		3.0059			
446.04	200		186			0.0000	3.7505	L	3.0059			
446.06	80					0.0000	3.7505	REVERSE CURVE	3.0059			

APPENDIX 85 — PAGE 9 — COLUMN 2

READ DOWN — WESTBOUND TRAFFIC | STATION | READ UP — EASTBOUND TRAFFIC

MILES	DISTANCE FT	HWY GRADE CROSSING	MAX MPH	DISTANCE FT	SPEED MPH	ET HOURS	TIME HOURS	STATION	TIME HOURS	ET HOURS	DISTANCE FT	SPEED MPH
446.08	150					0.0000	3.7505		3.0059			
446.14	300		186			0.0000	3.7505	R	3.0064			
446.24	950					0.0000	3.7505		3.0059			
446.43	1000					0.0000	3.7505		3.0059			
446.55	580					0.0000	3.7505		3.0059			
446.62	380					0.0000	3.7505		3.0059	0.0066	6122	A 165-156
446.62	30		233	1031	A 18	0.0010	3.7516	L	2.9993			
446.81	1000		233			0.0000	3.7516	L ALIQUIPPA	2.9993			
446.88	360		233			0.0000	3.7516	L	2.9993			
447.00	650			130	186	0.0001	3.7517		2.9993	0.0005	400	186
447.17	900			1246	D 18	0.0013	3.7530		2.9988	0.0010	900	186
447.19	150		186	100	186	0.0001	3.7532	R	2.9977	0.0001	100	186
447.31	600		186	600	186	0.0007	3.7538	R	2.9976	0.0007	600	186
447.38	400			400	186	0.0005	3.7543		2.9969	0.0005	400	186
447.55	900			900	186	0.0010	3.7553		2.9965	0.0010	900	186
447.57	100			100	186	0.0001	3.7556		2.9954	0.0001	100	186
447.62	250			250	186	0.0003	3.7557		2.9953	0.0003	250	186
447.73	600		186	600	186	0.0007	3.7564	L	2.9950	0.0007	600	186
447.76	150			150	186	0.0002	3.7566	WEST ALIQUIPPA	2.9943	0.0002	150	186
447.95	1000			1000	186	0.0011	3.7577		2.9942	0.0011	1000	186
448.14	1000			1000	186	0.0011	3.7589		2.9930	0.0011	1000	186
448.33	1000			1000	186	0.0011	3.7600		2.9919	0.0005	464	186
448.48	800			800	186	0.0009	3.7610		2.9913			
448.52	230		186	230	186	0.0003	3.7612	L	2.9913			
448.56	210			210	186	0.0002	3.7614		2.9913			
448.67	560		186	560	186	0.0007	3.7621	L	2.9913	0.0009	776	A 162-186
448.71	200		208	200	165	0.0002	3.7623	L	2.9904			
448.82	600		208	600	165	0.0007	3.7630	L	2.9904			
448.90	400			400	165	0.0005	3.7635		2.9904			
449.09	1000			185	165	0.0002	3.7637	HOPEWELL TOWNSHIP	2.9904			
449.10	100			1854	A 16	0.0024	3.7660		2.9904	0.0000		
449.27	900		229			0.0000	3.7660	L	2.9904			
449.33	300		229	296	172	0.0003	3.7664	L	2.9904			
449.43	500			400	D 17	0.0006	3.7669	REVERSE CURVE	2.9904			
449.46	200		162	200	162	0.0002	3.7671	R	2.9904			
449.56	480		162	480	162	0.0006	3.7677	R	2.9904			
449.65	500		162	400	162	0.0005	3.7682	R	2.9904			
449.67	50		162	0	162	0.0000	3.7682	R	2.9904			162
449.84	820			0	162	0.0000	3.7682		2.9904	0.0109	8107	A 117-162
450.03	1000					0.0000	3.7682		2.9795	0.0003	210	117
450.12	450					0.0000	3.7682		2.9792	0.0007	450	117
450.22	550		165			0.0000	3.7682	R	2.9785	0.0009	550	117
450.25	150		165	1429	162	0.0017	3.7698	R	2.9776	0.0002	150	117
450.25				1841	D 18	0.0025	3.7723		2.9773			
450.31	300		117	300	117	0.0005	3.7728	R	2.9773	0.0005	300	117
450.33	100			100	117	0.0002	3.7730		2.9768	0.0002	100	117
450.36	200		117	200	117	0.0003	3.7733	R	2.9767	0.0003	200	117
450.41	250			250	117	0.0004	3.7737		2.9764	0.0005	321	D 129-117
450.43	100			100	117	0.0002	3.7739		2.9759	0.0002	108	128
450.45	80		165	80	117	0.0001	3.7740	R	2.9757			
450.55	540			540	117	0.0009	3.7749		2.9757			
450.60	290			290	117	0.0005	3.7753		2.9757			
450.72	650			582	117	0.0009	3.7763		2.9757			
450.72				68	D 11	0.0001	3.7764		2.9757	0.0023	1490	A 115-128
450.79	350		115	350	115	0.0006	3.7770	L	2.9734	0.0006	350	115
450.98	1000		115	1000	115	0.0016	3.7786	L COLONA	2.9728	0.0016	1000	115
451.15	900		115	900	115	0.0015	3.7801	L	2.9712	0.0015	900	115
451.17	100			100	115	0.0002	3.7803		2.9697	0.0009	577	D 131-115
451.36	1000			1000	115	0.0016	3.7819		2.9688	0.0008	523	131
451.42	350			280	115	0.0004	3.7823		2.9680	0.0005	350	131
451.55	850		131	2214	A 11	0.0033	3.7856	L	2.9675	0.0001	66	131
451.56	80		131			0.0000	3.7856	L	2.9674	0.0000		
451.58	70		131			0.0000	3.7856	L	2.9674	0.0000		
451.74	850					0.0000	3.7856		2.9674	0.0000		
451.93	1000			36	131	0.0001	3.7857		2.9674	0.0000		
452.12	1000	XING 2	135	654	A 13	0.0007	3.7863	L 4TH STREET	2.9674	0.0000		
452.13	100		135	966	135	0.0013	3.7877	L MONACA	2.9674	0.0070	4022	A 95-131
452.31	900			651	135	0.0008	3.7885		2.9604	0.0007	380	95
452.42	600					0.0000	3.7885		2.9597	0.0012	600	95
452.42						0.0000	3.7885	OHIO RIVER BRANCH TO	2.9585	0.0000		
452.49	400				135	0.0000	3.7885		2.9585	0.0008	400	95
452.49				1349	D 13	0.0022	3.7907		2.9577	0.0000		
452.54	250		95	250	95	0.0005	3.7912	R	2.9577	0.0005	250	95
452.68	750		95	750	95	0.0015	3.7927	R	2.9572	0.0015	750	95
452.70	80		95	80	95	0.0002	3.7928	R	2.9557	0.0002	80	95
452.76	320			320	95	0.0006	3.7935		2.9556	0.0006	320	95
452.76						0.0000	3.7935	OHIO RIVER S BANK	2.9549	0.0000		
452.87	600			600	95	0.0012	3.7947		2.9549	0.0017	808	95
453.01	700			640	95	0.0009	3.7956		2.9532	0.0000		
453.01				248	A 95	0.0005	3.7960	OHIO RIVER N BANK	2.9532	0.0000		
453.06	300				96	0.0000	3.7960		2.9532	0.0000		
453.14	400		96		96	0.0000	3.7960	R	2.9532	0.0000		
453.25	600		96		96	0.0000	3.7960	R	2.9532	0.0000		
453.29	200		96	200	96	0.0004	3.7964	R BEAVER	2.9532	0.0000		96
453.37	400			400	96	0.0008	3.7972		2.9532	0.0073	2602	A 37-96
453.44	400		233	400	96	0.0008	3.7980	R	2.9460	0.0008	360	37
453.55	570			550	96	0.0011	3.7990		2.9451	0.0029	570	37
453.55						0.0000	3.7990		2.9422	0.0015	305	37
453.61	350					0.0000	3.7990		2.9407	0.0002	40	A 35-37
453.63	100		229	585	A 96	0.0010	3.8001	L	2.9405	0.0005	85	36
453.67	200		230	11	104	0.0000	3.8001	L BEAVER NEWSPAPERS	2.9400	0.0011	200	36
453.67				1386	D 98	0.0037	3.8038		2.9389	0.0000		
453.84	905		37	905	37	0.0048	3.8086	L	2.9389	0.0049	905	36
453.86	100			100	37	0.0005	3.8090		2.9340	0.0005	100	36
453.86	20		117	20	37	0.0001	3.8091	L	2.9335	0.0001	20	36
453.87	20			20	37	0.0001	3.8092		2.9334	0.0001	20	36
453.87	20		42	9	37	0.0000	3.8092	L	2.9333	0.0001	20	36
453.87	10			21	D 37	0.0001	3.8093	REVERSE CURVE	2.9331	0.0001	10	36
453.88	25		36	25	36	0.0001	3.8095	R	2.9331	0.0001	25	36
453.88	5		42	5	36	0.0000	3.8095	R	2.9330	0.0001	21	D 37-36
453.89	50		42	50	36	0.0003	3.8098	R	2.9328	0.0002	34	37
453.90	50	XING		50	36	0.0003	3.8100	SHARON ROAD	2.9327	0.0003	50	37
453.91	65			65	36	0.0004	3.8104	REVERSE CURVE	2.9324	0.0003	65	37
453.91	14		64	14	36	0.0001	3.8105	L	2.9321	0.0001	14	37
453.91	6			6	36	0.0000	3.8105	REVERSE CURVE	2.9320	0.0000	6	37
453.91	10		108	10	36	0.0001	3.8105	R	2.9320	0.0001	10	37
453.92	5	XING		5	36	0.0000	3.8106	TRACK IN 5TH AVENUE	2.9319	0.0000	5	37
453.93	100	XING		100	36	0.0005	3.8111	SHARON ROAD	2.9319	0.0005	100	37
453.95	100	XING		100	36	0.0005	3.8117	TO BUFFALO STREET	2.9314	0.0005	100	37
453.97	100			100	36	0.0005	3.8122	REVERSE CURVE	2.9309	0.0005	100	37
453.98	25	XING		25	36	0.0001	3.8123	BUFFALO STREET	2.9304	0.0001	25	37
453.98	36		42	36	36	0.0002	3.8125	L	2.9302	0.0002	36	37
453.99	40			40	36	0.0002	3.8127	REVERSE CURVE	2.9301	0.0002	40	37
453.99	16	XING		16	36	0.0001	3.8128	TUSCARAWAS ROAD	2.9299	0.0001	16	37
454.00	24		42	24	36	0.0001	3.8130	R	2.9298	0.0001	24	37
454.01	60			60	36	0.0003	3.8133		2.9297	0.0003	60	37
454.02	60			60	36	0.0003	3.8136		2.9293	0.0003	60	37
454.03	30		40	30	36	0.0002	3.8138	R	2.9290	0.0002	30	37
454.03	10	XING		10	35	0.0001	3.8138	WALNUT LANE	2.9289	0.0001	10	37
454.04	35			35	36	0.0002	3.8140	REVERSE CURVE	2.9288	0.0002	35	37
454.05	55		78	55	35	0.0003	3.8143	L	2.9287	0.0003	55	37
454.05	10		95	10	35	0.0001	3.8144	L	2.9284	0.0001	10	37
454.06	55		95	55	35	0.0003	3.8147	L	2.9283	0.0003	55	37
454.07	45	XING		45	35	0.0002	3.8149	SPRING LANE	2.9280	0.0002	45	37
454.09	105			105	35	0.0005	3.8154		2.9278	0.0005	105	37

READ DOWN WESTBOUND TRAFFIC — APPENDIX B5 STATION — READ UP EASTBOUND TRAFFIC

PAGE 10 COLUMN 1

MILES	DISTANCE	CROSSING	MAX MPH	DISTANCE	MPH	% GRADE	TIME	STATION	TIME	% GRADE	DISTANCE	MPH
454.09	25			25	36	0.0001	3.8156		2.9273	0.0001	25	37
454.10	35	XING	74	35	36	0.0002	3.8158	L DIVISION LANE	2.9272	0.0002	35	37
454.10	10		82	10	36	0.0001	3.8158	L	2.9270	0.0001	10	37
454.11	30			30	35	0.0002	3.8160		2.9269	0.0002	30	37
454.12	65			65	35	0.0004	3.8163		2.9268	0.0003	65	37
454.12	35		43	35	35	0.0002	3.8165	L	2.9265	0.0002	35	37
454.13	15	XING	44	15	35	0.0001	3.8166	L DIVISION LANE	2.9253	0.0001	15	37
454.13	30	XING		30	36	0.0002	3.8168	RC PA HWY 68	2.9262	0.0002	30	37
454.14	56		48	10	35	0.0001	3.8166	R	2.9260	0.0005	56	37
454.15	20		40	40	A 35	0.0002	3.8170	R	2.9258	0.0001	20	37
454.15	16		37	41	37	0.0002	3.8172	R	2.9257	0.0001	16	37
454.15	10		38	10	37	0.0001	3.8173	R	2.9256	0.0001	11	D 38
454.15	11		54	11	37	0.0001	3.8173	R	2.9255	0.0000		38
454.16	18			18	37	0.0001	3.8174	REVERSE CURVE	2.9256	0.0006	206	D 53
454.16	15		33	15	37	0.0001	3.8175	L	2.9247	0.0000		53
454.16	3	XING		3	37	0.0000	3.8175	RC BEAVER INDUSTRY	2.9247	0.0000		
454.19	10		63	10	37	0.0001	3.8176	R	2.9247	0.0000		
454.16						0.0000	3.8176		2.9247	0.0082	3169	D 14
454.16						0.0000	3.8176	RC VANPORT	2.9180	0.0016	1276	147
455.16	5253			1305	37	0.0174	3.8349	W BEAV MO	2.9140	0.0000		
455.29	700	XING 3.04		27	A 37	0.0001	3.8351	PRIVATE	2.9140	0.0018	1360	147
455.73	2300		147	418	A 38	0.0017	3.8368	L	2.9131	0.0025	1943	147
455.89	878			7301	A 53	0.0138	3.8506		2.9106			
456.01	600		166			0.0000	3.8506	L MPH50(TT)	2.9105	0.0000		
456.01				1861	A 13	0.0019	3.8526	REVERSE CURVE	2.9106			
456.16	800		166	27	146	0.0007	3.8532	R	2.9106	0.0025	1676	A 13
456.27	600			331	D 14	0.0004	3.8567		2.9081	0.0023	1675	137
456.44	900		137	2528	137	0.0006	3.8602	R	2.9058	0.0016	1127	A 13
456.80	1900			274	D 13	0.0004	3.8606		2.9042	0.0027	1868	130
456.97	900		130	900	130	0.0013	3.8619	R	2.9015	0.0009	510	A 12
457.16	978			1426	130	0.0021	3.8640		2.9005	0.0011	700	128
457.27	600			153	D 12	0.0002	3.8642		2.8994	0.0009	600	126
457.40	700		126	700	126	0.0011	3.8652	R	2.8985	0.0011	700	126
457.55	781		194	680	126	0.0010	3.8662	R	2.8975	0.0022	1713	D 16
458.16	3200			7661	A 12	0.0096	3.8762		2.8953	0.0003	254	166
458.76	3200	XING 6.16		1700	166	0.0020	3.8781	PRIVATE	2.8950	0.0000		
458.98	1172		166	753	166	0.0009	3.8790	R E INDUSTRY	2.8950	0.0000		
459.16	900				165	0.0000	3.8790		2.8950	0.0037	3101	A 15
459.55	2100	XING 2.80		500	D 15	0.0007	3.8796	WABASH STREET	2.8913	0.0000		
459.76	1085		159	1080	153	0.0010	3.8810	L W INDUSTRY	2.8913	0.0121	6115	A 97
460.16	2100	XING 8.08		1407	153	0.0001	3.8811	PRIVATE	2.8792	0.0007	1360	97
460.16				2053	D 15	0.0031	3.8842		2.8786			
460.31	800		97	806	97	0.0016	3.8858	L	2.8758	0.0016	606	97
460.65	1800			1360	97	0.0027	3.8885		2.8750	0.0010	560	D 11
460.86	1100		255	1032	A 97	0.0033	3.8917	L	2.8740	0.0026	2349	115
461.10	1283			0	115	0.0000	3.8917		2.8721	0.0005	315	115
461.16	300		115	0	115	0.0000	3.8917		2.8696	0.0036	2006	A 96
461.33	800		115	831	115	0.0014	3.8931	R	2.8680	0.0005	160	96
461.56	1300			584	115	0.0010	3.8940		2.8657	0.0024	1700	95
461.56				616	D 11	0.0011	3.8952		2.8633			
461.76	1100		96	1100	96	0.0022	3.8974	R	2.8633	0.0022	1100	96
461.76						0.0002	3.8974		2.8611	0.0016	895	D 12
461.93	900			900	95	0.0018	3.8991		2.8588	0.0033	6	123
462.15	1175		146	480	95	0.0009	3.9001	R	2.8562	0.0106	7085	123
463.16	5278		123	5011	A 85	0.0062	3.9063	R MICLAND	2.8459	0.0002	104	A 12
463.21	280		169	4342	123	0.0087	3.9120	R	2.8451	0.0056	3184	A 91
463.21				2194	A 12	0.0032	3.9152		2.8395			
463.21				98	132	0.0001	3.9152		2.8395			
463.95	3900			470	D 13	0.0008	3.9161		2.8395	0.0005	288	91
464.16	1100		122	130	122	0.0002	3.9163	L	2.8390	0.0094	1100	91
464.16				868	D 12	0.0017	3.9180		2.8296			
464.33	900		91	900	91	0.0019	3.9198	L MPH45	2.8296	0.0019	900	91
464.54	1100			1100	91	0.0023	3.9222		2.8278	0.0019	1077	D 15
464.54							3.9222		2.8258	0.0001	63	125
464.65	500		165	280	91	0.0005	3.9227	L	2.8257	0.0012	777	125
464.80	800			3318	A 91	0.0058	3.9285	MIDL TO USAC	2.8245	0.0000		
464.92	600		125			0.0000	3.9285	L	2.8245	0.0000		
465.02	500					0.0000	3.9285		2.8245	0.0000		
465.16	700		165			0.0000	3.9285	L	2.8248	0.0071	3962	A 67
465.71	2900			1457	125	0.0022	3.9308		2.8174	0.0042	1917	86
465.71				1134	D 12	0.0021	3.9328		2.8132			
465.88	800		86	800	86	0.0018	3.9346	L	2.8132	0.0021	961	A 70
466.03	900	XING 13.56		900	86	0.0020	3.9366	CHURCH STREET	2.8111	0.0008	282	70
466.10	676		132	676	86	0.0016	3.9381	R MPH45	2.8103	0.0018	676	70
466.19	200		132	34	86	0.0001	3.9381	R	2.8085	0.0005	200	70
466.23	200				86	0.0000	3.9381	MPH 45	2.8080	0.0006	200	70
466.23				366	D 86	0.0006	3.9388	REVERSE CURVE	2.8074			
466.35	631		70	631	70	0.0017	3.9407	L	2.8074	0.0017	631	70
466.35							3.9407	LITTLE BEAVER	2.8057	0.0066	3670	D 13
466.41	300			300	70	0.0008	3.9416	REVERSE CURVE	2.8001	0.0008	734	171
466.52	600	XING 1.166		600	70	0.0016	3.9432	R PRIVATE	2.7996	0.0000		
466.64	630		186	490	70	0.0012	3.9444	L CURVE	2.7990	0.0000		
467.15	2700			10665	A 70	0.0173	3.9617	OKWB OHIOPA LINE	2.7993	0.0000		
468.18	5832	XING 15.10				0.0000	3.9617	PUTNAM SATREET	2.7988	0.0028	3320	A 15
466.23	400					0.0000	3.9617		2.7956	0.0000		
468.28	256	XING 1.166				0.0000	3.9617	L RC MYRTLE STREET	2.7956	0.0000		
466.38	500	XING 1.196				0.0000	3.9617	R WALTERS STREET	2.7956	0.0000		
466.58	1100	XING 16.5				0.0000	3.9617	MPH50(TT) VIRGINIA 61	2.7965	0.0042	3067	A 14
466.77	1000		158	473	158	0.0006	3.9623	L	2.7912	0.0000		
466.96	1100		174	1102	158	0.0013	3.9636	L	2.7912	0.0000		
469.15	900			280	158	0.0003	3.9639		2.7912	0.0000		
469.64	2600			704	D 18	0.0009	3.9648	US 30	2.7912	0.0103	8447	A 90
469.83	1000		142	2187	142	0.0029	3.9677	L3e	2.7809	0.0009	418	90
469.91	400				142	0.0000	3.9677		2.7800	0.0028	1202	A 69
469.91				1709	D 14	0.0029	3.9706		2.7772			
470.00	500		90	500	90	0.0011	3.9718	R OKWB	2.7772	0.0007	240	69
470.10	500		138	500	90	0.0011	3.9727	R MPH50(TT)	2.7765	0.0014	500	69
470.16	320			130	90	0.0008	3.9730		2.7751	0.0009	320	69
470.22	300				90	0.0000	3.9730		2.7742	0.0008	300	69
470.22				400	D 90	0.0012	3.9741		2.7734			
470.25	200	XING 1.88		200	89	0.0000	3.9747	R PRIVATE	2.7734	0.0005	200	88
470.25						0.0000	3.9747		2.7729	0.0003	185	D 74
470.39	700			700	89	0.0019	3.9766		2.7726	0.0015	695	74
470.43	252	XING 1.74		252	89	0.0007	3.9773	R BROADWAY	2.7711	0.0006	252	74
470.43						0.0000	3.9773		2.7704	0.0002	89	D 75
470.51	400	XING 18.49		400	89	0.0011	3.9784	WASHINGTON STREET	2.7702	0.0008	311	76
470.62	600	XING 1.78		8	88	0.0000	3.9784	R MARKET STREET	2.7695	0.0015	600	78
470.62		XING 18.60				0.0000	3.9784	JACKSON STREET	2.7680	0.0019	347	D 11
470.74	600			238	A 88	0.0006	3.9790	E LIVERPOOL	2.7661	0.0001	59	112
470.85	600		113	716	74	0.0018	3.9809	R	2.7660	0.0000		
470.96	600			195	A 74	0.0005	3.9814	MPH35	2.7660	0.0000		
471.08	600		115	600	78	0.0020	3.9833	R	2.7660	0.0000		
471.15	400			2454	A 78	0.0049	3.9882		2.7660	0.0000		
471.23	300					0.0000	3.9882		2.7660	0.0040	2139	A 86
471.31	500		199	29	109	0.0001	3.9884	L	2.7620	0.0010	490	88
471.48	900			887	D 10	0.0012	3.9895		2.7610	0.0019	900	88
471.57	500		88	500	88	0.0011	3.9906	L	2.7591	0.0011	500	88
471.57							3.9906		2.7585	0.0016	838	D 11
471.68	600		117	600	88	0.0013	3.9919	L	2.7584	0.0001	62	116
471.89	1100			760	88	0.0016	3.9935		2.7583	0.0013	600	116
472.06	900		116	2839	A 86	0.0048	3.9984	L	2.7590	0.0015	900	116
472.10	215					0.0000	3.9984		2.7536	0.0001	34	D 11
472.67	3000			128	A 11	0.0002	3.9986		2.7535	0.0051	3101	117
472.80	1200		117	2885	117	0.0047	4.0033	R	2.7483	0.0019	1200	117
473.15	1337			1337	117	0.0022	4.0055	MPH 50(TT)	2.7464	0.0038	2344	117

READ DOWN WESTBOUND TRAFFIC — APPENDIX B5 STATION — READ UP EASTBOUND TRAFFIC

PAGE 10 COLUMN 2

MILES	DISTANCE	CROSSING	MAX MPH	DISTANCE	MPH	% GRADE	TIME	STATION	TIME	% GRADE	DISTANCE	MPH
473.42	1400			2907	117	0.0048	4.0103		2.7436	0.0014	833	A 110-117
473.76	1800		117	230	D 11	0.0004	4.0106	L	2.7412	0.0025	1565	110
473.93	900		110	900	110	0.0015	4.0127	L	2.7388	0.0015	900	110
474.14	1131		125	1131	110	0.0019	4.0141	L OKWB	2.7373	0.0008	460	D 124-110
474.29	800			229	110	0.0004	4.0145		2.7366	0.0000	4	124
474.47	900		310	2458	A 118	0.0037	4.0182	R	2.7365	0.0000		
474.60	700					0.0000	4.0182		2.7366	0.0053	3087	A 96-124
474.71	600		136			0.0000	4.0182	L	2.7313	0.0011	580	96
474.86	800			24	109	0.0000	4.0183	MPH 50(TT)	2.7302	0.0016	606	96
474.86				1028	D 12	0.0018	4.0201		2.7286			
475.10	1250		96	1250	96	0.0028	4.0230	R	2.7286	0.0025	1250	96
475.16	300			300	96	0.0006	4.0232		2.7261	0.0003	174	D 102-96
475.25	500			500	96	0.0010	4.0241		2.7258	0.0000	23	102
475.25				560	96	0.0011	4.0252		2.7257			
475.42	900		117	1007	A 96	0.0019	4.0271	R	2.7257	0.0031	1620	A 82-102
475.68	1380	XING 23.47		20	107	0.0000	4.0272	18TH STREET	2.7226	0.0031	1384	82
475.65				893	D 10	0.0014	4.0286	WELLSVILLE	2.7195			
475.85	900		80	900	82	0.0021	4.0306	L	2.7196	0.0021	906	82
475.99	700			700	82	0.0016	4.0322		2.7174	0.0009	451	D 99-82
476.12	700		147	680	82	0.0015	4.0338	L	2.7164	0.0018	946	99
476.15	200			1266	A 82	0.0026	4.0364		2.7146	0.0004	200	99
476.31	800					0.0000	4.0364		2.7142	0.0006	475	99
476.48	900		99	1824	99	0.0031	4.0395	L	2.7133	0.0026	1285	A 82-99
476.74	1400			451	D 99	0.0008	4.0403		2.7107	0.0031	1300	82
476.93	1000		82	1000	82	0.0023	4.0426	L	2.7076	0.0026	1134	82
477.17	1222			1222	82	0.0028	4.0456		2.7049	0.0034	1228	A 54-82
477.26	500			500	82	0.0012	4.0467		2.7015	0.0013	366	54
477.37	600		132	441	82	0.0010	4.0478	L	2.7002	0.0021	600	54
477.43	400				82	0.0000	4.0478	RC	2.6982	0.0014	400	54
477.45				359	D 82	0.0016	4.0493		2.6966			
477.58	700		54	700	54	0.0025	4.0518	R	2.6968	0.0025	700	54
477.58						0.0000	4.0518		2.6943	0.0004	136	D 78-52
477.71	700	XING 25.73		700	54	0.0025	4.0542	TT MPH 35 PRIVATE	2.6939	0.0017	564	82
477.87	800		82	690	54	0.0023	4.0565	L YELLOW CREEK	2.6921	0.0024	806	82
477.98	900			298	A 54	0.0010	4.0575	REVERSE CURVE	2.6897	0.0002	75	D 86-82
478.16	852		86	1202	82	0.0037	4.0612	R	2.6895	0.0006	1242	86
478.43	1400		98	141	A 82	0.0005	4.0617	R	2.6858	0.0040	1400	86
478.63	1150			2811	86	0.0081	4.0697		2.6819	0.0028	1305	D 76-70
478.73	495		117	2928	A 86	0.0083	4.0780	L	2.6791	0.0002	100	116
478.89	1400		117			0.0000	4.0780	LOW POST	2.6706	0.0036	2135	A 95-116
479.12	700					0.0000	4.0780		2.6751	0.0038	1866	95
479.31	1000			20	106	0.0000	4.0780		2.6714	0.0010	504	A 86-95
479.31				387	D 10	0.0007	4.0787		2.6703			
479.54	1200		95	1200	95	0.0024	4.0791	L	2.6703	0.0049	2208	86
479.77	1200			1012	95	0.0025	4.0812		2.6654	0.0016	487	A 77-86
479.77		XING 27.89		183	D 95	0.0004	4.0815	DANDOS CROSSING	2.6636	0.0000		
480.18	2190		86	2729	86	0.0058	4.0874	R	2.6636	0.0016	662	77
480.34	800		113	288	D 86	0.0006	4.0880	R	2.6621	0.0003	138	A 74-77
480.49	800		77	734	77	0.0018	4.0898	R	2.6618	0.0069	2591	74
480.49				86	D 77	0.0002	4.0900		2.6549			
480.64	800		74	800	74	0.0020	4.0920	R HAMMONDSVILLE	2.6549	0.0031	964	A 46-74
480.94	1500	XING 28.65		1380	74	0.0035	4.0955	WASHINGTON	2.6518	0.0003	86	46
481.07	700		129	185	A 74	0.0005	4.0960	L	2.6515	0.0026	700	46
481.18	575		76	788	78	0.0019	4.0979	L	2.6488	0.0003	575	46
481.18				504	D 78	0.0017	4.0995		2.6480			
481.35	900		48	900	48	0.0036	4.1031	L	2.6465	0.0006	500	48
481.40	225			225	48	0.0009	4.1040	REVC	2.6459	0.0001	39	D 50-48
481.40						0.0000	4.1040		2.6429	0.0007	106	50
481.61	1100		50	1100	48	0.0043	4.1083	R	2.6421	0.0042	1100	50
481.76	900			35	48	0.0021	4.1085		2.6379	0.0011	382	D 70-50
482.04	1400		76	72	A 48	0.0005	4.1087	L	2.6368	0.0053	1846	70
482.19	800	XING 30.04		800	50	0.0030	4.1118	EAST AVENUE	2.6315	0.0003	128	D 76-70
482.19						0.0000	4.1118		2.6312	0.0017	672	76
482.39	1000		76	403	50	0.0017	4.1135	R	2.6295	0.0025	1000	76
482.59	1100			727	A 50	0.0023	4.1158	IRONDALE	2.6270	0.0003	136	D 82-76
482.84	1300		116	1873	70	0.0043	4.1200	L	2.6267	0.0073	3141	82
483.06	1300		60	256	A 70	0.0008	4.1208	L	2.6194	0.0010	420	82
483.19	595			1505	76	0.0036	4.1245		2.6180	0.0016	600	82
483.19				326	A 76	0.0008	4.1253		2.6166			
483.25	300			2470	82	0.0057	4.1310		2.6166	0.0026	975	A 61-82
483.25				440	D 82	0.0012	4.1322		2.6141			
483.65	2100		61	2100	61	0.0065	4.1387	R	2.6141	0.0038	1215	61
483.68	145			145	61	0.0006	4.1391	REVC	2.6103	0.0005	145	61
483.98	1600		61	1600	61	0.0050	4.1441	L OKWB	2.6098	0.0050	1600	61
484.21	1200		101	1200	61	0.0037	4.1478	L	2.6049	0.0017	809	D 91-61
484.51	1800			150	61	0.0005	4.1483	NEW SALISBURY	2.6032	0.0049	2371	91
484.68	900		112	1577	A 61	0.0039	4.1522	R	2.5983	0.0014	668	91
484.81	700	XING 32.50		1463	91	0.0030	4.1553	PRIVATE	2.5969	0.0015	700	91
485.04	1200		91	1206	91	0.0025	4.1576	L OKWB	2.5954	0.0025	1200	91
485.20	865			865	91	0.0018	4.1595		2.5929	0.0011	562	91
485.37	900			505	91	0.0011	4.1605		2.5918	0.0041	1643	A 59-91
485.54	900		104	307	A 91	0.0006	4.1612	L OKWB	2.5877	0.0015	490	59
485.71	900			1075	95	0.0021	4.1634		2.5862	0.0028	905	59
485.71				910	D 95	0.0020	4.1654		2.5830			
485.96	1300		59	1300	59	0.0040	4.1695	R	2.5833	0.0040	1300	59
485.96						0.0000	4.1695		2.5793	0.0013	300	D 83-59
485.96						0.0000	4.1695		2.5776	0.0003	116	83
486.00	100			100	59	0.0006	4.1701	REVC	2.5776	0.0000	0	83
486.13	700		83	700	59	0.0022	4.1724	L OKWB	2.5776	0.0006	280	83
486.18	300			300	59	0.0010	4.1734		2.5769	0.0023	1245	D 124-83
486.71	2800			170	59	0.0005	4.1739		2.5747	0.0000	1	124
486.92	1100		120	1101	A 59	0.0029	4.1768	R	2.5747	0.0011	3985	A 90-124
487.20	1471			476	A 60	0.0010	4.1779		2.5735	0.0018	860	90
487.30	500			3118	A 90	0.0056	4.1834		2.5717	0.0011	500	90
487.30				48	121	0.0001	4.1835		2.5707			
487.30				950	D 12	0.0017	4.1852		2.5707			
487.47	900	XING 3[illegible]	90	900	90	0.0019	4.1871	L McGARY	2.5707	0.0082	3906	90
487.94	2500	XING 35.71		1369	90	0.0029	4.1900	NOLAN	2.5625	0.0026	1202	A 69-90
488.18	1267		158	2786	A 90	0.0052	4.1952	R	2.5598	0.0027	991	69
488.20	100			97	158	0.0050	4.2002		2.5569	0.0003	100	69
488.32	1700	XING 38.11		1344	D 11	0.0027	4.2029	PRIVATE	2.5562	0.0009	308	A 61-69
488.73	1100		89	1247	89	0.0034	4.2063	R	2.5557	0.0060	1984	61
488.79	300	XING 38.90			89	0.0000	4.2063	REV C HAYTI	2.5497	0.0002	68	A 59-61
488.75				193	D 90	0.0004	4.2068		2.5495			
489.15	1900		61	1900	61	0.0059	4.2127	L ROGERS	2.5495	0.0040	1269	59
489.17	100			65	61	0.0002	4.2129	RC	2.5455	0.0003	100	59
489.17				38	D 61	0.0001	4.2130		2.5451			
489.18	100		59	100	59	0.0003	4.2133	R OK	2.5451	0.0003	100	59
489.46	1431		59	1431	59	0.0046	4.2179	R OK	2.5448	0.0046	1431	59
489.46		XING 32.38				0.0000	4.2179	MAIN STREET	2.5402	0.0004	148	D 67-56
489.55	500			500	59	0.0016	4.2195	RC SALINEVILLE	2.5398	0.0010	352	67
489.70	800		87	800	59	0.0025	4.2221	L	2.5388	0.0023	800	67
489.87	900			60	59	0.0002	4.2223		2.5365	0.0010	401	D 85-67
490.04	900		117	301	A 59	0.0008	4.2232	L	2.5355	0.0031	1392	85
490.19	800			929	67	0.0028	4.2260		2.5324	0.0044	1988	85
490.48	1500			923	A 67	0.0023	4.2283		2.5280	0.0024	972	A 86-85
490.61	700	XING 3[illegible]	85	2517	85	0.0056	4.2339	R FOUNDRY HILL RD	2.5256	0.0002	60	65
490.67	300			300	85	0.0007	4.2346	REVC	2.5254	0.0008	300	86
490.78	600		94	179	85	0.0004	4.2350	L	2.5246	0.0017	850	86
490.86	400			421	D 85	0.0011	4.2360	REVC	2.5228	0.0011	400	86
491.05	1000		86	1000	86	0.0029	4.2389	R	2.5217	0.0029	1000	86
491.07	116			116	86	0.0003	4.2392	REV C	2.5186	0.0006	258	D 78-86
491.18	600		78	600	86	0.0017	4.2409	L	2.5183	0.0013	508	78
491.39	1100		78	644	86	0.0018	4.2428	L MPH 35	2.5170	0.0027	1100	78
491.39		XING 39.33		469	A 86	0.0013	4.2440	DOBSON	2.5143	0.0009	421	D 93-78

APPENDIX B5 — PAGE 11 COLUMN 1

READ DOWN WESTBOUND TRAFFIC | STATION | READ UP EASTBOUND TRAFFIC

MILES	DISTANCE	HWY GRADE CROSSING	MAX MPH	DISTANCE	SPEED	ET HOURS	TIME HOURS	STATION	TIME HOURS	ET HOURS	DISTANCE	SPEED
491.43	205			595	76	0.0015	4.2456	REV C	2.5133	0.0001	37	93
491.56	750		101	1152	78	0.0029	4.2484	R	2.5132	0.0009	450	89
491.71	600		93	1091	A 78	0.0024	4.2508	R	2.5123	0.0016	800	93
491.88	900			157	93	0.0031	4.2536		2.5107	0.0022	1326	D 12
492.19	1600		186	1260	93	0.0024	4.2560	L	2.5085	0.0003	216	133
492.28	500		186	2401	A 93	0.0043	4.2606	L	2.5082	0.0034	2249	A 11
492.41	700			348	A 11	0.0006	4.2612		2.5048	0.0000		
492.57	600	XING 4	165	144	129	0.0002	4.2614	R HULL ROAD	2.5048	0.0011	706	117
492.68	800			104	D 12	0.0002	4.2616		2.5038	0.0011	654	117
493.17	2613		117	2610	117	0.0042	4.2658	L MPH 35(TT)	2.5026	0.0081	2787	A 98
493.32	786		117	898	117	0.0011	4.2669	L	2.4975	0.0012	560	88
493.47	800			3	117	0.0025	4.2694		2.4963	0.0017	800	88
493.47				872	D 11	0.0016	4.2710		2.4946			
493.79	1700		88	1700	88	0.0037	4.2747	R	2.4946	0.0034	1569	88
493.91	600			606	88	0.0013	4.2760		2.4912	0.0026	1071	A 69
494.02	600		133	600	88	0.0013	4.2773	R	2.4886	0.0007	260	69
494.17	800			882	88	0.0014	4.2787		2.4879	0.0031	1131	69
494.23	300				88	0.0000	4.2787		2.4848	0.0009	309	A 61
494.23				438	D 88	0.0011	4.2798		2.4839			
494.48	1300		69	1300	69	0.0036	4.2833	L	2.4839	0.0030	980	81
494.55	400			247	69	0.0007	4.2840	MPH 35	2.4809	0.0012	400	81
494.55				153	D 81	0.0004	4.2845	REVERSE CURVE	2.4797			
494.75	800		81	800	81	0.0025	4.2869	R	2.4797	0.0025	800	81
494.70						0.0000	4.2869	REVERSE CURVE	2.4772	0.0003	33	D 66
494.84	700		66	700	81	0.0022	4.2891	L	2.4769	0.0004	148	58
494.84						0.0000	4.2891		2.4765	0.0028	1369	D 11
494.95	600			600	81	0.0019	4.2910	SUMMITVILLE	2.4736	0.0000		117
495.01	316		117	65	81	0.0002	4.2912	L WEND 2 TRACKS	2.4736	0.0000		117
495.16	800			186	A 61	0.0005	4.2917	MPH 40(TT)	2.4736	0.0013	800	117
495.16	32			514	86	0.0015	4.2932		2.4724	0.0001	32	117
495.26	500		117	3861	A 86	0.0080	4.3012	R	2.4723	0.0008	500	117
495.26		XING 43.40				0.0000	4.3012	PRIVATE	2.4715	0.0026	2231	D 17
495.26						0.0000	4.3012		2.4686	0.0001	113	170
496.17	4800			8645	A 11	0.0115	4.3127		2.4684	0.0000		
497.17	5280			219	164	0.0003	4.3130		2.4684	0.0148	10178	A 11
497.88	3600	XING 46.30		1587	D 16	0.0026	4.3156		2.4536	0.0022	1360	117
498.10	1100		117	1100	117	0.0018	4.3174	L	2.4514	0.0018	1100	117
498.18	480			486	117	0.0007	4.3181		2.4496	0.0017	1182	D 14
499.17	5200			900	117	0.0015	4.3196		2.4479	0.0001	107	147
499.17				4020	A 11	0.0058	4.3254		2.4478	0.0073	5073	A 11
499.17				212	145	0.0003	4.3257		2.4405			
499.36	1000			1068	D 14	0.0016	4.3272		2.4405	0.0022	1360	115
499.51	800		115	500	115	0.0013	4.3285	L	2.4382	0.0036	2008	A 95
500.17	3480			1360	115	0.0022	4.3306		2.4346	0.0027	1360	95
500.17				1000	A 11	0.0016	4.3324		2.4318			
500.17				222	123	0.0003	4.3327		2.4318			
500.17				806	D 12	0.0016	4.3343		2.4318			
500.40	1200		95	1200	95	0.0024	4.3367	L KENSINGTON	2.4318	0.0024	1200	95
500.40						0.0000	4.3367		2.4295	0.0012	984	D 11
500.59	1000			1000	95	0.0020	4.3387	MPH 40(TT)	2.4283	0.0003	156	117
500.76	900		117	380	95	0.0007	4.3394	R	2.4280	0.0015	900	117
501.06	1600			2269	A 95	0.0040	4.3434		2.4265	0.0026	1600	117
501.16	524		117	401	117	0.0006	4.3441	R	2.4240	0.0008	524	117
501.19	186		117	185	117	0.0003	4.3444	R	2.4231	0.0003	185	117
501.19						0.0000	4.3444		2.4228	0.0045	3742	D 16
502.16	5100			1360	117	0.0022	4.3466	MP	2.4183	0.0013	1356	166
503.01	4600	XING 51.18		5866	A 11	0.0118	4.3584	BROGAN ROAD	2.4170	0.0030	4051	166
503.07	300	XING 5 202				0.0000	4.3584	L LIPPINCOTT ROAD	2.4131	0.0108	11149	A 16
503.16	476	XING 51.85		130	165	0.0001	4.3585	PRIVATE	2.4024			
504.17	5342	XING 52.28		5342	165	0.0061	4.3646	PRIVATE	2.4024	0.0000		
504.32	800					0.0000	4.3646		2.4024	0.0005		
504.42	500		330			0.0000	4.3646	R	2.4024			
505.13	3752			6608	A 16	0.0073	4.3718		2.4024	0.0006	460	165
505.30	900					0.0000	4.3719		2.4018	0.0010	960	165
505.60	1600	XING 5 165		353	166	0.0004	4.3723	R US HWY 30	2.4008	0.0018	1600	165
505.75	800			1191	D 16	0.0013	4.3736	E ROCHESTER	2.3990	0.0008	800	165
506.14	2073		165	2073	165	0.0024	4.3759	L	2.3980	0.0024	2073	165
506.20	300		165	300	165	0.0003	4.3763	L	2.3957	0.0006	300	165
506.34	714			714	165	0.0008	4.3771		2.3953	0.0019	1874	D 20
506.34		JCT 54.6		646	165	0.0007	4.3778	JCT MAHONING SEC	2.3934	0.0000	39	200
506.38	213					0.0000	4.3779	BAYARD	2.3933	0.0000		90
506.45	400	XING 5 286		11149	A 16	0.0115	4.3894	R BAYARD STREET	2.3933	0.0000		60
506.53	418					0.0000	4.3894		2.3933	0.0000		200
506.60	370						4.3894		2.3933	0.0007	200	200
507.53	4900	XING 56.68					4.3894	PRIVATE	2.3931	0.0046	4900	200
508.53	5278	XING 57.57					4.3894	PRIVATE	2.3886	0.0040	4216	200
508.64	600			380	200	0.0004	4.3897		2.3846	0.0000	0	117
508.92	1477		117	1472	200	0.0014	4.3911	R	2.3845	0.0063	6306	A 18
509.53	3200			3200	200	0.0030	4.3942		2.3783	0.0000	0	182
510.53	5280			4306	200	0.0041	4.3982		2.3783	0.0053	4840	A 16
510.70	900			1874	D 20	0.0019	4.4002	HUGLE RUN	2.3730	0.0016	1360	165
510.96	1074	XING 5 165		1074	165	0.0012	4.4014	L PRIVATE	2.3714	0.0012	1074	165
511.53	3300			1360	165	0.0016	4.4030	MP	2.3702	0.0007	654	D 17
511.58	300			3586	A 16	0.0040	4.4069		2.3696	0.0003	283	178
511.72	700		188			0.0000	4.4069	R REVC	2.3692	0.0040	3586	A 16
511.85	750		188	283	178	0.0003	4.4072	L REVC	2.3652	0.0005	477	165
512.02	883			894	D 17	0.0007	4.4080		2.3647	0.0010	883	165
512.36	1800		165	1800	165	0.0021	4.4100	L	2.3638	0.0021	1800	165
512.53	900			906	165	0.0010	4.4111		2.3616	0.0011	973	D 17
513.52	5236			480	165	0.0005	4.4116	PARIS AVE	2.3605	0.0002	159	184
513.52				5444	A 16	0.0059	4.4175		2.3604	0.0059	5444	A 16
513.86	1800			159	184	0.0002	4.4177		2.3545	0.0018	1360	165
513.86				873	D 16	0.0011	4.4187		2.3529			
514.03	900		165	900	165	0.0010	4.4197	R PARKS RD	2.3529	0.0010	900	165
514.53	2980			1360	165	0.0016	4.4213		2.3519	0.0017	1641	D 10
514.69	800			11149	A 16	0.0115	4.4328		2.3501	0.0004	411	196
515.30	3229		233			0.0000	4.4328	R	2.3497	0.0000		
515.54	1300	XING 64.88				0.0000	4.4328	PRIVATE	2.3497	0.0000		
516.51	5086	XING 65.08		3600	200	0.0034	4.4362	PRIVATE	2.3497	0.0101	9625	A 16
517.53	5383	XING 65.71		1375	200	0.0013	4.4375	PRIVATE	2.3397	0.0000		
517.70	900			1874	D 17	0.0019	4.4395	NKP RD	2.3397	0.0000		
517.87	1473		165	1473	165	0.0017	4.4412	R	2.3397	0.0000		165
518.52	2900			2800	165	0.0033	4.4445		2.3397	0.0178	12058	A 14
518.76	1364			540	165	0.0006	4.4451	LOUISVILLE	2.3219	0.0018	1360	147
518.76				824	D 16	0.0010	4.4461		2.3201			
519.16	2000		147	2000	147	0.0026	4.4487	L COTTAGE ST & NIMISI	2.3201	0.0026	2000	147
519.16						0.0000	4.4487		2.3176	0.0013	1089	D 17
519.16						0.0000	4.4487	ABUTS MAIN	2.3163	0.0001	50	170
519.38	2200			1360	147	0.0018	4.4505		2.3162	0.0000		
						0.0000	4.4505		2.3162	0.0015	1250	A 16
519.94	1900			2350	A 14	0.0029	4.4534		2.3148	0.0016	1360	165
520.15	1124		165	1514	165	0.0017	4.4551	R	2.3132	0.0003	237	165
520.51	1900			1360	165	0.0016	4.4567		2.3129			
520.51						0.0000	4.4567	FAIRHOPE	2.3129	0.0034	2656	A 15
520.51		JCT 89.0/96.3				0.0000	4.4567	MP89+ JCT FT W LINE	2.3095	0.0003	200	153
520.61	500	XING 98.58		500	165	0.0006	4.4572	BEK ROAD	2.3092	0.0006	500	153
				580	165	0.0007	4.4579		2.3086			
520.72	600	XING 97.13		580	D 20	0.0007	4.4586	BROADWAY	2.3086	0.0007	800	153
520.78	300		153	300	153	0.0004	4.4590	L	2.3079	0.0004	300	153
520.83	300		153	300	153	0.0004	4.4594	L	2.3075	0.0004	300	153
521.16	1753			1360	153	0.0017	4.4610		2.3071	0.0003	228	D 15
521.37	1100		158	1140	A 15	0.0014	4.4624	L	2.3068	0.0031	2825	158
521.62	1300			1713	158	0.0021	4.4645		2.3037	0.0023	2205	D 20
						0.0000	4.4645		2.3014	0.0002	183	200
522.62	5286			12867	A 15	0.0135	4.4780		2.3012	0.0040	4200	200

APPENDIX B5 — PAGE 11 COLUMN 2

READ DOWN WESTBOUND TRAFFIC | STATION | READ UP EASTBOUND TRAFFIC

MILES	DISTANCE	HWY GRADE CROSSING	MAX MPH	DISTANCE	SPEED	ET HOURS	TIME HOURS	STATION	TIME HOURS	ET HOURS	DISTANCE	SPEED
523.62	5280			0	200	0.0000	4.4760		2.2872	0.0060	5280	200
524.61	5236			2877	200	0.0027	4.4807		2.2922	0.0058	5236	200
524.92	1800			1600	200	0.0015	4.4822		2.2873	0.0015	1600	200
525.26	1800			1800	200	0.0017	4.4839	CANTON	2.2857	0.0017	1800	200
525.52	1400		200	1400	200	0.0013	4.4853	R	2.2840	0.0013	1400	200
525.64	610			610	200	0.0006	4.4858		2.2827	0.0006	505	200
525.76	634			634	200	0.0006	4.4864		2.2822	0.0000	0	95
521.37	1100		158	1140	A 15	0.0014	4.4824	L	2.3008	0.0031	2625	158
521.62	1300			1713	158	0.0021	4.4845		2.3037	0.0023	2205	D 200 158
						0.0000	4.4845		2.3014	0.0000	183	200
522.62	5288			12867	A 15	0.0135	4.4780		2.3012	0.0040	4200	200
523.62	5280			0	200	0.0000	4.4780		2.2972	0.0060	5280	200
524.61	5236			2877	200	0.0027	4.4807		2.2922	0.0060	5236	200
524.92	1600			1600	200	0.0015	4.4822		2.2873	0.0015	1600	200
525.26	1800			1800	200	0.0017	4.4839	CANTON	2.2857	0.0017	1800	200
525.52	1400		200	1400	200	0.0013	4.4853	R	2.2840	0.0013	1400	200
525.64	610			610	200	0.0006	4.4858		2.2827	0.0006	500	200
525.76	634			634	200	0.0006	4.4864		2.2822	0.0000	0	90
526.83	4600			4600	200	0.0044	4.4906		2.2822	0.0000	0	180
527.63	5306			5306	200	0.0050	4.4958		2.2822	0.0000		
527.63					200	0.0000	4.4958		2.2822	0.0146	14007	A 153-200
527.63		XING 103.57			200	0.0000	4.4958	MARYLAND AVENUE	2.2673	0.0000	0	131
527.92	1500	XING 1: 200		1500	200	0.0014	4.4972	R RAFF RD	2.2673	0.0000	0	131
528.63	3780			3780	200	0.0036	4.5008		2.2673	0.0000	0	
529.31	3600	XING 105.56		2000	200	0.0019	4.5027	FERRY DRIVE	2.2673	0.0000	0	153
529.47	600		233	0	200	0.0000	4.5027	L	2.2673	0.0064	9657	A 112-153
529.64	887			0	200	0.0000	4.5027		2.2579	0.0036	2788	112
529.79	800			70	200	0.0001	4.5028		2.2540	0.0147	800	112
529.79				4027	D 20	0.0049	4.5077		2.2392			
530.07	1500	XING 1: 112		1500	112	0.0025	4.5102	L GENOA STREET	2.2392	0.0019	1085	A 101-112
530.64	2991			2808	112	0.0048	4.5150		2.2373	0.0022	1168	101
530.67	192			0	112	0.0000	4.5150		2.2351	0.0004	187	101
530.87				344	D 11	0.0006	4.5156		2.2347			
530.84	1400		101	1400	101	0.0026	4.5182	R	2.2347	0.0025	1400	101
							4.5182		2.2321	0.0016	985	D 129-101
531.26	1700			1360	101	0.0026	4.5208		2.2306	0.0001	70	129
531.34	400		165	3502	A 88	0.0056	4.5264	L	2.2305	0.0055	3346	A 101-128
531.64	1800					0.0000	4.5264		2.2250	0.0007	383	A 97-101
531.87	1187					0.0000	4.5264		2.2243	0.0003	160	97
531.96	500		165			0.0000	4.5264	L	2.2240	0.0010	500	97
532.09	700			128	127	0.0002	4.5266		2.2230	0.0014	700	97
532.09				1087	D 12	0.0019	4.5285		2.2216			
532.40	1600	XING 1 97		1800	93	0.0033	4.5318	R HIGH STREET	2.2216	0.0013	680	97
532.64	1300			1300	93	0.0026	4.5344	REED	2.2203	0.0032	1482	A 75-97
532.76	600			113	93	0.0002	4.5346		2.2171	0.0055	2168	75
532.76				487	D 93	0.0011	4.5357		2.2116	0.0006	240	A 70-75
533.12	1909		75	1909	73	0.0050	4.5407	L	2.2110	0.0023	800	70
533.21	500			377	73	0.0010	4.5417	REVERSE CURVE	2.2087	0.0014	500	70
533.21				123	D 73	0.0003	4.5420		2.2073			
533.63	2200		70	2200	67	0.0062	4.5482	R	2.2073	0.0060	2200	70
533.70	400		70	400	67	0.0011	4.5494	R	2.2014	0.0011	400	70
533.70						0.0000	4.5494		2.2003	0.0024	1154	D 113-70
533.91	1100			1100	67	0.0031	4.5525	MASSILLON	2.1979	0.0018	1088	113
534.03	700		165	260	67	0.0007	4.5532	R	2.1961	0.0012	700	113
534.39	1800			1276	A 67	0.0031	4.5563		2.1949	0.0023	1360	113
534.58	1021		113			0.0000	4.5563	L	2.1926	0.0016	1061	113
534.59		B&O RR XING				0.0000	4.5563	MACE TOWER	2.1908			
534.59				2130	A 90	0.0040	4.5602		2.1908	0.0030		
534.66	400			2333	113	0.0039	4.5641		2.1906	0.0009	827	A 108-113
535.08	2200	XING 111.36		162	D 11	0.0003	4.5644	EARL ROAD	2.1889	0.0024	1360	108
535.40	1700		108	1700	108	0.0030	4.5674	L TT MPH 70/60/50	2.1875	0.0030	1700	108
535.40						0.0000	4.5674		2.1846	0.0009	544	D 124-108
535.51	584	XING 111.92		584	108	0.0010	4.5684	RC BOYDS CRP	2.1837	0.0002	150	124
535.63	700		124	700	108	0.0012	4.5697	R	2.1834	0.0009	800	124
535.82	900		124	96	108	0.0001	4.5698	R	2.1828	0.0014	900	124
535.82				1381	A 10	0.0023	4.5721		2.1812	0.0004	282	D 131-124
536.03	1103	XING 112.36		613	120	0.0013	4.5733	BEAUMONT AVENUE	2.1808	0.0012	841	131
536.46	2300		131	1811	A 12	0.0024	4.5758	L	2.1795	0.0033	2300	131
536.65	1000			1402	131	0.0021	4.5778		2.1782	0.0014	1000	131
536.69	187			167	131	0.0003	4.5781		2.1748	0.0003	187	131
536.85	1400		131	1400	131	0.0020	4.5801	R	2.1745	0.0020	1400	131
537.63	3700			1360	131	0.0020	4.5821		2.1725	0.0015	1144	D 158-131
538.43	4100	XING 114.66		5209	A 13	0.0068	4.5889	KENYON AVENUE	2.1710	0.0080	8686	158
538.62	1000		158	2231	158	0.0027	4.5916	L	2.1630	0.0012	1000	158
538.64	136			136	158	0.0002	4.5918		2.1618	0.0015	1358	D 185-158
539.64	5282	XING 115.19		1225	158	0.0015	4.5933	SR 93	2.1603	0.0009	917	185
540.40	4000	XING 116.15		1718	A 15	0.0020	4.5952	CLERMONT	2.1594	0.0063	5762	A 165-185
540.40				5175	A 16	0.0056	4.6008		2.1531	0.0016	1360	165
540.40		XING 116.73		245	183	0.0003	4.6011	ALABAMA AVE	2.1510			
540.40				919	D 18	0.0010	4.6021		2.1516			
540.55	800		165	800	165	0.0009	4.6030	L	2.1516	0.0009	800	165
540.65	511	XING 117.06		511	165	0.0042	4.6072	RAILROAD STREET	2.1505	0.0018	1669	D 197-165
541.65	5299			849	165	0.0010	4.6082	NORTH LAWRENCE	2.1489	0.0002	225	197
542.67	5377	XING 119.28		11149	A 16	0.0115	4.6197	ANDERSON ROAD	2.1488	0.0124	11703	A 158-197
543.39	3600	XING 119.84		2478	200	0.0023	4.6220	PRIVATE	2.1362	0.0016	1360	158
543.39				2205	D 20	0.0023	4.6244		2.1346			
543.66	1422	XING 1 158		1422	158	0.0017	4.6261	L PRIVATE	2.1346	0.0017	1422	156
543.68	98	XING 1 156		98	156	0.0001	4.6262	L MP 120 PRIVATE	2.1329	0.0001	98	156
544.66	5200	XING 120.90		1360	158	0.0016	4.6278	SR 94	2.1328	0.0021	1973	D 196-156
545.67	5297	XING 121.82		467	A 15	0.0006	4.6284	TANNERVILLE	2.1306	0.0001	89	196
545.67				8404	A 16	0.0088	4.6371		2.1306	0.0115	10876	A 180-196
546.39	3800			376	180	0.0004	4.6375		2.1190	0.0016	1360	180
546.39				1484	D 18	0.0016	4.6391		2.1174	0.0000		
546.56	900		160	900	160	0.0011	4.6402	R	2.1174	0.0006	484	180
546.67	569	XING 123.82		366	160	0.0004	4.6406	WALNUT	2.1169	0.0014	1138	A 155-160
546.96	1500	XING 123.96		231	D 15	0.0003	4.6409	MARKET STREET	2.1155	0.0017	1380	155
547.67	3789		155	3789	155	0.0046	4.6455	L	2.1138	0.0034	2749	155
547.67						0.0000	4.6455	ORRVILLE	2.1104	0.0030		
548.17	2600		165	1960	155	0.0017	4.6472	L	2.1104	0.0000		
548.68	2700			4060	A 15	0.0049	4.6521	[illegible]	2.1104	0.0090	6420	A 117-155
548.96	1500	XING 125.11		348	172	0.0004	4.6525	CROWN HILL	2.1015	0.0022	1360	117
548.96		XING 125.36		2331	D 17	0.0031	4.6556	SCHOOLHOUSE ROAD	2.0993	0.0000		
549.23	1400		117	1400	117	0.0023	4.6579	R	2.0993	0.0023	1400	117
549.46	1400		179	1380	157	0.0022	4.6600	R	2.0970	0.0024	1747	D 180-117
549.69	1028		180	7535	A 17	0.0087	4.6687	R	2.0946	0.0008	682	180
549.93	1300	XING 1 174				0.0000	4.6687	R MUSKRAT ROAD	2.0938	0.0001	94	D 162-180
550.19	1385	XING 1 162				0.0000	4.6687	R BACK ORRVILLE	2.0937	0.0030	2571	162
550.68	2600	XING 127.65		472	A 19	0.0046	4.6733	ROHRER ROAD	2.0907	0.0007	643	D 175-162
550.68				1573	A 16	0.0030	4.6763		2.0900	0.0003	286	175
551.69	5302	XING 128.03		1490	158	0.0018	4.6779	PRIVATE	2.0897	0.0055	5603	A 153-175
551.73	200	XING 128.72		796	D 16	0.0008	4.6786	APPLEGATE	2.0842	0.0009	723	153
552.68	5100		153	5100	153	0.0063	4.6851	L	2.0823	0.0000		153
552.99	1600			1360	153	0.0017	4.6867		2.0823	0.0000		
553.16	960		158			0.0000	4.6867	L MPH 50	2.0823	0.0000		
553.30	705	XING 129.57				0.0000	4.6867	CO RD 75	2.0823	0.0148	9377	A 79-153
553.49	1000		172			0.0000	4.6867	R	2.0675	0.0018	760	79
553.60	600			922	153	0.0011	4.6879		2.0657	0.0014	800	79
553.60				2518	D 16	0.0041	4.6920		2.0640			
553.70	518		79	518	79	0.0012	4.6932	L SMITHVILLE	2.0643	0.0012	518	79
553.91	1123		79	1123	79	0.0027	4.6959	L	2.0630	0.0033	1123	79
554.67	4000			1360	79	0.0033	4.6992		2.0597	0.0016	784	D 107-79
554.70	187			177	A 79	0.0004	4.6996	TT MPH 35	2.0581	0.0001	42	107
554.70				1904	A 82	0.0038	4.7034		2.0581	0.0010	2001	A 82-107
554.70				84	108	0.0002	4.7036		2.0570	0.0057	2470	82

APPENDIX B5

PAGE 12 COLUMN 1

MILES	DISTANCE, FT	HWY GRADE CROSSING	MAX MPH	READ DOWN WESTBOUND TRAFFIC: DISTANCE, FT	SPEED, MPH	ET, HOURS	TIME, HOURS	STATION	READ UP EASTBOUND TRAFFIC: TIME, HOURS	ET, HOURS	DISTANCE, FT	SPEED, MPH
554.70				662	D 1	0.0013	4.7049		2.0513	0.0010	430	A 74
555.01	1600		82	1600	82	0.0037	4.7086	R	2.0503	0.0002	90	74
555.25	1300			1117	82	0.0026	4.7112		2.0501	0.0033	1300	74
555.25				183	D 8	0.0004	4.7116		2.0466			
555.69	2296		74	2296	74	0.0059	4.7175	L	2.0412	0.0093	3628	74
555.84	800			800	74	0.0020	4.7195		2.0375	0.0006	310	A 67
556.14	1600		80	560	74	0.0014	4.7210	R	2.0367	0.0021	760	67
556.14				311	A 74	0.0008	4.7218		2.0345			
556.26	600			1049	80	0.0025	4.7242		2.0345	0.0017	600	87
556.26				280	D 8	0.0007	4.7250		2.0328			
556.65	2100		67	2100	67	0.0059	4.7309	R	2.0328	0.0059	2100	67
556.79	700		67	700	67	0.0020	4.7329	R	2.0269	0.0020	700	67
556.79						0.0000	4.7329		2.0249	0.0003	123	D 72
556.83	236			236	67	0.0007	4.7335	REVERSE CURVE	2.0246	0.0003	113	73
557.32	2600	XING 1	73	1124	67	0.0032	4.7367	L E END 3 TR PRIVATE	2.0243	0.0104	4005	73
557.55	1200			256	A 67	0.0007	4.7374		2.0139	0.0001	53	A 72
557.68	700		79	2580	73	0.0067	4.7441	L	2.0137	0.0012	442	72
557.68				307	A 79	0.0008	4.7449		2.0126			
557.82	700		79	996	79	0.0024	4.7473	L	2.0126	0.0018	700	72
557.86	218			155	D 76	0.0004	4.7477	REVERSE CURVE	2.0107	0.0006	218	72
557.95	500		72	500	72	0.0013	4.7490	R	2.0102	0.0013	500	72
557.95				1360	72	0.0036	4.7525		2.0089	0.0012	536	D 11
558.39	2300			1377	A 72	0.0031	4.7557		2.0076	0.0045	2235	94
558.39							4.7557		2.0031	0.0001	69	A 93
558.54	800		94	383	94	0.0007	4.7564	L	2.0030	0.0005	260	93
558.65	600			573	94	0.0012	4.7576		2.0025	0.0012	800	93
558.75	500			27	D 91	0.0001	4.7576	WOOSTER	2.0012	0.0010	500	93
558.98	1207	XING 1	93	1207	93	0.0025	4.7601	L LIBERTY AVENUE	2.0002	0.0025	1207	93
558.98		XING 135.24				0.0000	4.7601	SOUTH STREET	1.9978	0.0019	1134	D 12
559.18	1100			1100	93	0.0022	4.7623		1.9958	0.0000	16	128
559.26	400		128	260	93	0.0005	4.7628	R	1.9958	0.0005	350	128
559.56	1600			3885	A 93	0.0066	4.7694		1.9953	0.0004	270	D 12
559.69	666	XING 1	135	1109	A 12	0.0016	4.7710	L PRIVATE W ED 3 TR	1.9949	0.0028	1986	135
560.58	4700	XING 136.6		6760	A 13	0.0085	4.7795	PRAIRIE ROAD	1.9921	0.0021	1717	D 17
560.71	700					0.0000	4.7795		1.9900	0.0029	2917	173
560.83	600		173			0.0000	4.7795	R	1.9871	0.0000		120
560.83	35					0.0000	4.7795	REVERSE CURVE	1.9871	0.0000		120
560.95	600		173			0.0000	4.7795	L	1.9871	0.0000		120
560.95				1663	A 16	0.0019	4.7814		1.9871			
561.67	3800					0.0000	4.7814		1.9871	0.0000		120
561.67				832	A 17	0.0009	4.7823		1.9871			
561.67				1468	176	0.0016	4.7839		1.9871	0.0115	9241	A 13
562.52	4500			2084	D 16	0.0026	4.7864		1.9757	0.0020	1360	130
562.65	714		130	714	130	0.0010	4.7875	R KILLBUCK	1.9737	0.0010	714	130
563.41	4000			1360	130	0.0020	4.7895		1.9726	0.0039	3388	D 16
563.62	1100		185	7071	A 13	0.0090	4.7985	R FISH	1.9687	0.0176	1712	200
563.64	134	XING 140.10				0.0000	4.7985	MESSNER ROAD	1.9512	0.0016	1716	200
564.67	5405	XING 141.6		11149	A 16	0.0115	4.8100	WILLOW ROAD	1.9496	0.0000		
565.68	5352	XING 142.94				0.0000	4.8100	FORCE ROAD	1.9496	0.0000		
566.67	5232			1643	200	0.0016	4.8116		1.9496	0.0000		
567.05	2000	XING 143.3		200	200	0.0002	4.8118	CEMETERY ROAD	1.9496	0.0000		
567.28	1216		222	1216	200	0.0012	4.8129	L	1.9496	0.0000		
567.66	2000					0.0000	4.8129		1.9496	0.0235	20183	A 11
567.89	1200		235			0.0000	4.8129	L	1.9261	0.0010	620	116
568.02	700	XING 144.3		1807	200	0.0017	4.8146	WOOD	1.9251	0.0026	1521	A 10
568.02		XING 144.4		3800	D 2	0.0047	4.8193	ROBINSON	1.9224			
568.34	1700	XING 1	116	1700	116	0.0028	4.8221	R MARKET STREET	1.9224	0.0025	1360	102
568.53	1000	XING 144.6		1000	116	0.0016	4.8237	SHREVE MAIN STREET	1.9199	0.0019	1000	102
568.67	753		165	305	116	0.0005	4.8242	R	1.9180	0.0015	752	A 93
568.67				448	D 11	0.0008	4.8250		1.9166			
568.92	1300		102	1300	102	0.0024	4.8274	R	1.9166	0.0018	560	93
569.02	500			243	102	0.0005	4.8278	REVERSE CURVE	1.9148	0.0010	500	93
569.02				257	D 1	0.0005	4.8283		1.9138			
569.28	1378		93	1378	93	0.0028	4.8311	L	1.9138	0.0026	1378	93
569.28						0.0000	4.8311		1.9110	0.0001	27	D 94
569.54	1400		106	1360	93	0.0028	4.8339	L	1.9110	0.0028	1373	94
569.64	500			69	A 93	0.0001	4.8341	RC	1.9082	0.0010	500	94
569.67	200		94	200	94	0.0004	4.8345	R	1.9072	0.0004	200	94
569.98	1601	XING 1	94	1091	94	0.0022	4.8367	R BROWNS	1.9066	0.0032	1601	94
570.68	3700	XING 147.1		1360	94	0.0027	4.8394	PRIVATE	1.9036	0.0042	2893	D 1
571.51	4392	XING 148.1		11218	A 9	0.0160	4.8554	SCRUBYS	1.8994	0.0000	27	169
571.85	1800	XING 148.58				0.0000	4.8554	BIG PRAIRIE RTE 226	1.8994	0.0000		
572.02	687		165			0.0000	4.8554	L	1.8994	0.0000		
572.68	3500			1750	165	0.0020	4.8574		1.8994	0.0156	11576	A 10
572.73	277					0.0000	4.8574		1.8835	0.0001	60	102
572.98	1300		165	42	165	0.0000	4.8574	L	1.8834	0.0024	1300	102
572.98				2467	D 1	0.0035	4.8609		1.8810			
573.44	2400		102	2400	102	0.0045	4.8654	L RC	1.8810	0.0045	2400	102
573.44						0.0000	4.8654		1.8765	0.0008	482	D 11
573.49	300			300	102	0.0005	4.8659		1.8757	0.0007	403	117
573.68	1000		117	1000	102	0.0019	4.8678	R REVERSE CURVE	1.8751	0.0005	300	117
573.89	1100		135	60	102	0.0001	4.8679	R	1.8746	0.0004	255	A 11
573.97	400			1394	A 10	0.0024	4.8703	LAKEVILLE	1.8742	0.0078	4726	115
574.21	1300	XING 1	115		115	0.0000	4.8703	R GULLS ROAD	1.8664	0.0006	374	A 11
574.69	2500	XING 150.9		1360	115	0.0022	4.8726	STATE ROUTE 179	1.8657	0.0001	60	112
574.69				1739	A 11	0.0027	4.8753		1.8656			
574.69				1484	126	0.0022	4.8775		1.8656			
574.93	1300			563	D 1	0.0009	4.8784		1.8656	0.0022	1300	112
575.16	1217		112	1217	112	0.0021	4.8804	L	1.8634	0.0021	1217	112
575.69	2800			1366	112	0.0106	4.8910		1.8614	0.0016	1076	D 14
575.69						0.0000	4.8910		1.8598	0.0002	119	141
575.69				5232	A 11	0.0076	4.8987		1.8596	0.0008	6226	A 91
576.69	5281			277	146	0.0004	4.8990		1.8588	0.0054	2584	91
576.82	700			1912	D 14	0.0031	4.9021		1.8534	0.0013	602	A 82
577.37	2900	XING 1	91	2900	91	0.0060	4.9081	R TR 385	1.8521	0.0025	1074	82
577.43	286			56	91	0.0001	4.9082	REVERSE CURVE	1.8496	0.0007	286	82
577.43				228	D 91	0.0005	4.9087		1.8489			
577.69	1400		82	1400	82	0.0032	4.9120	L	1.8489	0.0032	1400	82
577.83	700			700	82	0.0016	4.9136		1.8457	0.0011	510	D 10
578.00	900		101	660	82	0.0015	4.9151	L	1.8447	0.0020	1090	101
578.00						0.0000	4.9151		1.8426	0.0025	1630	D 14
578.15	804		156	1448	A 82	0.0030	4.9181	L	1.8401	0.0001	97	146
578.15							4.9181		1.8400	0.0080	5436	A 11
578.70	2900	XING 156.2		152	101	0.0003	4.9184	TR 5	1.8320	0.0008	435	A 10
579.59	4700			5635	A 10	0.0085	4.9269		1.8312	0.0014	806	106
579.59				326	142	0.0004	4.9274		1.8298			
579.59				1183	D 14	0.0018	4.9291		1.8298			
579.69	554	XING 1	110	1936	110	0.0033	4.9325	R JEFFERSON	1.8298	0.0010	554	106
579.69				127	D 11	0.0002	4.9327	LOUDONVILLE	1.8288			
580.23	2848	XING 1	106	2848	106	0.0051	4.9378	R SPRING STREET	1.8288	0.0051	2848	106
580.23						0.0000	4.9378		1.8237	0.0024	1808	D 14
580.23							4.9378		1.8213	0.0003	216	149
580.71	2500			1360	106	0.0053	4.9431	CP MOHICAN	1.8210	0.0051	3714	A 12
581.54	4400			2752	A 1	0.0044	4.9475		1.8160	0.0020	1360	128
581.82	1503	XING 1	128	2782	128	0.0041	4.9516	L PRIVATE	1.8140	0.0022	1503	126
582.71	4700			750	128	0.0011	4.9527		1.8117	0.0004	270	D 13
582.71				3334	A 12	0.0046	4.9573		1.8114	0.0062	4430	135
582.71				120	147	0.0002	4.9575		1.8051			
582.71				498	D 14	0.0007	4.9581		1.8051			
582.86	1303	XING 1	135	1303	135	0.0018	4.9600	L E PERRY PRIVATE	1.8051	0.0018	1303	135
583.72	4000			1360	135	0.0019	4.9619		1.8033	0.0013	968	D 1
583.72				7334	A 13	0.0091	4.9710		1.8020	0.0016	1349	158
583.91	1000	XING 160.9		42	169	0.0000	4.9710	BRIDGE STREET	1.8004			158
584.72	4264	XING 1	172	528	D 1	0.0006	4.9716	R PRIVATE	1.8004	0.0016	1360	158
585.30	3100		158	3100	158	0.0037	4.9753	R PERRYSVILLE	1.7968	0.0037	3100	158

APPENDIX B5

PAGE 12 COLUMN 2

MILES	DISTANCE, FT	HWY GRADE CROSSING	MAX MPH	READ DOWN WESTBOUND TRAFFIC: DISTANCE, FT	SPEED, MPH	ET, HOURS	TIME, HOURS	STATION	READ UP EASTBOUND TRAFFIC: TIME, HOURS	ET, HOURS	DISTANCE, FT	SPEED, MPH
585.73	2260	XING 162.5		2260	156	0.0027	4.9781	PRIVATE	1.7951	0.0027	2260	156
586.51	4100	XING 162.8		3500	156	0.0042	4.9823	TOWNSHIP ROAD 1037	1.7924	0.0049	4100	156
586.73	1197		156		156	0.0000	4.9823	L	1.7875	0.0094	6808	A 119-156
586.83	500		156		156	0.0000	4.9823	L	1.7781	0.0006	380	119
587.02	1000			1575	156	0.0019	4.9841		1.7775	0.0016	1000	119
587.02				1722	D 15	0.0024	4.9865		1.7759			
587.32	1600		119	1500	115	0.0028	4.9892	L	1.7759	0.0025	1600	119
587.56	1360	XING 163.8		1360	115	0.0022	4.9914	COUNTY ROAD 1075	1.7734	0.0007	440	D 131-119
587.74	800	XING 1	158	2997	A 11	0.0045	4.9959	L PRIVATE	1.7727	0.0003	207	131
587.89	800		158			0.0000	4.9959	L	1.7724	0.0060	3683	A 101-131
588.40	2700			116	136	0.0002	4.9961		1.7664	0.0026	1360	101
588.40				1217	D 13	0.0019	4.9980		1.7639			
588.61	1104		101	1104	101	0.0021	5.0001	R	1.7639	0.0042	2232	101
588.74	700			700	101	0.0013	5.0014		1.7597	0.0018	912	A 90-101
589.25	2700			660	101	0.0012	5.0026		1.7579	0.0029	1380	90
589.25				1218	A 10	0.0022	5.0048		1.7550			
589.25				137	113	0.0002	5.0050		1.7550			
589.25				685	D 11	0.0013	5.0063		1.7550			
589.54	1500		90	1500	90	0.0032	5.0095	L	1.7550	0.0032	1500	90
589.54						0.0030	5.0096		1.7519	0.0013	715	D 104-86
589.74	1102			1102	90	0.0023	5.0118		1.7505	0.0006	304	114
590.09	1800			258	90	0.0005	5.0123		1.7499	0.0030	1800	114
590.24	800		114	2255	A 90	0.0042	5.0165	L	1.7469	0.0013	800	114
590.44	1067	XING 166.6		1154	114	0.0019	5.0184	PRIVATE	1.7456	0.0018	1067	114
590.67	1200		114	1200	114	0.0020	5.0204	R	1.7436	0.0020	1200	114
590.74	400			400	114	0.0007	5.0211		1.7418	0.0003	171	D 132-110
590.89	800			960	114	0.0016	5.0227		1.7415	0.0035	2176	119
590.89				641	A 11	0.0010	5.0237		1.7381			
591.03	700		119	2497	119	0.0040	5.0277	R	1.7381	0.0018	1093	A 110-119
591.58	2900	XING 167.6		302	D 11	0.0005	5.0282	PRIVATE	1.7363	0.0023	1360	110
591.65	400	XING 1	110	400	110	0.0007	5.0289	R PERRYSVILLE RD	1.7339	0.0007	400	110
591.74	457			457	110	0.0008	5.0297		1.7332	0.0008	457	110
591.79	296			295	110	0.0005	5.0302		1.7324	0.0005	295	110
591.96	900		110	900	110	0.0015	5.0317	L	1.7319	0.0015	900	110
591.96		XING 168.3			110	0.0000	5.0317	UNION STREET	1.7304	0.0012	781	D 132-110
591.96		XING 168.4			110	0.0000	5.0317	CHURCH	1.7292	0.0000	19	132
592.36	2100	XING 168.6		1360	110	0.0023	5.0341	HANLEY BRETZ	1.7291	0.0019	1300	132
592.74	2000		132	2937	A 11	0.0046	5.0386	R	1.7273	0.0029	2000	132
593.16	2200		166	1163	132	0.0017	5.0403	R	1.7244	0.0015	1059	132
593.44	1500			489	A 13	0.0007	5.0410	CP LUCAS	1.7229	0.0037	1974	A 118-132
593.57	693		135	1913	136	0.0027	5.0437	R	1.7192	0.0022	1360	118
593.57				631	D 13	0.0009	5.0446		1.7170			
593.71	700		118	700	118	0.0011	5.0458	R	1.7170	0.0032	2010	118
593.74	200			2766	118	0.0044	5.0502		1.7136	0.0002	130	A 117-118
594.24	2600			34	D 11	0.0001	5.0502		1.7138	0.0022	1360	117
594.67	2300		117	2300	117	0.0037	5.0540	L	1.7114	0.0037	2300	117
594.74	377	XING 171.5		377	117	0.0006	5.0546	PRIVATE	1.7077	0.0042	3399	D 192-117
595.74	5281	XING 171.9		983	117	0.0016	5.0562	SITTS ROAD	1.7035	0.0002	164	192
596.69	5000			14330	A 11	0.0181	5.0742		1.7033	0.0021	2080	A 186-192
596.74	279		186			0.0000	5.0742	L	1.7013	0.0000		
596.80	300	XZING	186			0.0000	5.0742	L RC ILLINOIS AVENUE	1.7013	0.0000	0	186
596.89	481		186			0.0000	5.0742	R TT MPH 70	1.7013	0.0126	10988	A 143-186
597.74	4500			62	184	0.0001	5.0743		1.6887	0.0019	1427	A 135-143
598.12	2000			1986	D 18	0.0023	5.0766		1.6867	0.0002	160	135
598.24	600		143	1373	143	0.0018	5.0784	R	1.6865	0.0007	523	135
598.35	600	XING 1	186	326	D 14	0.0004	5.0788	R 4TH STREET	1.6858	0.0000		135
598.46	600	XING 1	135	600	135	0.0008	5.0797	R 5TH STREET	1.6858	0.0000		135
598.73	1512	XING 175.0		1360	135	0.0019	5.0816	ORANGE STREET	1.6858	0.0067	5229	A 90-135
598.75				2010	A 13	0.0027	5.0843	MANSFIELD	1.6771			
599.47	3800	XING 175.3		4	146	0.0000	5.0843	SURREY ROAD	1.6771	0.0029	1360	90
599.47				1938	D 14	0.0031	5.0874		1.6742	0.0000		
599.47			90		90	0.0000	5.0874	B&O XING	1.6742	0.0000	RR CR	90
599.47		XING 175.57				0.0000	5.0874	MAIN STREET	1.6742	0.0035	2245	D 153-90
599.53	300	XING 1	233	300	90	0.0006	5.0880	L LONGVIEW AVENUE	1.6707	0.0003	229	153
599.94	2174			1060	90	0.0022	5.0903		1.6705	0.0000		
600.01	400		233	8664	A 90	0.0132	5.1035	L	1.6705	0.0005	400	153
600.25	1268					0.0000	5.1035		1.6700	0.0016	1268	153
600.59	1800	XING 1	153			0.0000	5.1035	L EMP STL MILL	1.6684	0.0017	1341	153
600.76	900	XING 177.22				0.0000	5.1035	PRIVATE	1.6667	0.0000		
601.39	3300			2377	A 15	0.0028	5.1063		1.6667	0.0000	0	
601.77	1987		165			0.0000	5.1063	L	1.6667	0.0122	8186	A 96-153
602.31	2900					0.0000	5.1063		1.6545	0.0027	1360	96
602.31				63	163	0.0001	5.1064		1.6518	0.0000		
602.31				2545	D 16	0.0037	5.1101		1.6515	0.0000		
602.52	1100		96	1100	96	0.0022	5.1123	R	1.6518	0.0022	1100	96
602.52						0.0000	5.1123		1.6497	0.0020	1204	D 132-96
602.64	593			593	96	0.0012	5.1135	REVERSE CURVE	1.6477	0.0001	89	132
602.96	1697		132	767	96	0.0015	5.1150	L	1.6475	0.0014	997	132
603.11	800			4262	A 96	0.0070	5.1220		1.6461	0.0021	1633	D 169-132
603.22	600		219			0.0000	5.1220	L	1.6440	0.0002	166	169
603.35	700	XING 179.57				0.0000	5.1220	LEPPO ROAD	1.6439	0.0000		
603.51	800		171			0.0000	5.1220	R	1.6439	0.0016	1041	A 165-169
603.75	1300	XING 180.5		6003	A 13	0.0061	5.1300	SPRING MILL	1.6420	0.0010	844	165
604.04	1510		165			0.0000	5.1300	L	1.6411	0.0000		165
604.78	3900			275	162	0.0003	5.1304		1.6411	0.0088	9927	A 131-165
605.29	2700			2700	162	0.0032	5.1335		1.6323	0.0036	2340	A 114-131
605.42	701		131	58	162	0.0001	5.1336	L	1.6286	0.0001	60	114
605.67	1300	XING 181.94		1943	D 16	0.0027	5.1363	PRIVATE	1.6286	0.0022	1300	114
605.78	600		114	600	114	0.0010	5.1373	L	1.6264	0.0010	600	114
605.96	927	XING 1	114	927	114	0.0015	5.1388	L ROCK ROAD	1.6254	0.0015	927	114
605.96						0.0000	5.1388		1.6239	0.0048	3961	D 200-114
606.79	4400	XING 183.8		1360	114	0.0023	5.1411	PRIVATE	1.6191	0.0004	439	200
607.72	4912	XING 184.2		7952	A 11	0.0240	5.1650	SHELBY ROAD	1.6187	0.0047	4912	200
608.80	5710			5710	200	0.0054	5.1704		1.6140	0.0054	5710	200
609.80	5295	XING 186.4		5295	200	0.0050	5.1754	HORNING ROAD	1.6086	0.0041	4336	200
610.81	5308			5308	200	0.0050	5.1805		1.6045			
611.81	5302			5302	200	0.0050	5.1855		1.6045			
612.15	1800			1800	200	0.0017	5.1872		1.6045			
612.26	554		200	554	200	0.0005	5.1877	R	1.6045			
612.43	900			900	200	0.0009	5.1886	CRESTLINE	1.6045			
612.47	200	OHD		200	200	0.0002	5.1888	SR 61	1.6045			
612.58	600		200	600	200	0.0006	5.1893	R	1.6045			
612.73	800			800	200	0.0008	5.1901		1.6045			
612.94	1100		200	1100	200	0.0011	5.1911	R	1.6045			
612.98	204			204	200	0.0002	5.1913		1.6045	0.0206	18220	A 130-200
613.06	400			400	200	0.0004	5.1917		1.5839	0.0000		
613.06					200	0.0000	5.1917		1.5839	0.0069	4147	A 94-130
613.81	4000			1000	200	0.0009	5.1926		1.5770	0.0009	440	94
613.81				29	200	0.0000	5.1927		1.5761	0.0005	238	A 91-94
614.12	1600			4571	D 20	0.0059	5.1986		1.5756	0.0014	882	91
614.23	600		94	597	94	0.0012	5.1998	L	1.5742	0.0012	600	91
614.24	78			61	D 94	0.0002	5.1999	REV CUR TT 40MPH	1.5730	0.0002	78	91
614.40	800		91	800	91	0.0017	5.2016	R	1.5728	0.0017	800	91
614.40					91	0.0000	5.2016		1.5711	0.0061	4852	D 200-91
615.08	3600			1360	91	0.0028	5.2044		1.5651	0.0000	48	200
615.21	700	XING 1	219	22605	A 91	0.0280	5.2324	R SR 598	1.5650	0.0000	0	200
615.38	900	XING 1	219		200	0.0003	5.2324	R CRESTLINE ROAD	1.5650	0.0005	500	200
616.21	4380	XING 192.80			200	0.0000	5.2324	BIDDLE ROAD	1.5646	0.0041	4380	200
617.19	5192	XING 193.86			200	0.0000	5.2324	BECK ROAD	1.5604	0.0049	5192	200
618.20	5288	XING 194.62			200	0.0000	5.2324	MAIN STREET	1.5555	0.0050	5288	200
619.20	5284	XING 194.71		1379	200	0.0013	5.2337	WESTERN AVENUE	1.5505	0.0050	5284	200
620.20	5289	XOING 195		5289	200	0.0050	5.2387	OLENTANGY ROAD	1.5456	0.0050	5289	200
621.16	5092	XING 195.87		5092	200	0.0048	5.2435	PRIVATE	1.5405	0.0048	5092	200
622.20	5470	XING 196.2		5470	200	0.0052	5.2487	PRIVATE	1.5356	0.0052	5470	200
623.20	5275	XING 197.0		5275	200	0.0050	5.2537	LWR LEESVILLE	1.5305	0.0050	5275	200

APPENDIX B5 — PAGE 13 — COLUMN 1

READ DOWN WESTBOUND TRAFFIC — STATION — READ UP EASTBOUND TRAFFIC

MILES	DISTANCE FT	HWY GRADE CROSSING	MAX MPH	DISTANCE FT	SPEED MPH	HOURS	TIME HOURS	STATION	TIME HOURS	HOURS	DISTANCE FT	SPEED MPH
624.19	5254	XING 197.6		5254	200	0.0050	5.2587	PRIVATE	1.5255	0.0050	5254	200
624.19		XING 197.76			200	0.0000	5.2587	PRIVATE	1.5205	0.0000		200
624.40	1100	XING 198.2		1100	200	0.0010	5.2587	PRIVATE	1.5205	0.0010	1100	200
624.58	965	XING 198.7		965	200	0.0008	5.2607	ALBRIGHT	1.5195	0.0009	965	200
624.77	1000			1000	200	0.0009	5.2616		1.5185	0.0009	1000	200
625.10	1700			1700	200	0.0016	5.2632	BUCYRUS	1.5176	0.0016	1700	200
625.19	500	US 201.20		500	200	0.0005	5.2637	SANDUSKY RIVER	1.5160	0.0005	500	200
625.40	1100		200	1100	200	0.0010	5.2647	R	1.5155	0.0010	1100	200
626.19	4186	XING 203.1		4186	200	0.0040	5.2687	McCRACKEN	1.5145	0.0040	4186	200
627.19	5286	XING 203.6		5286	200	0.0050	5.2737	SIMMS CR	1.5105	0.0050	5286	200
628.19	5280	XING 204.7		5280	200	0.0050	5.2787	PRIVATE	1.5055	0.0050	5280	200
629.21	5380	XING 204.9		5380	200	0.0051	5.2838	GLENVILLE CR	1.5005	0.0051	5380	200
630.21	5272	XING 205.9		5272	200	0.0050	5.2888	PRIVATE	1.4954	0.0050	5272	200
631.21	5286	XING 206.1		5298	200	0.0050	5.2938	STRIEB CR KN	1.4904	0.0050	5298	200
632.08	4583	XING 207.1		4583	200	0.0043	5.2982	MARION-MELMO	1.4854	0.0043	4583	200
633.21	5926	XING 207.7		5926	200	0.0056	5.3038	PRIVATE	1.4810	0.0056	5926	200
633.21		XING 208.13			200	0.0000	5.3038	COUNTY LINE	1.4754			
633.21		XING 208.60			200	0.0000	5.3038	PRIVATE	1.4754			
634.24	5440	XING 209.0		5440	200	0.0052	5.3089	NEVADA AYERS STREE	1.4754	0.0052	5440	200
634.24		XING 209.10			200	0.0000	5.3089	MORRISON STREET	1.4703			
634.24		XING 209.27			200	0.0000	5.3089	GOODBREAD STREET	1.4703			
635.24	5270	XING 210.8		5270	200	0.0050	5.3139	EDENVILLE ROAD	1.4703	0.0050	5270	200
636.24	5290	XING 211.6		5290	200	0.0050	5.3189	PRIVATE	1.4653	0.0050	5290	200
637.24	5289	XING 212.1		5289	200	0.0050	5.3239	DOUGLAS RD	1.4603	0.0050	5289	200
638.24	5269	XING 213.7		5269	200	0.0050	5.3289	PRIVATE	1.4553	0.0050	5269	200
639.24	5280	XING 213.80		5280	200	0.0050	5.3339	ROCK RUN ROAD	1.4503	0.0050	5280	200
640.24	5284	XING 214.3		5284	200	0.0050	5.3389	PRIVATE	1.4453	0.0050	5284	200
640.39	787	XING 2 200		787	200	0.0007	5.3397	L WILLIAMS CR	1.4403	0.0007	787	200
641.24	4500	XING 215.5		4500	200	0.0043	5.3439	PRIVATE	1.4395	0.0043	4500	200
641.24		XING 215.70			200	0.0000	5.3439	PRIVATE	1.4353	0.0000		200
641.24		XING 216.10			200	0.0000	5.3439	PRIVATE	1.4353	0.0000		200
641.73	2600			2600	200	0.0025	5.3464	UPPER SANDUSKY	1.4353	0.0025	2600	200
641.82	479	XING 2 200		479	200	0.0005	5.3468	L WHITE CR	1.4328	0.0005	479	200
642.24	2200	XING 218.9		2200	200	0.0021	5.3489	PRIVATE	1.4323	0.0021	2200	200
643.24	5281	XING 220.2		5281	200	0.0050	5.3539	PRIVATE	1.4303	0.0050	5281	200
644.22	5185	XING 220.7		5185	200	0.0049	5.3588	WILL CR	1.4253	0.0049	5185	200
644.22		XING 220.99			200	0.0000	5.3588	MIGRET CR	1.4204	0.0000		
645.22	5281	XING 221.3		5281	200	0.0050	5.3638	PRIVATE	1.4204	0.0050	5281	200
646.22	5281	XING 222.2		5281	200	0.0050	5.3688	GAMBER CR	1.4154	0.0050	5281	200
647.22	5263	XING 222.5		5263	200	0.0050	5.3738	PRIVATE	1.4103	0.0050	5263	200
648.22	5280	XING 223.2		5280	200	0.0050	5.3788	KRAUS CR	1.4053	0.0050	5280	200
649.22	5279	XING 223.8		5279	200	0.0050	5.3838	PRIVATE	1.4003	0.0050	5279	200
650.22	5282	XING 223.9		5282	200	0.0050	5.3888	PRIVATE	1.3953	0.0050	5282	200
651.22	5284	XING 224.2		5284	200	0.0050	5.3938	MAIN STREET SR	1.3903	0.0050	5284	200
652.22	5275	XING 224.3		5275	200	0.0050	5.3988	ALLEY STREET	1.3853	0.0050	5275	200
653.22	5279	XING 225.8		5279	200	0.0050	5.4038	LILES CR	1.3803	0.0050	5279	200
653.22		XING 226.84			200	0.0000	5.4038	HELLER CR	1.3753			
653.22		XING 227.20			200	0.0000	5.4038	PRIVATE	1.3753			
653.22		XING 227.50			200	0.0000	5.4038	PRIVATE	1.3753			
653.22		XING 227.80			200	0.0000	5.4038	PRIVATE	1.3753			
653.22		XING 227.94			200	0.0000	5.4038	ANGLING ROAD	1.3753			
653.22		XING 228.38			200	0.0000	5.4038	COUNTY LINE	1.3753			
653.22		XING 228.90			200	0.0000	5.4038	LOUISA STREET	1.3753			
654.23	5298	XING 228.9		5298	200	0.0050	5.4088	FORREST MARY STREET	1.3753	0.0050	5298	200
654.23		XING 229.21			200	0.0000	5.4088	GORMLY STREET	1.3703			
655.22	5275	XING 229.3		5275	200	0.0050	5.4139	MARTIN STREET SR	1.3703	0.0050	5275	200
656.22	5273	XING 229.7		5273	200	0.0050	5.4188	CAMPBELL STREET	1.3653	0.0050	5273	200
657.22	5278	XING 230.5		5278	200	0.0050	5.4238	BERLIN CR	1.3603	0.0050	5278	200
658.22	5289	XING 231.0		5289	200	0.0050	5.4288	PRIVATE	1.3553	0.0050	5289	200
659.22	5289	XING 231.7		5289	200	0.0050	5.4338	PRIVATE	1.3503	0.0050	5289	200
660.22	5276	XING 234.11		5276	200	0.0050	5.4388	GROAT ROAD	1.3453	0.0050	5276	200
660.51	1500	XING 234.5		1500	200	0.0014	5.4403	WYKES CR	1.3403	0.0014	1500	200
661.22	3771			3771	200	0.0036	5.4438	DUNKIRK	1.3389	0.0036	3771	200
662.22	5278			5278	200	0.0050	5.4488		1.3354	0.0050	5278	200
663.22	5280			5280	200	0.0050	5.4538		1.3304	0.0050	5280	200
664.22	5276			5276	200	0.0050	5.4588		1.3254	0.0050	5278	200
665.22	5280	XING 241.6		5280	200	0.0050	5.4638	TRAVERSE PK	1.3204	0.0050	5280	200
666.22	5277	XING 241.8		5277	200	0.0050	5.4688	PRIVATE	1.3154	0.0050	5277	200
667.22	5278	XING 242.6		5278	200	0.0050	5.4738	PETERSON CR	1.3104	0.0050	5278	200
668.22	5273	XING 243.6		5273	200	0.0050	5.4788	SCOTT XING	1.3054	0.0050	5273	200
669.21	5265	XING 244.6		5265	200	0.0050	5.4838	1 MILE XING	1.3004	0.0050	5265	200
670.24	5394	XING 245.5		5394	200	0.0051	5.4889	ADA JOHNSON STREET	1.2954	0.0051	5394	200
671.24	5281	XING 245.8		5281	200	0.0050	5.4939	MAIN STREET	1.2903	0.0050	5281	200
672.24	5276	XING 246.7		5276	200	0.0050	5.4989	GILBERT STREET	1.2853	0.0050	5276	200
673.25	5377	XING 246.8		5377	200	0.0051	5.5040	KLINGLER RD	1.2803	0.0051	5377	200
674.26	5303	XING 247.6		5303	200	0.0050	5.5090	ST PAUL RD	1.2752	0.0050	5303	200
674.26		XING 248.10			200	0.0000	5.5090	PRIVATE	1.2702			
674.26		XING 248.65			200	0.0000	5.5090	COUNTY LINE	1.2702			
674.26		XING 249.55			200	0.0000	5.5090	BENTLEY ROAD	1.2702			
674.26		XING 250.13			200	0.0000	5.5090	PEVEE ROAD	1.2702			
675.26	5289	XING 250.4		5289	200	0.0050	5.5140	PRIVATE	1.2702	0.0050	5289	200
676.26	5286	XING 250.7		5286	200	0.0050	5.5190	LAFAYETTE RD	1.2652	0.0050	5286	200
677.26	5297	XING 250.9		5297	200	0.0050	5.5241	SHRIDER ROAD	1.2602	0.0050	5297	200
678.27	5304	XING 251.3		5304	200	0.0050	5.5291	VINT RD	1.2551	0.0050	5304	200
679.29	5371	XING 251.6		5371	200	0.0051	5.5342	PHILLIPS RD	1.2501	0.0051	5371	200
680.27	5202	XING 252.2		5202	200	0.0049	5.5391	S HIGH STREET	1.2450	0.0050	5231	200
681.27	5298	XING 252.3		5298	200	0.0050	5.5441	CHURCH ST	1.2401	0.0063	6309	A 16
681.73	2400	XING 252.4		2400	200	0.0023	5.5464	WASHINGTON	1.2338	0.0014	1360	182
681.78	293	XING 2 182		293	200	0.0003	5.5467	R RUMBAUGH RD	1.2324	0.0003	293	182
682.28	2600	XING 253.6		2600	200	0.0025	5.5491	FISHER ROAD	1.2321	0.0010	1008	O 20
683.29	5336	XING 254.1		5336	200	0.0051	5.5542	McCLURE XING	1.2311	0.0052	5503	200
683.86	3000	XING 254.6		3000	200	0.0028	5.5570	COOL ROAD	1.2259	0.0000		200
683.86		XING 256.64			200	0.0000	5.5570	THAYER ROAD	1.2259			
683.86		XING 256.93			200	0.0000	5.5570	FETTER ROAD	1.2259			
684.31	2373	XING 257.8		2373	200	0.0022	5.5593	METZGER ROAD	1.2259	0.0000	0	
684.51	1100	XING 258.4			200	0.0000	5.5593	ROUSH ROAD	1.2259	0.0000	0	
684.61	500				200	0.0000	5.5593	LIMA	1.2259	0.0116	11149	A 16
684.61					200	0.0000	5.5593		1.2144	0.0001	75	165
684.70	485		200		200	0.0000	5.5593	R	1.2143	0.0006	485	165
684.70				1011	200	0.0010	5.5602		1.2137	0.0000		
684.85	800			1874	D 20	0.0019	5.5622		1.2137	0.0009	800	165
685.29	2300		165	2300	165	0.0026	5.5648	R	1.2128	0.0026	2300	165
685.40	576		165	576	165	0.0007	5.5655	R	1.2102	0.0007	576	165
686.29	4700	XING 262.9		1360	165	0.0016	5.5670	CABLE ROAD	1.2095	0.0019	1874	D 20
687.29	5298	XING 263.5		11149	A 16	0.0115	5.5786	HARTZLER RD	1.2076	0.0050	5298	200
688.29	5286	XING 264.0		2775	200	0.0026	5.5812	EASTTOWN ROAD	1.2025	0.0050	5286	200
689.31	5383	XING 264.6		5383	200	0.0051	5.5863	EAST ROAD	1.1975	0.0051	5383	200
690.31	5280	XING 265.2		5280	200	0.0050	5.5913	BATY ROAD	1.1924	0.0050	5280	200
691.31	5286	XING 265.8		5286	200	0.0050	5.5963	PRIVATE	1.1874	0.0050	5286	200
692.31	5265	XING 266.4		5265	200	0.0050	5.6013	GREENLAWN AVE	1.1824	0.0050	5265	200
693.31	5282	XING 266.6		5282	200	0.0050	5.6063	MAIN ROAD	1.1774	0.0050	5282	200
694.31	5281	XING 267.2		5281	200	0.0050	5.6113	WAPAK ROAD	1.1724	0.0050	5281	200
694.31		XING 267.40				0.0000	5.6113	PRIVATE	1.1674			
694.31		XING 267.65				0.0000	5.6113	KEMP ROAD	1.1674			
694.31		XING 267.90				0.0000	5.6113	PRIVATE	1.1674			
694.31		XING 268.10				0.0000	5.6113	PRIVATE	1.1674			
694.31		XING 268.83				0.0000	5.6113	SHUBB ROAD	1.1674			
694.31		XING 269.20				0.0000	5.6113	PRIVATE	1.1674			
695.32	5292	XING 270.4		5292	200	0.0050	5.6163	STATE RD	1.1674	0.0050	5262	200
696.32	5283	XING 270.8		5283	200	0.0050	5.6213	OLD DELPHOS	1.1624	0.0011	1164	200
697.32	5280	XING 271.2		6618	200	0.0062	5.6275	DEFIANCE TRA	1.1613	0.0115	11149	A 16
697.91	3100	XING 271.7		1874	D 20	0.0019	5.6294	BAUGH RD	1.1498	0.0016	1360	165
698.32	2172	XING 2 165		2172	165	0.0025	5.6319	L PELTIER RD	1.1482	0.0025	2172	165
698.51	1000	XING 2 165		1000	165	0.0011	5.6331	L LEHMAN RD	1.1457	0.0011	1000	165
698.51		XING 273.69				0.0000	5.6331	PRIVATE	1.1446	0.0003	246	D 17

APPENDIX B5 — PAGE 13 — COLUMN 2

READ DOWN WESTBOUND TRAFFIC — STATION — READ UP EASTBOUND TRAFFIC

MILES	DISTANCE FT	HWY GRADE CROSSING	MAX MPH	DISTANCE FT	SPEED MPH	HOURS	TIME HOURS	STATION	TIME HOURS	HOURS	DISTANCE FT	SPEED MPH
698.51							5.6331		1.1443	0.0001	104	170
698.72	1100	XING 274.3		1100	165	0.0013	5.6343	DELPHOS PIERCE & TRE	1.1442	0.0015	1290	A 165-170
699.08	1900	XING 274.3		1900	165	0.0022	5.6365	S MAIN STREET	1.1427	0.0016	1360	165
699.27	1000	XING 2 165		1000	165	0.0011	5.6377	R CANAL STREET	1.1412	0.0011	1000	165
699.32	284	XING 274.71		284	165	0.0003	5.6380	JEFFERSON STREET	1.1400	0.0019	1874	D 200-165
700.32	5277	XING 274.7		1076	165	0.0012	5.6392	S CLAY STREET	1.1381	0.0047	5016	200
701.32	5268	XING 274.9		11146	A 16	0.0130	5.6522	S BREDICK STREET	1.1333	0.0036	3660	200
702.32	5265	XING 275.0		3605	200	0.0034	5.6556	STATE STREET	1.1296	0.0050	5265	200
703.32	5289	XING 275.6		5289	200	0.0050	5.6606	BRICKNER RD	1.1246	0.0050	5289	200
704.32	5284	XING 277.2		5284	200	0.0050	5.6656	PRIVATE	1.1196	0.0050	5284	200
704.32		XING 277.68				0.0000	5.6656	BOOKEY RD	1.1146			
704.32		XING 278.69				0.0000	5.6656	CONVERSE RD	1.1146			
704.32		XING 279.71				0.0000	5.6656	MIDDLEPOINT	1.1146			
705.32	5281	XING 280.0		5281	200	0.0050	5.6706	MAIN STREET	1.1146	0.0050	5281	200
706.32	5261	XING 280.2		5281	200	0.0050	5.6756	ADAMS STREET	1.1096	0.0050	5281	200
707.32	5285	XING 280.3		5285	200	0.0050	5.6806	MASON STREET	1.1046	0.0033	5285	200
708.32	5282	XING 280.7		5282	200	0.0050	5.6856	DOG CREEK RD	1.0712	0.0010	1015	200
709.32	5261	XING 281.0		5261	200	0.0050	5.6906	RINGWALD RD	1.0703	0.0000		
710.32	5297	XING 282.32		8032	200	0.0076	5.6983	CHENOWETH RD	1.0703	0.0000		
711.23	4830	XING 283.34		2065	D 20	0.0022	5.7004	GAMBLE RD	1.0703	0.0127	12105	A 161-200
711.33	486	XING 2 161			161	0.0000	5.7004	L HOAGLIN CR	1.0576	0.0133	5406	A 103-161
711.57	1300	XING 2 161			161	0.0000	5.7004	L GILLAND RD	1.0443	0.0000		103
711.57		XING 285.86			161	0.0000	5.7004	MENDON RD	1.0443	0.0015	800	103
711.86	1500	XING 286.7		1040	161	0.0012	5.7016	VAN WERT WAYNE STRE	1.0428	0.0022	1100	A 90-103
711.86		XING 286.85		2245	D 15	0.0032	5.7049	VINE STREET	1.0407			
712.03	900	XING 2 103		900	103	0.0017	5.7065	R FRANKLIN STREET	1.0407	0.0008	360	90
712.20	900	XING 2 117		832	103	0.0012	5.7077	R CHESTNUT STREET	1.0399	0.0019	900	90
712.22	100	XING 287.04		368	D 10	0.0007	5.7084	RACE STREET	1.0380	0.0002	100	90
712.22		XING 287.11			90	0.0000	5.7084	N TYLER STREET	1.0378	0.0000	0	
712.22		XING 287.17				0.0000	5.7084	HARRISON STREET	1.0378			
712.22		XING 287.24				0.0000	5.7084	CHERRY STREET	1.0378			
712.22		XING 287.30				0.0000	5.7084	WALNUT STREET	1.0378			
712.22		XING 287.37				0.0000	5.7084	MARKET STREET	1.0378			
712.22		XING 287.49				0.0000	5.7084	WASHINGTON STREET	1.0378			
712.22		XING 287.55				0.0000	5.7084	JEFFERSON STREET	1.0378			
712.22		XING 287.67				0.0000	5.7084	N SHANNON	1.0378			
712.22		XING 287.74				0.0000	5.7084	WALL STREET	1.0378			
712.22			90			0.0000	5.7084	CR VAN WERT	1.0378	0.0000	0	90
712.22		XING 288.07				0.0000	5.7084	BURT STREET	1.0378	0.0061	4679	D 200-90
712.34	642	XING 288.37		642	90	0.0014	5.7098	FISHER AVENUE	1.0317	0.0000	0	200
713.33	5241	XING 289.0		718	90	0.0015	5.7113	JOHN BROWN RD	1.0317	0.0011	1204	200
714.33	5264	XING 289.59				0.0000	5.7113	PRIVATE	1.0306	0.0050	5264	200
715.33	5269	XING 290.18		22671	A 90	0.0251	5.7364	LIBERTY UNION	1.0256	0.0050	5269	200
716.33	5287	XING 291.27				0.0000	5.7364	RICHEY CR	1.0206	0.0050	5287	200
717.34	5287	XING 292.3		2989	200	0.0028	5.7422	MACE XING	1.0155	0.0050	5287	200
718.34	5284	XING 293.0		5284	200	0.0050	5.7472	ALT RT US 30	1.0105	0.0050	5284	200
718.34		XING 293.59				0.0000	5.7472	BERGNER RD	1.0055			
718.34		XING 294.71				0.0000	5.7472	MAIN ST CONVOY	1.0055			
718.34		XING 294.88				0.0000	5.7472	TULLY ST	1.0055			
719.34	5287	XING 295.4		5287	200	0.0050	5.7522	PRIVATE	1.0055	0.0050	5287	200
720.34	5288	XING 295.82		5288	200	0.0050	5.7572	PAYNE CR	1.0005	0.0050	5288	200
721.34	5286	XING 296.36		5286	200	0.0050	5.7623	SHANER CR	0.9955	0.0050	5286	200
722.34	5280	XING 296.94		5280	200	0.0050	5.7673	LARE CR	0.9906	0.0050	5280	200
723.34	5292	XING 297.1		5292	200	0.0050	5.7723	MENTZER	0.9855	0.0050	5292	200
724.34	5286	XING 297.6		5286	200	0.0050	5.7773	PRIVATE	0.9805	0.0050	5286	200
724.34		XING 298.12				0.0000	5.7773	MENTZER CURC	0.9756			
724.34		XING 298.86				0.0000	5.7773	CLEM XING	0.9756			
724.34		XING 299.46				0.0000	5.7773	SCHOOLHOUSE	0.9756			
725.34	5282	XING 300.2		5282	200	0.0050	5.7823	OH IN ST LINE RD	0.9755	0.0050	5282	200
726.34	5281	XING 301.5		5281	200	0.0050	5.7873	MORGAN RD	0.9705	0.0050	5281	200
727.35	5285	XING 301.8		5285	200	0.0050	5.7923	PRIVATE	0.9655	0.0050	5285	200
728.35	5285	XING 302.64		5285	200	0.0050	5.7973	LORTIE RD	0.9605	0.0050	5285	200
729.35	5294	XING 303.9		5294	200	0.0050	5.8023	OHIO STREET	0.9555	0.0050	5294	200
730.35	5289	XING 304.0		5289	200	0.0050	5.8073	MAIN STREET	0.9505	0.0050	5289	200
731.35	5297	XING 304.3		5297	200	0.0050	5.8123	WASHINGTON	0.9455	0.0050	5297	200
732.35	5274	XING 305.5		5274	200	0.0050	5.8173	SNYDER RD	0.9404	0.0050	5274	200
733.36	5282	XING 306.8		5282	200	0.0050	5.8223	HOFFMAN RD	0.9354	0.0050	5282	200
734.36	5281	XING 307.1		5281	200	0.0050	5.8273	GROTRAIN RD	0.9304	0.0050	5281	200
734.36		XING 307.71				0.0000	5.8273	WILSON RD	0.9254			
734.36		XING 306.39				0.0000	5.8273	PACKLER RD	0.9254			
734.36		XING 308.85				0.0000	5.8273	GARADOT RD	0.9254			
734.36		XING 308.37				0.0000	5.8273	HOUK RD	0.9254			
734.36		XING 309.71				0.0000	5.8273	WASHINGTON	0.9254			
734.36		XING 309.94				0.0000	5.8273	FRANKE RD	0.9254			
735.36	5281	XING 310.8		5281	200	0.0050	5.8323	PRIVATE	0.9254	0.0050	5281	200
736.26	5261	XING 310.9		5261	200	0.0050	5.8373	PRIVATE	0.9204	0.0050	5261	200
737.35	5287	XING 311.07		5287	200	0.0050	5.8423	TILLMAN RD	0.9154	0.0050	5287	200
737.35		XING 311.75			200	0.0000	5.8423	MINNICH ROAD	0.9104	0.0000		200
737.35		XING 312.50			200	0.0000	5.8423	PRIVATE	0.9104	0.0000		200
737.35		XING 312.80			200	0.0000	5.8423	PRIVATE	0.9104	0.0000		200
737.35		XING 313.09			200	0.0000	5.8423	PRIVATE	0.9104	0.0000		200
737.35		XING 313.25			200	0.0000	5.8423	PAULDING ROAD	0.9104	0.0000		200
738.36	5263	XING 313.44		5263	200	0.0050	5.8473	HARTZELL ROAD	0.9104	0.0050	5263	200
739.36	5283	XING 315.1		5283	200	0.0050	5.8523	ADAMS CENTER	0.9054	0.0050	5283	200
740.36	5288	XING 316.2		5288	200	0.0050	5.8573	MEYER ROAD	0.9004	0.0050	5288	200
741.37	5329			5329	200	0.0050	5.8624		0.8954	0.0050	5329	200
742.35	5192			5192	200	0.0049	5.8673		0.8904	0.0049	5192	200
743.22	4600			4600	200	0.0044	5.8717		0.8855	0.0044	4600	200
743.39	749		200	749	200	0.0007	5.8724	L	0.8811	0.0007	749	200
743.52	800			800	200	0.0008	5.8731		0.8804	0.0008	800	200
743.78	1400	XING 319.3		1400	200	0.0013	5.8745	FORT WAYNE	0.8796	0.0013	1400	200
744.01	1200			1200	200	0.0011	5.8756		0.8783	0.0011	1200	200
744.48	2470			2470	200	0.0024	5.8779		0.8772	0.0024	2470	200
745.37	4700			4700	200	0.0045	5.8824		0.8746	0.0045	4700	200
745.55	962		200	962	200	0.0009	5.8833	R	0.8704	0.0009	962	200
746.40	4500	XING 323.0		4500	200	0.0043	5.8875	THOMAS ROAD	0.8695	0.0043	4500	200
747.41	5349	XING 324.9		5349	200	0.0051	5.8926	HADLEY ROAD	0.8652	0.0051	5349	200
748.42	5322	XING 325.04		5322	200	0.0050	5.8977	BASS ROAD	0.8602	0.0050	5322	200
749.40	5193	XING 326.4		5193	200	0.0049	5.9026	PRIVATE	0.8551	0.0049	5193	200
750.40	5277	XING 328.14		5277	200	0.0050	5.9076	MAIN STREET	0.8502	0.0050	5277	200
751.40	5278	XING 328.36		5278	200	0.0050	5.9126	AROOLA COLUMBIA STR	0.8452	0.0050	5278	200
752.40	5290	XING 328.9		5290	200	0.0050	5.9176	PRIVATE	0.8402	0.0050	5290	200
753.42	5363	XING 329.0		5363	200	0.0051	5.9227	PRIVATE	0.8352	0.0051	5363	200
754.41	5224	XING 329.3		5224	200	0.0049	5.9276	PRIVATE	0.8301	0.0049	5224	200
755.41	5296	XING 329.5		5296	200	0.0050	5.9326	PRIVATE	0.8252	0.0050	5296	200
756.42	5297	XING 330.51		5297	200	0.0050	5.9376	COUNTY LINE	0.8202	0.0050	5297	200
757.42	5294	XING 331.5		5294	200	0.0050	5.9426	BEECHING ROAD	0.8151	0.0050	5294	200
758.42	5293	XING 332.67		5293	200	0.0050	5.9477	ROSE	0.8101	0.0050	5293	200
759.42	5293	XING 333.83		5293	200	0.0050	5.9527	MILL STREET	0.8051	0.0050	5293	200
760.43	5292	XING 334.7		5292	200	0.0050	5.9577	MOWERY ROAD	0.8001	0.0050	5292	200
761.42	5269	XING 336.9		5269	200	0.0050	5.9627	PAGES CR	0.7951	0.0050	5269	200
762.42	5290	XING 336.44		5290	200	0.0050	5.9677	OHIO STREET	0.7901	0.0050	5290	200
763.45	5434	XING 338.81		5434	200	0.0051	5.9728	SR 9	0.7851	0.0051	5434	200
764.44	5185	XING 338.91		5185	200	0.0049	5.9777	LINE STREET SR	0.7799	0.0049	5185	200
765.43	5203	XING 339.9		5263	200	0.0050	5.9827	POOR FARM CR	0.7750	0.0039	4157	200
766.44	5298	XING 341.3		5634	200	0.0055	5.9882	WOLF ROAD	0.7711	0.0066	6842	A 181-200
766.74	1900	XING 342.0		1062	O 20	0.0011	5.9893	PEARL CR	0.7645	0.0011	1026	181
767.02	1503	XING 3 181		1503	181	0.0016	5.9909	BUTLER CR 40	0.7634	0.0000	0	181
767.44	2200	XING 343.8		1360	181	0.0014	5.9923	L PRIVATE	0.7634	0.0000		
767.44				4759	A 18	0.0046	5.9971		0.7634	0.0000		
768.44	5250			250	165	0.0002	5.9973		0.7634	0.0000		
768.80	1900			3001	D 19	0.0035	6.0008		0.7634	0.0135	11089	A 132-161
769.08	1481		132	1481	132	0.0021	6.0030	R	0.7498	0.0020	1360	132
769.44	1900			1360	132	0.0020	6.0049		0.7478	0.0038	3011	D 200-132
770.35	4800			16048	A 13	0.0183	6.0233		0.7440	0.0032	3369	200

APPENDIX B5 — PAGE 14 — COLUMN 1

READ DOWN: WESTBOUND TRAFFIC · STATION · READ UP: EASTBOUND TRAFFIC

MILES	DISTANCE	ANY GRADE CROSSING	MAX MPH	DISTANCE	SPEED	ET HOURS	TIME HOURS	STATION	TIME HOURS	ET HOURS	DISTANCE	SPEED
770.45	488	XING	3 202			0.0000	6.0233	L SRD 5 CENTR ST	0.7408	0.0005	488	200
770.48	184	XING	3 202			0.0000	6.0233	L LARWILL DEPOT STRB	0.7403	0.0002	184	200
771.45	5100	XING	347.30			0.0000	6.0233	BRINKLEY ROAD	0.7401	0.0048	5100	200
772.45	5294	XING	348.81			0.0000	6.0233	BLACK CR	0.7353	0.0050	5294	200
773.45	5286	XING	349.3(	1603	A 15	0.0018	6.0251	PRIVATE	0.7303	0.0050	5286	200
774.45	5287	XING	348.8(	9048	200	0.0085	6.0336	PRIVATE	0.7253	0.0050	5287	200
775.45	5282	XING	350.3(	5282	200	0.0050	6.0386	MARKET STREET	0.7203	0.0050	5282	200
776.45	5283	XING	350.6	5283	200	0.0050	6.0436	PIERCETON FIRST ST	0.7153	0.0050	5283	200
777.45	5279	XING	352.5(	5279	200	0.0050	6.0486	VAN NESS ROAD	0.7103	0.0050	5279	200
778.45	5278	XING	353.4	5278	200	0.0050	6.0536	WOOSTER ROAD	0.7053	0.0050	5278	200
779.45	5293	XING	354.7	5293	200	0.0050	6.0586	SCHADE 250E	0.7003	0.0050	5293	200
780.46	5323			5323	200	0.0050	6.0637		0.6953	0.0050	5323	200
781.45	5233			5233	200	0.0050	6.0686		0.6903	0.0050	5233	200
782.17	3800			3800	200	0.0036	6.0722		0.6853	0.0034	3647	200
782.46	1500	XING	3 202	1500	200	0.0014	6.0737	L	0.6819	0.0000		
782.65	986	XING	3 202	986	200	0.0009	6.0746	L WARSAW	0.6819	0.0088	8724	A 17
783.16	2700	XING	360.5(	2700	200	0.0026	6.0772	JACKSON CR	0.6731	0.0000	0	90
783.46	1600	XING	361.7	1600	200	0.0015	6.0787	R BURKET PRIV	0.6731	0.0027	2425	A 16
784.46	5282	XING	362.0	3586	200	0.0034	6.0821	PRIVATE	0.6704	0.0014	1182	165
784.49	178	XING	362.2	1674	D 20	0.0019	6.0840	THOMAS CR 10	0.6690	0.0002	178	165
784.85	1900	XING	3 165	1900	165	0.0022	6.0862	PRIVATE	0.6688	0.0022	1900	165
785.46	3200	XING	363.7	1360	165	0.0016	6.0878	SELBY CR 525	0.6666	0.0019	1874	D 2(
786.47	5318	XING	364.5	11149	A 16	0.0115	6.0993	PRIVATE	0.6647	0.0063	4574	200
787.46	5245	XING	365.3	1254	200	0.0012	6.1005	ATWOOD PRAIRIE ST	0.6604	0.0050	5245	200
788.46	5291	XING	365.8	5291	200	0.0050	6.1055	HARRISON STREET	0.6554	0.0050	5291	200
789.46	5352	XING	366.8	5352	200	0.0051	6.1105	ROBINSON ROAD	0.6504	0.0051	5352	200
790.47	5220	XING	367.4	5220	200	0.0049	6.1155	SNYDER CR 87	0.6453	0.0049	5220	200
791.47	5291	XING	368.1	5291	200	0.0050	6.1205	PRIVATE	0.6404	0.0050	5291	200
792.47	5279	XING	368.7	5279	200	0.0050	6.1255	COOK CR	0.6354	0.0050	5279	200
793.46	5227	XING	369.2	5227	200	0.0049	6.1304	WALNUT STREET SR	0.6304	0.0049	5227	200
794.47	5337	XING	369.3	5307	200	0.0051	6.1355	ETNA GREEN	0.6254	0.0051	5337	200
795.48	5379	XING	370.9	5379	200	0.0051	6.1406	S BEECH ROAD	0.6204	0.0051	5379	200
796.47	5193	XING	372.9	5193	200	0.0049	6.1455	HARRIS STREET	0.6153	0.0049	5193	200
797.50	5439	XING	373.0	5439	200	0.0052	6.1507	BOURBON MAIN ST	0.6104	0.0052	5439	200
797.50		XING	373.17			0.0000	6.1507	BOURBON STREET	0.6052	0.0000		
797.50		XING	373.26			0.0000	6.1507	CENTER STREET	0.6052	0.0000		
797.50		XING	374.13			0.0000	6.1507	TSCHR ST SE	0.6052	0.0000		
798.48	5162	XING	375.1	5162	200	0.0049	6.1556	STOCKMAN ROAD	0.6052	0.0049	5162	200
799.47	5252	XING	376.2	5252	200	0.0050	6.1606	WYMER CR	0.6003	0.0050	5252	200
800.47	5270	XING	377.0	5270	200	0.0050	6.1656	COMPTON CR	0.5953	0.0050	5270	200
801.47	5279	XING	377.8	5279	200	0.0050	6.1705	HAWTHORNE ROAD	0.5904	0.0050	5279	200
802.47	5279	XING	378.6	5279	200	0.0050	6.1755	PRIVATE	0.5854	0.0050	5279	200
803.47	5283	XING	379.1	5283	200	0.0050	6.1805	S IRIS ROAD	0.5804	0.0050	5283	200
804.47	5277	XING	379.6	5277	200	0.0050	6.1855	PRIVATE	0.5754	0.0050	5277	200
805.48	5315	XING	380.2	5315	200	0.0050	6.1905	PRIVATE	0.5704	0.0050	5315	200
806.45	5156	XING	380.5	5156	200	0.0049	6.1954	PRIVATE	0.5653	0.0049	5156	200
807.47	5387	XING	381.0	5387	200	0.0051	6.2005	KING ROAD	0.5604	0.0051	5387	200
808.30	4400	XING	383.2	4400	200	0.0042	6.2047	PLYMOUTH LIBERTY S1	0.5554	0.0042	4400	200
808.48	944	XING	384.0	944	200	0.0009	6.2056	FIFTH ST	0.5512	0.0009	944	200
809.47	5216	XING	384.6	5216	200	0.0049	6.2105	OAK ROAD	0.5503	0.0049	5216	200
810.47	5280	XING	385.7	5280	200	0.0050	6.2155	PRIVATE	0.5454	0.0050	5280	200
811.47	5287	XING	385.9	5287	200	0.0050	6.2205	PRIVATE	0.5404	0.0050	5287	200
812.47	5293	XING	386.3	5293	200	0.0050	6.2255	ANGLING ROAD	0.5353	0.0050	5293	200
813.47	5269	XING	386.0	5269	200	0.0050	6.2305	QUINCE ROAD	0.5303	0.0050	5269	200
814.47	5281	XING	387.7	5281	200	0.0050	6.2355	REDWOOD ROAD	0.5253	0.0050	5281	200
815.47	5268	XING	388.1	5268	200	0.0050	6.2405	PRIVATE	0.5203	0.0050	5268	200
816.47	5298	XING	390.8	5298	200	0.0050	6.2455	DONALDSON UNION R	0.5154	0.0050	5298	200
817.48	5298	XING	393.5	5298	200	0.0050	6.2505	PRIVATE	0.5103	0.0050	5298	200
818.50	5411	XING	394.1	5411	200	0.0051	6.2557	SR 23	0.5053	0.0051	5411	200
819.48	5164	XING	395.5	5164	200	0.0049	6.2606	WALSH CR	0.5002	0.0049	5164	200
820.48	5285	XING	396.5	5285	200	0.0050	6.2656	750E	0.4953	0.0050	5285	200
821.48	5288	XING	398.0	5288	200	0.0050	6.2706	STARKE STREET	0.4903	0.0050	5288	200
822.48	5285	XING	398.5	5285	200	0.0050	6.2756	HAMLET JEFFERSON S	0.4853	0.0050	5285	200
823.48	5279	XING	399.1	5279	200	0.0050	6.2806	900E	0.4803	0.0050	5279	200
824.49	5301	XING	401.1	5301	200	0.0050	6.2856	300E	0.4753	0.0050	5301	200
825.49	5282	XING	401.8	5282	200	0.0050	6.2906	PRIVATE	0.4703	0.0050	5282	200
826.49	5282	XING	402.3	5282	200	0.0050	6.2956	PRIVATE	0.4653	0.0050	5282	200
827.49	5270	XING	403.3	5270	200	0.0050	6.3006	50E	0.4603	0.0050	5270	200
828.49	5287	XING	405.2	5287	200	0.0050	6.3056	100W	0.4553	0.0050	5287	200
829.49	5292	XING	405.7	5292	200	0.0050	6.3106	1400E	0.4503	0.0050	5292	200
830.49	5280	XING	406.2	5280	200	0.0050	6.3156	SR 39	0.4453	0.0050	5280	200
831.52	5422	XING	407.2	5422	200	0.0051	6.3207	LONG LANE	0.4403	0.0051	5422	200
832.47	5045	XING	407.7	5045	200	0.0048	6.3255	PRIVATE	0.4351	0.0048	5045	200
833.30	4400	XING	408.6	4400	200	0.0042	6.3297	THOMPSON STREET	0.4303	0.0042	4400	200
833.30		XING	408.74		200	0.0000	6.3297	OHIO STREET	0.4262	0.0000		200
833.49	986	XING	408.8	986	200	0.0009	6.3306	HANNA 450W	0.4262	0.0009	986	200
834.49	5281	XING	409.2	5281	200	0.0050	6.3356	PRIVATE	0.4253	0.0050	5281	200
834.49		XING	410.35		200	0.0000	6.3356	MP 410 SCHULL 600W	0.4203	0.0000		200
835.49	5285	XING	411.3	5285	200	0.0050	6.3406	FARRESTER ROAD	0.4203	0.0050	5285	200
836.49	5273	XING	411.8	5273	200	0.0050	6.3456	TILDON 750W	0.4153	0.0050	5273	200
837.66	6182	XING	412.3	6182	200	0.0059	6.3515	PRIVATE	0.4103	0.0059	6182	200
838.49	4367	XING	413.4	4367	200	0.0041	6.3556	KLEINE 900W	0.4044	0.0041	4367	200
839.49	5280	XING	414.7	5280	200	0.0050	6.3606	N MAIN STREET	0.4003	0.0050	5280	200
840.48	5272	XING	414.8	5272	200	0.0050	6.3656	WANATAH S ILLINOIS	0.3953	0.0050	5272	200
841.46	5251	XING	415.1	5251	200	0.0050	6.3706	S WASHINGTON	0.3903	0.0050	5251	200
842.48	5293	XING	415.1	5293	200	0.0050	6.3756	LINCOLN STREET	0.3853	0.0050	5293	200
843.48	5270	XING	415.4	5275	200	0.0050	6.3806	150W	0.3803	0.0050	5270	200
844.08	3200	XING	416.5	3200	200	0.0030	6.3836	OSBRN CR-CO	0.3753	0.0050	5270	200
844.19	570	XING	4 218	570	200	0.0005	6.3841	R FISHER 575E	0.3703	0.0005	570	200
844.48	1500	XING	419.0	1500	200	0.0014	6.3855	DOLSON 450E	0.3698	0.0002	234	200
844.48		XING	419.76		200	0.0000	6.3855	BICKEL 400E	0.3696			
844.48		XING	420.50		200	0.0000	6.3855	MP 420 PRIVATE	0.3696			
845.48	5280	XING	420.7	5280	200	0.0050	6.3905	MONTDALE ROAD	0.3696	0.0115	11149	A 16
846.47	5263	XING	421.0	4089	200	0.0039	6.3944	PRIVATE	0.3580	0.0008	960	165
846.61	700	XING	422.2	1874	D 20	0.0019	6.3964	CEMETERY ROAD	0.3573	0.0006	700	165
847.40	4200	XING	4 165	4200	165	0.0048	6.4012	R GREENWICH ST	0.3565	0.0048	4200	165
848.01	3194	XING	423.4	906	165	0.0016	6.4027	AXE STREET	0.3516	0.0032	3194	165
848.01		XING	423.6	1290	A 16	0.0015	6.4042	FRANKLIN STREET	0.3480			
848.01				296	170	0.0003	6.4046		0.3480			
848.01				246	D 17	0.0003	6.4048	VALPARAISO	0.3480			
848.16	486	XING	A 165	486	165	0.0006	6.4054	L WASHINGTON	0.3480	0.0006	486	165
848.48	2000	XING	423.8	1360	165	0.0016	6.4069	LAFAYETTE STREET	0.3474	0.0019	1874	D 2(
849.48	5279	XING	423.8	11149	A 16	0.0141	6.4210	NAPOLEON STREET	0.3455	0.0051	5405	200
850.48	5274	XING	424.5	44	200	0.0000	6.4211	STEVENS XING	0.3404	0.0050	5274	200
851.31	4400	XING	430.2	4400	200	0.0042	6.4252	PRIVATE	0.3354	0.0042	4400	200
851.48	892	XING	430.5	892	200	0.0008	6.4261	WHEELER PARK AVE	0.3312	0.0008	892	200
852.48	5296	XING	431.0	5299	200	0.0050	6.4311	GOTT RD 7TH	0.3304	0.0050	5299	200
853.49	5340	XING	432.3	5340	200	0.0051	6.4362	TREAGER RD	0.3253	0.0051	5340	200
854.48	5206	XING	433.0	5206	200	0.0049	6.4411	COUNTY LINE	0.3203	0.0049	5206	200
855.48	5291	XING	434.5	5291	200	0.0050	6.4461	HOBART ILLINOIS ST	0.3154	0.0050	5291	200
856.48	5289	XING	434.6	5289	200	0.0050	6.4511	LINDA STREET	0.3103	0.0050	5289	200
857.47	5212	XING	434.8	5212	200	0.0049	6.4560	CLEVELAND STREET	0.3053	0.0049	5212	200
858.48	5354			5354	200	0.0051	6.4611		0.3004	0.0051	5354	200
859.49	5316			5316	200	0.0050	6.4661		0.2953	0.0050	5316	200
859.78	1544	XING	4 117	1544	200	0.0015	6.4676	R LAKE PARK AVE	0.2903	0.0015	1544	200
860.48	3700	XING	435.8	3700	200	0.0035	6.4711	WISCONSIN STREET	0.2888	0.0035	3700	200
861.49	5320			5320	200	0.0050	6.4762		0.2853	0.0013	1384	200
861.69	1060			1060	200	0.0010	6.4772		0.2840	0.0000		
861.69		XING	437.18		200	0.0000	6.4772	LIVERPOOL RD E END LIVERPOOL TERMINAL				
862.49	4200	XING	438.0	4200	200	0.0040	6.4811	INDIANA AVENUE	0.2840	0.0000		
862.60	600	XING	438.7	600	200	0.0006	6.4817	VIRGINIA STREET	0.2840	0.0000		
862.60					200	0.0000	6.4817	I-65	0.2840	0.0000		
862.81	1122			1122	200	0.0011	6.4828		0.2840	0.0000		
862.81					200	0.0000	6.4828	I-80	0.2840	0.0000		
863.50	3800			3800	200	0.0034	6.4862		0.2840	0.0000		

APPENDIX B5 — PAGE 14 — COLUMN 2

READ DOWN: WESTBOUND TRAFFIC · STATION · READ UP: EASTBOUND TRAFFIC

MILES	DISTANCE	ANY GRADE CROSSING	MAX MPH	DISTANCE	SPEED	ET HOURS	TIME HOURS	STATION	TIME HOURS	ET HOURS	DISTANCE	SPEED
863.86	1897			1897	200	0.0018	6.4860		0.2840	0.0000		
863.86					200	0.0000	6.4860	WEST END LIVERPOOL TERMINAL				
864.30	2363	XING	440.28	2363	200	0.0022	6.4882	21ST STREET	0.2840	0.0281	22671	A 90-200
865.30	5283	XING	440.3(	500	200	0.0005	6.4887	BROADWAY	0.2559	0.0029	1360	90
865.30		XING	440.47			0.0000	6.4907	WASHINGTON S	0.2530	0.0000		60
865.30		XING	440.60			0.0000	6.4907	16TH STREET	0.2530	0.0000		60
865.30		XING	440.70			0.0000	6.4907	MADISON STREET	0.2530	0.0000		60
865.30		XING	440.84			0.0000	6.4907	JACKSON AVE	0.2530	0.0000		60
865.30		XING	441.00			0.0000	6.4907	HARRISON STREET	0.2530	0.0000		60
865.30		XING	441.14			0.0000	6.4907	15TH AVENUE	0.2530	0.0000		60
865.30		XING	441.28			0.0000	6.4907	13TH AVENUE	0.2530	0.0000		60
865.30		XING	441.65	84	200	0.0001	6.4906	GRANT STREET	0.2530	0.0000		60
865.30		XING	441.74	4679	D 20	0.0061	6.4969	10TH AVENUE	0.2530	0.0000		60
865.30		RR	90		90	0.0000	6.4969	TOLLESTON CSX CR	0.2530	0.0000		90
865.30		XING	441.90	1360	90	0.0029	6.4997	9TH AVENUE	0.2530	0.0064	3696	D 188-90
866.30	5298	OHD	442.02	19023	A 90	0.0263	6.5261	IHB RR	0.2476	0.0004	368	188
867.31	5280	XING	442.25			0.0000	6.5261	TAFT STREET	0.2472	0.0000		
867.51	1000	XING	442.75			0.0000	6.5261	5TH AVENUE	0.2472	0.0000		
867.79	1500	XING	4 202			0.0000	6.5261	R CSSB OVERHEAD	0.2472	0.0000		
868.30	2700	XING	444.10			0.0000	6.5261	CLARKE ROAD	0.2472	0.0000		
869.31	5325					0.0000	6.5261		0.2472	0.0000		
869.71	2110	OHD	445.40			0.0000	6.5261	EJ&E RR (CN)	0.2472	0.0239	18330	A 90-186
869.86	780	XING	445.5	255	190	0.0003	6.5263	PRIVATE	0.2233	0.0029	1360	90
869.99	661			4107	D 19	0.0056	6.5319		0.2204	0.0014	681	90
869.99			90	1360	90	0.0029	6.5347	CSX CLARKE JCT	0.2190	0.0000		90
870.31	1718			1100	A 90	0.0002	6.5389	JCT PRR-PC-NYC	0.2190	0.0008	106	D 95-90
870.50	1000				104	0.0000	6.5388	CONN TO CP 501	0.2188	0.0002	101	95
870.65	807		104	37	100	0.0001	6.5389	R	0.2186	0.0047	1029	A 60-95
870.65				1028	D 10	0.0004	6.5393	CP 501 EX-NYC	0.2138	0.0043	1360	60
870.67	80		60	80	60	0.0003	6.5396	JCT NYC	0.2096	0.0003	80	60
870.97	1600		110	1360	60	0.0043	6.5439	R		0.0051	1600	60
871.33	1900		251	3359	A 60	0.0073	6.5512	R	0.2043	0.0028	1247	D 110-60
871.65	1700			7541	A 11	0.0106	6.5618		0.2015	0.0000	0	110
872.29	3380					0.0000	6.5616		0.2015	0.0036	2666	D 174-110
872.62	1700		213			0.0000	6.5618	L	0.1980	0.0002	302	174
872.65	200			139	166	0.0002	6.5620		0.1978	0.0064	5606	A 152-174
873.05	2100			181	D 15	0.0002	6.5622		0.1914	0.0017	1360	152
873.07	115		152	115	152	0.0001	6.5623	L	0.1897	0.0001	115	152
873.11	200			200	152	0.0002	6.5626	REVERSE CURVE	0.1895	0.0002	200	150
873.14	125		152	125	150	0.0002	6.5627	R	0.1893	0.0002	125	152
873.21	400			400	152	0.0005	6.5632		0.1891	0.0004	368	D 160-152
873.23	125		166	125	152	0.0002	6.5634	R	0.1887	0.0003	245	160
873.29	300			300	152	0.0004	6.5638	REVERSE CURVE	0.1884	0.0000		
873.31	115		156	115	152	0.0001	6.5639	L	0.1884	0.0000		
873.65	1800			35	152	0.0000	6.5640		0.1884	0.0000		
874.07	2150		176	9667	A 15	0.0076	6.5716	R	0.1884	0.0000		
874.65	3100			861	176	0.0008	6.5725		0.1884	0.0000		
874.79	740			357	D 17	0.0003	6.5728		0.1884	0.0000		
874.83	190		173	190	173	0.0002	6.5730	L	0.1884	0.0217	11422	A 70-180
875.65	4350			757	173	0.0009	6.5738		0.1667	0.0034	1772	70
875.67	88			3871	D 17	0.0057	6.5796		0.1632	0.0002	88	70
875.70	172		70	172	70	0.0005	6.5800	L	0.1630	0.0005	172	70
875.75	244		70	244	70	0.0007	6.5807	R REVERSE CURVE	0.1625	0.0007	244	70
875.75		XING	451.7	1360	70	0.0037	6.5844	LAKE AVENUE	0.1619	0.0028	1428	D 121-70
875.75				3632	A 70	0.0071	6.5914	ATK HAMMOND WHITING	0.1590	0.0000	30	121
876.47	3831	XING	452.1	96	115	0.0002	6.5916	CALUMET AVENUE	0.1590	0.0049	2812	A 94-121
876.82	1800			644	D 11	0.0012	6.5928		0.1540	0.0057	7631	94
877.04	1200		94	1200	94	0.0024	6.5952	R REVERSE CURVE	0.1483	0.0001	89	A 93-94
877.27	1200		94	1675	94	0.0034	6.5986	L	0.1482	0.0017	815	93
877.37	500			27	D 94	0.0001	6.5986		0.1466	0.0005	235	A 90-93
877.53	850		93	988	93	0.0020	6.6006	L	1.4610	0.0023	1090	90
877.56	200			81	D 90	0.0002	6.6007	REVERSE CURVE	0.1438	0.0001	78	90
877.77	1100		90	1100	90	0.0023	6.6031	R	0.1436	0.0004	200	90
878.36	3100			1360	90	0.0029	6.6060		0.1432	0.0023	1100	90
878.49	711		141	3116	A 90	0.0056	6.6115	L	0.1409	0.0018	986	D 122-90
878.84	1800		132			0.0000	6.6115	L	0.1391	0.0002	130	122
878.86	150			16	121	0.0000	6.6115	REVERSE CURVE	0.1389	0.0085	4401	A 70-122
878.89	180		132	1429	D 12	0.0028	6.6143	R	0.1304	0.0008	236	70
879.08	985						6.6143		0.1296	0.0002	180	70
879.15	400	UG 454	70	1385	70	0.0037	6.6180	CALUMET RIVER UG	0.1294	0.0026	985	70
879.46	1600						6.6180		0.1267	0.0011	400	70
879.62	880			880	70	0.0023	6.6204		0.1257	0.0034	1837	D 132-76
879.85	1230		132	500	70	0.0014	6.6217	L REVERSE CURVE	0.1222	0.0009	623	132
879.94	500		132	5891	A 70	0.0107	6.6324	R	0.1213	0.0017	1300	132
880.07	700			5064	A 13	0.0066	6.6390		0.1196	0.0007	500	132
880.21	734		158			0.0005	6.6390	R	0.1189	0.0014	1106	D 158-132
880.44	1200					0.0000	6.6390		0.1174	0.0000	0	158
881.47	5471			10828	A 15	0.0114	6.6504		0.1174	0.0023	2147	D 196-158
882.46	5250					0.0000	6.6504		0.1152	0.0012	1200	196
883.44	5184					0.0000	6.6504		0.1140	0.0000		
884.44	5276			3065	194	0.0030	6.6534		0.1140	0.0000		
884.74	1600			3512	D 19	0.0043	6.6577		0.1140	0.0000		
884.80	300		117	303	117	0.0005	6.6582	L ENGLEWOOD	0.1140	0.0230	16659	A 117-196
885.45	3400			2302	117	0.0037	6.6619	METRA (CRIP) CHIC	0.0910	0.0073	3705	A 72-117
885.47	154			1247	D 11	0.0025	6.6644		0.0837	0.0032	1206	72
885.70	1200		72	1200	72	0.0032	6.6675	R	0.0806	0.0004	154	72
885.70				1360	72	0.0036	6.6711		0.0803	0.0032	1300	72
885.70				1994	A 72	0.0036	6.6746		0.0771	0.0017	786	D 102-72
885.70				270	98	0.0006	6.6754		0.0754	0.0001	81	102
886.42	3770			485	D 98	0.0009	6.6763		0.0753	0.0043	1583	A 81-99
886.48	340		81	340	81	0.0008	6.6771	R	0.071	0.003	1300	81
886.78	1580			1390	81	0.0032	6.6803		0.0677	0.0044	1873	81
886.86	400		82	596	A 81	0.0013	6.6815	L	0.0633	0.0000		
887.06	1060			811	86	0.0017	6.6833		0.0633	0.0003	117	A 79-81
887.06				273	D 90	0.0006	6.6840		0.0631	0.0031	1276	79
887.15	516		79	1433	79	0.0034	6.6874	L	0.0600	0.0013	544	A 86-79
887.37	1154			237	D 79	0.0006	6.6880	REVERSE CURVE	0.0587	0.0021	790	86
887.45	411		88	411	86	0.0011	6.6891	R	0.0566	0.0022	771	88
888.48	5486			1360	88	0.0036	6.6929		0.0544	0.0003	120	85
889.53	5506			10498	A 88	0.0181	6.7110		0.0541	0.0052	3124	D 161-88
890.13	3200			290	160	0.0003	6.7112		0.0489	0.0001	155	161
890.13				2044	D 16	0.0029	6.7142		0.0488	0.0124	8903	A 108-161
890.20	380		108	380	108	0.0007	6.7148	L REVERSE CURVE	0.0364	0.0041	2079	A 83-108
890.27	340		108	340	108	0.0006	6.7154	R	0.0323	0.0000		
890.40	700			700	D 10	0.0014	6.7168		0.0323	0.0007	300	83
890.47	370		83			0.0006	6.7177		0.0316	0.0024	1040	83
890.53	320		83	320	83	0.0007	6.7184	R REVERSE CURVE	0.0292	0.0003	148	83
890.59	310			310	83	0.0007	6.7191	S BR CHICAGO RIV	0.0289	0.0005	230	83
890.65	300		132	300	83	0.0007	6.7198	L	0.0284	0.0007	320	83
890.83	960	XING	458.3	750	83	0.0017	6.7215	LUMBER STREET	0.0276	0.0016	829	D 112-83
890.89	320		132	2657	A 83	0.0054	6.7269	R	0.0260	0.0002	115	112
890.99	520					0.0000	6.7269	REVERSE CURVE	0.0258	0.0000		
891.07	420		132			0.0000	6.7269	L	0.0258	0.0000		
891.20	677					0.0000	6.7269		0.0208	0.0008	3263	A 67-113
891.31	600		105			0.0000	6.7269	L	0.0190	0.0000		67
891.35	240			58	105	0.0001	6.7271		0.0190	0.0016	574	67
891.59	1250			1281	D 11	0.0027	6.7298		0.0174	0.0033	1076	A 38-67
891.70	550		67	1053	67	0.0030	6.7327	R	0.0141	0.0020	410	38
891.88	560			447	D 67	0.0016	6.7343		0.0120	0.0047	860	38
891.98	550		36	550	38	0.0027	6.7371	L REVERSE CURVE	0.0073	0.0029	503	38
892.07	473			1360	38	0.0068	6.7439	CHICAGO UNION STA	0.0044	0.0044	440	A 0-38
892.08	80						6.7439	EAST END PLATFORM	0.0000			
892.18	550		186	261	38	0.0013	6.7452	R				
892.33	750			212	D 38	0.0021	6.7473					
							6.7473	WEST END PLATFORM				

Appendix B6 - Improved Right of Way on Extensively Rebuilt Alignment; sufficient for High-Speed TOFC Operations.

Aerodynamically designed TOFC equipment operating at 120 mph maximum speed. Maximum allowable speed on curves as shown for high speed equipment only.

Minimum of FRA Class 7 track where speeds over 110 mph are possible

Other tracks classified per maximum possible speed.

E_a = 6 inches; wheel load 0.6 minimum per FRA 213.307d1

Assumed c.g. of moving equipment 108 inches above top of rail

Train length 11,000 feet: 200 platforms plus 20 locomotive units @ 6,000 hp each. Maximum braking/deceleration rate 5 mph/second.

20% speed reduction for reverse curves with 600 feet or less between curves.

Railroad grade crossings eliminated except where they do not affect allowable speed. Other highly effective improvements made to right of way.

Class 7 track where speeds over 110 mph are possible

Other tracks classified per maximum possible speed.

Highway grade crossing locations shown are approximate

Distance Jersey Meadows to Liverpool IN 854.62 miles

Westbound trip	9.0313 hours
Eastbound trip	9.0163 hours

Appendix B6 table available in online

Appendix B7 - Passenger on Improved Right of Way on Extensively Re-built Alignment per Appendix B6. Streamlined high-speed passenger train operating at 200 mph maximum speed.

Maximum allowable speed on curves as shown for high speed equipment only.

FRA Class 9 track where speeds over 180 mph are possible.

Other tracks classified per maximum possible speed.

Track re-built for maximum speed up to 200 mph. E_a = 8.635 inches, E_u = 4 inches

Note: for HS Freight (TOFC) E_a = 5.84 inches; for HS passenger if E_a = 5.84 inches E_u = 4 inches.

Wheel load 0.6 minimum per FRA 213.307d1

Assumed c.g. of moving equipment 48 inches above top of rail

Train length 1,360 feet: 10 non-articulated streamlined passenger cars, 8 locomotive units @ 10,000 hp each. Maximum train braking rate 5 mph/second.

20% speed reduction for reverse curves with 600 feet or less between curves.

Gradients shown in Appendix B2.

Highway grade crossing locations shown are approximate. See Appendix B2 and Appendix B3 for more detailed right of way information.

Distance New York Penn Station to Chicago Union Station 886.5 miles

Westbound trip	6.2707 hours
Eastbound trip	6.2576 hours

Appendix B7 table available in online

Appendix B8 - Improved Right of Way on Extensively Rebuilt and Partly Relocated Alignment-New right of way between Upper Sandusky and Orrville, and between Monaca and Lewistown (Pavonia Cutoff and Sam Rae Line)
Class 9 track where speeds of over 180 mph are possible
Other tracks classified per maximum possible speed. New Tunnel under Hudson River, 80 mph maximum speed. All other track built, rebuilt or replaced for minimum 147 mph speed. All railroad crossings and highway crossings eliminated or reconfigured to be acceptable to FSA at speeds traversed. Maximum speed 235 mph at appropriate locations with special high-speed passenger equipment.
E_a = 8.635 inches, E_u = 4 inches
Note: for HS Freight (TOFC) E_a =5.84 inches; for HS passenger if E_a = 5.84 inches E_u = 6.795 inches.
Wheel load 0.6 minimum per FRA 213.307d1
Assumed c.g. of moving equipment 48 inches above top of rail
Train length 1,360 feet: 10 non-articulated streamlined passenger cars, 8 locomotive units @ 10,000 hp each. Maximum train braking rate 5 mph/second. 20% speed reduction for reverse curves with 600 feet or less between curves.
Gradients shown in Appendix B2.
Highway grade crossing locations shown are approximate. See Appendix B2 and Appendix B3 for more detailed right of way information.
Travel Distance New York Penn Station to Chicago Union Station 792.12 miles

Westbound trip 3.9980 hours
Eastbound trip 3.9924 hours

Appendix B8 table available in online

About the Author

Terry L. Koglin graduated from the University of Wisconsin–Madison with a bachelor's degree in Mechanical Engineering. He worked for the Milwaukee Road (the Chicago, Milwaukee, St Paul and Pacific Railroad Company) when they still operated passenger trains. He worked as a consultant on many engineering projects on Amtrak's Northeast Corridor. He worked briefly on projects for other passenger railroads including Finnish National Railways, The Lindenwold High-Speed Line, New Jersey Transit, the Long Island Rail Road, and Canadian National Railways. He had professional engineering registrations in several states, but currently only his Pennsylvania license is active. He is a member of the American Railway Engineering and Maintenance-of-Way Association, and have retired from their Steel Structures committee. He has ridden on, inspected, and investigated high-speed and other passenger railroads in North America and Europe, including trains in Canada, Great Britain, France, Spain, Portugal, Italy, Switzerland, Germany, Luxembourg, Belgium, Netherlands, Denmark, Sweden, Finland and Russia. He has authored and presented papers on high speed rail for ASCE and IABSE.

Index

9 781606 509838

CPSIA information can be obtained
at www.ICGtesting.com
Printed in the USA
FFOW03n2228080517
35361FF